What Is A Chemical Element?

What Is A Chemical Element?

A Collection of Essays by Chemists,
Philosophers, Historians, and Educators

Edited by

ERIC SCERRI AND ELENA GHIBAUDI

OXFORD
UNIVERSITY PRESS

OXFORD
UNIVERSITY PRESS

Oxford University Press is a department of the University of Oxford. It furthers
the University's objective of excellence in research, scholarship, and education
by publishing worldwide. Oxford is a registered trade mark of Oxford University
Press in the UK and certain other countries.

Published in the United States of America by Oxford University Press
198 Madison Avenue, New York, NY 10016, United States of America.

© Oxford University Press 2020

CIP data is on file at the Library of Congress
ISBN 978-0-19-093378-4

1 3 5 7 9 8 6 4 2

Printed by Integrated Books International, United States of America

Contents

Foreword

It can come as a surprise to those not versed in the philosophy of science to discover how apparently shaky its roots can be. The classic deductive method supposes that there are some statements and concepts we can assume to be fundamental and precise, and that we can use these to construct theories and hypotheses amenable to experimental testing. But those concepts might, on closer inspection, prove rather less precise and secure than we had hoped or imagined. In biology this is true of "genes" and "species"; in physics there are several ways of thinking about the concept of "force," while the origin and nature of at least one of these—gravity—remain hazy and contentious. In chemistry it is possible to argue without resolution or consensus about what is meant by, for example, "element," "bond," and "molecule"—concepts that even school pupils are taught at the early stages of instruction in the subject.

Yet there is nothing truly problematic about such fuzziness surrounding fundamental concepts. On the contrary, the debates and ambiguities reflect the provisional nature of science, which is not so much a progressive elaboration of the implications of a few timeless truths, but an ever deeper inquiry into the nature of physical reality. Put simply, it works in both directions: starting from what we can observe, at a scale intermediate between atoms and cosmos, and venturing out into the extremes beyond perception and—at least so it can often feel—beyond intuition.

The idea of a chemical element is, then, one of those scientific concepts that exists first and foremost because it is good for thinking with. And so has it always been, since long before we had any notion of an element in its modern sense, before there was any reason to think of atoms as real and tangible objects, before even the idea of science as a self-consistent process of deductive, inductive, and abductive reasoning was adduced. The element is a way of simplifying the world: reducing the bewildering diversity of experience to some countably small number of underlying principles. No one supposes now that what Empedocles and Aristotle understood by an element bears much relation to what Boyle, Lavoisier, and Mendeleev understood it to mean (or that those latter three individuals had a shared definition either). But supposing the physical world to be made up of elemental substances was useful to them all.

Likewise, the fact that it is possible to compile a book like this, in which the definition of an element is interrogated and contested, does not imply that the discipline of chemistry is unstable at its base. The element is a "good enough" organizing principle for the daily practice of chemists. They know what they mean by it, and even if they do not always mean the same thing (from one day to another, and one chemist to another), any confusion does not take long to resolve. In this respect, the language and discourse of scientists is not so different from that of anyone else: it has rules, but they often work best when there is some latitude, some room for maneuver, some flux. We can use the concept of an element as effectively (and as loosely) as we can speak of "love" or "money," "life" (another of biology's usefully vague words) or "liberty."

A chemist working at the bench with solvents and retorts might, then, be forgiven for asking, what is the point? Why debate a word that works perfectly well for my colleagues and me? And indeed one answer can be "fair enough, there's no obligation." But there are more than enough justifications all the same. First, because it is interesting, in the same way that the history of the subject is interesting (and one can't discuss the meaning of "element" without discussing the history of the term). But also because it is always a little dangerous in science to overconfidently assume that because you can use a word or concept without being challenged, you and others know what it means and agree on that. And it's important to realize that there are very real consequences of ambiguity, as some of the contributions in this volume point out. In an age when "new elements" are made one ephemeral atom at a time, with little prospect of ever establishing their chemical behavior, it seems imperative to think about how to judge and codify such claims. What are our criteria for existence? Does element 119 "exist" even in 2020 (although at the time of writing no one has claimed to have made it)? Does even element 118 (which has been reported and verified) truly exist, except for a few fleeting moments? Does it exist right now, if its atoms do not?

What's more, if we are not united on what are the defining properties of a chemical element, then we have to confront uncertainties about how to arrange and structure these entities in the periodic table. We have to acknowledge and deal with the possibility that students will be confused and even misled. The terms and names used in science encode assumptions, often invisibly. To be a good chemist, it is not a requirement that you know the name "oxygen" falsely imputes the element's ubiquity in acids, or that sodium and nitrogen share similar chemical symbols because they share a linguistic root (natro/nitro). But sometimes implications lie dormant in etymologies and terminologies that, brought to light, not only can deepen our appreciation of the subject and its heritage but might also alert us to notions we might too readily take for granted. As the disputes between the oxygen and phlogiston theories of combustion showed, words and terminology can dictate thinking, and so are worth taking seriously. By defining our terms, we state what we think is important.

There's a very real danger in science today, when the literature and the broader boundaries of scientific knowledge are so unmanageably vast, that the working scientist becomes too preoccupied with the question "is this information useful to me?" If the issue of utility is simply about getting an experimental result, a paper published, a grant secured, science itself risks becoming a pinched, pragmatic enterprise. But perhaps more important, scientists risk doing their job less well, trapped as they are in the valley of the familiar. I believe that these contributions, as well as being entertaining, intriguing, and challenging discussions in their own right, can have the value to chemical practitioners of re-engaging them with the core of their craft: to the intersection of ideas and substance, of the microscopic to the macroscopic, of concepts to the objects they are supposed to describe (or might that be: to create?). These essays prompt more general questions: Why do I do what I do? How do I know what I know (and is it right)? Are words tools or traps?

The questions faced by the chemists of past times struggling to refine the concept of an element apply equally elsewhere in chemistry today: To what extent is a mechanism a convenient model rather than a description of events? Are there natural boundaries of categories in nature (is this a bond or not?), and if so, where? What makes particles indistinguishable, and what is their status if they are? Can we live with the tension of multiple, overlapping and sometimes conflicting definitions and theories, or must it all dovetail perfectly? It's fun to think about these things, but it's also useful, and sometimes it will be essential.

Philip Ball
London, 2019

Introduction

What is a chemical element? The question in the title of this book may sound trivial. The notion of elements is so central to chemistry that one might expect it to have been established once and for all. There is no doubt that the notion of a chemical element is a foundational one. In fact, any chemist—if asked about it—would be able to provide a definition of elements and argue about its connection to the most renowned achievement of chemistry, namely the periodic system. Scientists such as Lavoisier and Mendeleev are familiar to chemists precisely because of their work on the elements. Yet, their own conceptions of elements are substantially different, not to say incompatible. How can this be? The truth is that concepts evolve and the notion of element has a long-standing history: it is rooted far back in the past, in Greek philosophy, and it changed along with the development of our thought and knowledge of the material world. Contributions to the definition of the notion of element have come from both philosophy and science, in the past just as in present times. The list of contributors to the historical debate about the elements starts with Greek philosophers (such as Aristotle and Democritus) and includes philosophers, chemists, epistemologists, and educators, all of whom bring distinct perspectives to the discussion. The development of chemical and physical knowledge has seriously challenged the concept of element. History shows that the main attribute of Greek elements (which was their *elementarity*) seems to have been refuted as a result of the discovery of isotopes, radioactivity, and the inner structure of atoms. Nevertheless, the notion endures, as it serves specific purposes in the conceptual architecture of chemistry. It is, therefore, no surprise that the discussion about elements persists, involving historians of scientific thought, chemistry practitioners, philosophers of science, and chemical educators.

One of the main issues in the debate over elements is their dual conception, which is partly captured by the double definition proposed by the International Union for Pure and Applied Chemistry, where an abstract meaning (a species of atoms) coexists with an operational one (the simple substances bearing the elements' names). This latter recalls the famous definition by Lavoisier, according to which the element is the final attainment of chemical analysis. Nevertheless, neither of the two IUPAC definitions accounts for the philosophical aspect of the element that Mendeleev mentions in his writings and that he claims was crucial for the construction of his periodic table, that is, to designate

Eric Scerri and Elena Ghibaudi, *Introduction*. In: *What Is A Chemical Element?*. Edited by: Eric Scerri and Elena Ghibaudi (2020). © Oxford University Press.
DOI: 10.1093/oso/9780190933784.003.0001

what remains unchanged in a chemical reaction. The problem has been noticed by several scholars: among them, Fritz Paneth, an Austrian chemist and epistemologist. Paneth is the author of a seminal article, published in 1931, in which he designates the two meanings of elements as *einfacher Stoff* (elementary substance) and *Grundstoff* (basic substance), respectively. Paneth's work raised a debate that still continues, as the ambiguity of the concept of chemical element is far from being resolved. Moreover, it raises problems at several levels: epistemic, logical, and educational. The epistemic problem mainly relates to defining what the term "element" refers to. The use of a single term (element) for designating distinct entities (namely, the simple substance and what remains unchanged in a chemical reaction) generates a logical conflict that, in turn, causes problems in the educational context of chemistry.

The present book aims to provide an update to the current state of the debate on elements. It hosts contributions from different perspectives: the authors are historians of chemistry, philosophers of chemistry, and chemists with epistemological and educational concerns. All of them focus on the notion of chemical element and its related problems, defending positions that are in some cases complementary and in others conflicting.

Eric Scerri opens the book by exposing a number of open issues related to the elements that are matters of discussion among the community of the philosophers of chemistry, starting from the above-mentioned problem of the dual meaning of the term "element."

Some of the issues mentioned in Scerri's chapter are tackled in subsequent chapters. The boundaries between the domains of philosophy of science, history, epistemology, education, and the practice of chemistry are loose: this book gives account of these blended perspectives, as it contains essays with a historical-philosophical slant and epistemological essays written by philosophers and chemists.

Bernadette Bensaude-Vincent analyzes the concept of elements as non-decompounded bodies, traditionally credited to Lavoisier, arguing that this claim is somehow misleading. She also argues that the distinction between the "analytical" and the abstract sense of elements was both a precondition and a product of the construction of the periodic system.

Nathan Brooks agrees with Bensaude-Vincent regarding the relevance of Mendeleev's abstract notion of element in the development of the periodic system. In his article, he retraces the evolution of Mendeleev's notion of elements. He also highlights the role played by concepts and reasoning in organic chemistry, and by studies on isomorphisms, upon the work of Mendeleev.

Geoffrey Blumenthal, James Ladyman, and Vanessa Seifert bring the reader back to pre-Lavoisian chemistry; they propose a causal-descriptive theory of reference for chemical substances, and discuss reference to elements. Marina Banchetti-Robino focuses on a crucial figure in the history of elements: John Dalton, who succeeded in chemically reconciling "atomicity" and "elementarity," a groundbreaking conceptual step that paved the way to the later development of modern chemistry.

The modern meaning of element is a matter of discussion among philosophers of science. Joseph Earley criticizes the internal inconsistency of the IUPAC dual definition of chemical element, and argues against the identification of elements with the indecomposable constituents of substances. Robin Hendry discusses the challenges posed by the fleeting existence of some super-heavy elements with the IUPAC definition of element and argues about the legitimacy of calling them "chemical elements." Farzad Mahootian explores the notion of element in the light of Kant's and Cassirer's thought, as an opportunity to discuss the dialectic relationship between substance and function typical of modern chemical thought. Joachim Schummer focuses on the operational definition of element as substance that cannot be further decomposed by chemical analysis, pointing out that its adoption meant giving up explanation, the primary goal of natural philosophy. The discussion of the legacy of the operational definition of elements gives him the chance to argue about the relevance of philosophy for both chemistry and history of chemistry.

Epistemological concerns pertain to philosophers as much as to chemists operating in their research lab or teaching in classrooms, with the benefit of highlighting the different aspects of a common issue. Jean-Pierre Llored relies on mereological arguments to analyze the material and conceptual aspects of elements, and to discuss the functional role played by elements in chemistry.

Klaus Ruthenberg comments on the historical and systematic aspects of the dictum "All chemical substances are preparations," in order to underline the irreducible material character of chemistry and to address the substantial character of chemical elements. Guillermo Restrepo relies on formal logic and the tools of mathematical chemistry to argue about the ontology of the chemical element, defined in a relational perspective. Sara Hijmans proposes to reformulate Paneth's identification of the two aspects of the elements, in order to reappraise the connection between chemical operations and chemical theory played by this notion. Chemical education plays a pivotal role in "building up" the epistemic community of chemists; Elena Ghibaudi, Alberto Regis, and Ezio Roletto point out the problems raised at the educational level by the dual definition of the elements and take a stance for an abstract conception of the element, understood as a category identified by the atomic number.

Finally, readers wishing to learn more about the ongoing debates in the philosophy of chemistry may refer to the list of selected works at the end of this volume.

Eric Scerri   Los Angeles
Elena Ghibaudi   Turin
January, 2020

1

The Many Questions Raised by the Dual Concept of "Element"

Eric R. Scerri, Department of Chemistry and Biochemistry, UCLA, USA

Introduction

The question of the conceptual nature of the term "element" represents a rather unique opportunity to examine the relationship that currently exists between chemists and philosophers of chemistry. This is one of the few topics that I know of in which the philosopher of chemistry can successfully challenge the professional chemist into considering the limitations of the still rather prevalent, but unspoken, positivist view that continues to dominate chemistry. However, any advances gained in the course of such a challenge very soon meet some serious obstacles that the philosopher of chemistry must seek to overcome.

Allow me to set the scene of a conversation between our intrepid philosopher of chemistry and his or her colleague from a department of chemistry. The philosopher of chemistry begins to inform the chemist about the dual nature of the concept of an element. On one hand, says the philosopher, an element is the familiar substance as defined by Lavoisier, namely the last stage of chemical decomposition, which has yet to be broken down any further by chemical means. This much the chemist is happy to concur with, even if expressing a little frustration at the philosopher announcing such a seemingly obvious truism.

Feeling a little more confident, the philosopher then takes on the task of explaining the other sense of the concept to her chemistry colleague. An element can also be regarded, says the philosopher, as an abstract bearer of properties that lacks chemical properties per se. At this point, not surprisingly perhaps, the chemist begins to feel a little uneasy about this metaphysical and paradoxical sounding notion. How can an element be the bearer of properties and yet not possess any properties?

The philosopher now needs to think quickly and reach for a few examples to try to justify her claim as to the dual nature of elements. The philosopher reminds the chemist that when we point to any particular element on the periodic table, let us say carbon for example, we do not mean any particular isotope, be it graphite or diamond or even C_{60}. The chemist readily concedes to the

Eric R. Scerri, *The Many Questions Raised by the Dual Concept of "Element"* In: *What Is A Chemical Element?*. Edited by: Eric R. Scerri and Elena Ghibaudi (2020). © Oxford University Press.
DOI: 10.1093/oso/9780190933784.003.0002

philosopher that one is referring to the abstract or more general notion of carbon that is not one of its allotropes. The allotropes of any particular element are the simple substance versions of the element, while the general concept of "carbon" refers to the element as what might be termed the more philosophical sense of element. At this point the chemist might begin to take more notice, contrary to his customary habit of thinking that philosophers of chemistry have nothing to contribute to scientific understanding.[1]

Encouraged even further by signs of progress, the philosopher now begins to point out another kind of example to drive home the distinction, while still using carbon as an example. When we point to the element carbon on the periodic table, the philosopher asks, are we referring to ^{12}C, ^{13}C, or ^{14}C? Once again the professional chemist is obliged to concede that the philosopher may have a point. What is being referred to is something of an average atom of the element carbon, not any particular isotope.[2]

At this point the chemist might readily agree to all that the philosopher of chemistry has carefully explained but may still conclude that these are trivial points that everybody is aware of, and that they are not worthy of making a fuss over. After all, the chemist says, regardless of whether we are speaking of allotropes or isotopes the failure of the periodic table to include such details does not necessitate that we think of elements in an abstract sense. In other words, if the periodic table does indeed fail in this respect, then so much the worse for the periodic table perhaps.

One more attempt on the part of the philosopher of chemistry

The philosopher might then try an alternative argument in order to convince her chemistry colleague of the need for an abstract conception of an element. She might ask the chemist to consider the reaction of sodium and chlorine to form sodium chloride,

$$2\ Na + Cl_2 \longrightarrow 2\ NaCl,$$

---

[1] One such chemist is the prominent textbook author and chemistry popularizer Peter Atkins who habitually claims that philosophy of science has nothing to offer to the study or practice of science.
[2] The atomic weight of most elements, as depicted on any periodic table, is literally weighted averages over the weights and abundances of the various isotopes that an element possesses. Exceptions to this state of affairs include mono-isotopic elements like fluorine, manganese, and gold.

in which a poisonous metal combines chemically with a poisonous gas to form a substance that is not only non-poisonous but, in fact, is essential for life. The philosopher might invite the chemist to say how or where the metal sodium and the gas chlorine are present in sodium chloride, since they are clearly not present in the form of simple substances. To the philosopher this is taken as a further motivation for appealing to elements as an abstract, but very useful, concept. Nevertheless, the chemist may not share this view and might merely shrug her shoulders and claim that the elements are now present as ions, while appealing to a microscopic explanation.

I believe that chemists are generally completely unaware of the distinction that is made in the dual sense of the concept of an element and that it has been regarded as very important by some chemists over many years. While a few of the chemists who have devoted attention to this issue have included Mendeleev, Urbain, and Paneth, their views are no doubt considered outmoded by contemporary chemists. In spite of all of this denial, many chemists do often concede to being troubled by the current Gold Book definition of the term "element," an issue to which I turn to next.

The Meaning of Element According to IUPAC's Gold Book

The Gold Book that is published by the International Union of Pure and Applied Chemistry (or IUPAC) contains a large amount of information concerning nomenclature and definitions especially pertaining to inorganic chemistry.[3] The entry for "element" reads as follows:

- a species of atoms; all atoms with the same number of protons in the atomic *nucleus*;
- a pure *chemical substance* composed of atoms with the same number of protons in the atomic *nucleus*. Sometimes this concept is called the elementary substance as distinct from the chemical element as defined above, but mostly the term chemical element is used for both concepts.

The first thing to notice is that the definition takes a dual form. Whether this duality corresponds to the more philosophical version of the duality that is the theme of the present book is another matter however. The first part of the definition requires little clarification and appears to be a straightforward appeal to microscopic components in order to identify any particular element with atomic

---

[3] IUPAC's Gold Book, s.v. "element." https://goldbook.iupac.org/html/C/C01022.html.

number, something that was introduced by Van den Broek and Moseley in the early years on the twentieth century.[4]

The second part of the definition if far more problematical however. The mention of a pure substance would seem to point in the direction of macroscopic samples of elements as simple substances, possessing what are normally considered to be properties such as color, melting point, and so forth. However, this apparent appeal to simple substances is immediately connected, rather uncomfortably, to the first definition and the number of protons in the nucleus of such elements. The final sentence creates even greater confusion by now drawing a distinction between what is simply termed "element" and perhaps intended to mean what philosophers refer to as the abstract element. Finally, we are further told that both concepts are generally referred to by the term "chemical element," which is of course begging the question, since this is the crux of the issue. Rather than acquiescing to the confusion caused by the dual usage of the term, one might have hoped for a definition that distinguishes between the two concepts as clearly as possible. The Gold Book definition appears to raise the issue at least tangentially but does not cast much light upon it.[5]

To summarize the Gold Book definition, in a way to connect it with the philosophical literature on "element," would perhaps be to say that for chemists the abstract sense of element is identified only microscopically, via its atomic number, whereas the element as a simple substance is identified with being a pure chemical substance. It is worth noting that according to the Gold Book, simple substances do not seem to deserve a definition that is independent of the microscopic one given in the first definition.[6]

An Appeal to Authority

One final attempt by our intrepid philosopher of chemistry could take the form of an appeal to authority, namely the work of Dmitri Mendeleev, the leading discoverer of the periodic table (Figure 1.1), who wrote a great deal about the nature of elements.[7]

---

[4] For Van den Broek, see Scerri 2016. For Moseley as well as Van den Broek see MacLeod et al. 2018.

[5] Interestingly, Mahootian disagrees on this point and believes that the Gold Book definitions accord well with the philosophical dualism under discussion (Mahootian 2013).

[6] A more charitable interpretation of the Gold Book definition is that it only identifies simple substances as substances and does not confuse the issue by also referring to the abstract element as a "substance." This point is taken up later in the present article in the context of the work of Earley and his response to Post's translation of the term Grundstoff.

[7] Whether contemporary chemists would be swayed by anything Mendeleev wrote is of course another matter.

In fact, the modern interest among philosophers of chemistry in the alternative, or abstract sense of "element," originates with the work of Mendeleev and an article on this subject by the philosopher-chemist Fritz Paneth writing in 1931. Paneth's article was translated into English by his son the physicist, turned historian and philosopher of science, and incidentally my own PhD advisor, Heinz Post (Paneth 1962).

Here are a few relevant quotations from Mendeleev on the dual sense of the concept of element:

> It is useful in this sense to make a clear distinction between the conception of an element as a separate homogeneous substance, and as a material but invisible part of a compound. Mercury oxide does not contain two simple bodies, a gas and a metal, but two elements, mercury and oxygen, which, when free, are a gas and a metal. Neither mercury as a metal nor oxygen as a gas is contained in mercury oxide; it only contains the substance of the elements, just as steam only contains the substance of ice, but not ice itself, or as corn contains the substance of the seed but not the seed itself. (Mendeleev 1891, 23)

The final parts of this quotation suggest that Mendeleev is recalling the Aristotelean distinction between actuality and potentiality.[8]

Elsewhere, Mendeleev argues firmly that the periodic table is primarily a classification of elements in this second abstract sense, not in the sense of simple bodies. Paneth emphasizes Mendeleev's point when writing:

> Thus, in terms of the distinction introduced here, we may refer only to a natural system [periodic system] of basic substances not of simple substances. (Paneth 1962, 152)

Mendeleev also argues that elements in the more abstract sense are characterized just by their atomic weights and that they have no "properties" in the usual sense of the term. In previous publications I have suggested that this is why Mendeleev could go beyond his competitors in making his predictions of eka-boron, eka-aluminum, eka-silicon, eka-manganese, which when found were named scandium, gallium, germanium, technetium (Scerri 2007).

---

[8] I am grateful to my co-editor, Elena Ghibaudi, for reminding me of this aspect of the discussion on element.

Why Did Only Mendeleev Make Significant Predictions?

Mendeleev's advantage over his competitors like Lothar Meyer lay in his philosophical approach to chemistry, for it allowed him to arrive at insights his less philosophically minded contemporaries could not have entertained. Mendeleev realized that abstract elements were to be regarded as more fundamental than simple substances. The explanation of why "elements" persist in their compounds was to be found in abstract elements and, as a consequence, if the periodic system was to be of fundamental importance, it would primarily have to classify the abstract elements. Stated otherwise, since abstraction is essential for generalization, Mendeleev was naturally led to placing greater importance on the abstract conception of elements rather than the more concrete one of elements as simple substances.

The predictions Mendeleev made were thus conceived of with the abstract elements in mind. If the available observational data on simple substances pointed in a certain direction, these features could be overlooked while believing that the properties of the more fundamental abstract elements might be different from what had been observed up to that point in the form of a particular "simple substance." Of course, any prediction would eventually need to be realized by the isolation of a corresponding simple substance, since "elements," in the more subtle sense of the term, are beyond observation, apart from the measurable property of atomic weight.

Because he was attempting to classify abstract elements, not simple substances, Mendeleev was not misled by any nonessential chemical properties. For example, the elements in the halogen group (fluorine, chlorine, bromine, and iodine) appear to be rather different from one another when one focuses on them as isolable simple substances, since they consist of two gases, a liquid, and a solid, respectively. The similarities among the members of the group are more noticeable when considering the compounds each one forms with sodium, for example, all of which are crystalline white powders. The point is that in these compounds, fluorine, chlorine, bromine, and iodine, are present not as simple substances but in a latent, or essential, form as basic substances.

Mendeleev's view of the elements allowed him to maintain the validity of the periodic law even in instances where observational evidence seemed to point against it. Such convictions may have resulted from the deeply held belief that the periodic law applied to the abstract elements as basic substances and that this law was as fundamental and equal in status to Newton's laws of mechanics (Scerri 2007). Had he been more of a positivist, Mendeleev might easily have lost sight of the importance of the periodic law and might have harbored doubts about his predictions and corrections involving known elements.

Reihen	Gruppe I. — R^2O	Gruppe II. — RO	Gruppe III. — R^2O^3	Gruppe IV. RH^4 RO^2	Gruppe V. RH^3 R^2O^5	Gruppe VI. RH^2 RO^3	Gruppe VII. RH R^2O^7	Gruppe VIII. — RO^4
1	$H=1$							
2	$Li=7$	$Be=9,4$	$B=11$	$C=12$	$N=14$	$O=16$	$F=19$	
3	$Na=23$	$Mg=24$	$Al=27,3$	$Si=28$	$P=31$	$S=32$	$Cl=35,5$	
4	$K=39$	$Ca=40$	$-=44$	$Ti=48$	$V=51$	$Cr=52$	$Mn=55$	$Fe=56,\ Co=59,\ Ni=59,\ Cu=63.$
5	$(Cu=63)$	$Zn=65$	$-=68$	$-=72$	$As=75$	$Se=78$	$Br=80$	
6	$Rb=85$	$Sr=87$	$?Yt=88$	$Zr=90$	$Nb=94$	$Mo=96$	$-=100$	$Ru=104,\ Rh=104,\ Pd=106,\ Ag=108.$
7	$(Ag=108)$	$Cd=112$	$In=113$	$Sn=118$	$Sb=122$	$Te=125$	$J=127$	
8	$Cs=133$	$Ba=137$	$?Di=138$	$?Ce=140$	—	—	—	— — — —
9	$(-)$	—	—	—	—	—	—	—
10	—	—	$?Er=178$	$?La=180$	$Ta=182$	$W=184$	—	$Os=195,\ Ir=197,\ Pt=198,\ Au=199.$
11	$(Au=199)$	$Hg=200$	$Tl=204$	$Pb=207$	$Bi=208$	—	—	— — — —
12	—	—	—	$Th=231$	—	$U=240$	—	— — — —

Figure 1.1. Mendeleev's periodic table of 1871 showing predicted elements with atomic weights of 44, 68, 72, and 100. These elements were all later discovered and called scandium, gallium, germanium, and technetium, respectively.
Source: *Zhurnal Russkeo Fiziko-Khimicheskogo Obshchestva* (1871), 3, 25–56.

Mendeleev expressed his general philosophical views on the nature of science when he wrote of the relationship between "matter, force, and spirit." He claimed that contemporary philosophical problems stemmed from a tendency to search for one unifying principle, while he favored three basic components of nature: matter (substance), force (energy), and spirit (soul). Everything was composed of these three components, and no one category could be reduced to any of the others. According to Michael Gordin, Mendeleev's use of "spirit" is an appeal to the modern notion of essentialism, or that which is irreducibly peculiar to the object in question. Gordin also claims that Mendeleev's position is metaphysical, and that it distances him from the "companionship of positivists" (Gordin 2004, 228) .

Mendeleev's Mistake

I think it fair to say that Mendeleev also sowed a certain amount of confusion regarding the question under discussion when he wrote,

> The central idea that aided me in undertaking the study of the periodic law of the elements, consists primarily in the absolute distinction between an atom [of, e.g., the element carbon] and a simple body [such as diamond or graphite]. (Mendeleev 1899, 193)

All that I am going to say [about the periodic law] must be understood as relating to atoms . . . and not simple bodies. (Mendeleev 1899, 193)

The problem that such passages raise has been noted by Paneth who wrote,

The reason why this distinction [two senses of "element"] has been so little noticed seems to be, on one hand, that the terms used by Mendeleeff are not very appropriate, and that, on the other hand, by coupling them to the pair of concepts, molecule and atom, he seems to have missed the essential point. (Paneth 1965, 57)

It is hardly possible in chemistry to introduce a contrast between elements and simple bodies, as the definition of element since Lavoisier is based on the simple body. It seems to me to be even less apt simply to equate the terms element/atom and simple body/molecule, respectively for apart from the fact that there are simple bodies whose molecules are single atoms, [e.g., Ar] molecules and atoms belong indubitably to one and the same group of scientific concepts, while the essential difference between element and simple body in the Mendeleeffian sense of the words, lies in their belonging to quite different spheres in epistemology. (Paneth 1965, 57)

The confusion is compounded further by the well-known fact that Mendeleev was very resistant to considering atoms realistically, or even in the usefulness of atomic theory in general, as evidenced in many passages from his writings:

There is a simplicity of representation in atoms, but there is no absolute necessity to have recourse to them. [Only] the conception of the individuality of the parts of matter exhibited in chemical elements is necessary and trustworthy. (Mendeleev 1891, 219)

Mendeleev repeatedly argues that the atom should be treated as a convention and not realistically as when he writes,

One may replace the term atomic weight by elementary weight in order to avoid recourse to the atom, which in any case is purely conventional. (Mendeleev 1968, 694)

The atomic hypothesis seems to me to be useless, if only because it does not lead to any general law and because it is not assured of a solid basis. (Mendeleev 1968, 733)

Paneth's View

In some previous articles I have attempted to summarize Paneth's dual view of elements and will only present a highly condensed view below (Scerri 2000; 2002; 2003; 2005; 2006; 2009a; 2012). In the 1920s, Paneth drew on the metaphysical essence of elements as basic substances in order to save the periodic system from a major crisis that it was facing. Over a short period of time many new isotopes of the elements had been discovered, which meant that the number of "atoms" or most fundamental units suddenly seemed to have multiplied. The question was whether the periodic system should continue to accommodate the traditionally regarded atoms of each element or whether it should be restructured to accommodate the newly discovered isotopes that might now be taken to constitute the true "atoms." Paneth responded that the periodic system should continue to accommodate the traditional chemical atoms and not the individual isotopes of the elements. One way to interpret Paneth might be that he regarded isotopes as simple substances that were characterized by their atomic weights, while basic substances were characterized by the more fundamental quantity of atomic number. This would be consistent with the view that simple substances are not primarily represented on the periodic table and similarly neither are individual isotopes of the elements:

A few key quotations from Paneth's article will suffice to recall his views,

I suggested that we should use the term "basic substance" whenever we want to designate that which is indestructible in compounds . . . and that we should speak of a "simple substance" when referring to the form in which such a basic substance, not combined with any other, is presented to our senses. (Paneth 1965, 65)

We cannot ascribe any particular qualities to an element as a basic substance, since it contributes to the production of an infinite variety of qualities which it exhibits both when alone and in combination with other basic substances. (Paneth 1965, 65)

With the concept of simple substance, we may remain within the realm of naive realism. When we are concerned with the basic substance, however, we cannot disregard its connection with the transcendental world without getting involved in contradictions. (Paneth 1965, 66)

Paneth's Mistake

However, I believe that Paneth too is guilty of sowing confusion on an already complicated debate when he writes,

> I have preferred to speak of basic substance and simple substance as different aspects of the chemical concept of element. (Paneth 1962, 155)

This statement could be interpreted to mean that "element" is the fundamental entity, while basic substance and simple substance are somehow derivative, which would surely be incorrect (figure 1.2).

I have previously proposed a three-way relationship for the purpose of clarifying this particular issue (Scerri 2012). The basic substance sense of "element" does not reside at the same epistemological level as simple substance. Instead it subsumes it, in the same way that it subsumes simple substance when it is present in combination with other simple substances. The scheme shown in figure 1.3 ensures that this is the case, but at the expense of now having three kinds of element to contend with!

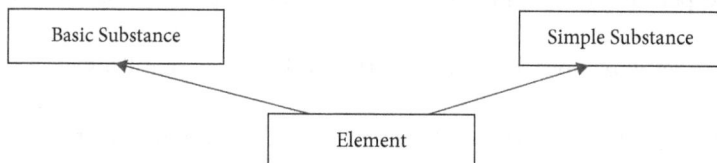

Figure 1.2. A representation of what appears to be Paneth's incorrect view on the relationship between basic substance, element, and simple substance.
Source: E. R. Scerri, Found Chem 14 (Springer Nature, 2012): 69.

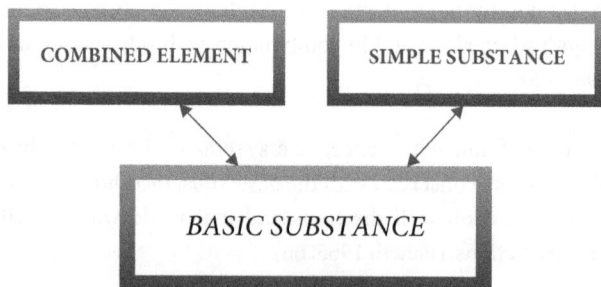

Figure 1.3. Proposed three-way relationship between combined element, basic substance, and simple substance.

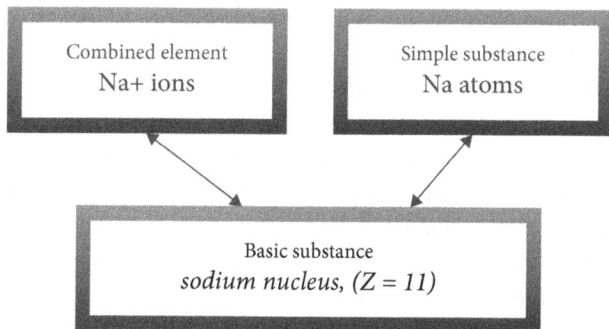

Figure 1.4. Proposed microscopic interpretation of the three-way definition of "element."

Microscopic Interpretation of the Dual or Even Three-way Sense of "Element"

Even if some contemporary chemists accept the philosophical duality concerning the concept of element, they tend to view the situation in a microscopic fashion instead of macroscopically. This section is intended to examine the possibility of a microscopic interpretation of the distinctions that have been discussed between the abstract sense of element, elements as simple substances, and in the sense mentioned above, namely combined elements.

Since the nucleus is the part of the atom that remains completely unchanged in the course of chemical reactions it would seem like the natural choice for the microscopic counterpart of the element as a basic substance. Similarly, the fact that the identity of elements resides in the nucleus rather than in the accompanying electrons would seem to further support the view that the nucleus should be identified with the element as a basic substance (figure 1.4).

As to the element as a simple substance, this could be identified with the entire neutral atom, meaning the nucleus with its attendant number of orbiting electrons.[9] This then leaves the notion of a combined element, as introduced in the previous section, which in its microscopic form must surely be an ion of any element in question, at least in the case of ionic bonding. Such a simple identification cannot be made in the case of covalent compounds, although perhaps

---

[9] This view represents a rather reductionist approach of course. In addition, the reader could object to the terminology since a true "substance" may not be reducible to just its fundamental particle components. Nevertheless, I have retained Post's terminology in discussing the main issue of there being at least two conceptions of the term element.

Richard Bader's notion of atoms in molecules could be co-opted to play such a role (Bader 1994).

On the other hand, such a simplistic set of identifications might run into serious problems if one were to attempt to make quantitative statements or even to carry out calculations on atoms of any particular element. The problem lies in the fact that a quantum mechanical approach to calculating the properties of elements and compounds is conducted through the time-independent Schrödinger equation, which in turn deals with the interaction between nuclei and electrons in any particular atomic or molecular case.

In this respect we would need to modify our earlier proposal so that elements as basic substances would be identified with the entire atom instead of just the nucleus. Moreover, restricting the characterization of atoms to just their number of protons and electrons severely underdetermines the chemical and physical behavior of atoms. Speaking more generally, it is not the mere number of electrons nor how they occupy various kinds of orbitals that causes these atoms to behave in a certain manner. After all, how could just a change in number lead to the observed subtle qualitative differences and in some cases quite abrupt changes that occur on moving from solid gold (element 79) to liquid mercury (element 80) for example?

The notion of electronic configuration per se does not therefore stand up to close scrutiny as anything but an oversimplified toy example of what causes chemical behavior, assuming that one wishes to speak of causation in the first instance.[10] A modern microscopic account of elements as simple substances must therefore do a good deal more than merely stating the number of elementary particles in the atoms of any particular element.

The cause of the properties of any particular atom lies in the complex dynamics that exist between a particular number of electrons and the protons in the nucleus in question. Consider, for example, the difference between an atom of hydrogen and one of helium. In the case of hydrogen there is just one contributing potential energy term, that of an attraction between the nucleus and the electron, namely—e^2/r where e represents the charge on the electron and r the distance between the nucleus and the orbiting electron.[11] In the case of the helium atom there are two electrons and as a result there are two terms to the one that are present in hydrogen.[12] But in addition there is an all-important electron-electron repulsion term given by $+ e^2/r_{12}$.

---

[10] A number of articles have discussed the role of causation in chemistry, including Stemwedel 2004.

[11] The electron has a charge of—e while the nucleus of hydrogen has a charge of + e. The product of the two charges gives—e^2, where the resulting negative charge denotes an attraction between the two particles.

[12] Another difference is the nuclear charge on the helium nucleus which is + 2e.

$$\text{Hydrogen}: \text{Potential Energy} = -e^2 / r$$

$$\text{Helium}: \text{Potential Energy} = -2e^2 / r_1 - 2e^2 / r_2 + e^2 / r_{12}$$

To consider the change from the hydrogen atom to that of helium as due to a change from one to two electrons, or from configuration $1s^1$ to $1s^2$, is to fail to recognize the electron-electron repulsion term. Chemistry is not just about a particular number of electrons but about the forces within any particular atom, which do not correlate precisely with the mere number of electrons. Expressed in different terms, a stipulation of the elementary particles that are present in an atom, or even its electronic configuration as a causal or explanatory device, would represent a gross oversimplification of the facts. A deeper explanation lies in the dynamics among the forces, which is captured in the time-independent Schrödinger equation of the atom in question.[13] This is why rigorous explanations in chemistry and physics require quantum mechanics.

This is also why, generally speaking, professional chemists might agree in saying that quantum mechanics provides chemical explanations while electronic configurations are a useful but crude generalization. Electronic configurations of atoms are more useful in chemical education than in "real chemistry." When a chemist wishes to predict the behavior of an element or a compound or the bond angles within a molecule, or indeed any number of properties, she reaches for her quantum mechanical computer program. Knowing the electronic configuration of the atom or molecule is of rather limited use, apart from making qualitative comparisons with other elements.

So if one really wishes to discuss the element as a basic, or even a simple substance, in microscopic terms, this requires a knowledge of the elementary particles present in an atom along with the use of the time-independent Schrödinger equation. This suggests that any attempt to identify the dual sense of the concept of element as was carried out in figure 1.5 is doomed to failure.

A Curious Dilemma in Chemistry

In the following section I would like to raise some issues that I do not believe have ever been discussed in the context of debates concerning the dual nature of an element. I am referring to some simple points that will no doubt have

---

[13] There are other approaches that do not appeal to Schrodinger's wave equation, such as the density functional approach to calculations.

1	2	3	4	5	6	7	8	9	10	11	12	13	14	15	16	17	18	19	20	21	22	23	24	25	26	27	28	29	30	31	32
H 1																															He 2
Li 3	Be 4																										B 5	C 6	N 7	O 8	F 9	Ne 10
Na 11	Mg 12																										Al 13	Si 14	P 15	S 16	Cl 17	Ar 18
K 19	Ca 20																Sc 21	Ti 22	V 23	Cr 24	Mn 25	Fe 26	Co 27	Ni 28	Cu 29	Zn 30	Ga 31	Ge 32	As 33	Se 34	Br 35	Kr 36
Rb 37	Sr 38																Y 39	Zr 40	Nb 41	Mo 42	Tc 43	Ru 44	Rh 45	Pd 46	Ag 47	Cd 48	In 49	Sn 50	Sb 51	Te 52	I 53	Xe 54
Cs 55	Ba 56	La 57	Ce 58	Pr 59	Nd 60	Pm 61	Sm 62	Eu 63	Gd 64	Tb 65	Dy 66	Ho 67	Er 68	Tm 69	Yb 70	*Lu 71*	Hf 72	Ta 73	W 74	Re 75	Os 76	Ir 77	Pt 78	Au 79	Hg 80	Tl 81	Pb 82	Bi 83	Po 84	At 85	Rn 86
Fr 87	Ra 88	Ac 89	Th 90	Pa 91	U 92	Np 93	Pu 94	Am 95	Cm 96	Bk 97	Cf 98	Es 99	Fm 100	Md 101	No 102	*Lr 103*	Rf 104	Db 105	Sg 106	Bh 107	Hs 108	Mt 109	Ds 110	Rg 111	Cn 112	Nh 113	Fl 114	Mc 115	Lv 116	Ts 117	Og 118

Figure 1.5. Thirty-two-column periodic table with lutetium and lawrencium in group 3. All elements are shown in order of increasing atomic number.

occurred to anybody who has taught chemistry at anything from high school to more advanced levels.

The fact that the identity of an element resides in its nucleus, or its atomic number, would seem to be a clear-cut reference to elements as basic substances. This situation is completely consistent with the idea that the nucleus does not change in the course of chemical reactions. In general philosophy, discussions of personal identity are tied to one or more features that do not change with time or whatever happens to the individual in question (Olson 2002). So too it would seem that discussion of the identity of elements must be connected with the unchanging nucleus. On the other hand, the manner in which an element reacts is closely associated with the electrons orbiting the nucleus of an atom. Electron arrangements are susceptible to change, without a corresponding change occurring to the identity of an atom. For example, the loss or gain of electrons results in the formation of positive or negative ions but without changing the identity of the atom in question.

Another duality that is encountered in the study or teaching of chemistry is that the amount of an element that reacts is associated with the atomic weight of an element, which in turn is almost entirely due to the nucleus.[14] I am referring to the topic of stoichiometric calculations in which students learn to calculate how a compound is obtained on reacting together two elements, for example. However, *how* an element reacts to form compounds, that is to say the qualitative rather than the quantitative question, is addressed by reference only to the electrons possessed by an atom or more specifically the electron configurations of atoms. To give just a couple of examples, elements with one outer-shell electron such as sodium and potassium, react by losing their outermost electrons, whereas elements whose atoms have seven outer electrons, such as chlorine or bromine, react by the gain of one electron. These points from the field of chemical education represent yet another form of duality concerning the behavior of elements that deserves to be examined further in the context of the general debate that this book addresses itself to.

---

[14] The virtual insignificance of electrons when carrying out stoichiometric calculations arises because of the very large mass discrepancy between protons and neutrons on one hand and electrons on the other. Whereas the mass of the electron is approximately 9.1×10^{-31} kg, that of the proton and neutron are approximately 1.67×10^{-27} kg or a ratio of 1835 to 1. Nevertheless, stoichiometric calculations do not neglect electron masses since these small quantities are included in the relative atomic weights that appear on periodic tables or tables of atomic weights.

A Direct Application of the Concept of the Abstract Element to Resolving the Group 3 Question

The question of precisely which elements should be placed in group 3 of the periodic table has been debated from time to time with apparently no resolution up to this point (Clark and White 2008; Jensen 1982, 2008; Scerri 2009b). This issue is of considerable importance for chemists, physicists, as well as students of the subject. The present author is currently the chair of a IUPAC working group that was convened in 2015 to consider the question of group 3 of the periodic table.[15]

There is a popular and mistaken belief that IUPAC supports the traditional periodic table with lanthanum and actinium in group 3. This view has been refuted by Jeffrey Leigh who made it clear that IUPAC has not traditionally taken a view as to the correctness of one or another periodic table and that there indeed is no such thing as an officially approved IUPAC periodic table (Leigh 2009).

Since the advent of quantum mechanics and the determination of the electronic configurations of atoms there has been a general belief that such an approach should settle any remaining questions having to do with the details of the periodic table. The early determination of the configurations of the elements ytterbium (70) and lutetium (71) seemed to indicate the following electronic configurations for their atoms.

$$Yb\left[Xe\right]\ 4f^{13}5d^{1}6s^{2}$$

$$Lu\left[Xe\right]\ 4f^{14}5d^{1}6s^{2}$$

As a consequence, it was believed that lutetium should mark the end of the lanthanide series. Moreover, the discovery of lutetium occurred at about the same time as the discovery of several other rare earth elements, which led to its generally being regarded as a rare earth element (Evans 1996).

In 1937 Meggers and Scribner (Meggers and Scribner 1937) published an article in which they reported that contrary to earlier observations, the configuration of ytterbium should be assigned as

$$Yb\left[Xe\right]\ 4f^{14}6s^{2}$$

---

[15] IUPAC working group for group 3of the periodic table, https://iupac.org/projects/project-details/?project_nr=2015-039-2-200.

If ytterbium possesses 14 f-electrons, rather than 13 as formerly believed, it can genuinely be thought of as the final rare earth element. Consequently, the next element, lutetium, can be regarded as a d-block element, thus placing it under scandium and yttrium in group 3. In the years following this discovery a few books and published periodic tables incorporated the newly assigned configuration of ytterbium but refrained from discussing the possible reassignment in the placement of lutetium.

The first published statement that these configurations provided grounds for regarding lutetium as a d-block rather than as an f-block element comes from the well-known book on quantum mechanics by Landau and Lifshitz (Landau and Lifshitz 1959, 245n):

> In books on chemistry, lutetium is also placed with the rare earth elements. This, however is incorrect, since the 4f shell is complete in lutetium.

This simple notion was then rediscovered separately by a number of authors, working in different sub-disciplines, although none of these proposals seemed to have any impact on the way in which the periodic table was presented (Hamilton and Jensen 1963; Matthias et al. 1967).

In 1982 a chemist, William Jensen, published a widely cited article in which he reviewed previous evidence and made perhaps the first concerted plea, to the chemical community, for periodic tables to be changed so that lutetium would replace lanthanum, and lawrencium would replace actinium in group 3 (Jensen 1982). It would seem that this article has not convinced many authors since the majority of periodic tables have remained unchanged and the debate has continued. Although Jensen's article represents a major step toward the reassignment of the elements lutetium and lawrencium to group 3, the proposal suffers from some limitations that have resulted in its not having the impact that it might have had (Scerri and Parsons 2018).

A Conclusive Argument in Favor of Sc, Y, Lu, Lr

What is required in order to settle the membership of group 3 seems to be a categorical argument. Such an argument immediately becomes available if one turns to consider the abstract concept of an element and in particular the feature that characterizes each element, namely atomic charge. Whatever interpretation is placed on this concept, there is general agreement that the one characteristic of an abstract element is its atomic number.

Here then is the proposed approach to resolving the group 3 question that relies mainly on using atomic number. It consists of two simple and uncontroversial requirements. The first is to present the periodic table in a thirty-two-column-long format rather than the more frequently displayed medium-long or eighteen-column format. The long form is a more correct representation of the periodic system, given that it incorporates the f-block into the main body of the periodic table. Meanwhile, the more familiar eighteen-column table features the f-block as a disconnected footnote, a format that has survived for pragmatic reasons. The thirty-two-column table is rather wide with the result that it becomes difficult to represent on wall charts or even on the printed page. However, it is generally accepted that this is merely a pragmatic issue.

The second and central requirement is that the elements be presented in such a fashion so that they show a smooth increase in atomic number as one progresses through the periodic table from left to right across each period. If these two recommendations are followed, it becomes quite clear that group 3 should contain lutetium and lawrencium rather than lanthanum and actinium. As can be seen in figure 1.5, this arrangement makes for a smooth and regular sequence in the atomic numbers of all the elements. On the other hand, if one insists on retaining lanthanum and actinium in group 3, the sequence of increasing atomic number becomes highly anomalous as highlighted in figure 1.6.

Open Questions Concerning the Two Senses of Element

According to the account that was given in the introduction to this article it might seem that philosophers of chemistry agree about the two senses of the concept of elements. In fact, nothing could be further from the truth. Perhaps the only point that current philosophers of chemistry can agree upon is the need for a dual concept, while almost every other aspect seems to raise profound disagreement. In the remainder of the present article I will not presume to resolve any of the current disagreements, but only to spell out the differences of opinion in the hope of fostering future discussions among ourselves. Of course not a word of such disagreement needs to reach the professional chemist since we would do well to "clean up our act" before being taken seriously by the chemical community. This is particularly important given that professional chemists may already be predisposed to the notion that no two philosophers can agree on anything whatsoever.

Here is a quick list of what I take the controversial issues to be, many of which are addressed in the present volume.

1. Are the two senses of element co-extensive or is one contained within the other?
2. To what extent did Lavoisier abandon the abstract sense of elements while promoting his more positive sense of elements as simple substances?

H 1																															He 2
Li 3	Be 4																									B 5	C 6	N 7	O 8	F 9	Ne 10
Na 11	Mg 12																									Al 13	Si 14	P 15	S 16	Cl 17	Ar 18
K 19	Ca 20															Sc 21	Ti 22	V 23	Cr 24	Mn 25	Fe 26	Co 27	Ni 28	Cu 29	Zn 30	Ga 31	Ge 32	As 33	Se 34	Br 35	Kr 36
Rb 37	Sr 38															Y 39	Zr 40	Nb 41	Mo 42	Tc 43	Ru 44	Rh 45	Pd 46	Ag 47	Cd 48	In 49	Sn 50	Sb 51	Te 52	I 53	Xe 54
Cs 55	Ba 56	Ce 58	Pr 59	Nd 60	Pm 61	Sm 62	Eu 63	Gd 64	Tb 65	Dy 66	Ho 67	Er 68	Tm 69	Yb 70	Lu 71	*La* 57	Hf 72	Ta 73	W 74	Re 75	Os 76	Ir 77	Pt 78	Au 79	Hg 80	Tl 81	Pb 82	Bi 83	Po 84	At 85	Rn 86
Fr 87	Ra 88	Th 90	Pa 91	U 92	Np 93	Pu 94	Am 95	Cm 96	Bk 97	Cf 98	Es 99	Fm 100	Md 101	No 102	Lr 103	*Ac* 89	Rf 104	Db 105	Sg 106	Bh 107	Hs 108	Mt 109	Ds 110	Rg 111	Cn 112	Nh 113	Fl 114	Mc 115	Lv 116	Ts 117	Og 118

Figure 1.6. Thirty-two-column periodic table with lanthanum and actinium in group 3. Atomic numbers no longer increase in a regular fashion, implying that lanthanum and actinium are misplaced.

3. What properties, if any, other than atomic number (or atomic weight in Mendeleev's time), do elements in the more general abstract sense possess?
4. Just how abstractly should Mendeleev's sense of element be regarded?
5. What influences caused Mendeleev to attach such great importance to the abstract sense of element?
6. To what use did Mendeleev put his sense of element in the course of discovering his version of the periodic table?
7. The terminological question of whether Paneth's Grundstoff, that has been translated as basic substance, is indeed a form of substance.

1. How Are the Two Senses of Element Related to Each Other?

In order to avoid the pitfalls mentioned in point 7, I have proposed using the terms "element-1" and "element-2," respectively, for Paneth's einfacher Stoff and Grundstoff, respectively (Scerri 2003). One way to consider these two senses would seem to be that they are co-extensive and that they belong to the same epistemological level. A moment's thought however reveals that this cannot be so. Perhaps the more concrete aspect of "element" represents a subset of the more general and deeper concept of element-2 (Grundstoff)? Conversely, it might also make sense to regard the matter from the opposite point of view. Perhaps element-2 can be thought of as subsisting within element-1 in the sense that the essence of an element might be contained within the more tangible and sensible aspect of element, or element-1, to remain within my proposed terminology. Alternatively, the conceptual nature of element-2 would seem to preclude any such inclusion or the converse view that it, element-2, includes element-1 within itself.

I do not propose to resolve such mereological issues here but would point to the work of Rom Harré and Jean-Pierre Llored, one of whom is represented in the current volume, and both of whom have written extensively on mereology in the context of the philosophy of chemistry (Harré and Llored 2013; Llored, this volume).

2. To What Extent Did Lavoisier Abandon the Abstract Sense of Elements While Promoting His More Positive Sense of Elements as Simple Substances?

A quick historical sketch of the development of the two concepts of element-hood can be described as follows. In the beginning the Greek philosophers thought of

elements as abstract bearers of properties though devoid of any tangible properties per se. As is well known, this view may have had considerable philosophical
influence for centuries to come but it was not conducive to modern science in
view of it having an overly metaphysical character.

Scientific chemistry is generally believed to have begun to take shape at
the hands of Antoine Lavoisier, one of whose major contributions was to promote the view of elements as the final stage of chemical decomposition. Such
elements could be isolated and each one possessed numerous properties that
could be experimentally observed. To a first approximation this move would
seem to represent a step in the direction of a positivist philosophy. Of course
Lavoisier did not abandon the abstract concept of element as the bearer of
properties or as a principle. As many authors point out, Lavoisier's famous list
of simple substances (element-1) contains a number of principles such as heat
and light.

The next decisive step in the evolution of the concept of an element
appears to have been taken by Mendeleev, or at least he is the author who
wrote on this theme in some detail. Moreover, Mendeleev begins his
Principles of Chemistry by carefully pointing out the need for a dual concept
of element. One comes away from these passages with the distinct impres-
sion that Mendeleev believed that the abstract concept is more fundamental.
Indeed, as noted earlier, he goes so far as to claim that his periodic classification of the elements is predicated upon the abstract sense of element rather
than elements as simple substances. Although a few more reflective chemists
such as Urbain and Soddy continued to write about the more abstract sense
of element, the increasingly positivistic attitude of twentieth-century science
has all but eradicated the abstract notion of element. One of the few accounts
that openly discusses the two senses of element, written by the philosophically inclined inorganic chemist Fritz Paneth in the 1930s, was only translated into English in 1962.

Many questions arise from this over-simplistic account that I have just
sketched out. For example, did Lavoisier originate the notion of an element
as a simple substance? How much if anything of this view is due to earlier
French chemists who influenced Lavoisier, a subject that has been explored by
Bernadette Bensaude, a contributor to this volume (Bensaude, this volume). We
also need to ask about the extent to which Lavoisier himself held a dual notion
of elements or whether he favored elements as simple substances but was simply
unable to shake off the remnants of the more abstract metaphysical view. On
this question we can look to Marina Banchetti and Geoffrey Blumenthal who
have conducted studies on Lavoisier and the chemical revolution and who we
are fortunate to also have represented in this volume (Blumenthal, this volume;
Banchetti, this volume).

3. What Properties, if Any, Other than Atomic Number (or Atomic Weight in Mendeleev's Time), Do Elements in the More General Abstract Sense Possess?

Mendeleev believed that an abstract element is also a material constituent, as he puts it. However, the precise meaning of this view remains to be elucidated.

> Mendelejeff emphasises very forcefully the difference "between the concept of simple body as a single homogeneous substance, and as the material constituent, not perceptible to the senses, of a composite body." (Paneth 1962)

Mendeleev also believed that such an element has one distinguishing attribute, namely its atomic weight, but no properties as such, a view that appears to be highly paradoxical. Why should the possession of an atomic weight not count as a property of the element in question?

Paneth's updated concept of abstract element takes account of Moseley's discovery that atomic number is a better ordering criterion for the elements that atomic weight is. According to Paneth, abstract elements are identified through their unique atomic number, but again this is not to be regarded as a property of the abstract element since one of their main attributes is that they lack the properties that are typical of elements as isolated simple substances. As in Mendeleev's account, this position would seem to raise questions that have yet to be settled.

4. How Abstract Is Abstract?

The notion of an abstract element raises the obvious question of the degree of abstraction that is intended. This issue is further complicated by some authors like Paneth who invoke the term "transcendental" when referring to elements. At one end of the spectrum it could mean completely abstract in the Kantian sense of a thing in itself, which must remain hidden from view (Ruthenberg 2009; 2010; Mahootian 2013). Stated otherwise this degree of abstraction could be taken to mean a metaphysical view, in the literal sense of lying beyond the physical, or even outside the realm of the physical world.

According to Klaus Ruthenberg, who has made a study of the Kantian influence on Paneth,

> In modern philosophical terms, simple substances are *observables*, basic substances are *nonobservables*. Here is seen a hint of the Kantian *phenomena* (the observables) and *noumena* (the non-observables). Paneth claimed: "to

understand the change of properties of substances we require transcendental hypotheses." Hence, he expanded his former merely phenomenological point of view and added the realm of transcendental ideas to his philosophy of chemistry. (Ruthenberg 2009)

Mendeleev is known to have been firmly opposed to any such a literal form of metaphysics. Indeed Mendeleev devoted considerable effort to criticizing the interest in spiritualism and seances that were gripping the attention of European society during Victorian times.[16] As for Paneth, he is known to have been sympathetic to Kantian philosophy, although it was a variety of Kantianism that was espoused by the German philosopher Eduard von Hartmann and termed by him as transcendental realism.

In the contemporary debate Robin Hendry has insisted on interpreting the notion of abstract elements in a physicalist manner. Hendry maintains that the abstract sense of element must be located in space and time and argues for a robust form of abstractness.[17]

5. What Caused Mendeleev to Attach Such Great Importance to the Abstract Sense of Element?

Mendeleev's deep commitment to the abstract sense of elements appears to come from nowhere at least according to the present author. How did Mendeleev arrive at such a view? What led him to reinterpret Lavoisier's placing of almost all the attention on simple substance. We know very little about why Mendeleev adopted the view that lies at the heart of many other articles in the present volume.

For greater clarification we would need to rely on historian scholars who are able to read the primary Russian literature. One thinks of Michael Gordin and Nathan Brooks, the latter of whom is represented in this volume. We are also reminded of the fact that Russian Mendeleev scholars have yet to join the debate that we have been conducting within the philosophy of chemistry community. One thinks of authors like Igor Dmitriev or Eugene Babaev, the latter of whom has attended some of the meetings of the International Society for the Philosophy of Chemistry (ISPC). We can hope that there may be a rapprochement between these communities as a result of the international conference being organized in St. Petersburg to celebrate the sesquicentennial of Mendeleev's article of 1869.

[16] I do not intend to imply an identity between spiritualism and metaphysics but merely bring up this point in wanting to open up the debate further.
[17] Robin Hendry has made this point at more than one meeting of the International Society for the Philosophy of Chemistry.

6. Do Basic Substances Have Properties or Not?

How can Mendeleev claim that the periodic table is primarily about elements as basic substances since they are supposed to have no properties, except increasing atomic weights? The recognition of periodicity or chemical similarity implies comparing properties of simple substances as well as those of the elements in combined form. In any case it seems to rest on observable properties. Elements as basic substances are supposed to have no properties. Could it be that Mendeleev did not in fact base his periodic system as firmly on the more abstract notion of property-less elements as he seems to claim he did? On the contrary, Mendeleev's recognition of chemical periodicity seems to depend on the simple substance interpretation of elements as well as the properties of elements in their combined form especially since Mendeleev frequently referred to trends in valency on moving across the periodic table.

This question is clearly related to just how abstractly Mendeleev's sense of element should be regarded and whether the distinction should be regarded macroscopically or microscopically.

7. The Terminological Issue

According to Paneth's influential article on the subject that was written in 1931, the two senses of element are referred to *einfacher Stoff* and *Grundstoff*. However, the most widely known version of this article was translated into English with the two key terms as "simple substance" and "basic substance." The chemist-philosopher Joseph Earley has objected to this translation (Earley 2005; 2009):

> The main thrust of Paneth's paper was that . . . the designation "element" properly belongs to what he calls Grundstoff—an ultimate constituent of a chemical substance—rather than to stable substances that cannot be decomposed (einfache Stoffe). . . . On this basis, a more appropriate (though less literal) translation of Paneth's einfacher Stoff into English would use the term "elementary substance" rather than "simple substance" or "element" for stable materials containing only one element. . . . The word "element" would then be available for exclusive use as an English translation of "Grundstoff." (Earley 2009, 75)

Some languages are better suited to making the distinction. For example, in the French language elements as simple substances are known as "substances simples" whereas elements as basic substances are known as "elements." In the English language both senses are given the name "element," which causes much confusion. Unfortunately, much of the recent debate about the dual concept of the term "element" has of course been conducted in the English language. Some work remains to be carried out to resolve, or agree on, a common terminology that would enable more fruitful discussion of "the element question" among contemporary philosophers of chemistry. Finally on the question of translation and terminology, readers should see the very interesting analysis by Boyce (Boyce, 2019).

Conclusion

We have reviewed the multifaceted discussions concerning a most fundamental question in chemistry, namely the nature of elements. We have found a shifting emphasis between an abstract notion (Greek philosophers) to a more concrete notion (Lavoisier) and back to an abstract notion (Mendeleev and Paneth). We have tried to reconcile the views of philosophers of chemistry with those of working chemists and chemical educators. In addition, we have applied the abstract concept of element, as characterized by atomic number, to resolving the question of which elements to place in group 3 of the periodic table.

When all is said and done, much remains to be clarified. Perhaps we are more confused than when we first undertook this question but hopefully we can better appreciate the depth in the field of chemistry that is still often thought to lack any serious philosophical questions.

References

Bader, R. F. W. 1994. *Atoms in Molecules: A Quantum Theory.* New York: Oxford University Press.

Boyce, C., Mendeleev's Elemental Ontology and Its Philosophical Renditions in German and English, *HYLE-International Journal for Philosophy of Chemistry*, 25: 49–70.

Clark, R. W., and G. D. White. 2008. "The Fly-Leaf Periodic Table." *Journal of Chemical Education* 85: 497.

Earley, J. E. 2005. "Why There Is No Salt in the Sea." *Foundations of Chemistry* 7:85–102. https://doi.org/10.1023/B:FOCH.0000042881.05418.15.

Earley, J. E. 2009. "How Chemistry Shifts Horizons: Element, Substance, and the Essential." *Foundations of Chemistry* 11:65–77.

Evans, C. H. 1996. *Episodes from the History of the Rare Earth Elements.* Berlin: Springer.

Gordin, M. 2004. *A Well-Ordered Thing.* New York: Basic Books.

Hamilton, D. C., and M. A. Jensen. 1963. "Mechanism for Superconductivity in Lanthanum and Uranium." *Physical Review Letters* 11:205–207.

Harré, R., and J.-P. Llored. 2013. "Molecules and Mereology." *Foundations of Chemistry* 15:127–144. https://doi.org/10.1007/s10698-013-9181-5.

Jensen, W. B. 1982. "The Positions of Lanthanum (Actinium) and Lutetium (Lawrencium) in the Periodic Table." *Journal of Chemical Education* 59:634–636.

Jensen, W. B. 2008. "Response to the Fly-Leaf Periodic Table." *Journal of Chemical Education* 85:1491.

Landau, L. D., and E. M. Lifshitz 1959. *Quantum Mechanics*. London: Pergamon.

Leigh, J. 2009. *Chemistry International* 31, no. 1 (January–February). http://www.iupac.org/publications/ci/2009/3101/1_leigh.html.

MacLeod, R., R. G. Egdell, and E. Bruton, eds. 2018. *For Science, King and Country: The Life and Legacy of Henry Moseley*. London: Unicorn.

Mahootian, F. 2013. "Paneth's Epistemology of Chemical Elements in Light of Kant's Opus Postumum." *Foundations of Chemistry* 15:171–184.

Matthias, B. T., W. H. Zacharisen, G. W. Webb, and J. J. Englehardt. "Melting-Point Anomalies." 1967. *Physical Review Letters* 18:781–783.

Meggers, W. F., B. F. Scribner. 1937. "Arc and Spark Spectra of Ytterbium." *J. Research National Bureau of Standards* J19:651.

Mendeleev, D. I. 1891. *Principles of Chemistry*, vol. 1. 1st English ed. Translated by G. Kamensky. London: Longmans, Green & Co.

Mendeleev, D. I. 1899. *Rev. Gen. Chim. Pur. Appl. 1*: 211. Translated in *Mendeleev on the Periodic Law, Selected Writings, 1869–1905*. Edited by W. B. Jensen, 33. Mineola, NY: Dover, 2002.

Mendeleev, D. I. 1968. *Izbrannye lektsii po khimii*. Izd-vo Vyshaya shkola. [A book of selected lectures in chemistry edited by Professor Makarenya.] Leningrad/Moscow.

Olson, E. T. 2016. "Personal Identity." In *Stanford Encyclopedia of Philosophy*. https://plato.stanford.edu/entries/identity-personal/.

Paneth, F. A. 1931. "Über die erkenntnistheoretische Stellung des chemischen Elementbegriffs." *Schriften der Königsberger Gelehrten Gesellschaft, Naturwissenschaftliche Klasse*, Heft 4. Halle: Max Niemeyer Verlag.

Paneth, F. A. 1962. "The Epistemological Status of the Chemical Concept of Element." *British Journal for the Philosophy of Science* 13:1–14, 144–160. Reprinted in *Foundations of Chemistry* 5 (2003): 113–145.

Paneth, F. A. 1965. In *Chemistry & Beyond: Collected Essays of F.A. Paneth*, edited by Dingle, H. and Martin, G.R. New York: Wiley.

Ruthenberg, K. 2009. "Paneth, Kant, and the Philosophy of Chemistry." *Foundations of Chemistry* 11:79–91.

Ruthenberg, K. 2010. "The Kantian Response in Paneth's Philosophy of Chemistry." *Kant Studien* 101: 465–479.

Scerri, E. R. 2000. "Naive Realism, Reduction and the 'Intermediate Position.'" In *Of Minds and Molecules*, ed. N. Bhushan and S. Rosenfeld. New York: Oxford University Press, 51–72.

Scerri, E. R. 2003. "Response to Vollmer's Review of Minds and Molecules." *Philosophy of Science* 70:391–398.

Scerri, E. R. 2005. "Some Aspects of the Metaphysics of Chemistry and the Nature of the Elements." *Hyle—International Journal for Philosophy of Chemistry* 11:127–145.

Scerri, E. R. 2006. "On the Continuity of Reference of the Elements, A Response to Hendry." *Studies in History and Philosophy of Science* 37:308–321.

Scerri, E. R. 2007. *The Periodic Table, Its Story and Its Significance*. New York: Oxford University Press.

Scerri, E. R. 2009a. "The Dual Sense of the Term "Element, Attempts to Derive the Madelung Rule and the Optimal Form of the Periodic Table, if any." *International Journal of Quantum Chemistry* 109: 959–971.

Scerri, E. R. 2009b. "Which Elements Belong to Group 3?" *Journal of Chemical Education* 86:1188.

Scerri, E. R. 2012. "What Is an Element? What Is the Periodic Table? And What Does Quantum Mechanics Contribute to the Question?" *Foundations of Chemistry* 14:69–81.

Scerri, E .R. 2016. *A Tale of Seven Scientists and a New Philosophy of Science*. New York: Oxford University Press.

Scerri, E. R., and W. Parsons. 2018. "What Elements Belong in Group 3 of the Periodic Table?" In *Mendeleev to Oganesson: A Multidisciplinary Perspective on the Periodic Table*, ed. E. R. Scerri and G. Restrepo, 140–151. New York: Oxford University Press.

Stemwedel, J. D. 2004. "Explanation, Unification, and What Chemistry Gets from Causation." *Philosophy of Science* 71, no. 5: 1060–1070.

2

From Simple Substance
to Chemical Element

*Bernadette Bensaude-Vincent, Université Paris 1
Panthéon-Sorbonne, France*

It is often assumed that a metaphysical notion of elements prevailed until Antoine Lavoisier introduced the modern notion of chemical element as simple substances on the basis of empirical tests of decomposition. The history of chemistry is simplistically divided up into a prescientific era ranging from Empedocles or Aristotle to Ernst-Georg Stahl and a scientific era, where experimental evidence together with quantitative methods enabled chemists to overthrow the metaphysical notion of elements as universal constituents of all material bodies in favor of the notion of elements as simple substances (Freund 1968; Siegfried 2002).

This standard positivistic account not only carries biased historical views, it also helps perpetuate a familiar confusion among the basic concepts of chemistry between simple substance and chemical element. I will argue that the notion of chemical element was clarified and stabilized by Mendeleev. In his effort to build up a classification of elements, he made clear distinctions between the basic concepts of chemistry that allowed him to successfully accommodate all the known elements and to allocate places for elements to be discovered.

The epistemological principle guiding the historical analysis of this paper is that concepts are much more than linguistic units but are indispensable for exchanging information in scientific practices. Many scientific concepts rather come into being as solutions to theoretical puzzles. They consequently need to be understood in their context of emergence in order to identify their connection with the issue at stake.

Using this contextual approach, the first section analyzes Lavoisier's notion as presented in the *Elements of Chemistry* (1789) in relation to his revolutionary ambition and points out the ambiguities introduced by this definition. The second section tries to assess the impacts of John Dalton's atomic hypothesis on the notion of chemical element: to what extent did it remove the ambiguities introduced by Lavoisier's notion? The third section considers the conceptual distinctions introduced by Mendeleev in the course of his attempt at classifying

Bernadette Bensaude-Vincent, *From Simple Substance to Chemical Element* In: *What Is A Chemical Element?*. Edited by: Eric Scerri and Elena Ghibaudi (2020). © Oxford University Press.
DOI: 10.1093/oso/9780190933784.003.0003

chemical elements. In analyzing the epistemic status of Mendeleev's notion of chemical element I argue that the concept of element is both a precondition and a product of the construction of the periodic system.

Advantages and Limitations of Lavoisier's "Elementary Chemistry"

The context that inspired Lavoisier's concept of element as simple substance is clearly indicated at the onset of his *Traité élémentaire de chimie* published in 1789. "When I began the following work my only object was to extend and explain more fully the memoir which I read at the public meeting of the Academy of Sciences in the month of April 1787, on the necessity of reforming and completing the Nomenclature of Chemistry" (Lavoisier 1799). The work on the construction of a new chemical language allegedly turned into the project of writing a textbook. Moved by the ambition to make chemistry accessible to beginners, Lavoisier set out to construct an *elementary* treatise of chemistry, in the dual meaning of the term: a treatise providing the most rudimentary notions of chemistry and concerned with the basic components or elements. These linguistic and didactic concerns drew Lavoisier's attention to the *Logic* of Etienne Bonnot de Condillac whom he credited with theorizing the strong link between language and science. He retained the supremacy of the analytic method referring to both the movement from compound to simple and from simple to compound. He also retained the key idea of composition as association or addition. Just as complex ideas form in the child's mind by association of simple sensations, a well-constructed language should proceed by association of simple words and science by the association of facts.

The *Method of Chemical Nomenclature* (1787), which forged the names of compounds on the basis of their composition by apposition of the names of the simple substances that combine to form them, encouraged Lavoisier to focus the whole of chemistry on the issue of composition. In his *Elements of Chemistry* he consequently attempted to frame the entire system of chemistry along this analytic logic. He took "analysis" in the dual meaning given by Condillac to this term (Albury 1972; Bensaude-Vincent 2010). The simple-to-compound axis provided an open-ended perspective of further and further decompositions: "Thus as chemistry advances toward perfection in dividing and subdividing, it is impossible to say where it is to end and those things we at present suppose simple may soon be found quite otherwise" (Lavoisier 1799, 247).

Condillac also provided a mine of resources to support Lavoisier's revolutionary ambition to build up a system of chemistry from scratch. In his *Traité* he aimed at making chemistry "elementary" by reconstructing chemical knowledge on a tabula rasa. He thus expected to alleviate the learning of chemistry, to make it accessible to beginners who would gradually proceed from the simple to the complex, according to Condillac's pedagogy, and quickly acquire a sound knowledge of chemistry. Consequently, Lavoisier discarded all reference to the chemical tradition that he viewed as an accumulation of errors and prejudices. As typical exemplars of such errors, the theories of elements would not even be mentioned:

It will no doubt be a matter of surprise that in a treatise upon the elements of chemistry, there should be no chapter on the constituent and elementary parts of matter; but I may here observe that the fondness to reducing all bodies in nature to three or four elements, proceeds from a prejudice which has descended to us from the Greek philosophers. The notion of four elements, which, by the variety of their proportions, compose all known substances in nature, is a mere hypothesis, assumed long before the first principles of experimental philosophy had any existence. In those days, without possessing facts, they framed systems (Lavoisier 1799, xxi).

Thus depreciating the ancient theories of elements Lavoisier ignored that the theory of four elements that prevailed in the mid-eighteenth-century was based on operational considerations rather than metaphysical foundations, in particular in Pierre-Joseph Macquer's *Elements of the Theory and Practice of Chemistry*:

In whatever way we attempt to go further [in the analysis of bodies], we are always stopped by substances in which we can produce no change, which are incapable of being resolved into others. . . .To these substances we may [. . .] give the title of Principles or Elements. . . .Of this kind the principal are Earth, Water, Air, and Fire. (Macquer 1758, vii.1, 1–2)

While chemists assumed that there were many kinds of earth they also acknowledged that their incapacity to decompose further the other three elements did not imply a radical impossibility to decompose them in the future.

Lavoisier introduced his famous definition of elements in a polemical tone as an antidote to the alleged metaphysical tradition although it was by large a rhetorical claim since he was not the first one to develop an operational notion of elements:

All that can be said upon the number and nature of elements is, in my opinion, confined to discussions entirely of metaphysical nature. The subject only furnishes us with indefinite problems, which may be solved in a thousand of different ways, not one of which in all probability, is consistent with nature.

I shall, therefore, only add to this subject that if, by the term *elements*, we mean to express those simple and indivisible atoms of which matter is composed, it is extremely probable we know nothing at all about them; but, if we apply the term *elements* or *principles of bodies*, to express our idea of the last point which analysis is capable of reaching, we must admit, as elements, all the substances into which we are able to reduce bodies by decomposition. Not that we are entitled to affirm that these substances which we consider as simple may not themselves be compounded of two or even a greater number of more simple principles; but since these principles cannot be separated, rather since we have not hitherto discovered the means of separating them, they act with regard to us as simple substances, and we ought never to suppose them compounded until experiment and observation has proved them to be so. (Lavoisier 1799, xxiii)

While Lavoisier fully endorsed the compositional definition of element initiated by previous generations of chemists, he radically discredited the alternative definition of elements as ultimate constituents of material bodies. Moreover, Lavoisier overlooked the notion of elements as instruments (when they are in a free state), largely spread by Gabriel-François Rouelle's courses that he, like many contemporary chemists, attended (Rappaport 1960; 1961). He equally discarded the notion of affinity as being too complex for beginners. He thus deprived chemical elements of the attractive force, which accounted for their elective combinatorial power. Affinity, a core notion in eighteenth-century chemistry (Kim 2003), allowed to predict the outcome of chemical operations and to formulate general rules of combination derived from the "*tables des rapports*" (affinity tables). In other words, in his effort to reorganize the system of chemistry exclusively along the lines of analytic logic, Lavoisier drastically reduced the number of attributes of the rich notion of elements forged by many generations of chemists. His ideal was to confirm the composition of chemical substances through analysis and synthesis, conducted with careful weighing of the inputs and outputs in chemical changes. In addition, he provided chemists with a simple criterion for discriminating simple bodies from compound ones by using analytic tools and a precision balance. And he convinced chemists that the number of elements could not be determined a priori.

However, this reorganization of chemistry was at the cost of its explanatory power.[1] Lavoisier's impoverished notion of element proved too weak a pillar to reorganize the entire system of chemistry. Unsurprisingly, in the

---

[1] From a different perspective, French philosopher François Dagognet (1969) pointed out the epistemic roadblocks of Lavoisier's analytic logic. Dagognet characterized his Condillacian ideal of bringing chemical bodies, names, and ideas into a fixed and transparent relation as a *verbal* mode of representation that was superseded by the *pictorial* mode of representation of Laurent's graphic formulae and Mendeleev's periodic table.

course of his textbook, Lavoisier surreptitiously reintroduced some of the attributes of the former notion of elements. For instance, in the list of thirty-three simple substances provided at the beginning of part 2, he titled the first category (which included light, caloric, oxygen, nitrogen, and hydrogen) "simple substances belonging to all the kingdoms of nature which may be considered as the chemical elements of bodies" (Lavoisier 1799, 245). Not only did Lavoisier implicitly reintroduce a distinction between simple substances and elements as universal constituents, he also explicitly endorsed the notion of principles as vehicles of properties that they transmit to the compounds they form when they are combined. Just as affinity, this function of principles was central in eighteenth-century chemistry, especially as the basis of the distinction between mixts and aggregates, often used to distinguish chemistry from physics. In aggregates the properties of the compounds are intermediate between the properties of the component parts, while in mixts new properties result from the combination of the ingredients. Lavoisier used the terms "elements" and "principles" interchangeably in the definition quoted above. And in his *Elements*, caloric is clearly described as the principle of heat conferring elasticity since it is responsible for the gaseous state of "elastic fluids." Oxygen, the principle of combustion, conferred acidity (as expressed in its name). Although Lavoisier defined the notion of element in a purely negative manner, he could not dispense with the positive functions assigned to chemical elements, even for teaching chemistry at the elementary level. In retaining only the negative criterion of undecomposition, he had deprived the concept of all the positive features that could account for the strong individual character of chemical substances. In particular, he could no longer account for the impossibility to isolate elements such as light and caloric that used to be attributed to their strong affinities. Here lies a first paradox in Lavoisier's definition of chemical elements. It provided a firm ground for building up an "elementary chemistry" but was far too weak to offer the explanatory power of the traditional polysemic notion of chemical elements that prevailed in mid-eighteenth-century chemistry.

The Impact of Chemical Atomism

Not long after the publication of Lavoisier's textbook and the adoption by most European chemists of the new language of chemistry based on a compositional paradigm, John Dalton's hypothesis of atoms seemed to complete Lavoisier's compositional chemistry. Clearly the atomic weight provided the positive individual attribute missing in Lavoisier's notion of element, and it was a quantified and measurable attribute. At least this is taken for granted by a number of historians. John R. Partington, for instance, emphasized the complementarity between Lavoisier's definition and the laws of proportions (Richter's law of fixed proportions and Dalton's law of multiple proportions).

> The concept of element seemed capable of a very simple statement: an element is a form of matter composed of identical atoms of a given mass. Apart from the circumstance that the atom was purely hypothetical and not capable of being an object of experiment, this definition seemed quite satisfactory to chemists in the nineteenth century. (Partington 1948, 116)

In fact, in the early nineteenth century few chemical textbooks bridged the two notions and gave the definition forged by Partington. Chemists were generally reluctant to introduce hypothetical notions among the basic concepts presented in the introductory chapters of textbooks. Most of them, in particular Jakob Berzelius, embraced Lavoisier's compositional paradigm and stressed the importance of stoichiometric laws but they were content with the notions of simple and compound substances.

Dalton's atomic hypothesis has opened up an avenue of experimental research. Chemists were busy determining the atomic weight value of every simple substance. For this well-defined purpose they used not only the laws of proportions, they also mobilized various measuring techniques (gravimetric, volumetric, calorimetric, goniometric), and theoretical resources ranging from the study of gases, the theory of heat, to crystallography. In the 1820s they had at their disposal an arsenal of laws. Dulong's and Petit's law of specific heats (1919) states that the heat capacity per weight multiplied by the presumed relative atomic weight of the element was close to a constant value for various elements. Gay-Lussac's law (1809) stated that there is a simple ratio between the volumes of gas entering in combination and also with the volume of the resulting compound. Based on Gay-Lussac's law, Avogadro's law (1811) states that under the same condition of volume and pressure, equal volumes of gas contain the same number of molecules. This law required a hypothetical conceptual distinction between atoms and molecules' (in modern terms). Avogadro conjectured that molecules could split to enter into combination. For instance water resulted from the combination of a half-molecule of oxygen and two half-molecules of hydrogen. This conjecture raised a lot of skepticism among chemists until the 1860s because the idea of constituent molecules made of two elementary molecules (or atoms) of the same nature was inconsistent with the prevailing theory of electrochemical combination (Brooke 1981; Fischer 1982). In this theoretical framework promoted by Berzelius, combinations resulted from the attraction between two opposite electric charges. It was consequently impossible that two atoms of the same nature could unite to form one molecule. As a result most chemists gave up

Avogadro's law and rather used other less conjectural resources to determine the atomic weight values of simple substances.

As this well-defined objective raised an intense research activity, it undoubtedly increased the amount of information about chemical elements. While most chemists used atomic weights in their research and teaching activities,[2] they tacitly assumed that each element was characterized by this positive and quantitative attribute, although they deeply disagreed on the meaning and significance of atomic weights. In any case, they did not feel any pressure to revise or complete Lavoisier's definition of chemical element.

Hence the striking contrast in the first half of the nineteenth century between the rapid advances of knowledge about chemical elements and the relative disinterest for refining the concept itself. In the golden age of positivism, conceptual work was not highly valued by scientists. It was, rather, disqualified as a source of endless metaphysical disputes, in contrast with the cumulative gain of knowledge provided by experimental work. Only dictionaries provided a ground for fine-grained conceptual analysis. For instance, in a monumental *Dictionnaire des sciences naturelles* edited by a consortium of famous French scientists between 1816 and 1830, chemist Michel-Eugène Chevreul proposed a thorough conceptual clarification:

> In a simple body, there are only atoms of the same nature. In a compound body, one reckons as many atoms of different natures as there are elements. . . . In the system of atoms, one accounts for the difference of nature between simple substances by the different density, figure, and arrangement of the atoms that constitute them; and the difference between compound bodies by 1) the proper nature of elementary atoms, 2) by the proportion in which they are combined, 3) by their arrangement or the way their respective faces are presented. (Cuvier et al. 1816–1830, 14:517–519, my translation)

This exceptional attempt at completing the notion of simple substance in light of the atomic hypothesis does not, however, take place in the entry "element," which stuck to the criterion of un-decomposition. It occurred in the course of the generic entry "body (chemistry)" [*corps (chimie)*]. It was authored by one of the few chemists who embraced the crystallographer's approach to atoms and consequently adopted Avogadro's hypothetical distinction between atoms and molecules (Mauskopf 1976; Rocke 1984).

----

[2]   A number of textbooks emphasized the gap between "empirical equivalent weights" and "hypothetical atomic weights" in order to prevent speculations about atoms. However the determination of equivalent-weight values involved as many hypotheses as the determination of atomic weight values (Rocke 1984; Kounelis 2000).

While most chemists relying on Lavoisier's definition of elements were reluctant to embark into speculations about the forms and figures of molecules,
they were inclined to speculate about the number of elements. They were even
encouraged in this direction by Lavoisier's injunction to endlessly test the simplicity of simple bodies on the one hand, and on the other hand by the success
of his analytic chemistry. His program of analysis has been galvanized by powerful analytic techniques such as the voltaic pile, which enabled chemists to isolate dozens of simple substances in the early decades of the nineteenth century,
followed by spectral analysis in the 1860s. The number of simple substances almost doubled from thirty-three in Lavoisier's *Elements of Chemistry* (1789) to
sixty-three in Mendeleev's first volume of *Principles of Chemistry* (1868). The
inflation of the population of simple substances deeply affected the teaching of
chemistry and created an urgent need for ordering the indefinite multiplicity of
elements.

Such tensions intrinsic in the analytic chemistry built on Lavoisier's and
Dalton's concepts are not alien to the tremendous success of William Prout's hypothesis that the multiple simple bodies emerged from one single, primary element: hydrogen. Formulated in the early nineteenth century this speculative
hypothesis got unexpected support from Dalton's choice of hydrogen as the
conventional unit in his system of atomic weights. It also fostered the precise
determination of the atomic weights of elements because they had to be whole
multiples of H = 1. When the hypothesis was threatened by unwelcome decimals
(such as Cl = 35.5) Prout ventured that all elements were whole multiples of an
unknown primordial element whose atomic weight was a fraction of that of hydrogen (Brock 1985). The hypothesis was thus rescued but became hardly fal-
sifiable. It nevertheless continued to stimulate not only precise determinations
of the atomic weights but also attempts at classifying elements on the basis of
their atomic weights. Underlying was the expectation to find markers in arithmetic regularities for tracing the genealogy of elements from the unknown hypothetical ancestor. Such attempts started as early as 1817 with Johann Wolfgang
Döbereiner's series of "triads"; they multiplied in the mid-century and continued
long after Mendeleev published his periodic system (Van Spronsen 1969).

The significant impetus that Prout's daring hypothesis gave to nineteenth-
century chemical research may be seen as the symptom of a second paradox
at the core of Lavoisier's system of chemistry. While the concept of element as
simple substance was central to reorganize chemistry along the simple/com-
pound axis, at the same time it generated doubts about the simplicity of chemical
elements. Simplicity was the unique criterion but it was questionable so that the
ontological status of element remained shaky, uncertain, and fragile. Being elements "for us," Lavoisier's elements had no objective mode of existence. Therefore,
instead of reinforcing the individuality of Lavoisier's chemical elements with

the positive attribute provided by Dalton's atomic weights, chemical atomism encouraged the quest for a reduction of the multiplicity of elements.

Mendeleev's Conceptual Network

A number of historians have emphasized the strong connection between Mendeleev's periodic system and his textbook *Principles of Chemistry* since it was in the course of writing this textbook that he discovered the periodic law (Leicester 1948; Graham 1983; Bensaude-Vincent 1986; Brooks 2000; Kaji 2002). It is clear that Mendeleev, like many authors of textbooks, was in need of a general scheme, a kind of system, to order the vast amount of knowledge accumulated about an ever-increasing number of chemical substances. It remains, however, to be shown that the pedagogical constraints also channeled his attention toward the basic concepts of chemistry and what the real impact of conceptual clarifications was on Mendeleev's periodic system. For, while Mendeleev was obviously not the first chemistry professor who tried to find an order among the multiplicity of chemical substances, he was the only one who cared about providing a clear definition of the objects to be classified.

Masanori Kaji found evidence of an early attempt at conceptual distinctions in the lecture notes of Mendeleev's course at St. Petersburg in fall 1867.

> It is necessary to distinguish the concept of a simple body from that of an element. A simple body substance, as we already know, is a substance, which taken individually, cannot be altered chemically by any means produced up until now or be formed through the transformation of any other kinds of bodies.
>
> An element, on the other hand, is an abstract concept: it is the material that is contained in a simple body and that can, without any change in weight, be converted into all the bodies that can be obtained from this simple body. (quoted in Kaji 2002, 6)

This early definition presents three noticeable features. First, Mendeleev is not content with repeating the conventional definition of simple substance as un-decomposed ones. Instead he used the phrase "altered chemically," which implies that they have an identity or individuality (something that Lavoisier omitted). He also completed the definition with an additional criterion— "cannot be formed through the transformation of other kinds of bodies"— which is more questionable since as residues of analytic operations, simple substances can be formed out of other bodies. Still this first attempt clearly demonstrates a concern with refining the definition of the basic concepts of chemistry.

Second, Mendeleev already stressed the contrast between the concrete and the abstract modes of existence of simple substances and elements, respectively. Elements are nowhere to be seen, they have no tangible reality; they are accessible only by the mind. Mendeleev insisted that for being pure abstraction, elements are nevertheless material entities. At first glance this definition looks like an oxymoron since abstract entities, being creatures of the intellect, are usually deprived of material properties. Had Mendeleev written "real and abstract," it could have been concluded that he adopted a realistic position in the medieval debate between realism and nominalism. Abstract concepts forged by the mind are more than just words or linguistic entities, they refer to a conceptual reality. Or in the more recent philosophical debate about "natural kinds" one could say that elements exist as "chemical kinds," or specific entities embracing a collection of material individuals (Harré 2005; Hendry 2005; Sharlow 2006). But Mendeleev used the phrase "material and abstract" instead of "real and abstract." Whatever the floppiness of the concepts he and his translators used, it is hard to believe that Mendeleev mistook material for real, especially because he repeated this conjunction of material and abstract in a number of publications. Moreover, Mendeleev's elements are not amenable to "chemical species" because he insisted that they are individuals. We will see that this conceptual puzzle can be clarified if we take into account the support that the periodic system brings into Mendeleev's conceptualization. However, it is clear from the start that for Mendeleev, elements, despite their abstract status, are not metaphysical entities. They belong to the solid ground of positive science. They are accessible through experimental and quantitative procedures using gravimetric. Through chemical changes, they circulate from simple bodies to compounds or from compounds to compounds. Their materiality is assessed on the basis of the conservation of matter in chemical changes. While Mendeleev took over Lavoisier's favorite weighing techniques to characterize elements, at the same time he reconnected with the mid-eighteenth-century notion of element as hidden entities that cannot be isolated.

Third, this early distinction between simple substance and element includes no mention of atoms. The striking absence of any reference to the atomic theory can be linked with Mendeleev's previous works of "indefinite compounds." As Kaji (2002) noticed, Mendeleev's early research on non-stoichiometric compounds (such as alloys and solutions) led him to point out the limitations of atomic theory and concluded that Mendeleev was extremely reserved on atoms. However, the lack of reference to atoms in this early definition may be simply due to Mendeleev's pedagogical effort to define the most elementary notions prior to introducing the atomic hypothesis.

In fact, Mendeleev later improved this early distinction with a clear reference to the atomic theory. Remarkably he formulated this conceptual distinction at the onset of the 1871 article published in *Annalen der Chemie und Pharmacie*, thus suggesting that it was a preliminary step, or even a *passage obligé*, on the way to the periodic law:

> Just as the words "molecule," "atom," and "equivalent" were used interchangeably, even as recently as the time of Laurent and Gerhardt, so now the terms "simple substance" and "element" are often confounded with one another. However these terms must be sharply distinguished in order to avoid confusion in chemical philosophy. A simple body is something material, a metal or a metalloid, endowed with physical and chemical properties. The idea of a simple substance corresponds to that of a molecule made of one . . . or more . . . atoms. A simple body is able to display itself in the form of isomeric and polymeric modifications and it is only distinguished from a compound body by the homogeneity of its material parts. But in opposition to this, the term "element" designates those material particles that form simple bodies and compounds and determine the manner in which they behave in terms of their physical or chemical properties. The word "element" calls to mind the idea of an atom. Hence carbon is an element, but coal, diamond and graphite are simple bodies. (Mendeleev 1871, 42)

Again three major points can be outlined in this well-known quotation. The first one concerns the relation between atoms and elements. Mendeleev suggests that the distinction between atoms and molecules came prior to—and maybe inspired—the distinction between elements and simple bodies. This presentation is in line with his personal experience at the Karlsruhe Congress in 1860 where Stanislao Cannizzaro's pamphlet convinced many participants to adopt the system of atomic weights based on Avogadro's distinction between atoms and molecule reformulated by Charles Gerhardt (Cannizzaro 1911; Rocke 1984). Mendeleev fully endorsed the atomic theory, although he was aware of its limitations. Not that he believed in the real existence of atoms,[3] but he assumed that the distinction between atoms and molecules was the key to understanding chemical combinations. For him, as for most chemists of his time, atoms were not structural units of matter, but units of chemical combination (Nye 1984). Mendeleev characterized them by their capacities to bind with others, that is, by their atomicity. He shaped his view of atoms as *binders* or relational entities through his first textbook of organic chemistry published in 1861. He was

---

[3] Remarkably, the champions of the atomic hypothesis (Gerhardt and August von Kekulé in particular) adopted essentially positivist epistemological positions, and rejected the ontological interpretation of their molecular formulae. For Mendeleev the real existence of atoms remained an open question.

essentially concerned with issues of valences, degrees of affinity, and saturation. Later in 1868 he used Gerhardt's type theory as a guide to structure the chapters of *Principles of Chemistry* with a first chapter on monovalent hydrogen, then a chapter on divalent oxygen, then one on trivalent nitrogen, followed by a chapter on tetravalent carbon. Mendeleev considered them as "typical elements," and used them as exemplars, or models, for ordering the others.

A second noticeable aspect in the above quotation is that Mendeleev was not content with emphasizing the contrast between abstract elements and concrete simple substances; he redefined the notion of simple body. Instead of the usual criterion of un-decomposition he mentioned the capacity "to show itself under isomeric and polymeric modifications and it is only distinguished from a compound body by the homogeneity of its material parts." Curiously he revived the criterion of homogeneity used in the Aristotelian tradition to distinguish "real mixts" from "apparent mixts." In thus transferring to simple bodies a characteristic of compounds, he suggested that the distinction between simple and compound that constituted the main pillar of Lavoisier's system was secondary. It was pushed to the back of the stage for the benefit of the dichotomy between simple/compound bodies and elements. Simple and compound bodies have no explanatory power, only the elements can account for their properties and behaviors.

> Everybody understands that in all changes in the properties of simple substances, something remains unchanged, and that in the transformations of elements into compounds, this material something determines the characteristics common to the compounds formed by a given element. (Mendeleev 1869, 33)

Simple bodies were treated as *explananda* and the notion of element as the *explanans*. Since simple bodies disappear in chemical combinations they cannot be responsible for the conservation of matter, a function assigned to elements. Mendeleev substituted Lavoisier's analytic program for an attempt to account for the empirical data gathered on concrete simple and compound bodies by reference to the abstract notion of element. This conceptual shift marks the end of the compositional paradigm based on the central operation of analysis (decomposition and recomposition) and the advent of a new paradigm, a true *elemental chemistry* distinct from Lavoisier's *elementary chemistry*. In response to Lavoisier's view that "chemistry advances towards perfection in dividing and subdividing" Mendeleev assumed that

> the principal goal of modern chemistry is to extend our knowledge of the relations between the composition, the reactions and the qualities of simple and compound bodies on the one hand, and on the other hand, the intrinsic

qualities of elements which are contained in them; so as to be able to deduce from the known character of an element all the properties of its compounds." (Mendeleev 1871, 42)

But how could the abstract notion of element gain such an explanatory power? The abstract entity itself derived its power from the network of basic concepts woven by Mendeleev through his parallel between the two dichotomies atom/molecule and element/simple body. This analogy is the third and most important advantage of Mendeleev's reconceptualization of elements. On the one hand, it provided a precise meaning to the notion of "atomic weight." Clearly the atomic weight is not of the property of simple substances.

The magnitude of the atomic weight, according to the actual, essential nature of the concept, which does not refer to the momentary state of simple body but rather belongs to a material portion of it—a portion that it has in common with all its compounds. The atomic weight does not belong to coal or to the diamond but rather to carbon (Mendeleev 1869, 39).[4]

Neither does it belong to atoms themselves (they are beyond our weighing capacities). It belongs to the element. It is its most constant attribute, the marker of its identity. Mendeleev was confident in this identity card of elements to the point that he assumed it would remain unchanged whatever the future advances in science:

The concept of atomic weight (considered as the smallest part of an element contained in a molecule of its compounds) will remain without change, whatever modifications the theoretical ideas of chemistry may undergo. It is true that the expression atomic weight implies the hypothesis of the atomic structure of matter, but, then we are not discussing terminology here, but rather a conventional concept. (Mendeleev 1871, 44)

On the other hand, the analogy between the two couples of concepts helped delineate the turf of chemistry. Chemists, in Mendeleev's view, are not directly concerned with the properties of atoms and molecules. They struggle with a multiplicity of material entities at the macroscale, and to master them they need to rely on an abstract entity. Whatever the advances in atomic science, elements will remain the core notion of chemistry.

---

[4] There is a striking contrast with an earlier and rather confusing definition given in Mendeleev's 1861 textbook on organic chemistry: "the minimum quantity of an element in the compound molecules of the element" (quoted in Kaji 2002, 5).

Thus the conceptual clarifications introducing the papers on the periodic law are another major contribution of Mendeleev to the discipline of chemistry.[5] Indeed it is not the brilliant flash of the discovery of the periodic law. Nothing like a eureka! It proceeds from Mendeleev's years of modest and painstaking reflections on the meaning of the terms that he used in his daily teaching and research practices. This reflexive attitude involves a number of philosophical engagements (Bensaude-Vincent 1986). In placing all the burden of explanatory power on the abstract notion of elements, Mendeleev made a philosophical decision. This choice in turn determined Mendeleev's emphasis on completeness and his ontological choice of the individuality of elements. Such crucial philosophical decisions exemplify the feedback of the periodic system upon the initial definition of element formulated by Mendeleev in the course of its construction.

Mendeleev's Philosophical Commitments

Unlike many of his contemporaries Mendeleev did not consider the periodic law as a pragmatic rule for ordering elements. He firmly claimed that it was a natural law. In the 1871 paper, he gave a lengthy description of all the consequences that he derived from it. He used it to disclose unknown properties of elements, to predict unknown elements, to correct the values of atomic weights, to complete knowledge of the forms of compounds. He was convinced that the periodic law could account for all observations related to all known elements, explain unexplained phenomena, and lead to the discovery of new ones. In addition to its explanatory power it has a heuristic power: it increased our knowledge, and also shed new light on all theoretical issues at stake in chemistry.

So striking was Mendeleev's confidence in the power of the periodic law that many commentators did not hesitate to revere him as a genius or a prophet (Dagognet 1969). Without minimizing the merit of Mendeleev's predictions and corrections, it is important to emphasize the epistemic choice that guided him because it is conditioned by his work on the basic concepts of chemistry. Although Mendeleev repeatedly expressed rough positivistic claims that science had to be rooted in "solid facts," observations, and experiments, he did not hesitate to go against empirical evidence, when he corrected the atomic-weight values

[5] The paper on the discovery of the periodic law in a French journal in 1899 systematically used the terms "atom" instead of "element." "The central idea that aided me in undertaking the stud of the periodic law of the elements consists precisely in the absolute distinction between an atom and the simple body" (Mendeleev 1899, 4). This conceptual mistake is so surprising that one can suspect an intervention of the journal editor, Dr. Quesneville, or the translator of Mendeleev's Russian text. In the context of longstanding controversy over the atomic theory in France, it is not impossible that Mendeleev has been instrumentalized by the atomists against Marcellin Berthelot and his followers.

of indium and uranium for instance, or when he left vacant spaces for elements still to be discovered. Among the spectrum of epistemic values embedded in scientific pursuit (Longino 2008), Mendeleev's top priority was neither empirical evidence nor accuracy. Even more than empirical evidence, he prioritized completeness. This epistemic value embraces two specific demands: first to account for all phenomena where chemical elements are involved and, second, to account for all observed aspects of those phenomena. Mendeleev was guided by a strong urge for eradicating all exceptions or anomalies that could affect the regularity expressed by the periodic law. Although he was well aware that chemical periodicity is approximate since the elements do not recur exactly after certain intervals, Mendeleev refused to present the periodic law as an approximate law. A natural law cannot tolerate a fringe of exceptions.

Mendeleev was able to put such strong epistemic demands on the periodic law because it was based on a robust conceptual network. In particular, the shift from simple substances to elements proved crucial. Had he relied on Lavoisier's definition, Mendeleev could never have predicted the existence of unknown elements. Simple substances being just concrete residues of analytic operations are literally unpredictable. They only come into existence when they are empirically isolated. They have a factual existence, whereas the existence of abstract elements can be known in advance and their properties can be anticipated, induced from those of neighbor elements.

In addition, the precision of Mendeleev's predictions of the properties of eka-boron, eka-silicon, and eka-aluminium is exemplar of the feedback of the periodic system on the individuality of elements. In fact, the success of the predictions based on the periodic law reinforced his ontological conviction of the irreducible pluralism of elements. For him the individuality of chemical elements became as fundamental an aspect of nature as Newton's gravitation:

> Kant said that there are in the world "two things, which never cease to call for the admiration and reverence of man: the moral law within ourselves and the stellar sky above us." But when we turn our thoughts about the nature of the elements and the periodic law, we must add a third subject, namely: "the nature of the elementary individuals that we discover around us." Without them the stellar sky itself is inconceivable; and in the atoms we see at once their peculiar individualities, the infinite multiplicity of individuals, and the submission of their seeming freedom to the general harmony of nature. (Mendeleev 1879, 159)

Following the discoveries of gallium, scandium, and germanium, which confirmed his prediction, Mendeleev increasingly tended to think of new elements as gaps in the system, as though the periodic law gave them a right to come into existence. Conversely Mendeleev denied this right to candidate elements that had no vacant place waiting for them in the system. Mendeleev was not the only one to closely associate the "right to be an element" with vacant spaces in the periodic system. When Lord Rayleigh and William Ramsay announced the discovery of argon in the summer of 1894, a number of chemists including Raffaello Nasini, James Dewar, and Marcellin Berthelot objected that argon with an atomic weight close to 39.9 could not be an element because there is no room for it in the periodic system. Mendeleev (1895), in turn, ventured that the so-called element might be N_3, an allotropic form of nitrogen with atomic weight about 42. But soon after the discovery of helium Ramsay realized that argon and helium could fit nicely in the system thanks to the addition of a new group of inert elements.

Mendeleev's conviction that elements were true individuals whose existence depended on the regularity of the periodic law grew stronger and stronger as he tested the heuristic power of his system. He consequently expressed strong doubts about the existence of electrons. Mendeleev's belief in the individuality of elements grew stronger and stronger as he tested the heuristic power of the periodic law. He questioned the discovery of radioactive elements after a visit to Paris in 1902 where he discussed the topic with Henri Becquerel, Pierre Curie, and Marie Curie. Not only he had left no vacant space for electrons or radium but his deep faith in the individuality of elements was threatened by J. J. Thomson's electron model of atoms and the interpretation of radioactivity as resulting from atomic transmutation. He consequently questioned the nature of radioactivity through a brave and desperate hypothesis in *An Attempt at a Chemical Conception of Ether* (Mendeleev 1904). He conjectured that ether was a chemical so light that it was imponderable as an element so he assigned to it a place in the periodic system above the lightest known element: hydrogen. However, in order to more or less keep the regularity of the increase of atomic weights in the periodic system he added another element, between ether and hydrogen: coronium. He estimated that this hypothetical element identified in the spectral analysis of the sun by astrochemists would have an atomic weight around 0.4. Mendeleev thus expected to save the integrity of elements, to extend the periodic system to more unknown elements at the frontier of materiality, and to reconcile chemistry with electromagnetism and mechanics (Bensaude-Vincent 1982; Kragh 1989; Gordin 2004, 217–227).

It is clear, and already well-known, that Mendeleev never shared the enthusiasm of his colleagues for Prout's hypothesis, which enjoyed a considerable success throughout the nineteenth century (Brock 1985). He never sought to reduce the plurality of elements to one single primary matter. To govern the crowd of chemical elements he searched unity in a general law rather than in matter itself. In the decades following the publication of the periodic system he became

extremely critical of the hypothetical primary matter and of the hypothesis of
the composition of elements based on a parallel between the periodicity of organic radicals and of elements (Mendeleev 1879, 158–164). Mendeleev strongly
protested against the uses of the periodic system by some colleagues as a decisive argument in favor of the quest for a unique element or a reduction of their
number. His metaphysical conviction that elements were true individuals had
been reinforced as new elements were empirically discovered that either confirmed or jeopardized the periodic system.

Whatever the future developments on the nature and structure of atoms,
chemical elements would remain individuals, as long as their individuality was
guaranteed by their atomic weight. Therefore, Mendeleev insisted that the periodic law did not express a continuous mathematical function. He rejected the
idea that intermediate elements could be inserted between two neighbor elements such as silver and cadmium for instance: "According to the very essence of
the periodic law there can be none" (Mendeleev 1879, 157). The periodic system
conferred a new kind of individuality on the concept of element. An element is
not just an individual in the sense of a singularity identifiable by its unique behavior and properties. It is also *individualized* by its position in the network of the
periodic system. It is defined by its own properties as much as by its place in the
matrix formed by the groups and periods of the periodic table.

> As a general rule when we study the properties of elements, bearing in mind
> practical conclusions, it is necessary to give equal attention to both the general
> properties of the group to which a given element belongs and to its individual
> properties. (Mendeleev 1871, 43)

Mendeleev's view of individuality as a place assigned in a network has
been confirmed by the adoption of the term "isotope" (same location). This
term coined by Frederick Soddy in 1913 saved both the periodic system and
Mendeleev's view of elements as chemical individuals from the critical situation
due to the production of radioelements. These elements challenged Mendeleev's
notion since they had different atomic weights and threatened the periodic
system since there were no blank spaces for incorporating them. As long as the
new substances had different atomic weights and different properties they deserved to be treated as individual elements and the periodic system should be
subverted, as Kazimierz Fajans argued. By contrast, Friedrich Paneth and his colleague George de Hevesy argued that despite the variation in their atomic weights
these elements shared the same chemical identity. Mendeleev's choice to place
the core of chemical identity in elements rather than in atoms was renewed at
the cost of a slight modification in the enunciation of the periodic law expressing
a dependency on atomic numbers rather than on atomic weights. Hence the

official definition of a chemical element adopted in 1923 by the International Union of Pure and Applied Chemistry as a species of atoms having the same number of protons in their atomic nuclei. The number of neutrons in the atomic nucleus does not affect the chemical identity of the element, which is only related to its position among other elements in the periodic system.

Finally, the role of the periodic system in the process of individuation of chemical elements helps clarify the chemical puzzle of their abstract-material mode of existence. Although they have no empirical existence, chemical elements are true materials in the sense that they are identifiable by their location in the periodic system. Let's use an anachronism to venture a twenty-first-century interpretation of the abstract/material mode of existence of elements. As the periodic system rules the material world and delineates the scope of potential exchanges among materials, let's assume that it constitutes the "chemosphere" (an equivalent of the biosphere). As members of the chemosphere, elements can be substituted in material changes for other members of the system. For instance, H can be substituted for Cl. They are equivalents in material interactions. This equivalence is the signature of their abstract mode of existence but it is limited to the members of the material chemosphere. It means that a chemical element such as carbon, for instance, cannot be substituted for an amount of money, as it is the case in the system of financial compensation known as carbon trading. A chemical element is certainly abstract but it cannot be exchanged for something that does not belong to the chemosphere defined by the periodic system. It should not be treated as a general equivalent precisely because the periodic system circumscribes the material world.

In conclusion, the notion of element as simple substance prevailed with the reform of the chemical language and the triumph of a compositional paradigm in chemistry. But given its concrete and empirical mode of existence Lavoisier's element has a limited explanatory potential and no predictive power. By contrast, Mendeleev's notion of chemical element as an abstract and material entity proved extremely powerful for explaining and ordering chemical phenomena, as well as predicting unknown elements.

The case of Mendeleev exemplifies the value of conceptual work in experimental sciences such as chemistry. As a professor of chemistry, Mendeleev certainly valued clear definitions as didactic tools. At the same time, he demonstrated their heuristic power in the process of construction of his periodic system. While in contemporary practices of chemistry most concepts are defined through discussions between peers in international conferences, conceptual work tends to be viewed as a matter of mere convention rather than as a support of invention. Clarifying the basic concepts is considered a minor activity useful for communication and for teaching purposes. Mendeleev's achievement demonstrates that conceptual work can be a creative activity, of high epistemic value. This case also demonstrates the importance of metaphysical views in chemistry. In fact, Mendeleev's view of elements as true individuals was just a metaphysical conviction, which gradually became a central dogma while he tested the heuristic power of the periodic law against candidate elements that could either confirm or jeopardize the system. This kind of interplay between Mendeleev's metaphysical views, his system, and experimental discoveries proved extremely fruitful since it contributed to the invention of the notion of "isotopes" and a subsequent redefinition of elements that saved the periodic system.

References

Albury, W. 1972. "The Logic of Condillac and the Structure of French Chemical and Biological Theory (1780–1800)." PhD diss., Johns Hopkins University.

Avogadro, A. 1991. "Essai d'une manière de déterminer les masses relatives des molécules élémentaires des corps et les proportions selon lesquelles elles entrent dans ces combinaisons." *Journal de physique, de chimie et d'histoire naturelle* 73 (1811): 58–79. Reprinted in *Les atomes: Une anthologie historique*, edited by B. Bensaude-Vincent and C. Kounelis. Paris: Presses Pocket.

Bensaude-Vincent, B. 1982. "L'éther, élément chimique: Un essai malheureux de Mendeleev en 1904." *British Journal for the History of Science* 15: 183–187.

Bensaude-Vincent, B. 1986. "Mendeleev's Periodic System of Chemical Elements." *British Journal for the History of Science* 19: 3–17.

Bensaude-Vincent, B. 2010. "Lavoisier lecteur de Condillac." *Dix-Huitième Siècle* 42: 49–65.

Bensaude-Vincent, B. 2012. "Le système périodique en perspective historique." *Comptes-rendus de l'Académie des sciences*, series 2 C 15, no. 7: 546–552.

Brock, W. 1985. *From Protyle to Proton: William Prout and the Nature of Matter.* Boston: Adam Hilger.

Brooke, J. H. 1981. "Avogadro's Hypothesis and Its Fate: A Case Study in the Failure of Case Studies." *History of Science* 19: 235–273.

Brooks, N. M. 2000. "Mendeleev's *Principles of Chemistry* and the Periodic Law of Elements." In *Communicating Chemistry: Textbooks and Their Audiences 1789–1939*, edited by B. Bensaude-Vincent and A. Lundgren, 295–310. Canton MA: Science History Publication.

Cannizzaro, S. 1911. *Sketch of a Course of Chemical Philosophy.* 1858. Chicago: Alembic Club Reprint N°18.

Cuvier, G., S. Lacroix, A. Fourcroy, M.-E. Chevreul et al. 1816–1830. *Dictionnaire des sciences naturelles.* 71 vols. Strasbourg-Paris: F. G. Levrault.

Dagognet, F. 1969. *Tableaux et langages de la chimie, Essai sur la representation.* Paris: Vrin.

Fischer, N. 1982. "Avogadro, the Chemists and Historians of Chemistry." *History of Science* 20: 241–276.

Freund, I. 1968. *The Study of Chemical Composition: An Account of Its Method and Historical Development.* New York: Dover Publications.

Gordin, M. D. 2004. *A Well-ordered Thing: Dmitrii Mendeleev and the Shadow of the Periodic Table.* New York: Basic Books.

Graham, L. 1982. "Textbook Writing and Scientific Creativity: The Case of Mendeleev." *National Forum* 22-23 (Winter).

Harré, R. 2005. "Chemical Kinds and Essences." *Foundations of Chemistry* 7: 7-30.

Hendry, R. F. 2005. "Lavoisier and Mendeleev on the Elements." *Foundations of Chemistry* 7: 31-48.

Kaji, M. 2002. "D. I. Mendeleev's Concept of Chemical Elements and the Principles of Chemistry." *Bulletin for the History of Chemistry* 27: 4-16.

Kim, M. G. 2003. *Affinity, That Elusive Dream*. Cambridge, MA: MIT Press.

Kounelis, C. 2000. "Atomism in France: Chemical Textbooks and Dictionaries." In *Communicating Chemistry: Textbooks and Their Audiences 1789-1939*, edited by B. Bensaude-Vincent and A. Lundgren, 207-231. Canton, MA: Science History Publication.

Kragh, H. 1989. "The Aether in Late Nineteenth Century Chemistry." *Ambix* 36: 49-63.

Lavoisier, A. 1799. *Elements of Chemistry*, 4th ed. Translated by Robert Kerr and William Creech. Edinburgh.

Leicester, M. 1948. "Factors Which Led Mendeleev to the Periodic Law." *Chymia* 1: 67-74.

Longino, H. 2008. "Values, Heuristics and the Politics of Knowledge." In *The Challenge of the Social and the Pressure of Practice: Science and Values Revisited*, edited by M. Carrier, D. Howard, and J. Kourany, 68-86. Pittsburgh: University of Pittsburgh Press.

Macquer, P.-J. 1758. *Elements of the Theory and Practice of Chemistry*. Translated by A. Reid. London.

Mauskopf, S. M. 1976. "Crystals and Compounds." *Transactions of the American Chemical Society*, 66 (part 3): 5-82.

Mendeleev, D. I. 1869. "The Correlation of the Properties and Atomic Weights of the Elements." Paper read before the Russian Society of Chemistry on March 6, 1869. English translation in *Mendeleev on the Periodic Law: Selected Writings*, edited by W. Jensen. Cincinnati: Dover, 2002.

Mendeleev, D. I. 1871. "The Periodic Law of the Chemical Elements." Originally published in *Annalen der Chemie und Pharmacie* 8 (supplement band) (1871): 133-229. English translation in *The Chemical News* 40 (1879): 231-232, 243-244, 255-256, 267-268, 279-280, 291-292, 303-304; *The Chemical News* 41 (1880): 2-3, 27-28, 39-40, 49-50, 61-62, 71-72, 83-84, 93-94, 106-108, 113-114, 125-126. Reprinted in *Mendeleev on the Periodic Law: Selected Writings*. Edited by W. Jensen. Cincinnati: Dover, 2002.

Mendeleev, D. I. 1895. "Professor Mendeléef on Argon." *Nature* 151: 543.

Mendeleev, D. I. 1899. "The Periodic Law of Chemical Elements." Faraday Lecture. 1879. *Journal of the Chemical Society* 55: 634-656. Reprinted in *Mendeleev on the Periodic Law. Selected Writings*, edited by W. Jensen. Cincinnati: Dover, 2002.

Mendeleev, D. I. 1901. "Comment j'ai découvert le système périodique de éléments." *Revue générale de chimie pure et appliquée* 1: 211-214, 510-513; 4: 533-546. English translation in *Mendeleev on the Periodic Law: Selected Writings*, edited by W. Jensen. Cincinnati: Dover, 2002.

Mendeleev, D. I. 1904. *An Attempt towards a Chemical Conception of the Ether*. Longmans, Green & Co: London.

Nye, M. J. 1984. *The Question of the Atom from the Karsruhe Conference to the First Solvay Conference*. Los Angeles: Tomash Publishers.

Partington, J. R. 1948. "The Concept of Substance and Chemical Elements." *Chymia* 1: 109-127.

Rappaport, R. 1960. "G.-F. Rouelle: An Eighteenth-century Chemist and Teacher." *Chymia* 6: 68-101.

Rappaport, R. 1961. "Rouelle G.-F. and Stahl: The Phlogistic Revolution in France." *Chymia* 7: 73–102.

Rocke, A. J. 1984. *Chemical Atomism in the Nineteenth Century: From Dalton to Cannizzaro.* Columbus: Ohio State University Press.

Scerri, E. R. 2007. *The Periodic Table, Its Story and Its Significance.* New York: Oxford University Press.

Sharlow, M. F. 2006. "Chemical Elements and the Problem of Universals." *Foundations of Chemistry* 8: 225–242.

Siegfried, R. 2002. *From Elements to Atoms: A History of Chemical Composition.* Philadelphia: Transactions of the American Philosophical Society.

Van Spronsen, J. W. 1969. *The Periodic System of Chemical Elements: A History of the First Hundred Years.* Amsterdam: Elsevier.

3

Dmitri Mendeleev's Concept of the Chemical Elements Prior to the Periodic Law

Nathan M. Brooks, Department of History,
New Mexico State University, USA

In the combined second and third issue of the first volume of the *Journal of the Russian Chemical Society* (Zhurnal Russkago khimicheskogo obshchestva) published in March 1869, Dmitri Ivanovich Mendeleev published his first account of the periodic law of the Chemical Elements (Mendeleev 1869).[1] In this article, Mendeleev sketched out his reasoning for asserting that the elements were ordered in a periodic fashion according to atomic weights. Mendeleev's arguments in this article were rather general and he did not provide much concrete experimental data to support his hypotheses. While his initial article did not elicit much attention from other chemists at the time, even inside Russia, Mendeleev was convinced that he had discovered a fundamental law of nature and he began to gather more data and conduct experiments to verify his assertions. Over the next two years, Mendeleev assiduously worked to shore up his theory and publicize it to a wider audience, culminating in an 1871 summary, in an influential German journal, where he included some predictions about undiscovered elements. It seems likely that these predictions attracted chemists' attention to the periodic law once the predicted elements were discovered and found to have properties close to those predicted by Mendeleev.[2]

The traditional interpretation of Mendeleev's discovery of the periodic law was proposed by B. M. Kedrov, who argued that Mendeleev arrived at the initial concept of the periodic law suddenly, while he was working on his textbook *Principles of Chemistry*. Mendeleev then over the course of one day developed

---

[1] I will use the terms "periodic system" and "periodic law" interchangeably instead of the more common "periodic table," because these two terms better express the complexities and interrelationships revealed by the Periodic System.

[2] For a discussion of Mendeleev's predictions see Brush 1996; and the individual chapters in Kaji, Kragh, and Palló, 2015.

Nathan M. Brooks, *Dmitri Mendeleev's Concept of the Chemical Elements Prior to the Periodic Law* In: *What Is A Chemical Element?*. Edited by: Eric R. Scerri and Elena Ghibaudi (2020). © Oxford University Press.
DOI: 10.1093/oso/9780190933784.003.0004

that insight into the first table of the elements (Kedrov 1958).[3] He quickly wrote up his first paper about the periodic system in which he elaborated his reasons for a periodic system based on the atomic weights of the elements. Kedrov's interpretation became the canonical interpretation in the Soviet Union and gradually spread to other parts of the world. Despite its dominance, Kedrov's interpretation has some glaring deficiencies, including arguments that it was based on inferences about documents that do not actually show what Kedrov says they show, as well as an emphasis on Mendeleev's use of "chemical solitaire" (Kedrov 1958).[4]

More recently, I. S. Dmitriev has proposed an alternative and quite sophisticated explanation of Mendeleev's path to the periodic system, arguing that Mendeleev developed his ideas over a longer period of time (from late 1868 to late 1870) in several stages until he arrived at a unified system for describing the periodicity of the elements (Dmitriev 2004a). In fact, Dmitriev believes that we should not see the period between 1855 and 1867 as a "path" taken by Mendeleev toward the periodic law. Instead, Mendeleev's work during these years gave him the necessary foundation upon which he could base his concept of the periodic law.

While Kedrov, Dmitriev, and others have focused their attention on Mendeleev's development of the periodic law, van Spronsen drew attention to six scientists, including Mendeleev, who have been credited with discovering the periodic law (van Spronsen 1969).[5] These predecessors of Mendeleev published incomplete or partial tables of elements, some aspects of which were close to the initial table drawn up by Mendeleev.

But why did Mendeleev discover the concept of periodicity and not any of the forerunners identified by van Spronsen? Bernadette Bensaude-Vincent explained this peculiarity by arguing that Mendeleev adopted a different definition of element than did most of these forerunners (Bensaude-Vincent 1986).[6] Most scientists before Mendeleev, like Lavoisier, saw elements as bodies that had not been able to be decomposed. Mendeleev, on the other hand, saw elements as individual, unique simple bodies having different properties. One consequence of this definition of element was that Mendeleev was opposed to Prout's hypothesis. This allowed him to avoid being sidetracked by a search for precise mathematical relationships between the different elements. Bensaude-Vincent based

---

[3] While this volume is the fullest and most detailed account of his argument, Kedrow published shorter versions more than a decade prior to this.

[4] Some of these flaws in Kedrov's analysis are discussed in Trifonov 1990.

[5] Also see Benfey 1992–1993.

[6] Bensaude-Vincent argues that Mendeleev developed his theoretical idea about the chemical elements before he arrived at the concept of periodicity, while Kedrov believes that Mendeleev developed his abstract definition of element after his initial conception of the periodic law.

her argument on evidence from Mendeleev's *Principles of Chemistry* and his first paper on the periodic law, but did not investigate Mendeleev's earlier views of elements. In this paper, I intend to examine Mendeleev's scientific work prior to the *Principles of Chemistry*, including his ideas about elements in the years leading up to his first paper on the periodic law. I will show that there is some evidence that Mendeleev had settled upon his abstract definition of element prior to initiating his work on *Principles of Chemistry*. However, in addition to this, from his very earliest scientific work, Mendeleev was searching for general laws of nature. Therefore, through his preparations for lectures in general chemistry in the 1860s and working on *Principles of Chemistry*, Mendeleev was primed to develop his abstract notion of chemical elements, which became fundamental to his elaboration of the periodic law.

Mendeleev was born on January 27, 1834, in Tobolsk, a small city in Siberia that served as an administrative and commercial center for the region.[7] His father was the director of the local gymnasium in the school system who lost his sight and died during Mendeleev's childhood years. Mendeleev's mother then shouldered the entire responsibility for raising the family, including one stint operating a glass factory. Mendeleev was an indifferent student, earning average or below-average marks in most of his classes, along with a few higher ones. However, his mother aspired for him to continue his education and after his graduation from the gymnasium in 1849, they set off for Moscow to gain admission to Moscow University. Unsuccessful in that endeavor, Mendeleev and his mother then went to St. Petersburg.

There, after much effort and multiple petitions and interviews, Mendeleev finally succeeded in being admitted as a state-supported student at the Main Pedagogical Institute, from which his father had graduated many years previously. This was an unusual institution for Russia, as it was "closed," meaning that students lived on its grounds and the discipline was strict. Looking back at his education at the Main Pedagogical Institute toward the end of his life, Mendeleev recognized how important the institute's structure was for his educational growth, especially as his mother, sister, and other close relatives died during his first years at the institute, while Mendeleev himself was quite ill for long periods of time. Another peculiar feature of the institute was that its curriculum was on a two-year cycle. Mendeleev and several other students who were admitted the same year were required to start at the mid-point of the introductory classes as this was the off year for the cycle. Mendeleev initially struggled with his classes and was even forced to start over with the introductory classes after one year. However, he soon righted himself and became one of the

---

[7] The most detailed account of Mendeleev's early life is Mladentsev and Tishchenko 1938. Also see Figurovskii 1983.

top students at the institute, focusing his attention on biology, chemistry, and physics. The faculty at the institute was outstanding, as it drew professors from St. Petersburg University and the Imperial Academy of Sciences, among other higher educational institutions of the capital. Once he had established himself at the institute, Mendeleev at first specialized in biology, even submitting several papers for publication that were based on field research he had conducted along with the biology professor. Growing frustrated when these papers were rejected for publication, he quickly switched his attention to chemistry and mineralogy, conducting analyses of several minerals from Finland during 1853–1854. Mendeleev presented the results of some of these investigations at a session of the Russian Geographical Society and they resulted in two publications in Russian scientific journals.

Mendeleev graduated with high honors (Gold Medal) from the Main Pedagogical Institute in 1855 following a series of examinations in various subject and his presentation of a dissertation for the candidate degree. One of his professors, A. A. Voskresenskii, highly praised one of Mendeleev's written examinations, calling it the work of "such a talented and prepared scholar" (Mladentsev and Tishchenko 1938, 96). Mendeleev's dissertation, "Isomorphism in Connection with Other Aspects of Crystal Forms in Relation to Composition," was an outgrowth of his mineralogical analyses conducted earlier.[8] Toward the end of his life, Mendeleev looked back upon this work: "At the Main Pedagogical Institute, it was a requirement to write a dissertation on some topic. I selected isomorphism, because I was interested in this due to my first publications and the subject seemed to me to be important in a natural scientific respect. The writing of this dissertation attracted me most of all to the study of chemical relationships. By that it determined much" (Shchukarev and Valk 1951, 43–44). Mendeleev here expresses his belief that this early scientific work on isomorphism helped prepare him both scientifically and intellectually for formulating his periodic system.[9]

This candidate dissertation contained no experimental work and was largely a literature review of isomorphism, the phenomena where different substances crystallize in the same form, first discovered by Eilhard Mitscherlich in 1819.[10] Isomorphism was soon put to use in developing the atomic theory and determining atomic weights using the concept that the same crystallographic forms

---

[8] This dissertation was published in two parts in *Gornyi Zhurnal*, no. 8 (1855) and no. 9 (1856). It was also published as a separate booklet in 1856. This separate booklet was republished in Mendeleev 1937, t.1, 7–137.

[9] Dmitriev believes that this early work did not directly assist Mendeleev in discovering the periodic law, but rather helped him refine his ideas once he had grasped the overarching principles. See Dmitriev 2004b, 9–89.

[10] For information about isomorphism, see Rocke 1984.

have analogous formulas. Mendeleev was attracted to isomorphism as he saw it as a law that could provide valuable information about atoms and molecules. "The reason that compels bodies to take specific geometrical forms is unknown to us, but it is conceivable for us to search for a means to show how the composition and form [of bodies] relate by means of external phenomena that we can determine. The correspondence of form and composition would be very easy to understand if we could know the laws of the interior structures of bodies" (Mendeleev 1937, t.1, 15).

In this work, Mendeleev was concerned with determining a classification system that would unite all matter into one. Following recent work by August Laurent, Mendeleev stated that "all organic and inorganic bodies are composed in one manner, according to one law" (Mendeleev 1937, t.1, 100). However, he criticized some beliefs of Haüy, noting that "[his] mistake consisted of seeing too tight a connection" between crystalline forms and chemical composition. "The sharpness of general laws, as found in Haüy's work, does not correspond to the mobile and flexible laws of nature" (Mendeleev 1937, t.1, 25). Still, Mendeleev found isomorphism to be useful in showing similarities in chemical composition, with isomorphic substances having similar properties. Mendeleev provided a table in which he grouped different elements by properties. This table arranged the elements into eleven different categories, which quite closely followed the groupings in the later periodic table (Mendeleev 1937, t.1, 37). What Mendeleev was trying to show in his dissertation was that isomorphism could help illuminate the similarities between different elements that in his periodic table would later be located in nearby groups to one another. Thus, he noted the similarities between Al and Si (Mendeleev 1937, t.1, 117). This showed that—using the language of the periodic table—isomorphism could indicate not only vertical similarities between elements but also horizontal ones as well.

While Mendeleev did not conduct his own experimental work for this dissertation, it is notable for its analysis of a mass of information, in different languages, and the evident assuredness of the author in its presentation. He performed extensive calculations using the data gleaned from the wide range of literature he consulted for this dissertation. In addition, he clearly followed and approved of the work of Gerhardt and Laurent, which would become even more clear in the subsequent years. Furthermore, the dissertation showed his attention to organic chemistry and his attempts to meld inorganic and organic chemistry into one unified system. All of these features were important steps toward the periodic law. There is another aspect about this work that we should note. Mendeleev was only twenty-one years old at this time and was only a few years removed from his decision to concentrate on the study of chemistry. His ability to comprehend and digest contemporary chemical theories while formulating his own views would continue to be a feature of Mendeleev's scientific career.

Mendeleev's candidate dissertation and his high marks earned him the right to continue his education, instead of the usual placement teaching at a secondary school. However, his poor health during his last years at the Main Pedagogical Institute convinced him to ask for a placement teaching in a secondary school in the south of Russia, preferably Odessa, which would allow him access to libraries and other cultural resources in that city. For some reason, he instead was sent in 1855 to Simferpol, in the Crimea, close by the front lines of the Crimean War, although he soon managed to arrange a transfer to a different secondary school in Odessa than the one where he had originally planned to teach.

Mendeleev returned to St. Petersburg in April 1856 in order to take his examinations for the master's degree in chemistry at St. Petersburg University and to defend his master's dissertation in September of that year. Mendeleev's dissertation for the master's degree, "Specific Volumes," like his candidate dissertation, did not contain any of his own experimental investigations and was based on published literature, although it did contain extensive original calculations made by Mendeleev using published data (Mendeleev 1937, t.1, 139–311). This dissertation, again like his candidate dissertation, also was published in the *Mining Journal*, in 1856 (Mendeleev 1856).

Mendeleev's master's dissertation continued the discussion of several themes from his candidate dissertation. His adherence to the views of Laurent and Gerhardt was even more clearly visible in this work than in the previous one. He used Laurent and Gerhardt's version of the type theory consistently in this work, as he did in later years, which he had not done in his candidate dissertation. He also used their atomic weights. In addition, he often cited with approval the work of Avogadro.[11]

Mendeleev began his master's dissertation with a criticism of the dualist views of Berzelius: "All of this focuses one to give up the electrochemical theory [of Berzelius]. It was a necessary step in science, like the theory of the philosopher's stone and phlogiston" (Mendeleev 1937, t.1, 151). While throughout the dissertation Mendeleev often referred to and used the work of Gerhardt and Laurent for his calculations, he was not completely uncritical toward the views of the French chemists. "All of the principles of Gerhardt's system that we have imported, are—without a doubt—far from the last word in science, but they are, at least, at the highest levels [of science] that are accessible to us now" (Mendeleev 1937, t.1, 167). Mendeleev considered Gerhardt and Laurent's views as the basis for the "new" chemistry that he supported wholeheartedly. He saw the basis for this "new" chemistry to be the distinction between atoms, equivalents, and molecules. After criticizing the dualist theory, Mendeleev described in detail the

---

[11] For example, see the citations and discussions about Avogadro's work in Mendeleev 1937, t.1, 238, 240.

methods for calculating molecular weights on the basis of vapor pressure and the new understanding of atomic weights. Here he devoted special attention to the principles involved in these calculations. Finally, he provided a table of atomic weights for forty-nine elements that had been determined by previous experimentation (Mendeleev 1937, t.1, 257–260).

In the master's dissertation, Mendeleev also paid close attention to many of Avogadro's publications, calling them "wonderful principles" (Mendeleev 1937, t.1, 236). In particular, Mendeleev emphasized the differences between atoms and molecules, but he also distinguished between chemical atomism and physical atomism. "We need to define a molecule as a group of atoms, which determines its known physical properties. Therefore, we need to distinguish molecules in a chemical sense from physical molecules. Studying molecules by their physical properties does not add to pure chemistry in those cases where the agreement of chemical and physical properties are [sic] impossible; that is, where the weight of a chemical molecule is not equal to the physical weight" (Mendeleev 1937, t.1, 238). This is where Avogadro's work became important for Mendeleev. "The main contribution of Avogadro is that he turned attention to which others wanted to discard; that is, to small or large differences of volumes. Earlier only Kopp in his work on the isomorphism of bodies turned attention to this difference and introduced it in agreement with the study of forms. This is, in our opinion, the most natural path to the study of specific volumes of solid bodies. Their volumes are very diverse; we do almost everything when we determine from that and instead of the variations in volume. There, where there is no uniformity, is it natural to study the causes of differences and inequalities" (Mendeleev 1937, t.1, 238). Mendeleev also drew attention to Avogadro's support for two-volume atomic weights and formulas.

We can obtain, perhaps, a clearer understanding of some of Mendeleev's views about the "new" chemistry from the outline of topics Mendeleev wrote for lectures in a course in organic chemistry that he taught at St. Petersburg University in 1857–1858. The first few topics include: "1) Stoichiometry. About atoms and equivalents. 2) Volume relationships and chemical compounds. 3) Chemical particles (molecules). Atomic theory and chemical formulas" (Mladentsev and Tishchenko 1938, 142). Later in these lectures, Mendeleev planned to provide a summary of Gerhardt's ideas. The lectures concluded with an exposition of the relationship between "different physical properties and composition." Mendeleev did not publish these lectures, so we only have the outline and a few notes on which to base our analysis of his thinking at this time.

Following the defense of his master's dissertation, Mendeleev became a privatdozent at St. Petersburg University where he taught courses mainly in organic chemistry and wrote articles for various periodicals in order to sustain himself. His brilliant performance in his candidate examinations and dissertation had qualified him for a fellowship to study abroad, which had been suspended between 1848 and 1856, but he did not receive official permission for this until January 1859.[12] Mendeleev decided to go to Heidelberg University, intending to gain a place in Robert Bunsen's laboratory.[13] There was a relatively large and lively Russian student community in Heidelberg at this time and for some years after. Mendeleev became an integral part of one small group that retained ties even after the students returned to Russia and settled into scientific careers there.[14]

According to Mendeleev's own account, he decided not to work in Bunsen's laboratory as it was crowded and "there was nothing I needed in that laboratory, even the scales . . . were bad" (Mladentsev and Tishchenko 1938, 161). He used much of his fellowship funds to purchase delicate and sensitive equipment and to install a gas line in his student apartment, where he conducted his experiments. Mendeleev had decided to investigate "that special field which connects chemistry with physics and mechanics." To do this, he conducted very careful measurements of the capillary action of various organic substances, trying to discover a general law. He believed that by identifying the differences in the capillary action of different hydrocarbons in homologous series, he could determine this general law of molecular cohesion. "I studied the capillarity of many chemical compounds and their changes with heat. I improved the means of determining specific weight precisely. I helped especially to introduce more precision and ease of correction in weighing in air, and by these more precise methods I studied the density of many liquids. I also had to study the expansion of liquids at temperatures higher than boiling, and I attained general results in this area" (Mladentsev and Tishchenko 1938, 223). Thus, we can see that he was continuing along the lines of his earlier work on specific volumes and isomorphism, but now he was conducting his own original experiments not just relying on published data. He published a number of articles based on this research in both Russian and German chemistry journals, but was not able to decipher a general formula for capillary action that he could develop into a general law. Later, in 1870, he claimed priority for discovering the concept of the critical point, but as Gordin has shown, this is a debatable point (Gordin 2009a).

While Mendeleev was abroad on his two-year fellowship term, he attended the International Congress of Chemistry in Karlsruhe in September 1860. One of seven Russians at the congress, he was chosen to be a member of the committee that formulated the questions to be voted on by the congress. It is not clear why Mendeleev was chosen to be a member of this committee, since he was

---

[12] Russian State Historical Archive (RGIA), f. 733, op. 149, d. 826, l. 60 ob.
[13] Bunsen's laboratory was popular with other foreign students, as well. For example, see Brock 2013.
[14] For Mendeleev's experiences in Heidelberg and his Russian circle, see Gordin 2008 and 2009b.

quite young and had not yet established himself as a chemist with an international reputation, but he had had many conversations with prominent chemists during excursions to Paris and other places in the German lands while based in Heidelberg. Likely he had impressed one of the organizers of the congress during some of these conversations. Mendeleev was excited by the congress and a few weeks later wrote in a letter to a relative: "At the congress it was gratifying to see those principles, which all young Russian chemists follow, being adopted over all routine understanding which reigns among the mass of chemists" (Mladentsev and Tishchenko 1938, 172).

Mendeleev also wrote a long letter about the congress to his advisor Voskresenskii in St. Petersburg, who, with Mendeleev's approval, had it published in a local newspaper (Mladentsev and Tishchenko 1938, 250–258).[15] "The essential reason for the convocation of an international congress was the wish to clarify and if possible to bring into agreement the basic differences of language existing between the followers of the different chemical schools. It was M. Kekulé who first proposed many questions for decision: the question of the differences between molecules, atoms, and equivalents; the question of the values of atomic weights, i.e., whether to accept the atomic weights of Berzelius, corrected afterwards by Liebig and Poggendorf, and now accepted by a majority; further, the question of formulas and even, finally, of those forces which in the contemporary state of the science we must consider as the reason behind chemical occurrences" (Mladentsev and Tishchenko 1938, 250–252).

Mendeleev was most impressed by Cannizzaro, both at the congress itself and through a copy of the Italian chemist's famous 1858 paper on atomic weights that was distributed to participants at the congress. "During the discussions, the most determined and, without a doubt, most original and valuable opinion was expressed by . . . Cannizzaro. . . . [He displayed] that ardor, that vital energy, those fully composed convictions which acted so forcefully on those who heard him" (Mladentsev and Tishchenko 1938, 252–253). Mendeleev provided a detailed discussion of Cannizzaro's method for determining atomic weights and included a list of the atomic weights of some elements, including some that differed from the usually accepted values when calculated according to Cannizzaro's method. Mendeleev summarized Cannizzaro's conclusions about why scientists should adopt these new atomic weights. "Having adopted the new understanding about molecules, we cannot continue to hold on to the old Berzelian understanding of atoms. . . . [This is accepted by] all those who are working in the new direction, in England, France, Russia, Germany and Italy, all those who more or less use Gerard's atoms, since they are based on firm, unavoidable principles." It

---

[15] It is also reprinted in Mendeleev 1958, 660–669.

seems that Mendeleev did not see Cannizzaro's new methods for determining atomic weight values as revolutionary. "We have corrected only a few of Gerard's mistakes—and we will achieve more such results. Now these corrections of Gerardian atoms are long from being new, and there is no need to demand their general introduction, but we must keep them constantly in mind" (Mladentsev and Tishchenko 1938, 257–258). In this, Mendeleev's attitudes closely follow what Alan J. Rocke has described as "the Quiet Revolution," where Cannizzaro's presentations at the Karlsruhe Congress were not decisive in altering chemists' views in and of themselves, but that most chemists had gradually come to accept these ideas over the course of the 1850s (Rocke 1984; 1993).

Mendeleev returned to Russia from his time abroad in February 1861. He had attempted to obtain a one-year extension of his trip, but despite much effort in this regard, he was not successful. Since the academic term had started, he was not able to find employment in either a secondary or higher educational institution. Desperately in need of funds, especially since he had gone into debt while abroad in order to purchase laboratory equipment, he began to publish articles and translations for various publishers. In addition, most important, he began work on a new textbook of organic chemistry, which he rapidly finished in June 1861. This textbook, succinctly titled *Organic Chemistry*, was based largely on Gerhard's type theory and was received with much favor by more senior Russian scientists, despite the fact that it was so derivative. In fact, it won the Demidov Prize (worth 1,000 rubles) from the Imperial Academy of Sciences in 1862, greatly assuaging Mendeleev's financial conditions. In this work, Mendeleev employed to some extent Cannizzaro's ideas for determining atomic weights, defining them as "the minimum quantity of an element in the compound molecule of the element." He also explicitly distinguished molecules, radicals, and atoms. He defined molecules as "something divisible" while atoms were "an indivisible whole" (Mendeleev 1948, t.8, 45). However, he provided a list of atomic weights that he would be using in the textbook that differed significantly from Cannizzaro's atomic weights. In fact, this list of atomic weights even differed from the list he had included in his letter about the Karlsruhe Congress (Mendeleev 1948, 39).

While the textbook was based on Gerhard's type theory, Mendeleev extended what we might term the traditional type theory with his concept of a "theory of limits."[16] Also published as a separate article in 1861 (Mendeleev 1948, t.8, 23–27), Mendeleev believed that his theory of limits was the first theory to show how chemical properties of a substance were linked with its structure. However, this was an entirely different approach than that set forth by the Structural

---

[16] For a detailed discussion of Mendeleev's theory of limits, see Dobrotin 1956.

Theory of Organic Chemistry, which was just at that time in the process of being elaborated by Kekulé and A. M. Butlerov. Instead, Mendeleev developed new types based on homologous series such as C_nH_{2n+2} and on their capacity for saturation. This worked for basic hydrocarbons, but it soon broke down for oxygen-containing compounds. Indeed, the theory of limits was rather a muddle, but Mendeleev was greatly excited by its potential and he continued promoting it long after the Structural Theory was established as the dominant theory of organic chemistry. A second edition of Mendeleev's organic chemistry textbook came out in 1863 in which he tried to defend the theory of limits against the Structural Theory, but to no avail. Even Mendeleev's friends, such as A. P. Borodin, chided him about his extravagant claims for the theory of limits, saying that "your theory, it seems to me, is more of an empirical law rather than an actual theory" (Volkova 1940).

Mendeleev soon threw himself into teaching mainly general and inorganic chemistry at a wide variety of educational institutions in St. Petersburg, including St. Petersburg University, the Technological Institute, the Second Cadet Corps, and others. Note that these teaching duties involved teaching general chemistry rather than organic chemistry, which had primarily been the courses had had taught previously, and he thus began to shift his vision away from organic chemistry. These teaching duties occupied much of his time and, as he later recalled, he did not take steps to write his doctoral dissertation until he had obtained a permanent position that would require such a degree (Shchukarev and Valk 1951, 91). In addition, Mendeleev did pursue other, more practical and applied activities during these years. For example, he began consulting for an oil company and started a wide-ranging series of experiments in agronomy and agriculture at his country dacha. Finally obtaining a permanent teaching position in January 1864, when he was appointed a professor of chemistry at the Technological Institute in St. Petersburg, he quickly began work on a doctoral dissertation. This dissertation, "On Compounds of Alcohol with Water," was defended in January 1865 and he received the doctoral degree on February 1 of that year. Mendeleev selected this topic as an outgrowth of his interest in what he termed "indefinite compounds," ones that had constant physical properties, but variable compositions. Examples of these indefinite compounds included solutions, alloys, silicate compounds, and others. Mendeleev had studied silicate compounds in an 1856 work, "On the Structure of Silicate Compounds" (Mendeleev 1952, t.25, 108–228). Now he turned to a study of solutions as another example of these indefinite compounds (Mendeleev 1937, t.4, 1–152). Focusing his attention on varying solutions of alcohol and water, he attempted to determine whether these solutions were mechanical mixtures or chemical combinations. As in his first two dissertations, he provided an extensive review of the existing literature on the topic of solutions. He then described a method that could be used to determine the specific weights of various solutions. In the next parts of his dissertation, he examined the specific weights of different solutions of alcohol and water at various temperatures. Mendeleev admitted that these experiments were difficult for him to conduct, in particular, those using very little amounts of water. While he claimed that he "attempted to avoid making any kind of theoretical assumptions about the nature of solutions" in this study, he concluded that "definite [regular] chemical compounds form only one part of [the group of] indefinite chemical compounds" (Mendeleev 1937, t.4, 1–2). As he had in his previous work, Mendeleev wanted to derive general theoretical conclusions from this study. However, he was forced to admit that he was unable to resolve his initial questions about the exact nature of these compounds. In a list of "propositions" concluding the dissertation, he stated that "chemistry does not bestow causes to give a real meaning to the atomic method of studying molecular phenomena." Also, "the understanding of the limit of chemical compounds or the contemporary understanding of the atomicity of elements can be seen as the result of the Gerard's type ideas" (Mendeleev 1937, t.4, 151). This dissertation developed Mendeleev's views on the absence of specific boundaries between "definite" and "indefinite" compounds, as well as the absence of a sharp differentiation between oxygen and hydrogen. Mendeleev would further develop these concepts in the *Principles of Chemistry*. For example, in the first part of this textbook, which he was writing during the time he was developing the periodic law, he stated there that "compounds with indefinite compositions . . . speak against the atomic doctrine as much as definite chemical compounds speak in its support" (Mendeleev 1949, t.13, 337).[17]

As already noted, Mendeleev was very busy with teaching during the mid-1860s. In 1865, following the defense of his doctoral dissertation, Mendeleev was appointed ordinary professor of technical chemistry at St. Petersburg University. Two years later, in October 1867, he moved from the chair of technical chemistry to the chair of general chemistry at St. Petersburg University. This move to the chair of general chemistry was the primary impetus for him to begin writing his textbook on general chemistry, the *Principles of Chemistry*, which scholars generally agree was the key initial spark for his discovery of the periodic law.

Mendeleev's initial series of lectures in general chemistry in the fall of 1867 shows that while he retained some skepticism about the atomic theory, he firmly believed in distinct individual chemical elements. However, he also believed that the elements underwent some type of change when they combined to form molecules. Using the example of mercury oxide, Mendeleev emphasized that this

---

[17] This is a reprint of part 1 of the first edition of *Osnovy khimii*.

compound differed significantly from the individual elements of mercury and oxygen. "It does not consist of . . . mercury as a metal or of oxygen in its gaseous state, which combine to form red mercury oxide. [This compound] contains only the substance of the simple bodies, as steam contains the substance of ice, but is not the same as ice" (Mendeleev 1949, t.15, 381). Mendeleev used a similar idea as the key passage in his first paper on the periodic law: "Everyone understands that in all changes in the properties of simple substances, something remains unchanged and that, in the transformation of the elements into compounds, this material something determines the characteristics common to the compounds formed by a given element. In this regard, only a numerical value is known, and this is the atomic weight appropriate to the element. The magnitude of the atomic weight . . . is a quantity which does not refer to the momentary state of a simple substance but rather belongs to a material portion of it. . . . The atomic weight does not belong to coal or to diamond but rather to carbon" (Mendeleev 1869, 24–25).

Further in his 1867 lecture, Mendeleev stated: "It is necessary to distinguish the concept of a simple body from that of an element. A simple body . . . is one that up to now has not been changed chemically and has not been formed through the transformation of any kind of other body. An element is an abstract concept. It is the material that is contained in a simple body and can without any change in weight be transformed into all of the bodies which can be obtained from this body" (Mendeleev 1949, t.15, 381–382). Masanori Kaji has argued that Mendeleev's distinction here between "simple bodies" and "elements" mirrors the distinction made in 1861 in the textbook *Organic Chemistry* between "bodies" and "radicals." Kaji concludes that Mendeleev developed the concept of the weight of the elements as an invariable property "not by adherence to the concept of chemical atoms, but by seeking freedom from it, as the failures of the law of definite proportions seemed to demand." Thus, according to Kaji, Mendeleev arrived at the abstract idea of an element through his belief that the atomic theory had limitations (Kaji 2002).

However, Mendeleev developed a further refinement in his concept of element in the first part of *Principles of Chemistry*. Using the example of the different forms of carbon (carbon, charcoal, graphite, diamond), he explained that when an element is combined with another, it either gains or loses caloric.[18] This is also true for different forms of elements, such as those noted for carbon. Mendeleev hastened to note that his use of the term "caloric" was only a formal means of expression, "since caloric should not be considered as a material

---

[18] Note that in his first published article about the periodic law, Mendeleev does not use the concept of caloric when discussing elements and their atomic weights.

substance" (Mendeleev 1949, t.13, 489). He explained that, instead, caloric was an "oscillatory state of matter." Carbon has the base amount of caloric, while charcoal has more, or to put it a different way, charcoal has "a different tension of oscillations and a different time of oscillation." And "carbon can be seen as an atom of carbon mater, while charcoal as a collection of such atoms in one whole, in a particle and in a massive body." Furthermore, "The weight of a carbon atom must be 12, because that is the smallest amount of carbon that enters into its compounds, while the weight of charcoal particles, certainly, must be very large." He then underlined that a simple body differs from an element. "It must be emphasized that there is a difference between the abstract notion of carbon and the real understanding of charcoal as a substance or body" (Mendeleev 1949, t.13, 490).

As Bensaude-Vincent and others have argued, Mendeleev's use of an abstract definition of an element had far-reaching implications for his development of a system to order the elements. Most important, this concept emphasized that each element was individual and had specific properties aligned with it. This feature assisted him in reasoning that the compounds of similar elements would reflect their being composed of similar elements.

One final point should be made. We have already noted Mendeleev's use of concepts from organic chemistry, such as the type system, to use as a basis for his development of the periodic law.[19] He also used reasoning from organic chemistry to speculate that elements had an internal structure, which functioned similar to homologs in organic chemistry. This internal structure of atoms, along with their size and weight, determined their properties.[20] While this speculation was distinct from Prout's hypothesis, it does indicate that Mendeleev was not unalterably opposed to the idea of a subatomic structure for elements. Thus, we should be careful not to overemphasize Mendeleev's opposition to Prout's hypothesis by implying that he was also opposed to the idea of a subatomic structure of elements.

References

Benfey, O. T. "Precursors and Cocursors of the Mendeleev Table: The Pythagorean Spirit in Element Classification." *Bulletin for the History of Chemistry* 13–14 (1992–1993): 60–66.
Bensaude-Vincent, Bernadette. "Mendeleev's Periodic System of Chemical Elements." *British Journal for the History of Science* 19 (1986): 3–17.
Brock, W. H. 2013. "Bunsen's British Students." *Ambix* 60, no. 3: 203–233.

---

[19] For an extensive discussion of Mendeleev's use of the type system for the periodic law, see Gordin 2002.

[20] For an extensive discussion of this point, see Dmitriev 2004a, 150–153.

Brush, Stephen G. 1996. "The Reception of Mendeleev's Periodic Law in America and Britain," *Isis* 87: 595–628.

Dmitriev, Igor S. 2004a. "Scientific Discovery in *Statu Nascendi*: The Case of Dmitri Mendeleev's Periodic Law." *Historical Studies in the Physical Sciences* 34, part 2: 233–275. This is an English translation and condensed version of "Nauchnoe otkrytie in statu nascendi." In I. S. Dmitriev, *Chelovek epokhi peremen: Ocherki o D. I. Mendeleeve i ego vremeni.* St. Petersburg: Khimizdat, 2004, 90–207.

Dmitriev, Igor S. 2004b. "Ubi plurima intent, ili o "deistvitel'nykh zadatkakh I vyzovakh periodeicheskoi zakonnosti." In I. S. Dmitriev, *Chelovek epokhi peremen: Ocherki o D. I. Mendeleeve i ego vremeni.* St. Petersburg: Khimizdat.

Dobrotin, R. B. 1956. "Teoriia Predelov Mendeleeva." *Trudy Istituta istorii estestvoznaniia i tekhniki* 35: 143–148.

Figurovskii, N. A. 1983. *Dmitri Ivanovich Mendeleev, 1834–1907,* 2nd ed. Moscow: Izd. Nauka.

Gordin, Michael D. 2002. "The Organic Roots of Mendeleev's Periodic Law." *Historical Studies in the Physical and Biological Sciences* 32: 263–290.

Gordin, Michael D. 2008. "The Heidelberg Circle: German Inflections on the Professionalization of Russian Chemistry in the 1860s." *Osiris* 23: 23–49.

Gordin, Michael D. 2009a "Points Critical: Russia, Ireland, and Science at the Boundary." *Osiris* 24: 99–119.

Gordin, Michael D. 2009b. "Running in Circles: The Heidelberg Kruzhok and the Nationalization of Russian Chemistry." In *Global Science and National Sovereignty: Studies in Historical Sociology of Science,* edited by Grégoire Mallard, Catherine Paradeise, and Ashveen Peerbaye, 40–62. New York: Routledge.

Kaji, Masanori. 2002. "D. I. Mendeleev's Concept of Chemical Elements and *The Principles of Chemistry.*" *Bulletin for the History of Chemistry* 27: 4–16.

Kaji, Masanori, Helge Kragh, and Gábor Palló, eds. 2015. *Early Responses to the Periodic System.* New York: Oxford University Press.

Kedrov, B. M. 1958. *Den' Odnogo Velikogo Otkrytiia.* Moscow: Izd. Sotsial'no-Ekonomicheskoi Literatury.

Mendeleev, Dmitri I. 1856. "Udel'nye obemy." *Gornyi zhurnal,* kn. 7 (1856): 1–104; kn. 8 and 9: 160–388.

Mendeleev, Dmitri I. 1869. "Sootnoshenie svoistv s atomnym vesom elementov." *Zhurnal Russkago khimicheskogo obshchestva* 1, no. 2–3: 60–77. Reprinted in *Mendeleev on the Periodic Law: Selected Writings, 1869*–1905, edited by William B. Jensen, 18–37. Mineola, NY: Dover Publications, 2002.

Mendeleev, Dmitri I. 1934–1954. *Sochineniia,* t. 1–25. Moscow-Leningrad: Izd. Akademiia Nauk SSSR.

Mendeleev, Dmitri I. 1958. *Periodicheskii zakon: Osnovnye stat'i.* Moscow: Izd. Akademii Nauk SSSR.

Mladentsev, M. N., and V. E. Tishchenko. 1938. *Dmitri Ivanovich Mendeleev, ego zhisn' i deiatel'nost',* t. 1. Moscow-Leningrad: Izd. Akademii Nauk SSSR.

Rocke, A. 1984. *Chemical Atomism in the Nineteenth Century: From Dalton to Cannizzaro.* Columbus: Ohio State University Press.

Rocke, A. 1993. *The Quiet Revolution: Hermann Kolbe and the Science of Organic Chemistry.* Berkeley: University of California Press.

Shchukarev, S. A., and S. N. Valk, eds. 1951. *Arkhiv Mendeleeva, t. 1: Avtobiograficheskie materialy, sbornik dokumentov.* Leningrad: Izd. Leningradskogo gosudarstvennogo universiteta imeni A. A. Zhdanova.

Trifonov, D. N. 1990. "Versiia-2: k istorii otkrytiia periodicheskogo zakona D. I. Mendeleeva." *Voprosy istorii estestvoznaniia i tekhniki*, no. 2: 25–36; no. 3: 20–32.

van Spronsen, J. W. 1969. *The Periodic System of Chemical Elements: A History of the First Hundred Years*. Amsterdam: Elsevier.

Volkova, T. V. 1940. "Pis'ma A. P. Borodina k D. I. Mendeleevu." *Uspekhi khimii* 9, no. 9: 1070.

4

Referring to Chemical Elements and Compounds

Colorless Airs in Late-Eighteenth-Century Chemical Practice

Geoffrey Blumenthal, James Ladyman, and Vanessa Seifert,
University of Bristol, UK

Introduction

How do we refer to chemical substances, and in particular to chemical elements? This question relates to many philosophical questions, including whether or not theories are incommensurable, the extent to which past theories are later discarded, and issues about scientific realism.[1] This chapter considers the first explicit reference to types of colorless air in late-eighteenth-century chemical practice.

Reference to a colorless air was usually initiated by using descriptions of a method of production and an observable property. This was done prior to naming, and independently of theories of the inner constitution or dispositions of the air. In general, more than one name was given to each type of air. In several cases a researcher who adhered to one theory wrote to another researcher, using more than one name for a specific substance *interchangeably*, even though terms within those names were part of the conceptual framework of different theories. This paper argues that reference to a gas by one chemist was generally intended to give others epistemological, methodological, and practical access to the gas; such reference included definite descriptions, which were chosen to refer uniquely or as uniquely as was practicable, to enable a user of the descriptions to know whether they were referring uniquely, and to enable a user of the descriptions to know whether a specific substance was what the descriptions described. Such reference underlaid the interchangeable use of names for substances using

---

[1] Recent work on these subjects includes Häggqvist and Wikforss 2017; Havstad 2017; Bursten 2014; and Hendry 2016.

Geoffrey Blumenthal, James Ladyman, and Vanessa Seifert, *Referring to Chemical Elements and Compounds* In: *What Is A Chemical Element?*. Edited by: Eric R. Scerri and Elena Ghibaudi (2020). © Oxford University Press.
DOI: 10.1093/oso/9780190933784.003.0005

terms from different theories, and hence the continuity of reference in chemical practice.

This chapter proposes a causal-descriptive theory of reference for chemical substances, by considering how particular types of air were referred to in late-eighteenth-century chemical practice as well as the problems of existing theories of reference for natural kinds. Implications for debates about incommensurability and realism are also noted.

The next section describes how specific participants referred to and distinguished between types of air. The third section gives generalizations concerning these cases and discusses reference to elements. The last section is a philosophical commentary that examines the descriptive, pure-causal, and causal-descriptive theories of reference. The paper concludes that a version of the causal-descriptive theory of reference fits best how late-eighteenth-century chemical practice referred to substances.

Some Actual Occurrences of Reference to Types of Colorless Air in Late-Eighteenth-Century Chemistry

In the primary literature of late-eighteenth-century chemistry there are the beginnings of reference to a number of newly discovered types of air and their compounds. Six examples of colorless air are considered here, each of which were given several names. The first names that were given to each of the six colorless airs were "nitrous air," "acid air," "fixed air," "dephlogisticated air,"[2] "phlogisticated air,"[3] and "inflammable air."[4] One more example is considered here, that of a green gas, whose first name can be translated as "dephlogisticated acid of salt."[5] These examples show a variety of things about reference, and are discussed in the order above. This section concentrates on selected information directly from the primary literature. Generalization and interpretation are in subsequent sections.

The first of our airs was found by Stephen Hales (1727), who dissolved Walton pyrites in spirit of nitre.[6] He described that when this air mixed with common air, there resulted a turbid red mixture, in which part of the common air was absorbed.[7] Hales did not identify this as a separate type of air, while Henry

---

[2] These four types of air are now known, respectively, as "nitric oxide," "hydrogen chloride" (gas), "carbon dioxide," and "oxygen."

[3] The usage of this term settled in 1785 on the air now known as "nitrogen."

[4] This initially indicated the air now known as "hydrogen," but the ambiguity of the name led to confusion between this air and other types of inflammable air.

[5] Now known as "chlorine."

[6] In modern terms, he reacted iron disulphide with nitric acid.

[7] As quoted in Priestley 1772, 210.

Cavendish (1766, 145), who dissolved metals in nitrous acid, only identified that the resulting air was not inflammable. After reading Hales and talking to Cavendish, Priestley (1772, 210) dissolved metals in spirit of niter and found the air, he explored its properties, and he named it "nitrous air."

The second air was found by Cavendish (1766, 157–158), by dissolving copper in spirit of salt.[8] He was engaged in investigating the production of inflammable air, and only noted that this air was not inflammable. Citing Cavendish's publication, Priestley (1772, 234) repeated Cavendish's method of production, by collecting the air over mercury. Priestley undertook further experiments, and found that the air was more simply generated from the acid alone. If water was introduced over the mercury, it rapidly absorbed the air and became acid. At that time, Priestley titled his work as being on "the air procured by means of spirit of salt," and he named it "acid air" two years later (1774, 143). Among its later names was "marine acid air" to distinguish it from "vitriolic acid air" (e.g., Priestley 1777, 298).

The third air was found when Joseph Black (1755) experimented on magnesia alba concerning which he cited Hoffman's publication that gave two methods of preparation from naturally occurring substances.[9] Black (1755, 6) repeated one of these methods of preparation. Black (1755, 14) found that when heated, magnesia alba lost more than half of its weight. He referred to Hales's (1727) publication that showed that air was released from substances, and in this case Black (1755, 19) found that the original magnesia alba differed from the magnesia usta that remained after heating, chiefly in that the former contained a considerable quantity of air.[10] Black (1755, 18) showed that this type of air could be separated from alkaline substances by dissolution in acids or by heating, and that magnesia alba could be decomposed and recomposed. He referred to the air he found as "fixed air," and in doing so appropriated and narrowed the meaning of the term that Hales had used for any air that was contained in a substance. Cavendish (1766, 141) specifically referred to Black and stated that he adopted Black's narrowed usage of the term, when undertaking more experiments on it. Priestley referred to several previous authors in his (1772) section on fixed air. The name "fixed air" later caused problems that were not due to the descriptions of the air, when researchers assumed that the air in metallic calces was "fixed air."[11] Examples are Lavoisier in his *Opuscules Physiques et Chymiques* (1774) and Priestley's *Experiments and Observations on Different Kinds of Air* (1774, 193), in

---

[8] Now known as hydrochloric acid, the aqueous solution of hydrogen chloride gas.

[9] Magnesia alba is now known as "magnesium carbonate."

[10] Magnesia usta is now known as "magnesium oxide."

[11] In Black's usage, "fixed air" meant the air now known as carbon dioxide, but the air that was fixed in metallic calces is now known as oxygen.

which Priestley heated mercury calx and released from it a quantity of air that he identified as the purest fixed air he had ever produced.

The fourth air was identified by Priestley (1775b, 34) who heated mercury calx[12] and found that in the air that had been released from the calx, a candle burned more vigorously than in common air. He used these two descriptions when he (1775b, 36) informed Lavoisier and other French chemists verbally in October 1774 about the new type of air.

While these two descriptions sufficed for basic communication, it was necessary to clarify issues concerning impurities in the new type of air. Priestley (1775b, 49; cf. Lavoisier 1780c, 327) noted that it was necessary to extract "fixed air" from a calx before the new "pure air" was emitted.[13] Lavoisier (1780b, 67) also clarified that this "pure air" was never obtainable entirely free from impurities of "mofette," which was Lavoisier's current name for the gas that Priestley had called "phlogisticated air." Also, Lavoisier (1778, 523–525) clarified in more detail the difference between the two types of air, by using lists of descriptions. The descriptions for "fixed air," which were mostly based on the work of Joseph Black, stated that "fixed air" (1) was susceptible to being combined with water by agitation, (2) caused animals placed in it to perish, (3) caused candles placed in it to be extinguished, (4) precipitated lime water, and (5) combined easily with fixed or volatile alkalis. The descriptions for "pure air," which were based on Priestley's work, were that it (1) was not susceptible to being combined with water by agitation; (2) did not precipitate lime water, but affected it in an almost imperceptible manner; (3) did not unite with fixed and volatile alkalis; (4) did not at all diminish the causticity of alkalis; (5) was capable of re-calcining metals; and (6) sustained the life of animals better than common air and caused candles to burn more brightly.

The descriptions that have been quoted so far did not involve hypotheses concerning the inner constitution or dispositions of such types of air. It is noteworthy that such hypotheses were not included in Priestley's original verbal communication nor in Lavoisier's lists. Priestley commenced producing such hypotheses alongside his detailed publication, while Lavoisier first used the descriptive terms "pure air" or "eminently respirable air" or "vital air," and coined the term "oxygen" three years later. Priestley (1775b, 48) hypothesized that "being capable of taking more phlogiston from nitrous air, it therefore originally contained less of this principle" and he accordingly named the new air as "dephlogisticated air."[14]

---

[12] In closed conditions resulting in a volume of air in the glass vial.

[13] "Pure air" was one of Lavoisier's names for the air that Priestley called "dephlogisticated air," and which Lavoisier later called "oxygen."

[14] However, he did not apply his concept of "dephlogisticated air" back into the experiment that he had done to create the air in the first place. If he had done so he would have found a problem, in that there was no free air present within the glass vial from which phlogiston could be absorbed (Blumenthal and Ladyman 2017a).

Priestley also noted that red precipitate was produced by a solution of mercury in spirit of niter,[15] and he now hypothesized that the peculiar nature of the new air was due to nitrous acid. He then speculated that since *mercurius calcinatus* was produced by exposing mercury to a certain degree of heat in common air, it must have "collected something of nitre...from the atmosphere" (1775a, 35). He speculated that "there is a regular gradation from dephlogisticated air, through common air, and phlogisticated air, down to nitrous air; the last species of air containing the most, and the first-mentioned the least possible phlogiston" (1775b, 392). He stated that "there remained no doubt in my mind ... that atmospherical air, or the thing that we breathe, *consists of the nitrous acid and earth*, with so much phlogiston as is necessary for elasticity, and likewise so much more as is required to bring it from its state of perfect purity to the mean condition in which we find it" (1775a, 55). Lavoisier (1779, 680) argued that it would be "dangerous" if Priestley's statement that atmospheric air consisted of nitrous acid and earth gained credit. Lavoisier's final name for the new air was based on a generalization from his (1779; 1780c) work that showed that pure air was present in nitrous acid, vitriolic acid and phosphoric acid, and from his (1781, 539) identification that more than one type of nitrous acid could be produced, and that the type with more pure air was more acid. He hypothesized that the new air could generate acidity in compounds, and so called it "oxygen." Scheele (1931) also gave the air the name "fire air."

Despite the existence of such disputed hypotheses, many examples show that reference to this type of air was stable and that communication about this air was easy in practice, despite the use of several different names for it. Lavoisier (1780b, 67) refers to "dephlogisticated air or eminently respirable air," giving the phlogistic term as well as his own term. Westrumb (1791) in a single paper used phlogistic terms such as "dephlogistisierte" as well as all of "Säurestoff" (oxygen), "Lebensluft" (vital air), and "reine Luft" (pure air). In the letter from Lavoisier's colleague Laplace to the phlogistian Deluc dated June 28, 1783 (Hahn 2013, v.1, 111–114), Laplace stated very briefly that he and Lavoisier had undertaken the combustion of dephlogisticated air and inflammable air by bringing together two currents, one of pure air and the other of inflammable air. This shows that Laplace used "dephlogisticated air" interchangeably with "pure air."

The fifth air was first identified and published by Daniel Rutherford in 1772 (Dobbin 1935). Rutherford referred to information received from Cullen and Black and from Hales's book, but does not seem to have taken Cavendish's (1766) paper into account. Rutherford identified experimentally that Black's "fixed air" differed from the air breathed out by animals (Dobbin 1935, 373), in that when

---

[15] Now known as nitric acid, HNO_3.

fixed air was removed from the latter, there still remained an unbreathable air, which was identical to air in which metals had been calcined (Dobbin 1935, 374). During the same period Priestley was also working on many of the same sorts of air that Rutherford did, with, in some cases, similar but many more and better-explored results. Priestley (1772, 228) identified that the reduction of air in cal-cination was greater than in any other of the cases he explored, including the air in which animals had breathed and died, so he hypothesized that in other cases there must have been "causes of addition." But despite all the distinctive descriptions he (1772) gave for the air resulting in different cases, he subse-quently tended to use the name "phlogisticated air" for both the air left over after calcination and the air breathed out by animals.

Lavoisier then used two lists of descriptions, of which individual descriptions had already been published by Priestley, in order to distinguish between these two types of air. Lavoisier (1780b, 187) first described the air left over after the calcination of mercury in closed conditions. This (1) never precipitated lime water, (2) extinguished candles, (3) caused animals placed in it to perish, (4) formed hardly any red vapors with nitrous air, (5) was not sensibly dimin-ished by nitrous air, and (6) had been diminished by about one-sixth relative to the original air within the experiment. Lavoisier (1780b, 188) second described the air that remained after a sparrow has been introduced into a specific volume of air in closed conditions, and had then died. This air (1) precipitated lime water, (2) extinguished candles, (3) was not diminished by nitrous air, and (4) had been diminished by only one-fortieth relative to the original air within the experi-ment. When he introduced a small amount of caustic alkali into such air, the amount of air was further diminished until it was about one-sixth of the original air within the experiment. The change to the caustic alkali was of the type which that substance underwent after absorbing fixed air. Accordingly, the air left over after breathing contained almost one-sixth fixed air.

The sense of "phlogisticated air" was later narrowed without this being un-derstood at the time. Cavendish (1785, 381) paraphrased Lavoisier's list for the air left over after calcination. Cavendish (1785) found that nitrous acid could not be produced in an experiment in which only phlogisticated air was present, and could only be produced in an experiment in which dephlogisticated air (and water) was also present.[16] Subsequently, Berthollet in his letter to Blagden on June 17, 1785,[17] stated that the Académie considered that Cavendish had combined dephlogisticated air with phlogisticated air to form nitrous acid (Sadouin-Goupil

---

[16] At that time, Cavendish attempted to maintain his view that nitrous acid plus phlogiston pro-duced nitrogen (see also Cavendish 1784).

[17] This letter is Royal Society reference number CB1/1/217, folio number b.126a, and was published by Sadouin-Goupil (1971).

1971, 92), and Cavendish adopted the same view in the title of his (1788) paper. This effectively provided another description for "phlogisticated air"—it was the air that when combined with dephlogisticated air formed (anhydrous) nitrous acid. In effect, it was with this additional description that the term was used in the *Nomenclature* (Guyton et al. 1787) and in Lavoisier's (1790) *Traité*, in which were listed the compounds of "phlogisticated air" under the Lavoisians' name of "azote," later named "nitrogen" by Chaptal. It was not understood at the time that this addition of the description of the combination of this air with dephlogisticated air to the previous list of descriptions had narrowed down the reference of the full list to the substance named "nitrogen."[18]

The sixth air was first discussed in print by Robert Boyle who obtained it by dissolving iron filings in dilute acids. Cavendish (1766, 144) produced this type of air by dissolving zinc, iron, or tin in diluted vitriolic acid or in marine acid, and he recognized it as a separate type of air, which he named "inflammable air." In this case, the description of the method of production of this air was sufficient to pick it out definitely. However, the name used by Cavendish was ambiguous and produced much confusion over the next thirty-five years as researchers did not always distinguish this inflammable air from other types of inflammable air.

The seventh air that is discussed in this paper was a green gas that was discovered by Scheele (1931, 20, 28–32) in 1774. His recipe for this air was clear— he reacted "manganese" with marine acid.[19] Scheele's name for the air has been translated as "dephlogisticated acid of salt." In Berthollet's letter to Guyton of May 4, 1785,[20] in which he discussed the contending theories in detail, he stated that his language would appear to Guyton to be "quite heterodox," in that he was using names from different theories interchangeably, including "aerated muriate," "dephlogisticated muriate," and "dephlogisticated marine acid." Additional antiphlogistic terms for this air were used, namely "oxygenated marine acid," "oxygenated muriatic acid," or "oxygenated muriate." Participants including Westrumb (1790; 1791) continued to use these terms and other variant names interchangeably to refer to this air without causing lack of understanding among those who were debating, as the sequence of papers shows. This was the case even though the inner constitution of this air remained a continuing subject of contention until after 1800.

---

[18] Previously, Lavoisier's (1780b, 187–188) and Cavendish's (1785, 381) lists had picked out an air that was roughly 98.8 percent nitrogen but about 1.2 percent argon, but the new usage picked out nitrogen.

[19] In modern terms, he reacted manganese dioxide with hydrochloric acid, the aqueous solution of hydrogen chloride, producing chlorine.

[20] This is the first in the collection of letters from Berthollet to Guyton that is in the collection that is housed in the dossier biographique Guyton de Morveau in the Archives of the Académie des Sciences.

These examples of the initial identifications, descriptions, and naming of types of air provide a resource for future discussions of reference in chemistry. In the next section, the seven examples of this section are used as the basis for some generalizations, interpretations, and points about elements.

Generalizations, Interpretations, and Points about Elements

The previous section shows that in the case of nearly all the examples, reference to a type of air started without the air being named and by using descriptions that tended to include a method of production of the air and at least one distinctive observable property related to the air.[21] The exception was "dephlogisticated acid of salt," concerning which Scheele presented the name prior to describing the qualities of the air.

In cases in which initial descriptions were deemed to be vague, more descriptions of distinctive observable properties were added as desired. This was usually done in passing. While Lavoisier's two pairs of lists of descriptions show the process particularly clearly, it is noteworthy that it was not common practice to produce such lists. In most of the examples, such descriptions of a type of air were stated prior to theories about the inner constitution of the air, or about the general dispositions of an air concerning reactions with other substances, being stated. That is, testable descriptions were used for reference to a substance, usually prior to untestable or only-partly-tested theories concerning the substance.

When names were added, they were generally less definite than the descriptions. In several instances this caused problems, as has been briefly seen in the cases of "fixed air," "phlogisticated air," and "inflammable air." In each of the seven examples of airs, more than one researcher gave a name to the same air, and many names accumulated for some airs, as has been seen in the cases of oxygen and chlorine. While some researchers routinely used terms and substance names belonging to a single type of theory, the examples show several other instances—involving Laplace, Berthollet, and Westrumb—in which a researcher adhering to one theory wrote to or concerning the work of a researcher adhering to another theory, using more than one name for a specific substance interchangeably, even though terms within those names formed part of the terminology of different theories.[22] For example, the substance name "dephlogisticated air" translated virtually exactly with other names for "oxygen," although "dephlogisticated" did not translate into an antiphlogistic term.

---

[21] Kuhn and Kitcher (1993, 97–98) seem to think that language like "dephlogisticated air" or "oxygen" was integral to Priestley's and Lavoisier's thinking from the outset, but this was not the case.
[22] Contrary to the claims of Kuhn (e.g., 1983, 1990).

There are a number of general views that all the researchers involved appear to have implicitly held, and that support all the points above. These general views were at least partly based on existing experimental evidence concerning substances, but were also partly general theories or assumptions. The first view was that investigation by experiment was of substances in the real world, and there did not need to be any doubt about this. The second view was that different types of substances were definitely distinguishable from one another, whether or not it was practicable to identify such distinguishing features with current experimental apparatus and expertise. The third was that samples of substances routinely included impurities, so that discussion concerning substances could take place without frequent explicit reference to the presence of impurities. The fourth view was that there could be many different names for and theories about a particular type of substance, which nevertheless each referred to that same particular type of substance.

So far, all these points have involved substances, and elements have not yet been considered. Prior to Lavoisier, the concept of a simple (undecompounded) substance existed, but chemists still considered that the basic constituents of matter were few and were usually some selection from among the Aristotelian elements or chymical principles. Emphasizing the latter point of view, Johann Juncker (1757, 132) argued that "the system which gives a particular and immutable form to each sort of thing, in order to establish an astonishing number of primordial entities" is an "ingenious frivolity." It was Lavoisier who applied the concept of simple (undecompounded) substances much more rigorously, although not fully consistently. Whether or not this concept together with then-current experimental techniques actually identified elements was not fully determined. Consequently, the new *Nomenclature* (Guyton et al. 1787) was framed in such a manner as to be deliberately open to alteration, to allow for future experimental findings concerning substances that were simple in Lavoisier's operational sense. In effect, the identification of a simple substance as an element was a hypothesis, which was provisionally used but was deliberately left open for possible alteration as a result of further work.

All the historical detail in the previous section and the generalizations in this section facilitate the philosophical discussion of the differing theories of reference in the next section.

Philosophical Commentary

There was no significant vernacular reference to colorless airs as distinct kinds of substances before they were identified in late-eighteenth-century chemistry, and this reduces the complications in understanding what occurred during reference to these airs. The implications of the history of chemical practice outlined in the above sections for theories of reference to natural kinds are as follows.

According to Putnam (1975) "meaning is not in the head." In the present context it is right that reference could not have been fixed directly by intension because, in the case of the first reference to a colorless air by a researcher, obviously that researcher's knowledge concerning that substance was very limited, so it was totally impracticable to refer using a relatively full indication of the intension of a term for the substance. In nearly all the examples, reference started by using a small number of the eventually possible descriptions of the substance. What generally occurred was that researchers intended to describe how to *pick out* a substance in the external world definitely. For practical purposes, it was a *disadvantage* to use more descriptions than were actually necessary in order to pick out a substance definitely. The small number of descriptions used to pick out the substance tended to include two very specific kinds of description. The first type described at least one method of obtaining the substance by a process involving one or more other substances. The second type described a distinctive property of the substance.[23]

In order to understand reference to colorless airs in chemical practice, it is necessary to note that reference-fixing could be scientifically useless or could be undecidable in a number of ways. Indeed, during this period, attempts at reference sometimes failed and the relevant theoretical terms had no referent. Obvious examples are when reference was to phlogiston, but other examples include Priestley's "principle of acidity" and Cavendish's reference to the water that was (claimed to be) present in calces as a result of the combination of absorbed dephlogisticated air and the phlogiston that had been in the metal. In some cases reference was achieved, but it was not possible then to know definitely that it had been. An example was when reference was to the oxygen that was the only constituent of calces other than metals.

By contrast, what was scientifically useful in the vast majority of cases was for reference to a substance to be definite, and to enable the listener or reader to know that a referent existed, to know how to pick out the referent, and to know how in practice to determine that a specific substance was the referent.[24]

---

[23] Devitt and Sterelny (1999, 84) argue that solid substances are observable while gases are unobservable. McLeish (2006, 178) also worries about unobservability. However, for experiments on gases in closed conditions, gases were effectively although "indirectly" observable. This usefully retains the distinction between observable and unobservable for that between, for example, oxygen and phlogiston.

[24] Consequently, in all cases there needed to be attributive and referential uses of the descriptions that were used (cf. Donnellan 1966), and in many cases researchers reading the initial published reference to the gas made both types of use of the relevant published descriptions. There is a philosophical debate concerning if such descriptions have referential meanings or if their use is merely pragmatic, in which the latter position is supported by many authors including Grice (1989) and Neale (1990, etc.), and the former position is supported by Devitt (e.g., 1996; 2004). The present paper does not take a position in this debate.

An example of the latter type of reference is that Priestley (1775b) told the French philosophers, including Lavoisier, that he had produced an air by reducing mercury calx, and that in the air a candle had burned with a much higher flame than in common air. This gave the French philosophers epistemological, methodological, and practical access to the type of air. The reference involved definite descriptions; the hearer was enabled to know that the referent exists, how to locate the referent, and that it was the referent. The listener or reader might not actually do any of these things, but he was in possession of all the information needed to do so.

Reference to the type of air was generally subsequently continuous, and this was knowable in all these ways by researchers due to the initial descriptions used in reference. Such reference underlaid the interchangeable use of names for substances using terms from different theories that is seen in the examples.

All this information can now be used to assess the existing theories of reference, the advantages that have been claimed for them, and the objections that have been raised to them. In doing so, it is helpful to consider Locke's theory of reference, which matches very well what happened in late-eighteenth-century chemistry. Locke (1689, v. 3, ch. 3, s. 17) argued that substances have "real essences" that we do not know, but from which follow their "nominal essences": those sensible qualities that serve us to distinguish them one from another. The examples show that any attempt to characterize an aspect of "real essence" was problematic, and that reference-fixing actually involved clearly observable properties. He (1689, v. 3, ch. 4, s. 15) argued that in the beginning of language, it is necessary to have the idea before one gives it the name. This is amply evident at the beginnings of reference to types of air. He (1689, v. 3, ch. 6, s. 18) argued that our nominal essences are not perfect collections of properties, but since we do not know real essences, it is impossible to know all ideas stemming from them.

The points that Locke did not bring out clearly are that reference was via parts of the "nominal essence" that served to pick out the substance relative to other substances, and did not need to involve any more of the "nominal essence" than was necessary to achieve this. Nevertheless, one way in which all subsequent theories of reference can be assessed is to identify the ways in which those later theories did not succeed in preserving the workable and crucial parts of Locke's theory.

Descriptivist Theories of Reference

It will be taken, from the work of Frege and Carnap, that in a descriptivist theory of reference, a single description, or most of a cluster of descriptions, expresses the sense (or intension) of the term, which then determines its reference (or extension). Due to the points made in the present paper, it can be seen that there are several immediate and major problems within a descriptivist theory of reference, in addition to those already proposed. First, during reference-fixing it is wholly impracticable to refer to more than a small number of the descriptions that will express the intension of the term reasonably fully. What actually happens is that a substance in the external world is picked out using a small part of the intension of the term. Moreover, as noted in the existing literature on descriptive theories of reference, when descriptions are added to the intension, this changes the determination of the extension of the word, and arguably implies changing the extension of the word. But in chemical practice, adding to the descriptions of what is known about a substance does not change the substance that is being referred to. A further detailed problem is that in the cases in which Lavoisier used lists of descriptions in referring to a substance, all the descriptions were intended to be part of the intension of the term, not just a selection of them.

However, the usually identified problems of description theories concerning natural kinds, which are mostly due to Kripke, appear to be somewhat less onerous or inapplicable in the cases of the examples of reference fixing to chemical kinds. The implicit "principled basis" of the selection of the descriptions used in reference was that such descriptions should pick out the substance definitely and give a user epistemological, methodological, and practical access to the type of air. The problem of "unwanted ambiguity," arising from variations in descriptions from person to person, was partly solved by the use of the previously published descriptions or by the statement of differences from previously published descriptions, as has been seen in several of the examples. In effect, and contrary to Kripke's line of argument, the names of types of air in several cases caused more problems than the associated descriptions.

The problem of "unwanted necessity," involving unusual environments producing anomalous members of a natural kind, was subject to control in the closed environments in which types of air could be investigated. The problem of the "lost rigidity" of descriptions that refer to different substances in different possible worlds, simply does not apply to chemistry, which (especially in the cases of these types of air) can only be based on research in limited environments in the actual world. Where researchers in chemistry were concerned, the problems of "ignorance and error" were effectively standard problems of expertise or of views in a science. The problem of the "regress" of descriptions, which is that any description uses other descriptions, was at least partly solved because the chain of derivation of substances tended to be short. For example, "dephlogisticated air" was first produced by reducing mercury calx, mercury calx was produced by calcining mercury over a continuous heat in open air, the mercury was produced from cinnabar.

The problem of "incompleteness," in which the intensions of a term for a type of air are never fully specified, was solved in practice by "singling out" an air, but this appears to show an unsolvable problem for a purely description theory of reference.[25] Moreover, the problems of ignorance and error in reference-*borrowing* appear not to be solvable in a descriptivist theory.

Causal Theories in the "Paradigm Case": Pure-Causal Theories

In causal theories of reference, the "paradigm case" (Devitt and Sterelny 1999, 88) of reference-fixing concerning a substance involves ostension and a sample, while reference-borrowing involves a causal chain transmitting the reference to those who have no contact with the referent. The extension of the term is then those examples of stuff that share the underlying essential nature of the samples, which is usually not known.[26] However, in Putnam's (1975) and Kripke's (1980) versions of a causal theory, this underlying essential nature concerns the inner structure of the substance.[27] According to such causal theories of reference, a term can be competently used when a user has no knowledge about what determines the reference of the term. It is this type of causal theory that has usually been taken as the basis for recent discussions of reference in chemistry.[28]

The standard problem with causal theories is the "qua" problem—a member of a natural kind is typically a member of many natural kinds, and in the paradigm case, it is not clear which natural kind is being picked out during ostension. In the light of the present paper, it is clear that there are several immediate and major problems of a pure-causal theory of reference, in addition to those that have been previously proposed.[29] In the case of reference to types of air, the sample could be taken to be one in the laboratory of the researcher who fixes the reference. But in this case, the "qua" problem with ostension seems particularly acute. Moreover, it is difficult to see how reference-borrowing could take place except in the case of other people who witness a repetition of the experiment that produces a sample

[25] Devitt and Sterelny (1999, 100) argue that description theories may work in the context of reference-fixing, but the present paper suggests that a causal-descriptive theory is necessary for reference-fixing as well as reference-borrowing.

[26] The "direct reference" theory states that the meaning of a name is simply its referent (e.g., Soames 2002, 5), while Devitt (e.g., 2012) argues that the meaning of the name is its causal mode of reference.

[27] However, as Devitt (e.g., 2012) pointed out, the commitment to meanings as causal modes does not rest on any particular theory of these modes.

[28] Häggqvist and Wikforss 2017; Havstad 2017; Bursten 2014; Hendry 2005; 2006a; 2006b; 2008; 2010; 2012; 2016. Some works also take into account description theories, but not causal-descriptive theories, for example, LaPorte 2004; Bird 2010.

[29] For example, Needham 2011 has proposed other problems for causal theories of reference.

like the first, and which is demonstrated to be so via an associated experiment that shows a specific effect (such as a candle burning higher than in an equivalent sample of common air). Yet in practice, reference-borrowing does not require such an elaborate procedure. However, the causal theory does not provide any other explanation of how reference-borrowing occurs in practice in the case of a person at a distance.

Also, the idea of a causal chain of reference to an original baptismal event does not match the cases of the actual examples in which communication using descriptions happened prior to the naming of many types of air. There is the further problem that in such a theory, reference to an element should be no more problematic than that to any substance, since both were implicitly due to the unknown inner structure. Yet in practice, reference to a substance was or could be relatively clear, whereas it was a hypothesis whether or not a substance was an element.

Hybrid Causal-Descriptive Theories

The considerations of the present paper suggest that a version of a causal-descriptive theory of reference can fit the actual examples of reference to substances in chemistry (Psillos 1999 proposes the idea of a causal-descriptive theory of reference). To do so it must include the following: First, a causal component according to which reference to substances in the external world involves causal relationships to them. Second, a descriptive component in which the substance is specified via descriptions sufficient to pick out the substance definitely in practice, and to enable the listener or reader to know that a referent exists, to know how to pick out the referent, and to know how it was practicable to determine that a specific substance was the referent. Ideally, such descriptions would be minimally complicated. In practice concerning chemical substances, this usually involved descriptions of methods of production and of clearly identifiable properties that were as non-theory-laden as practicable, and relatively simple descriptions were achieved.[30]

In such a causal-descriptive theory, continuity of reference is assured by the continuing applicability of the original (or developed) descriptions that picked out the substance definitely and that enabled a reader as previously indicated.[31]

---

[30] Pace LaPorte (2004, 144) who argues that dubbing ceremonies presuppose theory. Of course some minimal theory is presupposed in the sense that it is assumed that vessels can hold airs and so on, but crucially the idea is to avoid theory about which there is disagreement among correspondents. In particular, such minimal theory should be independent of claims about the inner constitution or dispositions of the type of air.

[31] LaPorte (2004, 134) acknowledges the possibility of full communication across a theoretical divide. We show how this can occur.

A particular sample is not required. The regress of the descriptions of substances appears to be short as indicated in the above section on descriptions. There is no role for a lengthy causal chain of reference because any reader of the descriptions used in reference could use the term competently and could identify the substance in practice if they wished. Hence, in many cases the causal chain would have only a single link. There is no "qua" problem. Implicit continuity in reference via the unknown inner structure of the substance is not required.[32] There is no problem concerning reference due to the initial vagueness of the intension of a term for a substance.[33] Such a theory matches the realization by Locke (1689, v. 3, ch. 3, s. 17) that while "nominal essences" followed from "real essences," the latter were not known. Such a theory matches how reference occurred in practice and solves the problem of why there can be multiple names for the same substance and yet there are no significant problems in communication about the substance.[34] There is no problem with the additional difficulty in identifying elements, when compared with identifying substances, in late-eighteenth-century chemistry.

Obviously there are implications of many of these points for theories of incommensurability and realism, which deserve further exploration.

References

Bird, Alexander. 2010. "Discovering the Essences of Natural Kind Terms." In *The Semantics and Metaphysics of Natural Kinds*, edited by Helen Beebee and Nigel Sabbarton-Leary, 125–136. Abingdon: Routledge.

Black, Joseph. 1898. *Experiments upon Magnesia Alba, Quick-Lime and Other Alcaline Substances*. Edinburgh: Alembic Club Reprints. First published 1755.

Blumenthal, Geoffrey, and James Ladyman. 2017a. "The Development of Problems within the Phlogiston Theories, 1766–1791." *Foundations of Chemistry* 19: 241–280. https://doi.org/10.1007/s10698-017-9289-0.

Bursten, Julia. 2014. "Microstructuralism without Essentialism: A New Account of Chemical Kinds." *Philosophy of Science* 81: 633–653. https://doi.org/10.1086/678043.

Cavendish, Henry. 1766. "Three Papers, Containing Experiments on Factitious Air." *Philosophical Transactions of the Royal Society of London* 56:141–184. https://doi.org/10.1098/rstl.1766.0019.

Cavendish, Henry. 1784. "Experiments on Air." *Philosophical Transactions of the Royal Society of London* 74: 119–153. https://doi.org/10.1098/rstl.1784.0014 2053-9207.

---

[32] This is contrary to the assumptions of Putnam (1975), Devitt and Sterelny (1999), and Stanford and Kitcher (2000) that such descriptions somehow must be to the inner structure of a sample.

[33] Contrary to LaPorte (2004).

[34] LaPorte points out the problems of description theories (2004, 115) and those of causal theories (2004, 118), but does not take into account that these problems need not occur in the case of a causal-descriptive theory.

Cavendish, Henry. 1785. "Experiments on Air." *Philosophical Transactions of the Royal Society of London* 75: 372–384. https://doi.org/10.1098/rstl.1785.0023 2053-9207.

Cavendish, Henry. 1788. "On the Conversion of a Mixture of Dephlogisticated and Phlogisticated Air into Nitrous Acid, by the Electric Spark." *Philosophical Transactions of the Royal Society of London* 78: 261–276. https://doi.org/10.1098/rstl.1788.0019 2053-9223.

Devitt, Michael. 1996. *Coming to Our Senses*. Cambridge: Cambridge University Press.

Devitt, Michael. 2004. "The Case for Referential Descriptions." In *Descriptions and Beyond*, edited by Marga Reimer and Anne Bedzuidenhout, 280–305. Oxford: Clarendon Press.

Devitt, Michael. 2012. "Still Against Direct Reference." In *Prospects for Meaning*, edited by Richard Schanz, 61–84. Berlin: de Gruyter.

Devitt, Michael, and Kim Sterelny. 1999. *Language and Reality*. Cambridge, MA: MIT Press.

Dobbin, Leonard. 1935. "Daniel Rutherford's Inaugural Dissertation." *Journal of Chemical Education* 12: 370–375. https://doi.org/10.1021/ed012p370.

Donnellan, Keith. 1966. "Reference and Definite Descriptions." *Philosophical Review* 75: 281–304. https://doi.org/10.2307/2183143.

Grice, Paul. 1989. *Studies in the Way of Words*. Cambridge, MA: Harvard University Press.

Guyton de Morveau, Louis-Bernard, Antione-Laurent Lavoisier, Claude-Louis Berthollet, and Antoine-François Fourcroy. 1787. *Méthode de Nomenclature Chimique*. Paris: Cuchet.

Häggqvist, Sören, and Åsa Wikforss. 2017. "Natural Kinds and Natural Kind Terms: Myth and Reality." *British Journal for the Philosophy of Science* 69, no. 4 (April 23). https://doi.org/10.1093/bjps/axw041.

Hahn, Roger. 2013. *Correspondance de Pierre Simon Laplace (1749–1827)*. Turnhout: Brepols.

Hales, Stephen. 1727. *Vegetable Staticks*. London: Innys.

Havstad, Joyce. 2017. "Messy Chemical Kinds." *British Journal for the Philosophy of Science*. https://doi.org/10.1093/bjps/axw040.

Hendry, Robin. 2005. "Lavoisier and Mendeleev on the Elements." *Foundations of Chemistry* 7: 31–48. https://doi.org/10.1023/B:FOCH.0000042886.65679.4e.

Hendry, Robin. 2006a. "Elements, Compounds and Other Chemical Kinds." *Philosophy of Science* 73: 864–875. https:// doi.org/10.1086/518745.

Hendry, Robin. 2006b. "Substantial Confusion." *Studies in History and Philosophy of Science* 37: 322–336. https://doi.org/ 10.1016/j.shpsa.2006.03.002.

Hendry, Robin. 2008. "Microstructuralism: Problems and Prospects." In *Stuff: The Nature of Chemical Substances*, edited by Ruthenberg, Klaus and Jaap van Brakel. Wûrzburg: Königshausen und Neumann.

Hendry, Robin. 2010. "The Elements and Conceptual Change." In *The Semantics and Metaphysics of Natural Kinds*, edited by Helen Beebee and Nigel Sabbarton-Leary, 137–158. Abingdon: Routledge.

Hendry, Robin. 2012. "Chemical Substances and the Limits of Pluralism." *Foundations of Chemistry* 14: 55–68. https://doi.org/10.1007/s10698-011-9145-6.

Hendry, Robin. 2016. "Natural Kinds in Chemistry." In *Chapters in the Philosophy of Chemistry*, edited by Grant Fisher and Eric Scerri, 253–75. Oxford: Oxford University Press.

Juncker, Johann. 1757. *Élémens de Chymie suivant les Principes de Becker et de Stahl*. 2nd ed. Translated by Jacques François Demachy. Paris: Hardy.

Kitcher, Philip. 1993. *The Advancement of Science*. Oxford: Oxford University Press.

Kripke, Saul. 1980. *Naming and Necessity*. Cambridge, MA: Harvard University Press.

Kuhn, Thomas. 1982. "Commensurability, Comparability, Communicability." In *Proceedings of the Biennial Meeting of the Philosophy of Science Association 1982*. no. 2: *Symposia and Invited Papers*: 669–688. https://doi.org/10.1086/psaprocbienme etp.1982.2.192452.

Kuhn, Thomas. 1990. "Dubbing and Redubbing: The Vulnerability of Rigid Designation." In *Minnesota Studies in the Philosophy of Science*, edited by C. Wade Savage, 298–318. Vol. 14 of *Scientific Theories*. Minneapolis: University of Minnesota Press.

LaPorte, Joseph. 2004. *Natural Kinds and Conceptual Change*. Cambridge: Cambridge University Press.

Lavoisier, Antoine-Laurent. 1774. *Opuscules Physiques et Chymiques*. Paris: Durand, Didot.

Lavoisier, Antoine-Laurent. 1778. "Mémoire sur la nature du principe qui se combine avec les métaux pendant leur calcination et qui augmente le poids." *Histoire de l'Académie Royale des Sciences, Année 1775, avec les Mémoires*, 520–526. Paris: Imprimerie Royale.

Lavoisier, Antoine-Laurent. 1779. "Mémoire sur l'existence de l'air dans l'acide nitreux, et sur les moyens de décomposer et de recomposer cet acide." *Histoire de l'Académie Royale des Sciences, Année 1776, avec les Mémoires*, 671–680. Paris: Imprimerie Royale.

Lavoisier, Antoine-Laurent. 1780a. "Expériences sur la respiration des animaux." *Histoire de l'Académie Royale des Sciences, Année 1777, avec les Mémoires*, 185–194. Paris: Imprimerie Royale.

Lavoisier, Antoine-Laurent. 1780b. "Mémoire sur la combustion du phosphore de Kunckel, et de la formation de son acide." *Histoire de l'Académie Royale des Sciences, Année 1777, avec les Mémoires*, 65–78. Paris: Imprimerie Royale.

Lavoisier, Antoine-Laurent. 1780c. "Mémoire sur la dissolution de mercure dans l'acide vitriolique, et sur la résolution de cet acide en acide sulfureux aériforme et en air éminement respirable". *Histoire de l'Académie Royale des Sciences, Année 1777, avec les Mémoires*, 324–328. Paris: Imprimerie Royale.

Lavoisier, Antoine-Laurent. 1781. "Considérations générales sur la nature des acides et sur les principles dont ils sont composés." *Histoire de l'Académie Royale des Sciences, Année 1778, avec les Mémoires*, 535–547. Paris: Imprimerie Royale.

Lavoisier, Antoine-Laurent. 1790. *Traité élémentaire de Chimie*. Paris: Cuchet, 1789. Translated as *Elements of Chemistry*. New York: Dover .

Locke, John. 1689. *An Essay Concerning Human Understanding*. London: Bassett.

McLeish, Christina. 2006. "Realism Bit by Bit: Part II. Disjunctive Partial Reference." *Studies in History and Philosophy of Science* 37: 171–190. https://doi.org/10.1016/j.shpsa.2005.07.010.

Neale, Stephen. 1990. *Descriptions*. Cambridge, MA: MIT Press.

Needham, Paul. 2011. "Microessentialism: What Is the Argument?" *Noûs* 45: 1–21. https://doi.org/10.1111/j.1468-0068.2010.00756.x.

Priestley, Joseph. 1772. "Observations on Different Kinds of Air." *Philosophical Transactions of the Royal Society of London* 62: 147–265. https://doi.org/10.1098/rstl.1772.0021.

Priestley, Joseph. 1774. *Experiments and Observations on Different Kinds of Air*. London: Johnson.

Priestley, Joseph. 1775a. "An Account of Further Discoveries in Air; in Letters to Sir John Pringle, Bart. P.R.S. and the Rev. Dr. Price, F.R.S." *Philosophical Transactions of the Royal Society of London* 65: 384–94. https://doi.org/10.1098/rstl.1775.0039 2053-9215.

Priestley, Joseph. 1775b. *Experiments and Observations on Different Kinds of Air*, vol. 2. London: Johnson.

Priestley, Joseph. 1777. *Disquisitions relating to Matter and Spirit*. London: Johnson. Page refs to Rutt, John T. (ed.) The Theological and Miscellaneous Works of Joseph Priestley, vol III. London (1818a).

Psillos, Stathis. 1999. *Scientific Realism: How Science Tracks Truth*. London: Routledge

Putnam, Hilary. 1975. "The Meaning of 'Meaning.'" In *Mind, Language and Reality*, 215–271. Cambridge: Cambridge University Press.

Sadouin-Goupil, Michelle. 1971. "La correspondence Berthollet-Blagden." In *Histoire de la Chimie depuis le XVIIIe siècle*, 91–97. Paris: Librairie Albert Blanchard.

Scheele, Carl. 1931. *The Collected Papers of Carl Wilhelm Scheele*. London: Bell.

Soames, Scott. 2002. *Beyond Rigidity*. New York: Oxford University Press.

Stanford, Kyle, and Philip Kitcher. 2000. "Refining the Causal Theory of Reference for Natural Kind Terms." *Philosophical Studies* 97: 99–129. https://doi.org/10.1023/A:1018329620591.

Westrumb, Johann Friedrich. 1790. "Neue Bemerkungen über einige merkwürdige Erscheinungen durch die dephlogistisierte Salzsäure." *Chemische Annalen* 1: 3–21, 109–129.

Westrumb, J. F. 1791. Bemerkungen über die Entzündung mehrerer Körper durch brennstoffleere Salzsäure; vom Hrn Professor Arbogast zu Strasburg. Übersetzt und mit einigen Erläuterungen versehen. *Chemische Annalen* 1: 10–31, 137–152.

5

The Changing Relation between Atomicity and Elementarity

From Lavoisier to Dalton

Marina P. Banchetti-Robino, Florida Atlantic University, USA

1. Introduction

Any freshman taking introductory chemistry will learn that an element is a basic substance that cannot be simplified and that an atom is the smallest amount of an element. Yet throughout most of the history of philosophy and early modern chemistry, it was not always clear how the concept of "atom" could possibly be related to the concept of "element," since these notions often implied completely different ways of conceptualizing fundamental reality. In fact, our current understanding of the relation between atoms and elements owes a great deal to the work of nineteenth-century chemist John Dalton, whose atomic theory provided a way of studying and measuring the properties of atoms and elements in a way that would allow chemists to finally understand the chemical relations between them. Dalton's pioneering work and the publication of his *New System of Chemical Philosophy* in 1808 would ultimately lead to Mendeleev's discovery of the periodic nature of valence and to the publication of his periodic table of the elements in *Zeitschrift für Chemie* in 1869.

Before I discuss the way in which Dalton succeeded in chemically reconceptualizing and reconciling "atomicity" and "elementarity," I will give a brief overview of the ways in which these two concepts were understood by natural philosophers prior to the chemical revolution. Although a detailed discussion of atomism, the theory of elements, and chymistry is beyond the scope of this paper, I will also address the development of chemical ontologies, the development of the epistemology of experiment, and the relation between the two, as well as the way in which these were tied to the abandonment of Aristotelian hylomorphism by early modern chymists. This discussion will be followed by an account of Lavoisier's reconceptualization of "elementarity" as an operational concept, which will serve as background for a detailed examination of Dalton's atomic theory. This discussion will help to reinforce the idea that Dalton's

Marina P. Banchetti-Robino, *The Changing Relation between Atomicity and Elementarity* In: *What Is A Chemical Element?*. Edited by: Eric R. Scerri and Elena Ghibaudi (2020). © Oxford University Press. DOI: 10.1093/oso/9780190933784.003.0006

reconceptualization of "atomicity" in relation to "elementarity" was truly groundbreaking and significant for the later development of modern chemistry, as well as for our understanding of the chemical constitution of material bodies.

2. "Atomicity," "Elementarity," and the Fundamental Nature of Material Reality

The concepts of "atomicity" and of "elementarity" have a long and venerable history both in philosophy and in chemistry and it would take us far from the scope of this essay to attempt a detailed history of these concepts. Thus the discussion here will be limited to a brief outline of these notions to make the point that, although "atomicity" and "elementarity" both served to elucidate the fundamental nature of substance, these notions were not regarded as interchangeable. In spite of this, it is interesting to note the way in which some early modern chymists attempted to theorize an operational conception of elementarity within the context of atomistic or corpuscularian conceptions of matter, thus anticipating the groundbreaking work of John Dalton. With this in mind, let us briefly examine the concepts of "elementarity" and "atomicity" from their classical roots to their appropriation by early modern chymists.

For both classical and early modern atomists, atomism was a thesis purporting to establish claims about the fundamental compositional nature of matter. They regarded atoms as all having the same properties of shape, size, and motion and as being indivisible or unsplittable, that is, as being mereologically simple and irreducible. Since such entities were considered unobservable in principle and their properties were not empirically determinable, the belief in their existence was founded more on metaphysical speculation than on sound empirical evidence. Thus, rather than being an empirical notion, the notion of "atom" was the idea of ultimate substances that grounded all material being.

Since *atomos* were understood as being indivisible, one of the debates that ensued as a result of the early modern revival of atomism was whether any physical particle could ever truly qualify as an "atom." At one extreme of this debate, we find René Descartes who argues that since the essence of matter is extension, all material bodies, no matter how small, are extended and therefore divisible in principle. Thus, no physical particle can be truly atomic. The only true atom, for Descartes, is the mathematical point because it is dimensionless. On the other extreme of this debate, we find Giordano Bruno, who develops a threefold understanding of "atomicity" to account for the different ways in which something can be considered mereologically minimal. Thus, for Bruno, the mathematical *minimo* is the Point, the physical *minimo* is the Atom, and the metaphysical *minimo* is the Monad. Most natural philosophers fell somewhere between these two extremes and several chose to postulate corpuscles, rather than atoms, in order to dispense with the notion of indivisibility. One can safely state that this debate over atomicity and indivisibility reflects the intimate relationship that persisted between metaphysics and natural philosophy in the seventeenth and early eighteenth centuries.[1]

In spite of efforts by Gassendi and other early modern atomists to conceptualize "atomicity" in a chemically viable manner, the revival of Democritean atomism would not be of great use to early modern chymistry since the creation of a chemical theory was never the goal or purpose of atomistic ontology. Rather, both classical and early modern atomism "aimed at an explanation of the riddle of the universe; it pretended to solve the difficult problem of change. The atomistic philosophers illustrated their doctrine by examples taken from observation, but their ideas were founded on metaphysical speculation, not on the data of experience. Theirs was a science of the mind unaided by science of the hands" (Hooykaas 1949, 65). We shall see presently how the "science of the hands" would require an ontology that was both compatible with and epistemically supported by chemical practice and experiment.

Just like the concept of "atomicity," the concept of "elementarity" has a long and venerable history. From classical Greek cosmology to early modern science before Lavoisier, "elementum" or elements were regarded as property-bearing first principles. One well-known early example of a theory of elements is Empedocles's theory of the four elements, which was later revised by Aristotle. In spite of some differences between the Empedoclean and Aristotelian theories, both philosophers regard the four elements and their associated properties as the first principles for the properties of all compound material bodies. However, for Aristotle, the four elements are neither compositionally fundamental nor mereologically irreducible. Instead, he considers amorphous matter as fundamental, while the four elements represent the ways in which this amorphous matter is manifested and organized in the world of existing things. Although many post-Aristotelian natural philosophers rejected his theory of the four elements, they still accepted the view that the properties of material bodies depend upon those of constituent elements or first principles.

Like atoms, elements were generally regarded as unobservable substances with no empirically determinable properties. For example, chymists like Geber (Newman 2006), Stahl (Stahl 1730), and Rouelle (Lehman 2009) considered that the elements or principles that formed the basic layer of their chemical ontologies

---

[1] For more detailed discussions of the relation between atomism, corpuscularism, and early modern chymistry, I refer the reader to the excellent work done by William R. Newman, Alan Chalmers, Lawrence Principe, Hiro Hirai, Antonio Clericuzio, Peter Anstey, and Ursula Klein, among other important historians and philosophers who have made significant contributions to this research.

were not experimentally accessible. Even when chymists like Macquer claimed that the substances in the basic layer of their chemical ontologies were experimentally accessible, they were either unable to access them in practice or they were making incorrect inferences from their experiments (Lehman 2009). One example of such incorrect inference is the case of Homberg's conclusion that his experiments on gold had identified its elementary constituents. For early modern chymists, the properties of individual material bodies were considered to be causally dependent on the nature of the elements that compose them, though not on the properties of such elements. In fact, the apparent properties of most compounds were extremely difficult to relate to those of their proposed constituents. Even before the time of Lavoisier, it was realized that, while the properties of substances had to be related in some way to the *natures* of their constituents, they did not match the *properties* that their individual constituents displayed when not combined with other substances.

Both the notions of "atomicity" and "elementarity" had significant, though not always successful, roles to play in the development of early modern chemical ontologies. However, when one considers the role that atoms and elements played in the context of early modern chemical ontologies, one must also consider the important role played by experiment in the development of such ontologies, as well as the significant role of the epistemology of experiment in the rejection of Aristotelian hylomorphism, which had dominated the theory of substance through the Middle Ages and well into the sixteenth century.

3. Aristotelian Hylomorphism and the Scholastic Theory of Substantial Form

In order to understand hylomorphism and the reasons for its eventual rejection, it is important to understand the role played by form in Aristotelian metaphysics. I frame this discussion of Aristotelian forms in mereological terms because form is a fundamentally mereological concept for Aristotle. Aristotle's hylomorphism has its roots in his theory of four causes, the two external causes being efficient (of first) cause and final cause and the two internal causes being material cause and formal cause. For Aristotle, forms are organizing principles that account for the unity, identity, order, and properties of things, whether simple wholes or composite wholes. Mereologically speaking, Aristotle is concerned with what makes composite wholes distinct from mere heaps. In other words, what is the something "over and above" the parts or "elements," to use Aristotle's term, that is present in the whole but not in the heap (Koslicki 2006). That "something" is form or essence, and it accounts for the unity that is present in wholes but that is absent in heaps.

One of the issues that Aristotle addresses in *Metaphysics* Z.17 is mereological hylomorphism, that is, the idea that the form is itself a part of the whole that it unifies. Aristotle rejects mereological hylomorphism because it would ensue in an unacceptable regress. Referring to the form of composite wholes, Aristotle says that it is the substance of each thing and the primary cause of its being (Aristotle 1984, 1041B32–33). To maintain mereological consistency, however, Aristotle also states that "both the form and the matter are proper parts of a matter/form compound" (Koslicki 2006, 727). So, to resolve the problem of potential regress, Aristotle takes the form of a composite whole to be ontologically distinct from the material parts or elements of that whole. Aristotle's concept of form is, therefore, distinctly metaphysical to the extent that "the unity of a substance is a product not just of its elements, but of some further unifying principle, the form, that is not itself an element" (Pasnau 2004, 33). Thus, the theory of form is a "top-down" theory of organization within Aristotelian metaphysics, in that forms are of a distinct ontological kind from the wholes that they unify. The elements that are part of the unified whole belong to the ontological category of matter, while the form belongs to the ontological category of immaterial principle, so that unified wholes are both mereologically and ontologically complex.

Scholastic philosophers undertook the task of clarifying the question of what form is exactly and twelfth-century metaphysicians proposed several answers to this question, the discussion of which is beyond the scope of this chapter. Suffice it to say that the Scholastic conception of substantial form was not a mereological notion as much as an attempt to understand the relationship between elements and mixed substances. The Scholastics inherited their conception of mixts from Aristotle but they modified the Aristotelian view in significant ways. For Aristotle, a genuine mixt is one in which "the ingredients act to change each other so that they cease actually to exist" (Wood and Weisberg 2004, 682). However, Aristotle adds that the ingredients within the mixt "continue to exist potentially, and [so] they can be separated out again" (Wood and Weisberg 2004, 682).

Medieval Scholastics, however, supplemented the Aristotelian conception of mixts with their own notion of substantial form. Unlike Aristotle, Scholastics did not consider the four elements to be fundamental. Instead, they believe that the primary substance is an undifferentiated prime matter. Scholastics then posited substantial forms as immaterial forms that were responsible for endowing fundamental matter with qualities, so that each of the four elements was generated and determined in its essential properties by the presence of a distinctive substantial form. This hylomorphic view impacts on the Scholastic theory of mixts. According to Scholastics such as Aquinas, when two substances are mixed, the substantial forms of the individual ingredients are destroyed and replaced by the substantial form of the homogeneous mixt, which determines the distinctive properties of the mixt. This view implies that the components of a genuine mixt cannot be recovered, since their substantial forms no longer exist. Thus, for Aquinas, true synthesis (*synkrisis*) precludes the possibility of analysis (*diakrisis*). It is precisely the notion that a genuine mixt cannot be analyzed that would incite early modern chymists to reject Aristotelian hylomorphism and the Scholastic theory of substantial form in favor of ontologies informed by chemical practice and experiment.

4. The Epistemology of Experiment and the Rejection of Hylomorphism

The development of an epistemology of experiment by early modern chymists significantly impacted their abandonment of Aristotelian hylomorphism and of the Scholastic theory of substantial form. An important figure in this transition away from the theory of substantial form is the experimental chymist Daniel Sennert, who rejects Aquinas's conclusion that true synthesis precludes the possibility of analysis, since this is clearly inconsistent with experimental evidence, and also rejects the idea that a homogeneous mixt is endowed with its own distinctive substantial form. Sennert is not simply reacting to problems endemic to the theory of substantial form but is also attempting to articulate the distinction that was later made explicit by Diderot and the French materialists between "la matière" and "les matières," that is, between theoretical and abstract prime matter and the diverse and concrete materials or substances of the chymist's laboratory practice (Pépin 2012). Sennert's (and the French materialists') concern over articulating a notion of substance that reflects the concrete experiences and practices of experimental chymists, rather than the abstract speculations favored by Scholastic metaphysics, foreshadows the later approach of Lavoisier who also rejects abstract metaphysical speculations about fundamental substance in favor of an empirical and quantitative approach to understanding elementarity.

Sennert believes that atomism can provide a better theory of mixture than can be provided by the theory of substantial form, although he does not reject hylomorphism entirely. It is worth noting that Sennert disagrees with the mechanistic aspects of classical atomism and develops a hybrid theory that we can call hylomorphic atomism, according to which the different types of atoms are endowed with distinctive immaterial substantial forms that define their essence and their distinctive properties. Sennert's theory is also an early version of a type of chemical structuralism. Although Sennert considers simple atoms to be fundamental particles, he also believes that the observable qualities of different substances are due to the structure according to which their atoms are arranged. So, the structural composition of atoms, along with their substantial forms, determine the specific properties of a substance. Sennert regards such corpuscular aggregations as non-fundamental atomic species that have specific chemical properties observable at the macro-level. There is an interesting interplay between essence and structure in Sennert's theory so that the properties of substances are due both to the substantial forms of their constituent atoms but also to the structure in which those atoms are composed in that specific substance. For him, chemical synthesis simply involves altering the structural arrangement of the fundamental corpuscles to form a new structural composition. When such microstructures are altered, the new structural arrangement of the mixt constitutes a different material species with different observable chemical properties, though the substantial form of the fundamental particles remains unchanged.

It is also important to note that Sennert's conception of "atomicity" is grounded in the limits attained by the analytical method of the laboratory. This notion of atomicity focuses on what cannot be achieved in the context of experimental practice and with the available methods of chemical analysis. "Atomicity" or "elementarity" understood in this way are "negative-empirical concepts." As William Newman has emphasized, the practice of defining either atomicity or elementarity operationally as the limits of analysis can be traced at least as far back as the thirteenth century, at which time it played an important role in Scholastic alchemy. However, the term "negative-empirical concept" was first coined by David Knight in 1967 to describe concepts that are defined as the limits of analysis (Knight 1967). It was then appropriated by Arnold Thackray in 1970 (Thackray 1970) and resurrected in 1993 by Bernadette Bensaude-Vincent and Isabelle Stengers in their discussion of Boyle's concept of element (Bensaude-Vincent and Stengers 1996). Bensaude-Vincent and Stengers explain that "the idea of a negative-empirical concept is . . . a purely epistemological notion. It represents a new type of argument that locates the authority of proof not within reason but within experimental practice" (Bensaude-Vincent and Stengers 1996, 51–52).

We see in Sennert an early attempt to theorize a chemically viable conception of elementarity within the context of a corpuscularian conception of matter, an attempt that was undertaken once more by several seventeenth-century natural philosophers. One of these attempts is made by Pierre Gassendi, who is arguably the strongest proponent of mechanistic atomism in early seventeenth-century France. From the point of view of chemical philosophy, his *Philosophia Epicuri syntagma* (1649) is significant because it marks the shift from the vitalistic corpuscularianism of van Helmont and Sennert to the sort of mechanistic atomism later defended by Walter Charleton, Robert Boyle, and Isaac Newton. Like Sennert, Gassendi believes that "the primordial atoms combine with one another to form compound corpuscles (Newman 2006, 191–192), which he calls "molecules" (*moleculae*). These molecules are stable, compounded corpuscles that cannot be further analyzed but that serve as intermediaries between indivisible atoms and tangible perceptible bodies. Since they are produced by chemical resolution, these molecules are in a certain sense "elementary" although they are not simple particles. However, it must also be stressed that for Gassendi the only true elements are atoms, since they are the only particles that are completely indivisible by natural means. Gassendi believes that there are several intermediary levels of compounded corpuscles between fundamental atoms and concrete bodies, and it is these molecules that compose the traditional chemical "elements" (sulfur, mercury, salt, earth, water). Thus, the notion of molecules endowed with chemical properties is compatible with his mechanistic atomism. Gassendi believes that textural alterations to molecules produce new qualities in substances and that such changes in qualities can be induced by chemical operations (Newman 2006, 192). Gassendi suggests that the molecules of chemical principles characterize the various species of bodies, depending on their proportions and composition. He finds it difficult, however, to distinguish homogeneous bodies with identical molecules from mixed bodies, especially when determining the nature of metals (Pinet 2004).

Gassendi's revision of classical Epicurean atomism had a great influence on the work of Robert Boyle, who attempts to bring the atomic and mechanical philosophies within the compass of experiment. Boyle agrees with Gassendi, against Descartes, that material corpuscles are impenetrable and indestructible by natural means, despite their theoretical divisibility. Boyle describes corpuscles as being indivisible by nature, although they are mentally and divinely divisible. He also claims that corpuscles form clusters or concretions of various sorts that affect the senses in various ways. Although Boyle does not adopt the Gassendian term "molecule," he does avail himself of the concept attached to this term. For him, "elementary" corpuscles, or elements, are corpuscular aggregations and are considered semi-permanent corpuscles because they cannot be further analyzed into smaller particles. The structure of these corpuscular aggregations accounts for the chemical properties of substances, which Boyle calls "essential properties." As Newman points out, such semi-permanent corpuscles can also be termed "chymical atoms" (Newman 2006). These chymical atoms are hierarchically secondary with respect to primary corpuscles, but they are primary with respect to mixed substances. When combined, they form primary mixtures that can, in turn, be combined to form different degrees of mixtures. To the extent that "chymical atoms" are semi-permanent and homogeneous with regard to their essential properties and they cannot be altered by chemical procedures, they can be considered as chemically "elementary."

In fact, Boyle's work signals a change in how elementary substances are conceptualized. To the extent that he accepts the conventional seventeenth-century definition of elements as "substances necessarily present in all bodies (Boas Hall, 27), Boyle rejects the existence of such substances and recognizes only "corpuscles [as the minutest portions of matter present in] all bodies, including those [bodies] regarded by others as elementary (Boas Hall 1968, 27). For him, what other chemists call elements are actually concretions of primary corpuscles. "Chymical atoms" are, for Boyle as for Sennert, the final products of chemical analysis. Boyle's concept of chymical atom is a negative-empirical notion because, although chymical atoms are corpuscular concretions, they are operationally atomic in that they resist any further chemical analysis. As Rocke has pointed out, this "operational criterion of elementarity gradually insinuated itself into the consciousness of chemists, so that by the time Lavoisier first clearly and unambiguously stated it in his classic *Traité élémentaire de chimie*, it could provoke but little controversy" (Rocke 1984, 4–5).

5. Antoine Lavoisier (1743–1794)

It is against this background that Antoine Lavoisier's own perspective regarding the nature of fundamental substances should be contextualized. For Lavoisier, the concept of fundamental substance is a suspect and speculative metaphysical notion and he rejects the epistemic and heuristic value of speculating about the grounding of all material reality. Lavoisier's concerns apply to the still-prevalent habit among the phlogistonists of his day of proposing inferences in terms of substances that were not experimentally accessible or at least not accessible in the supposed compounds in which they were thought to be present, such as phlogiston within metals.

In the preface of the *Traité Élémentaire de Chimie* (1789), Lavoisier makes a clear break with early modern chymistry's still-intimate relation with speculative metaphysics and explains the general principle that he proposes to apply to his chemical studies. He states, "When we first begin to undertake the study of a science, our relation to that science is analogous to that of children. . . . Just as in a child, it is ideas that are the product of sensation, it is sensation that gives birth to an idea, so it is for that individual who begins to undertake the study of the physical sciences: Ideas must only arise as a consequence, as an immediate result of, an experience or sensation"[2] (Lavoisier 1789, viii).

For Lavoisier, just as correct ideas can only arise from experience, a priori notions that are contrived by the imagination or by the faculty of reason unchecked can lead us into serious epistemic and scientific error. He claims that

---

[2] "Lorsque nous nous livrons pour la première fois à l'étude d'une science, nous sommes par rapport à cette science, dans un état très analogue à celui dans lequel sont les enfants. . . . De même que dans l'enfant l'idée est un effet de la sensation, que c'est la sensation qui fait naître l'idée; de même aussi pour celui qui commence à se livrer à l'étude des science physiques, les idées ne doivent être qu'une conséquence, une suite immédiate d'une expérience ou d'une observation."

"the only way to avoid such errors is to suppress or at least to simplify as much as possible our reasoning . . . which alone can lead us astray. . . .Convinced of these truths, I have imposed upon myself the rule of proceeding only from the known to the unknown, and of deducing no consequence that does not immediately derive from experience and observation"[3] (Lavoisier 1789, x–xi). Based upon these strictly empiricist principles of study, Lavoisier concludes that the positing of suspect metaphysical entities is scientifically untenable. Because these types of entities possess no empirically determinable features, their postulation contributes nothing to experimental knowledge. Such positing does not advance but, rather, hinders empirical knowledge about substances, their behavior, their interaction, and their transformation.

The relation between Lavoisier's new chemistry and Newtonian atomism championed by Laplace deserves comment. Lavoisier and Laplace's concepts of expansion, contraction, and change of state utilized Newton's indication that a gas composed of particles repelling one another by a force inversely proportional to distance would obey Boyle's law, which agreed with experiments. So the *physical* side of the new chemistry was corpuscular or particulate and, therefore, Newtonian to that extent. Consequently, Lavoisier must have had an implicit concept of "simple substances" as corpuscles or particles. However, Lavoisier's compositional chemistry relates to lack of comparable evidence about any *chemical* combination between heat and substances, and he specifically states that this was why no table of combinations had been developed for caloric, which in any case he defines as the "cause of heat."

In opposition to the existing chaotic collection of names of substances that had built up ad hoc over many centuries, Lavoisier proposes a reformed nomenclature for chemistry in which the names of compound substances reflect their elementary composition (Hendry 2012). The achievement of this nomenclature was to present and codify a simple, coherent, accurate, and fertile compositional chemistry that could not be matched by any view based on phlogiston. However, this proposed reform of the nomenclature of compound substances in terms of their elementary composition forces the question of what is to be regarded as an "element."

Lavoisier's conception of "element" is in no way associated with the ancient and early modern notion of principle or bearer of properties, though it remains associated with explanatory fundamentality and causal dependence. As will be seen, however, the explanatory fundamentality of Lavoisier's elements is understood as being provisional. His reconceptualization requires that "elements be isolable, and therefore more concrete and material than the 'principles' of

---

[3] "Le seul moyen de prévenir ces écarts, consiste à supprimer ou au moins à simplifier autant qu'il est possible le raisonnement . . . qui seul peut nous égarer. . . .Convaincu des ces vérités, je me suis imposé la loi de ne procéder jamais que du connu à l'inconnu, de ne déduire aucune conséquence qui ne dérive immédiatement des expériences & des observations."

alchemy and the phlogistonists" (Hendry 2005, 32). To clarify what he means by "element," Lavoisier explains that "if by the name of element, we mean simple and indivisible molecules that compose bodies, it is probable that we do not know them: if, on the contrary, we attach the name of element or principle of bodies to the idea of the last point at which analysis arrives, all of the substances that we have not yet been able to decompose by any means are, for us, to be considered elements"[4] (Lavoisier 1789, xii).

For Lavoisier, this analytical definition of element as a "simple substance" serves to supplant any a priori metaphysical speculation about the ultimate principles of which things are made (Hendry 2012, 66). However, as Robin Hendry points out, this analytical, pragmatic, and operational definition provides a criterion for deciding when a substance should be regarded as an element, but it does not tell us *what* the term "element" means (Hendry 2012, 66). Lavoisier's system is the culmination of the experimental program involving the investigation of the analysis and synthesis of chemical substances, and he also considers that chemical analysis goes hand in hand with weighing, which provides a way of regulating, shaping, and validating chemical experiments and theories. Together, these two techniques provide a way to replace the old ideas about elements with the new concept of simple substances with determinable and measurable volumes that can be converted into their corresponding weights. By 1789, Lavoisier uses analysis and weighing methods to produce a list of chemical elements that is shown in figure 1 below. He then subdivides this table into four general categories on the basis of their chemical properties, that is, acid-making elements, gas-like elements, metallic elements, and earthy elements.

Lavoisier understands that the limits of analysis of his time are probably only temporary and that future analytical methods might succeed in further decomposing what he calls chemical elements. He, thus, admits that the table of elements derived by this operational approach is provisional and entirely open to revision. He states, "We cannot assure that the substances that we regard as simple are not themselves composed of two or perhaps a greater number of principles. However, since these principles cannot be separated or, rather, since *we* have no means of separating them, they behave for us in the manner of simple substances, and we must not assume them to be composed until experience and observation prove otherwise"[5] (Lavoisier 1789, xii–xiii). Thus, although the

---

[4] "Si par le nom d'élémens, nous entendons désigner les molécules simple & indivisibles qui composent les corps, il est probable que nous ne les connoisons pas: que si au contraire nous attachons au nom d'élémens ou de principe des corps l'idée du dernier terme auquel parvient l'analyse, toutes les substances que nous n'avons encore pu décomposer par aucun moyen, sont pour nous des élémens."

[5] "Non pas que nous puissions assurer que ces corps que nous regardons comme simples, ne soient pas eux-mêmes composes de deux ou même d'un plus grand nombre de principes, mais puisque ces principes ne se séparent jamais, ou plutôt puisque nous n'avons aucun moyen de les séparer, ils agissent à notre égard à la manière des corps simples, & nous ne devons les supposer composes qu'au moment où l'expérience & l'observation nous en auront fourni la prévue."

TABLEAU DES SUBSTANCES SIMPLES.

	NOMS NOUVEAUX.	NOMS ANCIENS CORRESPONDANTS.
Substances simples qui appartiennent aux trois règnes, et qu'on peut regarder comme les éléments des corps.	Lumière.	Lumière.
	Calorique.	Chaleur. Principe de la chaleur. Fluide igné. Feu. Matière du feu et de la chaleur.
	Oxygène.	Air déphlogistiqué. Air empiréal. Air vital. Base de l'air vital.
	Azote.	Gaz phlogistiqué. Mofette. Base de la mofette.
	Hydrogène.	Gaz inflammable. Base du gaz inflammable.
Substances simples, non métalliques, oxydables et acidifiables.	Soufre.	Soufre.
	Phosphore.	Phosphore.
	Carbone.	Charbon pur.
	Radical muriatique.	Inconnu.
	Radical fluorique.	Inconnu.
	Radical boracique.	Inconnu.
Substances simples, métalliques, oxydables et acidifiables.	Antimoine.	Antimoine.
	Argent.	Argent.
	Arsenic.	Arsenic.
	Bismuth.	Bismuth.
	Cobalt.	Cobalt.
	Cuivre.	Cuivre.
	Étain.	Étain.
	Fer.	Fer.
	Manganèse.	Manganèse.
	Mercure.	Mercure.
	Molybdène.	Molybdène.
	Nickel.	Nickel.
	Or.	Or.
	Platine.	Platine.
	Plomb.	Plomb.
	Tungstène.	Tungstène.
	Zinc.	Zinc.
Substances simples, salifiables, terreuses.	Chaux.	Terre calcaire, chaux.
	Magnésie.	Magnésie, base de sel d'Epsom.
	Baryte.	Barote, terre pesante.
	Alumine.	Argile, terre de l'alun, base de l'alun.
	Silice.	Terre siliceuse, terre vitrifiable.

Figure 5.1. Antoine Lavoisier, table of simple substances from *Traité Élémentaire de Chimie* (1789) [Public domain]

properties of compound substances are considered ontologically dependent on the properties of the chemical elements that compose them, this dependence is epistemically provisional until these elements can be further analyzed into simpler substances to reveal a deeper level of ontological dependence.

It should be mentioned that Lavoisier's analytical definition is not entirely original with him. According to Paul Needham, this type of definition dates as far back as the writings of Aristotle, although the latter's conception of elements is obviously quite distinct from that of Lavoisier (Needham 2009, 149–164). What is original to Lavoisier is the combination of this type of definition with an open-mindedness about how many substances might be simple, as well as the elimination of the version of the elements and principles that were still present in Macquer's 1749 textbook. Although we note that Lavoisier's operational definition of element may seem similar to Boyle's notion of "chymical atom" as the undecomposed substance of the laboratory (Levere 2001, 81), there are fundamental differences between these two notions. Although Boyle accounts for his inability to analyze such substances by theorizing that their microstructure cannot be decomposed into their constituent primary corpuscles, Lavoisier refuses to theorize about what might be occurring at the micro-level to explain why a substance cannot be further analyzed.

There is no accessible way in Lavoisier's system to connect mereologically fundamental particles, such as corpuscles or atoms, with chemical elements. Restricting himself to the empirical facts means refraining from making any connections between atoms as fundamental particles and elements. According to Lavoisier, there is a profound philosophical distinction between them (Levere 2001, 81). As noted above, Lavoisier's system of chemistry depends on the consistent use of an experimental method that had not previously been systematically applied in chemistry, a method guided only by experimental evidence and by the facts of the senses (Levere 2001, 80). Although Cavendish, the most scrupulous of the phlogistonists, undertook quantitative investigations in terms of volume, these did not provide the information that would have been provided by measurements of weight and that would have, in fact, provided evidence against Cavendish's phlogistic views.

Eighteenth-century chemists like Lavoisier were also turning their attention to the importance of quantification and to the role of mathematics in chemistry. Therefore, besides the evidence of the senses, the most convincing chemical evidence for Lavoisier is what can be measured and built into a quantitative system (Levere 2001, 80). Elements, as simple substances that cannot be further analyzed, meet these two criteria. They have empirically determinable features, they can be weighted using the precision balance, and their volumes can be converted into the corresponding weights. For Lavoisier, weights are ways of regulating, shaping, and validating chemical experiments and theories. For him, chemical analysis and weighing go hand in hand. Together, these techniques give him a way to replace the old ideas about elements with the new concept of simple substances whose volumes and weights can be determined. For him, chemical theories based on anything other than experimental evidence and quantitative measurement are simply worthless, and that goes for every theory of the elements or of atoms that had been proposed by his predecessors (Levere 2001, 80).

Because Lavoisier's system of chemistry precludes connecting chemical elements with fundamental particles, the formerly intimate relation that had existed between metaphysics and early modern chymistry is decisively severed. Thus, following Lavoisier's pronouncements against speculative theorizing in chemistry, no atomic theory could garner credibility unless it could somehow disentangle "atomicity" from its association with metaphysics and render it into an empirical and quantitative notion. This disentanglement and transformation of the concept of "atomicity" would finally occur as a result of the development of chemical atomism by nineteenth-century chemist John Dalton.

6. John Dalton (1766–1884)

Dalton's *New System of Chemical Philosophy* (1808) affirms his commitment to Lavoisier's operational and analytical definition of elementarity, which had become conventional for chemists by the time we reach the nineteenth century (Boas Hall 1968, 21). In a manner similar to Lavoisier, Dalton states that "by elementary principles, or simple bodies, we mean such as have not been decomposed, but are found to enter into combination with other bodies. We do not know that any one of the bodies denominated elementary, is absolutely indecomposable; but it ought to be called simple, till it can be analyzed" (Dalton 1808, 222).

In spite of agreeing with Lavoisier regarding elementary principles, however, Dalton's meteorological studies on the properties and behavior of gases inspire a number of important questions that ultimately lead him to inquire about the compositional nature not only of gases but of all elementary substances. He asks, "Why do different gases have different solubilities in water?" "Why are light and elementary gases such as hydrogen and oxygen least soluble, while compound gases such as carbon dioxide are very soluble," and "Is solubility proportional to density and complexity?" These questions fuel Dalton's desire to ascertain the compositional nature of gases and of elementary substances but do so in a way that meets Lavoisier's strict empirical and quantitative requirements.

Dalton ultimately decides on the assumption, which he believes is supported by observations, that gases and chemical elements are composed of mereologically "ultimate particles" or atoms, and he conceptualizes atoms as dense spherical particles, each surrounded by a subtle fluid or "caloric" that prevents these particles from being drawn into actual contact with one another. He claims that this is proven by the observation that the bulk of a body may be diminished by abstracting some of its heat (Dalton 1808, 144). It is noteworthy that this is the material version of Lavoisier and Laplace's dual material-or-kinetic theory of heat. Dalton's theory of change of state does not take into account latent heat, which was first formulated by Black around 1761, following evidence from Mairan in 1749 and Cullen in 1756 on the cooling of solids when liquids evaporated from their surfaces, and it does not take into account the evidence from sources (including Martine in 1740) that specific heat did not depend on the weight of a substance.

Dalton regards chemical reactions as the shuffling and reshuffling of atoms into new clusters and, influenced by Newton's idea of forces of attraction, he writes, "Observations have tacitly led me to the conclusion which seems universally adopted, that all bodies of sensible magnitude, whether liquid or solid, are constituted of a vast number of extremely small particles, or atoms of matter bound together by a force of attraction, which is more or less powerful according to circumstances, and which . . . endeavours to prevent their separation, and is very properly called . . . *affinity*. . . .Besides the force of attraction . . . we find another which comes under our cognizance, namely, a force of repulsion" (Dalton 1808, 141–143).

Although the view expressed here may seem like a return to the speculative atomism that had been supplanted by Lavoisier, Dalton buttresses his chemical atomism by postulating that there are as many distinct atoms as there are distinct chemical elements. He assumes that atoms of the same element are similar in shape and mass but differ from atoms of other elements. He postulates that the properties of a compound substance are determined by the properties of its constituent elements and that the properties of each element are determined by the properties of the distinctive atoms of which it is composed. He thus reconceptualizes the notions of "atom" and of "element" by subsuming both under the concept of fundamentality understood as ontological dependence.

Dalton regards the antagonism between the forces of attraction and repulsion affecting atoms, as well as the quantities of heat surrounding each atom, as accounting for the differences between gaseous, liquid, and solid bodies (Dalton 1808, 144). He considers that gases consist of corpuscles that repel each other when exposed to heat and that one must differentiate these corpuscles not only by size and shape, as traditional atomists would have it, but also by weight (Bensaude-Vincent and Stengers 1996, 113). Regarding the solubility of gases, he concludes that "the circumstance depends upon the weight and number of the ultimate particles of the several gases: those whose particles are lightest and single being least absorbable, and other more according to their increase in weight and complexity" (Dalton 1805, 217–287).

One can see how this line of reasoning would then lead Dalton to pursuing an inquiry into the relative weights of chemical atoms (Leicester 1961, 155). In seeking to conduct such an inquiry, Dalton eliminates the intangibleness implied by metaphysical atomism and has fixed a determinable property to chemical atoms. Dalton believes that the relative weights of atoms can be determined by measuring the relative weights of elements that are isolated during the analysis of compounds. Thus, availing himself of the highly precise data provided by Lavoisier's research (Newman 2006, 221), Dalton proposes to use analysis to isolate the component elements of compound substances and to measure their relative weights. He explicitly states that "it is one great object of this work, to shew the important advantage of ascertaining *the relative weights of the ultimate particles both of simple and compound bodies, the number of simple elementary particles which constitute one compound particle, and the number of less compound particles which enter into the formation of one or more compound particles*" (Dalton 1827, 213).

In order to move from the measurement of the relative weights of elements to the relative weights of their constituent atoms, Dalton must make a number of assumptions both about chemical reactions and about chemical composition. One of these assumptions is the Law of Conservation of Mass, which had already been previously recognized by van Helmont. Although this law has been shown to be incorrect since the work of Einstein, it was considered "approximately true" or "empirically adequate" at the level of the chemical experimentation of the early nineteenth century. Agreeing with Lavoisier, Dalton concludes that atoms can neither be created or destroyed in chemical reactions. He states, "Chemical analysis and synthesis go no farther than to the separation of particles one from another and to their reunion. No new creation or destruction of matter is within reach of chemistry" (Newman 2006, 212). Dalton also assumes the Rule of Greatest Simplicity, so that chemical formulae are always to take the simplest form that is compatible with the empirical data. Therefore, Dalton postulates that "if there are two bodies, A and B, which are disposed to combine, the following is the order in which the combinations may take place, beginning with the most simple [combination]: namely, binary [AB], ternary [A^2B or AB^2], quaternary [A^3B or AB^3], etc." (Newman 2006, 213).

Dalton's theory also places Joseph Proust's Law of Definite Proportions on firm theoretical ground by assuming that compounds are made of combinations of different types of atoms in fixed proportions, regardless of the source or method of preparation. Regarding the composition of compound substances, Dalton formulates the Law of Multiple Proportions such that, when two elements combine to form two or more compounds, the ratios of the masses of one element that combine with the fixed mass of the other are simple whole numbers (Newman 2006, 213).

Armed with these postulates and with empirical data provided by chemical analysis and measurements using the precision balance, Dalton proceeds to calculate the relative weights of chemical atoms from experimental data about compound substances, using hydrogen as the fixed unit so that "the atomic weight of each element [is] the gravimetric proportion that combine[s] with a gram of hydrogen to form the most stable combination" (Bensaude-Vincent and Stengers 1992, 114). After having collected enough data on relative weights, Dalton moves on to create his famous tables of elements, several of which are shown in figures 2, 3, and 4 below. These tables grow larger and more complex over the course of twenty-four years as Dalton is able to identify more and more elements based on relative weights and atomic masses so that, by the time the second volume of the *New System of Chemical Philosophy* is published in 1827, the number of elements had grown to thirty-six.

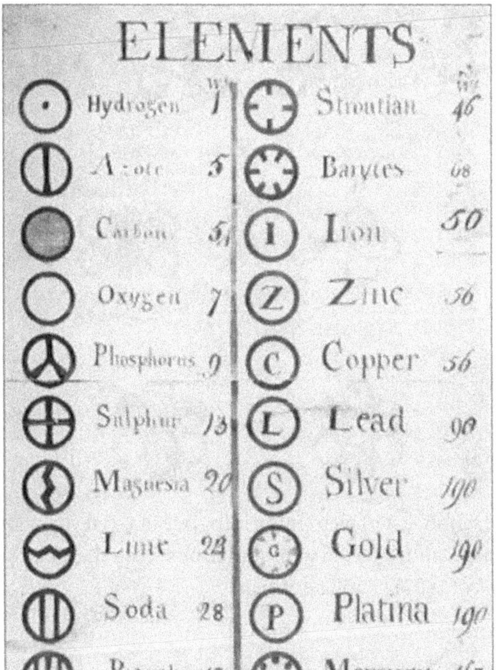

Figure 5.2. John Dalton, list of element symbols from the *New System of Chemical Philosophy* (1803) [Public domain]

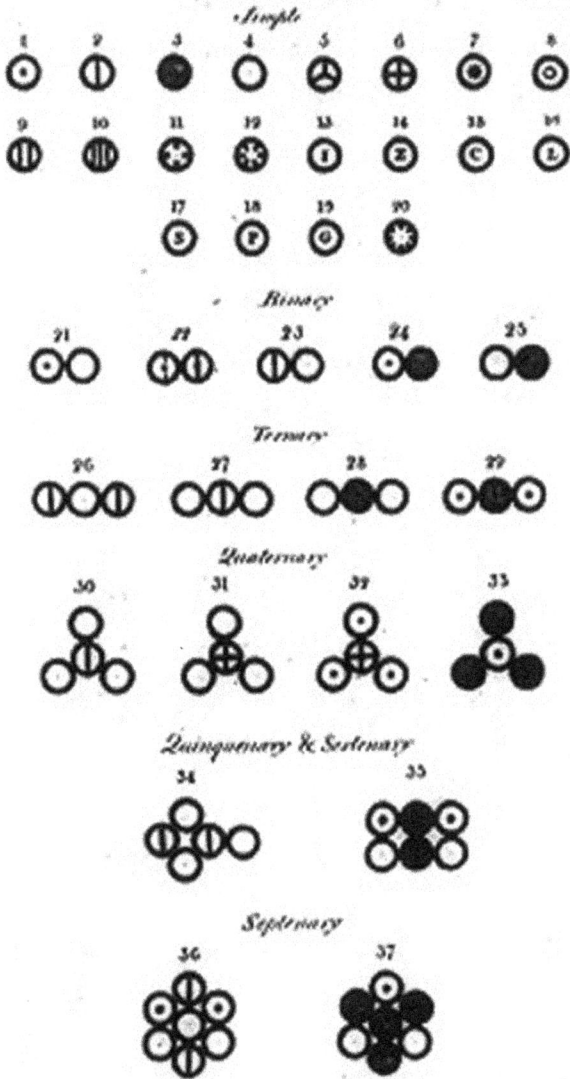

Figure 5.3. John Dalton, list of elements in single, binary, ternary, quaternary, quintenary, sextenary, and septenary combinations from the *New System of Chemical Philosophy* (1803) [Public domain]

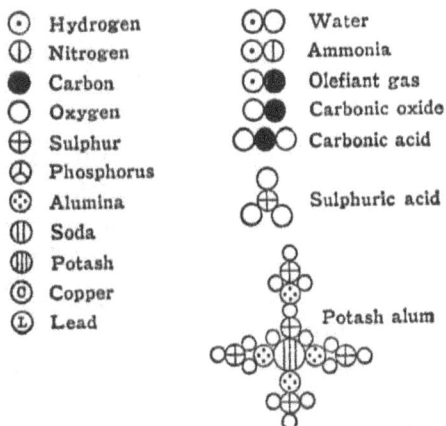

Figure 5.4. John Dalton, another list of elements and their various combinations from the *New System of Chemical Philosophy* (1803) [Public domain]

Thus, although compositional fundamentality does not necessarily overlap with explanatory fundamentality, Dalton succeeds in overlapping these conceptions of fundamentality in the context of his chemical atomism. Dalton's chemical atomism, as well as his laws of simplicity and proportionality, reconcile these two notions of fundamentality by showing that the chemical properties of elements are both mereologically and causally dependent upon the empirically determinable properties of their constituent chemical atoms.

7. Some Problems with Dalton's Chemical Atomism

As Alan Rocke explains, and as has been supported in this essay, there is a clear distinction between chemical atomism, which "forms the conceptual basis for assigning relative weights to elements and assigning molecular formals to compounds" (Rocke 1984, 10) and the physical atomism expounded by both Democritean and early modern atomists and clearly rejected by Lavoisier. Dalton himself embraced both theories, that is, the theory of chemical atoms as compositionally and mereologically fundamental and as explanatorily and causally fundamental. The clear relation between chemical atomism and physical atomism in his work generated much controversy among nineteenth-century chemists for whom physical atomism continued to be "far from universally accepted, since it made what seemed to many to be rather doubtful statements about the intimate mechanical nature of substances" (Rocke 1984, 10). For these reasons, many of Dalton's contemporaries continued to sympathize with

Lavoisier's view that all atomic theories were ultimately bound to be metaphysical (Knight 2012). "Few scientists distinguished between the two theories, and as a consequence every attack on physical atomism impugned, by association, the scientific worth of chemical atomism" (Rocke 1984, 10). Although several nineteenth-century chemists adopted a more pragmatic stance and chose to use chemical atomic theory as a convenient instrumental device without making any realist ontological commitments to the physical reality of atoms, other chemists rejected atomism altogether, whether physical or chemical. One example is the great physical chemist Wilhelm Ostwald, who held out against atoms into the twentieth century, believing that the facts of chemistry could be better accounted for thermodynamically (Rocke 1984, 76).

We also know that Dalton's rigid and arbitrary assumptions led to some erroneous conclusions about composition and weights. Later chemical discoveries showed that there are violations of Dalton's postulates. For instance, ferrous oxide, whose actual formula is non-stoichiometric ($Fe_{0.95}O$ so $Fe_{1-x}O$), rather than the more "ideal" stoichiometric formula (FeO), violates the Law of Definite Proportions, which requires a ratio of whole integers. Complex organic compounds like sucrose, whose formula is $C_{12}H_{22}O_{11}$, violate his Law of Multiple Proportions.

As well, Dalton conceptualizes the ontological dependence of properties on atomic mass and composition in a way that is too simplistic. Being unaware of the dependence of chemical properties on the structural arrangements of atoms, chirality, and other factors, Dalton was unable to predict and would have been unable to explain many types of chemical compounds later discovered. For instance, his chemical atomic theory could not have predicted or explained the existence of isomers of different types, such as constitutional isomers like butane and isobutane, which have entirely different melting and boiling points (since butane [C^4H^{10}]) melts at -138.4°C and boils at -0.5°C, and isobutane [C^4H^{10}] melts at -159.42°C and boils at -11.7°C), nor could it have predicted the existence or of stereoisomers, enantiomers, and diastereomers. He could not have predicted the discovery of isotopes, such as chlorine 35 and chlorine 37, which are the same element but have different masses and densities; of isobars such as argon atoms and calcium atoms, which are different elements but have the same atomic mass; or of allotropes, such as charcoal, graphite, and diamond, which are the same element (carbon) but entirely different properties.

8. Conclusion

In spite of its errors, Dalton's atomic theory did present a number of new and important concepts. It rendered intelligible the hundreds of quantitative analyses of substances recorded in the chemical literature. It provided a model for the long-standing assumption that compounds were formed from the combination of constant amounts of their constituents. It explained the discontinuity in the proportions of elements in compounds via the law of definite and multiple proportions. It suggested that the arrangement of atoms in a compound could be represented schematically in a way that indicated the actual structure of the compound, and it gave a precise quantitative basis to older and much more vague ideas of atoms.

There is a sense in which the relation between Dalton's chemical atoms and his elements is analogous to the relation between Friedrich Paneth's "basic substances" (*Grundstoff*) and his "simple substances" (*einfache Stoffe*), in the sense that Dalton's chemical atoms (like "basic substances") are non-observable and that elements (like "simple substances") are the form in which the properties of chemical atoms present themselves to our senses. This is where the possible analogy ends, however, and we should be careful not to stretch it too far. This is because Paneth regards "basic substances" as noumenal in the Kantian sense and, thus, as not having empirically determinable properties (Ruthenberg 2009, 83). For Dalton, on the other hand, chemical atoms have the empirically determinable and measurable property of weight and this is why he succeeds in pulling atoms from the realm of purely speculative metaphysics into the realm of concrete empirical science.

In spite of its flaws and errors, Dalton's chemical atomism represents the first major attempt at reconciling the empirical and quantitative requirements of modern chemistry as advocated by Lavoisier with the theory of discrete particles that compose material bodies and it constitutes one important step in our understanding of the nature of matter.

References

Aristotle. 1984. *The Complete Works of Aristotle: The Revised Oxford Translation*. Edited by Jonathan Barnes. 2 vols. Princeton: Princeton University Press.

Bensaude-Vincent, Bernadette, and Isabelle Stengers. 1992. *Histoire de la Chimie*. Paris: La Découverte. English translation: Bernadette Bensaude-Vincent and Isabelle Stengers, *A History of Chemistry*. Cambridge, MA: Harvard University Press, 1996.

Boas Hall, Marie. 1968. "The History of the Concept of Element." In *John Dalton and the Progress of Science*, edited by D. L. S. Cardwell, 21–39. Manchester: Manchester University Press.

Dalton, John. 1805. "On the Absorption of Gases by Water and Other Liquids." In *Memoirs of the Literary and Philosophical Society of Manchester*, 217–287.

Dalton, John. 1808. *A New System of Chemical Philosophy*. London: R. Bickerstaff.

Dalton, John. 1827. *A New System of Chemical Philosophy*. Manchester: George Wilson.

Hendry, Robin Findlay. 2005. "Lavoisier and Mendeleev on the Elements." *Foundations of Chemistry* 7: 31–48.

Hendry, Robin Findlay. 2012. "Antoine Lavoisier (1743–1794)." In *Philosophy of Chemistry*, edited by Andrea I. Woody, Robin Findlay Hendry, and Paul Needham, 63–70. Amsterdam: Elsevier.

Hookyaas, Reiger. 1949. "The Experimental Origin of Chemical Atomic and Molecular Theory Before Boyle." *Chymia* 2: 65–80.

Knight, David. 1967. *Atoms and Elements: A Study of Theories of Matter in England in the Sixteenth Century*. London: Hutchinson.

Knight, David. 2012. "John Dalton (1766–1844)." In *Philosophy of Chemistry*, edited by Andrea I. Woody, Robin Findlay Hendry, and Paul Needham, 71–78. Amsterdam: Elsevier.

Koslicki, Kathrin. 2006. "Aristotle's Mereology and the Status of Form." *The Journal of Philosophy* 103: 715–736.

Lavoisier, Antoine-Laurent. 1789. *Traité Élémentaire de Chimie*. Paris: Cuchet Libraire.

Lehman, Christine. 2009. "Mid-Eighteenth Century Chemistry in France as Seen through Student Notes from the Course of Gabriel-François Venel and Guillaume-François Rouelle." *Ambix* 56: 163–189.

Leicester, Henry M. 1961. *The Historical Background of Chemistry*. New York: John Wiley & Sons.

Levere, Trevor H. 2001. *Transforming Matter: A History of Chemistry from Alchemy to the Buckyball*. Baltimore: Johns Hopkins University Press.

Needham, Paul. 2009. "An Aristotelian Theory of Chemical Substance." *Logical Analysis and History of Philosophy* 12: 149–164.

Newman, William R. 2006. *Atoms and Alchemy: Chymistry and the Experimental Origin of the Scientific Revolution*. Chicago: The University of Chicago Press.

Pasnau, Robert. 2004. "Form, Substance, and Mechanism." *The Philosophical Review* 113: 31–88.

Pépin, François. 2012. *La philosophie expérimentale de Diderot et la chimie*. Philosophie, science et arts. Paris: Classiques Garnier.

Pinet, Patrice. 2004. "La philosophie de la matière de Galilée à Newton." *Revue d'histoire de la pharmacie* 50, no. 341: 67–82.

Rocke, Alan J. 1984. *Chemical Atomism in the Nineteenth Century: From Dalton to Cannizzaro*. Columbus: Ohio State University Press.

Ruthenberg, Klaus. 2009. "Paneth, Kant, and the Philosophy of Chemistry." *Foundations of Chemistry* 11: 79–91.

Stahl, Georg Ernst. 1730. *Fundamenta chymiae dogmaticae & experimentalis*. Translated by Peter Shaw as *Philosophical Principles of Universal Chemistry*. London: Osborn and Longman.

Wood, Rega, and Michael Weisberg. 2004. "Interpreting Aristotle on Mixture: Problems about Elemental Composition from Philoponus to Cooper." *Studies in History and Philosophy of Science* 35: 681–706.

Thackray, Arnold. 1970. *Atoms and Powers: An Essay on Newtonian Matter-Theory and the Development of Chemistry*. Harvard: Harvard University Press.

6

Origins of the Ambiguity of the Current Definition of Chemical Element

Joseph E. Earley, Department of Chemistry, Georgetown University, USA

Many of the papers in this collection refer to the *duality* or *ambiguity* of the current definition of chemical element. The International Union of Pure and Applied Chemistry (IUPAC) defines:

> Chemical element:
> 1. A species of atoms; all atoms with the same number of protons in the atomic nucleus.
> 2. A pure chemical substance composed of atoms with the same number of protons in the atomic nucleus. Sometimes this concept is called the elementary substance as distinct from the chemical element as defined under 1, but mostly the term chemical element is used for both concepts. (McNaught 1997)

This definition includes two distinct and incompatible concepts of element. In about 350 B.C.,[1] Aristotle provided a definition of an element, but pointed out that there was disagreement about an important aspect—whether or not an element was actually present in that of which it was an element.

> An element, we take it, is a body into which other bodies may be analyzed, present in them potentially or in actuality (which of these, is still disputable), and not itself divisible into bodies different in form. (Barnes 1984, p. 495.)

The duality of the IUPAC definition suggests that this ancient controversy still continues. Ghibaudi et al. (2013) reported that such lack of clarity has significant educational effects.

---

[1] In *On the Heavens*, 302a, line 15, for which Wikipedia gives this date.

Joseph E. Earley, *Origins of the Ambiguity of the Current Definition of Chemical Element* In: *What Is A Chemical Element?*.
Edited by: Eric R. Scerri and Elena Ghibaudi (2020). © Oxford University Press.
DOI: 10.1093/oso/9780190933784.003.0007

1. Concepts of Chemical Element

a. **The Ancient Four.** The earliest Greek philosophers had opinions about the ultimate constituents (τὰ στοιχεῖα, tà stoicheia) of natural things, and concerning the principles (sources) of observable change (ἀρχαί τὰ φύσις, archē tā physis) (Leclerc 1972). Aristotle summarized those suggestions (Barnes 1984) and settled on fire, air, earth, and water (understood as related to, but quite different from, the ordinary items with the same names), generally called "the four elements." These were considered to be ultimate components, and also property-bearers accounting for the observable characteristics of all items (four types of properties were taken to suffice). Versions of this approach were accepted for many centuries—mainly as convenient ways of speaking, rather than as useful understandings. After the collapse of the Roman Empire (O'Donnell 2008), what we now recognize as chemical activity (trade in salt, mining and smelting of metals, practice of medicine and pharmacy) continued, although on a reduced scale. Paracelsus (aka Theophrastus von Hohenheim) [ca.1494–1541], introduced chemicals and metals into medical practice, and taught that mercury, sulfur, and salt (understood as related to, but quite different from, the ordinary items with the same names) were ultimate constituents of things, thus effectively ending whatever four-element consensus may have existed.

b. **As Indecomposable Substances.** In 1750, Joseph Black [1778–1799] developed the analytical balance, greatly facilitating quantitative measurements in chemical research, and leading to "the chemical revolution" led by Antoine Lavoisier [ca.1743–1794]. That outburst of influential research largely operated with a different concept of elements, usually defined as "all the substances into which we are capable, by any means, to reduce bodies by decomposition" (Lavoisier 1793, xvii). Most of the hundreds of distinct chemicals known in the eighteenth century could be decomposed (analyzed) to yield a smaller number of "simple substances" that could not be broken down further. (These were designated "simple" substances to emphasize their distinction from "compound" substances.) The important restriction that applied to the four ancient elements, and also to the three Paracelsian ultimate constituents—that they were understood as related to, but quite different from, the ordinary items with the same names—did not apply to the indecomposable-substance elements of the chemical revolution, which were quite simply tangible substances. Identifying elements as specific, known materials avoided esoteric questions that accompanied earlier systems of elements.

c. **As Property-Determining Components.** In 1805, John Dalton [1766–1844] published a method for determining relative weights of the particles of gases. Applying related methods to available analytic data, especially including that pertaining to pairs of simple substances that corresponded to more than one compound (as MX_2, MX_3, and M_2X_3), Dalton came up with a theory (published in 1807) that compounds consisted of more than one sort of component, in fixed ratios by weight, while simples consisted of components of a single sort (Banchetti-Robino, 2020, this volume). On that basis, formulas (e.g., AX_3, B_2Y) of compounds, and minimum-combining-quantities for simple substances (atomic[2] weights), could be computed. Demetri Mendeleev [1824–1907] used such weights, W, along with much qualitative information on chemical reactivity, to develop his periodic table, published in 1869 (Bensaude-Vincent, 1986)—thereby bringing some degree of order to a dauntingly complex field of study. Mendeleev concluded that his results called for yet another different concept of chemical element:

Even as . . . the words: molecule," "atom," and "equivalent," [formerly[3]] were used, one for the other indiscriminately in the same manner, so now the terms "simple body" and "element" are often confounded, one with the other. They have, however, each a distinct meaning, which is necessary to point out, so as to prevent confusion of terms in philosophical chemistry. A simple body is something material, metal or metalloid, endowed with physical properties and capable of chemical reactions. The idea of molecule corresponds with the expression a simple body . . .But in opposition to this, the name "element" must be reserved for characterizing the material particles which form simple and compound bodies and which determine their behavior from a chemical and physical point of view; the word "element" calls to mind the idea of an atom: carbon is an element; coal, diamond, and graphite are simple bodies. (Mendeleev 1871, quoted by Bensaude-Vincent 1986)

For Mendeleev, elements are not independently existing bodies (chemical substances) as they were for many participants in the chemical revolution, but instead are components having characteristics that determine the observable attributes of chemical substances. The analysts of the chemical revolution had been explicit in stating that what they called elements were tangible substances with which chemists were quite familiar. Mendeleev's

[2] Here, atomic mainly meant indivisible; the word now has quite a different meaning.
[3] Brackets enclose words added for clarity.

proposal reversed this move. According to the Mendeleev definition, elements are actual components of the substances of which they are elements, but they are not independently existing substances.

d. **As Presumed Composites**: At the 1889 Annual Meeting of the American Association for the Advancement of Science, Professor Francis P. Venable lectured on "The Definition of the Element." He regarded the then-current definition of a chemical element as quite unsatisfactory, and was looking for a replacement. He said:

> At several times during the [nineteenth] century a wider vision has made it necessary to recast the definition of the elements to accord with increasing knowledge. It would seem as if another such period of change were [now] approaching. (Venable 1899, 274)

He credits Lavoisier with originating the definition of an element as a substance which could not be further decomposed—a concept he felt had been the basis of chemical progress during the nineteenth century—but also argued that the well-attested fact that some elements existed in several simple forms (allotropes) made it "quite apparent that the old definition would no longer hold good." Although Venable referred to "the Periodic System," he did not mention Mendeleev by name, or refer to his 1871 concept of a chemical element as a component rather than as a substance. He devoted half of his address to evaluating evidence that elements must somehow be composite, and to considering several suggestions (none plausible) as to what the as-yet-undiscovered components of elements might be. He found the then-recently-discovered noble gases puzzling because of their lack of electrical attraction or repulsion, but did not mention J. J. Thompson's then-recent determination of the charge of the electron. Clearly, discoveries made over the course of the next century identified many of the components of the elements that Venable had sought in 1899.

In 1902 Rutherford and Soddy showed that the stable materials produced by the decay of several radioactive elements all had the chemical properties of lead—but each turned out to have a slightly different atomic weight, depending on the original radioactive source. These were called "isotopes" of lead. After discussion, chemists agreed that since all the several isotopes of each element had seemingly identical chemical properties, they should all be considered as varieties of that element. At about the same time, it was reported that the characteristics of radiation (X-rays) produced (in evacuated tubes) when beams of electrons hit samples of various metals depended on the atomic number Z of the metal. These two results led to the conclusion that atomic number Z was somehow more fundamental than atomic weight W.

On June 30, 1921, the International Union of Pure and Applied Chemistry replaced the International Committee on Atomic Weights with an International Committee on Chemical Elements, partly because the discovery of isotopes had produced problems regarding the conception of a chemical element. In its first report (Aston et al. 1923), that new committee stated "a chemical element is defined by its atomic number" (p. 868).

2. Fritz Paneth on Two Concepts of Chemical Element

a. **Losses in Translation.** German chemist Friedrich Adolf "Fritz" Paneth (1887–1958, famous for demonstrating the reality of short-lived "free radicals") had been an important contributor to the discussion about isotopes and elements. In a German-language address delivered in Königsberg in 1931 (Paneth 1931), he compared the older (limit of analysis) and newer (property-determining component) notions of element. In translating that speech (Paneth 2003), British philosopher Heinz Post[4] used "simple substance" as the English translation of *einfacher Stoff* the established German name for a pure chemical substance composed of similar parts,[5] that is, the older concept of chemical element as analytical limit. Arguably, after the acceptance of Dalton's atomic approach and the identification of allotropes, the term "simple substance" should have been recognized as an anachronism, and replaced by "homoeomerous substance"—one having or consisting of many similar parts.

Post used "basic substance" to translate *Grundstoff*, the German word Paneth had used for the newer concept of element-as-component; this was a rarely used German word with no well-established meaning.[6] Post's translation of *Grundstoff* as basic substance is also regrettable since a main thrust of Paneth's paper is that a *Grundstoff* is not "a substance."[7] Stability

---

[4] Heinz Post was Fritz Paneth's son.

[5] In contrast with multi-component compounds, *Zusammengesetztestoffe*.

[6] The German language has two words, *Stoff* and *Substanz*, that are generally translated into English by the word *substance*. Johann Buchler (2004) pointed out that the German word *Stoff* does not carry as much philosophic freight as the English word *substance* does. Paneth used *Stoff* throughout his paper when referring to an ordinary chemical substance (*a* substance)—but switches to *Substanz* on two occasions where essence (*the* substance) is implied (Paneth 1931, 117, 118).

[7] Elsewhere (Earley 2006) I considered how the prominence of the term "substance" derives from compromises made by ancient grammarians (Luhtala 2005), and how substance has a meaning in chemistry which is related to, but quite different from, the meanings of that term in general English speech, and in philosophy.

over time (however brief) is taken as a centrally important aspect of whatever can properly be said "to exist" (Earley 1993). A second traditional requirement of substance is independent existence. The element carbon necessarily exists as part of more-inclusive wholes, as in carbon tetrachloride or in graphite. What Paneth designated *Grundstoff* is not a substance. In the present paper, the two German words *einfacher Stoff* and *Grundstoff* are left untranslated.

b. **Elements and Properties.** Paneth pointed out that an element properly so called (*Grundstoff*) does not have sensible properties. The chlorine in carbon tetrachloride is not green, gaseous, pungent, or toxic, but dichlorine—a homoeomeric substance—certainly is. The element chlorine has the characteristic atomic number of 17. That is the defining characteristic (Earley 2009) of that element—*the* substance of the element. We now recognize that the nuclei of the stable atoms with which we deal were assembled (Volk 2017, 42–56) in aging or exploding stars, long ago and far away. Recent research indicates that massive stars dominated this process in the early universe (Zhang 2018); although very massive stars no longer exist in our galaxy, traces of their former presence persist (Aoki 2014).

The atomic numbers of each of the component atomic types (elements) found in any chemical substance are permanent characteristics (properties) of long-persistent (stable) atomic nuclei. Color, pungency, toxicity, and the like are molecular-level properties that depend on which atomic centers are involved in a composite and also on the details of dynamic interactions among those centers—on structures of several sorts (Earley 2006).

Paneth does not explicitly call attention to the difficulty that the homoeomerous-substance notion of element is not consistent with elements being actually present in the substances of which they are said to be elements. (How can anything be an element of something without being a component of it?) However, he does raise that point indirectly. After recommending that allotropes of each element should be assigned diverse names, as had happened for carbon (coal, coke, graphite, buckminsterfullerene, diamond, etc.) Paneth remarked:

> However, I do not, of course, wish to suggest by this a linguistic reform impossible to carry out so late in the day, but only to point out what would seem to be indicated in the interest of logic. Much confusion could thus be avoided. Take, for instance, the following sentences, correct according to present chemical usage. "Copper oxide consists of copper and oxygen," "Copper is a metal; oxygen is a gas." Therefore, one must conclude copper oxide "consists" of a metal

and a gas, which is obviously nonsense. The fallacy here is based on the use of "copper and oxygen" in the sense of *Grundstoffe* in the first sentence and of *einfache Stoffe* in the second. Thus we have a *quarternio terminorum*[8] made possible by linguistic identity. (Paneth 2003, 132)

Throughout his lecture, and especially in this statement, Paneth made it clear that the correct current understanding of chemical element is *Grundstoff* (property-determining constituent) and not *einfacher Stoff* (separate, independently existing substance).

c. **Anti-transcendentalism.** Paneth maintained that *Grundstoff*, the modern concept of chemical element, is as "metaphysical" as the ancient notion.

> The concept of element must be taken in its naïve-realistic sense when meaning *einfacher Stoff*, but understood as transcendental when meaning *Grundstoff*." (Paneth 2003, 138)

Following philosopher Eduard von Hartmann (1842–1906), Paneth used "transcendental" to indicate lack of accessibility to ordinary direct human awareness, rather than in the far-wider sense used by Immanuel Kant (1724–1804), who wrote:

> I apply the term transcendental to all knowledge that is not so much occupied with objects as with the mode of our cognition of those objects, so far as this mode of cognition of these objects is possible a priori. . . .Supposing that we should carry our empirical intuition even to the very highest degree of clearness, we should not thereby advance one step nearer to the knowledge of the constitution of objects as things in themselves. (Kant [1781] 1952)

The understanding of chemical composition and structure (molecular, atomic, and nuclear) achieved during the twentieth century was well beyond anything that could have been imagined in the eighteenth century. A corresponding step forward, achieved by chemist C. S. Peirce and other pragmatist philosophers, was their realization that no description of reality can be so complete as to be adequate for every conceivable purpose: there is no "God's-eye view" (Earley 2015). Kant's eighteenth-century notion of the inscrutability of "things in themselves" incorrectly assumes that fully adequate descriptions are conceivable.

There is no need to resort to out-of-the-ordinary designations (metaphysical, transcendental, abstract) for Mendeleev's (property-determining component)

---

[8] "The fallacy of four terms," a logical error caused by assigning two referents to a single term.

concept of element. The atomic number Z, the element-determining character-istic, pertains to the stable nuclei of atoms not to properties that involve elec-tronic configurations of those atoms, which are usually of interest to chemists. This scale-difference has important consequences, but it is not in any sense mysterious.

3. Origins of Ambiguity

Like Venable and many others, Paneth linked Lavoisier's name with the def-inition of an element as homoeomerous substance. Although Lavoisier tol-erated the use of the name *"element"* to refer to the results of full analysis, he *did not endorse* that use of the name "element" or use it himself, preferring the designation "principles of bodies" (*"principes des corps"*). Although it has long been reported that Robert Boyle (1627–1691) originated the indecomposable-substance concept of element, Lavoisier *did not refer* to Boyle on that topic. Lawrence Principe, author of a recent (Principe 2000) Boyle biography, reports that Boyle followed Jan Baptist van Helmont (1580–1644) in being skeptical of then-current (Paracelsian) doctrine that all substances were composed of just three components and that Boyle also followed van Helmont in holding that all substances were made of *only one* sort of stuff, rather than from any number of "elements":

> Boyle, rather than defining chemical elements as we know them, doubted that there were any. (Principe 2011)

Antoine Lavoisier *shared* Boyle's view that there was no evidence for the exist-ence of a few basic components of all items. In the *Discours préliminaire* of his *Traité Élémentaire de Chimie*, Lavoisier wrote:

> The notion of four elements, which, by the variety of their proportions, com-pose all the known substances in nature, is a mere hypothesis, assumed long before the first principles of experimental philosophy or of chemistry had any existence... if, by the term elements, we mean to express those simple and in-divisible entities[9] of which matter is composed, it is extremely probable we know nothing at all about them; ...if we apply the term elements, or princi-ples of bodies, to express our idea of the last point which analysis is capable of

---

[9] "Entities" is substituted for Kerr's "atoms" as translation of <*molécules.*>

reaching,[10] then for us the elements are all the substances into which we are capable, by any means, to reduce bodies by decomposition. (Lavoisier 1793, xvii, translation by Robert Kerr 1790, modified as indicated.)

Lavoisier himself does not connect simple substances (limits of analysis) with the notion of elementarity, although he tacitly gives permission for others to do so, even though he clearly points out that this is not the traditional notion of element. The previously quoted definition of element as simple substance; "all the substances to which we are capable, by any means to reduce bodies by decomposition" is indeed present in this passage; citations of this definition given by Paneth, Venable, and the first part of this paper are correct, *as far as they go*. But by not including the first part of the sentence those citations give a *defective understanding* of Lavoisier's actual position. Why has that clarifying phrase generally been omitted? Lavoisier suggests:

The authority of these [ancient Greek] fathers of human philosophy still carry great weight, and there is reason to fear that it will even bear hard upon generations yet to come. (Lavoisier 1793, xiv–xv)

As Lavoisier had feared, during the eighteenth century the concept that everything is composed of only a few basic components was so widespread that, in spite of Lavoisier's own reservations, general acceptance of *indecomposable substance* as a definition of "chemical element" followed Lavoisier's success in quantitative analyses of important chemical substances.

Mendeleev had pointed out that the term element *"must be reserved"* for the newer concept of element-as-component. Paneth argues convincingly that the central and essential part of the concept of element is that of *Grundstoff*. Both authors clearly held that the name element should be assigned to *components* of specific stuffs rather than to substances. There may have been some justification, many decades ago, for Paneth, or for Lavoisier, to take a permissive (but not approving) attitude to the use of the word element as a more impressive term than "limit of analysis." If so, that justification lapsed long ago. More than ten generations after Lavoisier identified that concept as a residue of erroneous ancient ideas, the definition of element as indecomposable substance **should now be abandoned.**

---

[10] The phrase "for us, the elements are" substitutes for Kerr's translation "we must admit, as elements." The original text is: <...: *que si au contraire nous attachons au nom d'élémens ou de principes des corps l'idée du dernier terme auquel parvient l'analyse, toutes les substances que nous n'avons encore pu décomposer par aucun moyen, sont pour nous des élémens;* ... >.

4. *Cui bono?* Who benefits?

The *benefit* of the second part of the IUPAC definition of an element is that it allows old folks to keep talking as they are used to doing, even though their concepts are obsolete. With such slight benefit, why has this obsolete definition survived? Paneth suggested one reason, beginning by pointing out that two statements universally made in histories of chemistry are *quite wrong*. He claimed that it is *not correct* that

> . . . the four elements of Aristotle did not designate substances, but combinations of the fundamental properties, cold, warm, dry, wet: that Boyle was the first to mean by chemical element a substance that could not be decomposed further.

He then concludes:

> The fact that there is nevertheless general agreement with these [false] statements amongst the majority of chemists writing on this subject only shows how little importance they attach, in general, to the investigation of the concepts used in chemical teaching. (Paneth 2003, 114)

The most interesting aspects of chemistry are at the advancing frontiers of the science; most chemists assume that the basic terms of their field have been clarified long ago. In identifying Lavoisier with the undecomposed-substance concept of chemical element, Paneth has fallen into the same error (failing to closely examine the original sources of widely reported opinions) that he had criticized in the work of others. The lack of felt importance in clarifying meanings for fundamental chemical concepts, which Paneth identifies, is one possible reason for the long-term survival of an obsolete definition of element.

However, there is another (perhaps less unflattering) reason why chemists may have tolerated that long-superseded notion. The *language* used in scientific discourse continually changes, but linguistic modifications frequently lag behind the underlying conceptual or technological changes that occasion them. Such sluggishness is not specific to science: when language-communities split into seldom-interacting parts, some ways of speaking die out in the larger part but may survive in a smaller part, perhaps for centuries. For instance, the ordinary speech (*koiné*) of seventeenth-century Paris had distinctive features that have long been extinct in France—but some of them can still be identified in the French spoken in Québec (Wittmann 1997). Certainly, it is easier to switch from one explicitly recognized complex theory (or scientific technique) to another than to change unconsciously habitual ways of talking about ideas and methods.

The preservation of an obsolete definition of chemical element for nearly a century is an example of what John Dewey (2012) called "unmodern philosophy." Due to the unrecognized influence of prior habits of thought,

> ... traits which are genuinely new have been formulated in terms of old beliefs with which they are at war, instead of in their own terms. (Dewey 2012, 66)

During the nearly two millennia when four basic substances were generally considered to be property-determining constituents of all items, related assumptions became generally but unconsciously accepted. Dewey suggests that such notions have left residues that still influence interpretation of the conflicting ("at war") definitions of chemical element.

5. Composites and Properties

The reasons suggested above for the toleration of this ambiguity in a basic chemical concept seem rather weak; perhaps deeper influences are at work. Possibly, the ambiguity of the IUPAC definition of chemical element is connected with the widespread but incorrect idea that when the same names are applied to items as independent substances and as components of more inclusive entities, then the properties of the composite substances can be inferred from the characteristics their ingredients would have if they were independent materials. However it is invariably the case that inclusion in a coherence *modifies* each included entity; sometimes, parts maintain enough integrity to continue as recognizable components[11] of new items (for instance, atomic nuclei persist unchanged through chemical combinations that greatly modify how electrons are associated with those nuclei), but drastic modifications usually do occur (e.g., organisms normally *decompose* their food). The rather old-fashioned wording of the IUPAC definition ("the number of protons in the atomic nucleus") implies that certain protons are components of each nucleus, a view which prevailed in the late 1930s and early 1940s. Current concepts of nuclear structure hold that quarks and gluons rapidly move among hadrons (protons and neutrons) in nuclei, changing which hadrons are positively charged. Protons and neutrons may properly be regarded as *ingredients* of nuclei, but the massy components of nuclei are

---

[11] Components are ingredients that retain significant integrity in a composite. Items purchased are generally exempt from state sales tax if they are either ingredients or components of products that will eventually be sold. The difference between components and ingredients is irrelevant for sales-tax purposes; for other purposes, the difference is often significant.

hadrons; at any instant the collection of hadrons has a positive charge (Z), but how that charge may be localized is not obvious.[12]

This error seems to be connected with the regrettably widespread idea that features of composites *depend primarily on the characteristics of their ingredients*—the basis of the mistaken ancient notion of elements as *property-bearers*. Characteristics of composites are constrained by the identities of their ingredients, but those properties are mainly determined by *closure of networks of relationships* among components. Properties stay together in each chemical entity because networks of dynamic relationships among components generate *closed* sets of subsequent states (Earley 2016). Each such closure depends on continuance of less-inclusive closures involving the sub-parts of its components, and also may participate in (and partially be determined by) closures of more-inclusive relationship-networks of which it itself is a component.

Typically, characteristics of components are consistent with vast numbers of possible outcome-configurations: *structural* features, generally operating over time, *select* which of those possibilities become (or remain) actualized. Components impose necessary limitations, but those limitations *are not sufficient* to determine detailed outcomes. If any structure is a *criterion of selection* during the historical course of events through which an outcome arises (Earley 2012), that structure may partially *determine* the specific outcome.

The immensity of the number of galaxies in the universe began to be recognized in the 1920s; the complexity of the nuclei of atoms was clarified beginning in the first half of the twentieth century. We now recognize that all the items with which we deal are *composites* of less-extensive entities, and also themselves are *components* in more-extensive coherences (Earley 2013).

> Big fleas have little fleas upon their backs to bite 'em, And little fleas have lesser fleas, and so on, *ad infinitum*. And the great fleas, themselves, in turn, have greater fleas to go on; While these again have greater still, and greater still, and so on. (De Morgan 1915)

At any and every level of size, from sub-sub-microscopic to astronomically immense, entities interact with others; in favorable cases, *coherence* may be achieved so that more-inclusive new entities emerge. (The appendix of this paper proposes a formal statement of a criterion for this important class of event.) Each level of emergence involves novel closures of relationships that may ground relative-independence of causal properties of entities at different

---

[12] Elsewhere (Earley 2005), I argued that salt (crystalline, white, sodium chloride), although it is an ingredient of brine, *is not actually a component* of seawater.

levels. The closures of relationships that form and dissolve during chemical reactions are effectively independent of the details of the closures of relationships that account for the permanence of association of hadrons into stable atomic nuclei; this scale-difference accounts for the effects that Paneth called "transcendental." Since all such difference is fully understandable using ordinary reasoning, there is no need or justification for continuing a "dual nature" in the definition of a chemical element. The notion that all indecomposable constituents of substances are "elements"—usually attributed to Boyle or Lavoisier, but definitely *not* endorsed by either of those authors—should now, finally, be abandoned.

Appendix: A Formal Criterion of Emergence

Suppose we have a set of several agents $(x_1, x_2, x_3, \dots . x_i)$—let's call them "$X$, the set of xs." Each x may interact with other xs, and with itself. Also, consider an appropriate test agent y that has some property F when the xs do not interact in a significant way. If the xs do interact so as to generate a *closed set of states that continually recur* over a more or less extended period of time—indicated XI—then y may have different (other than F) properties ($\sim F$) due to interaction between y and x, the set of xs. If the latter condition prevails, it is legitimate to conclude that an emergent entity z exists, and that the xs are parts of z.

If y is an agent that, in the absence of significant interaction among the xs, has property F and XI indicates that the set of xs interact to yield a closed set of states,

$$\left(\exists x_1 \dots \exists x_i \right) \in X . \left[\left\{ \sim XI \supset \left(\exists y \left(F(y) \right) \right) \right\} \right]$$

then, when P is the part relation and the symbol z refers to an emergent entity of which the xs are parts,

$$\left(\exists x_1 \dots \exists x_i \right) \in X . \left[\left\{ XI . \left(\exists y \left(\sim F(y) \right) \right) \right\} \supset \exists z \left(XPz \right) \right]$$

That is to say, it would be legitimate to speak of the emergence of a new entity z, if and only if certain agents (the xs in this case) interact in such a way that some (any) test entity (y) suffers a change in its properties (F becomes $\sim F$) due to the closure of that interaction (XI). If such closure does occur, then the xs are correctly considered as parts of the emergent entity z—that is, XPz. Any such z might interact with other agents, of similar or different sorts, to generate yet-larger emergent entities, say, one of the ws. Also, each and every one of the xs, is itself an emergent entity made up of components, perhaps the us. Every one of these integrations (us, xs, zs, ws) can fail to persist, with consequences both for its constituents and for coherences of which it may be a component.

References

Aoki, W., N. Tominaga, T. C. Beers, S. Honda, and Y. S. Lee. 2014. "A Chemical Signature of First-generation Very Massive Stars." *Science* 345, no. 6199: 912–915. doi:10.1126/science.1252633.

Aston, F. W., G. P. Baxter, B. Brauner, A. Debierne, A. Leduc, T. W. Richards, F. Soddy, and G. Urbain. 1923. "Report of the International Committee on the Chemical Elements: 1923." *J. Amer. Chem. Soc.* 45, no. 4: 867–874. (Thanks are due to R. F. Hendry for locating this reference.)

Banchetti-Robino, Marina Paola. 2020. "The Changing Relation between Atomicity and Elementarity, from Lavoisier to Dalton." In this volume, *What is a Chemical Element?* edited by Eric Scerri and Elena Ghibaudi. New York: Oxford University Press, pp. 87–108.

Barnes, Jonathan, ed. 1984. *The Complete Works of Aristotle: The Revised Oxford Translation.* 2 vols. Princeton: Princeton University Press.

Bensaude-Vincent, Bernadette. 1986. "Mendeleev's Periodic System of the Elements." *The British Journal for the History of Science* 19: 3–17.

Buchler, Johan. 2004. "Chemistry Seen as Molecular Architecture and Notes on the German Term 'Stoff.'" In *Chemistry in the Philosophical Melting Pot (Dia-Logos 5)*, edited by Danuta Sobczyńska, Pawel Ziedler, and Ewa Zielonaka-Lis. Frankfurt: Peter Lang.

De Morgan, Augustus. 1915. *A Budget of Paradoxes*, vol. 2. Edited by David Eugene Smith. London: Open Court.

Dewey, John. 2012. *Unmodern Philosophy and Modern Philosophy*. Edited by Philip Deen. Carbondale: Southern Illinois University Press.

Earley, Joseph E. 1993. "The Nature of Chemical Existence." In *Metaphysics as Foundation*. Edited by Paul Bogaard and Gordon Treash, 272–284. Albany: State University of New York Press.

Earley, Joseph E. 2005. "Why There Is No Salt in the Sea." *Foundations of Chemistry* 7: 85–102.

Earley, Joseph E. 2006. "Chemical 'Substances' That Are Not 'Chemical Substances.'" *Philosophy of Science* 73: 841–852.

Earley, Joseph E. 2009. "How Chemistry Shifts Horizons: Element, Substance and the Essential." *Foundations of Chemistry* 11, no. 2: 65–77. doi:10.1007/s10698-008-9054-5.

Earley, Joseph E. 2012. "A Neglected Aspect of the Puzzle of Chemical Structure: How History Helps." *Foundations of Chemistry* 14, no. 3: 235–244. doi:10.1007/s10698-012-9146-0.

Earley, Joseph E. 2013. "A New 'Idea of Nature' for Chemical Education." *Science and Education* 22, no. 1: 1775–1786. doi:10.1007/s11191-012-9525-x.

Earley, Joseph E. 2015. "Pragmatism and the Philosophy of Chemistry." In *Philosophy of Chemistry: Growth of a New Discipline*, edited by Eric Scerri and Lee McIntyre, 73–90. New York: Springer.

Earley, Joseph E. 2016. "How Properties Hold Together in Substances." In *Essays in Philosophy of Chemistry*, edited by Eric Scerri and Grant Fisher, 199–233. New York: Oxford University Press.

Ghibaudi, Elena, Alberto Regis, and Ezio Roletto. 2013. "What Do Chemists Mean When They Talk about Elements?" *Journal of Chemical Education* 90: 1626–1631. doi:10.1021/ed3004275.

Kant, Immanuel. (1781) 1952. *Critique of Pure Reason*. Translated by J. M. Meiklejohn. Edited by R. Hutchins. Chicago: Britannica.

Lavoisier, Antoine Laurent. 1793. *Traité Élémentaire de Chimie, Présenté dans un Ordre Nouveau et d'après les Découvertes Modernes; avec Figures: Seconde Édition, Tome Premier*. Paris: Cuchet. Nineteenth Century Collections Online, accessed 21 June 2018, http://tinyurl.galegroup.com/tinyurl/6gEME7.

Lavoisier, Antoine Laurent, translated by Robert Kerr. 1790. *Elements of Chemistry in a New Systematic Order*. Edinburgh: William Creech.

Leclerc, Ivor. 1972. *The Nature of Physical Existence*. London: George Allen & Unwin.

Luhtala, Anneli. 2005. *Grammar and Philosophy in Late Antiquity: A Study of Priscian's Sources*. Amsterdam: John Benjamin.

McNaught, A. D., and A. Wilkinson, eds. 1997. *IUPAC Compendium of Chemical Terminology* ("The Gold Book"). Oxford: Blackwell Scientific Publications. doi:10.1351/goldbook. Last update: 2014-02-24; version: 2.3.3.

Mendeleev, Dmitri. 1871. "*Die Periodische Gesetzmässigkeit der Chemischen Element.*" *Ann. Chem. Pharm.* 8: 133–229.

O'Donnell, James. 2008. *The Ruin of the Roman Empire*. New York: Ecco.

Paneth, Fritz A. 1931. "*Über die erkenntnistheoretische Stellung des chemischen Elementsbegriffs.*" *Schriften der Königsberger Gelehrten Gesellschaft (Max Niemeyer)* 8, no. 4: 101–125.

Paneth, Fritz A. 2003. Translated by Heinz Post. "The Epistemological Status of the Chemical Concept of Element." *Foundations of Chemistry* 5, no. 2: 113–145.

Principe, Lawrence. 2000. *The Aspiring Adept*. Princeton: Princeton University Press.

Principe, Lawrence. 2011. "In Retrospect: The Sceptical Chymist." *Nature* 469: 30–31.

Venable, F. P. 1899. "The Definition of the Element." *Science* 10, no. 244: 274–282.

Volk, Tyler. 2017. *Quarks to Culture: How We Came to Be*. New York: Columbia University Press.

Wittmann, Henri. 1997. "*Le français de Paris dans le français des Amériques.*" *Proceedings of the International Congress of Linguists (Paris, 20–25 juillet 1997)*. Oxford: Pergamon (CD edition), 16.0416.

Zhang, Z., D. Romano, R. J., Ivison, P. P. Papadopoulos, and F. Matteucci. 2018. "Stellar Populations Dominated by Massive Stars in Dusty Starburst Galaxies across Cosmic Time." *Nature* 558: 260–263. doi:10.1038/s41586-018-0196-x.

7

The Existence of Elements, and
the Elements of Existence

Robin Findlay Hendry, University of Durham, UK

1. The Existence of Elements

It is sometimes said that chemistry is the science of synthesis, but analysis is just as important to chemistry, and was arguably more important during the early part of its historical development. From Lavoisier onward there was a research program involving the identification of new chemical elements—most of them metals—by separating them out from minerals and other naturally occurring substances, and then distinguishing them from other elements by virtue of their chemical and physical properties (see Hendry forthcoming b). The need to systematize the growing list of elements uncovered by this project, and the evident similarities between some of them led, by the 1860s, to the periodic table.

What does this have to do with the existence of elements? Since before Lavoisier's time, a commitment to the continued existence of elements within their compounds was an important strand running through chemical thinking (Hendry 2010; Hendry forthcoming b).[1] To anyone who is familiar with modern chemistry, this may seem a rather obvious assumption to make, but it is not something that could have been known a priori. Consider, for instance, an alternative assumption, associated with Aristotle's view of the elements (Needham 2006; 2009); on this view the elements may be "used up" when they combine with other elements to form a mixt.[2] Since a mixt displays none of the essential

---

[1] Note that, in everything that follows, by "element" I will mean what F. A. Paneth (1962) called a "basic substance" rather than what he called a "simple substance." Given that "simple substance" can be defined in terms of "basic substance" (a simple substance is a substance composed of only one basic substance), there is nothing of interest, from a philosophical point of view, in the distinction. See also Hendry 2006.

[2] Following Paul Needham (2006; 2009) I use the archaic term "mixt" to indicate that Aristotle does not distinguish between compounds and homogeneous mixtures such as solutions. "Mixt" here means "compound or mixture."

Robin Findlay Hendry, *The Existence of Elements, and the Elements of Existence* In: *What Is A Chemical Element?*. Edited by: Eric R. Scerri and Elena Ghibaudi (2020). © Oxford University Press.
DOI: 10.1093/oso/9780190933784.003.0008

properties of its elements, they can at best be said to be potentially present, in the sense that they can be recovered from the mixt when certain operations are performed.[3]

At the beginning of the nineteenth century, John Dalton provided an explanation of the survival of the elements within their compounds, by positing a one-to-one relationship between the identity of the chemical elements and kinds of atoms; elements survive in their compounds because their atoms can survive the processes that constitute chemical change. But this can hardly be said to have settled the matter against Aristotle. First, Dalton's version of atomism was widely regarded as speculative and controversial until the twentieth century. Second, Dalton's atomism provides only the merest explanatory sketch of how elements might compose compounds, rather than a fully worked-out account of how chemical combination is possible. Thus, for instance, Dalton's theory explains the empirical fact that elements sometimes combine with other elements in fixed proportions by supposing that atoms combine into microscopic bodies whose composition reflects that of the corresponding compound, but it offers no account of *how* atoms can do this (for criticisms of atomism, see Needham 2004 and Chalmers 2009). Third, many aspects of Dalton's thinking about his atoms seem odd, even arbitrary and fantastical. Thus, for example, Dalton supposed that elements are individuated by their atomic weights, but to make that quantity empirically accessible he had to make assumptions about how many atoms of each element were contained in each molecule of a compound substance. Having no reason to do otherwise, he arbitrarily chose to assume that each molecule of (for instance) water contained one atom each of hydrogen and oxygen (see Needham 2004 and Chalmers 2009). This is a familiar phenomenon. To take another case, Lavoisier is known as the discoverer of oxygen, but his compositional theories were based on a table of simple substances that also includes caloric, the matter of heat, which he introduced to enable his overall compositional system to occlude any explanatory role for phlogiston (see Lavoisier 1790, 175), and he argued on inductive grounds that oxygen confers acidity on its compounds (1790, 64–65). My own view is that we should approach cases like this in piecemeal fashion; when engaging with scientific systems of the past, we should decide for ourselves where we agree and disagree with them, and must neither ignore the disagreements, nor regard them as reasons to reject or contextualize the system as a whole.

Whatever one thinks of the modern successors to Dalton's theory, one can hardly fault those chemists who steered well clear of it. Nevertheless, it seems to have been an important source of inspiration for many subsequent chemists,

---

[3] How far modern chemistry rules out Aristotle's view is a complicated question (see Needham and Hendry 2018). For a wide-ranging discussion of potential versus actual existence of chemical substances, see Earley 2005.

and it constituted a substantial break with previous versions of atomism. Robert Boyle also advanced corpuscularian explanations of chemical change, but these are even more sketchy than Dalton's. Boyle's atomism fails even to give any account of how many different kinds of atoms there are, or how they are related to the number and variety of chemical elements. In contrast, Dalton's atoms are substance specific (see Klein 1994 and Chalmers 2009).

The nineteenth-century research program of seeking out new elements could be understood as a search for things that are given by nature; things that actually exist, although they may be extremely rare. Mendeleev's periodic table placed that search within a classificatory system, but Mendeleev himself saw the elements as "chemical individuals" (Mendeleev 1889, 640), individuated by their atomic weights, which were connected to their chemical behavior by his periodic law. Throughout the nineteenth century there were intermittent speculations about the underlying nature of the elements, attempts to see them as members of a series generated by deeper structure. For a number of reasons Mendeleev resisted such speculations, as he did the closely connected possibility of transmutation between elements (Gordin 2004, chapter 8). Mendeleev's view was overturned by the combined effect of three discoveries whose individual significance was disputed at the time: radioactivity as involving transmutations between the elements, isotopy as involving differences in the atomic weights of atoms of the same element, and nuclear charge as an alternative physical quantity with which to individuate atomic species. There was some debate on how close were the chemical properties of different isotopes, and consequently how isotopy should be accommodated in the periodic table (see van der Vet 1979; Kragh 2000), but in 1923 the International Committee on Chemical Elements, appointed by the International Union of Pure and Applied Chemistry (IUPAC), enshrined nuclear charge as the determinant of the identity of the chemical elements (see Aston et al. 1923). The overall result of these changes was that the elements ceased to be "chemical individuals" whose essential property is their atomic weight. They came instead to be seen as members of a series generated by a (quantized) physical property, namely nuclear charge. The possibility of transmutation between them implied that they could be created and annihilated; they no longer need be given by nature.

2. The Existence of Superheavy Elements

The chemical elements present a number of interesting questions about the nature of discovery. To decide on priority—which of two disputed claims to a discovery should be accepted—requires an account of what counts as discovery. The awards of Nobel Prizes have turned on just such issues. In a recent book, Helge Kragh (2018) explores how these issues become even more complicated in the context of research into superheavy elements (SHEs). In an accompanying paper (Kragh 2017), he argues that SHEs put pressure on the very concept of a chemical element. The term "superheavy element" is somewhat vague, as Kragh points out, but typically refers to "transactinide elements with Z ranging from 103 to 120" (2017, 9). What is so special about SHEs? They are very heavy compared to other nuclear species; they typically do not occur naturally; they tend to be very unstable, and have correspondingly short lifetimes; for any given space in the periodic table, such elements might be merely possible existences, rather than actual existents.

I think that Kragh poses an interesting philosophical challenge, which is to answer the following question: how is it possible that some sample of a nuclear species has been brought into existence, yet this does not entail that there exists a chemical element corresponding to this species?[4] Answering this question is the task of this section. Here are four possible answers:

1. A physical species has been brought into existence, but this doesn't itself count as a chemical element.
2. The new chemical elements have been synthesized (and are therefore artificial), so they cannot count as true chemical elements, which must be naturally occurring.
3. Establishing the existence of an atomic species is not enough to establish the *actual* existence of a chemical element, only the mere possibility of one.
4. What the evidence tells us is that a member of a nuclear species has been brought into existence, but this does not live long enough to ground the actual existence of a chemical element.

There may well be other possibilities, but for the remainder of this section I will investigate each of these proposals in turn.

2.1. Physics Versus Chemistry

SHEs might be argued to fall outside of chemistry, being (in some sense to be specified) physical rather than chemical entities. As Kragh puts it, research into SHEs involves the "formation of new atomic nuclei and relies crucially on advanced accelerator and detection technology," and so might be considered "a branch of nuclear physics rather than chemistry" (2017, 17). However, insofar

---

[4] I leave aside the issue of what pressure such "exotic" species as positronium might put on the 1923 IUPAC definition (Kragh 2017, 14–15). A plausible response to this issue is to regard the elements as a series generated by adding protons to nuclei, a series to which positronium does not belong.

as it is the business of IUPAC (rather than IUPAP, The International Union of Pure and Applied Physics) to confirm the discovery of SHEs, and to confer the naming rights for them, they seem to be the business of chemistry, at least from an institutional point of view. Moreover, it seems unsatisfactory for such institutional decisions to settle ontological questions, and as Kragh points out "the distinction between physics and chemistry in modern SHE research is in some way artificial as workers in the field rarely consider themselves as either physicists or chemists" (2017, 18).

It could be that those who work in the field of SHE research are atypical in where they see the boundary between chemistry and physics: such views are likely to vary across different sub-disciplines of chemistry. Other chemists might not see SHEs as "proper" chemical elements for a number of reasons. Here's one reason why they should not see things that way. Following the discovery of the distinction between atomic number and atomic weight, and their assimilation of this distinction, chemists came to see the periodic table as one whose spaces were individuated by the (quantized) values of nuclear charge. Some chemists might have resisted this change, but as we have seen IUPAC, the collective voice of chemistry itself, came down in favor of atomic number (or nuclear charge) as the defining property of elements. Once that move was made, the mathematical structure of the periodic table effectively corresponds to that of the natural numbers. Since there are endless natural numbers, there is no reason why the spaces in the table, each corresponding to a distinct chemical element, should not also be endless. Then it seems that the element with atomic number 298 is as much a part of the periodic table as the element with atomic number 8. The idea that there could be a limit to the endlessness of the spaces in the periodic table then raises the question of when SHEs cease to be chemical elements. What could provide the basis of a boundary between chemical elements proper, and the SHEs? What significance would such a boundary have? Why should we think that it did any more than reflect the (historically contingent) fact that the discipline of chemistry began by exploring the lighter elements? The sheer heaviness of the SHEs is surely not pertinent.

I would argue that an attempt to answer this kind of question through appeal to a principled distinction between chemistry and physics is wrongheaded. During the twentieth century, chemistry developed an ever more intimate relationship to physics, which was forged by three kinds of development. First, substances came to be identified by their physical structures at the molecular scale, a fact that is reflected in IUPAC's systematic nomenclature for chemical substances. Second, there emerged theories of structure and bonding, including quantum mechanics, central to which were physical interactions between physical entities such as electrons and nuclei. Third, there emerged physical methods for investigating and individuating chemical substances and their reactions, including X-ray crystallography and IR, UV, mass, and NMR spectrometry. By the end of the twentieth century, these methods were fully integrated into chemical practice, along with the traditional "chemical" methods. One might say that the distinction between chemical and physical methods had become obsolete: an irrelevance to modern chemistry. Whether or not one thinks that these achievements effect the reduction of chemistry, they do commensurate chemistry and physics.[5] Connections were forged between the central theories and concepts of the two sciences in ways that changed both of them deeply. Prior to these developments (some historians and philosophers might wish to say) chemistry and physics inhabited quite different worlds, but the two sciences were then brought into the very same world. Any time after the late twentieth century, it is just archaic to worry about an important distinction between the "chemical" and the "physical" as such. It appeals to some long-past (and possibly imagined) classical essence of chemistry. Of what relevance would such a distinction have to present-day science?

2.2. Synthesis Versus Discovery

One might worry that SHEs challenge the definition of a chemical element because they are products of the laboratory: synthetic substances rather than naturally occurring ones. For instance, Kragh asks,

> Can one reasonably claim that superheavy elements exist in the same sense that the element oxygen exists? After all, they are created in the laboratory and not discovered in nature such as has been the case with most elements. (2017, 8–9)

One might challenge the assumption that chemistry is not concerned with the artificial: the discipline has always involved itself with making new substances, an activity that came into its own during the nineteenth century, when organic chemistry began to specialize in the production of useful artificial substances, artificial dyes being a nineteenth-century example (Brock 1992, chapter 8). Manufacturing new elements is rather different from manufacturing new compounds, however. One might view chemistry as a discipline that takes the elements as things whose existence is given by nature, and investigates how they combine to form other substances. Part of this project would be investigating just which elements nature

---

[5] Some scientists and philosophers regard these theories as effecting a reduction of chemistry to physics, or as evidence that the subject matter of chemistry (that is, what it studies) is in some important ontological sense "no more than" the subject matter of physics. My own view is that this is too strong an interpretation, because the evidence is compatible with looser ontological relationships, such as the (strong) emergence of the chemical from the physical (see Hendry 2019).

does give us, but manufacturing new elements would fall outside it. That would be nuclear chemistry-cum-physics, at best an interdisciplinary borderland between chemistry and physics. I do not think this distinction between things that are "given in nature" and those that are artificially created can do the work required of it in the current context, for a number of reasons.

First, it is unclear how the distinction between what is created and what is discovered is supposed to work, because experimental discoveries typically involve technological achievements. For instance, Jed Buchwald gives a detailed account of Heinrich Hertz's discovery of electric waves (including radio waves), a process that involved him inventing a device that was capable of emitting them (Buchwald 1994). Second, the distinction between what is naturally occurring and what is artificial must be parochial, because it must be made relative to some particular environment and the processes that produce and consume substances in that environment. From a global point of view, any substance that the laws of nature allow to be made in some enviroment is, in a sense, natural. Third, it is not clear what is ontologically significant about the distinction, because the existence of an artificial element is still existence. Last, the distinction does not seem to have bothered past chemists, even where it applies to elements. The existence of technetium was first established by analyzing the residues on molybdenum plates from a cyclotron, at a time when it was an open question whether it exists naturally in detectable quantities. The very name "technetium" reflects its supposed origin in human activity (Zingales 2005).

2.3. Actual Versus Merely Possible Existence

One might instead worry that the SHEs to be found in the far reaches of the periodic table are merely possible rather than actual existences. A motivation for this worry would be as follows: chemists are able to write down full structural formulae that are allowed by valence rules. This does not mean that such structures actually exist, or that there is a chemical substance that corresponds to each one; perhaps the structures are too strained, or so reactive that their existence is short lived. Similarly, just because there is a place in the periodic table corresponding to a particular value of nuclear charge doesn't mean that the element to occupy it exists.

To explore this further, we need to make some further distinctions. First, there are structures that can be made, but have not in fact been made, versus structures that cannot be made. This is not the right distinction to shed light on the problem of SHEs, because there IUPAC accepts that the relevant nuclear species have actually been made. There is something occupying the place—a nuclear species—and we need to know why it doesn't count as a chemical element. Second, in the case of chemical compounds one should distinguish structures that can be made, but are so unstable or reactive that they exist only for a very short time, from structures that cannot be made at all, or are so short lived that there could be no evidence for their existence. In organic chemistry there can also be short-lived structures for whose existence there can be evidence; reactive intermediates are important in understanding the pathways of reactions, and they may even be detectable directly. Once again, this is not the problem with SHEs such as Oganesson, which (IUPAC accepts) do exist long enough to be detected.

One might say that what is actually created is a nucleus rather than a chemical element. Thus Kragh points out the SHEs are typically created only as nuclear species, but "an isolated atomic nucleus has no chemistry" (2017, 15). This applies to α-particles, which can readily be converted into helium atoms by the addition of electrons. However, it would seem more correct to say that α-particles do have a chemistry (that of helium), which is determined by the stable electronic structure that would result by adding electrons. This does not require that that electronic structure actually be possessed. Moreover, when discussing the relative abundance of elements, chemists usually understand this to mean the abundance of the corresponding nuclei. This reflects the IUPAC definition, according to which the names of the elements apply in abstraction from their states of chemical combination, and also in abstraction from their states of aggregation. Thus "helium" applies to anything with a nuclear charge of two, no matter whether it exists as part of a sample of helium gas, a plasma in which the electrons and nuclei are dissociated, or as a completely isolated α-particle. Now the key thing about the α-particle is that there is an electronic configuration it *would* have if electrons were added to it. Perhaps this might fail for SHEs.

2.4. Stability and Lifetime

Having investigated three other ways to explain how a nuclear species might exist without a corresponding chemical element existing, I should present a more plausible proposal that would explain some of the worries about SHEs as elements; in a nutshell, it concerns the different requirements for stability that apply to nuclear species and chemical elements. The stability of any physical thing (in a given physical environment) is closely related to how long it lives (in that environment); the lifetime of a nuclear species is a very different thing to the lifetime of the corresponding chemical element. Nuclear species can have very short lifetimes, so one might accept (on good experimental evidence) that a nuclear species has been brought into existence without accepting that there is a chemical element corresponding to that species. Before I present the proposal in any further detail, however, I will do some stage-setting.

Philosophers and scientists often claim that things can exist at different "levels." This, for instance, is one way to understand relationships between different sciences. Might this be the way to understand the relationship between an element and its characteristic nuclear species? I think not. The notion of a level that is involved in such claims is often unclear. Some philosophers even question whether the idea that reality itself is "layered" is fully intelligible (see for instance Heil 2003; 2012). At most, they argue, it may be that such levels or layers arise only from the different perspectives from which different sciences view reality. If that is so, then layers and levels are not really ontological, but instead reflect different systems of knowledge, and the concepts that are applied within them. It would be better to substitute talk of different "levels" with talk of different scales. I will do that in what follows, for two reasons. The first reason is that scales (whether of energy, length, or time) are crucial to understanding existence in chemistry and physics, as I hope the following discussion will establish. Second, the scales of things are not of human creation. Scales are part of how things are, not of how they are perceived, conceived, or described. That last claim may sound odd, but it is important to see that "scale" can mean two different things. Taking size as an example, it might mean the very different sizes of things (for instance, the different scales of a hillock and a mountain); on the other hand, it might mean the different ways that these sizes can be mapped onto the number line (using yards and meters, for instance). When I say that scales are not of human creation, I only mean that the differing sizes of things are not of human creation. Of course yards and meters are of human creation.

Philosophers have tended not to give due consideration to scale,[6] but as I have already said, scale is crucial to understanding existence in chemistry. Why? First, existence in chemistry requires stability, which is an energetic notion: to be stable is (broadly speaking) to be of lower enough energy than other relevantly accessible states.[7] Quite obviously, it is possible for something to be stable at one energy scale but not at another; as every chemist knows, if you heat something up then eventually it may cease to exist. Second, when chemists concern themselves with existence, they are typically concerned with the existence of structures, and structure is a scale-relative notion (see Hendry 2016, forthcoming a). Third, the ability of a structure to interact with the rest of physical reality depends on the timescale on which it exists.

So here is what I take to be a plausible answer to the central question of this section: the existence of an SHE is typically established when it exists long enough

---

[6] Honorable exceptions include Batterman 2001; 2013; Ladyman, Ross et al. 2007; McGivern 2008; McGivern and Rueger 2010; and Thalos 2013.

[7] The phrase "relevantly accessible states" is intended to cover both kinetic and thermodynamic stability.

to be detected in a high-energy experiment in which nuclei of some kind are bombarded with particles of some other kind (typically other nuclear species). The energy scales at which these interactions occur (gigaelectron volts, or GeV) are very high compared to those at which the interactions governing chemical reactions typically occur (that is, interactions between electrons and nuclei). This entails that the processes involved in the nuclear reactions occur at much shorter timescales than chemical processes. A nuclear species may come into existence and then decay before it could possibly have time to acquire a stable electronic structure. If the lifetime of a nuclear species is in the femtosecond range, while the time it would take for electrons to relax into a stable electronic structure corresponding to a neutral atom might be of the order of picoseconds or nanoseconds, then the existence of the nucleus may be thousands or millions of times too brief to allow the formation of a neutral atom with a determinate chemistry. This is an important difference: if a nuclear species cannot—owing to its own instability—live long enough to have a stable and determinate electronic structure, then in some important sense it cannot have a chemistry. If it cannot have a chemistry, then it cannot be a chemical element.[8]

It is important to distinguish two possible versions of this proposal. On the stronger view, suggested by comments from Denys Wilkinson (quoted by Kragh [2017, 15]), the ground-state electronic structure of a neutral atom is essential to it; it is part of what makes it the particular element it is. While the nuclear species exists, the element doesn't because nothing has the property that would be essential to it. The weaker version is that SHEs might fail to correspond to chemical elements proper because they do not live long enough for the addition of electrons to result in an atom of the corresponding element with a determinate electronic structure, since by the time the electrons settle down into a stable electronic structure, the nucleus will have decayed into something else. On this weaker view, the electronic structure of a chemical element need not be essential to it, but SHEs might fail to be elements because they cannot have any stable electronic structure at all, since they are too short lived.

I prefer the weaker view for two reasons. First, it is weaker, and yet still strong enough to give the required result, that "although a chemical element is defined by its atomic number, not everything with an atomic number is an element" (Kragh 2017, 15). Second, it seems a coherent possibility that an element might have had (under different physical laws) a different ground-state electronic structure to the one it actually has. For that to be a genuine possibility, the ground-state electronic structure has to be metaphysically contingent, in which case it cannot be essential. To give a related example, chemists sometimes say that

---

[8] This proposal develops along more clearly ontological lines the lifetime requirement that Wapstra (1991, 883) places on evidence required to establish the existence of a new element.

water's boiling point is anomalously high. One way to interpret such comments is that water would boil at a much lower temperature if it behaved more like the hydrides of other elements in oxygen's group (S, Se, and Te). But this requires that water could have a different boiling point, that is, that its actual behavior is not metaphysically necessary. This still leaves open that water's boiling point is nomologically necessary, that water's boiling point is determined to be what it is by the laws of nature (see Hendry and Rowbottom 2009 for further discussion of this example).

The key point about the above reasoning, aside from the focus on timescales, is that different kinds of thing may have different criteria for existence. What it takes for a nuclear species to exist is different from what it takes for a chemical element to exist. One might expand on this rather abstract point in one of two ways. From an epistemic point of view, one might see it as expressing a pragmatic constraint on evidence for existence; I can have reasons to believe that something exists only if it exists long enough to be detectable. Detectability, of course, is a moving barrier; the invention of a new experimental device may shift it significantly. I think a more robustly physical point of view is required to do justice to the difference between elements and their constituent nuclei. From this point of view one might say that an object of a particular kind can be said to exist only if it is in principle possible for it to exert the causal powers that are characteristic of members of that kind. A diatomic molecule cannot exist unless the restoring force between the constituent atoms is strong enough to bring them back to their equilibrium positions. Hence it must be able to vibrate, and so interact with radiation at appropriate frequencies. A nuclear species can be said to exist when it interacts with its surroundings, but it takes a much longer life for an element to acquire what is required of it: a stable electronic structure, and the chemistry that that structure makes possible.

3. The Elements of Existence

In this final section I would like to relate the approach developed in the last section—according to which different kinds of thing may have different criteria for existence—to questions of the existence of composite objects more generally. One might imagine that philosophers spend their time debating what existence is, but they more often take it to be a basic or even unanalyzable notion, instead proposing criteria for when something can be said to exist. First-order logic, for instance, offers no account of what existence is, but merely represents it formally using the existential quantifier. This supports the widespread view that existence is neither a predicate nor a property. It is also the source of Quine's (1980) criterion for ontological commitment: to be is to be the possible value of a bound variable. This criterion provides neither an account of what existence is, nor of what exists; to answer the latter question we have to decide which theories we believe, and then render them in a canonical formulation that can be formalized in first-order logic. Deciding which theories we believe requires that we apply standards for theoretical commitment; in other words, what kinds of evidence we need for a theory in order to take it seriously enough to play a role in answering ontological questions.

Quine's criterion of ontological commitment may seem difficult to apply to chemical theories, simply because they are often presented in the form of diagrams rather than as sets of statements that could be translated into first-order logic. Nevertheless, one can think in broadly Quinean terms about how to determine the existential commitments of such theories, and what kinds of evidence are needed to support them. By this I mean (what I take to be) the following truisms: (1) that it is reasonable to acquire ontological commitments by becoming committed to scientific theories; (2) that it is reasonable to acquire commitments to scientific theories on the basis of the evidence we have for them; and (3) that the ontological commitments we acquire by becoming committed to a scientific theory should be closely related to the distinctive inferences we use that theory to make.

In a now-classic discussion, Peter van Inwagen (1990) raises what he calls the "Special Composition Question" (hereafter SCQ). On a common-sense view, depending on how they are related to one another, groups of objects sometimes form further things—composite objects—and sometimes they don't. For instance, a blade and a handle, suitably attached, form a knife. In contrast, a particular blade and handle that have never been so attached do not form anything at all. To take a scientific example, consider a proton and an electron that are so distant from each other that effectively they do not physically interact. The proton and the electron are just that: a proton and an electron. We can consider them as forming a "physical system," but any description of the composite system will reflect their separateness, and will be analyzable into a description of two separate physical systems. Bring them together, however, and they will interact strongly enough for the proton to capture the electron, and the physical properties will be transformed. Here we have a hydrogen atom: since the proton and the electron still exist, a further thing has come into being. The key point is that the union makes a difference, in the sense that we would fail to account properly for the composite system's behavior unless we treat it as a composite system. One might say that a composite object has come into existence.[9]

---

[9] The idea that in order to be considered to exist something must bear causal powers is, of course, an old one: applications of this idea follow Plato's Eleatic stranger (see, for example, Colyvan 2001, chapter 3) or, in the context of emergence, Samuel Alexander (see, for example, Kim 1998, 119; 2005, 159).

Following van Inwagen, one might pose the SCQ as follows: consider a plurality of things (the xs). When is it the case that there is a y such that the xs compose y? (1990, 30) There are three general kinds of answer to the SCQ: "never,"
"always," and "sometimes, depending on the circumstances." These are known
as compositional nihilism, compositional universalism, and the moderate answer. The moderate answer—that the xs sometimes compose a y, and that sometimes they do not, depending on the circumstances —would seem to reflect the
common-sense view of composition outlined above. In fact, it covers a great variety of different positions, depending on whether it is claimed that there is just
one criterion of composition covering all categories of composite object, or many
(for many different kinds of thing). Thus, for instance, one might think that the
conditions under which a blade and a handle form a knife (e.g., attachment) are
different to those under which a proton and an electron form a hydrogen atom.
Alternatively, one might argue that there is a single such criterion for every case
of composition. Both the relationship between the blade and the knife and that
between the proton and the electron might be claimed to fall under some more
abstract term, such as, for instance, "bonding." So there are as many "moderate
positions" as there are proposed criteria for composition, and we can add more
by considering theories according to which there are different criteria of composition for different categories of composite object.

First let us consider compositional nihilism and universalism. Some
philosophers have found it hard to accept the common-sense view, preferring
these more systematic alternatives.

3.1. Nihilism

Many philosophers who address the SCQ defend nihilism, or variants of it,
against the alternatives (see, for instance, Sider 2013). Nihilists sometimes say
that we (philosophers attempting to answer ontological questions) should be
wary of unthinkingly endorsing the commitments of common sense, which in
other cases can be overturned by science. But this seems unfair to composite
objects, which arise not only in our common-sense view of the world, but right
across the sciences: consider atomic nuclei, atoms and molecules, cells, multi-
cellular organisms, and other entities at various scales all the way up to such as-
tronomical objects as planets, stars, and galaxies. In rejecting the existence of
composite objects, one is rejecting the apparent commitments of both common
sense and science. The relative epistemic standing or ontological trustworthiness
of common sense and science are not relevant to the discussion.

There are two tasks involved in defending nihilism. One is to defend it against
an obvious, and seemingly crushing, objection: that nihilism is incompatible with

the obvious fact that in science and everyday life we encounter many composite objects. The other is to establish that nihilism is preferable to the alternative kinds of position on composition (universalism and moderate answers). One way to do both these jobs is to argue that nihilism has all the benefits of an apparent commitment to composite objects, but without the cost of real commitment to them. Nihilists attempt to avoid conflicts with the apparent commitments of common sense and science using paraphrase. The idea is this: apparent commitments to composite objects can be explained away by replacing statements involving references to composite objects (such as chairs) with paraphrases such as "simples arranged chair-wise." "Simples arranged chair-wise" is understood not to involve commitment to chairs. Via obvious substitutions, the same strategy can be applied to methane molecules, benzene molecules, methane, benzene, and of course cabbages, queens, and kings. A "simple" is whatever there is that is mereologically simple (that is, has no parts). If this strategy is successful, nihilism would appear to be preferable to either universalism or moderate answers, as it has the benefits of their commitments but on the basis of a simpler and more economical ontology.

There are three questions about this strategy: (1) whether use of the phrase "simples" implies a problematic (because possibly false) claim about the existence of a class of fundamental entities (i.e., lacking parts); (2) whether phrases containing the phrase "simples arranged chair-wise" genuinely provide paraphrases of sentences containing the word *chair*, but lack commitment to composite objects; and (3) whether, even if one accepts a positive answer to (2), there is any genuine reason to prefer nihilism to universalism, or some brand of moderate answer to the SCQ. I shall set aside (1), because scientific theories are often developed successfully on the basis of merely hypothetical commitments. The chemical elements themselves provide an example.

My objections to nihilism's paraphrase "simples arranged chair-wise" concern not the simples, but the notion of an arrangement. What is an arrangement supposed to be? Unless some positive characterization is given, the nihilist's claim that sentences about chairs can be replaced by sentences about simples arranged chair-wise cannot be assessed. We don't know what the theory is. The nihilist might insist that the notion of an arrangement is something we clearly grasp: here are building blocks arranged to form a cross (arranged cross-wise), here they are arranged to form the letter *R* (arranged *R*-wise). Fair enough, but that notion of an arrangement is no help in describing things that are constantly in motion, such as flocks of birds and atoms. It is therefore an inadequate basis for understanding sciences that engage with the Heraclitean world we find at the molecular scale, in which such arrangements are fleeting (see Hendry forthcoming a). The nihilist has given only the appearance of constructing a paraphrase, an appearance that vanishes as soon as we give the word *arrangement* any thought. The nihilist might then say that the term is meant simply as a placeholder: "arrangement" should be understood to stand for whatever relation would do the job of replacing statements about chairs. But then how do we know that paraphrases involving the phrase "simples arranged chair-wise" are free of commitments to chairs? It is common to reject skeptical alternatives to scientific theories on the grounds that they are not genuine alternatives: the hypothesis that the world behaves as if electrons exist is not a genuine alternative to the hypothesis that electrons exist, because it does not have its own theoretical machinery for making predictions of worldly phenomena. It is a parasitic pseudo-theory. One can summarize the point as a dilemma: either (1) "arrangement" should be understood in terms of the static spatial relationships that are entered into by medium-sized physical objects, or (2) it is understood as a placeholder that allows the nihilists to help themselves to the benefits of composite objects but without the ontological cost of commitment to them. If (1), then it is inadequate for replacing statements about some composite objects. If (2), then an argument needs to be given that the paraphrases have no commitments to composite objects. Until then it is not clear that it is a distinct alternative to accepting commitment to composite objects. I fear that whatever currency compositional nihilism has results from the failure to distinguish between (1) and (2). My objection to nihilism is not that it is false. It is that, as presented, it is a form of words rather than an answer to the SCQ.

3.2. Universalism

Another very general approach to answering the SCQ is mereological universalism, which can be understood by analogy with versions of set theory according to which the existence of some things automatically entails the existence of a set containing them. In just the same way, the existence of some xs automatically entails the existence of an object that is composed of them. The problem with universalism is that alongside its commitment to (for instance) methane molecules, it is committed to the existence of many unimportant objects including, for instance, the object composed of my left arm and the Eiffel Tower. Universalists are committed to interpreting all such contrasts in pragmatic terms, that is, in terms of why we care about the existence of some composite objects (e.g., methane molecules) and not others (the object composed of my left arm and the Eiffel Tower). But some such reasons present themselves as ontological: methane demands our attention and commitment because it impinges on our experience by exerting causal powers that its components do not have. The object composed of my left arm and the Eiffel Tower then falls away because it does not so impinge. Applying the causal criterion mentioned earlier, one might

EXISTENCE OF ELEMENTS, AND ELEMENTS OF EXISTENCE   139

as well say that for all practical purposes it doesn't exist, even though my left arm and the Eiffel Tower do.

On the face of it, universalism's commitment to the existence of composites independent of any physical relationships among their parts suggests that it must be unhelpful in answering *contrastive* existence questions, which are common in chemistry: why does species A exist, but not species B, which is similar to it in some relevant way? Thus, for instance, one of the most basic applications of quantum mechanics allows us to show why the hydrogen molecule H_2 exists, but not the helium molecule He_2 (see, for instance, Atkins 1986, 379). The universalist sees this discussion as one that does not involve existence. Both H_2 and He_2 exist, and quantum mechanics gives us an account of why we are interested in the former, but not the latter.

3.3. Moderate Answers

Moderate answers to the SCQ identify specific conditions under which a group of objects act together so as to form a composite object. For an answer to the SCQ to be a moderate one, such conditions must not apply universally, so that there are some groups of objects to which they do not apply, and which do not therefore form composite objects. One can distinguish simple answers to the SCQ, which identify a single condition for the existence of composites, from disjunctive answers that allow that there may a (possibly open-ended) list of such conditions. The problem with simple answers is that they face easy counterexamples: For physical entities, one might suppose that bonding or fastening them together would form a new object, but this fails for people. We would not say that you and I would form a new object even if we were attached at the hip. On the other hand, van Inwagen (1990, section 7) criticizes disjunctive answers (which he calls "Series-style answers") for their ad hoc character (for a Carnapian response, see Thomasson 2007, chapter 7).[10] The conclusion of section 2 was that atomic nuclei and chemical elements have different criteria for existence, even though the latter are individuated in terms of the former. Generalizing the point, it should be no surprise that the criteria for existence of different kinds of a thing are quite diverse. Material objects, as studied by the physical sciences, are composed from their parts in a range of different ways, but one thing is clear: they cannot survive change in their parts. Within this class we can make distinctions based on the variation of existence conditions with scales (of energy, length, or time). Elements, and their characteristic atoms, considered as composite entities, provided just such a case. This is why it is possible to synthesise an SHE atom without synthesising the SHE itself.[11]

---

[10] There are real problems with the way van Inwagen construes the terms in which he frames his candidate simple answers, but discussing them would be too much of a digression.

[11] James Ladyman and Don Ross (Ladyman, Ross et al. 2007; Ladyman 2017) have previously criticized the debate over the SCQ, and defended the idea of scale-relative ontology. I have two worries

Living organisms relate to their matter in an entirely different way: they constantly absorb and release matter of different kinds, but this is no threat to their diachronic existence.[12] Artifacts are yet different: Their existence may sometimes depend on our intentions rather than any physical relationship between their parts. Think of the bicycle in my garage, which I took apart during the summer, and have been meaning ever since to reassemble. Should there be any hope that anything other than a disjunctive answer to the SCQ could work? Perhaps all this is not so much to answer the SCQ as to ignore it. If that is the case, there seems to be no more adequate response.

Acknowledgments

I am most grateful to Helge Kragh for sending me the paper that inspired me to write this one, and also to an anonymous referee for comments on an earlier draft.

References

Aston, F. W. et al. 1923. "Report of the International Committee on Chemical Elements." *Journal of the American Chemical Society* 45: 866–874.
Atkins, P. W. 1986. *Physical Chemistry,* 3rd ed. Oxford: Oxford University Press.
Batterman, Robert. 2001. *The Devil in the Details: Asymptotic Reasoning in Explanation, Reduction and Emergence.* Oxford: Oxford University Press.
Batterman, Robert. 2013. "The Tyranny of Scales." In *The Oxford Handbook of Philosophy of Physics,* edited by R. W. Batterman, 255–286. Oxford: Oxford University Press.
Brock, William H. 1992. *The Fontana History of Chemistry.* London: Fontana Press.
Buchwald, Jed Z. 1994. *The Creation of Scientific Effects: Heinrich Hertz and Electric Waves.* Chicago: Chicago University Press.
Chalmers, Alan. 2009. *The Scientist's Atom and the Philosopher's Stone: How Science Succeeded and Philosophy Failed to Gain Knowledge of Atoms.* Dordrecht: Springer.

about their approach to these issues: one is terminological, and one (I think) more substantive. The terminological worry is that "scale-relative ontology" suggests meta-ontology; we must construct different ontologies for different scales. I would prefer to emphasize the ontological point that existence itself can be dynamic and scale-relative. The more substantive worry is that Ladyman and Ross attempt to commensurate physics and the special sciences using a single notion (that of a "real pattern") that (to my mind) invites an ontologically non-serious stance (non-seriousness being perfectly compatible with scientific realism). I think that we naturalists should avoid approaching all the special sciences with a single template, and try to express relationships between different sciences using their own concepts and theories.

[12] One might hope that a causal criterion for existence might unify at least the existence conditions for physical objects and living organisms. However, I think that the Eleatic Principle and Alexander's Dictum (see above) are best understood as providing sufficient conditions for when something should be considered to exist. So construed they are epistemic rather than ontological principles.

Colyvan, Mark. 2001. *The Indispensability of Mathematics.* Oxford: Oxford University Press.

Earley, Joseph E. 2005. "Why There Is No Salt in the Sea." *Foundations of Chemistry* 7: 85–102.

Gordin, Michael. 2004. *A Well-Ordered Thing: Dmitrii Mendeleev and the Shadow of the Periodic Table.* New York: Basic Books.

Heil, John. 2003. *From an Ontological Point of View.* Oxford: Oxford University Press.

Heil, John. 2012. *The Universe as We Find It.* Oxford: Oxford University Press.

Hendry, Robin Findlay. 2006. "Substantial Confusion." *Studies in History and Philosophy of Science* 37: 322–336.

Hendry, Robin Findlay. 2010. "The Elements and Conceptual Change." In *The Semantics and Metaphysics of Natural Kinds*, edited by Helen Beebee and Nigel Sabbarton-Leary, 137–158. London: Routledge.

Hendry, Robin Findlay. 2016. "Structure as Abstraction." *Philosophy of Science* 83: 1070–1081.

Hendry, Robin Findlay. 2019. "Emergence in Chemistry: Substance and Structure." In *The Routledge Handbook of Emergence*, edited by S. C. Gibb, R. F. Hendry, and T. Lancaster, 339–351. London: Routledge.

Hendry, Robin Findlay. Forthcoming a. "Structure, Scale and Emergence." *Studies in History and Philosophy of Science.*

Hendry, Robin Findlay. Forthcoming b. "Elements and (First) Principles in Chemistry." *Synthese.* DOI https://doi.org/10.1007/s11229-019-02312-8.

Hendry, Robin Findlay, and Darrell Rowbottom. 2009. "Dispositional Essentialism and the Necessity of Laws." *Analysis* 69: 668–677.

Kim, Jaegwon. 1998. *Mind in a Physical World.* Cambridge, MA: MIT Press.

Kim, Jaegwon. 2005. *Physicalism, or Something Near Enough.* Cambridge, MA: MIT Press.

Klein, Ursula. 1994. "Origin of the Concept of Chemical Compound." *Science in Context* 7: 163–204.

Kragh, Helge. 2000. "Conceptual Changes in Chemistry: The Notion of a Chemical Element, ca. 1900–1925." *Studies in History and Philosophy of Modern Physics* 31B: 435–450.

Kragh, Helge. 2017. "On the Ontology of Superheavy Elements." *Substantia* 1: 7–17.

Kragh, Helge. 2018. *From Transuranic to Superheavy Elements: A Story of Dispute and Creation.* Berlin: Springer.

Ladyman, James. 2017. "An Apology for Naturalized Metaphysics." In *Metaphysics and the Philosophy of Science: New Essays*, edited by Matthew Slater and Zanja Yudell, 141–161. Oxford: Oxford University Press.

Ladyman, James, and Don Ross, with David Spurrett and John Collier. 2007. *Every Thing Must Go: Metaphysics Naturalized.* Oxford: Oxford University Press.

Lavoisier, Antoine. 1790. *The Elements of Chemistry.* Edinburgh: William Creech. Translation of *Traité Élémentaire de Chimie* by Robert Kerr. Paris, 1789.

McGivern, Patrick. 2008. "Reductive Levels and Multi-scale Structure." *Synthese* 165: 53–75.

McGivern, Patrick, and Alexander Rueger. 2010. "Hierarchies and Levels of Reality." *Synthese* 176: 379–397.

Mendeleev, D. I. 1899. "The Periodic Law of the Chemical Elements." *Journal of the Chemical Society* 55: 634–656. Reprinted in *Classical Scientific Papers—Chemistry*, 2nd ser., edited by D. M. Knight. London: Mills and Boon, 1970.

Needham, Paul. 2004. "Has Daltonian Atomism Provided Chemistry with Any Explanations?" *Philosophy of Science* 71: 1038–1047.

Needham, Paul. 2006. "Aristotle's Theory of Chemical Reaction and Chemical Substances." In *Philosophy of Chemistry: Synthesis of a New Discipline*, edited by Davis Baird, Lee McIntyre, and Eric Scerri, 43–67. Dordrecht: Springer.

Needham, Paul. 2009. "An Aristotelian Theory of Chemical Substance." *Logical Analysis and History of Philosophy* 12: 149–164.

Needham, Paul, and Robin Findlay Hendry. 2018. "Aspects of the Concept of Potentiality in Chemistry." In *Handbook of Potentiality*, edited by Kristina Engelhard and Michael Quante, 375–400. Berlin: Springer.

Paneth, F. A. 1962. "The Epistemological Status of the Chemical Concept of Element." *British Journal for the Philosophy of Science* 13: 1–14, 144–160.

Quine, W. V. 1980. "On What There Is." In *From a Logical Point of View*, 2nd ed., rev. Cambridge, MA: Harvard University Press.

Sider, Theodore. 2013. "Against Parthood." In *Oxford Studies in Metaphysics*, vol. 8, edited by Karen Bennet and Dean Zimmerman, 237–293. Oxford: Oxford University Press.

Thalos, Mariam. 2013. *Without Hierarchy: The Scale Freedom of the Universe.* Oxford: Oxford University Press.

Thomasson, Amie L. 2007. *Ordinary Objects.* Oxford: Oxford University Press.

van der Vet, Paul. 1979. "The Debate between F.A. Paneth, G. von Hevesy and K. Fajans on the Concept of Chemical Identity." *Janus* 92: 285–303.

van Inwagen, Peter. 1990. *Material Beings.* Ithaca, NY: Cornell University Press.

Wapstra, A. H. 1991. "Criteria That Must Be Satisfied for the Discovery of a New Chemical Element to Be Recognized, Being the Report on Phase (I) of Operations of the Transfermium Working Group of IUPAC and IUPAP." *Pure and Applied Chemistry* 63: 879–886.

Zingales, Roberto. 2005. "From Masurium to Trinacrium: The Troubled Story of Element 43." *Journal of Chemical Education* 82: 221–227.

8

Kant, Cassirer, and the Idea
of Chemical Element

Farzad Mahootian, Global Liberal Studies, New York University, USA

1. The Official IUPAC Definition

The concept of element is fundamental to modern chemistry and yet it embodies an apparently persistent ambiguity that has remained unresolved for the nearly one hundred years since it was made official by the International Union of Pure and Applied Chemists (IUPAC) in 1923. IUPAC's Gold Book is the current compendium of definitions and protocols for chemical methods and nomenclature. The Gold Book defines "chemical element" as follows (emphasis added):

1. *A species of atoms; all atoms with the same number of protons* in the atomic nucleus.
2. *A pure chemical substance composed of atoms* with the same number of protons in the atomic nucleus. Sometimes this concept is called the elementary *substance as distinct from the chemical* element as defined under 1, but mostly the term chemical element is used for *both* concepts.

This dual definition appears awkward and troublesome for a number of reasons, and yet it has withstood the test of time. This chapter presents a take on why this definition has the form it does, how it arose, and why it endures. Following the introductory overview, there will be two historical sections, one on Immanuel Kant whose imprint on modern philosophy and science persists in various forms, including the genesis and interpretation of the IUPAC definition (see Paneth 2003; Scerri 2000; van Brakel 2006; Ruthenberg 2009; Mahootian 2013). The other section is on Ernst Cassirer, founder of the Marburg school of neo-Kantian thought, who wrote about chemistry both before and after the IUPAC definition above. We conclude with a comparison of Cassirer's understanding of chemistry, first with the mathematical chemistry of Guillermo Restrepo, and then with Joachim Schummer's conceptual analysis of the "chemical core of chemistry" (Schummer 1998).

Farzad Mahootian, *Kant, Cassirer, and the Idea of Chemical Element* In: *What Is A Chemical Element?*. Edited by: Eric R. Scerri and Elena Ghibaudi (2020). © Oxford University Press.
DOI: 10.1093/oso/9780190933784.003.0009

In definition one, the word *atom* is used to designate both a species, and a member of the species—this ambiguity is relatively common and often easy to sort out by context. But in the second definition there is a reference back to the first, and a vaguely qualified "sometimes . . . but mostly," that effectively amounts to sanctioning equivocation. What strikes me first is the strange allowance for polysemy. Scientists are supposed, by philosophers and laypeople alike, to maintain tightly controlled, nearly literal technical definitions of their foundational terms, aren't they? Isn't the ideal a precise, unambiguous single meaning, not a pair or range of possible meanings? After all, this isn't poetry, it's science! This is IUPAC's Gold Book—and one is immediately struck again by apparently conflicting currents of thought. The very title of the IUPAC book of protocols is poetical: the metaphorical gold in Gold Book deliberately invokes chemistry's ancestor, alchemy, as (it seems) must all references to the history and language of chemistry. Faulting IUPAC for tolerating a loosely held together definition of a term that resides at the conceptual foundations of its discipline is perhaps to misunderstand IUPAC's function, and the way chemistry works in practice, as well as the logical limits of definition itself.

The duality at the heart of IUPAC's definition reappears in various systematic approaches to chemistry. Schummer's idea of the "chemical core of chemistry" (Schummer 1998) highlights chemical reaction i.e., the dynamic interactivity and relationality of chemistry. Here, relationality is both ontological and epistemological (Bernal and Daza 2010), for both chemical substances and the science of chemistry partake of this characteristic— the form of this science matches its content in this respect. Restrepo (this volume) shows how the inherent *looseness* of the IUPAC definition enables the needed dynamic and developmental space for the prolific discipline of chemistry, which actively invents its own subject matter in a never-ending flow of new materials. While physics, biology, and other fields are engaged in the manufacture of their facts and subject matter, this productive tendency characterizes chemistry far more than other fields of natural science— certainly the rate of generation of new chemical substances far exceeds that of the material productivity of any natural science (Schummer 1997). While the subject matter of physics involves matter, chemistry's subject involves materials. The distinction between matter and material is noted as early as Plato and Aristotle.[1]

---

[1] Both philosophers maintained different versions of a property-less substratum which underlies the transformation of substances and gives rise to the properties of elementary "bodies," or "kinds" or "qualities"; both distinguish the substratum from elements, and both describe a more or less articulate account of how the latter arise from the former. For Plato the process is mathematically definable; for Aristotle it is the interplay of contraries in (non-mathematically defined) proportions. Following suit with most histories of chemistry, Needham and Hendry (2018) sideline Plato and instead take on the labor of formalizing the logic of Aristotle's interplay of contraries to compensate for its lack of precision. By contrast, Plato relies on the precise geometry of regular solids to explain the emergence and transformations of materials.. The mischaracterization and relative neglect of Plato's matter theory in the historiography of chemistry is an oddity—and the subject of a separate article, in process.

A similar duality is at play in IUPAC's definition where the tension between the two definitions of element signals the range of possible abstractions in chemistry. Indeed, Paneth blames the historical misunderstanding between Boyle and Spinoza on the ambiguity of the term, and endorses the effort to keep but strictly distinguish both definitions (Paneth 2003, 136).

Paneth had some influence on the development of the IUPAC definition (Scerri 2000, 65–66), so why is that concern with maintaining distinction not reflected in the final definition? Why is there a "sometimes . . . but mostly" in this definition? Why draw attention to a difference just before diminishing it? Why does IUPAC leave that hanging? Is it a sign that this issue is not really an issue? Certainly, this is mostly true in practice: most chemists just don't care because it seems not to affect their work—at least not in the short run. Do philosophers care (beyond the few who discuss it in the meetings and journals of the philosophy of chemistry)? Does it matter? Does it make a material difference? Paneth's first detailed epistemological treatment makes the case for distinguishing two senses of element. To keep them distinct, he dubbed one *Grundstoff* and the other *einfacher Stoff*. These terms were translated (by his son, philosopher of science Heinz Post) into English as "basic substance" and "simple substance," to the chagrin of many (e.g., Earley 2009; this volume). For Paneth, *Grundstoff* is transcendental in the sense of being unobservable, irreducible "essence," whereas *einfacher Stoff* refers to the perceptible stuff that was at one time extracted from various parts of the Earth, processed and isolated in varying degrees of purity, and finally bottled for storage in chemical stockrooms.[2] Paneth sought to clearly distinguish them to prevent chemists and students of chemistry from mixing up and misapplying them. As the existence of this volume indicates, that effort was not effective. Paneth pointed to the necessary *Schwanken* (translated as oscillation, but could also mean vacillation) between the two meanings of element. Paneth implies that each sense of element performs a function that the other can't, and that they are inconvertible therefore both are needed, not by choice but by necessity, thus leaving them in a state of tension.

---

[2] Paneth's original article was published in German in 1931, in English (in two separate issues of the *British Journal for the Philosophy of Science*) in 1962, reprinted in a single volume in 2003. Philosophical interest in chemistry reignited in the mid- to late-1990s and continues into the present volume.

2. Kant's Philosophy and Its Relevance to the Science of Chemistry

2.1. Kant on the a priori Organization of Experience; Constitutive Versus Regulative Principles

This section undertakes an exposition of Kant's philosophy, his initial rejection of chemistry as a science, and the dramatic shift of his philosophy in response to the growth of chemistry late in his life. Ruthenberg's (2009) detailed article on the influence of Kant on Paneth points out that the latter's use of the term "transcendental" follows that of the neo-Kantian von Hartmann, who called himself a transcendental realist, as opposed to Kant's original transcendental idealism. Unfortunately, von Hartmann's interpretation of transcendental is, in my opinion, considerably less nuanced than Kant's. This resulted in Paneth's relatively blunt idea of transcendental as simply imperceptible in principle. The impact of this derivative form of the concept is to diminish its clarity and usefulness. In this section, I recur to Kant's philosophy, drawing chiefly from his *Critique of Pure Reason* (Kant 1998) and *Opus Postumum* (Förster 1998) in order to analyze its influence on Cassirer's neo-Kantian blend of history and philosophy of science.

Immanuel Kant lived in a society basking in the glow of successful expansionism, born of the marriage of science, technology, finance, and politics. Few questioned the growing success of science, but David Hume, an older contemporary of Kant, rejected the assumption of the unity of nature as mere habitual belief, unjustifiable by reason or experience. Hume posed the problem of induction as the weakness at the core of science, and thereby "roused" Kant from his "dogmatic slumbers." Kant realized that if perception is purely passive, there would be no way to justify any knowledge of causal connections, or any necessary connections among the objects of perception. Kant held that we contribute the forms and structures of perception and conception—notably, space, time, substance, and causality—to our experience of the world. Our very act of consciousness is the source of the necessary and universal structures of any possible experience. So, to look for the ground of order and structure in the world of sense experience, we must shift the direction of our gaze back to ourselves. This 180° shift is what Kant referred to as his "Copernican revolution."

While the contents of experience are known only after we experience them ("a posteriori"), the conditions, or form in which they appear can be known before experience ("a priori"). Kantian philosophy argues that acts of consciousness condition every possible experience, that our experience is partly constructed by the act itself, partly discovered in the content of our observations, measurements, and experiments. Our acts of consciousness contribute the form of experience, its continuity; the content of experience is to be discovered, sometimes confirming and sometimes conflicting with expectations, but always conditioned by space, time, substance, and causality. Kant derived the conditions for experience by what he called "transcendental deduction" in order to explain the continuity of "phenomenal experience."

Kant claimed that the unity of nature results from the unity of experience, that is, the universally consistent, logically necessary continuous application of the forms of perception and categories of understanding. Kant called this the constitutive function of reason, which he distinguished sharply from its regulative function. The latter guides and regulates the legitimate applications of the forms and categories of experience. In its regulative function, reason saves consciousness from believing the inevitable illusions created by overextending the constitutive ordering principles of perception and understanding beyond their legitimate usage. Just as visual perception can't help falling prey to optical illusions, so too cognition can't help falling prey to cognitive illusions. Furthermore, just as *understanding* an optical illusion frees us from actually believing what we see but can't prevent us from experiencing the illusion, so too reason frees our understanding from actually believing a cognitive illusion but can't prevent us from thinking it. Reason functioning in its constitutive mode occasionally makes illusions to which it naturally falls prey; reason functioning in its regulative mode identifies these as illusions, thus freeing the mind. These two terms, constitutive and regulative reason, are central to Kant's thinking, so when he changes their interrelationship in his later thought, the results are profound.

Perception, understanding, and reason work together in Kant's system to ground the universality and necessity of phenomenal experience. Objects of experience are organized by acts of consciousness, so every object in the flow of experience is the effect of not just one chain of causes and effects, but an interlinked net of chains. The interconnected community of moving objects constitutes the universe of human experience. Finally, in addition to being conditioned by the "forms of perception" (space and time), experience is ordered by the "categories of understanding" (causality and substance). Kant argues that through the category of substance, consciousness unifies sets of properties and renders them recognizable as persistent individual objects. The object, as the integration of several sets of properties, coexists with other objects in a causal network that spreads across space with every moment of time. In summary, space, time, substance, property, causality, and community are necessarily and universally constitutive of human experience.

Reason's constitutive function grounds the continuity of experience, but Kant points out that the systematic unity projected by regulative principles into the field of perception is only a rule. It is the intelligible unity of a set of perceptions as a singular object; it is not one more perception in that set.

Accordingly, the regulative principle will constrain thinking to organize space, time, and causality in a manner consistent with rules that it discovers through experimental interactions within the field of objects. Science is possible at all because observations and measurements embody the empirical correspond-ence to ideal mathematical, constitutive, principles. Kant distinguishes these carefully from principles that are "merely" regulative. However, a contrasting view of regulative principle is found elsewhere as Kant notes, "[This] method for seeking out order in nature . . . is a legitimate and excellent regulative prin-ciple of reason, which, however . . . only points the way toward systematic unity" [CPR A668, B696]. To summarize: it is a "mere" rule in reference to *perceiving* and *understanding* actual objects, but an "indispensable and excellent principle" for *thinking* about those objects in that it seeks the deepest sense of unity that an object embodies –and this is exactly the sense of unity that the idea of chemical element conveys.

The relevance of the regulative-constitutive distinction for the idea of chem-ical element is now apparent. The regulative principle of reason is a methodolog-ical, or epistemic imperative that guides reason to the worthiest goal (systematic unity) while simultaneously acting as a preventive measure against unwar-ranted epistemological and ontological claims, specifically, it circumvents the unwarranted certainty that the unity in question is an actual thing. So, for Kant, "element" is a regulative idea of pure reason, not a thing.

> Admittedly, it is hard to find pure earth, pure water, pure air, etc. Nevertheless, concepts of them are required (though as far as their complete purity is con-cerned, have their origin only in reason) in order appropriately to determine the share that each of these natural causes has in appearance. (CPR A646/ B673)[3]

On the largest scale, perhaps the most general idea of reason is the unity of na-ture: the working assumption underlying and regulating the conduct of all lab-oratory experimentation. The *continuity* of experiences (and thus, experiments) is guaranteed by reason's constitutive principles. The *unity* of all nature, past, present and future—which is impossible to experience—is an *idea* proposed by reason to regulate and integrate any possible experience (including experiments).

---

[3] Granted, in CPR Kant is referring to earth, air, fire, and water as elements and he makes no direct reference in this text to "elementary substances" or "chemical substances." However, as we note in the following subsection, in his later work Kant employs exactly those terms, and it is arguably clear from the context that he intends for matter and materials to be taken as regulative ideas. As discussed below, though in CPR Kant did not speak of chemical elements, in OP he distinguished "matter" from "specific materials," explicitly referring to oxygen and hydrogen as *stoichiea*, the Greek term for elements.

2.2 Kant's Later Thought

With his *Metaphysical Foundations of Natural Science* (MF) of 1786, Kant developed a metaphysic of Newtonian physics—a philosophical synthesis of Newton's mathematical synthesis of space, time, matter, and gravity. Kant succeeded in establishing the metaphysical ground for the dynamics of matter and bodies on the cosmic scale.[4] Matter on the macroscopic scale behaved just as Newton's laws predict, but microscale matter was quite another matter, so to speak. Both MF and the first edition of CPR denied chemistry the status of a proper science. However, laboratory chemistry was making rapid progress by the late 1700s:[5] Lavoisier's laboratory shook the faith of phlogiston theorists with his discovery of oxygen. Kant, who was an avid follower of the science, switched his allegiances from phlogiston to oxygen, and softened his harsh assessment of chemistry. This is evidenced in his second edition of CPR, and in his unpublished late work.

In his translator's notes on Kant's *Opus Postumum* (OP), Eckart Förster notes, "In 1796, he requested 'two lectures' from his friend and colleague, the professor of medicine, Carl Gottfried Hagen, 'in which he [i.e., Hagen] conducted all the experiments on which Lavoisier bases his theory, and the doctrine of the composition of different bodies according to it.'" Kant requested that Hagen perform another chemical experiment in 1800, and went as far as to commission an apparatus presumably to measure ether in order to track its involvement in the Lavoisian experiments: "Wasianski also had to build for Kant an instrument to measure the electricity of the air (electrometer); much to Kant's disappointment, it did not function as planned" (Förster 1993, note 90).

Kant began to reconsider the nature of matter and its importance to metaphysics in the late 1790s and proposed the "Transition" project to bridge the gap his system left between metaphysics and physics: "there is a gap to be filled between the metaphysical foundations of natural science and physics; its filling is called a Transition from the one to the other." (OP 21: 482). His shift in thought was quite dramatic. Friedman (1992) argues that Kant reconfigured his system to accommodate the new chemistry; McNulty (2016) makes a strong counter-argument. Friedman believes that Kant required principles that are both regulative and constitutive to fill this gap, creating a paradoxical "intersection" of "two, formerly entirely independent domains" (Friedman 1992, 262). "Paradoxical" may be too mild a term to capture the about-face that Kant pulls when he says that "the form of the system

---

[4] Kant proposed the nebular theory of solar system formation in 1755, based in part on Newtonian dynamics, and prior to his transcendental philosophy.

[5] Fourteen new elements were discovered between Kant's CPR (1780) and his death (1804). During this time, crossovers from physics to chemistry and biology were stimulating the common imagination. In the field of electricity, Leyden jars made people's hair stand on end at European salons, and Galvani's experiments with frogs were setting the intelligentsia abuzz.

of physics" requires "regulative principles which are at the same time constitutive" (OP 22, 241 in Friedman 1992).

A key goal of the Transition project was Kant's "transcendental deduction of ether," which reconceives his idea of space. The resultant conception of the ether is a "*realized* or *hypostatized* space" (Friedman 1992, 310, emphasis in original). Kant draws attention to the radical implication of the new concept: "now the principle of the possibility of all experience is (the realization of) space itself as an individual [*einzeln*] object of the senses (i.e., of empirical intuition)" (quoted in Friedman 1992, 311n138). This is indeed the very opposite of the transcendental deduction of space as the "form" of, as opposed to an object of, sense perception. In his CPR, Kant had made the sharp distinction between constitutive and regulative reason (see section 2.1, above) as one of the foundations of his critical philosophy:

> I accordingly maintain that *transcendental ideas never allow of any constitutive employment* . . . On the other hand, they have an excellent, and indeed indispensably necessary, employment, namely, that of *directing the understanding towards a certain goal upon which the routes marked out by all its rules converge, as upon their point of intersection.* This point is indeed a mere idea, a *focus imaginarius.* . . . Hence arises the illusion that the lines have their source in a real object lying outside the field of empirically possible knowledge—just as objects reflected in a mirror are seen as behind it. (CPR A644–5, emphasis added)

If we were to go along with Kant's Transition project, space would cease to be the transcendental form of empirical intuition and would instead be the all-pervasive, all-encompassing stuff that mediates interactions (e.g., electrical, magnetic, energetic) among bodies. Indeed, some of Kant's nineteenth- and twentieth-century proponents seem to have followed him through the looking glass in pursuit of the idea that Kant struggled to establish in his OP. The *Naturphilosophen*, the post-Kantians and neo-Kantians followed Kant's cue, and to various degrees and in different directions they hypostasized the systematicity and purposiveness of regulative reason, making significant contributions to nineteenth-century chemistry and physics in the process.[6]

McNulty's thorough (2016) analysis of Friedman and his critics concludes that the Transition project may be seen less as a departure from Kantian philosophy and more as a specific extension of it. This view gains further support if we consider how in its constitutive mode, reason is involved in the discovery of mathematical laws according to which objects move and interact (CPR A178–9/ B221–2). These "laws of nature" prepare the stage, as it were, for reason in its regulative mode to seek a unity of systematic interconnections. Beyond its unifying

---

[6] M. Friedman, "Kant-*Naturphilosophie*-Electromagnetism," in Friedman and Nordmann 2006.

vision, regulative reason's proclivities to purposiveness enable us to effectively intervene in practice with decisions of moral consequence, whether such consequential actions are taken under laboratory conditions, or in other aspects of our practical lives. In Kant's philosophy, the possibility of a moral universe is based on universalizable moral principles enacted in a natural world ordered according to mathematical causal laws. Thorndike (2018) argues that the Transition project is continuous with the transcendental idealism laid out in the *CPR* and that the merger of constitutive and regulative principles is a transcendental guarantee for morality in Kant's philosophy (Thorndike 2018, 111). In short, Thorndike argues that the Transition project provides further grounds for the possibility of taking morally principled, effective action despite necessarily limited knowledge.

In this connection, we must consider Kant's puzzling statement that the method of critical philosophy is the very same as the method of chemistry even though he does not count chemistry as a science. During his critical period, he considered chemistry as, at best, a set of rationally organized empirical rules for dealing with materials and their associated forces. What was the basis of his comparison of chemistry and critical philosophy? Kant mentions this in two sections of the CPR. In a footnote to the preface to the second edition he notes that when

> Stahl changed metals into calx and then changed the latter back into metal by first removing something and then putting it back again, a light dawned on all those who study nature. They comprehended that reason has insight only into what it itself produces according to its own design. (Bxii–xiii)

Though Stahl is replaced by Lavoisier in Kant's later writings, the methodical separation of chemical analysis, followed by a new synthesis remains the core method of critical philosophy:

> This experiment of pure reason has much in common with what the chemists sometimes call the experiment of reduction, or more generally the synthetic procedure. The analysis of the metaphysician separated pure a priori knowledge into two very heterogeneous elements, namely those of the things as appearances and the things in themselves. The dialectic once again combines them. (CPR, Bxxi)

In his final mention of chemistry in the *Critique of Pure Reason*, Kant tells us plainly that the "purity" of pure reason is not given, it is laboriously produced.

> It is of the utmost importance to isolate cognitions that differ from one another in their species and origin, and carefully to avoid mixing them together with

others with which they are usually connected in their use. What chemists do in analyzing materials . . . the philosopher is even more obliged to do. . . .Hence human reason has never been able to dispense with a metaphysics . . . though it has never been able to present it in a manner sufficiently purified of everything foreign to it. (A842 / B870)

The chemical analog of philosophical method, the paradigm reaction that Kant had in mind, is Stahl's formation of calx from metal and the subsequent restoration of the metal from calx. Lavoisier chose this very same reaction (as the "crucial experiment" of the phlogiston debates) to demonstrate the measurable existence of oxygen as the gas responsible for forming calx and rust. Oxygen can be captured by passing water through a sufficiently hot gun barrel containing iron rings. As mentioned earlier, Kant had both of those experiments performed for him and, with a little help from his friends, was trying to get involved in the chemistry game. Like some contemporary philosophers of chemistry, Kant got into chemistry for philosophical reasons.

Kant was very much a transitional figure (and his choice of "Transition" as the title for his OP project is quite apt). He continues to use the quasi-alchemical paradigm of dissolution and separation, and his acceptance of Lavoisier's methods, findings, and chemical theories is not fully dephlogisticated. Kant's investment in experimental investigation late in life is also suggestive of his mode of thought. Contrary to the usual portrayal of him as a dry rationalist, the spirit of experiment is apparent in the sometimes wildly speculative thrust of the OP. What Kant learned from chemistry was the procedure of analysis and synthesis. This is just what critical philosophy does in its transcendental analyses of perception, understanding and reason, and their subsequent synthesis for the sake of moral action and aesthetic judgment.

3. Cassirer's Philosophy of Chemistry

Kant's *Transition* project was meant to address shortcomings of his critical philosophy. One result of this project was the reconfiguration of the relationship between regulative and constitutive uses of reason. In the context of the lively debate[7] about what exactly Kant was trying to do in OP, an examination of Ernst Cassirer's philosophy of science presents fruitful lines of inquiry. Despite Cassirer's deep engagement with the cultural and intellectual history and

[7] Most early work on the OP has yet to be translated into English. Van Brakel (2006) has an overview of the major commentators on Kant's late work, especially as it relates to chemistry. A number of more recent books and articles on the OP have appeared in English, including those from Hall, McNulty, Thorndike, Friedman, and so on.

philosophy of science, very little is to be found about him in the literature of the philosophy of chemistry, beyond a few short references to him in the "prehistory" of that area of study.

Cassirer's view of the Kantian a priori as thoroughly regulative puts the OP's claims in an ambivalent light because at first glance, the regulative function of reason seems to have lost ground in OP's hypostasization of space, especially if one clings to Kant's original distinction of constitutive and regulative principles. However, one might instead argue that by erasing the line between regulative and constitutive functions in the case of space, Kant actually extends the domain of the regulative function into the constitutive, opening the latter to more than merely mathematical organization by amplifying the demands of reason[8] for systematicity. Cassirer further generalizes Kant's later thought by using the history and philosophy of chemistry as a paradigm case of scientific concept-formation, in which the growth of knowledge is not a steady progress toward the "truth" but a process of transition from substance-based to relational concepts. Even more important, in this transition, substance is not discarded but rather understood for what it is: a symbol—or more prosaically, a placeholder—for the complex intersection of relations that are progressively revealed with advanced empirical investigation. Cassirer's discussion of chemical atomism is instructive in this respect. As we shall see in the concluding sections of this chapter, his philosophy of science is consistent with current work in mathematical chemistry without succumbing to any variety of formalist reduction.

Cassirer's account is more nuanced than, but still consistent with, Kant's— indeed, Cassirer quotes the famous *focus imaginarius* passage of the CPR on regulative versus constitutive principles (see 2.2, above). Cassirer elaborates his own view on the basis of this passage, so it is worth quoting in full:

> Thus the atom of chemistry is an "Idea," in the strict meaning Kant gave this term, in so far as it possesses "a most admirable and indispensably necessary regulative use, in directing the understanding to a certain aim, towards which all the lines of its rules converge and which, though it is an idea only (*focus imaginarius*) . . . that is, a point from which, as lying completely outside the limits of possible experience, the concepts of the understanding do not in reality proceed [just as objects reflected in a mirror are not in reality behind it], serves nevertheless to impart to them the greatest unity and the greatest extension." (Cassirer 1923, 210–211)

---

[8] Some Kant scholars identify the source of OP's merger of constitutive and regulative reason as unresolved issues in his critical philosophy.

This passage culminates a section of *Substance and Function* titled "The 'regulative' use of the concept of the atom," where Cassirer presents in the most direct fashion his ideas about the appropriate relationship between regulative ideas and empirical ones (i.e., those arising from the constitutive function of reason). The passage ends as follows:

> This function remains as a permanent characteristic of the concept of the atom, although its content may completely change; thus e.g., the atom of matter becomes the atom of electricity, the electron. *Precisely this sort of change shows that what is essential in the concept does not consist in any material properties, but that it is a formal concept, that can be filled with manifold concrete content according to the state of our experience.* (Cassirer 1923, 211, emphasis added)

Shortly after Cassirer published this in 1910, the concept of isotope was introduced and the basis of the definition of element began to shift from atomic weight to atomic number, that is, the number of protons in the nucleus. By 1923, when Cassirer's book was translated into English, the IUPAC definition that opened this chapter became official. Nothing in the history of chemistry during the intervening thirteen years had prompted Cassirer to change anything. His philosophical work followed the ideal he had stated for scientific concept-formation: he defined an open formal concept that continues to be "filled with manifold concrete content according to the state of our experience." This is also demonstrated in his later work on Einstein's relativity:

> Concepts do not gain their truth by being copies of realities presented in themselves, but by *expressing ideal orders* by which the connection of experiences is established and guaranteed. The "realities," which physics affirms, have no meaning beyond that of being ordering concepts. They are not *grounded* by pointing out a particular sensuous being, that "corresponds" to them, but by being recognized as the *instruments of strict connections* and thus of thoroughgoing *relative* determinateness of the "given." (Cassirer 1923, 319, emphasis added)

Concepts, in Cassirer's sense, enable the development of new technologies, new experiments, and new theories that move the frontiers forward. The emphasis on the "thoroughgoing *relative* determinateness of the 'given'" means that at any given point in the history of science, determinateness of scientific knowledge is only partially met because determinateness is contingent upon the mathematics, methods, and instrumentation available at the time.

In the case of contemporary chemistry, though we have had partial success, we do not have the ideal mathematical relations to generate the series of elements of the periodic table such that for any element every physical and chemical property (heat of fusion, electronegativity, ionization energies, valences, reactivity, and so on) is derived purely from quantum theory. To expect such results is to misconstrue chemistry. Rather what we can reasonably expect is that,

> the originally confused factual material is organized; it is no longer unrelated, but is arranged around a fixed central point. When we ascribe to one and the same subject the observations on vapor density, on heat capacity, on isomorphism etc., they thereby enter into true conceptual relation. (Cassirer 1923, 209)

Indeed the unifying power of the concept is its chief purpose: "The atom functions here as the conceived unitary center of a system of coordinates, in which we conceive all assertions concerning the various groups of chemical properties arranged" (Cassirer 1923, 208). His use of mathematical language here and elsewhere is not merely metaphorical—*Substance and Function*, he notes, emerged from his study of mathematics (Cassirer 1923, xii). Elsewhere he states that "to express the relation between philosophy and science in a blunt and paradoxical way, one may say, The eye of philosophy must be directed neither on mathematics nor on physics; it is to be directed solely on the connection of the two realms" (Cassirer, in Mormann 2015, 35).

In the following section, we examine how mathematical chemistry returns from an excessive reliance upon—some might say a reduction to—quantum physical chemistry back to Mendeleev's relational way of thinking, but newly armed with updated mathematical techniques and digital firepower. Even at its best, mathematical chemistry does not take the place of empirical findings and computer-assisted chemical intuition. While the former changes from age to age, contingent on the advancement of sensor technology and theory, the latter is tied closely to what Schummer has called the "chemical core of chemistry." As we shall see, Schummer's "chemical core" may function as a regulative check on two potential infinities: the limitless application of the purely relational concept of chemical element, and the limitless world of conditional experimental findings.

Interestingly, the victory of quantum over classical physics in the microcosmic domain represents the development of a regulative idea in physics that was recognized as such at its inception: Bohr's principle of complementarity. As discussed below, complementarity played a regulative role with respect to classical physics concepts by restricting their formerly universal applicability (i.e., under classical physical assumptions), and, paradoxically, rendering these concepts functional in a domain for which they were not conceived. This moment in the history of science also demonstrates Cassirer's general approach to intellectual history: we learn more about a science from its process of concept-formation than from its concepts. The concept of quantum complementarity not only clarifies the concepts of physics, it explicates certain normative standards for applying existing concepts to newly discovered domains, domains for which they were not devised. Bohr's concept of complementarity is a paradigm case of Kant's regulative use of reason. This is one of those wonderful moments in the history of thought when both philosophy and science win. A somewhat longer consideration of this point appears in Mahootian 2013 (178–180), and will be the subject of a future treatment.

4. Rumors of Chemistry's Dematerialization Are Somewhat Exaggerated

With any application of the regulative principle of reason, there is a limitation of attainable possibilities in the phenomenal field, but there is always a gain in another aspect of that field. The principle of complementarity's regulative adjustment of fundamental physical concepts limits their constitutive application, and strongly confirms Cassirer's claims that the entities of physics and chemistry are progressively dematerialized in the development of science. When a particle's position, for example, is no longer a universally applicable parameter, that particle has lost its absolute materiality: its determinateness is, as Cassirer says, only relative. This can be said of everything we designate as an entity. "All that the 'thing' of the popular view of the world loses in properties, it gains in relations; for it . . . is connected inseparably by logical threads with the totality of experience" (Cassirer 1923, 166).

The dissolution of things into relations gains support from James Ladyman's ontic structural realism (OSR), which defends a view of relations having precedence over individual relata by reconceiving individuals (Ladyman 2014, 32). Mormann notes that

> the contemporary structural realists French and Ladyman comment that this "structural dissolution of physical objects leads to a blurring of the line between the mathematical and the physical" (2003). This may well be true, but . . . it should be noted that Cassirer does not contend that physical and mathematical concepts are identical. (Mormann 2015, 51)

Mormann concludes his article on Cassirer's philosophy of science striking a note of uncertainty with respect to its relation to OSR and other forms of structuralism:

> contemporary versions of structuralism [including OSR] . . . in no way claim to have settled the issue of determining definitively Cassirer's place in the

landscape of the many versions of structuralism presently available. (Mormann, 2015, 61)

The advent of chemical informatics and mathematical chemistry would seem to further bolster the structuralist trend by driving chemistry toward purely structural characterization and away from its historical focus on the stuff of the world. How far can this go?

Beginning with Aristotle, the opponents of idealism and its contemporary variants deny the possibility that structures of any kind—mathematical, logical, and so on—could ever result in things. "Abstracting from thing to relational structures is fine," they might say, "but never the other way around!" Logical, geometrical, and algebraic models, group theory, graph theory, and other such devices employed, no matter how sophisticated, will never suffice. The misconception of such critics is that idealists like Cassirer are actually trying to make "its" (things, individuals, electrons) out of bits (data, information, equations, relational networks, etc.). Rather, these "bits" are analogous to Kant's regulative ideas: rules for understanding complex behaviors, such as chemical reactions; see Cassirer (1923, 319), above.

Our contact with the known world takes place through the ever-expanding medium of human-designed objects and materials that literally take the shape (albeit imperfectly) of our ideas. If this makes us sound like the Demiurge of Plato's *Timaeus*, so be it. Our concepts always express ideal orders of connection among our experiences, whether or not they do so in the specific ways we intend, and whether or not we are even aware that they do. Of course, the more we know about our own processes of concept-formation, the more deliberately we can participate in these processes. This would make us less like the president of a corporation that makes widgets and more like one that sells services. The best version would be that of an artist in the sway of the beautiful complexities of their medium, yet still codirecting the work of art in gentle, but insistent and endless dialogue with their medium. This is, in somewhat metaphorical terms, what Cassirer has in mind for the progress of concept-formation in science. In less metaphorical terms, it is an intimate interplay of the symbolic and the sensory, mediated by whatever conceptual and technological tools are at hand, under a watchful eye that is more attentive to processes of concept-formation than to the ultimacy of its products. The processual approach to relational ontology results in coupled discoveries that articulate evermore comprehensive relations among our ideas even as they reveal more detail about the materials that mediate that knowledge. The concluding two sections of this chapter will add specific detail to the speculative balance expressed here between process and product, relations and relata.

5. A Chip Off the Old Eggshell: Cassirer and Mathematical Chemistry

From the relations embodied in a well-ordered fragment, it is possible to re-construct the whole of which the fragment is a part. The simplest case to im-agine is that of a rigid sphere, any fragment of which embodies the plan of the whole. The sphere is the simplest case; the complex chemical analog of the sphere is the periodic table of elements, every member of which is itself a network of relations. This is Cassirer's idea of the chemical element as a web of relations in which each element implicates the whole network, as noted in section 3, above: "The atom functions here as the conceived unitary *center of a system of coordinates, in which we conceive all assertions concerning the various groups of chemical properties arranged*" (Cassirer 1923, 208, emphasis added). This notion proposes the possibility of deriving everything about the elements, their every property, natural distribution, their positions in any possible periodic table.[9] To accept this proposal is to commit to a research program that seeks ever-greater generality that may translate into a broad array of applications. As such, this research program must be supported by vast amount of experimental legwork. Experiments based upon concepts of the highest levels of generality require the tightest control over experimental parameters.

In this penultimate section, I provide brief overview of mathematical chemistry's methods and accomplishments and only touch on chemical infor-matics. It is important to note at the outset that though they are similar, drawing on sophisticated relational concepts and take advantage of ever-more powerful information technologies, mathematical chemistry and chemical informatics are different disciplines. Of the two, informatics has been around since the 1950s and is well established as a discipline. That is only recently the case with mathe-matical chemistry (Restrepo 2016).

Guillermo Restrepo and his collaborators have sought similarities among sets of various properties within different groupings of elements in the peri-odic table. This work originated with Villaveces who proposed that "the set of chemical substance is a topological space wherein relationships such as belonging to a class, neighbourhood or hierarchy of classes, etc. are more im-portant than differential relationships." Villaveces continues in this vein and issues a call to go back to Mendeleev: "This seems to be the mathematical struc-ture underlying the periodic table and which would explain that predictions were possible when considering the whole set of chemical elements along

---

[9] Scerri notes over seven hundred graphical displays since its inception. See "Axiomatized Yet?" in Scerri 2008.

with their equivalence classes" (Restrepo 2018, 85). Restrepo proposes "not focusing on the 'periodic law' but on the relations of classes and on the internal ordering of classes." That is, "the periodic system—its structure and the different functions (not necessarily periodic) that can be found between the system and a desired property." The big picture here is that the periodic law is *not* a law: it pertains to some (e.g., ionization energies) but not to all properties of the elements; furthermore, the periodic system, not the periodic table, is the prize here—periodic tables, like the basis sets of chemical informatics, are driven by specific needs.

Mathematical chemistry has found previously unexpected or undiscovered relationships among a select group of elements based on data about a select group of properties that may be of interest to a given research community, for example, metallurgists specializing in the chemistry of transition metal alloys. Leal, Restrepo et al. (2012) found that the noble gases and the elements near them on the table are quite stable with respect to their properties because it makes no difference what set of properties you start with, the resulting arrangement among them is invariant—the elements keep the same neighbors. However, it is a completely different story with the transition metals: a variety of different possible arrangements arise depending on which properties are chosen as the basis for similarity studies. These are new groupings whose order may be suggestive to metallurgists in ways that a standard periodic table cannot be.

The theoretical approach of mathematical chemistry remains pragmatic. Restrepo argues that chemistry has a wealth of orders only some of which are represented by the periodic table, and asks whether the ones that are represented are "actually total orders, or just mappings of a partial order into a total order?" (Restrepo and Klein 2011). By using different mathematical models to group compounds according to a select set of properties, different neighborhoods emerge that are not immediately obvious from merely structural studies of compounds. Interestingly, Restrepo has extended the metaphor of neighbors and neighborhoods that have similar preferences and so forth, with social network theory and gotten encouraging preliminary results by representing concrete social relationships among neighbors in the context of the periodic system (Restrepo 2018, 92). Imagining the periodic system as a neighborhood of neighborhoods recalls the fact that all properties of elements and compounds are dispositional and hence relational. In this view, individual substances are fully immersed in a network of relations in a world of networks:

> The study (Leal et al. 2012) is influenced by Schummer's (1998) idea that chemistry is rooted on the relational character of substances (Bernal and Daza 2010), which combined with concepts from category theory (Bernal et al. 2015) and social network analysis (Restrepo 2017) leads to the idea that a chemical

substance (not necessarily an element) is not only characterized by properties measured upon the isolated substance, but, most important, by the other
substances with which the substance in question is related, for example by
chemical reactivity. (Restrepo 2018, 92)

We have seen that Cassirer's view of chemistry as a science whose growth
is guided by a relational idea of element is consistent with a growing body of
studies in mathematical chemistry (Restrepo et al. 2006). Both views ground the
possibility that a plurality of periodic relations—and thus periodic tables—can
be generated by selectively and systematically analyzing relations among various
sets of chemical properties.

Chemical informatics is closely related to mathematical chemistry, but has
been shaped by its ongoing associations with the pharmaceutical and biomedical
industries. The majority of existing chemical informatics databases are based on
structural features of molecules and designed for structural search.[10] Databases
designed for chemical properties and other relational qualities are generally
fewer (Warr 2011, 567), and property prediction remains as one of the major
challenges for chemical informatics. As the author of one review article put it,
"It has to be realized that the major task of chemistry is not so much to produce
chemicals but to produce properties, properties that happen to be attached to
chemicals" (Gasteiger 2006, 57). The suggestion is to shift to the pursuit of what
chemicals *do*, rather than what they *are*.

The shift signaled by both mathematical chemistry and Cassirer seems to be
that the destiny of substance is to be replaced by function and relation. This is
especially the case if one is predisposed to expelling the outdated concept of substance from science and philosophy once and for all. If that is the case, then why
is Cassirer's historical study of scientific concept-formation titled *Substance and
Function*, rather than *From Substance to Function*? Schummer's idea of the core
of chemistry sheds some light on that question.

7. Conclusion: Cassirer and the Chemical Core of Chemistry

In his 1998 article, "The Chemical Core of Chemistry: A Conceptual Approach,"
Joachim Schummer notes that the chemical core of chemistry comprises "the

---

[10] For example, Jean-Louis Reymond notes, "This 'chemical space project' led to the chemical
universe databases (GDBs, generated databases) enumerating molecules following criteria for size,
chemical stability, and synthetic feasibility" (Reymond 2015, 722).

whole complex—from the systematical investigation of chemical properties, over the classificatory networks of chemical substances and substance classes, to the chemical sign language" (Schummer 1998, 158). Schummer's relational view of chemistry is systematic and thoroughgoing, and yet maintains a place for chemical substance. We conclude with a comparison between the views of Cassirer and Schummer on chemistry and the idea of chemical element.

Schummer closes his article with two long quotations from Cassirer's *Philosophy of Symbolic Forms* that affirm the gist of his conceptual analysis of chemistry's chemical core. The quotations are powerful statements of Cassirer's confidence in the symbolic tools of chemistry to capture the essential form of scientific knowledge. The following is Schummer's own translation of Cassirer's, *Philosophie der symbolischen Formen*, vol. 3, p. 513:

> In general, the scientific value of a [structural] formula is not only that it unites given empirical facts, but that it *lures out*, so to speak, new facts. It puts forward problems about relations, connections, and formation of order, which precede immediate observation. Thus, it becomes one of the most outstanding means of what Leibniz has called the 'logic of invention', *logica inventionis* (Schummer 1998, 158).

Cassirer's *Substance and Function*, written thirteen years earlier, is consistent with this analysis but, as I will show, offers more direct support for Schummer's key conclusions. For example, Schummer argues for chemistry's relational property-based network structure, but one in which "pure substances" play a key role "despite serious problems of defining them by empirical and theoretical means."

> Pure substances, though artificially produced and definable only in operational terms ... perfectly fulfil the chemical requirement of distinct substances. ...Not only are pure substances the basic chemical species; they also form the nodes of the chemical network which is already a basic kind of chemical classification. (Schummer 1998, 145)

Furthermore, "since every substance class is defined with reference to another class, all we need is an anchor point" and pure substances fulfill that function because chemical classification "is essentially based upon similarities of chemical properties [of substance], and, consequently, it retains [the] peculiar nature of a relational network." Chemical classification in turn provides a "sophisticated set of concepts enabling us to grasp the diversity of entities and phenomena at all; and to provide for predictions of properties and of new entities, including the ways to find or make them" (Schummer 1998, 146).

Schummer's relational network, whose logical connections become increasingly sophisticated with ongoing empirical investigation, is reminiscent of Cassirer's "atom-as-regulative-idea." The similarity goes further: Cassirer also retains a function for substance within his vision of science as a system of laws of ever-increasing generality. As noted earlier in this chapter, for Cassirer, "The atom functions here as the conceived unitary center of a system of coordinates, in which we conceive all assertions concerning the various groups of chemical properties arranged." But the similarity of this statement with Schummer's concept of the relational property-based network increases as we examine the passage further. Not only do relational properties play a similar role, so do the atoms:

> The diverse and originally heterogeneous manifolds of determinations gain a fixed connection when we relate them to this common center. The particular property is <u>only apparently connected with the atom as its absolute "bearer,"</u> in order that the system of relations can be perfected. In truth, we are concerned not so much with relating the diverse series to the atom, as rather with *relating them reciprocally* to each other through the mediation of the concept of the atom. (Cassirer 1923, 208; italics in original, underlining added)

The "atom" for Cassirer seems more of a placeholder than a thing. Regarding the increasingly precise determinations of atomic weight he speculated that "in the 'absolute' atomic weights all possible relations would be expressed."[11] But Cassirer hastens to add that "the real positive outcome of chemical knowledge here is in the systematic analysis of these relations" (Cassirer 1923, 209), not the atoms. Yet atoms and substances, the "thing-like" bearers Cassirer refers to in the passage below, continue to play a key role for him—and that role seems remarkably the same as what Schummer (1998, 145) assigns to them:

> In fact, it is justified and unavoidable, that science should condense a wealth of empirical relations into a single expression, into *the assumption of a particular thing-like "bearer."* The critical self-characterization of thought, however, must analyze this product once more into its particular factors, although it conceives this product as *necessary for certain purposes of knowledge.* This is

---

[11] Soon after Cassirer wrote this, the trend of precision resulted not in "absolute atomic weights," but rather in the discovery of isotopes whose rapid proliferation in turn resulted in the shift of the periodic table's organizing principle from atomic weight to atomic number. This follows the spirit of Cassirer's projection.

done because critical thought is not directed forwards on the gaining of new objective experiences, but backwards on the origin and *foundations of knowledge*. (Cassirer 1923, 210, emphasis added)

In a similar vein, Schummer notes that although the identity of classical chemical substances is not absolute, but operationally defined, "it is only because our chemical species *per definition* retain their identity during purification, that we are able to connect single facts of chemical relations with each other to build *a systematic network structure of chemical knowledge*" (Schummer 1998, 142, emphasis added). It is for the sake of a coherent process of chemical investigation that we must tolerate the apparent arbitrariness of the system, however, "what seems to be, at first glance, an arbitrary and coarse approach, turns out to be *a necessary requirement for systematical chemical research*" (Schummer 1998, 142, emphasis added). Whereas the nodes in Schummer's network are operationally defined "pure substances," in Cassirer's system the nodes are atoms; in both cases, an epistemological imperative justifies the acceptance and employment of these arrangements.

From this vantage point, as we look back at the IUPAC definition of chemical element we note that, whereas in the first part of the definition element is replaced by "atom," in the second definition it is replaced by "pure substance." Furthermore, in light of the analogy between Schummer's and Cassirer's epistemic imperatives, IUPAC's apparently arbitrary dualism of modern atomism based on atomic number, on the one hand, and an antiquated concept of pure substance, on the other, seems less arbitrary and less dualistic. Instead, there appears to be a reciprocal relationship between the two parts of the definition, a coupling of two irreducible, but necessary aspects of chemical practice. In (Mahootian 2013), I proposed that the two parts of the IUPAC definition exist in a complementary relationship. The present analysis lends further support to that proposition.

So, what of the question posed at the end of the previous section as to why the title of Cassirer's book is *Substance and Function*, rather than *From Substance to Function*? My view, which is consistent with Cassirer's general philosophy,[12] is that the relationship between substance and function is like an ongoing conversation: it must flow in both directions at any point in its history. Indeed, Gregory Moynahan *reverses* the direction proposed in my question to show that in the process of concept-formation, function is often *construed* as substance *until* it is recognized as an example of a mathematical series, i.e., as a function. In relation

---

[12] This is best developed in his main work, *Philosophie der symbolischen Formen*, (1923-29) translated into English as *The Philosophy of Symbolic Forms*, in 3 volumes, 1955-57.

to chemical element, Moynahan notes that Cassirer's project in *Substance and Function* concerns

> the constructed nature of "substance" *from* "function," as well as the role of series and group in forming such definitions. On the one hand, a chemical element or compound literally is an example of a series *when it is categorized, respectively*, in the periodic table or in relation to related compounds. (Moynahan 2013, 134; emphasis added)

It is clear that Cassirer's title signifies a dialectical project. This difficult but familiar mode of thought is discernible as ascent and descent in the divided line and cave images of Plato's *Republic* (Book VI–VII). It is also celebrated as the core of critical philosophy which Kant astutely framed on the analogy of chemical analysis and synthesis. Kant's choice is remarkable in that it simultaneously invokes the alchemist's quest for purity and points beyond it to the controlled and conscious process of arriving at pure reason in order to recombine it with the elements of daily life for the sake of evermore refined human creativity.

References

Cassirer, E. 1923. *Substance and Function and Einstein's Theory of Relativity.* Translated by W. C. Swabey and M. Swabey. New York: Dover.

Cassirer, E. 1955. *The Philosophy of Symbolic Forms.* Translated by Ralph Manheim. New Haven: Yale University Press

Cassirer, E. 1957. *Determinism and Indeterminism in Modern Physics.* Translated by R. T. Benfey. New York: Dover.

Doxiadis, A. 2009. *Logicomix. An Epic Search for Truth.* New York: Bloomsbury.

Earley, J. 2009. "How Chemistry Shifts Horizons: Element, Substance, and the Essential." *Foundations of Chemistry* 11: 65–77. doi: 10.1007/s10698-008-9054-5.

Förster, E., and M. Rosen. 1993. *Kant's Opus Postumum.* New York: Cambridge University Press.

French, S., and J. Ladyman. 2003. "Remodelling Structural Realism: Quantum Mechanics and the Metaphysics of Structure." *Synthese* 136: 31–56.

Friedman, M. 1992. *Kant and the Exact Sciences.* Cambridge, MA: Harvard University Press.

Friedman, M., and A. Nordmann. 2006. *The Kantian Legacy in Nineteenth-century Science.* Boston: MIT Press.

Gasteiger, J. 2006. "Chemoinformatics: A New Field with a Long Tradition." *Anal. Bioanal. Chem* 384: 57–64.

Kant, I. 1998. *Critique of Pure Reason.* Translated by Paul Guyer and Allen W. Wood. New York: Cambridge University Press.

Ladyman, J. 2014. "Structural Realism." *Stanford Encyclopedia of Philosophy*. PDF Winter 2016 edition.

Leal, W., G. Restrepo, A. Andres Bernal. 2012. "A Network Study of Chemical Elements: From Binary Compounds to Chemical Trends." *MATCH Communications of Mathematical and Computational Chemistry* 68: 417–442.

McNulty Michael Bennett. 2016. "Chemistry in Kant's Opus Postumum." *HOPOS: The Journal of the International Society for the History of Philosophy of Science*, 6(1), 64.

Mahootian, Farzad. 2013. "Paneth's Epistemology of Chemical Elements in Light of Kant's Opus Postumum." *Foundations of Chemistry* 15:171–184.

Mormann, T. 2015. "From Mathematics to Quantum Mechanics—On the Conceptual Unity of Cassirer's Philosophy of Science (1907-1937)." In *The Philosophy of Ernst Cassirer: A Novel Assessment*, edited by J. Tyler Friedman and Sebastian Luft. Berlin: De Gruyter.

Moynahan, Gregory. 2013. *Ernst Cassirer and the Critical Science of Germany, 1899–1919*. New York: Anthem Press.

Needham, P., and R. Hendry. 2018. "Aspects of the Concept of Potentiality in Chemistry." In *Handbook of Potentiality*, edited by K. Englehardt and M. Quante, 375–400. New York: Springer.

Paneth, F. A. (1962) 2003. The Epistemological Status of the Chemical Concept of Element. *Br. J. Philos. Sci.* 13 (1962): 1–14, 144–160. Reprinted in *Foundations of Chemistry* 5: 113–145.

Restrepo, G. 2016. "Mathematical Chemistry, a New Discipline." In *Essays in the Philosophy of Chemistry*, edited by Eric Scerri and Grant Fisher. Oxford: Oxford University Press.

Restrepo, G. 2018. "The Periodic System: A Mathematical Approach." In *Mendeleev to Oganesson: A Multidisciplinary Perspective on the Periodic Table*, edited by Eric Scerri and Guillermo Restrepo, 80–103. New York: Oxford University Press.

Restrepo, G., and R. Harré. 2015. "Mereology of Quantitative Structure-Activity Relationships Models." *HYLE-International Journal for Philosophy of Chemistry* 21: 19–38.

Restrepo, G., and D. J. Klein. 2011. "Predicting Densities of Nitrocubanes Using Partial Orders." *J. Math. Chem.* 49: 1311–1321.

Restrepo, G., E. J. Llanos, and H. Mesa. 2006. "Topological Space of the Chemical Elements and Its Properties." *J. Math. Chem.* 39: 401–416.

Reymond, J-L. 2015. "The Chemical Space Project." *Accounts of Chemical Research* 48: 722–730.

Ruthenberg, K. 1997. "Friedrich Adolf Paneth (1887–1958)." *HYLE-International Journal for Philosophy of Chemistry* 3: 103–106.

Scerri, E. 2000. "Realism, Reduction and the 'Intermediate Position.'" In *Of Minds and Molecules*, edited by N. Bhushan and S. Rosenfeld, 51–72. New York: Oxford University Press.

Scerri, E. 2005. "Some Aspects of the Metaphysics of Chemistry and the Nature of the Elements." *HYLE-International Journal for Philosophy of Chemistry* 11: 127–145.

Scerri, E. 2008. *Collected Papers on Philosophy of Chemistry*. London: Imperial College Press.

Schummer, J. 1997. "Scientometric studies on chemistry I: The exponential growth of chemical substances, 1800–1995." Scientometrics Volume 39, Issue 1, pp 107–123.

Schummer, J. 1998. "The Chemical Core of Chemistry I: A Conceptual Approach." *HYLE—International Journal for Philosophy of Chemistry* 4: 129–162.

Thorndike, O. 2018. *Kant's Transition Project and Late Philosophy*. New York: Bloomsbury.

van Brakel, J. 2006. "Kant's Legacy for the Philosophy of Chemistry." In *Philosophy of Chemistry*, edited by D. Baird, E. Scerri, and L. McIntyre, 69–91. Boston: Springer.

Warr, W. 2011. "Representation of Chemical Structures." *WIREs Comput. Mol Sci* 1: 557–579.

9

The Operational Definition
of the Elements

A Philosophical Reappraisal

Joachim Schummer, js@hyle.org

In memory of Rom (Horace Romano) Harré (1927–2019)

1. Introduction

The definition of elements to be reappraised from a philosophical point of view in this chapter is well known among chemists and historians of chemistry. It plainly says:

> An element is any substance that we, at the current state of our art, cannot decompose further by chemical analysis.

Lavoisier was perhaps the most forceful advocator of that definition.[1] However, from an epistemological point of view, it does not matter who first formulated the definition, for what reasons, in what context, and if such formulations were consistent with other views by the respective author. It is sufficient to acknowledge that the definition was accepted by the vast majority of chemists around 1800, but even the exact date is largely unimportant.

The reason for the neglect of such details is not motivated by disinterest in the history of chemistry, but justified by the epistemological status of definitions. A definition is not to be confounded with a discovery, a hypothesis, or a theory, achievements for which we can frequently give credit to an individual. In contrast, a definition defines the meaning of a term, here "element," which serves communicational purposes. Individuals can suggest a new definition and provide arguments pro and con its adoption, but only a community can agree upon a definition by convention.

---

[1] "Nous nous contenterons de regarder ici comme simples toutes les substances que nous ne pouvont décomposer; tout ce que obtener en dernier résultat par l'analyse chimique. Sans doute un jour ces substances, qui sont simple pour nous, seront décomposées à leur tour...mais notre imagination n'a pas dût devancer les faits, & nous n'avons pas dû en dire plus que la nature ne nous enapprend" (Lavoisier 1787, 17–18, quoted from the original).

Joachim Schummer, *The Operational Definition of the Elements* In: *What Is A Chemical Element?*. Edited by: Eric R. Scerri and Elena Ghibaudi (2020). © Oxford University Press.
DOI: 10.1093/oso/9780190933784.003.0010

Almost all commentators of the "chemical revolution" have discussed the definition, whether it should be considered part of that "revolution," and if Lavoisier consistently applied it in his own works, which does not concern us here. They have used various terms, such as "empirical definition," "analytical definition," or "operational criterion," but only a few have called it an "operational definition," which it actually is from an epistemological point of view.

An operational definition defines a general term, here "being elemental," by reference to one or more operations. For instance, modern physicists define "time" by reference to measurement operations involving a clock. Empirically working psychologists and social scientists work hard to operationalize their concepts by reference to measurements. Operationalism, that is, the view that all basic concepts of science should ideally be operationally defined, was first elaborated by Physics Nobel Laureate Percy Bridgman (1927). Although it received harsh criticism by logical positivists who mistook it as a general semantic theory, it arguably plays an important methodological role in most empirical sciences.

The operational definition of elements, as formulated above, is special in two regards. First, it defines a kind of material entities, elements, in such a way that they can be produced in the laboratory by following the definition: if you take any substance and apply all available methods of chemical analysis, you end up with elements per definition. That is because the definition refers to experimental operations of chemical analysis rather than to measurement operations. (The definition also implicitly refers to measurements of equivalent masses that allow one to decide on empirical grounds if a chemical reaction is actually a decomposition or a synthesis.) Second, the definition refers to the current state of the art, which may change over time, such that what was once considered an element is no longer an element, but not the other way round. The meaning of "element" is thereby bound to the limits of human capacities, acknowledging that this is not something given but changing in an unpredictable manner. All assignments of the elemental status are thus provisional and contingent on the current laboratory practice.

If one considers that in the entire history of natural philosophy before, elements had been the central theoretical entities for explanations, not only in chemistry but also in medicine, mineralogy, and most of the sciences, the adoption of the operational definition with all its oddities and contingencies appears almost crazy. And yet, all of modern science has been built on the operationally defined elements, both experimentally and theoretically, and quite successfully so.

In the following,[2] I first point out the radical disruption that the adoption of the operational definition implied for chemistry and natural philosophy. Against

---

[2] Much of this chapter draws on Schummer 1996.

the background of the traditional role of elements in natural philosophy (section 2.1), the main disruption consisted in giving up explanation, the primary goal of natural philosophy, because the new elements had to be discovered first of all (section 2.2). Then I compare the operational turn in chemistry with several well-discussed "revolutions," including the Kantian, relativistic, and quantum revolutions in physics, which similarly modified our understanding of fundamental concepts of natural philosophy, such as time, space, and causation (section 2.3). Section 2.4 offers some explanation of why most historians of science have neglected the radical disruption and its significance for science.

Complementary to section 2, the subsequent section emphasizes continuities of the concept of elements across the operational turn, by using a threefold epistemological framework (section 3.1) that the eminent philosopher of science Rom Harré developed in 1986. If one considers all three roles or functions that elements have played with varying emphasis throughout history—explanation, classification, experimental accessability—the operational turn perfectly meets the needs of experimental access (section 3.2) and classification (section 3.3) at the temporary expense of explanation. With the exception of that temporary period, conceptual tensions arising from reconciling all three functions in one concept have always been obvious and caused numerous debates (section 3.4), which I illustrate by medieval debates on elements in compounds and by IUPAC's current definition of chemical elements.

In conclusion I discuss the legacy of the operational definition and the importance of philosophy for both chemistry and the history of chemistry.

2. Discontinuity

2.1 Principles of Nature

The notion that the material variety and dynamics of our world are somehow based on fundamental principles has been central to all natural philosophy, from early ancient Greek, Chinese, and Indian philosophy to modern science. To be sure the particular ideas have greatly varied, for instance, whether these principles have material, nonmaterial, or processual qualities; whether there is only one principle (e.g., Thales), or infinitely many (Anaxagoras), or an ordered set of a few principles (Plato, Aristotle); and whether they were called *stoicheia* (Greek), *elementa* (Latin), *xing* (Chinese, Taoism), *mahābhūta* (Sanskrit and Pāli, Buddhism), or elementary particles (modern physics). Regardless of their differences, all natural philosophies have shared the combined ontological and epistemological assumptions that the world is based on stable or recurrent principles and can be best understood (explained, predicted, controlled) by referring to such principles. These assumptions have distinguished natural philosophy from all other approaches to understanding the world, including theistic religion, craft, and mere description, be it qualitatively as in natural history or quantitatively as in applied mathematics. And they became a model for other scientific approaches, including geometrics, for which Euclid first introduced his definitions and axioms as "elements."

How did natural philosophers arrive at their principles? Those who favored material principles mostly developed the characteristic properties of their principles by analogy from material experience and supported them by explanations. The analogy reasoning is already obvious on the surface level of their terminology. Although they usually emphasized that the principles should not be confounded with ordinary materials, they used terms such as "fire," "air," "water," "earth," "metal," "wood," "sulphur," and "mercury." The more sophisticated approaches, such as by Aristotle and the Buddhist mahābhūta doctrine, developed their principles from a systematics of material properties. For instance, Aristotle (De gen. et cor., II. 2–3) selected two pairs of opposite properties, cold-warm and liquid-solid, such that all properties are tangible and the first pair has active effects on other materials (expanding or shrinking them) and the second one passive properties (being more or less ductile). With that he could redefine the four classical elements (called fire, air, water, earth) by the four possible binary combinations of active and passive properties, which allowed him and his followers to make explanations of what we would today largely call thermodynamic, mechanical, and chemical properties and interactions.

The principles or elements of nature were not only used in armchair philosophy and supported by a few explanations of everyday life phenomena. Aristotle, for instance, widely used them for explanations, from the structure and dynamics of the cosmos, to the physiology of biological organisms, to processes of chemical crafts. However, their most important, and most popular, use was probably in medicine, both in China and the West. For instance, already Hippocrates employed, perhaps co-created, the classical four elements doctrine in his physiology, which Galen would develop into his theory of the four humors, that is, four fundamental fluids of the body, corresponding to the four elements. Assuming that the four humors are in a certain balance in a healthy body, he explained many diseases (as well as a variety of human temperaments) by their imbalance, to be cured by medicines that could restore the balance. The humoral theory was the most important theoretical basis in Greek, Roman, Islamic, and European medicine at least until the end of the eighteenth century.

Also most alchemists and chemists accepted the classical four elements as their basic principles up to the late eighteenth century. However, they supplemented them by higher-level principles, such as "sulphur," "mercury," and phlogiston, to explain chemical phenomena, including combustion and calcination, which all showed radical changes of properties. As "property-conferring principles" they were supposed to survive all chemical transformation, but would change their appearance from being latent in some combinations to openly displaying their properties in others. Apart from occasional references to divine help and astrological influence, which was favored by theological authorities such as Thomas Aquinas, the general explanatory approach of material properties did not change much since antiquity. The properties of materials were explained by the principles that they were supposed to be composed of and had inherited from other materials by earlier chemical transformations. When the simple combination of the properties of the principles did not suffice, the usual assumption, in accordance with Aristotle, was that the principles combined to form new substantial forms. As the number of known chemical transformation increased, references to new substantial forms grew, which in modern terms correspond to ad hoc hypotheses to save an explanatory approach.

2.2 Epistemological Disruption

Historians of chemistry usually call the principles of premodern chemistry "metaphysical principles," but it is not so clear what that means, particularly if one considers Aristotle's inference of his elements from tangible properties. If postulating a theoretical entity for explanatory purposes were metaphysics, then all of classical natural philosophy and most of today's theoretical sciences would be branches of metaphysics. It is more likely that historians just adopted the term "metaphysical" from contemporary chemists who favored an operational approach in chemistry. And that was probably the most radical departure from the received understanding of science ever since. It not only broke up with metaphysics, it also gave up, at least temporarily, the idea that the main goal of science (natural philosophy) is explanation.

Remember that elements had before been conceptually developed for explaining material properties and transformations, not only in chemistry but also in biology, medicine, meteorology, and many other fields of science. If you now define elements as those substances that resist any kind of separation according to the contemporary state of the art, you are not replacing one set of principles with another one. Instead you abolish the explanatory basis of all these sciences. At the beginning you do not even know what the new elements might look like, because the definition gives no hint at all. Once you have found a substance that resists any separation effort, you first need to study its properties. How can you make use of these empirical properties for developing explanatory approaches? You can never be sure whether that substance remains an element or not, because further improvement of separation techniques might take it apart. And would not a scientific approach require that the complete set of elements, that is, all explanatory factors, are known before you should dare any serious explanation?

The overall discovery of the new elements was a very slow process that included numerous research programs (Schummer 1997). After an early period of many discoveries their number has almost linearly grown since about 1808 (1800, 27; 1808, 39; 1850, 55; 1900, 81; 1950, 98; 2000, 114). Of course later discoveries brought about only rare and even artificially made elements. However, in the early nineteenth century the most widely spread elements, including the alkali metals, alkaline earth metals, and halogens (except chlorine), were unknown. What kind of chemistry is possible on such a limited elemental basis? Chemists focused on discoveries, both of new elements and their combinations, and postponed explanation.

What appears like a crazy move has a deeper reason in philosophical ontology. Indeed, it redefines the *ontological* concept of "substance," that which persists through all possible changes, on experimental grounds. If by definition the elements cannot be destroyed by any chemical means, then they necessarily persist in all our chemical transformations.

2.3 Similarities and Differences to Scientific "Revolutions"

Was there any other event in the modern history of science comparable to the epistemological disruption in chemistry? If we believe our historians and philosophers of science, all the big discontinuities or "revolutions" have been theory changes, one theory replacing another one. In endless debates since the 1960s they have discussed if in all these cases the explanatory potential of the competing theories can either be compared or not, which implies that no single discontinuity consisted in temporarily giving up explanations at all. Moreover, all big revolutions concerned only very specific fields of science, with little to no immediate impact on others, such as astronomy (Copernican revolution), mechanics (relativistic, quantum mechanics), biology (Darwinian evolution), geology (tectonics), and so on. In contrast, the epistemological disruption in chemistry strongly influenced most other sciences who relied on the age-old concept of elements.

At the risk of provoking harsh criticism from philosophers, I would like to point out some epistemological similarities to the relativistic, quantum, and Kantian revolutions, all being favorite topics in philosophy. The core of the operational definition consists in relating elements to human capacities, here to the experimental capacities of taking chemical substances apart. In all the other three examples we can identify the same move (although there exist alternative interpretations that try to avoid that).

Both the special and general relativity theories can be derived from classical mechanics, and in fact have frequently been described and taught so in the tradition of Ernst Mach, if one requires that all properties in mechanics are strictly related to human measurement capacities, rather than being conceived of from a God-eye's view. The measurement of time and length of some object moving at a distance is bound by the signal speed of light, which was empirically found to be constant. If you include that so-called relativistic factor into the formalism of classical mechanics for different observers, thereby forgoing a God-eye's view, you end up with special relativity theory. Similarly, there exists no physical measurement apparatus that can distinguish between gravitational force and acceleration (or between gravitational mass and inert mass). Once you replace gravitational force with acceleration in the formalism of classical mechanics, you end up with general relativity theory.

According to the most influential Copenhagen interpretation of quantum mechanics, particularly in Heisenberg's version, quantum mechanics followed a similar move, best expressed in Heisenberg's uncertainty principle. It claims that for certain pairs of physical properties, such as location and momentum or time and energy of a particle, there is a limit of precision at which they both can be known simultaneously. Whereas others have later taken that principle as an axiom that requires no interpretation, Heisenberg himself considered the limit being posed by any possibly accurate measurement that would necessarily interact with the particle. In that tradition, physics is no longer striving for a God-eye's view but for a generalized human view that needs to take into account the limits of human capacities.

In a certain sense physicists followed the third and much earlier example of Kant, who had developed his "Copernican Revolution" in epistemology around the time of the "chemical revolution" in his *Critique of Pure Reason* (1781). Kant argued that true scientific knowledge ("synthetic statements *a priori*") cannot be derived from mere sense perception nor from metaphysical assumptions about the world as such, but only from understanding the fundamental capacities of the human mind. For instance, space, time, and causality are not something given or to be inferred from perceptions. Instead the human mind necessarily constructs our sensible world such that it has a certain spatial, temporal, and causal order. By investigating the perceptual and intellectual capacities in detail, Kant hoped to proof the a priori truth of Euclidean geometry and Newtonian mechanics, which turned out to be wrong, however, by Riemannian geometry and relativistic mechanics.

All four examples have in common that they redefine central concepts of natural philosophy by relating them to human capacities and their limits: experimental measurement in physics, cognitive capacities in epistemology, and experimental separation in chemistry. Yet, while the first three cases are widely considered landmark revolutions in the history of science and philosophy, the chemical turn to operationally defined elements has never been recognized as an epistemological revolution despite the fact that all of modern chemistry, and almost all the sciences, nowadays refer to the chemical elements that have first been identified by the operational definition.

Furthermore, the historical move from a God-eye's view to one that is specified by human conditions is not restricted to science. Since Renaissance humanism that has been a central feature of modernity in most branches of Western culture, including art (e.g., central perspective, aesthetics modeled after the human body), ethics (justification of moral rules by rational principles instead of divine order), law (natural law based on human nature), and politics (legitimation of power by democratic procedures instead of divine order). It would be more appropriate to understand the chemists' operational turn of the elements of nature in this broader cultural context rather than discussing it as an appendix of some contemporary theory change.

2.4 Why Was the Disruption Neglected?

To be sure, historians and philosophers of chemistry have all been aware of and mentioned the change from "metaphysical principles" to operationally defined elements. However, they usually did so as if just one set of elements was replaced by another one, and they did hardly acknowledge the radical epistemological disruption in natural philosophy. There are several reasons for the neglect that are all related to the so-called "chemical revolution" and Lavoisier's role therein.

Lavoisier has rightly been considered the most influential proponent of the operational definition of elements during his time. However, he did not consistently apply it in his own scheme. Instead he included, for instance, imponderables such as light and heat (*calorique*) that would not meet the operational definition, suggesting that the definition was not of central importance to him. For instance, the decomposition of light by diffraction in prisms was well known and accepted since the seventeenth century through works by Francesco Grimaldi. The more Lavoisier was seen as the hero of the New Chemistry, the more one could copy his mixed attitude toward the operational definition.

Second, historians of chemistry have dealt with the operational definition mostly by answering who-did-what-first questions. For instance, some argued that not Lavoisier but Boyle would have first formulated the definition, while others rightly pointed out that Boyle had rejected elements and chemical principles altogether in his *Skeptical Chymist* (Davis 1931). Some tried to give Joachim Jungius the credit, although his complex ontology was entrenched in medieval scholastics and had little connection to experimental practice (Meinel 1982). Others have argued that the operational definition was already used in mid-eighteenth-century mineral chemistry for pragmatic reasons, particularly by Torbern Bergman (Oldroyd 1975). All these historiographical debates on whom should credit be given have moved epistemological questions to the background.

Third, Lavoisier's own engineering of a "chemical revolution" was tailored against the received phlogiston theory to promote his own oxygen theory. Oxygen, which literally means "acid generator," actually met the operational definition and at the same time took a unique explanatory role in his acid theory, according to which all acids contain the acidic principle oxygen. It thus appeared that the new elements could immediately replace the old principles in chemical explanations. However, Lavoisier's acid theory was very soon given up because of massive counter-evidence: many basic substances such as potash were found to be composed of oxygen, and acids such as muriatic acid (HCl) contained no oxygen. The short life of the acid theory thus covered the fact that the new elements could at first not serve in explanations.

Fourth, from the 1960s onward, the "chemical revolution" was usually framed in terms of a change of competing theories that are structurally similar, such as classical and relativistic mechanics. For instance, while the phlogiston theory described combustion as the release of phlogiston from the burning substance into the air, the oxygen theory described it as a reaction of that substance with oxygen from the air. However, that framework made people ignore that the two "theories" are of entirely different epistemological kinds. It is one thing to describe a chemical transformation in terms of the exchange of one or the other element, and quite another thing to provide an explanation for why the transformation occurs at all. While the phlogiston theory explained the combustibility of all substances by their containment of phlogiston, the oxygen theory had no explanation to offer for combustibility or any other kind of chemical property for more than a century, before explanatory theories emerged in terms of chemical bonding energies. Thus, a misleading philosophical framework, taken from debates in the philosophy of physics, hided the radical epistemological disruption.

Fifth, Lavoisier was quick to incorporate the newly recognized elements into a chemical classification and nomenclature of hitherto unseen systematics, such that the identity of compounds was now based on their new elemental composition, which became extremely successful. While that has been recognized comparatively recently as being part of the "chemical revolution" (Siegfried and Dobbs 1986), it shifted the attention even more away from explanation. As will be shown below (section 3.1), classification and explanations are distinctly different epistemological functions of elements.

Sixth, two decades after the "chemical revolution," Dalton's atomism offered a new theoretical approach in which atoms corresponded to the elements. That suggested that the new elements could almost immediately take on the role of explanatory entities. However, Dalton's original atomism, while explaining or reformulating the laws of definite and multiple proportions, did not explain chemical properties of compounds. It took many decades before the approach gained any explanatory potential worth mentioning, such that most chemists considered it just a formalism to describe relative equivalent masses of the respected elements. Many twentieth-century historians and philosophers of science, who were enthusiastic about the much later success of atomism in structural chemistry, have overlooked that long period. And they equally tend to overlook that all of modern chemistry, and most of modern science, is based on the turn from the received principles to operationally defined elements.

Overall, the historiographical focus on the "chemical revolution" and its various interpretations have made historians and philosophers of chemistry neglect the epistemological disruption caused by the adoption of the operational definition of elements that occurred at about the same time. There is no hero in adopting a definition. The international chemical community did it sometime around 1800. But the exact dates and individuals do not matter for its epistemological understanding and appreciation.

3. Continuity

3.1 An Epistemological Framework for Understanding the Elements

A more general reason for the little appreciation is the lack of an epistemological framework that allows one to understand what epistemological roles elements have played in the history of science. A framework is not to be confused with a theory; it does not depict anything. It is a conceptual tool or scheme that can be applied to any historical period in order to grasp its epistemological particularities and to identify continuities and discontinuities.

The framework I am suggesting is borrowed from Rom Harré's pioneering work *Varieties of Realism* (Harré 1986). Although he himself developed it for understanding particle physics, despite his earlier background in chemical engineering before he became one of the most influential philosophers of science, his ideas can easily be generalized and applied to the entire history of elements (Schummer 1996, chapter 4).

In his book Harré presented three versions of realism: (1) for ordinary world experience, (2) for theoretical entities, and (3) for abstract mathematical structures. The second one, called "reference realism," transformed the contemporary language-focused debate (when do theoretical terms in science refer to real entities?) into a question of scientific practice: when is it reasonable to believe that a theoretical entity exists such that we should actually start a research program to experimentally identify it? To answer that question, Harré developed a sophisticated scheme that allows assessing the overall scientific context. What matters in the present context is that he distinguished between three different epistemological roles or functions of theoretical entities, all of which should ideally be considered at the same time.

The received philosophy of science has usually considered only the first role of theoretical entities, their explanatory potential within a certain theory. Many even went as far as to define theoretical entities just by that, such that two theories that happen to use the same theoretical term, say "electron," refer to different entities. However, Harré added a second function, their ontological role in a classification system. While classification came to the awareness of most philosophers of science only through particle physics, it has always played a pivotal role in most sciences other than mechanics, including chemistry, biology, mineralogy, and so forth. Scientific classifications frequently employ theoretical entities, the most prominent historical case being the classical elements or principles that allowed ordering the realm of substances by their supposed composition of the elements. The third role, which is a central requirement of Harré's "reference realism," is that the theoretical entities are conceptualized such that they potentially belong to the observational world. Even if we do not yet exactly know how to do it, there should be a possible way to get direct experimental access to these entities.

If we apply this threefold framework to the entire history of chemistry and alchemy, it turns out that there have always been tensions between the three roles, up to the present day. Many theoretical conceptions of the elements or principles have focused on explanation, at the expense of classification and direct access. Others have highlighted classification, neglecting explanations and experimental access. And there were, even long before the operational definition, approaches that emphasized experimental access.

Rather than analyzing the entire history of elements here, I will use the framework to point out three aspects: (1) the continuity of the third role, (2) the importance of the operational definition for classification, and (3) conceptual tensions arising from difficulties in reconciling all three roles.

3.2 Experimental Access

The threefold framework allows us to recognize the operational definition of elements as an extreme concept that takes the third role, experimental access, as the defining characteristics of elements. However, other historical conceptions of elements have taken that role also into account, even though they put more emphasis on the other roles, such that the framework helps us to see more continuity in the history (Schummer 1996, 101–165). Let us take a brief look at four examples.

In Aristotle's scheme the four elements are considered real material substances (composed of matter and form, in his ontology) with tangible properties. They therefore seem to perfectly meet Harré's third criterion for theoretical entities, direct accessibility, even if one does not yet know exactly how to do that. But Aristotle himself was skeptical for reasons consistent with his own account. The elements can not only move around, mix with one another, and build compounds, they can also convert into one another. Any attempt to isolate them by experimental methods would have to apply material means—for instance, fire in distillation—which would result in transformation rather than in isolation.

The experimental tradition of alchemy and chemistry was more optimistic than Aristotle was. For instance, Christian Gottlob Gmelin, who based his 1780 chemistry textbook on the Aristotelian elements, provided even a version of the operational definition (Gmelin 1780, 36):

> Simple bodies in the chemical sense are those which can be no further decomposed into unlike particles by chemical artifices, they are called by another name "elements."

The second example, classical atomism, literally includes an operational definition in its name. An atom (from Greek *átomos*: the indivisible, uncuttable) is a material body that cannot further be divided. In the Christian tradition, that limit of division usually marked the distinction between human and divine capacities of division. For mechanically minded philosophers, it was the limit of cutting a piece of matter by an imagined small knife. Whatever the specific account, classical atoms were conceived as material entities by a hypothetical operational criterion, even though direct access was considered to be restricted by their supposed smallness. However, by making the operational criterion hypothetical beyond human reach, atomism did not encourage experimental approaches of isolating individual atoms.

My third example is the sulphur-mercury theory from Arab and Latin medieval alchemy. Both "sulphur" and "mercury" were usually considered higher-level principles composed of the classical four elements, and in turn were the essential components of many materials. Particularly detailed were views about metals whose characteristic properties alchemists explained by their different proportions of the two principles. Rather than being "metaphysical principles," or identical with the common substances of the same name, "sulphur" and "mercury" were thought to be material substances that can be isolated by experimental methods (Newman 2014). Unlike the Aristotelian elements and the atoms, these chemical principles encouraged developing for the first time in history sophisticated experimental laboratory techniques for taking substances apart and combining them anew (*solve et coagula*, as it was called in Latin), which became the model of the modern chemical laboratory for analysis and synthesis, and, for that matter, of laboratory science in general. Thus, the notion that "sulphur" and "mercury" are potentially real and pure substances, from which one could even produce highly valued substances such as gold, enabled the development of the idea of laboratory science in the sense of performing controlled, reproducible, and theoretically guided operations.

Fourth, many later commentators have ridiculed phlogiston as a fancy detour of science that postulated an imagined chemical principle for explanatory reasons only without any experimental foundation (Harré's first criterion), echoing Lavoisier's own rhetoric. However, eighteenth-century "phlogistonists," that is, almost all chemists by then, had long given up speculative natural philosophy, as the philosopher of chemistry Elisabeth Ströker (1982) convincingly argued on the basis of the original, mostly Latin, texts. Already Johann J. Becher, on whose theory of various "earths" Georg Ernst Stahl developed the generalized phlogiston theory, tried to base his principles on experimental grounds. Before the so-called "chemical revolution" chemists identified phlogiston, which by then had gained tremendous explanatory success for such diverse fields as combustion, calcination, breathing, and meteorological cycles, with experimentally identifiable substances, including what we would today call hydrogen, carbon, and energy. One could even argue, as Chang (2012) has done, that the phlogiston theory pre-formulated modern redox theory, making phlogiston a predecessor of today's electrons. Whatever the interpretation and its specific role in explanations, phlogiston was from its very beginning thought to be a chemical substance that could possibly be isolated and analyzed in pure form in the laboratory.

In all four examples, experimental accessiblity is an operational criterion of elements, at least on hypothetical grounds, that illustrates the continuity of Harré's third role in the history of chemistry.

3.3 Classification

The simplest form of a systematical classification takes two independent properties and their opposites (A, non-A, B, non-B) and combines them in a table of four classes (A and B, A and non-B, B and non-A, non-B and non-A). In some sense, Aristotle's scheme of fundamental properties (solid, fluid, cold, warm) corresponds to that account. However, the central idea of natural philosophy has always been the classification of matter not by properties but by elemental compositions, such that Aristotle defined the elements by primary matter bearing each a binary combination of fundamental properties. Indeed, the notion of elemental entities is logically tied to a classificatory approach based on elemental composition, that is, every account of elements implies a classification based on composition, ideally in quantitative terms.

The main historical problem of classifying matter in terms of elemental composition was insufficient knowledge about composition. The composition could not simply be inferred from observational properties. As long as elements were theoretical entities postulated for explanatory purposes, the only way of inference was from successful explanations, such that classification and explanation were epistemologically tied together, despite being epistemologically different functions. For instance, determining whether a substance contained phlogiston or not could only be assessed from the explanation of certain chemical properties, such as combustibility, which the phlogiston theory explained. On the other hand, an ever-increasing repertoire of experimental techniques of analysis, such as various forms of distillation, allowed taking substances apart and provided direct access to components. Much of the history of alchemy and early chemistry is about reconciling the explanatory and experimental approaches to elemental composition, without success.

Against that background we can appreciate the operational definition of elements as a move that perfectly combines classification and experimental access, at the expense of explanation. If elements are those substances that by definition resist any separation technique, then the experimental separation of any substance ultimately provides its elemental composition by experimental access to the elements. Moreover, if the analysis is performed quantitatively in terms of relative elemental masses, every substance can be classified based on the quantitative composition of elements, which previous accounts could only dream of.

The new definition of elements was not the only operational solution to classificatory problems in chemistry (Schummer 1996, 170–181). Instead, the old hierarchy of materials, from elements to compounds, to homogeneous mixtures and heterogeneous mixtures, that Aristotle had already developed on partly observational and partly theoretical grounds, could be redefined by reference to experimental techniques. Mechanical separation techniques—such as cutting, grinding, and sorting—decided if the material was a heterogeneous mixture or not; thermodynamic separation techniques (distillation, crystallization) decided if it was a homogenous mixture or not; and chemical analysis decided if it was a compound or an element. Of course, the operational definitions did not always provide simple and decisive results, as with azeotrope mixtures and many presumed elements that turned out to be compounds by more sophisticated chemical techniques. However, those substances could not undermine the classificatory approach because they thereby just moved one step up the hierarchy.

It should be noted that the operational definitions of elements and compounds worked safely only on the basis of a reliable and uniform system of relative equivalent masses, which was developed only during the nineteenth century by tremendous collective efforts of the chemical community. Whether a chemical transformation was actually an analysis rather than a synthesis did not depend on the absolute masses of educts and products, but on their relative equivalent masses, the determination of which employed a large set of mutually correcting experimental techniques (Schummer 1996, 185–203), including volumetric and gravimetric measurements of combining volumes and masses in chemical reactions; thermodynamic measurements such as the decrease of melting points, the increase of boiling points, osmotic pressure, specific heat capacities; and analogy reasoning such as by isomorphism of crystals and chemical similarities of elements that eventually led to the periodic system.

Against that background it would be misleading to use the terms "empiricism" or "positivism" for describing the chemical approach to classification. The most appropriate term would be "experimentalism."

3.4 Tensions: Debates on Elements in Compounds

One major strength of Harré's threefold framework is that it allows us to identify and understand tensions within the concept of elements throughout history. Such tensions occur when the three roles or functions cannot smoothly be integrated into one concept. The notion of elements then tends to disintegrate into two or three concepts.

In one of the rare philosophical papers by a twentieth-century chemist, Fritz Paneth (1931) argued that modern chemistry has actually two concepts of elements, which he called *einfache Stoffe* (simple substances) and *Grundstoffe* (basic substances). The first concept refers to material bodies that meet the operational definition; the second one is used when we think of a compound being composed of elements to explain the compound's properties from the elements. Obviously the split results from obstacles to reconciling the two roles, experimental access and explanation. Note that classification can be achieved by both concepts: operationally by chemical analysis and synthesis, and representationally by referring to the constituents of a compound.

Paneth was not the first to notice the double meaning. As Scerri (2007, 117) and Hooykaas (1947) have pointed out, already Mendeleev observed the inconsistency and distinguished between simple substances and abstract elements. To some extent, also Lavoisier's terminological vacillation in his *Traité élémentaire de chimie* (1789) between "*principes*" and "*substances simples*" for his elements expresses the double meaning, because of his efforts to take some elements, particularly oxygen, as explanatory principles. In the period between Lavoisier and Mendeleev, who tried to reintroduce an explanatory account with his periodic system, the double meaning was largely absent, because explanatory ambitions were temporarily given up for the elements that still had to be discovered. However, in the past six or seven centuries that was rather an exception.

The main tension became visible in endless debates about whether elements are constituents in compounds, and, if so, how one could explain that most compounds radically differ in their properties from the elements they were presumably built of and classified by. The issue has been a major explanatory challenge to chemistry and its precursors. Much of the debate during the ancient, early modern, and modern periods has been documented (e.g., Duhem 2002; Hoykaas 1947), such that I confine myself to a few remarks on the medieval period that is lesser known and then look at IUPAC's solution.

Aristotle himself argued that in true compounds, elements are only in the state of potentiality (*De gen. et cor.*, I. 10). The two leading Islamic commentators, Avicenna and Averroes, developed—each with a sophisticated metaphysical apparatus—opposite views (for detailed references and discussion, see Lasswitz 1890, 239–254; and Maier 1943, chapter 1). In much simplified terms, Avicenna held that the elements as such are preserved in compounds and that only their properties combine into new substantial forms, whereas Averroes thought that the elements lose their identity in compounds (unlike in mixtures) to build new substantial forms. Almost all prominent Latin philosophers who dealt with natural philosophy debated the issue and sided either with Averroes or Avicenna or developed a middle way, including Albert the Great, Thomas Aquinas, Roger Bacon, Duns Scotus, and William of Ockham. They all engaged in what came to be a central issue of chemistry, although none of them could convincingly solve it.

Starting with Lasswitz (1890) modern commentators have frequently argued that the issue was eventually solved by modern atomism. However IUPAC still upholds a double definition of "chemical element" based on atomism in its Gold Book:[3]

(1) A species of atoms; all atoms with the same number of protons in the atomic nucleus.

[3] IUPAC, *Compendium of Chemical Terminology*, 2nd ed., comp. A. D. McNaught and A. Wilkinson (Oxford: Blackwell Scientific Publications, 1997), s.v. "chemical element," https://goldbook.iupac.org/html/C/C01022.html.

(2) A pure chemical substance composed of atoms with the same number of protons in the atomic nucleus.

While definition 2 highlights the role of experimental access by purification resulting in a collection of atoms, definition 1 defines elements not by a collection but by a species of atoms that might serve in explanations and classificatory representations such as structural formulas. The phrase "all atoms with the same number of protons in the atomic nucleus" suggests that the number of neutrons is an accidental property of elements, such that isotopes belong to the same atomic species and element in both definitions. But what about electrons? IUPAC defines "atom" as the

> smallest particle still characterizing a chemical element. It consists of a nucleus of a positive charge (Z is the proton number and e the elementary charge) carrying almost all its mass (more than 99.9%) and Z electrons determining its size.[4]

Hence, the number of electrons, Z, is an essential property of the atom and the element, such that an ion is, according to IUPAC's definition, not an element. It follows that elements do not exist in ionic crystals because the electrons can no longer be assigned to the nuclei as in elements, and similar reasoning can be developed for covalent and metallic bonds, or any other theory of chemical bonding. Since compounds are not composed of elements, they cannot be classified according to elemental constituents. Obviously IUPAC, I assume unknowingly and unwillingly, sides with Averroes, despite the double definition.

If, on the other hand, and contrary to IUPAC, one considers the number of electrons, like the number of neutrons, an accidental property of atoms, then one could define atoms and, thereby, chemical elements as proton aggregates, regardless of neutrons, electrons, energy and spin state exchange with the surroundings, and so on. Atoms/elements in that sense can without contradiction be said to exist in compounds, and their specific configuration may be used for classificatory purposes. That corresponds to Avicenna's view. However, all modern chemical explanations refer to the electrons of atoms, which would, strictly speaking, be ruled out. One could try a middle way, as did Albert and Thomas, for instance,

---

[4] IUPAC, *Compendium of Chemical Terminology*, 2nd ed., comp. A. D. McNaught and A. Wilkinson (Oxford: Blackwell Scientific Publications, 1997), s.v. "atom," https://goldbook.iupac.org/html/A/A00493.html.

by distinguishing between core electrons and outer shell electrons, but that only blurs the issue rather than solves the tension convincingly.

In sum, from the Middle Ages to today's chemistry, all three roles or functions of elements could not be reconciled in one consistent approach. The operational definition took experimental access as the defining feature of elements and achieved classification by experimental analysis, but at the price of giving up explanations. Once the explanatory role of elements was revived, conceptual tensions arose anew, of which IUPAC's definitions are a telling example.

4. Conclusion

4.1 The Legacy of the Operational Definition

Definitions cannot be true or false, they are the undisputed conventional parts of science, that is, the scientific community decides whether they accept it or not. Once accepted the community can at any time change their minds and drop the definition; or the community splits such that one group accepts it and the other one drops it. In general, definitions specify the meaning of terms which enables more precise communication and avoids misunderstandings. Operational definitions, for example, the definition of "time" by clock measurement, do not essentially differ in that regard, but they relate concepts to established and shared laboratory practices and try to cut off theoretical and metaphysical connotations as far as possible, which makes the definition acceptable across metaphysical views and theory changes.

However, the operational definition of elements is very special in two regards. First it defines not some property but elements, that is, the fundamental entities that all explanatory and classificatory approach in natural philosophy is supposed to refer to. Thus, any definition of elements shapes the entire conceptual and theoretical apparatus. Second, the operational definition of elements does not only define a term, it provides laboratory rules for literally producing those entities to which the term "element" refers. All of modern chemistry, both experimental and theoretical, has been built on studying those entities, their properties and compounds, their systematization and theoretical conceptions from Daltonian atomism to quantum chemistry. Moreover, almost all sciences—including physics, medicine, biology, and mineralogy—have adopted these chemical elements in their experimental and theoretical frameworks as the unquestionable material basis or starting point. When nuclear physics began to study subatomic particles, they did so not on the basis of Aristotelian elements but on the basis of the operationally defined and produced chemical elements; that is, without the operational definition today's nuclear and particle physics would not exist.

Theoretical scientists might be inclined to take atoms in their models as given entities, but the properties of the atoms that they take for granted came to be known only by studying those pieces of matter that met the operational definition. Thus the operational definition of elements became materialized in science in a unique way, both literally by providing a new material basis for science, and conceptually by integrating the new entities in theoretical frameworks.

One could argue that, once the operational definition was materialized, it could have been abandoned, like a tool that was temporarily useful to redirect science into a new direction. However, the operational definition was challenged several times, particularly through the discovery of electrons and isotopes (Kragh 2000). That caused debates on whether the electron is an element and whether isotopes are different elements or not, because their experimental isolation and separation actually met the operation definition, unless the techniques of separation would be better specified. In these debates, chemists had to rethink the definition and employ the usual criteria for accepting a definition, including its usefulness. In both cases their negative decisions reflect the contemporary chemical perspective of usefulness: electrons could not be isolated and experimentally employed in the same way as the other elements; isotopes do hardly differ in their chemical properties from one another such as the other elements. They could have decided, and perhaps will do so in the future, otherwise. The examples illustrate that definitional problems can come up anew at any time, such that it is better to be aware of the conceptual basis on which chemistry historically rests.

4.2 The Usefulness of Philosophy for Understanding Science

Epistemological reasoning might not be the main strength of scientists and historians of science, but some acquaintance with epistemology is certainly an advantage in both fields. The history of elements provides ample evidence.

To be sure, some time ago historians of science made frequent use of the methodological theories by Popper, Kuhn, Lakatos, and others in case studies that were meant to support or criticize ideas about scientific progress. However, using the historiography of science in support of a philosophical theory and using an epistemological framework for analyzing a historical period of science are two different things. The former takes epistemology as a theory that claims truth; the latter employs epistemological concepts as tools for better understanding the epistemic practices of science. Only the latter approach, philosophy as a toolbox or skills, serves historiography.

That was illustrated in section 3 by using Harré's threefold framework. By considering all three functions of elements—explanation, classification, and experimental access—we can analyze both the continuities and discontinuities in the

186 WHAT IS A CHEMICAL ELEMENT?

entire history of elements. The turn to the operational definition of elements then appears as a radical disruption, by temporarily giving up explanation in favor of perfectly meeting experimental access and classification. That was arguably the most radical and most influential disruption in science ever since, because it broke up with the entire tradition of natural philosophy and reoriented not only chemistry but most of the sciences toward a new material basis on which all subsequent ideas of science have built. One may call it a "revolution," but that term has long been watered down by numerous case studies on marginal events in science. Its scientific impact dwarfs all the other so-called revolutions, such as the Kantian, relativistic, and quantum revolutions, which have been vividly debated in philosophy despite their limited impact on specific fields.

Although the operational turn follows a similar epistemological pattern as these local revolutions, the history and philosophy of science has hardly acknowledged its scientific importance and epistemological significance. Whereas philosophers of science have focused on physics and neglected chemistry, historians of chemistry would rather debate the so-called "chemical revolution" and Lavoisier's role therein. The person-centered historiographical approach tends to overlook the broader epistemological dimension of the operational turn in chemistry, that it redefined central concepts of natural philosophy by relating them to human capacities and their limits, and its place in the wider cultural history of modernity.

Moreover, Harré's framework allows us to identify and understand conceptual tensions and debates that arose from insufficient integration of all three roles into one account, from the Middle Ages up to the present day. Of course one needs detailed knowledge of scholastic metaphysics to understand the medieval debates. But only little training in logic is required to understand that current chemistry seems to favor the view of Averroes. I assume that most chemists agree to IUPAC's definitions of "chemical element" and "atom" cited above. However, they would probably disagree about the claim that elements are not constituents of compounds, although that exactly follows from these definitions. As conceptual tensions have reached a historical maximum, philosophy can help build a more consistent basis of chemistry.

References

Bridgman, Percy W. 1927. *The Logic of Modern Physics*. New York: Macmillan.
Chang, Hasok. 2012. *Is Water H2O? Evidence, Realism and Pluralism*. Dordrecht: Springer.
Davis, Tenney L. 1931. "Boyle's Conception of Element Compared with That of Lavoisier." *Isis* 16, no. 1: 82–91.
Duhem, Pierre. (1902) 2002. *Mixture and Chemical Combination*. Edited and translated by Paul Needham. Dordrecht: Kluwer.

Gmelin, Christian Gottlob. 1780. *Einleitung in die Chemie.* Nuremberg: Raspe [online http://reader.digitale-sammlungen.de/de/fs1/object/display/bsb11274713_00048.html].

Harré, Rom. 1986. *Varieties of Realism: A Rationale for the Natural Sciences.* Oxford: Blackwell.

Hooykaas, Reyer. (1947) 1983. "The Law of the Conservation of Elements." In *Reyer Hooykaas: Selected Studies in the History of Science,* 121–143. Coimbra: University of Coimbra.

Kragh, Helge. 2000. "Conceptual Changes in Chemistry: The Notion of a Chemical Element, ca. 1900–1925." *Studies in History and Philosophy of Modern Physics* 31B, no. 4: 435–450.

Lasswitz, Kurd. 1890. *Geschichte der Atomistik,* vol. 1. Hamburg & Leipzig: Voss.

Lavoisier, Antoine. 1787. "Nomenclature chimique." In *Méthode de Nomenclature chimique,* edited by Guyton de Moreau, Antoine Lavoisier, Claude-Louis Berthollet, and Antoine de Foucroy, 1–25. Paris: Cuchet.

Lavoisier, Antoine. 1789. *Traité élémentaire de chimie.* Paris: Cuchet.

Maier, Anneliese. 1943. "Die Struktur der materiellen Substanz." In *An der Grenze von Scholastik und Naturwissenschaft.* Essen: Essener Verlagsanstalt, 7–40.

Meinel, Christoph. 1982. "Der Begriff des chemischen Elements bei Joachim Jungius." *Sudhoffs Archiv* 66, no. 2: 313–338.

Newman, William R. 2014. "Mercury and Sulphur among the High Medieval Alchemists: From Rāzī and Avicenna to Albertus Magnus and Pseudo-Roger Bacon." *Ambix* 61, no. 4: 327–344.

Oldroyd, David R. 1975. "Mineralogy and the 'Chemical Revolution.'" *Centaurus* 19: 54–71.

Paneth, Fritz. 1962. "Über die erkenntnistheoretische Stellung des chemischen Elementbegriffs." *Sitzungsberichte der Köngisberger Gelehrten Gesellschaft / Naturwissenschaftliche Klasse* 8, no. 4 (1931): 101–125 (English trans. "The Epistemological Status of the Chemical Concept of Element." *British Journal for the Philosophy of Science* 8: 1–14, 144–160).

Scerri, Eric. 2007. *The Periodic Table: Its Story and Its Significance.* New York: Oxford University Press.

Schummer, Joachim. 1996. *Realismus und Chemie.* Würzburg: Königshausen & Neumann.

Schummer, Joachim. 1997. "Scientometric Studies on Chemistry I: The Exponential Growth of Chemical Substances, 1800–1995." *Scientometrics* 39, no. 1: 107–123.

Siegfried, Robert, and Betty Jo Dobbs. 1968. "Composition, a Neglected Aspect of the Chemical Revolution." *Annals of Science* 24: 275–293.

Ströker, Elisabeth. 1982. *Theoriewandel in der Wissenschaftsgeschichte. Chemie im 18. Jahrhundert.* Frankfurt: Klostermann.

10

Substance and Function

The Case of Chemical Elements

*Jean-Pierre Llored, Ecole Centrale of Casablanca, Morocco;
Linacre College, UK*

Introduction

Substances, atoms, and *elements* are keywords upon which most chemists'
practices have revolved, independent of the period of the history of chemistry
at stake, even if we do acknowledge that their meaning and relative importance
have very often been different for specific periods of history or within a particular period as well. Such words have also been keywords of our metaphysics
from the early Greek philosophy onward; words that have often been used in
order to explain the persistence of "things" through change (Wiggins 2001).
In his book *The Same and Not the Same*, the Laureate of the Nobel Prize in
Chemistry, Roald Hoffmann, asserts: "If there is anything central to chemistry,
it is change" (Hoffmann 1995, xv). So we can easily understand why this centrality of change in chemistry offers the temptation to slip into drawing hasty
conclusions as to the ways chemists are influenced by metaphysical approaches.
The changes of substances that occur within a chemical transformation are
partly explained by changes in composition, using chemical formulas in which
elements are rearranged. The same elements are used by chemists to write the
formulas of both the chemical reagents and products, but the chemicals in the
reactor are not the same because a transformation has just occurred. In the title
of Hoffmann's book, *The Same and Not the Same*, the conjunction "and" is thus
particularly interesting. Hoffmann does not write the same *or* not the same, as it
is expressed by the principle of non-contradiction according to which contradictory statements cannot both be true in the same sense at the same time, but
he does write "the same *and* not the same" because of the chemical change(s)
and of the importance, for instance, of isotopic modification in a molecule. But
chemists are seeking out something that persists through the transformation in
order for them to explain change.

In §43 of *Philosophical Investigations* (1953), Wittgenstein famously
wrote: "For a *large* class of cases—though not for all—in which we employ the

Jean-Pierre Llored, *Substance and Function* In: *What Is A Chemical Element?*. Edited by: Eric R. Scerri and Elena
Ghibaudi (2020). © Oxford University Press.
DOI: 10.1093/oso/9780190933784.003.0011

word 'meaning' it can be defined thus: the meaning of a word is its use in the language." In that same book, he also claimed: "What we do is to bring words back from their metaphysical to their everyday use" (§116). Wittgenstein believed that there are general misconceptions about meaning whose debunking is of direct relevance for the method of asking how words are actually used in certain domains of human activities. In order to clarify the meaning of the word *substance*, Timmermans (1928) and van Brakel (2000a; 2000b; 2008; 2012) have carried out many investigations about the uses and definitions of this word in many different domains of chemistry. Van Brakel concludes:

> "Substance" is a proto-scientific concept (not defined in science) and identification (isolation, separation) depends on specific properties that are considered relevant and must be known, "substance" will remain a pragmatic notion and ambiguous situations may arise as to whether to count a sample as one, two, or three substances (cf. non-stoichiometric compounds, inclusion compounds, racemic species, tautomers, etc.). (van Brakel 2012, 224)

According to van Brakel, the notion of substance will remain pragmatic. By "pragmatic," we understand "the set of practical effects that can be deduced from it [the word at stake], by all the members of a community, when this [term] is used in a certain context" (Bächtold 2008, 845). James (1995, 71) asserted that by virtue of the properties we ascribe to a substance, this later has a "steering function": knowing these properties helps us "anticipate" the possible observable "consequences" issued from the operations made on the substance at stake (1997, 246). Pragmatic perspectives can be flexibly varied and multiplied through new relevance aspects, new problems, and new methods. Moreover, when referring to a chemical substance using everyday language (e.g., "carbon dioxide") or a chemical formula (e.g., CO_2), chemists are implicitly referring to a set of properties. A chemical substance thus also has an economy of thought function in that it concisely embodies the knowledge chemists have of the observable effects likely to occur when they act upon the substance. As Dewey (1993, 129) asserted: "A chemical substance is represented not by enumeration of qualities as such, but by a formula which provides a synoptic indication of the various types of consequences that will result." In the same vein, in his well-known definition of lithium, Peirce (1931–1958, CP 2.330)[1] confidently endorsed the idea that lithium can be defined as a set of instructions aimed at permitting not only the identification but also the production of a specimen of lithium. This definition is clearly provisional insofar as the word *lithium* will acquire new meanings

---

[1] The reference in the text is given by volume and paragraph number.

as we learn more about the thing or stuff to which it refers. Peirce remarked: "The peculiarity of this definition is that it tells you what the word lithium denotes by prescribing what you are to do in order to gain a perceptive acquaintance with the object of the word" (Peirce1931–1958, CP 2.330.). In certain cases, the approach taken by ordinary language philosophy does not succeed in clarifying the meaning that a word has in a particular set of activities. For those cases and others, a pragmatic approach focused on both the function(s) played by a word and the consequences to which a group of practitioners expect when using this word could be of help to a philosopher of chemistry.

The investigation of the meaning of the word *element* leads us to similar conclusions. The answer to the question "What exactly do chemists mean when they say that elements both underlie all these changes while being common to both reagents and products?" is anything but simple. As a matter of fact, as Ghibaudi et al. (2013, 1628) state: "the problem of unambiguously defining what is conserved in a chemical transformation has been a matter of discussion for a long time within the chemistry community." In fact, there is no consensus within the chemical community regarding a unique and clear definition of the word *element*. The multitude of different and often contradicting definitions leads to confusion among both students and professionals of chemistry. For instance, the International Union of Pure and Applied Chemistry (IUPAC 1997) proposes two separate definitions according to which the chemical element is defined as: (1) a species of atoms, that is, all atoms having the same number of protons in the atomic nucleus, or (2) a pure chemical substance composed of atoms with the same number of protons in the atomic nucleus. In some cases, the second definition refers to an "elementary substance," which is considered to be distinct from the chemical element as defined under 1, but in most cases the expression "chemical element" is used to denote both definitions.

This paper aims to analyze the two aspects of elements, material and conceptual, in order to further understand the functional role played by elements in chemistry. First, we will refer to how Mendeleev gradually built his periodic table. Second, we will both highlight and reflect upon the functional role of elements in chemistry. In conclusion, we propose that a functional approach is important for understanding current chemical practice, especially in nanochemistry and quantum chemistry. This approach also contributes to the analysis of different types of mereology that coexist in chemistry today.

Mendeleev and the Two Aspects of Elements

William Prout's assumption of the derivability of all the elements from hydrogen entailed the primacy of the atomic weight system as the criterion for classifying chemical elements. Following Prout's line, researchers such as Johann Döbereiner pointed out accurate and astonishing correlations between arithmetical relations and chemical analogies within sets of three chemical elements called "triads." He carried out his calculations in 1829 by using Berzelius's values of atomic weights. Leopold Gmelin then widened the correlations from triads to a larger "family" of elements. The proliferation of correlations and classifications occurred after 1850 when researchers started to discover new elements and integrated the new concept of "organic radicals" into mineral chemistry. Those strategies focused their interest on local arithmetic correlations to the detriment of global analogies between chemical properties and thus failed in classifying elements into a coherent scheme while contemporary spectroscopic methods provided scientists with new accurate correlations (Bensaude-Vincent 1986; 1989).

In 1860, the Congress of Karlsruhe aimed at clarifying and framing the situation using: (1) basic notions such as molecule, atom, and element; (2) atomic weight systems; (3) nomenclature; and (4) notations. For the first time, chemists coming from all over the world met together in order to both choose and define common methodologies and definitions. The Gerhardt-Cannizzaro's system of atomic weight thus became the official framework despite deep divisions between atomists and equivalentists. In line with this new standard, John Alexander Newlands proposed a law of octaves (1865) while William Odling set up a whole periodic system (1865). Those classifications neither allowed them to predict new elements nor did they pave the way for accurate correlations of atomic weights (Nye 1984; 1993). Mendeleev acknowledged that this congress suggested to him the idea of an overall periodicity of the elements depending on the increasing numerical value of their atomic weight. He believed in "chemical individuals," that is, in the "peculiar" and "infinite diversity of the elemental individualities" (Mendeleev 1889, 637–640). He was not searching for a proto hyle to think about the unity of the universe, but for a unique periodic law connecting the "multifarious *relations* of matter" and the "many-sided relations" that elements share with one another (Mendeleev 1889, 644–645). According to Mendeleev, a crucial distinction must be made between an observable simple body that displays chemical and physical properties, and a chemical element that "causes" those properties. This basic element possesses at least one attribute, namely, the atomic weight that served to distinguish it from the other elements, and that was used to order the elements in a unique and coherent sequence (Scerri 2005). Mendeleev stated (1891, 23):

> It is useful in this sense to make a clear distinction between the conception of an element as a separate homogeneous substance, and as a material but invisible part of compound. Mercury oxide does not contain two simple bodies, a gas and a metal, but two elements, mercury and oxygen, which, when free, are

a gas and a metal. Neither mercury as a metal nor oxygen as a gas is contained in mercury oxide; it only contains the substance of the elements, just as steam only contains the substance of ice, but not ice itself. . . .The existence of an element may be recognised without knowing it in the uncombined state, but only from an investigation of its combinations, and from the knowledge that it gives, under all possible conditions, substances which are unlike other known combinations of substances.

This quotation emphasizes that the combinations are first in order for Mendeleev to investigate elements. This relational approach is carried out by following a few principles derived from experiment. We can also notice the pragmatic meaning ascribed to elements insofar as it is what chemists can do with a particular element, under all possible conditions, that enables them to identify it. Mendeleev nevertheless adds (1891, 23):

Besides, many elements exist under various visible forms whilst the intrinsic element contained in these various forms is something which is not subject to change. Thus carbon appears as charcoal, graphite, and diamond, but the element carbon alone contained in each is one and the same.

Hoffmann's expression "The same and not the same" is exemplified in this last quotation. This carbon as a stuff is identified by criteria such as specific gravity and so on. The "element carbon" in compounds is "carbon" in another sense. But the two aspects of the chemical element are mutually dependent in Mendeleev's system. The possibility of considering something to be invariant cannot be detached from empirical observation, since the knowledge about each particular element is continually adjusted to experimental results. Notwithstanding their abstract aspect, elements are material as well. They are characterized by their atomic weights, which result from experiments. The element is thus employed within an experimental setting, which involves "both the head and the hand" to use Holmes's turn of phrase (Holmes 1989, 125). That which remains invariant cannot be envisaged without considering the operations used by Mendeleev in order to identify chemical regularities.

This approach enables Mendeleev to predict some properties and the existence of other elements such as the *eka*-iodine. Mendeleev brought chemical similarities and contrasts to the fore and paid attention to differences between elements. In doing so, he gradually construed a network of relations thanks to approximations and a careful study of available elements by means of trial and error. He defined an element from within its relations with other elements. For example, the reactions between alkali metals and halogens allowed him to infer that they shared the same valency even if they displayed different combinations with other elements. This contrast helped Mendeleev identify an interesting regularity regarding the differences of their atomic weights (Scerri 2007; Bensaude-Vincent and Stengers 1996). Mendeleev stated that his method consists of simultaneous interpolations within groups or columns as well as within periods or rows of the periodic table. He carried out his procedure by taking the average of the sum of the values of the four elements flanking the element in question. In doing so, he accurately calculated the atomic weight of the element selenium (Scerri 2001; 2005; 2012). He thus gradually co-defined elements by means of their relations with other elements. Some groups of elements share similarities and it is always possible for a chemist to foresee the kinds of reactions that are possible between elements belonging to two different columns. Bensaude-Vincent and Simon claim that "by assigning a place to each element, the periodic system anchors the unit-element in a whole network of material, chemical relationships" (Bensaude-Vincent and Simons 2008, 160). The element gradually got a functional role and became a "node" interlinked with other elements within a network of material transformations. The concept of element thus gained a strong explanatory and heuristic power. Elements can circulate and be exchanged while respecting the "conservation law" (Bensaude-Vincent and Simons 2008). Let us further investigate the role played by the notion of element in the periodic table and in chemistry at large.

On the Functional Role of Elements: Critical and Epistemic Insights

Following the line defended by Cassirer in the book *Substance and Function* (1910), but adapted to the concept of element instead of that of atom, we could assert that it is only the totality of all the different points of view that have been defended from Lavoisier to Mendeleev, mutually confirming and correcting one another that finally after many experiments gives rise to a unitary table of atomic weights, and thus lays the basis of a definite system of chemical formulas. As chemists apparently investigate the element itself in its manifold determinations, they at the same time place these different groups of circumstances in a new relation to one another. "The real positive outcome of chemical knowledge here is in the systematic analysis of these relations. This unification is rather directly productive; it produces a general schema for future observations, and indicates a definite direction for these" (Cassirer 1910, 209). The initially dispersed facts are now organized. Instead of coexisting in indifference, they are ordered around a center of precise reference, that is the element at stake. The unification thus construed plays an immediately functional and productive role; it sets up a total table applicable to the future observations and assigns to them a given direction. The element functions as "the conceived unitary center of a system of coordinates" (Cassirer 1910, 208) thanks to which chemists can classify the various groups of chemical properties. Chemists are concerned with relating the different series of experiments reciprocally to one another. To do so, they use the notion of element as a conceptual tool (Llored and Bitbol, 2013). Cassirer asserted:

> The totality of what is empirically known is condensed, as it were, to a single point [a place in the periodic table in our case], and from this point issue the different lines of direction, in accordance with which our knowledge advances into the unknown. Those manifolds already discovered and defined according to law function [the periodic "law" in our case] as a fixed logical unity in opposition to those manifolds newly to be discovered; and it is this unity of the fundamental point of connection, which renders possible our assumption of an ultimate identical subject for the totality of possible properties [the invariant element through transformation]. (Cassirer 1910, 208)

Cassirer goes on to say: "Science today collects the plurality of elements into a fundamental series, whose members succeed each other according to a definite principle, and then determine the individual properties of bodies as functions of their position in this series" (1910, 218), and "we penetrate no deeper into the absolute being of bodies by this means; but we grasp the rules of their systematic connection more definitely" (1910, 219).

According to Cassirer (1910, 210), it is "justified and unavoidable that science should condense a wealth of empirical relations into the assumption of a particular thing-like 'bearer.'" Such an assumption is as a necessary condition for achieving certain purposes of knowledge. This "thing-like bearer" is not a substance, but rather it displays the unitary function within the ongoing edification of the knowledge at stake. So the coexistence of the two aspects of elements stressed by Mendeleev, and which has then been considered as a "peculiar dual epistemological status" by Paneth (1962, 137–138), can be explained otherwise. Following Cassirer's line of reasoning, this double aspect of elements is the result of "two tendencies of thought"—as far as we are concerned, we prefer to refer to two modes of research practice rather than to "tendencies" of thought—which can never be directly united: the conditions of scientific production and that of critical reflection, in reference to Critical Philosophy. We cannot use *functions* to account for what we call the empirical reality and *at the same time* consider and describe them. The peculiar character of the two aspects of the elements rests on the tension and opposition remaining between these two standpoints. In the light of this, it can be understood that the chemical concept of element also shows a different form, according to the way we approach it. Element may appear as a fixed substantial kernel, a "simple substance," from which different properties can be successively ascribed to chemical bodies sharing the same atomic weight (Mendeleev)—or the same atomic number (Paneth)—, while, conversely, from the standpoint of the critique of knowledge, those "common properties" and their mutual relations *form* the real empirical data and *play a regulative function* in the construction of the periodic table. In this latter case, that of Critical Philosophy, the concept of "basic element," to use Paneth's turn of phrase, would be a "mere idea"—*focus imaginarius*—, in the strict meaning Kant gave this term in the *Critique of the Pure Reason* (1781, A644–46, B672–74), that is to say an idea which possesses an indispensably regulative use and in virtue of which our various concepts and empirical generalizations gain a collective unity. Following this line of thought, Cassirer concluded that:

> The chemical constitution-formula at first seems to offer a direct intuitive picture of the serial order and position of the atoms among themselves; but what it finally achieves is not such a knowledge of the ultimate, absolute elements of reality, but rather a general analysis of the bodies and materials of experience. The formula of a definite compound does not teach us to know it merely in its composition, but inserts it into various typical series, and thus refers to the totality of such structures, that can arise by substitution out of a given combination. The individual member becomes the representative of the whole group to which it belongs, and it can issue from the group by variation according to law of certain fundamental parts. Since the constitution-formula represents this connection, this formula is indeed the real scientific expression of the empirical reality of the body; for it means nothing else than the thorough-going objective connection, in which an individual "thing" or particular event stands with the totality of real and possible experiences. (Cassirer 1910, 214)

We will not enter into more details concerning the Neo-Kantian interpretation developed by Cassirer, nor will we discuss whether or not we are in agreement with him or the extent to which we could agree with him. That is not really the point here. We believe the reference to Cassirer is of interest in that it highlights the functional role of the elements in the open-ended development of chemical knowledge, and because it enables us to understand further the mutual dependence of, or at least the connection between, the two "aspects" of the word *element* in chemical discourses. As Hans-Jörg Rheinberger states (2009, 83–84): "The productivity of a complex research endeavor depends on its capacity for orchestrating a polyphonic texture of experimental operations within which the contingent, the unthought-of, the unprecedented can take on meaning." The functional role of the chemical elements in the periodic table first, and then in chemistry at large, is the result of a history, that is of an ongoing process. The more chemists learn about elements, the more important their functional role is. This functional role is growing along with the alteration of the ways in which chemists produce, compare, identify, and individuate chemical substances. This growing functional role is both the result of the historicity and the contingency of chemists' inquiry into material substances, and of the material substances themselves. The element connects theoretical and experimental practices, and what they mean is continuously reconstructed (van Brakel 2000a). It is precisely what Ursula Klein's notion of "historical ontology" of materials is about (Klein 2008). The experimental production and individuation of substances, and the way their chemical differences and similarities are scrutinized and ordered by Mendeleev and other chemists take part constitutively in the very definition and understanding of what elements are at a particular period of the history of chemistry.

In line with Rheinberger (1997) and Knorr-Cetina (1999), we would like to assert that in order for us to understand the connection between the two aspects of the chemical element, and to understand its functional role in the making of chemical knowledge as well, we have to scrutinize the "epistemic cultures" in which chemists previously worked or are still working, and to take into account the "epistemic history" of the chemical elements. Hasok Chang is typically doing this kind of work, and especially when he introduces the notion of "epistemic iteration" (Chang 2004; 2012; 2016). Chang refers to mathematical accounts of iteration, in which successive approximations build on preceding ones, and suggests that scientific research frequently displays a similar process, "in which successive stages of knowledge, each building on the preceding one, are created in order to enhance the achievement of certain epistemic goals. . . .The whole chain exhibits innovative progress within a continuous tradition" (Chang 2004, 226). Chang is perfectly aware of the fact that the iterative process in mathematics is used to approach a correct answer that can be verified in other ways, and that is not the case in chemistry for instance. Nevertheless, he emphasizes that scientists do have a number of criteria that they can use in order for them to deem whether an iterative process is progressive or not. They can consider whether successive changes to their system of knowledge increases its accuracy, consistency, scope, efficacy, unifying power, explanatory power, and testability, among many other criteria of scientific assessment. Chang's understanding of epistemic iteration in science is based on a coherentist approach of epistemic justification in which scientists move from a tentative starting point to an improved provisional ending point (Bonjour 1985).

As we have seen, most strategies of classification after 1850 focused their interest on local arithmetic correlations to the detriment of global analogies between chemical properties, and failed in classifying elements. Mendeleev perfectly knew what his colleagues tried to do before him, and thus chose to carry out his procedure by taking the average of the sum of the values of the four elements flanking the element under study. As a result, he accurately calculated the atomic weight of the element selenium and had been able to predict the existence of other elements. The way elements were discussed in the Congress of Karlsruhe, the epistemic cultures of his time, his reject of Prout's hypothesis, and his use of two interdependent aspects of the notion of elements partly explain why he succeeded in achieving a classification whereas others failed to do so. Before Knorr-Cetina, Rheinberger, Klein, and Chang, the French philosopher Gaston Bachelard ([1938] 2002) already proposed a dynamic understanding of the way science transforms itself through time; approach in which the conditions of application of a concept have to be incorporated into the very meaning of the concept in order to include its historical evolution. Bachelard stated that ([1938] 2002, [61] 69):

> we need to create a new word here, between intension and extension, in order to refer to this activity of inventive empirical thought. This word would have to be given a very particular dynamic sense. Indeed, in our view the richness of a scientific concept is measured in terms of its power of deformation. This richness cannot be attached to an isolated phenomenon that would be regarded as growing increasingly rich in characteristics, and therefore ever richer in intension. Nor can this richness be attached to a collection that would bring together the most heterogeneous phenomena and extend in a contingent way to new cases. An intermediate meaning will be achieved if enrichment in extension becomes necessary, and as co-ordinated as richness in intension. In order to include new experimental proofs, we must then deform our initial concepts, examine these concepts' conditions of application, and above all incorporate a concept's conditions of application into the very meaning of the concept....The traditional division that separated a theory from its application was unaware of this need to incorporate the conditions of application into the very essence of the theory.

According to us, the integration of the conditions of application into the understanding of the notion of elements is strongly relevant for analyzing the functional role of such elements because elements cannot be detached from the operations used by chemists to explore material substances. To conclude, we would like to widen the scope of this paper, and to insist on the fact that the functional role of chemical elements is more than ever important considering the current evolution of chemistry, and the possibility of developing different types of mereology for chemistry.

Perspectives: Chemical Heterogeneity, Elements, and Mereology

The material production and individuation of chemicals has enormously been expanded in current nanochemistry, solid-state chemistry, and materials science. New instrumentation and chemical devices enable chemists to explore temporal and spatial scales that have been completely unreachable until now. Chemists have gained an enlarged capacity to synthesize, scrutinize, and modify particle size and distribution; agglomeration state, shape, and crystal structure; chemical composition; surface area; surface chemistry; surface charge; porosity; and interfaces. A "science of particulars" arises and chemists are now able to generate and study multifarious details at the nanoscale (Llored 2013). Nonstoichiometric compounds are now legion. Chemists even contrive to combine organic and inorganic ingredients into the same hybrid body; thus, holding together types of chemistry that have always been incompatible hitherto. The heterogeneity that chemists are now able to produce is unprecedented. The French philosopher François Dagognet (1989, 166, my translation) once stated: "We must accept the continuous renewal and inexhaustible richness of what is extended."

This diversity is not solely a question of ingredients, quantities, and structure. It also depends on the devices and the instruments involved. As a matter of fact, the structure of a nanocompound can be different, depending on what surrounds it, the device, and the time employed for its precipitation. Though the compounds that are obtained from two different approaches can share the same chemical formula, they can be different bodies. Their composition is the same, but the environment, the procedure, and the time involved in the production of the compound have influenced the relatedness of their parts. Thus, the whole compound, its parts, and their environment are intertwined within a process. The mode of operation cannot be eliminated from the final product, because this mode determines the whole and its correlative parts. The structure of the crystals may also differ if the chemical device changes, and those crystals can even differ within the same particular chemical device, depending on their size, which itself depends on the environment. Therefore, for the first time in their history, chemists have to hold the composition, the structure, the parts, the whole compound, the environment, and the device together within the same coherent explanation. Chemical bodies, and especially nanocompounds, are context-sensitive, and the ways they act upon the world always depend on contexts. Chemical bodies are mutually defined within a network including operations, instruments, transformations, and other purified bodies: they are not simply predictable by considering the body in isolation! Behind the same formula, for instance, ZnO, different types of nanocrystals having different reactivity, structure, and (eco) toxicity may exist depending on acidity of the solvent in which the precipitation is carried out (Aimable et al. 2010). Things turn out to be far more complicated than before.

In the same vein, quantum chemists have contrived specific methods that enable them to propose more details about the evolution of the electron density inside a molecule. For instance, the Quantum Theory of Atoms in Molecules (QTAIM) uses the notion of "topological atoms" for describing the constituent of molecules; such atoms are non-overlapping entities leaving no gaps between them (Llored 2014). Paul Popelier, a prominent chemist of the field, has pointed out that each topological atom is endowed with properties it inherits from the molecule of which it is a part, and reflects the features of its particular chemical environment inside the molecule (Popelier 2000). Popelier uses the expression "molecular atom" in order to express this heterogeneity inside a molecule. He writes (2000, 49):

> There are literally many millions of molecular atoms because there are millions of molecules which all give rise to a set of constituent atoms. Nevertheless the sometimes bewildering shapes of atoms have been criticized as being contrary to chemical intuition. This should not be disconcerting, rather it could be interpreted as an expression of the richness of chemistry. Indeed, the amazing variety of atoms is a result of quantum systems cutting themselves into fragments, each leaving behind on the fragments detailed fingerprints of the total molecule. Is it possible, then, to find exactly the same atom more than once coming from different molecules?

Popelier then shows that it is not possible to cut an atom from one molecule and insert it exactly into a corresponding cavity of another molecule, that is, it is impossible to transfer perfectly an atom from one molecule to another—even if such transferability remains possible to some extent and within certain limits. For instance, supported by the faith that a model reveals the correct degrees of transferabilities, this information can be used to set up a library of atoms, enabling a rapid and accurate construction of large sets of atoms, such as proteins. Transferability of atoms is, according to Popelier's turn of phrase, "an unattainable limit," that is a heuristic guide for carrying out research programs only (Harré and Llored 2018; 2019).

Proliferation of both non-stoichiometric compounds and organic-inorganic hybrids; context-sensitivity of nanobodies; size-dependence of structures; molecular landscapes; topological atoms; and, above all, "molecular atoms"; partial transferability of topological atoms from a molecule to another are among many other examples to which we could have referred: the whole sets of concepts, instrumentations, methods, and of empirical generalizations that have been used by chemists so far are transformed in-depth. The ways chemists produce, individualize, develop models of, and analyze chemicals is deeply changing. The "historical ontology" of materials and molecules is changing, and with it the content and the meaning of classical concepts. Philosophers should scrutinize how concepts as atoms, molecules, and elements will be "deformed," to use Bachelard's turn of phrase, in order to study the current evolution of chemistry. At first glance, the functional role of concepts such as that of element will play an active role in the rationalization and the ordering of this unprecedented growing heterogeneity. But would it be useful to chemists to keep on referring to elements in the same way they previously did in order for them to address the increase of all those more and more individualized particulars? Will the functional role of elements be the same in order to answer questions raised by "molecular atoms"?

Last but not least, the aforementioned context dependence and size sensitiveness bring to the fore the necessity for philosophers to propose a relational concept of structure in which the wholes, the parts, and the surroundings are required, and not the whole and its parts only (Harré and Llored 2011; 2018; 2019). Which version of mereology should we use to express the context dependence of the structures of chemicals? Is a molecule a thing of which its parts are also things? A structured collective? Or is a molecule a set from which its constituents are subsets? The former would require the classical mereology of Lesniewski, the C-mereology in its functional form. The latter would require one to make sense of the mereologized set theory of Lewis, the S-mereology, in chemical contexts (Harré and Llored 2011). In that mereology there is no provision for distinguishing parts functionally with respect to their roles in the wholes of which they are parts. Is the grammar of chemistry classical mereology or mereologized set theory? How are we to analyze this alternative in the light of the recent ideas mentioned above?

The case for adopting set theory as "the mereology" for chemistry begins with the predictions by Odling and Mendeleev of the properties of elements yet to be discovered. At the time of their proposals only the intensions of the set of atoms of *eka*-iodine were present in chemical discourse. The set had a null extension for the users of the grammar appropriate to the situation as it then stood, since the set had no members, and might conceivably never have any. Obviously there cannot be a fusion or a sum of which there are no parts. To talk of *eka*-iodine in the grammar of classical mereology made no sense. It does seem to make sense in a discourse in which the parts of sets are subsets (Harré and Llored 2011).

Are hydrogen and oxygen atoms (ions) subsets of the water molecule set? Each water molecule would be a subset of the superset, the stuff water. However, what is the intension of the set of which two sets, a pair of hydrogen atoms and a singleton oxygen atom, are the subsets? The hydrogen atoms are members of the set of all hydrogen atoms, while the oxygen atom is a member of the set of all oxygen atoms. Does this have any advantage over the classical mereological grammar?

The first argument for the C-mereology depends on the possibility of a whole having emergent properties as a result of some structural invariants. A set only accidentally has structural properties because it is a conceptual object. A whole has structural properties because it is a material entity, with real relations between its parts.

The second argument for C-mereology depends on the criteria for class membership that is the intensionality component of the set concept. If H^+ and O^{2-} are subsets of the water molecule set what is their common property that makes them members of this set? It can only be that they are constituents of a water molecule. Hence the S-mereology treatment of chemical unity in multiplicity depends on a C-mereological understanding of the relation between atoms (ions) and the molecules of which they are parts (Harré and Llored 2011). The two types of mereology are thus mutually dependent. Understanding the relationships between atoms and the molecules to which they are parts requires a functional concept of element, one that enables chemists to take into account all the possibilities, that is, atoms, ions, isotopes, and maybe the new types of atoms defined by quantum chemists. Further investigations are thus required regarding the functional role of the concept of element in order for philosophers to analyze the kind(s) of mereology with which they will address the specificity of issue of chemical compounds.

References

Aimable, A. et al. 2010. "Assisted Precipitation of ZnO Nanoparticles with Narrow Particle Size Distribution." *Journal of the European Ceramic Society* 30: 591–598.

Bachelard, Gaston. (1938) 2002. *The Formation of the Scientific Mind.* Translated by M. McAllester Jones. Bolton: Clinamen. Originally published as *La formation de l'esprit scientifique.* Paris: Vrin.

Bächtold, M. 2008. "Interpreting Quantum Mechanics according to a Pragmatist Approach." *Foundations of Physics* 38: 843–868.

Bensaude-Vincent, B. 1986. "Mendeleev's Periodic System of the Elements." *British Journal for the History of Science* 19, no. 1:3–17.

Bensaude-Vincent, B. 1989. "Mendeleïev : Histoire d'une découverte." In *Eléments d'histoire des sciences,* edited by Michel Serres, 447–468. Paris: Bordas.

Bensaude-Vincent, B., and J. Simon. 2008. *Chemistry: The Impure Science.* London: Imperial College Press.

Bensaude-Vincent, B., and I. Stengers. (1993) 1996. *A History of Chemistry.* Cambridge, MA: Harvard University Press. First published in French.

Bonjour, Lawrence. 1985. *The Structure of Empirical Knowledge.* Cambridge, MA: Harvard University Press.

Cassirer, E. 1910. *Substance and Function and Einstein's Theory of Relativity.* Translated by W. Swabey and M. Sawbey. Chicago: The Open Court Publishing Company.

Chang, H. 2004. *Inventing Temperature: Measurement and Scientific Progress.* New York: Oxford University Press.

Chang, Hasok. 2012. *Is Water H2O? Evidence, Realism and Pluralism.* Dordrecht: Springer.

Chang, Hasok. 2016. "The Rising of Chemical Natural Kinds through Epistemic Iteration." In *Natural Kinds and Classification in Scientific Practice*, edited by Catherine Kendig, 33–46. New York: Routledge.

Dagognet, François. 1989. *Rematérialiser.* Paris: Vrin.

Dewey, John. (1938) 1993. *Logic: The Theory of Inquiry.* New York: Henry Holt and Company.

Ghibaudi, Elena, Alberto Regis, and Ezio Roletto. 2013. "What Do Chemists Mean When They Talk about Elements?" *Journal of Chemical Education* 90: 1626–1631.

Harré, Rom, and Jean-Pierre Llored. 2011. "Mereologies as the Grammars of Chemical Discourses." *Foundations of Chemistry* 13: 63–76.

Harré, Rom, and Jean-Pierre Llored. 2018. "Products, Procedures, and Pictures." *Philosophy* 93: 167–186.

Harré, Rom, and Jean-Pierre Llored, 2019. *The Analysis of Practices.* Newcastle upon Tyne: Cambridge Scholars Publishing.

Hoffmann, Roald. 1995. *The Same and Not the Same.* New York: Columbia University Press.

Holmes, Frederic. 1989. *Eighteenth-century Chemistry as an Investigative Enterprise.* Berkeley: University of California Press.

IUPAC. 1997. *Compendium of Chemical Terminology*, 2nd ed. (the "Gold Book"). Compiled by A. D. McNaught and A. Wilkinson. Oxford: Blackwell Scientific Publications.

James, William. (1907) 1995. *Pragmatism.* New York: Dover.

James, William. (1909) 1997. *The Meaning of Truth.* New York: Prometheus Books

Klein, Ursula. 2008. "A Historical Ontology of Material Substances, C. 1700–1830." In *Stuff: The Nature of Chemical Substance*, edited by Jaap van Brakel and Klaus Ruthenberg, 21–44. Würzburg: Königshausen & Neumann.

Knorr-Cetina, K. D. 1999. *Epistemic Cultures: How the Sciences Make Knowledge.* Cambridge, MA: Harvard University Press.

Llored, Jean-Pierre. 2013. *Chimie, chimie quantique et concept d'émergence: Étude d'une mise en relation.* PhD diss., Free University of Brussels.

Llored, Jean-Pierre. 2014. "Wholes and Parts in Quantum Chemistry: Some Mereological and Philosophical Consequences." *HYLE International Journal for the Philosophy of Chemistry* 20: 141–163.

Llored, Jean-Pierre, and Michel Bitbol. 2013. "From Chemical Practices to a Relational Philosophy of Chemistry." In *The Philosophy of Chemistry: Practices, Methodologies, and Concepts.* Edited by Jean-Pierre Llored, 385–415. Newcastle upon Tyne: Cambridge Scholars Publishing.

Mendeleev, Dimitri. I. 1889. "The Periodic Law of the Chemical Elements." *Journal of the Chemical Society* 55: 634–656.

Mendeleev, Dimitri. I. 1891. *The Principles of Chemistry*, vol 1. London: Longmans, Green and Co. First English edition from the fifth Russian edition.

Nye, M. J. 1984. *The Question of the Atom: From the Karlsruhe Congress to the First Solvay Conference, 1860–1911.* Los Angeles: Tomash.

Nye, M. J. 1993. *From Chemical Philosophy to Theoretical Chemistry: Dynamics of Matter and Dynamics of Disciplines, 1800–1950*. Berkeley: University of California Press.

Paneth, Friedrich. (1962) 2003. "The Epistemological Status of the Chemical Concept of Element." *Foundations of Chemistry* 5, no. 2: 113–145.

Peirce, C. S. 1931–1958. *The Collected Papers of C. S. Peirce*, vol. 2. Edited by C. Hartshorne and P. Weiss. Cambridge, MA: Harvard University Press.

Popelier, Paul. 2000. *Atoms in Molecules: An Introduction*. London: Prentice-Hall.

Rheinberger, H.-J. 1997. *Towards a History of Epistemic Things: Synthesizing Proteins in the Test Tube*. Stanford, CA: Stanford University Press.

Rheinberger, H.-J. 2009. "Experimental Reorientations." In *Going Amiss in Experimental Research*, Boston Studies in the Philosophy of Science, volume 267. Edited by G. Hon, J. Schickore, and F. Steinle, 75–90. Dordrecht: Springer.

Scerri, Eric. 2001. "The Periodic Table: The Ultimate Paper Tool in Chemistry." In *Tools and Modes of Representation in the Laboratory Sciences*, , vol. 222. Edited by Ursula Klein, 163–177. Boston Studies in the Philosophy of Science. Dordrecht: Kluwer Academic Press.

Scerri, Eric. 2005. "Some Aspects of the Metaphysics of Chemistry and the Nature of the Elements." *Hyle: International Journal for Philosophy of Chemistry* 11, no. 2: 127–145.

Scerri, Eric. 2007. *The Periodic Table: Its Story and Its Significance*. New York: Oxford University Press.

Scerri, Eric. 2012. "What Is an Element? What Is the Periodic Table? And What Does Quantum Mechanics Contribute to the Question?" *Foundations of Chemistry* 14: 69–81.

Timmermans, Jean. (1928) 1941. *Species in Chemistry*. London: Macmillan and Co. Translated by R. E. Oesper of the revised version *La notion d'espèce en chimie*. Paris: Gauthier-Villars & Cie.

Van Brakel, Jaap. 2000a. *Philosophy of Chemistry: Between the Manifest and the Scientific Image*. Louvain: Leuven University Press.

Van Brakel, Jaap. 2000b. "The Nature of Chemical Substances." In *Of Minds and Molecules*, edited by N. Bhusham and S. Rosenfeld, 162–184. Oxford: Oxford University Press.

Van Brakel, Jaap. 2008. "Pure Chemical Substances." In *Stuff: The Nature of Chemical Substances*, edited by J. van Brakel and K. Ruthenberg, 145–161. Würzburg: Königshausen & Neumann.

Van Brakel, Jaap. 2012. "Substances: The Ontology of Chemistry." In *Philosophy of Chemistry*, edited by Robin Findlay Hendry, Paul Needham, and Andrea Woody, 171–209. Vol. 6 of *Handbook of the Philosophy of Science*, Boston: Elsevier.

Wiggins, David. 2001. *Sameness and Substance Renewed*, 2nd ed. Cambridge: Cambridge University Press.

Wittgenstein, Ludwig. 1953. *Philosophical Investigations*. Translated by. G. E. M. Anscombe. Oxford: Blackwell.

11

Making Elements

Klaus Ruthenberg, Coburg University of Applied Sciences and Arts, Germany

1. Introduction

What we call a chemical element is, at any time, a convention. Obviously, these conventions can change. While Empedocles, Plato, and Aristotle, to mention the inventors of the four-element concept for example, decided to use a priori (and abstract) principles to characterize their elements air, water, fire, and earth, the "official" element definition of the twentieth century has, at least in the first place, a strong a posteriori character. If we consider the atomic number 79 as the decisive property (or essence) of gold, for example, we know (or presume) that someone reliable determined that measure or is at least able to do this. In everyday chemical practice, however, atomic numbers play only a minor role because usually no chemist really works with atoms and only a few chemists actually determine atomic numbers (except some specialists, as for example mass spectrometrists or X-ray fluorescence analysts).[1] Hence, it is even questionable whether atomic numbers should count as chemical properties. Gold, for example, is characterized chemically by empirically obtained knowledge about samples of that yellow, dense, and noble metal. These aspects are relational. They represent only a few out of a long list of possible items, and though not one of these descriptions is essential or all-embracing, the group of them makes the deal. We can suspect that those people who are dealing with samples of gold—experts or laymen—do know its character already well before the atomic number is determined. The differentiation between gold and other metals and even its purity could be inferred from tests without any reference to microphysical structure already very early (see, e.g., Bodemann 1845). In other words, we know about the nature of the chemical elements well before we have developed a sophisticated theory because of our actions and sensual interactions, manipulations, and experiments with material samples.

---

[1] Note that even in such cases an often very elaborate preparation of real stuff samples is inevitable. These preparation steps are usually suppressed in chemistry when it comes to the theoretical interpretation, but they should not be suppressed in the history and philosophy of chemistry.

Klaus Ruthenberg, *Making Elements* In: *What Is A Chemical Element?*. Edited by: Eric R. Scerri and Elena Ghibaudi (2020). © Oxford University Press.
DOI: 10.1093/oso/9780190933784.003.0012

This chapter is explicitly devoted to the empirical part of the description of chemical elements. I am taking seriously what chemists are really doing when they explore and identify chemical elements instead of discussing ready-made facts from atomic physics. I shall try to avoid prejudices of modernist or physicalist kinds. In order to do this, I divide this chapter into four sections and start with a short narrative about the former element, water. Second, I will criticize the dogma of the autarchy of atomistic compositionism in the history and philosophy of chemistry, referring to Wald and Ostwald. In section 4, I shall discuss the applicability of the expression "metachemistry," and in section 5 I will discuss the sometimes over-interpreted origin of the modern definition of chemical elements.

2. Water as an Element

Water, in the sense of river water, rain, drinking water, snow, and what have you, is without any doubt the most often addressed substance in human history. For many years now, several philosophers (of chemistry) accordingly have discussed its nature. In the twentieth century, the core of these debates circles around the question "Is water H_2O?" that is the controversy between essentialists and descriptivists (see, for instance, van Brakel 2000; Hendry 2006; Needham 2002; 2017; Ruthenberg and van Brakel 2008, and further references there).

In European antiquity, natural philosophers derived their concept of elements by abstracting principles from everyday experience. "Water" becomes the principle of liquidity; "earth" the principle of solidity, stability, and ruggedness; "air" what we would call the gaseous state; and "fire" was transformed into an active substantial entity capable of causing dramatic changes (something we today would call energy). Hence, we can call the a priori basis of the four-element concept a kind of principlism with respect to elements (in contrast to the later compositionism). There is no place for pure elements in this concept, and it can even go so far that—for example, in Plato's Timaeus—the principle of liquidity is applied to metals (most of which are solid at room temperature) because these are meltable.

Table 11.1 shows a rough comparison of main aspects of the four-element theory with Empedocles, Plato, and Aristotle.[2] Empedocles, who seems to be the first to use the concepts of four elements to describe natural phenomena, uses gods to characterize the elements, and "love" and "hate" to address their behavior. Plato describes them in a speculative atomistic manner using the regular

---

[2] I adapt this table from my Ruthenberg 2005.

Table 11.1 Comparison of the descriptions of the four elements in antiquity.

	Water	Fire	Air	Earth
Aristotle	cold/moist	warm/dry, heat	warm/moist	cold/dry, solid rock, salt, clay
Plato	icosahedron liquid meltable	tetrahedron warm light, flame, ember	octahedron light, ether, fog	cube solid, hard rock, salt, clay
Empedocles	Nestis rain, sea	Zeus (Hephaistos) bright, warm, sun	Aidoneus/Hera ether, sky	Hera/Aidoneus solid, durable, bones

polyhedra together with a bunch of quotidian experiences, among them melt-ability, combustability, evaporation, hardness, and density. Aristotle prefers pairs of the sensible qualities cold, warm, dry, and moist to characterize elements and abandons any atomistic interpretation.[3]

The "making of elements" with respect to this tradition from antiquity is a rational act: although quotidian experiences are the starting point of the reflection, there is no purpose to prove hypotheses empirically. Operational aspects remain rather superficial; only Plato offers a kind of explicit explanatory frame for stuff behavior. According to that speculative conceptional frame with roots in the Pythagorean thinking, there are two reaction mechanisms, the cutting and the breaking mechanism. Plato explains, for example, the impact of fire on water during evaporation by letting the sharp edges of the tetrahedral elementary particles rip the less sharp icosahedral elementary water particles such that air particles are the result. An example for the breaking mechanism would be the choking of an open fire by an excess of sand.

Nevertheless, all the mentioned drawbacks—at least from a modern perspective—should not entrap us to reject the whole approach of four elements as being wrong without further thought only because we are living in (or connected to) a world theoretically dominated by microphysical descriptions. First, it is still worthwhile to investigate the reasons for the extraordinary historical stability of the four-element concept in more detail, and second, there is enough evidence to claim that this historical stability reaches well into the present. Although scientists do not call water an element anymore there are still vivid traces of the fundamentality of the four-element concept, as, for example,

---

[3] Note that a closer look on the explanatory approaches of substance behavior reveals that Plato is not less comprehensive than Aristotle.

the concept of the state of aggregation. The purpose of this short section is indeed not to convince the reader of the actual equivalency of an earlier type of element definition to the modern one, but to emphasize the fact that our description of nature is conventional. As far as I am concerned, water, like at least air and earth as well, was legitimately denoted to be an element. Of course, these four elements are not elements in the sense Lavoisier or the IUPAC gave the entities of the same name. As mentioned above, they rather are a priori material principles to describe the very fundamentals of the material world. I take a quite simple quotidian conception of stuff as starting point and central concept that allows me to discuss a variety of different elemental descriptions from different historical horizons.[4] Hence, it is just a different perspective that separates the old from the modern approach, and not the question of inferiority or superiority with respect to a prevailing paradigm.[5] I will try to illustrate the relevance of the more empirical part of chemistry in the following section.

3. How to Begin Chemistry?

3.1 Another Take on Twin Earth

Surprisingly, it has been neither a chemist nor a philosopher of chemistry who introduced this metaphor into the general discussion about the naming and characterization of substances, which in certain circles went really far—some would say astray. Hilary Putnam (1926–2016), the influential American philosopher of mind, invented the thought experiment of two planets—identical except the microstructural composition of what we call water—to illustrate his idea of semantic externalism (Putnam 1975). He summarized his main conclusion with the claim that "meanings are not in the head," which was considered to be a convincing support of naturalism (or realism) by many analytical philosophers. Such approaches take microphysical descriptions for granted and

---

[4] With reference to an earlier version of the present text, a reviewer claims that "the four element doctrine is not comparable to modern science as—among other things—it claimed to fix the elements' number a priori." Of course, we can consider the (in part) operational step of Lavoisier as revolutionary progress on the one hand, but we must not forget how much (chemical) knowledge has been gained in the framework of the old doctrine, on the other. I do not claim that there is only one possible usage or meaning of "element." One main aspect of the earlier doctrines (I only refer to the European traditions here) was obviously not to find the ultimate building blocks of our substantial world, but to understand certain behavior of its inventory. Perhaps this simplified characterization may serve as a reminder of what the different element doctrines have in common.

[5] The prevailing paradigm in modern chemistry goes by the name of atomism. During the last 150 years, this paradigm led most chemists and some philosophers of chemistry to believe that chemistry is the science of atomic permutations. In contrast, I still prefer to define chemistry as the science of substance behavior.

Table 11.2 Finite Sequence of Components of the Normal Form Description of the Meaning of a Word—Here *Water*—after Putnam (1975, 269)

Syntactic Markers	Semantic Markers	Stereotype	Extension
Mass noun, concrete	natural kind, liquid	colorless, transparent, tasteless, thirst-quenching, etc.	H_2O (give or take impurities)

as most fundamental without discussion. Putnam proposes what he calls the normal form of description of the meaning of a word and suggests the above table of the most important components of this description for the example of water (Putnam 1975, 269).

Table 11.2 should be read as a sequence (or "vector") from the left to the right and refers to the competence of the actual speaker, except the extension, which represents the scientific essence. Putnam does not exclude other possible entries to that list, but he neglects other kinds of designators—rigid or not—like thermodynamic ones (melting, boiling, and triple points, etc.) and reaction behavior (the functions as solvent and reaction partner, carrier of acids/bases, etc.). The expression "H_2O" is mistaken—not only by Putnam—if it is considered as a microstructural statement. Philosophers like Joachim Schummer (e.g., 1996, 54–55), Avrum Stroll (1998, 37–73), and Jaap van Brakel (e.g., 2005) rightly criticize that microstructuralist monism,[6] but that criticism has had—so far—no visible impact on the otherwise esoteric debates in some closed circles of analytical philosophy.

Van Brakel is very clear about the Twin Earth debates:

Perhaps the most disturbing fact about this literature is the insistence of sticking with an example that simply makes no sense at all. (van Brakel 2005, 22)

The picture of Putnam's Twin Earth is odd with respect to at least two points. First, the assumed microphysical priority is questionable. It is an old materialist belief—we can call it the strong reductionist claim—that the submicroscopic world determines the macroscopic, but how? This is a claim that is meant for

---

[6] Stroll is mainly interested in the theory of meaning and thus somewhat neutral with respect to the essentialisms of Putnam and Kripke and even to the Twin Earth idea. Particularly Schummer and van Brakel, however, reject uncritical microstructural essentialism. Addressing the claim "water is H_2O" in a footnote to the passage mentioned here, Schummer says: "If philosophers of language or mind-body-theoreticians enthusiastically refer to this example, this might be forgiven. Of a philosopher of science, however, we should expect the minimal chemical knowledge that the expression "H_2O" is an empirical formula, which is the result of an elementary analysis." Some of the main aspects of Jaap van Brakel's point of view should become clear in the following.

a perfect future physics that is capable of explaining and predicting everything inferred from the knowledge about the ultimate tiny building stones. And the second point, which is even more pertinent for our present topic, is that there cannot be a water that has exactly the same manifest properties and capabilities, but at the same time a totally different composition (which Putnam called "XYZ"). The identification of chemical individuals successfully follows exactly the method that if a few properties are the same, all other properties are identical, too (see the section on Ostwald). Hence, the search for essences is not necessary on the one hand, and not the only one performed in the real world, on the other. Nevertheless, essentialists call themselves realists, neglecting the relevance of all the parts of human activity apart from the most recent scientific theories—like, for example, the entries to the three columns on the left of the extension column in table 11.1. Van Brakel writes:

> The problem of being dependent on the (contingent) context arises for all natural kind properties, both "manifest" and "underlying." The marking of boundaries at the microscopic or "real essence" level is just as natural or conventional as the marking of boundaries at the macroscopic or "phenomenal" level. (van Brakel 2005, 72)

Hence, there is no good reason to talk of natural kinds without accepting at least some human actions, among which there are conventional decisions.

In addition to the previous discussion, I want to add the picture of two remote material worlds or planets in a different way to illustrate another point. I borrow that idea from the Czech chemist Frantisek Wald (1861–1930) who used it to describe the motivation for his theoretical approach. In his book *Chemistry of Phases* (Chemie Fasí) from 1918 he wrote the following:

> Already over twenty years ago I baited some chemists by asking the question what they would do if they for example would be sent to the moon by Jules Verne, where there would be no factories for chemical reagents, as well as no single ready chemical preparation carrying a chemical sign, yes, perhaps even not one substance familiar to us from the life on earth. (Wald 2004, 110; my translation)

Wald calls the contemporary chemistry "synthetic," because it explains substances as combinations of elements, which are "preexisting." Where these elements come from (the practical and operational background) does not count as proper part of their theories. Just as in the current methods of analytical chemistry, in which there is a tendency to underestimate sample preparation and sample pre-treatment and to overestimate the very measurement steps, chemical narratives usually are told from an upside-down perspective. Wald, contrasting that customary perspective, just asks, "How to begin chemistry?" by taking the moon-journey thought experiment seriously. In the first attempt of his own theoretical approach, which he calls "analytic," he applies the phase rule of Gibbs to reconstruct the stoichiometric laws. In his final theoretical attempt, he consequently skips the phase rule because it presupposes pure substances and tries to start from phases. The latter he defines as "any part in a mixture in equilibrium that is recognised by the senses as a homogeneous part, differentiable from the others" (Wald 2004, 112; my translation).[7]

When Jules Verne published *De la Terre à la Lune* in 1865, it was indeed open to speculation whether one would find the same inventory of substances as on earth. It is obvious—though neglected in large parts of philosophy of chemistry—that chemical substances are preparations. Even the knowledge about what we call elements since Lavoisier can only be obtained by empirical work. Hence, to find out the nature of the material constituents of the moon (or Twin-Earth), the traveling researchers[8] would have to perform a bunch of systematic manipulations. These systematic manipulations, however, are a genuine part of chemistry. Although questions like, for example, "How do we know about oxygen?" are usually confined to class-rooms, they are fundamental for the history and philosophy of chemistry as well. A look at the periodic table is obviously not enough to answer this question.

The "making" of oxygen—to take up this example—is quite easy to describe. If we send electric current through a sample of water, two different gases occur at the electrodes (in a constant volume ratio). Once isolated, one of these gases might let a candle flame burn brighter, react with the other gas to form water, react also very lively with other stuff, and so on. Without the primary preparatory step, we would know much less about that Lavoisierian element. No doubt, however, that there are many more things to learn after its first manufacture (here by electrolysis). Among these, there are its boiling point, the atomic structure, the fact that oxygen comes in molecular pieces at usual temperature, all the imaginable spectroscopic data we can obtain, and so forth. We might systematize some of these facts. We can, for example, ascribe the ordinal number of 6 to oxygen and put it into main group 16 in the periodic table, but none of these pieces of knowledge alone can claim the status of an essence. What is natural here—if there is any need to subscribe to naturalism[9] at all—is the entire network

---

[7] This definition, by the way, comes close to my characterization of a substance given above.

[8] Note that the communication of inhabitants of both worlds would cause the problem of radical translation, particularly if there would be no possibility to travel.

[9] It seems that Bird and Tobin (2018) presuppose a naturalist stance when they write: "To say that a kind is *natural* is to say that it corresponds to a grouping that reflects the structure of the natural world rather than the interests and actions of human beings." In this chapter, I argue just for the

of knowledge, and how we reach certain knots of that network depends on practical success.[10]

The mentioned network of chemical knowledge forms, of course, the restriction frame for vaulting ambitions like the Twin Earth idea. In such net, we cannot just cut off single knots by will without risking paradox results. The intertwined bits of substantial knowledge of reality form a system of necessary relations. Again, if a substance—on our earth or Twin Earth—has all the manifest properties we know of water, then it can by no means have a different composition other than H_2O.[11] Hence, the first of two premises in the One-World-One-Science (OWOS) principle can be applied here.[12] OWOS is the beliefe that we are, first, living in one world and, second, that we gain the appropriate and true description of that world by (only) one science. I hope it becomes clear another time that with respect to chemistry the second premise does not hold. Chemistry is a paradigm case for a pluralistic natural science. "Manifest and underlying properties . . . are on equal footing" (van Brakel 2005, 72).

3.2 Chemistry without Atoms?

The idea to describe and perform chemistry without using atomistic and elementary concepts is (still) intriguing if not fascinating.[13] František Wald, however, did not succeed in his attempts to establish an "a-priori-chemistry." At about the same time—that is during the last three decades of the long nineteenth century—another complementary attempt was developed that irritated and even offended

---

opposite, because I hold that human interests and actions are inevitable for the description of (the structure of) nature. For a critical discussion of the notion of nature with respect to chemistry, see Janich 1996).

[10] Note that the network of "chemical" knowledge is a very old concept. In his Meteorologica IV, for example, Aristotle speaks about different substances like silver and gold and gives the following characterization (which complements what I have described in section 2, by the way): "All these mixed bodies are distinguished from one another, firstly by the qualities special to the various senses, that is, by their capacities of action. (For a thing is white, fragrant, sonant, sweet, hot, cold in virtue of a power of acting on sense.) Secondly by other more characteristic affections which express their aptitude to be affected: I mean, for instance, the aptitude to melt or solidify or bend and so forth, all these qualities, like moist and dry, being passive" (Aristotle 1931, 385a).

[11] Note that the compositional formula H_2O is inferred from a specific perspective. If we would change that perspective, for example if we would include ionic constituents, we would get different results that perhaps would contain information on H^+ and OH^-. We could even ask for polymolecular structures and again the results would be different. H_2O is only one quite abstract description out of a bunch of possibilities.

[12] Intriguingly, André Kukla (2010, 47–62) puts OWOS into the framework of the discussion on extraterrestrials.

[13] Cf. the so far most comprehensive account of Wald's work in van Brakel 2013.

many members of the mainstream scientific community. I shall give a brief critical account of that attempt in this section.

In his latest chemo-theoretical work *Prinzipien der Chemie* from 1907, Wilhelm Ostwald (1853–1932) offers a foundation of chemistry on a phenomenological-operational basis. Although he is sympathetic to it, he does not make the philosophical position of Ernst Mach his own.[14] In contrast to Mach's epistemology, Ostwald presupposes a pan-energetic world. According to his view, scientists investigate into energy-driven events. Instead of claiming otherwise, Ostwald's theory of science is not free of metaphysics. He wishes to discuss the foundational relations of chemical phenomena methodically and systematically without reference to unquestioned theoretical pieces—except the all-embracing energetics dogma. We can read the *Prinzipien* as a pioneering contribution to the philosophy of chemistry.

According to the classificatory definition Ostwald gives in the very beginning of the main text, chemistry is an inorganic natural science, devoted to non-living bodies (*Körper*) of the outer world (Ostwald 1907, 1). Intriguingly, the author neglects the parts of contemporary chemistry that also address (parts of) living objects. Ostwald defines bodies as areas of space which behave in a certain way different than the environment (Ostwald 1907, 1). This typical behavior is conceivable via properties of everyday experience, like color, gloss, shape, and weight. He then says:

> So everyday experience offers me the spatial connection of certain properties, and the sum of these experiences is put together in the notion of *body.* . . . Thus it is a natural law that certain properties stick together such that they cannot be moved from one place to another independently, but always move together. (Ostwald 1907, 1)

The stable coexistence of certain properties of objects—if not the existence of these objects—Ostwald calls a "natural law." He even considers notions (words, names, formulas) to be natural laws at other places in the text. All these natural laws are empirical, though, and Ostwald is well aware of the fact that they are not valid once and forever. The natural laws are signs along the way of explanation of facts or the realization of desired results, and furthermore:

> a natural law thus presents itself as the expectation of a correlation between possible experiences; this expectation is based on the realisation of that correlation in all observed cases. (Ostwald 1907, 4)

---

[14] Note that Wald explicitly referred to Mach, see Ruthenberg 2011.

Thus, falsification of these natural laws is possible, but more or less improbable. They cannot be just discovered in nature; they are assigned to nature in the framework of purposes. Another aspect is the following: The closer a natural law, such as the aforementioned "body law,"[15] is to commonplace experience, the smaller the (for him) unwelcome hypothetization and the favorable ontological or metaphysical economy is.

Ostwald quite classically differentiates typical (*arteigene*) and arbitrary (*willkürliche*) properties and gains an additional way to characterize chemistry and draw a line to physics, because chemistry is engaged in the typical or specific properties, whereas the arbitrary properties are assigned to physics (Ostwald 1907, 6). As examples of arbitrary or physical properties he mentions temperature changes, electrification, illumination with colored light, magnetization, quantity, and outer shape (Ostwald 1907, 6). One problem emerges when these assignments are compared to Ostwald's already mentioned opinion—given later in the book—that every science has its own sort of energy. The questions arise of where we have to localize the chemical energy and which properties we should assign to it, if all arbitrary properties are physical ones. Enthalpies of bonding, dissociation, hydration, combustion, and so forth, are physical properties according to Ostwald's point of view, but of crucial importance in chemistry. From that point of view, the entire research program of physical chemistry is a threat to the autonomy of chemistry, because chemical phenomena are treated with physical means and physical notions. Thus, how to represent typical properties in both that program and in the context of the *Prinzipien* remains unclear. As to his *Grundriss der Naturphilosophie*, written at about the same time as the *Prinzipien*, Ostwald's statements are not convincing, either. Within that book, he claims that even the single elements are independent kinds of energy (Ostwald 1919, 170). This appears to be consistent within the given energetical context, but what if we can only give an appropriate account on chemical substances via the application of physical means? The chemical kind of energy postulated by Ostwald is a mysterious entity.

Methodologically, Ostwald is significantly more convincing. He mentions silver as an example for which a group of the following properties cannot be taken away from this material or altered separately: the solid, typical metallic state, high electrical and thermal conductivity, reasonable chemical stability, and solubility in nitric acid. Then he executes the transition to the general notion of stuff (*Stoff*) that he derives from that of body:

---

[15] I tentatively suggest this term here. Ostwald does not explicitly apply the term *Körpergesetz*.

Bodies which [without respect to quantity and shape] are considered referring to their specific properties are called *stuff*. Thus the subject of chemistry is the stuff. The substances which show the same specific properties are called identical. (Ostwald 1907, 7)

In the last sentence Ostwald indicates the *stuff law*, which he introduces explicitly in the third chapter "Mixtures, Solutions, and Pure Stuff":

If some specific properties of two bodies are identical, then all other specific properties are the same, too. (Ostwald 1907, 74–75, emphasis in original)

For Ostwald, the stuff law is a natural law, too (and in fact one that even today is accepted as valid, although the naming of which has not yet gained currency). From this law an operational prescription can be derived for the identification of substances. If there is agreement over which properties are specific, and if at least two of these specific properties of two stuff samples are identical, these samples belong to the same species. In both the area of scientific education and chemical research this operational prescription was followed even before the time of its formulation (Ostwald 1907, 126) and today as canonical standard (measurement of melting point, boiling point, refraction constant, absorption and emission spectra of various sorts, etc.).

As already mentioned, Ostwald is aware of the preliminary and inductive character of the theoretical comprehension of the results of chemical research. Because there is always the possibility to discover new properties, and at any time new substances composing new properties can be synthesized (Ostwald 1907, 76), chemistry will never come to an end. However, Ostwald sees an opportunity to economize research—an expression first used by Mach—via the application of, for instance, the stuff law, because not all properties have to be measured to characterize or identify substances. As to the characterization or identification of newly synthesized and unknown substances, recourse to known samples is of course not possible. Hence, the application of the stuff law results in a relational comparison when it comes to identification. What happens if—and that is the usual case—one faces an unknown stuff and does not know whether it is pure or mixed? If this stuff is a liquid, which, for instance, contains two miscible components, then the properties are changing with the composition of the mixture in a certain area. If the boiling point is to be measured as a specific property, the composition of the mixture varies during the measurement in most cases. Thus, if mixtures are analyzed the stuff law does not apply without a pretreatment of the samples (Ostwald 1907, 83). We have to ensure that pure stuff is measured. Consequently, Ostwald pays considerable attention to mixtures and their separation techniques in the third chapter.

Substantial ("stuffy") species can exist in several states of aggregate (*Formarten*). Ostwald considers the change of a species from one state to another a chemical process proper:

> Following our general definition we have to consider this transition as a *chemical* event because the one stuff with its properties, a gas, vanishes, and the other, a liquid with new properties, comes into existence. (Ostwald 1907, 95)

During condensation of a gas, new specific properties emerge—Ostwald mentions the surface tension as an example—and because of this the liquid species strictly speaking is different to the gaseous. Hence, ice, liquid water, and water vapor should be regarded as different substances. If the transition to another state takes place at an invariant temperature (melting point, boiling point, other phase transition constants), and the composition of both substances remains the same, then these substances are two forms of the same species (Ostwald 1907, 98). Another natural law is applicable here that Ostwald calls the "law of continuity" (*Stetigkeitsgesetz*) and of which he claims that it is applied (even before in the *Prinzipien*) at various instances without being mentioned explicitly.

> The law of continuity says that those properties of a thing or of an event which stand in interdependency are steady or unsteady at the same time and the same place. (Ostwald 1907, 275)

As to water, for instance, this law means that in the moment of freezing the mechanical properties, specific volume, refraction of light, heat capacity, and all other specific properties take different values irregularly or—like for example solidity and hardness—become observable and definable at all (Ostwald 1907, 276). That there is a defined melting point for a pure stuff, to take one example, stands in close relation to the steadiness law, because the melting point refers to just this sort of specific unsteadiness.

We shall now look into the phase concept that Ostwald propagates as far more general than that of a substance in his Faraday Lecture. How does he treat the notion of phase in the *Prinzipien*? He introduces this notion in the fourth chapter "Transformation of States and Equilibriums," and (perhaps with water in mind) describes compression and cooling of a gas, leading to condensation and solidification. All possible stuffy variants of this species will occur during this operation sometime. After all, a homogeneous solid (*gleichteiliger Körper*) is built out of a homogeneous gas. Ostwald proceeds:

> The components of the mixtures which can be composed by such transformation are called phases. Thus, phases are the homogeneous substances

which occur in mixtures. They can be pure substances or solutions. (Ostwald 1907, 117)

Obviously this formulation is not convincing.

Of course, "components" means the homogeneous stuffy variants rather than single stuffy components. The second sentence of this citation is confusing, too. To describe homogeneous substances as phases, which are components of mixtures, turns upside down the sequence of phase and stuff as claimed by the author back in 1904, as it were. Although there is nothing wrong with the contents of the third sentence, the reader would wish a much clearer characterization of such a central notion in this very textbook.[16] In the following, Ostwald introduces the phase law of Gibbs via the example of a single component system and illustrates its properties and application area. Within this passage, he introduces both the important notions "allotropy" and "metastability." Referring to the allotropic forms of a substance and their transformation into each other, Ostwald speaks of an analogy to the change of aggregate states. However, these phase transformations are hindered sometimes and are sometimes even so slow that substances can exist at conditions in which they are not thermodynamically stable (after equilibrium has been reached). This kind of state—which, for example, is applicable to benzene and ozone at standard temperature—is still considered as metastable in modern science. Transformation of allotropic substances is accompanied by the formation of certain products. Ostwald describes this phenomenon in another natural law that he calls the "stage law" (*Stufengesetz*; Ostwald 1907, 142–145); of the possible products, the one is formed that is next on the stability scale rather than the most stable one in terms of thermodynamics.[17] Ostwald describes pure stuff as borderline cases of solutions (Ostwald 1907, 167). As to phase transitions (*Änderungen der Formart*), solutions vary in composition; a pure stuff does not. The purer a stuff becomes during a certain operation, the smaller the variability becomes, and eventually we get constant measurements. If the composition of the formed and the remaining phases during a phase transition is the same, this transition is called "hylotropic" (Ostwald 1907, 259). Hylotropy can occur for pure substances as well as for solutions and mixtures. The area of hylotropic stability of pure stuff, although restricted in general, is much wider than that of hylotropic solutions or mixtures that exhibit this property only at certain points (azeotropes, eutectics).[18] It should be mentioned that the convincing notion

[16] František Wald is, as we have seen already, much more careful with the definition of this notion in *Chemie Fasi* (Wald 2004, 112). A deep, critical discussion of the operational applicability of the notion of phase is provided by Schummer (1996, 175–180).

[17] In a well-known German university textbook this law is used for the description of polymorphic sulphur and referred to as *Ostwaldsche Stufenregel* (Holleman and Wiberg 1964, 190).

[18] Discussing chemical kinds, Paul Needham (2002, 214–218) gives a comprehensive sketch of this part of Ostwald's theories.

of hylotropy has found no positive reception—if any at all—by the chemical community.

If the field of thermal stability of pure stuff is left due to the application of extreme conditions, several other pure substances might be obtained. According to Ostwald these processes are chemical events in the narrow sense (Ostwald 1907, 263). If this method is applied repeatedly, substances are finally obtained that cannot be decomposed further: the elements. In introducing the latter, Ostwald calls these "simple" or "not-decomposed stuff" (Ostwald 1907, 265). He then gives the following definition:

> Accordingly, an element is a stuff which can never be transferred into another hylotropic stuff within the whole area of applicable energetic manipulations. (Ostwald 1907, 267)

Ostwald outlines—not without minor flaws—the argumentative strength of a methodical foundation of chemistry, which explicitly does not use the critical but prevailing picture of elements, which are just unquestioned natural kinds and ready to be used, as basis. He shows that elements have to be made or prepared rather than only to be discovered. Chemical elements are the result of mental and experimental efforts. Hence, the constructive aspect of this part of the chemical sciences becomes obvious by his analysis.

Concluding this section, I want to emphasize that both Wald and Ostwald implicitly and to a reasonable amount argue for the reimport of the old general theories of the Greek in that they use the phase concept as a refined general concept of stuffiness.[19]

4. Metaphysics or Metachemistry?

In the early twentieth century, another unusual attempt emerged to come to terms with the peculiarities of chemistry. In his *La Philosophie du non* from 1940, the French Philosopher Gaston Bachelard (1884–1962) differentiates metaphysics and metachemistry. He says:

> Metaphysics could have only one possible notion of substance because the elementary conception of physical phenomena was content to study a geometrical solid characterized by general properties. (Bachelard 1968, 45)

---

[19] Although I not exclusively address elements in this section, most of the general aspects of the discussion apply to both elements and compounds.

For a materialist metaphysics—and this is still the prevailing stance—everything is made of one matter.[20] If we turn our attentiveness to chemical substances and their transformations, the description becomes different, and the reductionist and physicalistic approach more and more unapplicable. At the same spot, Bachelard proceeds:

> Metachemistry will benefit by the chemical knowledge of various substantial activities. It will also benefit by the fact that true chemical substances are the products of technique rather than bodies found in reality. This is as much as to show that the real in chemistry is a realization. (Bachelard 1968, 45)

Like Wald and Ostwald before him,[21] Bachelard claims that substances—compounds as well as elements—are preparations.[22] Moreover, he supplements the famous statement of Berthelot that chemists are building their own field by making substances. At several points of his work Bachelard criticizes—sometimes ironically—the widespread belief of metaphysical philosophers "to define reality as a mass of irrationality" (Bachelard 1968, 47). Philosopher Bas van Fraassen gives a somewhat similar description—here explicitly of materialism:

> materialism . . . is not identifiable with a theory about what there is but only with an attitude or cluster of attitudes. (van Fraassen 2002, 59)

Among these attitudes, van Fraassen mentions a strong deference to actual scientific knowledge and an inclination to accept completeness claims for actual science.

According to Bachelard, the modern periodic table with its fundamental atomic numbers is in fact the result of striving for completeness, which brings his interpretation close to van Fraassen's attitudes of materialism. Many scientists and philosophers would perhaps subscribe to the claim that the periodic table is a genuine part of modern materialism. Bachelard, however, characterizes this situation differently:

---

[20] Cf. the essay of Alfred Nordmann (2006), on which I draw in this section. As to Bachelard, see also Chimisso 2001. My attempt to describe the difference of the notions matter and substance with respect to chemistry is Ruthenberg 2016.

[21] Bachelard sometimes mentions Ostwald. As far as I know, he was not aware of Wald's work.

[22] Similar statements can be found in other parts of his writings. In his discussion of an experimental hierarchy of substances, Schummer refers to a citation from "Le matérialisme rationnel" (1953), in which Bachelard claims that the purity of a substance is a human act, which conserves the main relativity of human activities (Schummer 1996, 180).

Hence, it is not less true that with the table of Mendeleev a metachemistry has been formed, and that the ordering and rationalizing tendency has led to more numerous, more deep-going successes. (Bachelard 1974, 85)[23]

As to Bachelard, metachemistry is formed by the negating transformation of metaphysics (*La Philosophie du non*), such that the science of substances is supported by non-chemical concepts like the atomic number—but without becoming just physics. This situation might be tempting for philosophers from the "realist" party with an inclination to natural kinds to apply this notion not only to chemical elements but to the periodic table as a whole.[24] As becomes clear from the above discussion, I would be very critical with that kind of interpretation, but that critique does not belong to the present chapter.

5. Misunderstandings

The Austrian (later British) chemist Friedrich Paneth (1887–1958) wrote perhaps the most frequently cited essay in modern philosophy of chemistry,[25] which was titled "Über die erkenntnistheoretische Stellung des chemischen Elementbegriffs." Moreover, his earlier papers were the basis for the predecessors of the International Union for Pure and Applied Chemistry (IUPAC) to define chemical elements as follows:

chemical element

1. A species of atoms; all atoms with the same number of protons in the atomic nucleus.
2. A pure chemical substance composed of atoms with the same number of protons in the atomic nucleus. Sometimes this concept is called the elementary substance as distinct from the chemical element as defined under 1, but mostly the term chemical element is used for both concepts. (IUPAC 1997)

There are at least two misunderstandings that emerged from the interpretation of Paneth's impact on the current (or official) definition of chemical elements,

---

[23] I here translate and cite from the German translation of his *Épistémologie* from 1971, which is a collection of selected texts from other publications.

[24] Note that it is useful to differentiate between the notions periodic law, periodic system, and periodic table. The perhaps most neutral expression is the latter, which I prefer in the present text. I thank Pieter Thyssen for leading my attention to that point.

[25] See several of the chapters of the present volume. Paneth gave the "Königsberg-lecture" in Kant's hometown on July 12, 1931.

which I would like to bring to the attention of the community of historians and philosophers of chemistry in this context.

The first misunderstanding refers to the role of atomism for the definition of elements. Paneth excludes atomism from the definition of chemical elements. In his dualistic scheme, the atomic number is the—though transcendental[26]—center of description and refers directly to the basic substance. The atomistic picture, hence, is unnecessary to understand basic nor simple substances:

> The atomic theory can . . . contribute enormously to . . . visualizing how the basic substances persist in simple substances and compounds; but the concept of basic substance as such does not in itself contain any idea of atomism. It was, after all, while explicitly rejecting atomism that Lavoisier carried this concept to victory; and also in more recent times, there were, and are, chemists who avoid the atomic theory but retain the elements, including, of course, elements in the sense of basic substances. (Paneth 1962, 155)

As can be inferred from the cited IUPAC definition, modern chemists tend to forget or neglect that fact. Unfortunately, however, this is true as well for some circles of the philosophy of chemistry.

The second misunderstanding of Paneth's interpretation refers to his explicit appreciation of the operational concept of Lavoisier. Already as a fresh and promising radiochemist (he finished his chemistry studies with a synthetical thesis and changed to that physicochemical field) he was interested in general—philosophical—questions:

> An element is a substance that cannot be dissected yielding simpler substances. Substances that fit to that definition are considered the same element if they, once mixed, cannot be separated by chemical means. Atoms are those constituents of matter to which the chemical analysis can proceed, which nevertheless stay intact during all chemical reactions. (Paneth 1916, 198, my translation)

---

[26] Paneth insists on an epistemological interpretation of the expression "transcendental" (Paneth 1931, 3–4). Whereas in the customary interpretation (which is still questionable) Kant held that the transcendental realm—the world behind our direct perceptions—is not accessible at all, Paneth claims that the world is, at least to a certain amount, knowledgeable by scientific efforts. He explicitly follows the "transcendental realism" proposed by the neo-Kantian philosopher Eduard von Hartmann (for a discussion of the Kantian influence on Paneth, see my Ruthenberg 2009 and 2010). According to Paneth, the basic elements, as well as the chemical radicals of his time, do belong to the transcendental realm; nevertheless, we can obtain knowledge about these entities epistemologically. In other words, we are able to shift borders. Hence, in a certain sense this interpretation is similar to Bachelard's "non-chemical" support.

Although he is a little too optimistic about how far chemical analysis can go in reality (analytical chemists are not working with atoms), and although he refined this statement in later articles (atomic numbers replaced atoms), he throughout held that the operational part of the realization of elements is significant. Unfortunately, that fact—which he pointed at again in his Königsberg lecture, about ten years after the declaration of the International Commission—was suppressed since the early 1920s and finally eliminated. Again the members of the scientific community were satisfied by a theoretical microphysical picture.[27] To make my point clear: It is of course true that Paneth had a main influence on the "official" element definition; it is also true, however, that he still emphasized the empirical part of that definition. Hence, there is a plurality here, not a reductionistic monism.

I close this section with a remark on the making of elements in a literal sense, which at the same time belongs to one of Paneth's favorite fields of interest: the elemental transmutation. During experiments on the small-scale measurement of helium, Paneth and Peters (1926) found traces of helium in their apparatus, which they first ascribed to a transformation of hydrogen into helium. Later they corrected their misunderstanding: they had found out that the glass was contaminated by helium from air, which was released by hydrogen.[28] What matters in the present context is the fact that the contemporary chemists—those of the 1920s—believed in the possibility of transmutation.[29] That belief prevails until today, although perhaps not with respect to a cold fusion concept. Most modern scientists will probably agree that the synthesis of a handful of atoms of a certain, elevated atomic weight in sophisticated large particle accelerators counts as proof for the existence of that element. At least this is the official procedure to get "artificial" elements listed in the periodic table. Obviously, several of these artificial elements do not exist in the sense many philosophers of chemistry— among them most of the scholars I mentioned before in this chapter—would subscribe to. Whether or not one likes this procedure, it doubtlessly is another support for the thesis of the conventional character of the notion of chemical element.

---

[27] Again, in chemistry atomic numbers are unobservables, that is, theoretical (or microphysical or transcendental) entities.

[28] For a short account of that story, see Ruthenberg 2015.

[29] Shortly after World War II, Paneth himself gave a synopsis of modern element transmutations titled "The Making of the Missing Chemical Elements" (Paneth 1947).

6. Modes of Making Elements (Conclusions)

It is tempting to deploy the philosophical stories of chemistry from the successful end, that is, from the side of the most recent empirical and theoretical achievements. As we have experienced in the last twenty years, this method lead to a widespread emphasis of a naturalization of microphysical descriptions. In large parts of the current philosophy of chemistry, the prevailing scientific pictures of chemistry are just mirrored without further ado.

I believe that this stance—we might call it materialist, naturalist, or realist—cannot be sufficient to obtain a fully blown philosophy of chemistry. In this chapter I therefore tried to emphasize the central empirical efforts in chemistry and their possible place in the philosophy of this science of substance. "Making of elements" can mean to invent abstract pictures derived from everyday experience, to search for the composition with or without the presuppositions of atoms and elements, or to run accelerators in order to form bigger atomic nuclei from smaller ones. Each version has its own philosophical attachment, but for all—and hopefully for this chapter, too—we can claim with Bachelard "cela suffit pour designer le reel en chimie comme une realization."

References

Aristotle. 1931. "*Meteorologica IV.*" In *The Works of Aristotle*, vol. 3, trans. E. W. Webster. Oxford: Clarendon Press.

Bachelard, Gaston. 1968. *The Philosophy of No: A Philosophy of the Scientific Mind.* New York: Orion Press.

Bachelard, Gaston. 1974. *Epistemologie—Ausgewählte Texte.* Frankfurt: Verlag Ullstein,.

Bird, Alexander, and Emma Tobin. 2018. "Natural Kinds." In *The Stanford Encyclopedia of Philosophy* (Spring), edited by Edward N. Zalta, https://plato.stanford.edu/archives/spr2018/entries/natural-kinds.

Bodemann, Theodor. 1845. *Anleitung zur berg- und hüttenmännischen Probierkunst.* Clausthal: Verlag der Schweigerschen Buchhandlung.

Chimisso, Cristina. 2001. *Gaston Bachelard—Critique of Science and the Imagination.* London: Routledge.

Hendry, Robin. 2006. Elements, Compounds, and Other Chemical Kinds. *Philosophy of Science* 73: 864-875.

Holleman, Arnold F., and Egon Wiberg. 1964. *Lehrbuch der anorganischen Chemie*, 57–70 th ed. Berlin: Walter de Gruyter & Co.

IUPAC. 1997 "Chemical Element." In *Compendium of Chemical Terminology*, 2nd ed. (the "Gold Book"). Compiled by A. D. McNaught and A. Wilkinson. Oxford: Blackwell Scientific Publications. XML online corrected version: http://goldbook.iupac.org (2006–) created by M. Nic, J. Jirat, B. Kosata; updates compiled by A. Jenkins. https://doi.org/10.1351/goldbook.

Janich, Peter. 1996. "Natürlich künstlich: Philosophische Reflexionen zum Naturbegriff der Chemie." In *Natürlich, technisch, chemisch—Verhältnisse zur Natur am Beispiel der Chemie*, edited by P. Janich and C. Rüchardt, 53–79. Berlin/New York: Walter de Gruyter.

Kukla, André. 2010. *Extraterrestrials: A Philosophical Perspective*. Lanham, MD: Lexington Books.

Needham, Paul. 2002. "The Discovery that Water Is H2O." *International Studies in the Philosophy of Science* 16: 205–226.

Needham, Paul. 2017. *Macroscopic Metaphysics: Middle-Sized Objects and Longish Processes*. Cham: Springer Nature.

Nordmann, Alfred. 2006. "From Metaphysics to Metachemistry." In *Philosophy of Chemistry: Synthesis of a New Discipline*, edited by D. Baird, E. Scerri, L. McIntyre, 347–361. Dordrecht: Springer.

Ostwald, Wilhelm. 1904. Elements and Compounds (Faraday Lecture). *Journal of the Chemical Society* 85: 506–522.

Ostwald, Wilhelm. 1907. *Prinzipien der Chemie*. Leipzig: Akademische Verlagsgesellschaft.

Ostwald, Wilhelm. 1919. *Grundriss der Naturphilosophie*, vol. 1, 3rd ed. Leipzig: Verlag von Philip Reclam jun.

Paneth, Friedrich. 1916. "Über den Element- und Atombegriff in Chemie und Radiologie." *Zeitschrift für Physikalische Chemie* 91: 171–198.

Paneth, Friedrich. (1931) 1962. "Über die erkenntnistheoretische Stellung des chemischen Elementbegriffs." *Schriften der Königsberger Gelehrten Gesellschaft* 8: 101–125. Citations refer to the English translation: *The British Journal for the Philosophy of Science* 13: 1–14, 144–160.

Paneth, Friedrich. 1947. "The Making of the Missing Chemical Elements." *Nature* 159: 8–10.

Paneth, Friedrich, and Kurt Peters. 1926. "Über die Verwandlung von Wasserstoff in Helium." *Die Naturwissenschaften*: 957–962.

Putnam, Hilary. 1975. "The Meaning of Meaning." In *Mind, Language and Reality*, 215–271. Cambridge: Cambridge University Press.

Ruthenberg, Klaus. 2005. "Empedokles' Naturdichtung aus chemischer Perspektive." *Antike Naturwissenschaft und ihre Rezeption* 15: 7–12. Trier: Wissenschaftlicher Verlag.

Ruthenberg, Klaus. 2009. "Kant, Paneth, and the Philosophy of Chemistry." *Foundations of Chemistry* 11: 79–91.

Ruthenberg, Klaus. 2010. "Das Kant'sche Echo in Paneths Philosophie der Chemie." *Kant-Studien* 101: 465–479.

Ruthenberg, Klaus. 2011. "Chemietheorie 'nach dem Vorgange von Mach'—ein phänomenalistischer Ansatz in den Annalen der Naturphilosophie." *Abhandlungen der Sächsischen Akademie der Wissenschaften zu Leipzig. Philologisch-historische Klasse* 82: 94–115.

Ruthenberg, Klaus. 2015. "Als das Unmögliche noch möglich erschien." *Nachrichten aus der Chemie* 63: 639–643.

Ruthenberg, Klaus. 2016. "Matter and Stuff: Two Sides of the Same Medal?" In *Contemporary Lines*, edited by A. Le Moli and A. Cicatello, 153–168. Vol. 2 of *Understanding Matter*. Palermo: New Digital Press.

Ruthenberg, Klaus, and Jaap van Brakel, eds. 2008. *Stuff: The Nature of Chemical Substances*. Würzburg: Königshausen & Neumann.

Schummer, Joachim. *Realismus und Chemie*. 1996. Würzburg: Königshausen und Neumann.

Stroll, Avrum. 1998. *Sketches of Landscapes: Philosophy by Example*. Cambridge, MA: MIT Press.

Van Brakel, Jaap. 2005. "On the Inventors of XYZ." *Foundations of Chemistry* 7: 57–84.

Van Brakel, Jaap. 2013. "František Wald's Empiricism." *HYLE—International Journal for Philosophy of Chemistry* 19: 161–183.

Van Fraassen, Bas. 2002. *The Empirical Stance*. New Haven: Yale University Press.

Wald, František. 2004. *Chemie Fasi*. Prague: Vydala Univerzita Karlova v Praze.

12

A Formal Approach to the Conceptual Development of Chemical Element

Guillermo Restrepo, Max Planck Institute for Mathematics in the Sciences, Leipzig, Germany; Interdisciplinary Center of Bioinformatics, Leipzig, Germany

Concept-Formation. How Difficult Is It?

Grouping together similar objects is a basic task of the human condition, which we also share with some primates. Concepts are the tools we use to navigate the enormous amount of information we are exposed to every day.[1] They are the mental devices that simplify the complexity of the word to make sense of it.[2] It is by far more efficient to learn something about birds in general than try to learn something about any single member of the concept "bird." Moreover, the lack of concepts would hinder making predictions. How could we say something about noble gases without the concept of noble gas? There was something new in argon that allowed estimating a family of elements.[3]

Some features concepts hold are (Rouvray 1997): They (1) arise from the need of structuring our world, (2) depend on the input perceptions, (3) involve mental abstraction, (4) are theoretical devices that cluster our world, and (5) are time-dependent as a consequence of continuous updates.[4] As Rouvray stated, even if concepts are soft-edged and unstable, they work efficiently for most practical

---

[1] According to psychologist Medin, a concept is a set to which some assertion or set of assertions might apply (Medin 1989). In general, the most salient feature of a concept is its closed structure, where objects or particulars of the concept are uniquely characterized by a set of attributes such that those attributes uniquely characterize the objects of the concept. We will see these details when discussing the approaches from formal concept analysis. We adopt the philosophical position that a concept is a universal, that is, a property that can be instantiated by more than one individual thing.

[2] In *Funes the memorious*, a short story by Jorge Luis Borges, he tells us about a boy (Funes) who is able to remember every detail of every single activity. However, he is incapable of abstraction, of generalizations; all in all, of concept-formation (Borges 2007).

[3] The very fact that argon was discovered on Earth with absence of compounds indicated that the attribute "non-reactivity" was shared by the family. This has changed, for Xe and other noble gases have formed stable compounds. This shows that concepts are not static; rather, that they evolve. The evolution of the concept of chemical element is the subject of this chapter.

[4] In general, concepts form a covering of our world, that is, they create a collection of classes that may overlap with one another because of sharing objects.

Guillermo Restrepo, *A Formal Approach to the Conceptual Development of Chemical Element* In: *What Is A Chemical Element?*. Edited by: Eric R. Scerri and Elena Ghibaudi (2020). © Oxford University Press.
DOI: 10.1093/oso/9780190933784.003.0013

purposes; nevertheless, problems arise when we try to lead them to crystal-clear definitions. Is this not the case when trying to define chemical element?

Rouvray has exemplified the problem of concept definition through the assessment of the concept of "bird." He claims that one typically starts with an exemplar, say a robin; then after being exposed to several other things, one finds a dove and similar attributes pop up, for example, robin and dove fly, build nests, and are quite small in size regarding several other things one is customarily exposed to. Still, more experiences come, for example the finding of a turkey, which is a bit bigger than the robin and the dove, and builds nests but cannot fly. What to do? We say that birds only build nests. However, later on we find cuckoos that do not make nests, but that lay their eggs in the nests of other "birds." Which are the attributes of our concept?

The very idea of trying to define a concept is difficult. In fact, "there is a growing trend to abandon altogether any idea that...concepts are possessed of a set of characteristic features that might be used to define them" (Rouvray 1997). Some of the problems of attaching attributes to concepts are that (1) not even experts can state which are the attributes,[5] and that (2) there are some objects that one knows belong more than others to the concept.

As concepts are difficult to glue together by their defining attributes, what keeps them so stable (Rouvray 1997)? By taking Oden and Lopes's (1982) results along with Medin's ideas (1989), Rouvray states that concepts are coherent entities, for beyond including attributes, they are held together by structural principles (Rouvray 1997). He claims that "the role played by similarity should embrace not only the attributes of the entities but also the relationships existing between them, including higher order relationships."[6] In other words, if "chemical element" is a concept, it is not because elements have melting point, density, and so on; rather because they are objects that relate to other objects in a similar fashion.[7] All in all, materials, in general, have a particular density under the

---

[5] This is seen, for example, with the case of the criteria that must be satisfied for the discovery of a new chemical element. The Transfermium working group of the International Union of Pure and Applied Chemistry (IUPAC) and the International Union of Pure and Applied Physics (IUPAP) was created to establish these criteria; however, in its first publication the group recognized that "it is not feasible to specify criteria, or combinations of criteria, that, in the words of its Terms of Reference "must be satisfied..." in order to achieve recognition of the existence of a new chemical element and that would cover all cases. Very few properties indeed, of which perhaps the only uncontentious example is the characteristic X-ray spectrum, unambiguously determined, are sufficient of themselves to establish the existence of a new element" (Jeannin 1991).

[6] Perhaps the idea of the structural principles and the high order relations is better exemplified with the concept of family. A family is not only the grouping of its members, who share a common surname (at least in most of the Western cultures) and some physical resemblance, but also cultural traditions and rituals, internal customs, a whole set of relations with uncles, nieces, grandparents, and so on.

[7] This is similar to the structuralist approach to language (Allison 1999), where a sign of the language is made of a phonetic signifier and a meaning or signified element. Therein, what brings specificity and generality to the signifier is its relation to the other signifiers of the language. Moreover,

conditions of experimentation. Why do not all materials belong under the concept of chemical element?

With this difficult panorama of concept-formation, let us take a brief look to the history of chemistry to see how chemists have developed the concept of chemical element.

The Evolution of the Concept of Chemical Element

If "chemical element" is a concept, we here explore its history following the idea that a concept consists of objects and attributes, which we have seen is shared by psychologists (Medin 1989).[8] We are particularly interested in determining whether the objects of the concept have been of the same ontological category, for example, materials, substances, and so on. We also analyze which attributes have been used as part of the concept.

The idea of element has been around since ancient Greek times, but mainly as a metaphysical construct. As our interest is on chemical elements, we find it appropriate to start with Lavoisier, who in the second half of the eighteenth century claimed that a chemical element is a substance that cannot be decomposed into other substances with the technological means available at the time of experimentation (Lavoisier 1965).[9] The definition requires substances and a single attribute, non-decomposability, and entails the possibility of change, for it allows enlarging the number of elements.[10] Historically, new technologies, for example, electrolysis, spectral analysis, and nuclear science have actually enlarged the set of elements.[11]

---

signs and their generality or specificity is given by the system of grammatical, phonemic, and syntactic rules applied upon them. All of these relations and rules constitute the high order relations mentioned by Rouvray (1997).

---

[8] This idea has been nicely formalized by Ganter and Wille (1999) in the so-called Formal Concept Analysis, which we explain latter.

[9] Lavoisier claimed: "If we apply the term *elements*,...to express...the last point which analysis is capable of reaching, we must admit, as elements, all the substances into which we are capable, by any means, to reduce bodies by decomposition. Not that we are entitled to affirm, that these substances... may not be compounded of two, or even of greater number of principles; but, since these principles cannot be separated, or rather since we have not hitherto discovered the means of separating them, they act with regard to us as simple substances, and we ought never supposed them compounded until experiment and observation has proved them to be so" (Lavoisier,1965, xxiv).

[10] If element e decomposes into $n > 1$ other substances, the number of elements increases by n-1 units. Lavoisier's definition is based upon substances and non-decomposability, which requires a definition of these two terms. We will discuss this further in the coming sections.

[11] An open question in chemistry and physics is estimating the largest number of possible elements. It is claimed that the heaviest element has atomic number 173 (Indelicato, Bieroń, and Jönsson 2011). Further information is found in Karol 2018.

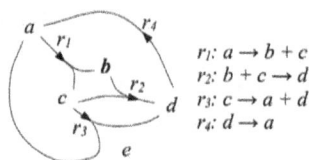

$$r_1: a \to b + c$$
$$r_2: b + c \to d$$
$$r_3: c \to a + d$$
$$r_4: d \to a$$

Figure 12.1. Reactions r_1 to r_4 depicted as directed hypergraphs, where for each reaction $r_i : S_i \to P_i$, substances in S_i (substrates) are connected through arrows to substances in P_i (products). Substance b (bold face) corresponds to a chemical element. There is an isolated substance, e, which is neither substrate nor product of any reaction and could be an element.

Given the central role of reaction for Lavoisier's approach and for the ensuing discussion, we define chemical reaction as follows: [12]

Definition 1. Let X be a non-empty set of substances; we call $r_i = (S_i, P_i)$ the *chemical reaction i*, where $S_i, P_i \subset X$. S_i and P_i are called the *substrates* and *products* of r_i, respectively.[13]

Reactions, and in general sets of reactions, can be modeled as networks (Schummer 1998), where the most suitable framework is that of directed hypergraphs (Klamt, Haus, and Theis 2009). In such a setting, a reaction is an ordered pair of sets, that is, the set of substrates leading to the set of products. Figure 12.1 exemplifies the model of few reactions in a directed hypergraph.

Figure 12.1 shows the particular structural feature of chemical elements in the network, according to Lavoisier's ideas. They are not decomposable substances, which may be substrates or products of reactions. Hence, substance b (figure 12.1) is a chemical element holding all three attributes, while e is a chemical element holding only non-decomposability. Examples of b are oxygen, hydrogen, and so forth; and of e are light and caloric, for Lavoisier (1965).[14]

---

[12] There are several formal definitions of chemical reaction in the scientific literature (Klamt, Haus, and Theis 2009; Restrepo and Stadler 2016), some of them only emphasize the relation between substrates and products, while others consider stoichiometric coefficients of the balanced reaction, for example.

[13] Note that S_i and P_i may share substances, for example, in autocatalytic reactions. However, most of the customary reactions assume $S_i \cap P_i = \varnothing$. In this setting, reversible reactions or tautomer formation are regarded as different reactions.

[14] In the Lavoisian context, light and caloric were regarded as substances. Note that a and c in figure 12.1 are not elements, for they decompose into b and c; and a and d, respectively. Likewise, d is not an element, for it decomposes into a and because it is synthesized from b and c.

The hypergraph setting for the Lavoisian account of chemical element leads to the following definition:

Definition 2. Let X be a non-empty set of substances endowed with a set of chemical reactions. We say that $s \in X$ is a *chemical element in the Lavoisian context* if:

- there is no reaction r_i such that $S_i = s$ (no decomposability).

Hence, a chemical element is a non-decomposable substance (figure 12.2a). About a century after Lavoisier's claim, there was a much larger number of substances, reactions, and properties of substances (Llanos et al. 2019), which led to a structure for the elements based on their chemical similarity and on their ordering by atomic weight (Restrepo 2018; Leal and Restrepo 2019): the periodic system of the elements. The system was possible because of the introduction of a stable system of atomic weights, which modified the concept of element to that of non-decomposable substances having an associated atomic weight, that is, a real positive number (figure 12.2a). Hence, the element went beyond the network and, for the first time, considered its projection into another space. An element was in fact characterized by the network and by a function to the real numbers (figure 12.2a). However, the role of the mapping was overemphasized to the extent that the network was somehow downplayed.

The corresponding definition of chemical element then changed to:

Definition 3. Let X be a non-empty set of substances endowed with a set of chemical reactions. We say that $s \in X$ is a *chemical element in the periodic system* context if:

- there is no reaction r_i such that $S_i = s$ (no decomposability); and

- there is a mapping $f(s) = w$, with $w \in \mathbb{R}$ (atomic weight).

Definition 3 also accounts for non-reactive elements (noble gases) discovered after the formulation of the periodic system. In the hypergraph setting they take the place of isolated substances in figure 12.1. That is, Lavoisian caloric and light are now replaced by argon and the other noble gases.

The observation of radioactivity around the turn of the twentieth century and the transmutation of some elements into others made chemists realize that the non-decomposability of the Lavoisian context was no longer generally applicable as an attribute characterizing elements.[15] Hence, the condition of not having a

---

[15] For example, ^{232}Th $\rightarrow$ ^{228}Ra + ^{4}He, where thorium decays to radium and helium. By the non-decomposability attribute Th would no longer be considered an element. As Th and Ra were actually

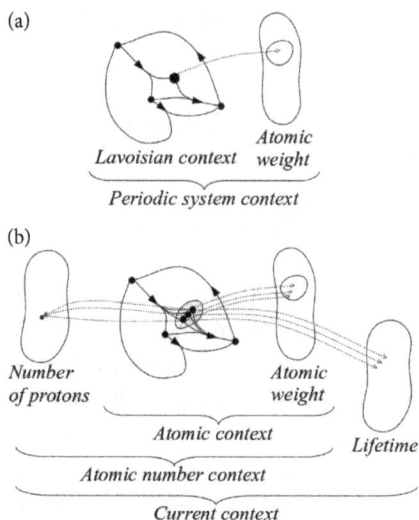

Figure 12.2. (a) The network on the left is a simplification of the network in figure 12.1, where substances are replaced by vertices (dots) and reactions are indicated by head arrows. For simplicity, substance e of figure 12.1 is not included. Nevertheless, all predicates upon elements are applied to e. The dotted arrow indicates the mapping between the chemical element (bold dot) and a subset of real values associated with the atomic weight of the element in the periodic system context*. (b) Still the same network as in (a), but now the chemical element has been refined to new species: isotopes, where each one has a corresponding atomic weight (dotted arrows heading to the right). The element is now a collection of isotopes (gray region). A second mapping is defined to the natural numbers (number of protons) and represented by arrows pointing to the left. The third mapping assigns a lifetime to each isotope.

*Note that a subset of $\mathbb{R}$ is indicated rather than a particular point in the space. This is to account for the discussion on whether atomic weights were whole numbers (Prout's hypothesis) or not and to account for the different atomic weights assigned to an element, most of which were close to one another.

reaction r_i such that $S_i = s$ was abandoned. However, the new nuclear reactions, where $S_i = s$ holds, could be incorporated into the network framework.[16]

considered chemical elements, the only possibility was to modify the attributes belonging to chemical elements.

[16] It is so, for $s \in X$ was now decomposed, or transmuted into other species of X. Another attribute associated with chemical elements that became important for the definition of chemical element was half-life, as we discuss later.

Knowledge about atomic structure led scientists to conclude that in several cases objects of the Lavoisian and the periodic system contexts were a collection of more than one species that undergoes the same chemical reactions and that can be differentiated by their atomic weight (figure 12.2b). The so-called isotopes challenged the periodic system definition of element based on atomic weight. The network and mapping structure shows (figure 12.2b) that the network structure remains even if new species show up, for now an element turns out to be a set of species that is structurally equivalent, that is, that behave in an equivalent way in the network. The mapping to real numbers was also important. It was found that isotopes have very close atomic weight values.[17]

The notion of equivalence in the network, that is, chemical equivalence of isotopes, can be defined as:

Definition 4. Let X be a non-empty set of chemical species.[18] For any $s \in X$, we call $\text{in}(s) = \{q \in X \mid q \in S_i \text{ and } s \in P_i\}$ and $\text{out}(s) = \{q \in X \mid q \in P_i \text{ and } s \in S_i\}$ the *input* and *output* of s, respectively. For all $s_i, s_j, \ldots \in X$, whose $\text{in}(s_i) = \text{in}(s_j) = \ldots$ and whose $\text{out}(s_i) = \text{out}(s_j) = \ldots$, we say that $s_i, s_j, \ldots$ are *chemical equivalent species.*

In Definition 4, $\text{in}(s)$ gathers those species that produce s through reactions. Likewise, $\text{out}(s)$ contains those species that are produced from s. In this atomic context, the definition of chemical element is then:

Definition 5. Let X be a non-empty set of chemical species endowed with a set of reactions. We say that $[s]$ is a chemical element in the atomic context if:

• $[s] = \{s_i, s_j, \ldots \in X \mid s_i, s_j, \ldots$ *are chemical equivalent species*$\}$ (chemical equivalence), and

• for every s_k in $[s]$ there is a mapping $g(s_k) = w_k$, with $w_k \in \mathbb{R}$ (atomic weight).

It was found that a common property of isotopes that are equivalent in the net-work is their number of protons (figure 12.2b). Therefore, a chemical element now included the mapping to the atomic number (number of protons), which is a natural number.[19] This new function became utterly important. There are no further natural numbers between consecutive naturals, but there are infinite between consecutive reals. This new mapping, once and for all, brought up cer-tainty. Any missing natural number indicated a missing element. Again, the role of the mapping, this time to number of protons, was overemphasized and the re-lational character of the element in the network was downplayed.

The concept of element was then modified to:

Definition 6. Let X be a non-empty set of chemical species endowed with a set of chemical reactions. We say that $[s]$ is a *chemical element in the atomic number context* if:

- $[s] = \{s_i, s_j, \ldots \in X \mid s_i, s_j, \ldots$ *are chemical equivalent species*$\}$ (chemical equivalence),

- for every s_k in $[s]$ there is a mapping $g(s_k) = w_k$, with $w_k \in \mathbb{R}$ (atomic weight), and

- for every s_k in $[s]$ there is a mapping $h(s_k) = Z_k$, with $Z_k \in \mathbb{N}$ (atomic number).[20]

As the twentieth century progressed, the as yet unobserved elements were syn-thesized through nuclear technologies. However, in most of the cases, the former amount of elements required for chemical analyses used to build up the chemical network was almost impossible to attain, as few atoms of the species were ever obtained, and their life spans were usually less than one second.[21] In light of these findings, the definition of element needed a further update.

The current requirement for an object to be called an element is "the exist-ence of at least a nuclide with an atomic number Z . . . existing for at least 10^{-14} s" (Jeannin 1991). The selection of that particular time "is chosen as a reason-able estimate of the time it takes for a nucleus to acquire its outer electrons" (Jeannin 1991).[22] To the detection of such a short lifetime species has to be

---

[19] Strikingly, this natural number corresponded to the ordinal number of the elements in the peri-odic system.

[20] By natural numbers we mean whole numbers greater than 0.

[21] Chemical studies of elements beyond Sg are rather dim, except for Hs, with some isotopes of half-lives on the order of seconds (Türler 2013).

[22] It is curious that the selection of the lifetime for the nuclide is based on chemical assumptions, where there is no room for chemistry. It is claimed that such a lifetime was selected as "it is not

added the determination of Z, which is normally attained through X-ray spectroscopy (Jeannin 1991) (figure 12.2b). This brought up a new mapping, now to the lifetime (figure 12.2b). Therefore, the current account of a chemical element includes the mapping to atomic number and to lifetime. The updated definition of element is as follows:

Definition 7. Let X be a non-empty set of chemical species endowed with a set of chemical reactions. We say that $[s]$ is a *chemical element in the current context* if:

- $[s] = \{s_i, s_j, \ldots \in X \mid s_i, s_j, \ldots$ *are chemical equivalent species*$\}$ (chemical equivalence),

- for every s_k in $[s]$ there is a mapping $g(s_k) = w_k$, with $w_k \in \mathbb{R}$ (atomic weight),

- for every s_k in $[s]$ there is a mapping $h(s_k) = Z_k$, with $Z_k \in \mathbb{N}$ (atomic number), and

- for every s_k in $[s]$ there is a mapping $l(s_k) = t_k$, with $t_k \in \mathbb{R}$ (lifetime).

Nevertheless, as Schwarz has pointed out, if what we are seeking is a definition of chemical element, allowing the possibility of chemistry as a criterion, then we need to have at least bonded atomic ensembles, for example, molecules, able to undergo chemical reactions.[23] This requires no less than 10^{-10} s.[24] The lack of chemistry for the new elements has made more notorious the historical overemphasis on the mappings from the network to other spaces. Therefore, the network as an attribute characterizing the elements, specifically the new ones, has become irrelevant for the definition of chemical element.

Thus, the concept of element has changed over time to fit experimental results. Its objects belong to different ontologies; from substances, or simple bodies, in Lavoisian times, to nuclides in contemporary terms. The attributes of the concept have also changed from "not decomposable" to "having lifetime $\geq 10^{-14}$ s," or if we want to allow for the possibility of chemistry, "having lifetime $\geq 10^{-10}$ s."

considered self-evident that talking about an 'element' makes sense if no outer electrons, bearers of the chemical properties, are present" (Jeannin 1991).

---

[23] Schwarz's ideas were part of a talk he gave at the Summer Symposium of the International Society for the Philosophy of Chemistry, held in Bristol in 2018.

[24] Data taken from Schwarz's presentation at the International Society for the Philosophy of Chemistry, held in Bristol in 2018.

What happens if we find an atom with a lifetime very close to 10^{-14} s or a bit lower—is that a chemical element? What if Z cannot be easily identified or if it is identified as an interval of possible values overlapping the values of other known elements? This has been considered by the criteria of IUPAC and IUPAP (Jeannin 1991). They say, "The exact value of Z need not be determined, only that it is different from all Z-values observed before, beyond reasonable doubt" (Jeannin 1991).[25] But problems lie not only on the side of the attributes, objects have had also several problems. For example, Lavoisier considered caloric as part of the object set (Lavoisier 1965, 175), Mendeleev tried to find a place for ether in the periodic system (Jensen 2005), and noble gases were also difficult to regard as chemical elements. Are the set of objects and of attributes always immediately crystal clear? Or only in hindsight?

Having seen the different shortcomings in concept-formation and the evolution, along with the problems on the construction of the concept of chemical elements, we explore new possibilities for the concept in the next section.

Seeking a Definition

A mathematical approach to the idea of concept was formulated in the 1980s by Ganter and Wille in the so-called Formal Concept Analysis (FCA) (Ganter and Wille 1999). As this approach is going to be used in the ensuing discussion, we introduce the basics of FCA, which involve the definition of a formal context that gathers the objects and the attributes of the universe (context) that we want to analyze.

> *Definition 8.* A *formal context* (G,M,I) consists of a set G of *objects*, a set M of *attributes*, and a *binary relation* $I \subseteq G \times M$. It is said that $x \in G$ has the attribute $y \in M$ if and only if $(x,y) \in I$.

Note that the objects of the binary relation are all possible couples where an object is related to an attribute.

The idea is that from the context, concepts arise. And to facilitate their finding, a closed relationship between objects and attributes needs to be defined, which is the key point of the formalism.

> *Definition 9.* Let (G,M,I) be a formal context. For any $O \subseteq G$ and $A \subseteq M$:

---

[25] The IUPAC/IUPAP document (Jeannin 1991) recognizes that "reasonable doubt" is "necessarily somewhat vague."

$O' = \{y \in M \mid (x, y) \in I, \text{ for all } x \in O\}$, that is, the set of attributes common to the objects in O.

$A' = \{x \in G \mid (x, y) \in I, \text{ for all } y \in A\}$, that is, the set of objects that have all attributes in A.

Definition 10. Let $K = (G,M,I)$ be a context. A *formal concept* in K is a pair (O,A) with $O \subseteq G$ and $A \subseteq M$ such that $O'=A$ and $A'=O$. O and A are called the *extent* and the *intent* of the concept.

Definitions 9 and 10 show that a formal concept is a couple of objects and attributes, but not any couple. It is a couple such that the objects O are characterized by the attributes A and those attributes A characterize the objects O. It is in this sense that the concept is a closed structure, a coherent pair of objects and attributes.[26]

In philosophical terms, one can situate FCA in the twentieth-century debates of philosophy of language (Lycan 1999). In such a framework, there are intensional and extensional definitions (Cook 2009). In the former, definitions are given by the attributes the objects defined and have to be accounted as a particular (object) of the universal (defined term). An intensional definition of *bachelor* is "unmarried man," which is both a necessary and sufficient condition for being a bachelor. Extensional definitions, in contrast, require the listing of the particulars of the defined universal. Hence, the extensional definition of bachelor would include the set of all unmarried men. Thus, FCA is based on the merging of intensional and extensional definitions, where a concept (O, A) results from the intensional definition of O' and the extensional one of A'.[27]

---

[26] There is indeed a closure operator $A \rightarrow A'$, and $O \rightarrow O'$ (Ganter and Wille 1999).

[27] FCA can also be framed in an Aristotelian setting, which is traditionally, but not explicitly, acknowledged by FCA researchers (Ganter and Wille 1999). In the *Categories*, Aristotle introduces four predicables (Shields 2016): accident, definition, proprium, and genus. By "accident" is meant what may or may not belong to a subject, which in the FCA formalism corresponds to the attributes of the found context (Definition 8). If the formal context is given, for example, by the objects (Bellerophon, Perseus, Pegasus), by the attributes (rational, animal, risible), and by the binary relation between objects and attributes, then rational, animal, and risible are accidents. Aristotle's definition entails what signifies a subject's essence, which is a formal concept (Definition 9). For example, *man* is defined as "rational animal," that is, all rational animals are men and all men are rational animals. Proprium is not in the essence of a subject but is unique to or counterpredictable of it. The proprium in the FCA setting accounts for all additional attributes of each object in the concept that are not part of the concept, that is, which are not essential for the concept. For instance, *risible* is not in the essence of *man*, although all and only men are risible. The genus corresponds to what is in the essence of subjects differing in species. As concepts in FCA are orderable by subsethood of their extensions (i.e., superhood of their intensions) (Ganter and Wille 1999), the genus corresponds to the most general concept, which in formal terms is the set of all objects of the context that share no attribute, that is $(G, \varnothing)$. However, according to several of Aristotle's interpreters (Shields 2016), the most general concept could also be that one gathering objects with at least a single common attribute. In

Coming back to the search of the concept of chemical element, we claim we need chemical objects and chemical attributes, which, as constituents of a formal concept, need to be interrelated. According to Schummer, chemical reactions are a fundamental source of chemical knowledge, which we take as the providers of objects and attributes for our attempt to define the concept of chemical element (Schummer 1998).

As seen, chemical reactions depend on the definition of substance, which, in terms of van Brakel, is impossible (van Brakel 2012). According to him, substance is a proto-scientific term not defined in science. Van Brakel, elaborating on the ontological status of substance and after discussing the manifold realizations of "chemical substances," for example, "components (in the sense of the phase rule), substances (identified in terms of separation methods)," and quasi-molecular species "(typically identified using spectroscopic techniques)" concludes that no single concept encompasses all realizations. He adds that "substance" will remain a pragmatical notion. As Schummer has found, when analyzing the ontology of chemistry, "substance" is an artificial concept needed as an anchor point to talk about matter transformation (Schummer 1998). It is what chemists, witnessing matter transformation around them and wanting to describe it, have found as the best option to address the issue. This situation of taking an anchor point to build up knowledge is typical in science; after all, sciences are not built up from scratch and there are necessary and fundamental terms lacking definition, at least if sciences are independent (Kline 1980). If the assumption of independence is regarded, every concept of a science must be defined in terms of another one inside the science, which leads to an infinite regress of definitions (Kline 1980); therefore, the need of undefined concepts. As Kline discusses it, "Aristotle in the *Organon*, Pascal in *Treatise on the geometrical spirit* and Leibniz in *Monadology* have emphasized the need for undefined terms" (Kline 1980). Putting it simply, chemical substances are not like animals in a zoo, waiting for the visit of the chemist, but rather pragmatic inventions of chemists trying to make sense of their zoo.

To consider not only the traditional chemical substances, but the other objects of the current chemical ontology, as discussed by van Brakel (2012) and by Restrepo and Harré (2015), we consider chemical species as the ontological objects that matter scientists have related and relate in their experiments. Note (1) that by matter scientists we include, besides chemists and alchemists, current physicists and engineers working in material sciences, biochemists, geologists, environmental scientists, botanists, marine biologists, and so forth. In general, the category includes all those who have been or are interested in matter

---

our context example, *animal* is in the essence of both *man* and *Pegasus*, therefore a suitable genus is
$$\left(\left\{Bellorophon, Perseus, Pegasus\right\}, \left\{animal\right\}\right) \cdot$$

interaction, independent of any disciplinary affiliation.[28] (2) By the very broad set of ontological objects manipulated by matter scientists we mean electrons, quarks, fermions, bosons, nuclei, staffs, molecules, materials, quanta, photons, phonons, and so on. (3) There is a temporal scale running across the definition and it is justified to consider and to stress that, over the history, objects may be discovered to be collections of other objects or simply dropped out for different scientific reasons from the set of objects.

After the above intermezzo on chemical species, we can come back to chemical reactions, where chemical species are related. It is by conducting chemical reactions that chemistry progresses and it is through them that its theories come up and its practices make sense. Note that now a chemical reaction involves the different instances of chemical species—for example, photochemical reactions—where the photon is a chemical species; reactions involving inclusion complexes, addition compounds, and so on; where the strong attachment to chemical bond is not any more decisive to call an object a chemical species. Hence, we are generalizing the idea of chemical reaction to that of *chemical relation*. The generalization goes beyond being able to obtain a stable stuff we can store in bottles or even the energetic constraints favoring one molecular configuration over some others. The level of detail meant by chemical relation encompasses traditional chemical reactions, including tautomerism and the several other flavors of relations between species matter scientists have found or theorized upon.[29] An important aspect of chemical relations is that they must be detected experimentally, which bridge these high-level relations to the available theoretical and technological means of a particular time.

We think there are mainly two attributes characterizing the concept of chemical element: (1) relational attributes encoded in chemical relations, and (2) a measured atomic number.

The atomic number is needed as a means of regarding chemical species of traditional interest for chemistry. This condition rules out phonons, photons, and so on. In general, the combination of relational attributes and of objects with atomic number makes that species with short lifetimes be necessarily excluded, therefore regarding only species of chemical interest.

Thus, we claim that the concept of chemical element entails chemical species as objects; the attributes of the concept are (i) detected chemical relations, and (ii) experimental measurement of atomic number. Hence, a chemical element

---

[28] Note the change from matter transformation to matter interaction. The former presupposes "generation," while the second "discovery." A transformation of a substance may lead to either a known substance or a new one. An interaction presupposes knowledge of the interacting species, at least potentially. This potential set of chemical species spans the so-called chemical space (van Brakel 2012; Llanos et al. 2019).

[29] See, for example, van Brakel 2012 for some of these relations.

is a chemical species, for which we have experimental evidence of its interaction with other chemical species.

Noble gases are then regarded as chemical elements, for we have obtained compounds for some of them, in the traditional sense, and also have experimental evidence of their interactions with other chemical species, for example, forming van der Waals complexes (Tanjaroon and Jäger 2007). Superheavy elements are also covered by the definition, as their chemical relations have been explored experimentally through the attachment of some of their atoms to surfaces (Ball 2019). The definition is time dependent, for it relies on the technological conditions of each epoch and on their associated theories and identified chemical species of the time. Hence, when we claim that flevorium is a chemical element, it is because it lasts enough allowing us to detect its interactions with other chemical species, for example, bonds with gold surfaces (Yakushev et al. 2014) and because its atomic number has been determined experimentally. A phonon, in contrast, in spite of being a chemical species and of having relationships with other species, lacks atomic number; therefore is not considered an element.

Conclusion and Outlook

After exploring the evolution of chemical element as a concept, we found that the very concept mirrors the changing ontology of chemistry, which initially was stuff-based, afterward substance-based, and currently based on chemical species that includes not only traditional substances one can store in bottles, but quasi-molecular species, nuclides, and other contemporary constructs. We showed that in such a dynamical ontology, while the concept of element adjusted to the changes, nevertheless it departs from traditional chemistry to the extent of including as chemical elements species unable to undergo chemical interactions, for the simple fact that they do not last enough.[30]

Facing and overcoming the difficulties of concept building, we made a broad attempt of defining the concept of chemical element by looking for the closed set of objects and attributes proper to chemical elements. This required expanding the definition of chemical species to non-traditional chemical entities and the selection of attributes stressing the relational character of the chemical species. Likewise, we expanded the definition of chemical reaction to that of chemical

[30] It is very difficult to state who is departing from whom: Are chemical elements departing from mainstream chemistry? Or is it chemistry departing from the science of exploring matter transformations? A possible way to solve this question is by a historical exploration of the kinds of ontological objects chemists have reported in the literature and how those objects are treated and reported by other disciplines, say physics. This also involves considering disciplinary changes. Of particular interest in such a study would be the so-called non-stoichiometric compounds, which have been on the borderline between chemistry and physics.

relation, which entails interactions of chemical species. Hence, traditional chemical reactions are a particular kind of chemical relation. We claim that a chemical element is a chemical species with experimental atomic number and with detected relations with other chemical species.

This situation of building up the edifice and then looking back to ask for its foundations is not only proper to chemistry and it should not seem counterintuitive. The history of science, for example, the history of mathematics, shows a similar trend. In mathematics, geometry was devised as well as calculus and then analysis, but it was after such an edifice was constructed that mathematicians were able to ask themselves about the meaning of number, number systems, their properties, and relationships (Kline 1980).

Acknowledgments

We are indebted to Joachim Schummer and José Luis Villaveces for motivating with thoughtful discussions over the years several of the ideas here presented. We also thank Jeff Seeman for his comments upon a draft of this chapter. We are especially indebted to Farzad Mahootian for his careful reading of an advanced draft and for his multiple suggestions to improve its readability. This document is dedicated to the memory of José Luis Villaveces.

References

Allison, David. 1999. "Structuralism." In *The Cambridge Dictionary of Philosophy*, 2nd ed., edited by R. Audi, 882–884. Cambridge: Cambridge University Press.

Ball, Philip. 2019. "Extreme Chemistry: Experiments at the Edge of the Periodic Table." *Nature* 565: 552–555.

Borges, Jorge Luis. 2007. *Labyrinths, Selected Stories and Other Writings*. New York: New Directions Publishing Corporation.

Cook, Roy T. 2009. *A Dictionary of Philosophical Logic*. Edinburgh: Edinburgh University Press, 155.

Ganter, Bernhard, and Rudolf Wille. 1999. *Formal Concept Analysis: Mathematical Foundations*. Berlin: Springer.

Indelicato, Paul, Jacek Bieroń, and Per Jönsson. 2011. "Are MCDF Calculations 101% Correct in the Super-heavy Elements Range?" *Theoretical Chemistry Accounts* 129: 495–505.

Jeannin, Y. P. 1991. "Criteria that Must Be Satisfied for the Discovery of a New Chemical Element to Be Recognized." *Pure and Applied Chemistry* 63: 879–886.

Jensen, W. B. 2005. *Mendeleev on the Periodic Law, Selected Writings, 1869–1905*. Mineola: Dover.

Karol, Paul J. 2018. "Heavy, Superheavy . . . Quo Vadis?" In *Mendeleev to Oganesson: A Multidisciplinary Perspective on the Periodic Table*, edited by Eric Scerri and Guillermo Restrepo, 8–42. New York: Oxford University Press.

Klamt, Steffen., Utz-Uwe Haus, and Fabian Theis. 2009. "Hypergraphs and Cellular Networks." *Plos Computational Biology* 5: e1000385.

Kline, Morris. 1980. *Mathematics: The Loss of Certainty.* Oxford: Oxford University Press.

Lavoisier, Antoine. 1965. *Elements of Chemistry.* New York: Dover.

Leal, Wilmer, and Guillermo Restrepo. Forthcoming. "Formal Structure of Periodic System of Elements." *Proceedings of the Royal Society A: Mathematical, Physical and Engineering Sciences,* 2019, vol. 475.

Llanos, Eugenio J., Wilmer Leal, Duc H. Luu, Jürgen Jost, Peter F. Stadler, Guillermo Restrepo. Forthcoming. "The Exploration of the Chemical Space and Its Three Historical Regimes." *Proceedings of the National Academy of Sciences of the United States of America,* 2019, vol. 116, 12660–12665.

Lycan, William G. "Philosophy of Language." 1999. In *The Cambridge Dictionary of Philosophy,* 2nd ed., edited by R. Audi, 673–676. Cambridge: Cambridge University Press.

Medin, Douglas L. 1989. "Concepts and Conceptual Structure." *American Psychologist* 44: 1469–1481.

Oden, Gregg C., and L. L. Lopes. 1982. "On the Internal Structure of Fuzzy Subjective Categories." In *Recent Developments in Fuzzy Set and Possibility Theory,* edited by R. R. Yager, 75–89. New York: Elsevier.

Restrepo, Guillermo. 2018. "The Periodic System: A Mathematical Approach." In *Mendeleev to Oganesson: A Multidisciplinary Perspective on the Periodic Table,* edited by Eric Scerri and Guillermo Restrepo, 80–103. New York: Oxford University Press.

Restrepo, Guillermo, and Rom Harré. 2015. "Mereology of Quantitative Structure-Activity Relationships Models." *HYLE—International Journal for Philosophy of Chemistry* 21: 19–38.

Restrepo, Guillermo, and Peter F. Stadler. 2016. "Assessing Greenness of Chemical Reactions and Synthesis Plans through Posetic Landscapes." *ACS Sustainable Chemistry and Engineering* 4: 2191–2199.

Rouvray, Dennis H. 1997. "Are the Concepts of Chemistry All Fuzzy?" In *Concepts in Chemistry: A Contemporary Challenge,* edited by Dennis H. Rouvray, 1–16. Taunton: Research Studies Press.

Schummer, Joachim. 1998. "The Chemical Core of Chemistry I: A Conceptual Approach." *HYLE—International Journal for Philosophy of Chemistry* 4: 129–162.

Shields, Christopher. 2016. "Aristotle." *The Stanford Encyclopedia of Philosophy* (Winter), edited by Edward N. Zalta, https://plato.stanford.edu/archives/win2016/entries/aristotle/.

Tanjaroon, Chakree, and Wolfgang Jäger. 2007. "High-resolution Microwave Spectrum of the Weakly Bound Helium-pyridine Complex." *The Journal of Chemical Physics* 127: 034302.

Türler, A., and V. Pershina. 2013. "Advances in the Production and Chemistry of the Heaviest Elements." *Chemical Reviews* 113: 1237–1312.

van Brakel, Jaap. 2012. "Substances: The Ontology of Chemistry." In *Philosophy of Chemistry,* edited by Robin F. Hendry, Paul Needham, and Andrea I. Woody, 190–229. Vol. 6 of *Handbook of the Philosophy of Science.* Amsterdam: Elsevier.

Yakushev, Alexander et al. 2014. "Superheavy Element Flerovium (Element 114) Is a Volatile Metal." *Inorganic Chemistry* 53: 1624–1629.

13

Chemical Elements and Chemical Substances

Rethinking Paneth's Distinction

Sarah N. Hijmans, Université Paris-Diderot, France

Introduction

In order to illustrate the way in which written chemical reactions translate to spectacular transformations of matter, students of chemistry can be shown practical demonstrations, such as reactions involving manganese that show a solution changing color as manganese changes its oxidation state (Pearson 1988). The demonstration is a dilute solution of $KMnO_4$ contains permanganate ions that give it a bright purple color. The addition of NaOH followed by $NaHSO_3$ turns the initial solution green; the MnO_4^- ion has been transformed into MnO_4^{2-}. The resulting solution can be turned a neutral grey by adding HCl; the mixture now contains purple MnO_4^- ions, green MnO_4^{2-} ions, and orange MnO_2 precipitate in suspension. The addition of more HCl turns the sample red-orange, indicating that all of the MnO_4^{2-} has disappeared. One can then make MnO_4^{2-} reappear by adding more NaOH and $NaHSO_3$, and mixing. The green color of the manganate ions will reappear mixed with the bright orange of the manganese dioxide in suspension. During the course of the demonstration, the mixture turns from purple, to green, to gray, to red-orange, and then back to green.

The circularity of this series of reactions exemplifies a remarkable aspect of chemistry: in some cases, chemists can bring about consecutive transformations and end up with the substance with which they started. This leads to the notion that something is conserved in the reaction: parts are separated and reconfigured in such a way that the initial substance can eventually be retrieved. These changes of matter are represented as changes in composition, using chemical reactions in which symbols are rearranged into different chemical formulas. The initial and final states are thereby linked through the identities of their components. Manganese, a transition metal, is capable of exhibiting multiple states of oxidation, giving rise to ions of different colors. To a seasoned chemist

Sarah N. Hijmans, *Chemical Elements and Chemical Substances: Rethinking Paneth's Distinction* In: *What Is A Chemical Element?*. Edited by: Eric R. Scerri and Elena Ghibaudi (2020). © Oxford University Press.
DOI: 10.1093/oso/9780190933784.003.0014

these transformations might seem unremarkable, but a student may be left wondering, How can this spectacular range of colors correspond to one component? How does the rearrangement of the symbols in the chemical reactions translate to macroscopic transformations? How is it possible for elements to change properties?

As Ghibaudi et al. (2013, 1628) state: "the problem of unambiguously defining what is conserved in a chemical transformation has been a matter of discussion for a long time within the chemistry community." When analyzing chemical textbooks, they found a variety of different, often contradicting definitions of the chemical element. Even the International Union of Pure and Applied Chemistry (IUPAC 1997) proposes two separate definitions. A chemical element is defined as:

1. A species of atoms; all atoms with the same number of protons in the atomic nucleus.
2. A pure chemical substance composed of atoms with the same number of protons in the atomic nucleus. Sometimes this concept is called the elementary substance as distinct from the chemical element as defined under 1, but mostly the term chemical element is used for both concepts.

Thus, there is no consensus within the chemical community regarding a single, clear definition of the concept of element. Such ambiguity surrounding the concept of element leads to confusion among students of chemistry. Most students have trouble switching between macroscopic and microscopic levels, as well as between operational and theoretical propositions (Ghibaudi et al. 2013, 1628–1629). Chemists correlate change in observable properties with change in composition in terms of elements (Siegfried 2002, 7). However, the connections between the change of color, smell, density, melting point, and other properties that can be observed in the laboratory on the one hand, and the rearrangement of the symbols of elements on paper on the other, are anything but straightforward.

These questions have led the chemist Friedrich Paneth (1962) to propose a dual meaning for the chemical element. In this paper I argue that we should reinterpret Paneth's identification of two aspects of the concept of element in order to understand the way in which the chemical element provides a connection between chemical operations and chemical theory. First I will analyze Paneth's distinction between two meanings of the concept of chemical element—translated as "basic substance" and "simple substance" —and show that neither of them individually is sufficient to specify the concept of chemical element. Then, based on a brief analysis of the history of chemistry, I will propose a way of rethinking this distinction in order to understand the different aspects of this complex chemical concept.

Two Meanings of the Chemical Element

In his 1931 speech on the epistemology of the chemical element, Paneth analyzed the past and current meanings of the concept, stating that two separate meanings coexist. According to Paneth (1962, 137–138), this "peculiar dual epistemological status" results from the fact that chemists characterize elements by their observable properties, such as color, smell, and so on, while also admitting their presence as components of compounds. Often, these compounds do not share the element's properties and this implies the element is present within the compound without manifesting the properties by which that element is usually identified. Manganese, for example, is characterized as a metal with a melting point of 1246 degrees Celsius on the one hand, but it is also said to be present in MnO_2 on the other hand—even though the latter is an orange salt rather than a metallic solid. How can we say manganese is present within manganese dioxide if this compound has none of the qualities of isolated manganese?

According to Paneth, the solution to this paradox lies in the dual nature of the concept of element. He states that the element is considered to be a substance, characterized by its properties, and at the same time a metaphysical entity that subsists even without those qualities. He thus identifies two meanings of the concept of element: on the one hand the material that cannot be decomposed, characterized by its properties, which Paneth is said to have called the "simple substance," and on the other hand the entity that does not manifest any qualities. The latter is what Paneth is said to have called the "basic substance," which according to him underlies both the simple substances and the compounds.

Paneth's basic substance has no intrinsic properties other than its atomic number, which is its defining characteristic. In the case of manganese, the metal with the melting point of 1246 degrees Celsius, it is a simple substance that ceases to exist when MnO_2 is formed; the basic substance, however, is preserved. According to Paneth (1962, 136), throughout history, the element has undergone an "oscillation" between the two meanings of the term: at times the simple substance had more importance, and at others the metaphysical meaning was more dominant. He states that, despite this oscillation, the two meanings of the term "element" have been used in parallel for many decades. In this section, I will analyze Paneth's distinction between these two aspects of the chemical element, and address some issues surrounding this notion.

Elements as Simple Substances

The term "simple substance," H. R. Post's translation of Paneth's "*einfacher Stoff*," is used to identify materials composed of only one element, which are also referred to as "elementary substances" or "simple bodies." In contrast to compounds, simple substances cannot be decomposed any further. They are chemical substances, and therefore empirical and macroscopic: they can be touched, weighed, smelled, and so on, and are characterized by a number of observable properties. Dihydrogen, for example, is a simple body that can be identified based on the fact that at standard temperature and pressure, it is a colorless, odorless, highly flammable gas. Thus, in Paneth's terms, the simple substance is "that form of occurrence . . . in which an isolated basic substance uncombined with any other appears to our senses" (Paneth 1962, 130).

Yet, Earley (2009, 68) reminds us that, though simple substances are "simple" in the sense that they cannot be decomposed into several components, "they definitely are not 'simple' in the sense of having no internal structure." Indeed, many elementary substances exist as diatomic molecules; for these bodies, molecular orbitals can be established by combining the electronic orbitals of two of the same atoms. Others exist in the form of crystals, as metals with outer shell electrons flowing freely between them, or as ions. Even though these substances are identified as simple, they consist of molecules or other multi-atomic systems,[1] and are characterized in the same way as compounds. In this sense, they are composed like any other type of substance: elemental composition itself does not suffice to explain their properties, since structure, particle interactions, and even milieu also contribute to their macroscopic properties.

The internal structure and the observable properties of the simple bodies are not present in compounds; they are produced during their decomposition. Rather than a mere separation of parts, analysis is a process during which new macroscopic substances are formed, with an internal structure and properties that were not present in the compound. Different ways of producing a substance composed of only one element result in different properties: whereas amorphous carbon can be produced by combustion of organic substances, the decomposition of small hydrocarbons by high heat leads to the production of graphite. In both cases, elemental carbon is isolated, but the resulting properties such as hardness and conductivity vary depending on the process of isolation and the resulting arrangement of the atoms, even though they have the same elemental composition.

Thus, chemical analysis is not just a decomposition analogous to taking apart Lego pieces; it is a chemical transformation that results in entirely different substances. As Harré (2015, 113) says, the assumption that empirical analysis, which is a chemical transformation, yields the parts that were present in a compound substance is a fallacy. Rather, we should say that compounds have the

---

[1] Noble gases such as helium and neon primarily occur in the form of monoatomic gases. Still, the atoms in such a gas interact through Van Der Waals forces, giving rise to properties that belong to the macroscopic substance as a whole.

capacity of yielding simple substances only when put through analytical processes that transform those very compounds and their parts (Harré and Llored
2011). In other words, simple bodies are not conserved during chemical analysis and synthesis; they are produced through decomposition and they disappear
when they combine to form a compound.

Indeed, in the same way that the properties of elementary substances are
produced through chemical analysis, they also disappear during chemical synthesis. In other words, elementary substances are not present as such within the
compound, whereas elements are said to be present as constituents. This is the
paradox that Paneth points out. For instance, according to Paneth (1962, 129) a
chemist would describe "the element sulphur" as "a substance of pale yellow
colour, without taste or smell, insoluble in water, etc." All of these properties
are lost as it combines with oxygen to form "the gaseous, colourless, pungently
smelling sulphur dioxide." Nevertheless, no chemist would deny that sulphur
dioxide contains sulphur. This is the key to the distinction between elementary
substances and abstract elements: sulphuric acid contains the element sulphur
but it does not contain the substance sulphur.

As Jacob (2001, 36–37) states, there are two kinds of analysis and synthesis: the
theoretical kind, which is performed on symbols through the formulation of reaction equations, and the empirical kind, which is an experimental procedure
that is performed on chemical substances. Whereas theoretical composition and
decomposition rearranges symbols without modifying them, the experimental
operations of synthesis and analysis transform chemical substances. The simple
substance disappears to give rise to a compound, but the presence of the element
as a constituent is indicated by its symbol in the chemical formula.

For instance, by taking apart the formula "NaCl" to show the presence of "Na"
and "Cl," a chemist knows that sodium and chlorine were used to produce the salt.
He or she can "confidently enter the laboratory with the clear expectation that
his/her melted salt will yield sodium and chlorine, but not potassium or bromine.
He/she also 'knows' that the compound named by 'NaCl' consists of sodium and
chloride ions and will react with silver nitrate to form a white precipitate of silver
chloride" (Jacob 2001, 38). In order to make these kinds of predictions, and to
interpret and represent the transformations of matter as changes in composition,
chemists base themselves on a notion of element that is stable throughout chemical transformations. For any type of explanation of chemical phenomena based
on composition, they need a constituent that exists as a part of compounds, and
the simple substance does not fulfill this requirement.

Therefore, it is important to clearly distinguish between stable elements
and elementary substances. Elementary substances are not the components of
matter, and they are themselves composed of elements. Yet, the contradiction
that Paneth points out results from the fact that chemists don't always make this

distinction; they use the word "element" to refer to simple substances as well. This is why Paneth identifies two distinct meanings for the term "element": the elementary substance and the element as a stable component of matter. However, this use of the term "element" only results from a confusion caused by the fact that many elements carry the same name as (one of) their corresponding simple substances. Allotropy shows that the terms "element" and "simple substance" are not used in exactly the same way.

Though they are all simple substances, there is usually no more than one allotrope that is referred to as an element: the one that carries the same name as the element. For example, diatomic oxygen gas, also referred to as "oxygen," is often characterized as an element because it carries the same name as element number 8, oxygen. However, one would rarely hear anyone refer to "the element ozone," even though ozone is a simple substance just as dioxygen. Likewise, it is clear that "the element carbon" does not refer to any simple substance since none of the allotropes of carbon carry the same name as the element. It is rare to hear someone speak of the chemical substance carbon, or of the element graphite, for example, since it is very clear that "carbon" refers to the sixth element of the periodic table and graphite to a chemical substance. These examples underline the importance of a clear distinction between elements and simple substances in order to avoid confusion.[2] As Ghibaudi et al. (2013, 1630) say, "The element is a name, a symbol, an atomic number, and a position within the table."

With these issues in mind, Earley (2009, 75–76) proposes to translate "einfacher Stoff" exclusively as "elementary substance" and to reserve the term "element" for the English translation of "Grundstoff"—now translated as "basic substance." This would introduce more clarity and consistency into chemical language, and solve the paradox resulting from the use of one term in two contradictory meanings. However, this would imply identifying the element as the basic substance: a metaphysical, abstract entity that does not exist in the form of a substance. What exactly is this abstract notion of basic substance and is it sufficient to specify the concept of chemical element?

Elements as Basic Substances

As was stated above, the notion of simple substance, according to which the element is a chemical substance, fails to capture the fact that the element is used to refer to a stable constituent of matter. As opposed to the simple substance, the abstract or metaphysical meaning of the element that Paneth is said to have

---

[2] The IUPAC recommends making a clear distinction between elements and their corresponding simple substances (Ghibaudi et al. 2013, 1629).

called the "basic substance"[3] fulfills this requirement. Since it is not defined by a number of observable qualities, the basic substance doesn't appear and disappear with these qualities; it is a component that remains stable throughout chemical transformations. The only characteristic of Paneth's basic substance is its atomic number, which is inaccessible to human perception.

Indeed, the basic substance, according to Paneth (1962), is the "transcendental principle underlying the phenomena" (p.129). Paneth uses "transcendental" in an epistemological sense, meaning "beyond the sphere of consciousness" (note 3), thereby implying that the abstract element is impossible to see, touch, smell, or even detect. Beyond the epistemological claim of inaccessibility to human perception, Paneth makes the ontological claim that the basic substance exists in a separate reality: in a "quality-less, objectively real sphere of nature" (p.130) or "a transcendental world devoid of qualities" (p. 130). Characterizing the element in this way implies that its meaning is fixed independently of scientific practice and completely unchangeable by it.[4]

The element is a concept, and in that sense it is theoretical—however, this is not the same as it being abstract or inaccessible; it has developed through a process of continuous scientific inquiry and therefore it exists thanks to human activity, not independent of it. Van Brakel (2000, 69) argues against this kind of reification of meaning, saying that all theoretical entities in science have to be seen as "products of communicative interaction, continuously reconstructed." Indeed, though his analysis of the concept of element is interesting and valuable in many ways, Paneth uses "unnecessary and unhelpful" philosophical terminology (Hendry 2006, 324) that complicates his discussion of chemistry by introducing questions relating to the metaphysics of perception.

Therefore, Paneth's notion of basic substance, too, is insufficient to grasp the meaning of the concept of chemical element. Furthermore, besides the issues concerning each of the definitions taken separately, the combination of the two is impossible: while Paneth aims to solve the paradox of the subsistence of the elements within compounds, he creates a new contradiction. How can the element be both devoid of qualities and characterized by them? How is it possible that in a simple substance, the element is present both as simple substance and as basic substance, but in a compound only as a basic substance? How can these substances exist separately from each other but still be the same element? How

---

[3] The term "basic substance" is Post's translation for the German *Grundstoff*. This translation, as Earley (2009, 69) explains, is quite problematic because the basic substance is not *a substance*, in contrast to the simple substance.

[4] For example, Paneth (1962, 128) says that "the 'metaphysics' of the basic philosophic idea of the concept of element is not touched" by Lavoisier's new chemistry. In other words, he states that even in insisting on the meaning of the element as simple substance, Lavoisier left the meaning of the element as a basic substance unchanged.

Simple Substance $\longleftarrow$ Element $\longrightarrow$ Basic Substance

Figure 13.1. Paneth's (1962) conception of the duality of the concept of chemical element
Source: F. Paneth, *Foundations of Chemistry* 5 (Springer Nature, 2003):113.

come the element is purely empirical when uncombined, and metaphysical when it exists as part of a compound?

Some of the above issues are solved by Scerri's (2012) analysis of the meanings of the concept of element. Scerri adopts the distinction between "basic substance" and "simple substance," but refrains from using the term "transcendental" when characterizing the basic substance (though he keeps the meaning of a metaphysical or fundamental entity that underlies the empirical phenomena). He explains that the two meanings should be placed on different epistemological levels: the simple body can be physically isolated whereas the abstract element cannot be directly observed by human experience. Scerri therefore says that we should assign the more fundamental, epistemologically profound meaning of the element to the basic substance since it exists "at a deeper level of nature" (p. 74). The interpretation of the two meanings as existing on different epistemological levels enables him to avoid making any ontological claims about the basic substance existing in an inaccessible reality.

Furthermore, Scerri adds a third meaning to the concept of element, which corresponds to the way in which the abstract element exists as part of a compound. Both simple and compound substances are macroscopic substances in which the abstract element exists as a constituent, but only the simple substance is named by Paneth as an instantiation of the basic substance. Scerri therefore concludes that there must be another empirical instantiation of the basic substance, which he names the "combined simple substance" (Scerri 2012, 74 figure 13.2). He uses this term to refer to the way in which an element is characterized as part of a compound substance, in order to make a clear distinction between the empirical forms of the element on the one hand, and the abstract element on the other.

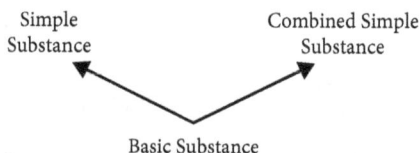

Simple   Combined Simple
Substance   Substance

Basic Substance

Figure 13.2. Scerri's (2012) conception of the three meanings of the chemical element.

Scerri places the basic substance on an epistemological level that underlies the empirical simple substance and combined simple substance, since the basic substance subsists in both empirical forms but is not an empirical body itself. This solves part of the contradictions resulting from Paneth's proposal (see figure 13.1); rather than the basic substance subsisting and the simple substance appearing and disappearing, Scerri's analysis shows a stable fundamental entity (the basic substance) that manifests itself in variable empirical forms (simple or combined simple substance) that can be transformed into one another. His schematization (see figure 13.2) illustrates the idea that Paneth did not aim to distinguish between the way an element exists in a simple substance and the way it exists as a compound, but rather between an empirical, macroscopic substance on the one hand and a stable, abstract element on the other. This emphasizes the idea that the element does not exist either as a basic substance or as a simple substance, but that the basic substance is instantiated either in a simple substance or as part of a compound.

Though Scerri's proposition offers an interesting point of view and an improvement on Paneth's thesis, it does leave some issues unsolved. Scerri follows Paneth in proposing multiple, mutually exclusive meanings for a single concept, which leaves the question of how the element can be multiple things at once. Furthermore, he adopts the term "basic substance" and characterizes it as being metaphysical, in the sense of existing beyond observation and independent of human endeavor (Scerri 2005).[5] This raises some questions about the relationship between the abstract and the concrete: how do chemists study and access metaphysical entities? What exactly does it mean to say that the basic substance underlies the simple substance, and what is the relation between different levels of nature? How can an element be both devoid of properties and characterized by them?

A possible solution to these issues, as proposed by Hendry (2005, 46; 2006, 324–325), is to see the relation between the simple and basic substance as logical, rather than metaphysical; in other words, the "abstract" term refers to the element "in abstraction from any particular state of chemical combination" (Hendry 2006, 324), and specific elementary substances are simply exemplifications of the basic substance. This eliminates metaphysical questions, as well as any logical contradictions between the simple and basic substance, since it allows for a way of including observable properties in the constitution of a more generic concept of element, rather than defining the element as being characterized by its properties on the one hand, and without any properties on the other.

---

[5] According to Scerri (2005) the metaphysical view of a basic substance as existing beyond observation coincides with a second metaphysical interpretation, which is the view of the basic substance as a natural kind.

Because of the issues addressed in this section, I argue that Paneth's proposition for distinguishing between the simple and basic substance does not solve the contradiction that he rightly points out. Considering the requirement of stability on the one hand and the description based on observable properties on the other, it seems these two aspects should be united in a single concept, rather than separated into two contradicting notions. Hendry's proposition for a logically abstract notion of element seems to do just that. But does the idea of an element in abstraction from any state of combination enable a reinterpretation of the concept without leading to a contradiction? This would mean we can abandon the idea of two separate meanings in favor of a single term that includes the properties of simple and compound substances in a stable concept of element. In order to find out, we should look at how the theoretical, microscopic, macroscopic, and operational have related to one another at different points in history. How has the concept evolved as part of chemical practice?

Theory and Experiment

Paneth (1962) bases his account of the chemical element on a historical analysis, in order to show the use of the term "element" both as a stable constituent and an elementary substance. According to Earley (2009, 66–67), Paneth mainly aimed to disprove the idea that Lavoisier completely eliminated metaphysical ideas from chemistry. Indeed, Paneth argues that the metaphysical view was only hidden, but never completely discarded by Lavoisier's definition, and that a metaphysical idea of the element persists in modern chemistry. According to his analysis, the two meanings exist in parallel and they vary in importance in chemistry at different points in history.

In this section I will argue that the development of the concept of element between Lavoisier and Mendeleev shows that different meanings of the term overlap and change, and we cannot identify two static and distinct meanings of the element that coexist in chemistry. There seems to be a single concept of element, which unites experimental and theoretical knowledge, and which corresponds to Hendry's interpretation of Paneth's basic substance. The oscillation that Paneth observes is not a process in which one meaning is temporarily replaced by another, but rather a process of abstraction during which theoretical knowledge becomes more important in the determination of the element. Finally, I will propose a possible reinterpretation of Paneth's distinction that takes these considerations into account.

A Brief History of the Chemical Element

In its most basic form, the idea of element entails simply that all of matter comprises a (finite) number of qualitatively different constituents. Such an idea of the element dates back to ancient Greek philosophy, and most notably Aristotle, who believed that matter comprised four elements (Hooijkaas 1933, 18–27). His ideas remained important in chemistry for a long time, and elements therefore continued to be thought of as metaphysical principles that could not be isolated. However, this changed around the beginning of the eighteenth century when composition took an increasingly important place in chemistry (Chang 2011; Klein 1994; Siegfried 2002), and chemical transformations came to be seen as the rearranging of stable material components.

In the late eighteenth century, the law of the conservation of weight offered a way to distinguish between composition and decomposition and thus substances could be identified as "compound" or "simple" (Chang 2012, 37–41). The simple substances were provisionally identified as the chemical elements by Lavoisier, who reluctantly accepted but did not endorse the provisional criterion of elementhood as the current endpoint of chemical analysis. Using the means available to him at the time, it was impossible to know what the true elements would be; he therefore proposed to eliminate all metaphysical speculation regarding the ultimate constituents of matter and to identify the operationally indecomposable substances as the elements—at least until they were experimentally proven to be composed (Lavoisier 1793, xvii–xviii). In doing so, he is generally considered to have "tied the concept of element down to the operational procedures of chemical analysis (by which Lavoisier meant decomposition)" (Chang 2016, 37).

During the following century, chemistry developed rapidly. New elements were discovered and an atomic weight was determined for each of them. Basing his views on the more extensive knowledge on the element available at his time, Mendeleev could insist on a more abstract aspect of this concept, stating that its only unchangeable quality was its atomic weight (Scerri 2007, 116). In doing so, he insisted on recognizing a positive identifying criterion for each individual element, rather than seeing them as indivisible only until proven otherwise. The atomic weight of each element was the ordering principle of his periodic table, first published in 1869, but he based himself on the properties of both simple and compound substances for the idea of periodicity (Scerri 2012, 72–73). Mendeleev distinguished between elements and simple substances, characterizing elements as both an abstract concept and a "material but invisible part" contained in both compounds and simple bodies (Mendeleev 1889, cited in Scerri 2007, 115; Mendeleev 1867, cited in Kaji 2003, 197).

At first glance, it may seem as though Mendeleev's view constitutes a rupture in the conception of the chemical element, and a return to a pre-existing notion of a stable element that was explicitly absent from Lavoisier's work. However, these is no clear distinction between two separate meanings of the element during this time. First, the conception of the element as being stable was not hidden by Lavoisier's definition—it was implicit but present in his work and that of others. Lavoisier added some substances to his list of elements that he never isolated. He also relied on the notion of element as a stable constituent in identifying oxygen as the principle of acidity in his system of chemistry (Hendry 2005, 35–42). Similarly, Berzelius (1813; 1814) had a theory of oxidation according to which the proportion of oxygen in a substance determined its electrochemical properties.

Second, the ever-present notion of a stable element did not refer to a metaphysical object to which no properties were attributed. During the early nineteenth century the element was seen as a stable material constituent of matter, and this combination of materiality and existence in compounds constituted a link between abstract and empirical conceptions of the element (Scerri 2007, 114). Elements were thought to combine as discrete units of matter, and thus they were seen as stable because they carried a certain quantity of matter that was not modified. Yet, these portions of matter were units of chemical substances, characterized by a certain reactivity and properties rather than size and shape (Klein 2001, 15–17). In short, the idea of element was neither completely empirical nor completely abstract; it was an intermediary notion that united these aspects into a single concept.

Therefore, the oscillation that Paneth observes is not the replacement of one meaning with another, but rather the gradual shift of the concept of element from a primarily operational notion to a more stabilized generally applicable idea. The case of aluminum exemplifies this development. On Lavoisier's list of elements we find *alumine* (alumina, now known as aluminum oxide; Lavoisier 1793, 192), which he thought to be a simple substance. With the development of electrolysis in the early 1800s, previously indivisible substances (among which several oxides) could be decomposed, which led to the discovery of new elements. Davy, who had already discovered barium, magnesium, and calcium using electrolysis (Davy 1808), predicted by analogy that alumina would also be divisible. Thus, he relied on a notion of stable element, and a generalization of experimentally observed properties of compounds and simple substances in order to make his prediction. He named the future element aluminum, which was finally successfully isolated in 1825 by Ørsted.

Thénard (1813, 229) included aluminum in his textbook as part of a group of metals that had not yet been isolated, but of which the existence was generally accepted, and attributed the failure to decompose alumina to aluminum's great affinity for oxygen. Thus, not only did he assume the existence of aluminum in a compound, he also attributed observable properties to it. In 1869, Mendeleev left a gap beneath aluminum in his periodic table. He predicted that this gap would be filled by *eka*-aluminum, which would have intermediate properties between its surrounding elements (aluminum, indium, zinc, and *eka*-silicon [now known as germanium]). Gallium, which was discovered fifteen years later, matched his predictions quite well (Scerri 2007, 131–137). Mendeleev thus insisted on a theoretical notion of the element but he also based this on knowledge of the observable properties of aluminum in every form of combination.

In short, as chemical knowledge developed, an increasingly generic notion of element was formed that could refer to an element in any state of combination. During this development, there was not an opposition between the stable element on the one hand and a simple substance on the other—both observable qualities and theoretical reasoning played a role in the constitution of the element. Gradually, the operational criterion of indivisibility lost importance and more and more predictions could be made that surpassed or even contradicted operational results. Eventually, the theoretical relations of periodicity that existed between elements sufficed in order to accurately predict the atomic weight, properties, and chemical behavior of unknown elements as part of a classification. Today, prediction goes even further beyond operational results: chemists are capable of artificially producing new elements based on empty spaces in the periodic table. Rather than predicting a natural element that will later be discovered, chemists can produce new elements based on theory.

Rethinking Paneth's Distinction

Looking at the history of chemistry, it is hard to identify two distinct notions of element, one stable and devoid of properties and the other macroscopic. The element is both a stable portion of matter and a conceptual tool used to describe chemical transformations; it is characterized by its macroscopic properties as part of both simple and compound substances; both experimental and theoretical considerations play a role in its determination. In that sense there is a certain duality to the concept, but the term "element" does not have two distinct meanings: it refers to the element in all forms of chemical combination. In other words, the simple substance, combined simple substance, and the basic substance are included in the concept of element without mutually excluding one another.

The contradictory use of the term as referring to both substances and stable ingredients of matter, pointed out by Paneth, can be explained by the fact that the concept of element enables chemists to connect different levels of abstraction

Theoretical

↓

Element

↑

Empirical

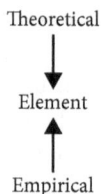

Figure 13.3. Proposition for an alternative view of the duality of the chemical element.

from experimental results. The establishment of the representation of composition, which on the one hand corresponds to operational results and on the other enables a theoretical interpretation of what happens during a reaction, is essential to systematic chemical knowledge.[6] For this, the concept of a constituent of matter is required. Therefore, the element allows chemists to pass from completely operational to more theoretical considerations.

So is there a way in which we can rethink Paneth's duality of the element? I argue that there is, if we see the element as having two aspects rather than two separate meanings. Paneth (1962, 133) himself speaks of the two meanings as "aspects" of the concept of chemical element, when referring to the fact that the abstract and the empirical have been closely linked at certain moments in the history of chemistry. This implies that he does not want to adopt too strong a distinction between the simple and basic substance. If we take this notion of "aspects" further and see them as belonging to one and the same concept, as if they were two sides of the same coin, it might be possible to rethink the duality of the concept of element in a way that is truer to the history of chemistry.

Figure 13.3 depicts a schematization of a new way of interpreting the duality of the chemical element, which might solve some of the issues with Paneth's interpretation. In this figure, the notions of basic and simple substance have been replaced with theoretical and empirical aspects of the concept; "empirical" was chosen so as to include both the simple substance and Scerri's combined simple substance. I have followed Scerri in placing the aspects on different epistemological levels, and I have followed Hendry in seeing these different levels as logical, rather than metaphysical abstraction. Since the element is never purely theoretical nor purely empirical, I propose to interpret it as that which connects the two: according to this interpretation, the element is neither a chemical substance, nor an abstract entity, yet both theoretical and empirical considerations

[6] According to Jacob (2001, 32), "It is at this interface between the manipulation of substances and the manipulation of symbols that simple operations (such as mixing, burning) become describable and generally reproducible, become part of science."

contribute to the meaning of the concept. Thus, the concept of element is a single concept, referring to the constituents of matter, of which the meaning has changed together with chemical practice, and which allows chemists to pass between theory and experiment.

The element might not be the only chemical concept that lies at the crossroads of theory and experiment. Indeed, chemists are constantly engaged in the co-production of macroscopic, observable changes on the one hand and theoretical interpretations and representations on the other (Van Brakel 2000, 98). They do not think only in terms of propositions: they also handle substances in the laboratory and think with their hands and their senses (Knight 1998, 6). Thus, other chemical concepts might exhibit a similar duality as the chemical element. Paneth himself argues that the duality and the questions regarding persistence apply to many chemical concepts, most notably to radicals. So what makes the element such a special concept? Perhaps not its uniqueness in terms of combining theory and experiment but its centrality in chemistry, the fact that it is so closely linked to composition—and thus its essentiality for the systematic establishment of chemical theory. Perhaps, by understanding the role of the element a bit better, we might understand more about how chemical knowledge develops.

References

Berzelius, Jöns Jacob. 1813. "Experiments on the Nature of Azote, of Hydrogen and of Ammonia, and upon the Degrees of Oxidation of Which Azote Is Susceptible." *Annals of Philosophy* 2:357–368.

Berzelius, Jöns Jacob. 1814. "Essay on the Cause of Chemical Proportions, and on Some Circumstances Relating to Them: Together With a Short and Easy Method of Expressing Them." *Annals of Philosophy* 3: 51–62.

Chang, Hasok. 2011. "Compositionism as a Dominant Way of Knowing in Modern Chemistry." *History of Science* 49: 247–268.

Chang, Hasok. 2012. *Is Water H2O? Evidence, Realism and Pluralism.* Dordrecht: Springer.

Chang, Hasok. 2016. "The Rising of Chemical Natural Kinds through Epistemic Iteration." In *Natural Kinds and Classification in Scientific Practice*, edited by Catherine Kendig, 33–46. New York: Routledge,.

Davy, Humphry. 1808. "Electro-chemical Researches on the Decomposition of the Earths; with Observations on the Metals Obtained from the Alkaline Earths, and on the Amalgam Procured from Ammonia." *Philosophical Transactions of the Royal Society* 98: 333–370.

Earley, Joseph E. 2009. "How Chemistry Shifts Horizons: Element, Substance, and the Essential." *Foundations of Chemistry* 11: 65–77.

Ghibaudi, Elena, Alberto Regis, and Ezio Roletto. 2013. "What Do Chemists Mean When They Talk about Elements?" *Journal of Chemical Education* 90: 1626–1631.

Harré, Rom. 2015. "Mereological Principles and Chemical Affordances." In *Philosophy of Chemistry: Growth of a New Discipline*, edited by Eric Scerri and Lee McIntyre, 107–120. Dordrecht: Springer.

Harré, Rom, and Jean-Pierre Llored. 2011. "Mereologies as the Grammars of Chemical Discourses." *Foundations of Chemistry* 13: 63–76.

Hendry, Robin Findlay. 2005. "Lavoisier and Mendeleev on the Elements." *Foundations of Chemistry* 7: 31–48.

Hendry, Robin Findlay. 2006. "Substantial Confusion." *Studies in the History and Philosophy of Science* 37: 322–336.

Hooijkaas, Reijer. 1933. *Het Begrip Element en zijn Historisch-Wijsgeerige Ontwikkeling*, Utrecht: Schotanus & Jens.

IUPAC. 2006–. *Compendium of Chemical Terminology*, 2nd ed. (the "Gold Book"). Compiled by A. D. McNaught and A. Wilkinson. Oxford: Blackwell Scientific Publications, 1997. XML on-line corrected version: http://goldbook.iupac.org.

Jacob, Claus. 2001. "Analysis and Synthesis: Interdependent Operations in Chemical Language and Practice." *HYLE—An International Journal for the Philosophy of Chemistry* 7, no. 1: 31–50.

Kaji, Masanori. 2003. "Mendeleev's Discovery of the Periodic Law: The Origin and the Reception." *Foundations of Chemistry* 5: 189–214.

Klein, Ursula. 1994. "Origin of the Concept of Chemical Compound." *Science in Context* 7, no. 2: 163–204.

Klein, Ursula. 2001. "Berzelian Formulas as Paper Tools in Early Nineteenth-century Chemistry." *Foundations of Chemistry* 3: 7–32.

Knight, David. 1998. Introduction to *The Development of Chemistry 1789–1914*, vol. 1. Edited by David Knight, 1–16. Routledge: London.

Lavoisier, Antoine-Laurent. 1793. *Traité Elémentaire de Chimie, Présenté dans un Ordre Nouveau, et d'après les Découvertes Modernes*, vol. 1. 2nd ed. Paris: Cuchet.

Paneth, Friedrich. (1962) 2003. "The Epistemological Status of the Chemical Concept of Element." Reprinted in *Foundations of Chemistry* 5, no. 2: 113–145.

Pearson, Robert S. 1988. "Manganese Color Reactions." *Journal of Chemical Education* 65, no. 5: 451–452.

Scerri, Eric. 2005. "Some Aspects of the Metaphysics of Chemistry and the Nature of the Elements." *HYLE—International Journal for Philosophy of Chemistry* 11, no. 2: 127–145.

Scerri, Eric. 2007. *The Periodic Table: Its Story and Its Significance*. Oxford: Oxford University Press.

Scerri, Eric. 2012. "What Is an Element? What Is the Periodic Table? And What Does Quantum Mechanics Contribute to the Question?" *Foundations of Chemistry* 14: 69–81.

Siegfried, Robert. 2002. *From Elements to Atoms: A History of Chemical Composition*. Philadelphia: American Philosophical Society.

Thénard, Louis-Jacques. 1813. *Traité de chimie élémentaire, théorique et pratique*, vol. 1. Paris: Crochard.

Van Brakel, Jaap. 2000. *Philosophy of Chemistry: Between the Manifest and the Scientific Image*. Louvain: Leuven University Press.

14

The Dual Conception of the Chemical Element

Epistemic Aspects and Implications for Chemical Education

Elena Ghibaudi, Alberto Regis, and Ezio Roletto,[1] Dept. of Chemistry, University of Torino, Italy

El vanadio se vestía de lluvia
para entrar a la cámara del oro,
afilaba cuchillos el tungsteno
y el bismuto trenzaba
medicinales cabelleras.
Minerales, P. Neruda

A few years ago, Martin Goedhart published a paper titled "The Chemical Element, an Unnecessary Concept?" (Goedhart 1999). The question sounds surprising, as the notion of element stands among the foundational concepts of chemistry and certainly cannot be banned from chemical textbooks. In fact, Goedhart's provocative title was aimed at drawing attention to the following issues:

- What is the notion of chemical element for?
- How ought this notion be taught?

The first question conceals the risk of taking the notion of chemical element as a trivial tool for classifying distinct kinds of atom. If so, one might wonder why chemical classifications shouldn't be restricted to empirically accessible simple substances rather than rely on the more problematic microscopic realm.

The second question stems from the first: chemists assign a dual meaning to the notion of chemical element, as the same term designates either macroscopic simple substances and that (whatever it may be) which is classified in the

---

[1] Prof. Ezio Roletto passed away in December 2018 during the development of this manuscript. This work is dedicated to him and to his commitment to the improvement of chemical education in Italy.

Elena Ghibaudi, Alberto Regis and Ezio Roletto, *The Dual Conception of the Chemical Element: Epistemic Aspects and Implications for Chemical Education* In: *What Is A Chemical Element?*. Edited by: Eric R. Scerri and Elena Ghibaudi (2020).
© Oxford University Press.
DOI: 10.1093/oso/9780190933784.003.0015

periodic table. This results in deep misunderstandings among students and is inherently contradictory. Several authors have pointed out the inconsistency of this duality (the element as indecomposable substance and as what persists throughout chemical changes) when reported to chemical reactions (e.g., Earley 2009; Hendry 2005; Scerri 2012; Jensen 1998; Luft 1997):

> Indecomposable substances disappear during the reaction. While elements are conserved. So when elements are defined as indecomposable substances— which is very usual in schoolbooks—this is contradictory with the law of element conservation. This ambiguity of the element concept was already mentioned by Mendeleev. (Goedhart 1999, 55)

This quote highlights two relevant aspects of the notion of element: (1) the relation between the chemical element and chemical changes; and (2) the fact that Mendeleev, while compiling the periodic table, clearly had in mind the ambiguity of the term "element." What did he classify? The periodic table is certainly not a classification of simple substances: Mendeleev's concept of chemical element went well beyond the mere empirical level (Earley 2009; Hendry 2006a; Scerri 2007; Vihalemm 2003).

Let's focus on the relation between the chemical element and chemical changes, as it is directly related to the issue of what the notion of chemical element is for.

Chemistry textbooks often present the chemical element as a mere classificatory tool that allows us to discriminate between different kinds of atom, with the periodic table as the somehow obvious outcome of such approach.

This is a very unfortunate choice as it overlooks the meaning of the concept of chemical element as well as the value of the amazing theoretical construct represented in the periodic table.[2]

Reducing the notion of element to a means for distinguishing one atom from the other implies reducing the periodic system (and its representation through the periodic table) to a mere collection of information about chemical and physical properties that vary more or less regularly along with the atomic number. These properties would either refer to atoms or substances, depending on the nature of the chosen feature, and the periodic table would be no more than an organized presentation of heterogeneous data.

This is clearly untrue. In addition, such a system would not account for the explanatory and predictive role that the periodic system plays in chemistry, or

---

[2] In the text, we will refer to the periodic system as a system of thought, and to the periodic table(s) as the representation(s) generated from that system.

its endurance, faced with the challenging discoveries of the internal structure of atoms and of isotopes.

Such a reductive view has several drawbacks in chemical education. In fact, not only does it hamper understanding of the difference between the chemical element (taken as an abstract concept) and the simple substance. Even worse, it hampers understanding of the need for such a distinction.

Chemistry deals primarily with the transformation of substances. Hence it faces the problem of defining what is modified and what is conserved throughout a chemical change. A primary need in chemistry is the identification of the factor of continuity in a system undergoing a chemical transformation. Chemistry must also provide an explanation for the mass conservation law, at the microscopic level.[3]

The epistemic problem of defining what persists unchanged in a chemical transformation was clear to Mendeleev who, according to Scerri, acknowledged the question of "how, if at all, the elements survive in the compounds they form when they are combined together" (Scerri 2007, xvi).

The element Mendeleev had in mind when he undertook the search for a consistent classification represented precisely those factors that guarantee the continuity of a chemical system, before and after a chemical change: "Mendeleev realized that abstract elements were to be regarded as more fundamental than simple substances" (Scerri 2007, 117).

In the educational context, this point is rarely stressed; it is rather overlooked. It should, on the contrary, be highlighted as soon as the issue of chemical transformation is introduced:

> A prerequisite for the introduction of "element" is that students understand the conservation of elements as a characteristic of chemical reactions. (Goedhart 1999, 55)

So, to get back to the opening questions of this work, we can say that the notion of element addresses the need for defining what survives in a chemical change.

Once the purpose of this notion is clarified, the problem of finding a proper definition of chemical element still remains. It is somehow paradoxical that the need for this notion goes along with the difficulty of grasping the actual meaning of the term "element."

Both the history and the longstanding debate about the concept of chemical element—among chemists, philosophers, and chemistry educators—points to a genuine difficulty. In fact, even the International Union of Pure and Applied

---

[3] The mass conservation law in physics refers to the undistinguished matter, but in chemistry it implies the conservation of elements.

Chemistry (IUPAC) proposes a definition that keeps together two clearly conflicting views. In fact, according to the IUPAC, a chemical element is:

1. A species of atoms; all atoms with the same number of protons in the atomic nucleus.
2. A pure chemical substance composed of atoms with the same number of protons in the atomic nucleus. Sometimes this concept is called elementary substance as distinct from the chemical element as defined under 1, but mostly the term chemical element is used for both concepts. (IUPAC 1997)

The difficulty in converging to a single, unambiguous, and consistent definition of chemical element stems from a number of different reasons. The philosophical burden associated with both terms "substance" and "element" is not the least of these reasons, as several philosophers have pointed out in their works (Paneth 1931; Earley 2009; Hendry 2006a; Harré 2005; Sharlow 2006; Mahootian 2013).

Even if the question has little impact on the everyday activity of chemistry practitioners, whose expertise and knowledge allow them to discern whether the term "element" is meant to indicate a simple substance or a more abstract conception, the same is not true for chemistry students. Chemistry classes are the places where chemical knowledge is primarily conveyed and inherited by new generations of chemists; chemical education assures the continuity and the dissemination of chemical knowledge. It has a pivotal role in "building up" the epistemic community of chemists. In this respect, the unambiguous definition of foundational chemistry notions is not an irrelevant issue. So, which answer are we able to provide to the question "what is an element?"

If the notion of chemical element is meant to identify what survives a chemical change, it follows that its identification with *any* material body is troublesome (Restrepo 2018). Let's examine three possible cases:

1. *Element vs. simple substance*: This identification places the notion of chemical element at the macroscopic level. Two main drawbacks are that (1) a simple substance is clearly not conserved in a chemical process (e.g., dioxygen in the synthesis of water); (2) an element may correspond to several simple substances (e.g., Carbon vs. graphite, diamond and fullerene). This latter issue was already pointed out by Perrin. While discussing the presence of the element "oxygen: in oxygenated compounds such as water, carbon dioxide, or sugar—he underlined that:

it is not quite correct to speak of this particular elementary substance as "oxygen." Clearly we might just as well call it "ozone," since oxygen and ozone can be completely transformed into each other. One and the same substance, *to which a distinct name should be given*, appears to us, according to circumstances, sometimes in the form "oxygen" and sometimes in the form "ozone." (Perrin 1916, 8, emphasis added)

This is a convincing argument against taking "chemical element" and "simple substance" as synonyms. It also implies that the periodic table cannot be a classification of simple substances, otherwise it would include a single case for each allotrope. Scerri remarks: "Because he was attempting to classify abstract elements, not simple substances, Mendeleev was not misled by nonessential chemical properties" (Scerri 2007, 118). Similarly, Paneth insists on the relation between the periodic table and a notion of element distinct from that of simple substance:

If it is maintained that the elements are not present in their compounds in actuality (as basic substances), but only potentially (as simple substances), then the fact that the properties of their compounds can form a basis for the most profound systematics of the chemical element becomes inexplicable. (Paneth 1931)

2. *Element vs. atom*: In the atomic-molecular perspective that characterizes modern chemistry, the temptation of identifying the element with the atom is strong. Nevertheless, the chemical element is preserved in compounds, whereas atoms are not preserved as such, in that their structure is perturbed upon formation of chemical bonds: what is preserved is their chemical identity. Further, the periodic table does not list atoms; the atomic weight associated with each element is a weighted average calculated over the population of atoms with the same atomic number. This implies that, except for elements that are monoisotopic, there are no atoms that possess the atomic weight reported in the table.

3. *Element vs. nucleus*: The identification of the chemical element with a further microscopic entity, the nucleus, has been put forward. Although the atomic nucleus is certainly an invariant in chemical processes, the discovery of isotopes has definitely torn down the arguments in favor of the identification of the element with the nucleus. The finding of isotopes was a challenge for the periodic table and fostered a fierce debate regarding whether each isotope deserved its own case in the table or not (Paneth 1931; Soddy 1918). The consequences for the periodic table would have been similar to those mentioned for the inclusion of allotropes as representatives of single

elements. Urbain remarks that the fact of sharing common chemical properties and the same atomic number was judged final by the *Commission
International des Eléments Chimiques,* that stated the atomic number as the
necessary and sufficient requirement for identifying the element: "these
isotopes that display the same atomic number, and their *ensemble* do represent, according to the currently admitted definition, one single and same
chemical element" (Urbain 1925, our translation)

Therefore, the identification of the chemical element with either macroscopic
or microscopic material bodies does not seem admissible. A major drawback
of this identification would be to get lost in a wealth of inessential details.
This would definitely obscure the theoretical and rational content of the periodic system that, on the contrary, concurs to make the theoretical construct
of chemistry logically consequent. Actually, were the periodic table a mere
organization of disparate data, it would have likely been replaced by other
classifications along with the increase of chemical-physical knowledge. Its endurance proves that the periodic system is a system of thought strong enough
to consistently incorporate experimental breakthroughs that could have hardly
been imagined at the time of its formulation (such as the discovery of the
atomic structure).

So far, we have argued that the chemical element can neither admit a dual definition nor be identified with a material body; yet, the problem of how to qualify
this notion remains open.

Seeking a way out of this problem, we now examine the main outcomes
of the philosophical and educational debate over the concept of chemical
element.

A Few Milestones of the Epistemic Debate over the Chemical Element

Philosophy of chemistry has paid a good deal of attention to the notion of chemical element. Two main issues have been tackled: the different perspectives
(empirical vs. metaphysical) adopted by various chemists toward the notion of
chemical element, and the discussion over the philosophical nature of this notion (chemical kind vs. essence).

Regarding the first issue, two conceptions are often set one against the
other: Lavoisier's versus Mendeleev's.

Lavoisier's definition of chemical element is reported in this renowned passage of the *Traité élémentaire de Chimie:*

If, by the term elements, we mean to express those simple and indivisible atoms of which matter is composed, it is extremely probable we know nothing at all about them; but, if we apply the term *elements, or principles of bodies,* to express our idea of the *last point which analysis is capable of reaching* we must admit, as elements, all the substances into which we are capable, by any means, to reduce bodies by decomposition. Not that we are entitled to affirm that these substances we consider as simple may not be compounded of two, or even of a greater number of principles. (Lavoisier 1789, xvii, emphasis added)

This definition has been thoroughly analyzed by many authors. Hendry points out that Lavoisier's perspective is analytical (Hendry 2005): Lavoisier aims at affirming the prevalence of empirical evidence on a philosophical thought that presumes to establish the number of chemical elements a priori. With his definition of *element*, he fixes a criterion for discriminating between simple substances and compounds. It has been claimed that, in doing so, Lavoisier rejects the metaphysical conception of the element (Scerri 2007, 31). Nevertheless, Bensaude-Vincent and others convincingly demonstrated that the idea of elements as principles (namely, oxygen as the principle of acidity) is still present in the *Traité* (Bensaude-Vincent 1986; Siegfried and Dobbs 1968; Hendry 2005; 2006a). It cannot be ignored that even the above-mentioned definition refers to elements as "principles of bodies." In fact, Lavoisier was surely aware of the problem of defining what "(i) is a component of other substances, (ii) can survive chemical change, and (iii) whose presence can explain the chemical and physical behaviour of its compounds" (Hendry 2006a). The simple substance obtained as the endpoint of the analysis does not, as such, fulfill these requirements.

Mendeleev's conception of the element was dictated by different concerns compared with Lavoisier's. Mendeleev was primarily concerned with the three items listed above, that converge on the problem of defining the factor of continuity in a system undergoing chemical changes. Answering this question clearly requires going beyond the empirical and analytical level: an abstraction process is needed and Mendeleev's philosophical mind was drawn to thinking in these terms (Scerri 2007, 114):

No matter how properties of simple bodies may change in the free state, *something remains constant,* and when the element forms compounds, *this something is material existence* and establishes the characteristics of the compounds, which include the given element. In this respect we know only one constant peculiar to an element, namely, *the atomic weight.* The size [magnitude] of the atomic weight, by the *very essence of matter,* is common to the simple body and all its compounds. Atomic weight belongs not to coal or diamond but to carbon. (Mendeleev 1869, cited by Kaji 2002, emphasis added)

Strangely enough, according to Mendeleev, the chemical element is abstract, but it is also strictly associated with the empirical experience: the element is a material, but invisible, part of a compound, endowed with a measurable property: atomic weight. As such, it cannot be identified with the corresponding simple substance:

> It is necessary to distinguish the concept of a simple body from that of an element. A simple body, as we already know, is a substance, which, taken individually, cannot be altered chemically by any means produced up until now or be formed through the transformation of any other kinds of bodies. An element, on the other hand, is an abstract concept, it is the material that is contained in a simple body and that can, without any change in weight, be converted into all the bodies that can be obtained from this simple body. (Mendeleev 1867/68 cited by Kaji 2002, emphasis added)

So, unlike Hendry (2005), who claims that Mendeleev's primary goal was to define "a criterion of sameness and difference for elemental substances," we believe that this is rather a *consequence* of Mendeleev's thought, whose primary aim was to establish the nature of what remains unchanged within a system undergoing a chemical transformation. We rather agree with Scerri's claim that Mendeleev's philosophical view was highly responsible for the achievement of the periodic table (Scerri 2007, 118).

Let's examine further definitions, proposed by other scientists and philosophers.

Kaji reports that Meyer, the co-discoverer of the periodic table, had quite a distinct attitude toward the element as compared to Mendeleev:

> Whereas Mendeleev discarded the atom and relied solely on the refined concept of a chemical element, *Meyer embraced the atom and even supported the speculation of Prout's hypothesis of a primordial matter* (hydrogen) as the building block of the elements. *This prompted Meyer to underestimate his findings* and prevented his having full confidence in his discovery of 1869. (Kaji 2002, emphasis added)

According to Kaji, Meyer—unlike Mendeleev[4]—was an early atomist. Meyer's physicalist approach to the element might have diminished the impact of his table of elements compared with Mendeleev's.

---

[4] Whether Mendeleev was an atomist or not has been a matter of discussion (Scerri 2007; Hendry 2005; Kaji 2003). At the time of the compilation of the periodic table he did not seem to fully embrace the atomistic theory, but his attitude seems to have changed toward the end of his career, as in 1899 he wrote: "The central idea that aided me in undertaking the study of the periodic table consists precisely in this absolute distinction between an atom and a simple body" (Mendeleev 1899).

Dmitriev recalls that "Mendeleev, while working on the Principles, was not only seeking teaching principles allowing a great deal of chemical information to be contained in a rather simple scheme, but was also solving a chemical problem. *He was seeking a rational system of inter-elemental dependencies*" (Dmitriev 2004, emphasis added). The abstract conception of the element was functional to this aim.

A big challenge to the definition of element and to the periodic table was issued by the discovery of isotopes. The problem was so relevant that Soddy, who had proposed to call isotopes the species characterized by the same nuclear charge and different atomic weights, in 1918 wrote:

> I am not much concerned with definitions, but I think the Chemical Society might safely offer a prize of a million pounds to any one of its members who will shortly and *satisfactorily define the element and the atom* for the benefit of and within the understanding of a first-year student of chemistry at the present time. (Soddy 1918)

In this lecture delivered before the Chemical Society in London, Soddy recalls the story of the notion of element, stating that—thanks to Dalton— "The element was first atomised and then the atom was made the central conception of the theory of the ultimate constitution of matter." Since then, "the atom and the element became synonyms, related as the singular to the plural," until the discovery of isotopes. After such a step, the element must be thought differently as "unique chemical and spectroscopic character is the criterion, not of a single kind of atom, but rather of a single type of external atomic shell." He concludes that "the elementary and even the homogeneous character has departed from the conception of the chemical element, but the conception remains, and, whatever we choose to call it, will remain." Soddy's view of the element is clearly atomistic (the material basis for the element is the atom); the discovery of isotopes does not require the rejection of this view, it rather imposes a refinement, as the determinant of the element becomes the nuclear charge and the related electron shell. According to Soddy, the notion of element is a chemical notion, so one must not be diverted by the fact that the element is associated with an isotopic mixture (i.e., it is neither simple nor elementary in a physical sense). All that matters is that each element is "unique in chemical character"[5] and "all this, of course, does not in the least affect or minimise the practical importance of the conception of the chemical elements as understood before these discoveries."

---

[5] The chemical equivalence of isotopes remains a debated issue among chemists. See Weisberg 2010; Needham 2008; Hendry 2010; Scerri 2005.

In 1923, the International Committee on Chemical Elements (1923), comprising outstanding scientists of the time—including F. W. Aston, F. Soddy, and G. Urbain—gave the new official definition of chemical element, as a result of the discovery of isotopes, and decreed that positions in the periodic table should correspond to atomic number:

> *A chemical element is defined by its atomic number.* This number represents *the excess of positive over negative charges in the constitution of the atomic nucleus;*[6] theoretically, the atomic number represents also the *number of electrons* which rotate round the central positive nucleus of the atom. Each atomic number also represents the place occupied by the element in the Mendeleev table." (Aston et al. 1923, emphasis added)

The reference to atomic number as the excess of positive over negative charges inside the atomic nucleus sounds astonishing nowadays, but it mirrors the debate over the structure of atoms that was ongoing at the time. In fact, in 1928, Paine reports that the discovery of isotopes "has enabled scientists to count with certainty the exact number of positive and negative particles inside every nucleus" (Paine 1928).

In 1925, Urbain—who had been part of the commission that decreed the new definition of chemical element—published a book titled "*Les notions fondamentales d'élément chimique et d'atome*" (Urbain 1925). His epistemological introduction points out the need for keeping hypothesis and "positive facts" well apart from each other, and goes on to analyze the notion of chemical element. After pointing out that terms such as "simple substance" and "element" are often used as synonyms, he underlines the necessity to keep them sharply distinct: "The simple body is a given substance, a *chemical species*" (emphasis in original). On the other hand, the substance shared ("*substance commune*") by chemical species dioxygen and ozone "we cannot know it objectively. It is a pure idea and defies any positive description" (Urbain 1925, our translation). Like Soddy, Urbain discusses the challenges faced by chemistry's definition of simple substance and of element with the discovery of atomic structure, radioactivity, and isotopes. What Lavoisier used to take as simple bodies are no longer simple, either because they contain mixtures of isotopes or because even the simplest chemical species, that is, the atom, is now known to be composite. Urbain shifts the discussion to the epistemic ground by remarking the conventional character

---

[6] This apparently odd statement reflects the fact that, before Chadwick's discovery of neutrons in 1932, the atomic nucleus was thought to be a mixture of protons and electrons, with protons in excess over electrons (Friedlander 1981, 20–21).

of categories, related with the aims and needs of each scientific discipline. His conclusion is very clear:

> What, in practice, makes a distinction between chemistry and physics is the routine of laboratory practices imposed by the arbitrary requirements of compartmentalization of classical education. (Urbain 1925, our translation)

Seeking for clarifying the notion of element, Urbain underlines that this notion is related to what is shared by a simple body and its derivatives. After examining the different methods applicable for characterizing elements, and their drawbacks, as well as the parameters that can identify an element, Urbain comes up to the choice of atomic number as the crucial and unique feature that allows to define elements unambiguously, in agreement with the recent conclusions reached by the International Committee. All in all, Urbain remarks that the chemical element is an idea, a category built up by chemists for the inherent purposes of their discipline. The distinctive feature of this abstract category is atomic number. In making this statement, he gets rid of the ancient idea of "underlying substance," although he remarks:

> The philosophical sense of elements will nonetheless remain, despite the use of this word by thousands of experts. For the vast majority of educated men, the element idea will always imply that of irreducibility. (Urbain 1925, our translation)

A further, fundamental contribution on the concept of chemical element comes from Friedrich Paneth, the radiochemist who authored a foundational paper that has fostered a good deal of secondary literature on the subject (Scerri 2012; Hendry 2006b; Ruthenberg 2009; Mahootian 2013; Ghibaudi 2013).

Paneth (1931) remarks on the distinction between the simple substance and the chemical element, and suggests the use of specific terms for each of them. The former is "that form of occurrence in which an isolated basic substance uncombined with any other, appears to our senses" and should be named "einfacher Stoff" subsequently translated as "simple substance." The latter is "the indestructible substance present in compounds and simple substances" and should be named "Grundstoff," subsequently translated as "basic substance." Paneth's basic substance is invariant in chemical transformations and accounts for the chemical properties exhibited by simple substances and the compounds wherein it is contained.

According to Paneth, "The fundamental principle of chemistry that elements persist in their compounds refers to the quality-less basic substances only" as the basic substance designates "transcendental principles underlying the phenomena": Paneth's basic substance belongs to "a transcendental world devoid of qualities."

In taking this stance, Paneth put himself in the stream of Greek philosophy, and was influenced by von Hartmann's view of the Kantian distinction between noumenal and phenomenal realms, as Ruthenberg (2009) and Mahootian (2013) have argued. We will not enter into this philosophical debate, except to stress that Paneth's basic substance belongs to the same formal reality as Mendeleev's element. Besides, just as Mendeleev contradictorily assigns a measurable property to the abstract element, the atomic weight, Paneth does the same with Grundstoff, whose distinctive property is atomic number, that is, nuclear charge. In both cases, despite its belonging to an abstract, or to use Paneth's own words, transcendental world, the element is endowed with a measurable property.

In the process of analyzing the definition of element proposed by different scientists, Paneth goes even further. He criticizes the following statement: "A chemical element is a substance all of whose atoms have the same nuclear charge" because it requires the preparation of the simple substance as a necessary condition. Curiously, it is only in a footnote of his paper that he reports this formulation "A chemical element is the class of all atoms of equal nuclear charge," where the chemical element is defined with complete generality as basic substance and the purely formal character of this notion clearly comes out.

Regarding the way the basic substance persists in compounds, the following statement might suggest that Paneth, despite his insistence on the transcendental nature of the element, thought of elements in terms of atoms or atomic components:

The atomic theory of Rutherford and Bohr enables us to visualise particularly vividly how we are to understand the persistence of an element in its compounds; namely, as the unchanged presence of all atomic *nuclei* of the simple substance. (Paneth 1931, emphasis in original)

Nevertheless, this was certainly not the case as, later, he explicitly rejects Hell's statements: "the concept of element coincides with that of atoms" and "the atoms are the true elements of bodies" (Hell 1878, 1).[7] In fact, Paneth believed that:

the atomic theory can, it is true, contribute enormously to—indeed, may be necessary for—visualising how the basic substance persists in simple substances and compounds. But the concept of basic substance as such does not in itself contain any idea of atomism. (Paneth 1931)

---

[7] This issue has been raised and discussed by Scerri 2012.

Despite its meticulous epistemic analysis of the notion of element, Paneth does not fully resolve the ambiguity addressed by proposing a specific terminology. In fact, at the end of his paper, he admits that—due to its established use—the term "element" may still refer to both *Grundstoff* and *einfacher Stoff*: "I have preferred to speak of simple substance and basic substance as different aspects of the chemical concept of element." And, in the very last lines of his paper: "As has been shown, this concept must be taken in the naive-realistic sense when meaning 'simple substance,' but has to be understood as transcendental when meaning 'basic substance'" (Paneth 1931).

Several researchers have analyzed and further developed Paneth's thought on the chemical element. Among them, we will recall Scerri (2012) who raises the issue of how a transcendental basic substance, devoid of inherent properties, may be investigated and characterized:

> Paneth's insistence that the periodic system only classifies elements as basic substances invites the obvious question of how we might learn about these elements, especially as they are said to have no properties. (Scerri 2012)

Despite this criticism, Scerri supports Paneth's view regarding the distinction between basic and simple substance. Nevertheless, according to Scerri, element, simple substance, and compounds share a "three-way relationship whereby element as basic substance underlies both simple substance and combined element." Hence he suggests distinguishing between the underlying "element" (taken as Paneth's "basic substance") and two kinds of real substances wherein the element is actually found: simple substance and "combined simple substance" (Scerri 2012).

Earley (2009) also focuses on terminology, pointing out that "Post's translation of Grundstoff as 'basic substance' is more problematical than his use of the designation 'simple' for elementary substances." The problem lies in using the term "substance" because, according to Earley "the main thrust of Paneth's paper clearly implies that a Grundstoff is not 'a substance'" as the term "substance" is understood and employed in chemistry and philosophy.

Ruthenberg's (2009) analysis of Paneth's concepts of basic versus simple substance aims at clarifying Paneth's philosophical position that—according to Ruthenberg—cannot be ascribed to Kant's transcendental idealism, but rather to von Hartmann's transcendental realism. In the conclusion of his paper, Ruthenberg points out:

> First, *the basic substances are non-observables and rather concepts than concrete objects.* They are not bearers of properties. Second, in contrast to the simple

substances, *the concept of basic substance carries a considerable amount of metaphysics* (which is a prerequisite for good science). Third, there are borderline cases which suggest that the concepts of some entities can pass the interface between the transcendental and the empirical world due to the progress in scientific inquiry. (Ruthenberg 2009)

The use of terms such as "transcendental" or "metaphysical" with regard to the chemical element has raised a good deal of discussion. Different viewpoints have been expressed. Hendry refuses the recourse to transcendental as misleading: "One way to interpret the talk of transcendence is that the elements inhabit a supersensory noumenal world. But that seems unhelpful: the sodium in common salt (which remember is sodium the element), inhabits the same ordinary sensible world as common salt" (Hendry 2006b).

Earley (2009), commenting on Paneth, points out that as a follower of philosopher Eduard von Hartmann, Paneth used the term "transcendental" only to indicate lack of accessibility to ordinary human perception. In fact, Paneth's conception of element "does not seem consistent with Kant's phenomena/noumena distinction—if the second Kant quotation given earlier is interpreted to hold that the essential natures of things are forever unknowable ('transcendental' in some absolute sense)" (Earley 2009).

Ruthenberg (2009) plainly states that metaphysics is a "prerequisite for good science." Mahootian, on his side, maintains that "in chemistry the relation obtains not between specific sets of concepts, but rather between transcendental and empirical explanatory frameworks" (Mahootian 2013).

We conclude this survey (which is not comprehensive of the countless literature on the subject), with the following points:

- Different conceptions of the term "element" coexist in chemical knowledge. They can be ascribed to two distinct approaches: analytical-empirical (the element as the final term of analysis) versus the abstract-philosophical (the element as the invariant in a chemical change).
- The adoption of a fully empirical stance toward the element does not, by any means, suppress the need for abstraction. Abstraction is unavoidable for addressing the question of what persists in chemical changes and accounts for the chemical-physical behavior of its compounds. Whether or not this abstraction requires metaphysical or transcendental conceptions is a matter of discussion. It remains that abstraction is a the constitutive dimension in construction of scientific knowledge.

The Chemical Element from an Educational Perspective

We will now focus on the educational aspects of the element: how ought the notion of element be taught in order to avoid fueling misconceptions and logical contradictions?

A number of educational papers are devoted to this issue (Luft 1997; Jensen 1998; Nelson 2006; Roundy 1989; Agudelo Carvajal 2015 and references therein). Let us examine some relevant contributions in the field that reflect a variety of epistemic positions of their authors regarding the notion of element.

In the years that followed the pronouncement of the international committee, the debate over the nature of the chemical element was intense, especially because of the discovery of isotopes, whose immediate consequence was the acknowledgment of the element as an inhomogeneous mixture. In 1928, Conant writes that "for all chemical purposes a mixture of isotopes is a constant, definite element" (Conant 1928).

In 1937, Menschutkin (1937) maintains that the chemical element is a principle, albeit stressing its relationships with atoms. He even pinpoints the Karlsruhe conference of 1860 as "the date when atoms were brought into relation with principles, not with simple substances." His definition of element is "based on the resolutions passed by the International Union in 1923" and reads as follows:

> Each group of atoms and their ions having the same atomic number form one aggregation. Each such aggregation is *one chemical element*, which can be defined thus: *a chemical element is a principle, all atoms and ions of which have the same atomic number.* (Menschutkin 1937, emphasis added)

Several attempts at re-conceptualizing the chemical element can be found in the *Journal of Chemical Education* in those years. For example, in 1941 Payne states that the element may be thought "as a substance of the second (or third) order of complexity"[8] (Payne 1941). This provokes Weiner's reaction, who rather suggests that "a chemical element is a pure substance" that complies with a list of spectroscopic requirements (Weiner 1941).

Elements' Misconceptions

More recent contributions are concerned with the cognitive obstacles to the comprehension of the notion of element and the need for avoiding misconceptions

---

[8] The reader may refer to the text of Payne's article for a clarification of the concept of "complexity level."

(Barker 2004; Kruse and Roehrig 2005; Laugier and Dumont 2003; 2004; Papageorgiou and Sakka 2000; Taber 2002; Vogelezang 2015). A recent survey on textbooks and teachers' conceptions carried out by Agudelo Carvajal (2015, 329–330) shows a widespread habit to present the element as simple substance, whereas a more abstract conception is less common. Rather oddly, this material conception goes along with the insistence on the chemical element as an invariant of chemical changes.

Laugier remarks that the difficulty of teaching the notion of element (and chemistry, in general) "lies in the necessity for switching between macroscopic and microscopic; between the observable and the model; between the concrete and the abstract" (Laugier 2004). According to this author, the notion of element "introduced at the macroscopic level cannot be fully defined at this level. It is not a strictly macroscopic and empirical concept" (Laugier 2003, our translation). On the other hand, Stains (2007) denounces the risk of trivializing the concept: "Students think of element as a kind of atom" and Schmidt (1998) points out that the term "element" is often used as a synonym for "atom of an element." Khanfour-Armalé and Le Marechal (2008) flip the perspective and argue that "the duality of element can be exploited in chemistry teaching, as a means to achieve the comprehension of the chemical element as the invariant of a chemical transformation."

Vogelezang (2015) argues that the concept of element can be built stepwise, through various conceptualization levels, as something that cannot be created or decomposed ("nonsynthesizable, nondecomposable substances") in a chemical change.

Talking of conceptualization levels, Kruse and Roehrig (2005) criticize the fact that the atom is usually taught as "the smallest part of the element that has all the same [physical and chemical] properties of the element" because this "suggests a common confusion between the term "element," which has both macroscopic and microscopic connotations, in relation to the atom, which has a microscopic connotation only." Unfortunately, in saying so, they seem to refer to the element as both simple substance and atom, thus fostering further confusion.

Reference to Chemical Change

Most authors maintain that the element has to be introduced with a sharp reference to chemical changes and the idea of conservation: "the element must be conceived as what is conserved in the chemical reaction" (Martinand 1993, our translation). It is also remarked that "the peculiarity of the element, as compared to principles of conservation of mass or energy, is that it does not designate a physical quantity, but an abstract entity" (Khanfour-Armalé and Le Marechal

2008, our translation). In fact, the chemical element is an invariant in chemical changes, but—unlike those cited above "it is not a physical quantity, and as such it is neither measurable nor calculable" (Khanfour-Armalé and Le Marechal 2008, our translation).

An interesting remark on conservation comes from De Vos and Verdonk (1987), who believe that "the presence of the element copper in copper sulfate is not a conservation but a theoretical construct. Inevitably, something is to be added to the conservation in order to achieve understanding."

Finally, we note that Soddy remarks that a univocal definition of element is crucial for qualifying the different kinds of transformations—in fact, a chemical change preserves uniquely the element, a physical change preserves substances as well, whereas a nuclear change does not even preserve the element (Soddy 1918, 5).

Abstract or Material Character?

According to Schmidt (1998), the discovery of atomic structure induced a shift of meaning of the term "element," that is now to be placed at the microscopic level: "In the Periodic Table the term *element* is used as a synonym for *atom of an element*." Strangely enough, this author also maintains that "elements are substances consisting of atoms of one kind," a definition that mixes up the macro (substance) and micro (atoms) levels.

A similar definition—that raises similar problems—comes from Kolb (1977): "We might best define an element as a substance all the atoms of which have the same atomic number, or the same number of protons."

Myers (2012) takes a clear stance for the identification of the element with atoms: "Only the atomic form of each element should be considered as the element and most other forms should be called elementary substances." He also remarks without hesitation that "the symbol 'C' in the periodic table means the atom and not its various forms."

On the opposite side, Khanfour-Armalé and Le Marechal (2008) insist on the abstract character of the element: "This notion needs to be built as a category, based on the idea of conservation" and remark that "a chemical element is a category, a class, a case of the periodic table. It cannot display chemical properties." According to their view "between the chemical element and the atom, there is thus a relation of hyperonymy as fruit and apple, or seat and chair" (our translation).

Roundy (1989) recalls: "Each element is defined by its atomic number (or number of protons in the nucleus), whether it is isolated or combined." He also specifies that "we use the same symbol for the element when it is pure as when it is combined" as "in compounds and ions, the nuclei are unchanged. The properties of the elements and their electronic structures are changed." This definition suggests the identification of the element with the nucleus, as an invariant of chemical changes; but, in this perspective, talking about pure or combined elements sounds odd.

Nelson (2006) remarks that a correct understanding of chemistry requires a progression from the macroscopic level toward the concepts and models of the microscopic level. Based on this premise, that complies with the psychology of learning, he introduces first the macroscopic "Elementary substance" as "a substance that does not undergo chemical decomposition into, and cannot be made by chemical combination of, other substances." The "Element" is then "a basic type of matter existing as elementary substances that can be interconverted without change in mass." This last specification stems from the need to take allotropy into account. The different terminology implies that these are clearly distinct concepts. Nelson's basic type of matter recalls Paneth's basic substance: Nelson's basic type of matter exists as elementary substance just like Paneth's simple substance is the way a basic substance appears.

Luft (1997) goes definitely abstract: "element" is an immaterial entity, devoid of physical or chemical properties, characterized by a symbol and the atomic number, that establishes its position in the periodic table. The element is understood as the root of a specific chemical species and common feature to his atoms, molecules, ions, and isotopes.

Finally, according to Jensen, the notion of element belongs to neither the molar nor the molecular level of discourse but to the electrical level, due to his relation to the nucleus. Notwithstanding, it is a formal concept:

> The term "element" is in fact a *descriptor* for a particular kind of nucleus, or more accurately, for a particular "class" of nuclei, all of which have the same atomic number. Hence the formal definition: *Element: A class of nuclei, all of which have the same atomic number.* (Jensen 1998, emphasis added)

Jensen's definition highlights the formal and systematic character of the notion of element, that was crucial for the building up of the periodic system, but it does not fully resolve the problem of relationship between the abstract element and the variety of its material counterparts.

Lexical Choices

Several authors have remarked on the relation between the ambiguity of the definitions of chemical element and the polysemy that characterizes the term "element" (Agudelo Carvajal 2015, 337) that, in turn, stems from the historical evolution undergone by this term. According to Luft (1997), from Boyle on, the use of the term "element" instead of "simple body" is the result of a sort of "lexical laxity" dictated by an established habit that is still present.

The IUPAC (1997) sanctions a double meaning (simple substance and species of atoms) for the term "element," but contradictorily recommends a distinct nomenclature for simple substances and elements (e.g., dioxygen is the simple substance and oxygen is the element).

The problem of lexical ambiguity is particularly relevant in the educational context.

A number of lexical ways out from this ambiguity have been proposed. Paneth (1931) suggests to discriminate between "basic substance" and "simple substance"; Nelson (2006) recommends the use of "element" and "elementary substance"; Scerri (2012) discriminates between "simple substance," "combined simple substance," and "basic substance," suggesting that the latter belongs to a distinct conceptual level compared with the former two—the "basic substance" is what we should call the "element."

Earley (2009) criticizes the use of "simple substance" instead of "elementary substance"; he argues that "elementary substance" and "element" are the best terminological choice for pointing out the difference between Paneth's *einfacher Stoff* and *Grundstoff*, respectively.

Conclusion

In a previous work (Ghibaudi 2013) we pointed out that the lexical heterogeneity stems from the coexistence of distinct conceptualizations of the element. We argued that the chemical element is an abstract notion, the conceptual tool that allows chemists to designate the invariant throughout a chemical change. As such, it should not be confused with a substance—that is never preserved in chemical transformations—or any material body. Hence, terms like "basic substance" are misleading, as they suggest the identification of the element with some sort of (quality-less and somehow mysterious) substance (Earley 2009; Hendry 2006b). The element is a unifying chemical concept that designates an abstract category, identified by the atomic number (Urbain 1925; Luft 1997; Khanfour-Armalé and Le Marechal 2008):

> An element is a name, a symbol, an atomic number, and a position within the [periodic] table; a chemical element does not exhibit macroscopic properties. (Ghibaudi 2013).

Table 14.1 Epistemic Relations between Element, Simple Substance, Chemical Species

Formal level	Element (abstract category identified by Z)
Physical level (Macro)	The simple substance corresponding to a given element
Physical level (Micro)	Any chemical species characterized by Z

All chemical entities (nuclei, atoms, ions, simple substances, etc.) sharing the same atomic number are designated by the same symbol and belong to the same element, without that element being identical to any of them. The element, understood as a category, lies on a distinct, epistemological level as compared to material entities (table 14.1).

This resolves the contradiction (already found in Mendeleev) that lies in conceiving the element as an abstract (or even transcendental) notion and assigning to it a material feature such as the atomic weight. The measurable feature belongs to material entities that are gathered under the abstract category of the "element."

From the educational viewpoint, it is then crucial to build the concept of element as an abstract notion, as this avoids stumbling into logical contradictions regarding chemical changes.

References

Agudelo Carvajal, Carlos Guillermo. 2015. "*La functión de la Tabla Periódica en la enseñanza della Química. Clasificar o aprender.*" PhD diss., Universidad Autonoma de Barcelona.

Aston, F. W., G. P. Baxter, B. Brauner, A. Debierne, A. Leduc, T. W. Richards, F. Soddy, and G. Urbain. 1923. "Report of the International Committee on the Chemical Elements: 1923." *J. Amer. Chem. Soc.* 45, no. 4: 867–874.

Barker, Vanessa. 2004. *Beyond Appearances: Students' Misconceptions about Basic Chemical Ideas.* London: Royal Society of Chemistry.

Bensaude-Vincent, Bernadette. 1986. "Mendeleev's Periodic System of Chemical Elements." *British J. Hist. Sci.* 19: 3–17.

Conant, James. 1928. "Atoms, Molecules and Ions." *J. Chem. Educ.* 5: 25–35.

De Vos, Wobbe, and Adri Verdonk. 1987. "A New Road to Reactions. Part 5: The Elements and Its Atoms." *J. Chem. Educ.* 64: 1010–1013.

Dmitriev, Igor. 2004. "Scientific Discovery in Statu Nascendi: The Case of Dmitrii Mendeleev's Periodic Law." *Historical Studies in Physical Sciences* 34: 233–275.

Earley, Joseph. 2009. "How Chemistry Shifts Horizons: Element, Substance, and the Essential." *Found. Chem.* 11: 65–77.

Friedlander, Gerhart, Joseph W. Kennedy, Edward S. Macias, and Julian Malcolm Miller. 1981. *Nuclear and Radiochemistry*, 3rd ed. New York: Wiley.

Ghibaudi, Elena, Alberto Regis, and Ezio Roletto. 2013. "What Do Chemists Mean When They Talk about Elements?" *J. Chem. Educ.* 90: 1626–1631.

Goedhart, Martin. 1999. "The Chemical Element: An Unnecessary Concept?" In *Chemiedidaktik im Wandel—Gedanken zu einem neuen Chemieunterricht: Festschrift für Altfrid Gramm*, edited by Helke Sumfleth, 49–65. Münster: LIT Verlag.

Harré, Rom. 2005. "Chemical Kinds and Essence Revisited." *Found. Chem.* 7: 7–30.

Hell, Carl. 1878. *Neues Handwörterbuch der Chimie*, vol. 3. Brauschweig: Vieweg and Sohn.

Hendry, Robin. 2005. "Lavoisier and Mendeleev on the Elements." *Found. Chem.* 7: 31–48.

Hendry, Robin. 2006a. "Elements, Compounds, and Other Chemical Kinds." *Phil. Sci.* 73: 864–875.

Hendry, Robin. 2006b. "Substantial Confusion." *Studies in History and Philosophy of Science* 37: 322–336.

Hendry, Robin. 2010. "Entropy and Chemical Substance." *Phil. Sci.* 77: 921–932.

IUPAC. 1997. *Compendium of Chemical Terminology* (the "Gold Book"). Oxford: Blackwell Scientific Publications. http://goldbook. iupac.org/.

Jensen, William. 1998. "Logic, History, and the Chemistry Textbook II: Can We Unmuddle the Chemistry Textbook?" *J. Chem. Educ.* 75: 817–828.

Kaji, Masanori. 2002. "D. I. Mendeleev's Concept of Chemical Element and the Principles of Chemistry." *Bull. Hist. Chem.* 27: 4–16.

Kaji, Masanori. 2003. "Mendeleev's Discovery of the Periodic Law: The Origin and the Reception." *Found. Chem.* 5: 189–214.

Khanfour-Armalé, Rita, and Jean-François Le Marechal. 2008. "Construire une catégorie grâce à une analogie: Cas du concept d'élément chimique." *Didaskalia* 32: 117–147.

Kolb, Doris. 1977. "What Is an Element?" *J. Chem. Educ.* 54: 696–700.

Kruse, Rebecca, and Gillian Roehrig. 2005. "A Comparison Study: Assessing Teachers' Conceptions with the Chemistry Concepts Inventory." *J. Chem. Educ.* 82: 1246–1250.

Laugier, Andrè, and Alain Dumon. 2003. "Obstacles épistémologiques et didactiques à la construction du concept d'élément chimique : Quelles convergences?" *Didaskalia* 22: 69–97.

Laugier, Andrè, and Alain Dumon. 2004. "The Equation of Reaction: A Cluster of Obstacles Which Are Difficult to Overcome." *Chem. Educ. Res. Pract.* 5: 327–342.

Lavoisier, Antoine. 1789. *Traité Élémentaire de Chimie, I, Discours Préliminaire.* Paris: Cuchet.

Luft, Robert. 1997. *Dictionnaire des Corps Purs Simples de la Chimie.* Nantes: Cultures et Techniques.

Mahootian, Farzad. 2013. "Paneth's Epistemology of Chemical Elements in Light of Kant's Opus Postumum." *Found. Chem.* 15: 171–184.

Martinand, Jean-Louis. 1993. "Histoire et didactique de la physique et de la chimie: Quelles relations?" *Didaskalia* 2: 89–99.

Mendeleev, Dimitri. 1867/68. *Lektsii po Obshchei Khimii* 1867/68g., Lecture V, St. Petersburg, reported in Works 15: 381–382.

Mendeleev, Dimitri. 1869. "Sootnoshenie Svoistv s Atomnym Vesom Elementov." *Zh. Russ. Khim. O-va.* 1, nos. 2/3: 60–77.

Mendeleev, Dimitri. 1899. *Rev. Gen. Chim. Pure Appl.* 1: 211–214, cited in D. Mendeleev, *On the Periodic Law: Selected Writings*, edited by William Jensen. Mineola: Dover, 2005.

Menschutkin, Boris. 1937. "Historical Development of the Conception of Chemical Elements." *J. Chem. Educ.* 14: 59–61.

Myers, Rollie. 2012. "What Are Elements and Compounds?" *J. Chem. Educ.* 89: 832–833.

Needham, Paul. 2008. "Is Water a Mixture? Bridging the Distinction between Physical and Chemical Properties." *Studies in History and Philosophy of Science* 39: 66–77.

Nelson, Peter. 2006. "Definition of Element." *Chem. Educ. Res. Pract.* 7: 288–289.

Paine, Phoebe. 1928. "The Nature of Matter." *J. Chem. Educ.* 5: 1135–1147.

Paneth, Fritz. (1931) 1962. "Über die erkenntnistheoretische Stellung des chemischen Elementbegriffs." *Schriften der Königsberger Gelehrten Gesellschaft* 8, no. 4, Halle: Max Niemeyer. Translated by Heinz Post as "The Epistemological Status of the Chemical Concept of Element." *Brit. J. Phil. Chem.* 13: 1–14, 144–160. Reprinted in *Found. Chem.* 5 (2003): 113–145.

Papageorgiou, George, and Despina Sakka. 2000. "Primary School Teachers' Views on Fundamental Chemical Concepts." *Chem. Educ. Res. Pract.* 1: 237–247.

Payne, Eric. 1941. "What Is an Element?" *J. Chem. Educ.* 18: 195.

Perrin, Jean. 1916. *The Atoms.* New York: Van Nostrand. Originally published as *Les atomes.* Paris: Felix Alcan, 1913.

Restrepo, Guillermo. 2018. "The Periodic System: A Mathematical Approach." In *From Mendeleev to Oganesson*, edited by Eric Scerri and Guillermo Restrepo, 80–103. New York: Oxford University Press.

Roundy, Willard. 1989. "What Is an Element?" *J. Chem. Educ.* 66: 729–730.

Ruthenberg, Klaus. 2009. "Paneth, Kant, and the Philosophy of Chemistry." *Found. Chem.* 11: 79–91.

Scerri, Eric. 2005. "Some Aspects of the Metaphysics of Chemistry and the Nature of the Elements." *HYLE—Int. J. Phil. Chem.* 11: 127–145.

Scerri, Eric. 2007. *The Periodic Table: Its Story and Its Significance.* New York: Oxford University Press.

Scerri, Eric. 2012. "What Is an Element?" *Found. Chem.* 14: 69–81.

Schmidt, Hans-Jürgen. 1998. "Does the Periodic Table Refer to Chemical Elements?" *School Sci. Rev.* 80: 71–74.

Schwartz, Eugen. 2010. "Theoretical Basis and Correct Explanation of the Periodic System: Review and Update." *Int. J. Quantum Chem.* 110: 1455–1465.

Sharlow, Mark. 2006. "Chemical Elements and the Problem of Universals." *Found. Chem.* 8: 225–242.

Siegfried, Robert, and Betty Dobbs. 1968. "Composition: A Neglected Aspect of the Chemical Revolution." *Annals of Science* 24: 275–293.

Soddy, Frederick. 1918. "The Conception of the Chemical Element as Enlarged by the Study of Radioactive Change." *J. Chem. Soc.* 115: 1–26.

Stains, Maryline, and Vicente Talanquer. 2007. "A2: Element or Compound?" *J. Chem. Educ.* 84: 880–883.

Taber, Keith. 2002. *Chemical Misconceptions: Prevention, Diagnosis, and Cure. Classroom Resources.* London: Royal Society of Chemistry.

Urbain, Georges. 1925. *Les notions fondamentales d'élément chimique et d'atome.* Paris: Gauthier-Villars.

Vihalemm, Rein. 2003. "Are Laws of Nature and Scientific Theories Peculiar in Chemistry? Scrutinizing Mendeleev's Discovery." *Found. Chem.* 5: 7–22.

Vogelezang, Michiel, Berry Van Berkel, and Adri Verdonk. 2015. "An Empirical Introduction to the Concept of Chemical Element Based on Van Hiele's Theory of Level Transitions." *Science Education* 99: 742–776.

Weiner, S. 1941. "What Is an Element? Letter." *J. Chem. Educ.* 18: 296.

Weisberg, Michael, and Paul Needham. 2010. "Matter, Structure, and Change: Aspects of the Philosophy of Chemistry." *Philosophy Compass* 5, no. 10: 927–937.

Appendix

Selected References

Baird, David, Eric Scerri, and Lee McIntyre, eds. 2006. *Philosophy of Chemistry: Synthesis of a New Discipline*. Boston Studies in philosophy of science. Dordrecht: Springer.

Bencivenga, Ermanno, and Alessandro Giuliani. 2014. *Filosofia chimica*. Rome: Editori Riuniti.

Bensaude-Vincent, Bernadette. 1998. *Éloge du mixte: Matériaux nouveaux et philosophie ancienne*. Paris: Hachette.

Bensaude-Vincent, Bernadette. 2008. *Matière à penser. Essais d'histoire et de philosophie de la chimie*. Paris: Presses universitaires de Paris Nanterre.

Bensaude-Vincent, Bernadette, and Simon Jonathan. 2008. *Chemistry: The Impure Science*. London: Imperial College Press.

Berson, Jerome. 2003. *Chemical Discovery and the Logicians' Program*. Weinheim: Wiley-VCH.

Bhushan, Nalini, and Stuart Rosenfeld. 2000. *Of Minds and Molecules: New Philosophical Perspectives on Chemistry*. New York: Oxford University Press.

Brandas, Erkki, and Eugene Kryachko, eds. 2003. *Fundamental Perspectives in Quantum Chemistry: A Tribute Volume to the Memory of Per-Olov Löwdin*. Dordrecht: Kluwer Academic Publishers.

Chalmers, Alan. 2009. *Scientist's Atom and the Philosopher Stone: How Science Succeeded and Philosophy Failed to Gain Knowledge of Atoms*. Boston Studies in the Philosophy of Science. Dordrecht: Springer.

Chamizo, José Antonio, ed. 2007. *La Esencia de la Química*. Mexìco City: UNAM.

Chamizo, José Antonio, ed. 2010. *Historia y Philosofía de la Chimica*. Mexico City: Siglo veintiuno editores.

Chang, Hasok. 2012. *Is Water H_2O? Evidence, Realism and Pluralism*. Boston Studies in Philosophy of Science. Dordrecht: Springer.

De Landa, Manuel. 2015. *Philosophical Chemistry: Genealogy of a Scientific Field*. New York: Bloomsbury.

Earley, Joseph, ed. 2003. *Chemical Explanation: Characteristics, Development, Autonomy*. Annals of the New York Academy of Sciences. New York: New York Academy of Sciences.

Hendry, Robin, Andrea Woody, and Paul Needham, eds. 2012. *Philosophy of Chemistry*. Handbook of Philosophy of Science. Amsterdam: Elsevier.

Hettema, Hinne. 2017. *The Union of Chemistry and Physics: Linkages, Reduction, Theory Nests and Ontology*. European Studies in Philosophy of Science. New York: Springer.

Hoffmann, Roald. 1995. *The Same and Not the Same*. New York: Columbia University Press.

Janich, Peter, ed. 1994. *Philosophische Perspektiven der Chemie*. Mannheim: BI-Wissenschaftsverlag.

Janich, Peter, and Nikolaos Psarros, eds. 1998. *The Autonomy of Chemistry*. Würzburg: Würzburg Königshausen & Neumann.

Jensen, Knud, J., and Anita Kildebaek, eds. 2008. *Aspects of the Philosophy of Chemistry*. Copenhagen: Danish Society for the History of Chemistry.

Knight, David. 1978. *The Transcendental Part of Chemistry*. Folkestone: Dawson.

Knight, David. 1992. *Ideas in Chemistry*. London: Athlone Press.

Kovac, Jeffrey, and Michael Weisberg, eds. 2011. *Roald Hoffmann on the Philosophy, Art, and Science of Chemistry*. New York: Oxford University Press.

Laidler, Keith. 1993. *The World of Physical Chemistry*. Oxford: Oxford University Press.

Llored, Jean-Pierre, ed. 2013. *Philosophy of Chemistry: Practices, Methodologies and Concepts*. Cambridge: Cambridge Scholars Publishing.

Lombardi, Olimpia, and Ana Rosa Perez Ransanz. 2012. *Los Multiples Mundos de La Ciencia*. Mexico City: Siglo veintiuno editores.

Needham, Paul. 2017. *Macroscopic Metaphysics*. Dordrecht: Springer.

Nye, Mary Jo. 1993. *From Chemical Philosophy to Theoretical Chemistry*. Berkeley and Los Angeles: University of California Press.

Paneth, Friedrich. 1964. *Chemistry and Beyond: A Selection of the Writings of the Late Professor F. A. Paneth*. Edited by Herbert Dingle and G. R. Martin. New York: Interscience Publishers.

Pépin, François. 2011. *Les matérialismes et la chimie Perspectives philosophiques, historiques et scientifiques*. Paris: Éditions Matériologiques.

Primas, Hans. 1983. *Chemistry, Quantum Mechanics and Reductionism: Perspectives in Theoretical Chemistry*. Heidelberg: Springer.

Psarros, Nikolaos. 1999. *Die Chemie und ihre Methoden*. Weinheim: Wiley-VCH.

Psarros, Nikolaos, and Kostas Gavroglu. 1999. *Ars mutandi: Issues in the Philosophy and the History of Chemistry*. Leipzig: Leipziger Universitätsverlag.

Psarros, Nikolaos, Klaus Ruthenberg, and Joachim Schummer, eds. 1996. *Philosophie der Chemie: Bestandsaufnahme und Ausblick*. Würzburg: Königshausen & Neumann.

Rocke, A. 1993. *The Quiet Revolution: Hermann Kolbe and the Science of Organic Chemistry*. Berkeley and Los Angeles: University of California Press.

Rouvray, Dennis H., and R. Bruce King, eds. 2004. *The Periodic Table: Into the 21st Century*. Philadelphia: Research Studies Press.

Ruthenberg, Klaus, and Jaap van Brakel, eds. 2008. *Stuff: The Nature of Chemical Substances*. Würzburg: Königshausen and Neumann.

Scerri, Eric. 2007. *The Periodic Table: Its Story and Its Significance*. New York: Oxford University Press.

Scerri, Eric. 2008. *Collected Papers on Philosophy of Chemistry*. London: Imperial College Press.

Scerri, Eric. 2009. *Collected Papers on the Periodic Table by Eric Scerri*. London: Imperial College Press.

Scerri, Eric. 2011. *The Periodic Table: A Very Short Introduction*. New York: Oxford University Press.

Scerri, Eric. 2013. *A Tale of Seven Elements*. New York: Oxford University Press.

Scerri Eric. 2016. *A Tale of Seven Scientists and a New Philosophy of Science*. New York: Oxford University Press.

Scerri, Eric, and Grant Fischer, eds. 2016. *Essays in the Philosophy of Chemistry.* New York: Oxford University Press.

Scerri, Eric, and Lee McIntyre, eds. 2015. *Philosophy of Chemistry: Growth of a New Discipline.* Dordrecht: Springer.

Scerri, Eric, and Guillermo Restrepo, eds. 2018. *Mendeleev to Oganesson: A Multidisciplinary Perspective on the Periodic Table.* New York: Oxford University Press.

Schummer, Joachim. 1996. *Realismus und Chemie. Philosophische Untersuchungen der Wissenschaft von den Stoffen.* Würzburg: Königshausen and Neumann.

Sobczynska, Danuta, Pawel Zeidler, and Ewa Zielonaka-Lis, eds. 2004. *Chemistry in the Philosophical Melting Pot.* Frankfurt: Peter Lang.

Soentgen, Jens. 1997. *Das Unscheinbare: Phänomenologische Beschreibungen von Stoffen, Dingen und fraktalen Gebilden.* Berlin: Akademie Verlag.

Van Brakel, Jaap. 2000. *Philosophy of Chemistry: Between the Manifest and the Scientific Image.* Leuven: Leuven University Press.

Villani, Giovanni. 2001. *La chiave del mondo. Dalla filosofia alla scienza: l'onnipotenza delle molecole.* Naples: CUEN.

Villani, Giovanni. 2008. *Complesso e organizzato: Sistemi strutturati in fisica, chimica, biologia ed oltre.* Rome: Franco Angeli.

Index

For the benefit of digital users, indexed terms that span two pages (e.g., 52–53), may, on occasion, appear on only one of those pages.

Philosophy of Symbolic Forms, 161,
163–64n14
reason and, 153–54
on regulative vs. constitutive
principles, 153–54
Substance and Function, 161,
163–64, 193–94
causal and causal-descriptive theories of
reference, 3, 70, 81–83
causality and causal relationships
of elements in compounds, 89–90, 96–97,
105–6, 120–21
existence through, 135n9, 138–39, 140n12
in Kant's philosophy, 146, 147, 173
Cavendish, Henry, 70–72, 73–75,
74–75n18, 78, 99
Chang, Hasok, 179, 196–97
Charleton, Walter, 93–94
chemical elements. *See* elements
chemical reactions. *See* reactions
chemistry
causation and, 16, 16n10
education in, 1–2, 3–4, 19
historical divisions of, 32, 38, 43, 44, 110,
125–26, 167–68, 171, 172–76, 238n30
Kant's relevance to, 146–52
linguistic changes in, 118–19
mathematical techniques in, 99–100, 153,
154–55, 158–60
operationalism and, 168–69, 173–74
Ostwald and, 212–17
philosophy's relationship with, 5–7, 5–6n1, 8,
9–11, 45, 185–86, 222
physics' influence on, 127–29, 128–29n5
quantum mechanics and, 17, 128–29,
155–56, 199
Chevreul, Michel-Eugène, 38
chlorine, 10, 19
classification
atomic weights and, 39, 196–97
continuity of, 259
education and, 258
Harré's epistemological framework and, 169,
177, 179–81
Lavoisier and, 175–76
operationalism and, 160–61
Paneth and, 181
periodic system and, 9–10, 25, 32–33, 40, 57,
126, 181–82, 261
colorless airs
descriptions of, 76
examples of reference to, 70–76
general views on, 77

names for, 76
theories of reference and, 69–70, 77–83
communication across terminologies, 73, 75,
76. *See also* causal and causal-descriptive
theories of reference
complementarity, 155–56
composites, 26, 90–91, 112–13, 114, 119–21,
134–39, 266–67
compounds and compound bodies. *See also*
mixts; substances
Aristotle on, 124–25, 124–25n2, 182
covalent, 15–16
Dalton on, 102–3, 106–7, 125
electron configurations and, 19
elements in, 89–90, 96–97, 105–6, 120–21,
175–76, 182–84, 186, 248, 252, 253, 259,
260–62, 268
heterogeneity and, 198–99
historical perspectives of, 251
Mendeleev on, 9, 43–44, 63–65, 191–92
mereology and, 200–1
non-stoichiometric, 189, 198,
199–200, 238n30
operationalism and, 181
simple substances in, 244–45, 248
Conant, James, 271
concepts
in Cassirer's philosophy, 153–56, 157, 163–64
chemical element's evolution and,
227–34, 250
chemical reactions and, 228f, 228
education and, 272, 275–76
experimentation and, 193–94, 195–96,
197, 250–55
Formal Concept Analysis and,
227n8, 234–35
formation of, 225–27
Condillac, Etienne Bonnot de, 33–34, 35–36n1
Congress of Karlsruhe (1860), 60–62, 191,
196–97, 271
consciousness and Kant, 146–47
conservation of mass. *See* Law of Conservation
of Mass
continuity, law of, 215
core of chemistry, 144–45, 160–64
coronium, 47
corpuscles, 88–89, 92–95, 96, 99,
101–2, 125–26
critical point, 60
crystals and crystallography, 37–38, 56–57,
128–29, 180–81, 183, 198–99, 244
Cullen, William, 73–74, 100–1
Curie, Marie & Pierre, 47

Praise for *Extending MFC Applications with the .NET Framework*

"Using Microsoft .NET in a new application is easy. Taking advantage of .NET in existing MFC applications is the real challenge! Tom Archer's book provides the information developers need to begin profiting from .NET today without a costly up-front rewrite of existing software."

—Mark Burhop
Systems Architect
Integraph Corporation

"To anyone using MFC who hasn't decided whether or not to jump onto the .NET bandwagon—do it! This book clearly illustrates the major productivity advantages and cost benefits when combining the .NET classes with existing MFC applications."

—Brian Delahunty
CodeProject.com

"To anyone looking to augment their existing MFC code base and knowledge with the powerful .NET classes—providing such capabilities as disconnected data, in-memory database (IMDB), regular expressions, and data encryption—Tom Archer's book has it all."

—Erik Westermann
Lead Architect
Eidenai Innovations

Extending MFC Applications
with the .NET Framework

Mike! Thanks for the
Scholman book. I hope
you enjoy this one
as well!

Tom Archer

Extending MFC Applications with the .NET Framework

Tom Archer
Nishant Sivakumar

✦ Addison-Wesley

Boston • **San Francisco** • **New York** • **Toronto** • **Montreal**
London • **Munich** • **Paris** • **Madrid**
Capetown • **Sydney** • **Tokyo** • **Singapore** • **Mexico City**

About the Cover:

Bridging Two Worlds
By Tom Archer and Justin Dunlap

Two bold entities
One lit—known and understood
One dark—mystery

Two worlds set apart
Both with so much to offer
How best to combine?

This book—like a bridge—
Bringing both worlds together
Joining light and dark

Library of Congress Cataloging-in-Publication Data

Archer, Tom.
    Extending MFC applications with the .NET framework / Tom Archer, Nishant Sivakumar.
        p. cm.
    ISBN 0-321-17352-X (pbk.)
        1. Application software—Development. 2. Microsoft foundation class library. 3. Microsoft .NET Framework.
    I. Sivakumar, Nishant. II. Title.
    QA76.76.A65A69 2003
    005.268—dc22   2003020714

ISBN 0-321-17352-X
Text printed on recycled paper
1 2 3 4 5 6 7 8 9 10—CRS—0706050403
First printing, December 2003

To my loving, dear wife, Krista.
I don't know where I'd be without you
and have no desire to find out.
I love you, Special K.
—Tom

To my parents and my dearest sister, Seetha,
who's always been by my side through
joys and worries alike.
—Nish

Contents

Foreword

"Yes!" That was how I felt when I started reading the proofs of this book. This is the first book I have seen that explains very clearly how Visual C++ developers can use the .NET Framework *right now* to increase their productivity in writing new MFC applications and enhancing existing ones. It addresses very successfully the mechanics of how you start using .NET APIs from your MFC applications. It explains where the feature our team called IJW (*It Just Works*), really just works and where you need to coax it across some pitfalls. But this book dispenses with that in short order and then goes straight to the real meat of selecting specific API sets from the vast breadth of the .NET Framework and explaining how to use them from your MFC code.

Just reading through the table of contents and the first few chapters it becomes clear immediately that Tom Archer is writing from the ideal vantage point of both having a deep understanding of the technology and the direct hands-on experience of using it in the real world, on a real project, with real deadlines, and on a team with real people facing all the pressures and deadlines of the typical modern development project.

What you won't find in here is the umpteenth overview of the .NET Framework, or even a book just focused on explaining the .NET Framework from the viewpoint of a C++ or MFC developer. What you will find is exactly which parts of the .NET Framework you can use *now* to increase your productivity—which parts are clearly superior in the .NET Framework compared to equivalent (or in some case non existent) Win32, COM, or MFC functionality. This book simply couldn't have been written a year ago because it is so deeply steeped in real-world usage. The author's experience with all the relevant parts of the technology—C++, MFC, Win32, the CLR, and the .NET Framework—clearly shows through in every single page. And so does his talent for clear, unambiguous explanations.

Because WinFX, the new API for the Longhorn OS, is a managed API set, using the techniques in this book ideally positions you to make your MFC applications great *Longhorn* applications. Starting now to use the

.NET Framework for specific technologies like remoting, file access, cryptography, and XML will not only jump-start your experience, but also will later make it easier to integrate the new Longhorn features into your application.

Designing the new releases of Visual C++ to help make you, the C++ developer on the Windows platform, ideally positioned to take advantage of the new platform functionality, even with a major shift like .NET and Longhorn, is what our team is all about and what I personally spend almost all my waking moments on. This book is all about how that works in the real world, with the current product. It both validates where we got it right and shows you how to make it work in the places where we haven't yet reached the ideal solution. It makes the hard value judgments in the context of your MFC applications as to which parts of the .NET API are beneficial to use now and which parts don't offer compelling advantages over the traditional ways of implementing the functionality.

I have already been using the material in this book in guiding the future evolution of Visual C++. This is a book that is going on my shelf of books that I use daily. I think it will go on yours as well.

Ronald Laeremans
Group Program Manager, Visual C++
Microsoft Corporation

Preface

Assumptions and Goals

This book assumes that you are an experienced MFC developer who, while not ready to completely migrate your code to .NET, is interested in seeing how .NET can make you more productive. As an MFC developer, you're infinitely more comfortable with the document/view and dialog model of creating user interfaces and are simply much more productive with this development environment. In addition, you might have a large amount of source code that you're not willing or able to migrate to .NET right now—if ever. However, you're still intrigued by some of the .NET classes and how they might complement or even replace various MFC counterpart classes. That is specifically who this book targets. As opposed to the many .NET books that seem to take the attitude that it will be all or nothing for developers, this book is more realistic in realizing that many developers (especially those using C++ and MFC) will want to use .NET only where there is a clear and obvious advantage.

In addition, while I do go into some detail on the various .NET classes that I illustrate throughout this book, the book is not a ".NET class library book." There are hundreds of .NET types and classes, and that simply isn't the focus here. My objective is to show you a few select classes that I feel would benefit most MFC developers; see, for example, the XML chapter.

XML has become the standard for exchanging data in many organizations. However, in order to use XML from MFC, you're probably going to be using the Microsoft IE (Internet Explorer) XMLDOM via COM (Component Object Model). While this is workable solution—I've written several articles and book chapters on how this is done—using the XMLDOM just doesn't compare to the very elegant and powerful set of .NET XML classes. Therefore, a complete chapter is devoted to the .NET XML namespace,

covering everything from reading/writing XML files and traversing XML documents to querying XML data using XPath.

As a result of this approach, when finished with this book, you will know how to integrate the power of the .NET Base Class Library (BCL) into your existing—and future—MFC applications, which will ultimately make you a more productive—and more marketable—programmer.

Technical Matters

While it's sometimes helpful to think of Visual C++ .NET as containing two separate C++ compilers—one for managed (.NET) code and one for unmanaged code—technically, the Visual C++ .NET product contains only one compiler that has two distinct parts:

- **ISO Standard C++**, an ISO-compliant C++ compiler that can be used (with the inclusion of the appropriate libraries) to generate native (x86) executables and libraries for projects such as console applications, Windows applications, Windows services, and so on. When combined with the MFC class library, this is what pays the bills for most of the intended readers of this book (and its authors).
- **Managed Extensions**, which are the language constructs available for mixing managed C++ syntax with your standard C++ code. The Managed Extensions allow you to define and use managed types (garbage-collected reference type, value types, scoped enums, delegates, etc.) and use managed runtime facilities such as reflection and security.

Therefore, there's only one compiler, and you can think of Managed Extensions as somewhat like other Microsoft C/C++ extensions, such as __declspec, except obviously more extensive, as it has the responsibility of exposing all of the CLI features: garbage collection, reflection, and so on. You might even think of Managed Extensions as a superset language containing Standard C++ as a subset. So now we know that a single compiler can generate both native and managed code for Standard C++. What about mixed code applications? Let's look at two techniques for mixing native and managed C++ code: COM Interop and IJW (It Just Works).

There are actually two types of COM Interop. There is the COM Callable Wrapper (CCW), which allows you to create a "pure" native application (with no .NET build-time dependencies) that accesses .NET code via COM, and the Runtime Callable Wrapper (RCW), which allows you to access native code from managed applications. Therefore, one direction I could have taken in this book would have been to write MFC applications and then access the various selected .NET classes via CCW. However, CCW has many drawbacks that inevitably make it unappealing. First, CCW is not very efficient in terms of execution speed. Second, CCW cannot use the entire CLR type system—method parameters and properties are restricted to automation types. Finally, many .NET types are not exposed to COM, so accessing them would be very difficult or impossible.

IJW refers to the collection of mechanisms that allow managed code to call native functions, compile unmanaged types in MSIL, use unmanaged types in managed type method signatures, provide native entry points to MSIL methods, and so on. However, IJW is not only about managed code calling native functions. It's also about exposing native entry points to managed functions and basically doing all the plumbing necessary to get C++ code to work under .NET.

If you're a bit confused by the last paragraph, that's perfectly understandable. After all, it now sounds like I'm talking about writing .NET applications that use native code when the book's topic promises to be about the opposite—using .NET to augment MFC applications! As it turns out, although at first blush this does seem to be backwards, and the designers of IJW probably intended it for .NET applications that need to occasionally call a native function, it turns out that it works just fine for a .NET application that is 90% native. As a result, since IJW has full access to the .NET class hierarchy and type system and is much more efficient than CCW, this is the technique used throughout the book.

Conventions Used in This Book

All programmers have their own pet terms and coding styles. Therefore, let me clarify a few terms you'll see throughout the book:

- **"Problem domain"**—I first picked up this term many years ago while using the Coad/Yourdon Object-Oriented Analysis and Design

methodology. "Problem domain" is a generic term that refers to the set of problems to be solved.

- **"Consumers" and "clients"**—These terms are used interchangeably to represent any code that uses a class or type. You'll also see this term used from the perspective of the MFC application being a consumer, or client, of the .NET class library or a specific .NET class.

- **"Server"**—This term is used to refer to a piece of code—typically a class—that is used by a client or consumer.

- **"Arguments" and "parameters"**—Like most programmers, I use these terms interchangeably when referring to the values passed to a method.

- **"Function" vs. "method"**—The object-oriented programming purists will tell you that you're supposed to use the term *method* instead of *member function*. However, having been an MFC developer for more than 10 years, I still find it difficult to refer to a C++ function as a method. Therefore, you'll see both terms used, depending on the context. If I'm referring to native C++ code, I use the term *function* and if I'm referring to .NET code, I use the term *method*.

- **Naming conventions**—This particular book presented a unique challenge when it came to naming conventions. I've been using Visual C++ and MFC since version 1 and have always used Hungarian notation for the main reason that the MFC source code did, and I wanted to maintain consistency in my own code bases. However, the generally accepted naming convention with .NET programming is a mixture of camel and pascal casing. Mix the two development environments of MFC and .NET, and you can see the problem. In the end I tried to localize my .NET code to standalone functions, but this wasn't always possible. As a result, you will often see a combination of the two distinct naming conventions.

- **Breaking lines of code**—Unfortunately, there are some times when the format of a book requires some odd line breaking of source code. This is regrettable—and visually unappealing—but unavoidable.

Demo Applications

System Requirements

I'm a big believer in "hands-on" training, which is why there are several step-by-step, tutorial-like demo applications in each chapter. In order for you to get the most out of this book, I highly recommend that you work your way through these demos. Most are intentionally very simple in scope in order to focus on the chapter's topic, but coding the demo will often help you to solidify what you've learned in the chapter's text.

Here's what you'll need to work through this book's demos:

- **Coding the demo applications**—You'll need Visual Studio .NET (this book was written using Visual Studio 2003 and the .NET Framework 1.1).
- **Running the demo applications**—You'll need to execute the applications on a machine equipped with the .NET runtime.

About the Demos

Here a few notes specifically related to this book's demos and chapter code snippets.

- I've included a Visual Studio .NET Custom AppWizard with this book that automates the creation of MFC/.NET projects. The wizard is located in the Tools folder of the CD-ROM and includes a readme.txt file that explains installation and usage. Note that the chapter's step-by-step instructions for creating the various demos do not refer to this wizard because some people will want to manually create and configure their projects. However, once you are comfortable with creating MFC/.NET mixed mode projects, I would suggest using the wizard, as it does save you a few steps and prevents needless recompiles when you manually create the project and forget to include a needed namespace or assembly reference in your code.
- When you run the demos provided on the book's CD-ROM, you will see that the dialogs and views will occasionally contain additional controls not seen in the chapter's figures or mentioned in the step-by-step instructions. The reason is that the chapter focuses on what you need to code in order to illustrate a given technical point or technique, while the additional controls on the finished demo are

simply there to either make the final UI more appealing or to provide additional instructions for running the demo.

- For the same reason noted above, some demos will have code that is not mentioned in chapter, so the chapter's code and the CD-ROM code won't always match up 100%. As an example, several demos use a list view control, and therefore each of these demos includes functions to allow for full-row selection and the auto-sizing of columns to match the control's text. However, to explain those functions or to even include them in the step-by-step instructions for something like a demo whose focus is regular expressions would be to take the focus away from the main point of the demo.
- Most of the demos have a button labeled Contact Me that allows you to easily access my Web site in order to view the most current version of the book's demos and a current errata list for the book, as well as to contact me.

Author Feedback

I'm always eager to hear from readers with constructive criticisms (and, I hope, ways of making the next edition even better) or questions about using what you've learned in these pages. To that extent, I can be reached through my company's Web site: www.ArcherConsultingGroup.com. Along with this book's updated demo applications, I will also maintain current errata for this book on my site.

If you do submit a question, critique, or anything that requires feedback from me, I would only ask for a bit of patience. Due to the success of my books and training seminars, my schedule gets very hectic and I travel quite a bit. Having said that, I'll respond as quickly as time permits, which typically translates to within a couple of business days.

Acknowledgments

Many people help take a book from the germ of an idea to completion, and it's important to recognize their individual contributions to the team. The book is much better as a result of the following contributors.

- **Sondra Scott**—Without Sondra's truly caring and compassionate support, this book would never have been completed. Thanks so much for all your help, Sondra!
- **Nishant Sivakumar**—Known affectionately on the CodeProject Web site simply as Nish, he contributed several chapters and demos to this book, and his contributions are greatly appreciated.
- **Erik Westermann**—I'd like to give many heartfelt thanks to my dear friend and fellow author, Erik—not just for helping on the book, but also for being someone I can depend on in a crunch. At the very last minute, I made some rather significant changes to the text, and Erik jumped in literally on a second's notice and provided some incredible insight and technical help. The end result is what you hold in your hands now.
- **Mark Burhop** and **Dmitri Riz** were the technical editors assigned to this book from Addison-Wesley. It's widely accepted in the technical publishing industry that Addison-Wesley only hires the best, and you guys are shining examples of that. You were both incredibly constructive and helpful with your critiques and advice and helped to shape much of what became the final result.
- **Brian Delahunty** wrote and donated the "MFC.NET" Visual Studio Custom AppWizard. This wizard enables you to automate the process of making the required project settings and including the desired .NET namespaces and assembly references for a combined MFC and .NET project with a few clicks of the mouse. The wizard is located in the Tools folder of the CD-ROM and includes a `readme.txt` file that explains installation and usage. Great work, Brian!
- **Don J. Plaistow** is a Perl and Regular Expressions guru who was kind enough to help me with the Regular Expressions chapter—specifically the expression used in parsing text files for e-mail addresses.

- **Mauricio Ritter** kindly donated his time and database expertise and helped solve one especially tricky problem—inserting relational data using the `DataSet` and `DataAdapter` classes—with the ADO. NET chapter.
- **James Johnson** acted as a technical advisor to the book and came up with a couple of eleventh-hour resolutions to difficult problems. He's one of the most knowledgeable people I know in the area of .NET and Managed C++, and his help was indispensable.
- **Anson Tsao** was my main contact on the Visual C++ team. He tirelessly answered many of my questions, resolved doubts, and even helped to shape the direction of this book in many ways.

Technical Editors

Erik Westermann is a seasoned developer and author with more than 12 years of professional programming experience who has written books and published articles both in print and on the Web. Erik is an active member of the developer community and is affiliated with the IEEE Computer Society and the ACM. He and Tom Archer have formed a new company—Windmill Publishing—whose aim is to aid first-time and unknown authors get the book *they want* to write published. Erik can be reached via Erik@WindmillPublishing.com.

Mark Burhop is a senior software engineer in Cincinnati, Ohio, and also part of the adjunct faculty at the University of Cincinnati. He has more than 15 years of experience developing software in the computer aided engineering (CAE) and product data management (PDM) fields using technologies ranging from Microsoft .NET to C++ and Java. As an author, he is currently focused on showing how to write high-performance C# applications. More information on Mark and his work can be found at http://www.markburhop.com.

Dmitri Riz has been designing and developing software long enough to remember the times when the complete set of development tools would fit on a single floppy. Currently he enjoys playing with his kids, riding his motorcycle, and, in his free time, writing code for Information Strategies, Inc. Dmitri holds several Microsoft certifications (MCSD, MCDBA, and MCSE). He can be reached at dmitrir@infostrat.com.

About the Authors

Tom Archer has been the lead programmer on three award-winning applications, has authored ten programming books, and has written countless magazine and online articles. He owns the Archer Consulting Group, which specializes in training courses, management and project consulting, and contract programming. Serving international clients such as IBM, AT&T, Equifax, and VeriSign, the Archer Consulting Group has offices in Atlanta, GA and Toronto, Canada. To learn more about what they can do for your company, be sure to visit them on the Web at www.ArcherConsulting Group.com.

Nishant Sivakumar (Nish) is a Microsoft VC++ MVP who began programming at the age of 13 and hasn't stopped since. Starting with GWBASIC and Assembler in 1990, he has gradually moved on to C++ and in the last five years has been actively using Visual C++ and MFC. Over the past couple of years he has also been extremely captivated by .NET and the Managed Extensions to C++. Nish considers having written a romantic comedy at the age of 20 as one of his greater achievements in life and is an avid lover of science fiction. In his free time, he can be found on the CodeProject forums participating in lively discussions and helping his fellow developers.

Combining MFC and Managed Extensions

Introduction

Beginning with Visual C++ .NET, Microsoft added the Managed Extensions to its C++ compiler, so that the C++ language can be used for developing managed applications that run on top of the .NET Common Language Runtime (CLR). Managed Extensions are the language constructs that allow you to mix managed (.NET) syntax with your unmanaged, or native (non-.NET) code. These Managed Extensions consist of additional keywords, preprocessor directives, and compiler options that allow—among other things—access to the .NET Base Class Library (BCL), the ability to define and use managed types (garbage-collected reference types, value types, scoped enums, delegates, etc.), and access to managed runtime facilities such as reflection and .NET security. In addition to being able to produce .NET-only applications, Visual C++ .NET is the only .NET language that enables programmers to freely mix managed and unmanaged code within the same application—even within the same source file and function. This unique capability is the focus of this book.

As promised in the Preface, this book is not a Managed Extensions book. Rather, the focus is on mixing MFC and .NET code. Therefore, the goal of this first chapter is to cover just enough Managed Extensions and .NET basics to ensure that you get the most out of the following chapters without repeating a lot of basics in each chapter. We begin by examining the canonical "Hello World" application in order to understand the key differences between a Managed Extensions application and an MFC application. This part includes broad overviews of such .NET topics as *assemblies, manifests,* and *references.* Next I provide a basic demo application illustrating

how to mix MFC and .NET code. The following part introduces some important mixed-mode programming issues and how to deal with them. Once that is done, we'll delve into the subject of working with .NET types, focusing on their definition and usage in your MFC applications. This part covers the topics of defining and using managed types, boxing and unboxing, managed and unmanaged pointers, managed arrays, and destructors.

Writing a Managed Extensions Application

It's always helpful when learning a new programming language or development environment to either examine any provided sample applications or generate a bare-bones application to study. That way, you can see how a working application's source code is formed in terms of syntax, include files or declarations that are needed, necessary compiler or linker options, and so on. Therefore, at this point, let's create a simple Managed Extensions console application:

1. Start the Visual Studio .NET AppWizard by selecting New Project from the File menu.
2. Once the New Project dialog is displayed, select the Visual C++ Projects project type and then click the Console Application (.NET) icon in the Templates list view.
3. Name the project `FirstMEApp` and specify the desired location of your new project.
4. When you click the OK button, the AppWizard will generate the project and its required files.

As with any Visual Studio .NET project, the Solution Explorer will contain a listing of all the projects defined in the current solution as well as a listing of the files that make up each project. However, in the case of a Managed Extensions application, there are some extra bits you might not be familiar with (see Figure 1–1).

The first thing to note is the References section. References are used to indicate which .NET libraries (and their defined namespaces and types) will be used by the source code. References are added to a project in one of two manners: via the Add References dialog and via the `#using` directive. For the references to show up in the Solution Explorer and for the references to be recognized by the various Visual Studio .NET designers,

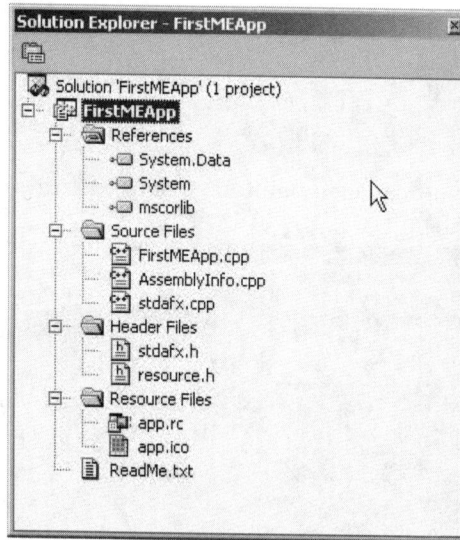

Figure 1–1 In order to provide access to the .NET class library, Visual Studio .NET automatically inserts several references and key files into AppWizard-generated Managed Extension projects.

they must be added via the Add References dialog (Figure 1–2). Any component specified here will be passed to the C++ compiler using the /FU compiler switch at build time. (We'll cover the #using directive shortly.)

The second item of note is the AssemblyInfo.cpp file. As the name indicates, this file contains information about the current module's assembly. A brief explanation of the .NET definition of an assembly and how it relates to this file follows.

All .NET binaries that use the .NET CLR consist of an *assembly* or a group of assemblies. Therefore, when you created this Managed Extensions console application, you actually created a .NET assembly and not the standard Windows Portable Executable (PE) that you're accustomed to creating when writing MFC Windows applications. There are three main types of binaries that you can create: executable (EXE), dynamic-link library (DLL), and netmodule. The first two are self-explanatory, and the netmodule is new to Windows developers. A netmodule is really just a DLL with an extension of .netmodule that does not contain a *manifest* (I describe what a manifest is in the following paragraph). As a result it doesn't belong to an assembly and must be added to an assembly (typically by using the .NET Assembly Generation tool).

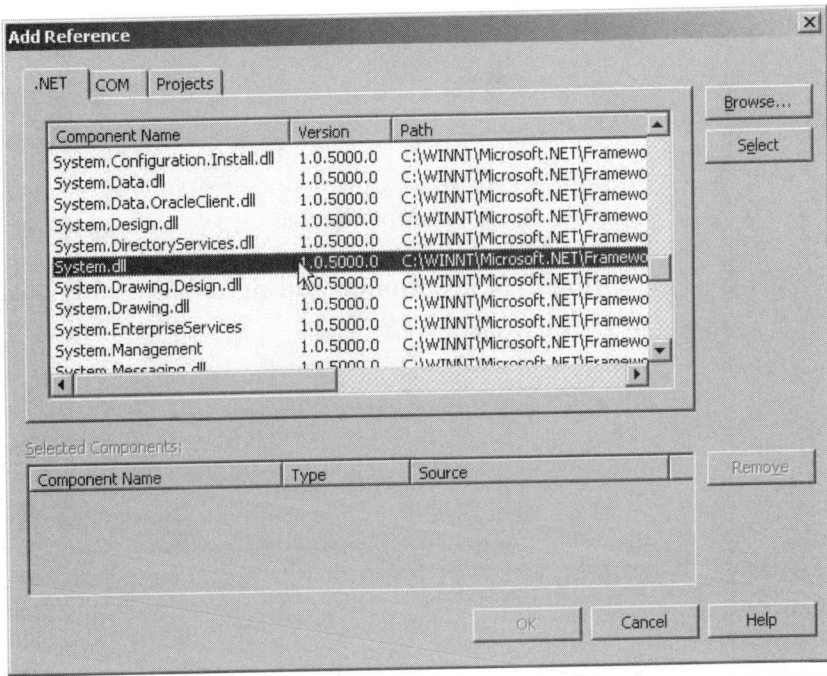

Figure 1-2 Adding .NET references to a Visual Studio .NET project works in much the same way as adding Component Object Model (COM) references to a Visual Basic project.

Now let's talk about manifests. If you compile a standalone application (such as our `FirstMEApp`) or a DLL, the manifest will be incorporated into the resulting binary. This is called a *single-file assembly*. A *multifile assembly* can also be generated, with the manifest existing as either a standalone entity within the assembly or as an attachment to one of the modules within the assembly. The manifest is essentially a repository of information about what assemblies the current assembly links to, what security permissions your application requires, and information about what assemblies constitute the application.

So what, then, is the definition of an assembly, and what is it used for? Well, that depends on your orientation. From a client's perspective, an assembly is a named, versioned, self-describing (via the manifest) collection of modules, exported types and, optionally, resources. From the assembly creator's standpoint, an assembly is a means of packaging related modules, types, and resources, and of exporting only what a client should use. That said, note that it is the manifest that provides the level of indirection

between the implementation details of the assembly and the assembly's client. A few of the most integral elements stored in an assembly's manifest are listed below:

- **Assembly name:** Textual name of the assembly.
- **Versioning information:** Contains four distinct parts that constitute the version number, including a major and minor version number in addition to a revision and build number.
- **(optional) Shared name and signed assembly hash:** Pertains to the deployment of assemblies.
- **Files:** List of all files contained in the assembly.
- **Referenced assemblies:** List of all external assemblies that are directly referenced from the manifest's assembly.
- **Types:** List of all types in the assembly with a mapping to the module containing the type's implementation.
- **Security permissions:** List of security permissions that are explicitly refused by the assembly.
- **Custom attributes:** As with types, custom attributes are stored in the assembly's manifest for quicker access during *reflection*. Reflection is the ablity to query, or discover, type information at runtime.
- **Product information:** Includes fields such as company, trademarks, product name, and copyrights.

With so much information crammed into this file, it must be important to the overall running of a .NET application, right? Absolutely. In fact, assemblies play an integral part in many of the strongest selling points of .NET, such as language interoperability, reflection, the end of "DLL Hell," and versioning, to name just a few.

Now that we've had a quick look at assemblies, let's return to the AssemblyInfo.cpp file.

```
#include "stdafx.h"

#using <mscorlib.dll>

using namespace System::Reflection;
using namespace System::Runtime::CompilerServices;

//
// General Information about an assembly is controlled through
// the following set of attributes. Change these attribute values
```

```
// to modify the information associated with an assembly.
//
[assembly:AssemblyTitleAttribute("")];
[assembly:AssemblyDescriptionAttribute("")];
[assembly:AssemblyConfigurationAttribute("")];
[assembly:AssemblyCompanyAttribute("")];
[assembly:AssemblyProductAttribute("")];
[assembly:AssemblyCopyrightAttribute("")];
[assembly:AssemblyTrademarkAttribute("")];
[assembly:AssemblyCultureAttribute("")];


//
// Version information for an assembly consists of the
// following four values:
//
//       Major Version
//       Minor Version
//       Build Number
//       Revision
//
// You can specify all the value or you can default the
// Revision and Build Numbers by using the '*' as shown below:

[assembly:AssemblyVersionAttribute("1.0.*")];


//
// In order to sign your assembly you must specify a key to
// use. Refer to the Microsoft .NET Framework documentation
// for more information on assembly signing.
//
// Use the attributes below to control which key is used for signing.
//
// Notes:
//   (*) If no key is specified, the assembly is not signed.
//   (*) KeyName refers to a key that has been installed in
//       the Crypto Service Provider (CSP) on your machine.
//       KeyFile refers to a file which contains a key.
//   (*) If the KeyFile and the KeyName values are both specified, the
//       following processing occurs:
//       (1) If the KeyName can be found in the CSP, that key is used.
//       (2) If the KeyName does not exist and the KeyFile does
//           exist, the key in the KeyFile is installed into the
//           CSP and used.
```

```
//    (*) In order to create a KeyFile, you can use the
//        sn.exe (Strong Name) utility.
//        When specifying the KeyFile, the location of
//        the KeyFile should be
//        relative to the project directory.
//    (*) Delay Signing is an advanced option; see the
//        Microsoft .NET Framework documentation for more
//        information on this.
//
[assembly:AssemblyDelaySignAttribute(false)];
[assembly:AssemblyKeyFileAttribute("")];
[assembly:AssemblyKeyNameAttribute("")];
```

As you can see, this file contains a number of *attributes,* which are denoted by the text within square brackets. While attributes are a fantastic addition to the language, they aren't essential to get started with mixing MFC and .NET code. For now, it's enough to know that attributes are a means of allowing the programmer to insert textual annotations that describe—or help to define the behavior and runtime use of—types and methods. These values can then be queried (reflected) at runtime by client code.

As you can see from the file above, most of the default attributes that are created for you pertain to versioning and assembly information. Visual Studio .NET inserts these values automatically into the `Assembly Info.cpp` file for you and defaults them to blank. Obviously, you can modify these values to suit your particular application.

Quick Note about the Placement of Attributes

While Visual Studio .NET automatically generates the `AssemblyInfo.cpp` file as a convenient location for your assembly-level attributes, you can define your attributes (custom or predefined) in any of your source files. However, I recommend placing all assembly-level attributes in this file as a matter of consistency.

One last note on assemblies concerns the ability to view the assembly's manifest. This can be done via a .NET utility called the Intermediate Language Disassembler (ILDASM). As all .NET code is compiled to MSIL (Microsoft Intermediate Language), this utility functions as a disassembler, and using reflection displays the contents of an assembly. Figure 1–3 shows the manifest for the compiled `FirstMEApp` application.

```
┌─ MANIFEST ─────────────────────────────────────────────────── _ □ ×
│ .assembly extern mscorlib
│ {
│   .publickeytoken = (B7 7A 5C 56 19 34 E0 89 )              // .z\U.
│   .hash = (DC 56 4C 0F 00 39 23 0D 65 34 2E 92 90 5C D3 5F  // .UL..9#.e4...\.
│               77 65 50 DE )                                 // weP.
│   .ver 1:0:5000:0
│ }
│ .assembly extern Microsoft.VisualC
│ {
│   .publickeytoken = (B0 3F 5F 7F 11 D5 0A 3A )              // .?_..
│   .hash = (98 D7 9D DC 4E 0E 1C 62 53 96 00 EF C8 DA 1E 07  // ....N..bS......
│               A2 84 80 5B )                                 // ...[
│   .ver 7:0:5000:0
│ }
│ .assembly FirstMEApp
│ {
│   .custom instance void [mscorlib]System.Reflection.AssemblyCopyrightAttribute:
│   .custom instance void [mscorlib]System.Reflection.AssemblyCompanyAttribute::.
│   .custom instance void [mscorlib]System.Reflection.AssemblyTitleAttribute::.ct
│   .custom instance void [mscorlib]System.Reflection.AssemblyTrademarkAttribute:
│   .custom instance void [mscorlib]System.Reflection.AssemblyProductAttribute::.
│   .custom instance void [mscorlib]System.Reflection.AssemblyDescriptionAttribut
│   .custom instance void [mscorlib]System.Reflection.AssemblyConfigurationAttribu
│   // --- The following custom attribute is added automatically, do not comment
│   //    .custom instance void [mscorlib]System.Diagnostics.DebuggableAttribute::.
│   //
│   .custom instance void [mscorlib]System.Reflection.AssemblyKeyNameAttribute::.
└──────────────────────────────────────────────────────────────────────
```

Figure 1-3 The assembly's manifest contains a lot of predefined as well as custom (user-defined) information about an application or library/module.

As you can see, the first two entries listed from the manifest are the externally linked assemblies for the `mscorlib` and `Microsoft.VisualC` libraries. Following those are assembly attributes that are defined in the `AssemblyInfo.cpp` file. Looking back at the external assemblies for a moment, you might be wondering where `mscorlib` came from, since it wasn't in the References section of the Solution Explorer. To understand it's origin, open the `FirstMEApp.cpp` file (listed below).

```
#include "stdafx.h"

#using <mscorlib.dll>

using namespace System;

int _tmain()
{
  Console::WriteLine(S"Hello World");
  return 0;
}
```

As an MFC developer, you're certainly accustomed to seeing the inclusion of the `stdafx.h` file. However, things get interesting after that with the use of the `#using` directive, which will be new to most C++ developers. The `#using` preprocessor directive imports metadata from a .NET Framework executable, library (both .DLL and .LIB), .OBJ, or netmodule. Basically, it's the same as using the Add Reference dialog except that here it's done at the source code level, and the information is not available to any of the Visual Studio .NET designers. Now you see how the `mscorlib.dll` library was referenced in the assembly's manifest despite not being in the project's References section. I should note also that `mscorlib.dll` is a mandatory library for interfacing to managed objects or code.

Next comes the `using` directive (not to be confused with the `#using` directive). Once you include a reference to a .NET library, you have source-level access to that library's public types. As all .NET types are nested in varying levels of namespaces, you can access these types using the standard C++ scope resolution operator (`::`). As an example, the BCL provides a `Path` class that performs operations on strings containing file or directory path information in a cross-platform manner. Using this class, you can combine paths, strip out directory names, isolate file names and extensions, and do much more with a variety of static methods. (You'll see this useful class throughout the book's demos.) As the `Path` class resides in the `IO` namespace, which in turn resides in the `System` namespace, you could write code like the following (which returns the directory portion of a string containing a qualified file name):

```
String* str =
    System::IO::Path::GetDirectoryName(S"c:\\tom\\test.txt");
```

As you can see, the call to the `GetDirectoryName` method of the `Path` class was fully qualified (including from the top-most layer, all namespaces). However, since the BCL hierarchy includes many nested layers of namespaces, typing in the entire qualified name of every single type or method name can get quite tedious. Therefore, the `using` statement is meant to reduce the amount of typing necessary.

As an example, in the `FirstMEApp.cpp` file, the `using namespace System;` line enables you to refer to any type defined in the `System` namespace without having to qualify its name. As a result, our use of the `Path` class could be shortened to `IO::Path::GetDirectoryName`. If you then add the directive `using namespace System::IO`, your call to

GetDirectoryName becomes the much more palatable Path::Get DirectoryName. Note that you can only refer to namespaces via the using directive. Therefore you may not write using namespace System::IO:: Path, because Path is a class (and not a namespace) that is defined within the System::IO namespace.

When the compiler encounters a type that is not fully qualified, it simply searches for the type declaration in each specified namespace (in the order defined). If the declaration is not found, a compile-time error is generated.

The next line is the _tmain function. I won't go into any detail here; as in any C++ console application, an entry point named main must be specified (_tmain is #define'd as the equivalent of main in the tchar.h include file). The last thing you see in the code is the call to the Console::Write Line method, which writes text to the output device à la cout or printf.

One final note worth mentioning here is that if you peek at the project settings (via the Solution Explorer), you can see that this application is compiled with the /clr compiler switch. This switch enables the use of the Managed Extensions to C++ and results in the generation of a .NET Framework EXE that will only execute on a machine that has the .NET Framework installed. Note that all Managed Extensions applications must have this switch specified (done automatically by AppWizard-generated projects).

In Visual Studio .NET 2003, there is a /clr:InitialAppDomain switch. The default /clr behavior has changed in version 1.1 of the CLR, where a transition from managed to native and back to managed code returns to the app domain that originated the call chain. In V1.0 of the CLR, the same call sequence results in the managed call executing in the first app domain created in the process. Since the new V1.1 CLR behavior is not backwards-compatible, Visual C++ 7.1 provides /clr:InitialAppDomain to compile code that will run in V1.0 of the CLR.

Now that you've written and explored your first Managed Extensions application, let's see how to define a project so that it allows you to mix MFC with .NET code.

Writing an MFC Application That Accesses .NET

As mentioned in the Preface, there are two means of interfacing native and managed code: COM Interop and the combination of Managed Extensions and IJW (It Just Works). For reasons spelled out in the Preface, the latter technique will be used throughout this book as you discover the myriad BCL classes that can greatly enhance your productivity as an MFC developer. However, before we get into all those BCL classes, you'll first need to see the steps required for mixing MFC and .NET code. Note that these steps remain constant regardless of the simplicity or complexity of the application being coded.

1. Create a new MFC Dialog–based application called `MFCAndDot Net`, keeping all the default AppWizard settings.
2. As mentioned, the `/clr` compiler switch must be specified for any Managed Extensions application. In order to set that switch, open the Project Properties dialog and select the Use Managed Extensions option, as shown in Figure 1–4.

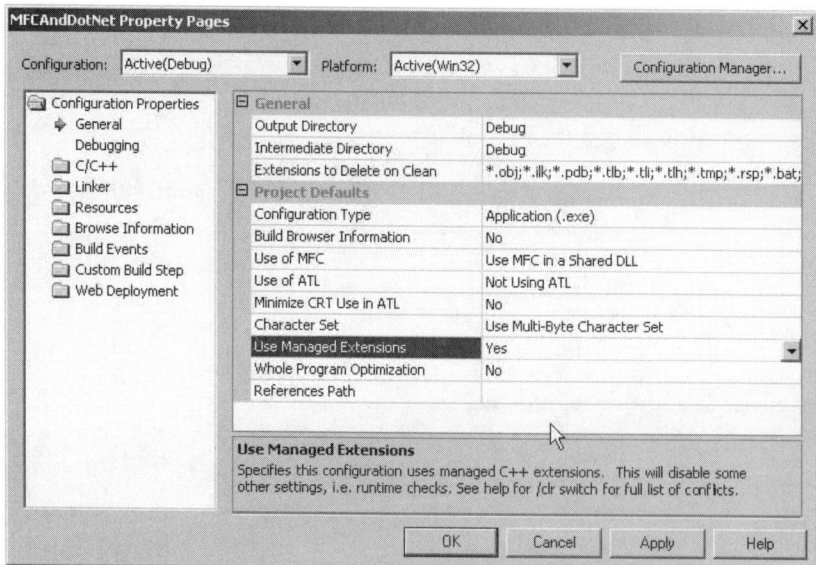

Figure 1–4 Specifying that the application should use Managed Extensions is as easy as flipping a (compiler) switch.

3. Some of the default debug-build compiler options for an MFC project conflict with the `/clr` compiler switch, and programmers using the initial releases of Visual Studio .NET would receive compiler errors after setting the `/clr` option for an MFC project. In subsequent releases of Visual Studio .NET (such as Visual Studio .NET 2003), the code for the Project Properties dialog box was modified so that these conflicting compiler switches are automatically changed when the `/clr` switch is selected (and applied). If you see these errors, you are running an older beta version of Visual Studio .NET, and you'll need to change your project settings manually each time you set the `/clr` switch. If that is the case, see Figures 1–5 and 1–6, which highlight the necessary changes.

4. Now build the project, and you'll find that it compiles and builds successfully without any errors or warnings. What you have done at this point is to build a .NET executable; however, so far it doesn't make any calls to managed code. In order to accomplish that, you need to add a couple of directives to your code to specify what BCL types you want to use. Since these directives are typically listed in the MFC precompiled header file, open the `stdafx.h` include file

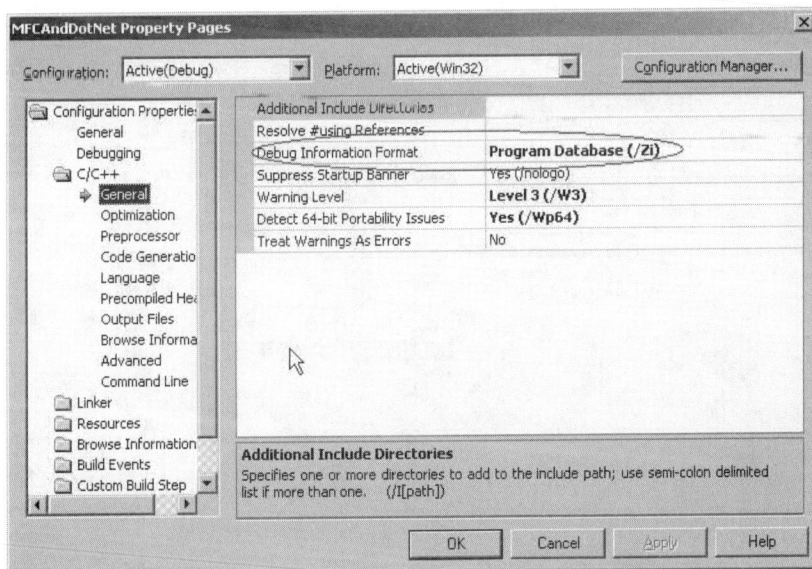

Figure 1–5 Project settings necessary for defining a mixed-mode (MFC and Managed Extensions) project

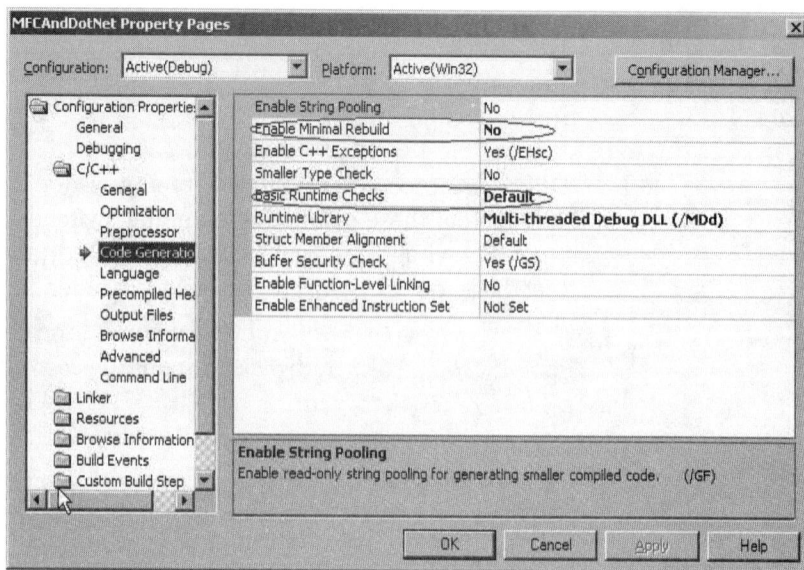

Figure 1-6 Additional project settings necessary for defining a mixed-mode (MFC and Managed Extensions) project

and add the following #using directives and using statements to the end of the file.

```
#using <mscorlib.dll>
#using <System.Windows.Forms.dll>
using namespace System;
using namespace System::Windows::Forms;
```

5. That's it! At this point, your MFC project is able to access managed code. Let's test that by adding a little user interface (UI) to this application. Start by using the dialog resource editor to add a list box control to the application's main dialog box.

6. Next, add a DDX (Dialog Data Exchange) control variable called m_lbxSysInfo.

7. Now modify the dialog's OnInitDialog method by adding the following code just before the return statement:

```
CString str;

str.Format(_T("User Domain: %s"),
            (CString)SystemInformation::UserDomainName);
m_lbxSysInfo.AddString(str);
```

```
str.Format(_T("Computer Name: %s"),
           (CString)SystemInformation::ComputerName);
m_lbxSysInfo.AddString(str);

str.Format(_T("Logged in User: %s"),
           (CString)SystemInformation::UserName);
m_lbxSysInfo.AddString(str);
```

At first glance it doesn't appear much different from typical MFC code, but as you look closer, you find that the `System::Windows::Forms::SystemInformation` class is being used to populate the list box with some basic information about the current machine and user. The `SystemInformation` class has several static properties—some of which are being used here to retrieve the needed information for this simple demo. (Properties are basically accessor functions that are referred to in code like member variables.)

8. Now build and execute your first MFC/.NET application. Your results should approximate what you see in Figure 1–7 (I've added a couple of static text controls and a Contact Me button to the CD's version that you see here).

One exciting issue to point out here is the ability to use MFC classes (such as `CString` and `CListBox`) with .NET BCL types (such as the `SystemInformation`) without having to write any special conversion code. Now you see why IJW stands for *It Just Works!*

MFC.NET Custom AppWizard

As you've seen in this part, the only two things you need to do in order to mix MFC and Managed Extensions is to set the Use Managed Extensions project property (which equates to the `/clr` compiler switch) and add the appropriate `#using` and `using` directives to your code.

To eliminate even those small chores, I've included a Custom AppWizard on the book's CD (located in the Tools folder)—donated to us by Brian Delahunty—that automatically adds the `/clr` switch to the compiler settings and presents a list of .NET modules and namespaces to be added to the new project. The selected module references are added to the compiler settings (via the `/FU` switch), and the selected namepaces are inserted into the project's `stdafx.h` file in the form of "using namespace" directives.

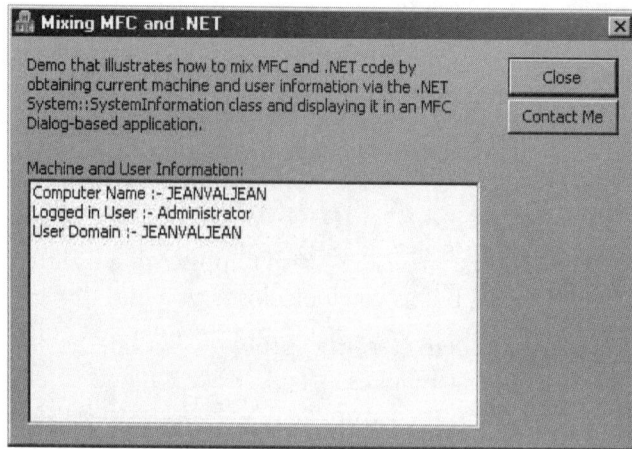

Figure 1-7 With Managed Extensions and IJW, you can even mix unmanaged and managed code and types in the same function!

Mixed-Mode Programming Issues

As with any new development environment, there are some caveats that you should be aware of when mixing managed and unmanaged code. In this part, we examine a few of the most common problems encountered by those new to mixed-mode programming:

- Avoiding inadvertent #define side-effects
- Allocating reference objects in MFC debug builds
- Using string literals and converting strings

Avoiding Inadvertent #define Side Effects

One of the first issues you're sure to run into is the problem caused when you attempt to use a function, operator, or type that has been #define'd to a different value than you expect. One example of that is in attempting to use the MessageBox::Show method:

```
using namespace System;

...
```

```
// Display a message box with the specified text.
MessageBox::Show(S"sample text");
```

Attempting to compile this code will yield the following error in an ANSI build (the error would state `MessageBoxW` in a Unicode build):

```
error C2653: 'MessageBoxA' : is not a class or namespace name
error C3861: 'Show': identifier not found, even with argument-
   dependent lookup
```

In this particular case, the problem is that the `WinUser.h` file contains a `#define` that maps the `MessageBox` name to either the `MessageBoxA` (marshals the passed string to ANSI) or `MessageBoxW` (marshals the passed string to wide-character format, or Unicode) function based on the defined character set option. Since default AppWizard-generated projects are ANSI-compliant (as opposed to Unicode-compliant), the precompiler will substitute `MessageBoxA` in the place of `MessageBox` resulting in a call to `MessageBoxA::Show(S"sample text")`. When the compiler then attempts to compile this code, it barks at seeing a function name followed by a scope resolution operator and lets you know that `MessageBoxA` is not a class, union, or namespace.

Therefore, we simply need a way to "undefine" `MessageBox` for the context in which we wish to use the .NET version. This is accomplished with the `pop_macro` and `push` macro directives in the following manner:

```
#pragma push_macro("MessageBox")
#undef MessageBox


...


MessageBox::Show(S"hello");


...


#pragma pop_macro("MessageBox")
```

Here the the `push_macro` directive has been used to preserve the current definition of `MessageBox` by pushing its value onto the precompiler's internal macro definition stack. `MessageBox` is then undefined (via the `#undef`

directive). Now the precompiler will not modify our call to `MessageBox::` `Show`. The `pop_macro` directive then restores the `#define`'d version of `MessageBox` by popping it off the compiler's internal macro definition stack.

Allocating Reference Objects in MFC Debug Builds

Another mixed-mode programming issue to keep in mind involves the use of the `new` operator when instantiating managed objects. Specifically, this problem appears in MFC debug builds. The following code attempts simply to allocate a `System::String` object.

```
// Allocate a managed object via the new operator.
String* str = new String(S"sample text");
```

This seemingly innocuous line of code can definitely catch the unwary off-guard, as it results in the following compiler error:

```
error C3828: 'System::String':
placement arguments not allowed while creating instances of managed
    classes
```

While at first this error message seems a bit obtuse, the hint as to the source of the problem lies in the word "placement." In the DLL chapter of my MFC book, *Visual C++ .NET Bible,* I explain how to use the `placement new` operator to enable the sharing of a single instance of a C++ object across multiple invocations of a DLL. MFC also uses the `placement new` operator at times—for example, to track memory allocation in Debug builds. What is happening here is that MFC redefines the `new` operator in the `afx.h` include file as a `placement new` operator. Since `String` is a reference type, and reference types are always allocated on the CLR heap—as opposed to the native C++ heap—this natually causes a problem in our attempt to allocate the `String` object above.

Luckily the solution is easy—and familiar. You simply use the same `push_macro` and `pop_macro` directives, as illustrated in our first name-collision example.

```
#pragma push_macro("new")
#undef new
String* str = new String(S"sample text");
#pragma pop_macro("new")
```

One helpful technique to understand errors like these is to turn on pre-processor output in the project settings. To do that, simply open the Project Settings dialog, locate the Preprocessor tab, and set the Generate Pre-processed File option to the desired setting—with line numbers or without. Here's an example of what you would see if attempting to instantiate a `String` object in an MFC debug build without first undefining the `new` operator:

```
String* str = new(".\\DefineTest.cpp", 177) String();
```

Visual Studio `push_macro` and `pop_macro` Macros

Since remembering to wrap your usage of conflicting macros with the appropriate pragmas can get a bit tedious, I've included a Visual Studio .NET macro on the book's CD (in the Tools folder) that allows you to select a word and, after clicking a single button, have the appropriate pragma directives inserted into the code for you. (The scope of the added pragmas is the current function.)

Using String Literals and Converting Strings

Converting strings between two distinct development languages is typically nothing short of painful, so I'll cover that here as, of course, you'll see lots of string conversions throughout the book's code snippets and demos.

String literals are composed of ASCII characters in standard C++ applications. However, managed types and methods are not designed to work with ASCII strings. To save programmers from having to cast every literal to a `String` object, the Managed Extensions introduces the `S` literal prefix:

```
String* s1 = "a string literal";   // standard literal

// both of these can be passed to .NET types and methods expecting
// a string object
String* s2 = L"a wide string literal"; // wide character literal
String* s3 = S"a wide string literal"; // .NET compatible wide literal
```

You should note that regular C++ wide-character string literals (prefixed by L) and managed string literals (prefixed by S) can technically be used interchangeably where `String` types are expected. However, there are two important issues to realize here.

First, while the documentation implies that the L and S prefixes work the same way when passing managed string literals, the S prefix is more efficient. The reason for this is that the S prefix simply resolves to an `ldstr` (load string) MSIL instruction, while the Unicode L prefix calls the `String(char*)` or `String(__wchar_t*)` constructor.

Second, managed string literals cannot be used where ANSI C++ string types are expected. To that extent, let's now talk a bit about string conversion in mixed mode-applications.

Fortunately, the conversion between the MFC `CString` and the .NET `String` classes is made extremely easy (and mostly transparent) through IJW. The following code illustrates how to convert a `CString` to a `String*`.

```
CString str1 = T("hello");
String* s1 = str1;
```

This works because the `String` class has a constructor that takes an `LPCTSTR` value, and `CString` has an implicit `LPCTSTR` operator. Converting a `String*` to a `CString` is equally easy. The following works because `CStringT` has a constructor taking `String*` when compiled with `/clr`. It takes care of whether `CString` is defined for ANSI or Unicode.

```
String* s2 = S"world";
CString str2 = (CString)s2; // ANSI build
```

Starting with Visual C++ .NET the `CString` constructor has an overload that accepts a `System::String` object, which makes things very easy.

Converting from `String*` to `__wchar_t*` is also possible, though it takes a little more effort, as shown here:

```
String* s = S"Hello";
const __wchar_t __pin * str = PtrToStringChars(s);
```

This requires some explaining. Let's start with the `PtrToStringChars` function. This function allows you to access the internal memory representation of a managed string instance. The returned value is the interior reference pointer to the first character of the `string` object. But we have one possible problem here. As the variable s represents a managed `string` object, how can be sure that the CLR won't move this object in memory, thus rendering our `__wchar_t` pointer invalid? That is where the `__pin` keyword comes in. It is specifically designed to help prevent just this sort of thing. Use the `__pin` keyword whenever you need to make sure that an

object will remain in memory even if it goes out of scope and thus becomes eligible for destruction via the .NET Garbage Collector (GC).

In addition, if what you need is a `char*`, you can then convert the `__wchart_t` value from above, as follows (new code is shown in bold):

```
using namespace System::Runtime::InteropServices;
...
String* s = S"Hello";
const __wchar_t __pin * str = PtrToStringChars(s);
char* str2 = (char*)(void*)Marshal::StringToHGlobalAnsi(str);
// Use str2
Marshal::FreeHGlobal(str2);
```

As you can see here, the `Marshal` BCL class is being used, as this class includes methods for allocating unmanaged memory, copying unmanaged memory blocks, and converting managed to unmanaged types, as well as other miscellaneous methods used when interacting with unmanaged code. In this example, the call to `Marshal::StringToHGlobalAnsi` is being used to copy the the contents of the `String` object (`str` variable) into the native heap, which in turn returns a pointer. The returned pointer can then be used as any other native `const*`. You need only remember to call the `Marshal::FreeHGlobal` to free the pointer when you're finished using it.

Regarding the `PtrToStringChars` function, there will be times when you will receive a less-than-obvious error message stating that you're attempting to pass it incorrect parameters when you don't have any explicit calls in your code to that function. As usual, an example will better illustrate this issue. In the following example, I've allocated an `ArrayList` object (the `ArrayList` and other collection types are covered in Chapter 11) and added to it a `String` object. I then allocate a `CString` object and attempt to format it using the first value in the `ArrayList` object.

```
using namespace System::Collections;
...
#pragma push_macro("new")
#undef new

  ArrayList* al = new ArrayList();
  al->Add(S"apples");
```

```
CString strTemp;
strTemp.Format("First value of array is %s", (CString)a1->Item[0]);
```

```
#pragma pop_macro("new")
```

However, if you were to attempt to compile this code, the compiler will generate an error similar to the following. (This very long error message has been reformatted for the book, and the key words are bold.)

```
C:\Program Files\...\Vc7\atlmfc\include\cstringt.h(875):
error C2664: 'PtrToStringChars' :
cannot convert parameter 1 from 'System::Object __gc *' to
'const System::String __gc *'
```

As noted above, there is no explicit call to PtrToStringChars, so the question is, "Where is it coming from?" A little spelunking into the MFC source code (cstringt.h) reveals that CStringT (the templatized class that implements the new MFC CString class) overrides the assignment operator (=) and calls the PtrToStringChars function.

Our debugging lives get a bit easier now that we know the source of the call the compiler is referring to. As the error message states, the value being passed to the PtrToStringChars function is an Object and not the required String type. While I did pass a String object to the ArrayList::Add method, it was automatically upcast to an Object, as that is what ArrayList objects contain. Therefore, in order to compile the code successfully, I need to modify the code as follows:

```
...
strTemp.Format("First value of array is %s",
               (CString)a1->Item[0]->ToString());
...
```

Unicode and Mixed-Mode Programming

While I've allowed the AppWizard to default the character set used for this book's demos—partly due to habit and partly due to the fact that I know most fellow MFC developers do as well—the Visual C++ team does recommend that mixed-mode applications be built using Unicode, as not having to continually convert string values between ANSI and Unicode is more efficient.

The point to all this is that when you see this error message about an invalid parameter to `PtrToStringChars`, the most common cause is that you're attempting to convert a type to a `CString` that can't be automatically (implicitly) converted.

Working with .NET Types

Now that you've seen the various mixed-mode programming issues to be aware of, let's look at the basics of working with .NET types. Specifically, this part covers how to define and use both reference and value types, the concepts of boxing and unboxing, using managed and unmanaged pointers, handling .NET arrays, and the proper destruction of .NET objects.

Defining and Using Managed Types

As you know by now, the Managed Extensions part of the Visual C++ .NET compiler allows you to write C++ applications that declare and use BCL types. To that extent, there are two distinct managed types that are supported by the Managed Extensions: reference types and value types. In order to interact with the .NET types from your MFC applications, you must be acquainted with the specifics of their differences—especially with regard to their definition and consumption.

Defining Reference Types with the __gc Keyword

Reference types are allocated on the CLR heap, and their lifetimes are managed by the GC. Thus, you should never explicitly call the `delete` operator nor otherwise take responsibility for the cleanup of objects of this type.

In order to define a reference type, you need simply to prefix the desired class or struct name with the __gc keyword (where "gc" stands for "garbage-collected"). In addition, note that all classes defined as __gc are implicitly derived from the BCL `System::Object` class and do not support multiple inheritance—although they may implement any number of interfaces. (Interfaces are basically abstract classes and provide the COM-like interface programming model.) Let's take a look now at how to declare a reference type.

If you've programmed in other .NET languages—such as C#—you've come to understand the rule that classes are reference types and structs are value types. However, with Managed Extensions whether a class or struct is a reference or value type is determined by your use of the __gc (reference) or __value (value) keyword when defining the type. Therefore, in the context of Managed Extensions, the terms "__gc type" and "reference type" are synonymous, as are the terms "__value type" and "value type."

```
__gc class Person
{
public:
  Person()
  {
    name = S"Anonymous";
  }

protected:
  String* name;

public:
  void SetName(String* name) { this->name = name; }
  String* GetName() { return this->name; }
};
```

The preceding code snippet defines a reference class named Person that looks like any other C++ class except for the aforementioned __gc prefix to the class name.

Additionally, take special note of the fact that the name member is defined as a pointer to the String type as opposed to an embedded String object. The reason for this is simple: the String class is a reference type; and therefore, you can only be declare a pointer to it and instantiate it via the new operator (which will allocate it properly on the CLR heap). Attempts to declare an instance of a reference type on the stack (as happens when you attempt to embed an object member into another type) result in compile-time errors like the following:

```
error C3149: 'System::String' :
illegal use of managed type 'System::String'; did you forget a '*'?
```

Once it is declared, consuming a reference type is very similar to using any C++ class, as shown in the followng code snippet:

```
#pragma push_macro("new")
#undef new
  Person* person = new Person();
  person->SetName(S"Tom Archer");

  AfxMessageBox((CString)person->GetName());
#pragma pop_macro("new")
```

As you can see in the client code above, the `Person` reference type must be instantiated using the Managed Extensions `new` operator. When you use Managed Extensions, the compiler can tell if the type being allocated is a native or managed type and perform the appropriate object construction. Also note that although the object was dynamically allocated, a memory leak will not occur, since the CLR will handle the object's removal—it will get queued for release once there isn't any code referencing it.

Defining Value Types with the `__value` Keyword

Value types are defined by prefixing the type name with the `__value` keyword and are generally used to hold primitive data values (`bool`, `char`, `int`, etc.). Value types can be created either on the heap or on the stack as required by the client. Note that I'm referring to the C++ (native) heap here (as opposed to the CLR heap, where reference types and reference structs reside). However, there are a couple of rules to remember regarding the dynamic (native heap) allocation of value types:

- You must use the `__nogc` keyword.
- Pointers to value types cannot be declared within other types.
- Value types cannot be allocated on the heap if they contain internal reference types.

The biggest advantage in defining value types is that the code does not incur the performance and resource overhead of automated garbage collection. As a result, value objects are much more efficient than their reference-type counterparts. In addition, note that although all value types implicitly derive from the same base `System::ValueType` class, these derived classes are sealed, which means that you cannot further derive your own types from existing value types.

Now let's look quickly at how a value type would be defined and instantiated.

```
__value struct OrderDetailLine
{
public:
 OrderDetailLine()
  {
     sku = S"111-006-116";
     qty = 42;
     unitPrice = 1.60;
  }

protected:
   String* sku;
   int qty;
   double unitPrice;

public:
   String* GetSku() { this->return sku; }
   int GetQuantity() { return this->qty; }
   double GetUnitPrice() { return this->unitPrice; }
};
```

As you can see, there's really not much difference from a code-definition standpoint between a reference type and value type—just a different keyword. In other words, the major differences lie in how they're allocated and managed by the runtime. The following code illustrates the client consumption of the Address type.

```
OrderDetailLine detailLine;

CString str;
str.Format(_T("SKU = %s\nQty = %ld\nPrice = $%.2f"),
           (CString)detailLine.GetSku(),
           detailLine.GetQuantity(),
           detailLine.GetUnitPrice());

AfxMessageBox(str);
```

Here, the object is created on the stack, and it will be destroyed when it goes out of scope, just as any stack-allocated object would.

Defining Managed Members within Unmanaged Classes

While you can define a value type as a member of an unmanaged class or struct, the compiler will not allow you to define a reference type as a member of an unmanaged class or struct. Therefore, the following code yields a compiler error.

```
class CNativeClass
{
public:
  // results in error C3265: cannot declare a
  // managed 'm_string' in an unmanaged 'CNativeClass'
  String* m_string;
};
```

To overcome this, use the gcroot type-safe wrapper template declared in the vcclr.h header file. The gcroot template allows you to embed a reference-object pointer in an unmanaged class and treat it just as if it were the underlying type. It is worth mentioning here that creating a circular reference between a native object and a managed object will cause a memory leak, as with COM circular references. The reason for the leak is that the GC won't collect the managed object because the native object is holding a reference to it.

```
// You can also use the gcroot to define members in
// native structs as well
class CNativeClass
{
  gcroot <String*>m_string;
  void SetString(String* s)
  {
    m_string = s;
  }
};
```

Boxing and Unboxing

All reference objects ultimately derive from the System::Object top-level base class, which provides the capability to treat all reference types polymorphically and have a few basic and common methods, like ToString and GetHashCode. This single-rooted object hierarchy can lead to problems in

a mixed-mode application when a method signature demands an `Object` as a parameter and you wish to pass a value type. This is where boxing comes in. In its simplest form, boxing is the process in which a value type is wrapped within a reference stub so that it can be passed as an `Object` type.

In some .NET languages like C# and Visual Basic .NET, boxing is an implicit operation and automatically occurs when needed. Boxing can be (computationally) expensive because memory is allocated on the CLR heap and the contents of the value object are copied to the newly allocated memory each time a value type is boxed. As a result—and because we C++ developers have always demanded far greater control over our development environment than programmers that use other languages—the Visual C++ design team decided to give us that level of control via the __box keyword. The downside, of course, is that we must explicity state when we want a boxing operation to occur.

Consider the following method:

```
void Show(Object* o)
{
   Console::WriteLine(o);
}
```

The `Show` method simply takes a pointer to an `Object` and writes out the string representation of the `Object` through the `Console::WriteLine` method. Now consider the following lines of code:

```
String* str = S"Hello World";
Show(str);
```

This works fine, as `str` is a `String` object and thus can be passed to a method expecting its base class (`Object`). Consider the following call to the `Show` method:

```
int foo = 100;
Show(foo); //This won't compile
```

If you try to compile that, you'll get a compiler error because `foo` is an `int` and thus a value-type object. As a result, this value type must be boxed into a reference type:

```
int foo = 100;
Show(__box(foo));
```

Now the code will compile, as foo is boxed into a completely distinct object on the CLR heap, and that object is passed to the Show method. This also means that if the Show method were to modify the incoming foo parameter, it would have no impact on the caller's foo variable because Show received a copy and not the original.

Unboxing refers to the process in which a reference-object pointer is obtained for a boxed value object using either the dynamic_cast casting operator or the __try_cast casting operator, and then the reference pointer is dereferenced to get back the boxed value object. The following code snippet should make things clearer:

```
int i = 99;
Object* o1 = __box(i);
int j = *dynamic_cast<__box int*>(o1);
j *= 3;
Show(__box(j));
```

Managed and Unmanaged Pointers

When you use the Managed Extensions to C++, you have two kinds of pointers: unmanaged pointers, which are pointers to unmanaged memory blocks in the native C++ heap, and managed pointers, which are pointers to managed memory blocks in the CLR heap. The managed memory blocks are automatically garbage-collected by the GC. Only managed pointers are Common Language Specification (CLS) compliant. Consider the following code:

```
class CNativeClass
{
   int i;
};

__value class ValueClass
{
   int i;
};

__gc class ReferenceClass
{
   int i;
};
```

```
...
#pragma push_macro("new")
#undef new

CNativeClass* pNativeClass = new CNativeClass(); //C++ heap
ValueClass* valueClass = __nogc new ValueClass(); //C++ heap

// CLR heap and thus garbage-collected
ReferenceClass* referenceClass = new ReferenceClass();

delete pNativeClass; //must delete
delete valueClass; //must delete

#pragma pop_macro("new")
```

As you can see, the unmanaged C++ class and the value class both got allocated on the native C++ heap, which means that we must manually delete those objects when necessary. On the other hand, the reference object is allocated on the CLR heap and thus gets automatically garbage-collected. Both the unmanaged and the value objects can be declared on the stack, but the reference object cannot be declared on the stack:

```
CNativeClass nativeClass;
ValueClass valueClass;
ReferenceClass referenceClass; //Won't compile - error C3149
```

In unmanaged C++, we can use a void* as a pointer to anything and everything (casting if necessary); similarly, in the managed world we can use an Object* as a pointer to any reference type, because all reference typees ultimately have Object at the top of their inheritance chain. Also note that managed pointers are automatically initialized to zero by the CLR.

At times, you will have an unmanaged pointer to a managed object—such as when passing a managed pointer to a function whose signature requires an unmanaged pointer. This scenario can be problematic regarding garbage collection: The GC maintains a list of all managed pointers to managed objects and updates those pointers as necessary when GC-controlled memory blocks are compacted for more efficient use. However, the GC has no way of tracking unmanaged pointers that point to managed objects. As a result, there is always the risk of the GC moving a managed object in memory; thereby invalidating any unmanaged pointer to that object. To remedy this situation, Managed Extensions provides the ability to pin a

managed pointer to a fixed memory location—so that the GC never moves it—using the __pin keyword (as noted earlier under "Using String Literals and Converting Strings"). The following code illustrates this:

```
__gc class Num
{
public:
  int m_num;
  Num()
  {
     m_num = 10;
  }
};
...
void Increment(int *n) { (*n)++; }



...



#pragma push_macro("MessageBox")
#undef MessageBox

#pragma push_macro("new")
#undef new

Text::StringBuilder* sb = new Text::StringBuilder();

//Here we pin the pointer so that
//the unmanaged function can use the
//managed pointer safely
Num __pin* num = new Num();
sb->AppendFormat(S"Before = {0}\r\n", __box(num->m_num));

Increment(&num->m_num);
sb->AppendFormat(S"After = {0}\r\n", __box(num->m_num));

MessageBox::Show(sb->ToString());

#pragma pop_macro("new")
#pragma pop_macro("MessageBox")
```

This example is obviously contrived, as you would probably not have a function dedicated solely to incrementing. The point, however, is to illustrate

The BCL `StringBuilder` Class

The `StringBuilder` class is used quite often in this book, but it is not covered explicity in any chapter. Since the `String` class is immutable, it doesn't lend itself well to demos where a diagnostic string needs to be continually modified and formatted throughout a given code snippet or function. Therefore, the `StringBuilder` class (which is mutable) is used in these situations.

how you would handle a situation where the function being called requires a pointer, and the pointer you have is managed by the CLR.

In a situation where you have control over the function's signature, you would use the __gc keyword in the function's signature to let the GC know that the value being passed to this function is in use, as shown in the following listing:

```
// Uses the __gc keyword so that the GC will take care
// in manipulating this variable in memory.
void Increment2(int __gc* n)
```

Managed Arrays

The BCL implements a powerful `Array` class in the `System` namespace that provides an easy and flexible means of working with large groups of any object type. The `Array` object supports single-dimensional arrays, multi-dimensional arrays, and even jagged arrays. Arrays are used frequently in this book in the passing of data to BCL class methods and the handling of return data from BCL class methods, so I'll cover the basics in this part.

Value Types in Arrays

Both reference types and value types can be embedded in other managed types (reference, value, or array). However, arrays of value types do not incur GC overhead, just as they don't when individually defined.

Single-Dimensional Arrays

Single-dimensional arrays can store either reference or value types, with syntax varying a little in each case. The following code snippet (which displays the letters A–E) illustrates how to declare and use a single-dimensional array

that contains instances of a reference type (String objects, in this case). The key points to note here are the syntax of the array definition and the fact that, like any reference object, the Array object has methods and properties (such as Length).

```
void DisplayManagedArrayOfReferenceObjects()
{
#pragma push_macro("new")
#undef new

    String* strArray[] = new String*[5];

    for (int i = 0; i < strArray->Length; i++)
    {
        // Uses the String(wchar_t, int) constructor
        strArray[i] = new String(L'A' + i, 1);
    }

    Text::StringBuilder* sb = new Text::StringBuilder();
    for (int i = 0; i < strArray->Length; i++)
    {
        sb->AppendFormat(S"[{0}] = '{1}'\r\n", __box(i), strArray[i]);
    }
    AfxMessageBox((CString)sb->ToString());

#pragma pop_macro("new")
}
```

Now let's take a look at using a single-dimensional array that holds value types.

```
void DisplayManagedArrayOfValueObjects()
{
#pragma push_macro("new")
#undef new

    int numArray __gc[] = new int __gc[5];

    for (int i = 0; i < numArray->Length; i++)
    {
        numArray[i] = i;
    }
```

```
Text::StringBuilder* sb = new Text::StringBuilder();
for (int i = 0; i < numArray->Length; i++)
{
   sb->AppendFormat(S"[{0}] = {1}\r\n", __box(i),
      __box(numArray[i]));
}
AfxMessageBox((CString)sb->ToString());

#pragma pop_macro("new")
}
```

In this case you can see that I have explicitly used the __gc[] keyword
to tell the compiler that the array is to be a managed array; otherwise the
compiler will simply create a normal unmanaged array of int types. In this
particular case, I could have simply used the Int32 type and avoided the
__gc syntax, as the compiler knows that In32 is a managed (value) type.
However, the point of this part is that when you are defining an array of
types that are not managed types, you need to use the __gc[] syntax.

Also note that while the numArray is a managed array, its members are
still value types, and thus they must be boxed in order to pass them to a
method that requires a reference object (such as the StringBuilder::
Format method).

Multidimensional Arrays

Managed multidimensional arrays are always rectangular arrays and should
not be confused with unmanaged arrays of arrays. Consider the following
function.

```
void MultiDimensionalArray()
{
#pragma push_macro("new")
#undef new

  String* multiString[,] = new String*[2,3];

  for (int i = 0; i < multiString->GetLength(0); i++)
  {
     for (int j = 0; j < multiString->GetLength(1); j++)
     {
        int k = i + j;
        multiString[i,j] = String::Format("{0} * {1} = {2};",
```

```
            __box(i),  __box(j),  __box(k));
    }
  }

  Text::StringBuilder* sb = new Text::StringBuilder();
  for (int i = 0; i < multiString->GetLength(0); i++)
  {
    for (int j = 0; j < multiString->GetLength(1); j++)
    {
      sb->AppendFormat(S"[{0}][{1}] = {2}\r\n",
                       __box(i), __box(j), multiString[i,j]);
    }
    sb->Append(S"\r\n");
  }
  AfxMessageBox((CString)sb->ToString());

#pragma pop_macro("new")
}
```

The syntax looks peculiar to most of us C++/MFC folks, because when defining a managed, multidimensional array, we specify only a single pair of brackets with each of the dimensional indexes separated by commas. Also note that to retrieve the length of multidimensional array, you must call the `Array::GetLength` method, passing it the desired dimension. Other than that, the code reflects what you've already learned about having to box value types when reference types are required—such as when the numeric equation is combined into a a `String` object for addition to the multi-dimensional `String` array.

Jagged Arrays

Jagged arrays are nonrectangular, multidimensional arrays (arrays of arrays) that allow you to define arrays where the number of dimensions do not match. As an example, you might have a two-dimensional array where each of the single-dimensional subarrays is of a different dimension. Unfortunately, while some .NET languages such as C# and Visual Basic .NET do directly support jagged arrays, this support is missing in the current incarnation of Managed Extensions. The following code illustrates how to simulate jagged arrays in C++.

```
void JaggedArray()
{
```

```
#pragma push_macro("new")
#undef new

  Array* jagged[] = new Array*[2];

  jagged[0] = new String*[4];
  jagged[1] = new String*[2];

  for (int i = 0; i < jagged->Length; i++)
  {
    for (int j = 0; j < jagged[i]->Length; j++)
    {
      jagged[i]->SetValue(String::Format(S"{0} + {1} = {2}",
        __box(i), __box(j), __box(i + j)),
        j);
    }
  }

  Text::StringBuilder* sb = new Text::StringBuilder();
  for (int i = 0; i < 2; i++)
  {
    for (int j = 0; j < jagged[i]->Length; j++)
    {
      sb->AppendFormat(S"[{0}][{1}] = {2}\r\n",
                       __box(i), __box(j), jagged[i]->GetValue(j));
    }
    sb->Append(S"\r\n");
  }
  AfxMessageBox((CString)sb->ToString());

#pragma pop_macro("new")
}
```

As you can see, all I did was define and add two single-dimensional arrays of different lengths (4 and 2) to another single-dimensional array. This works in Managed Extensions because—unlike native C++—when an `Array` object is defined it will hold any `Object`, so that I'm not restricted to only one type. Note that the `Length` property is used again, since the `GetLength` method is only for use with multidimensional arrays.

Passing Managed Arrays to Functions

Since managed arrays are essentially objects of type Array, it is possible to use them as arguments to functions. In fact, you'll see this quite often throughout the book. The following function takes a two-dimensional array as argument:

```
void MakeArrayUpper(String* twoDArray[,])
{
   for (int i = 0; i <= twoDArray->GetUpperBound(0); i++)
   {
      for (int j=0; j <= twoDArray->GetUpperBound(1); j++)
      {
         twoDArray[i,j] = twoDArray[i,j]->ToUpper();
      }
   }
}
```

This function is also type-safe in that it will only accept a two-dimensional array. In the unmanaged world we wouldn't get this level of type-safety because arrays are passed by pointer (to the address of the array's first element). If an incompatible array is passed to an unmanaged function, the chance of disaster is high (as it is whenever you make a mistake in passing pointers). Also note the use of the Array::GetUpperBound method (returns the dimension of the array) and String::ToUpper method (converts to uppercase all characters in the string).

Now let's see how to pass an array of value objects to a function:

```
void DoubleIntArray(int intArray __gc[])
{
   for (int i=0; i < intArray->Length; i++)
   {
      intArray[i] *= 2;
   }
}
```

The DoubleIntArray function takes a single-dimensional array of int values and uses a for loop to iterate through the array, doubling the value of each index. The point to note here, however, is the use of the __gc keyword. Keep in mind that arrays of value types are passed by reference. This might appear peculiar at first, but remember that while the array members

are value objects, the array itself is a reference object; hence the need to specify __gc[].

Returning Managed Arrays from Functions

In unmanaged C++, arrays are returned to client code via a pointer to the array's first element. As a result, no type-safety is involved, and the responsibility for properly handling the return array falls on the client code. With Managed Extensions, the exact array type and dimensions are specified in the function signature, so that client code is more protected from common errors. In the following function that populates some (very optimistic) dummy sales figures for a given 12-month period (with goals being in the first index and sales realized being in the second), you can see how to define a function as returning an Array object.

```
Int32 GetYearlySalesFigures() [,]
{
#pragma push_macro("new")
#undef new

   int goalIndex = 0;
   int realizedIndex = 1;
   Int32 salesFigures[,] = new Int32 [2,12];

   for (int i = 0; i < 12; i++)
   {
      salesFigures[goalIndex, i] = ((i+1) * 10);
      salesFigures[realizedIndex, i] = salesFigures[goalIndex, i] * 2;
   }

   return salesFigures;

#pragma pop_macro("new")
}

...

// Get the GetYearlySalesFigures function...
Int32 salesFigures[,] = GetYearlySalesFigures();

// Display multi-dimensional array values
for (int i = salesFigures->GetLowerBound(0);
```

```
                 i < salesFigures->GetUpperBound(0);
                 i++)
     {
       Console::Write(S"Row {0}: ", __box(i));
       for (int j = salesFigures->GetLowerBound(1);
              j < salesFigures->GetUpperBound(1);
              j++)
       {
         Console::Write(S"[{0}]={1} ", __box(j), __box(salesFigures[i,j]));
       }
       Console::WriteLine();
     }
```

As you can see, the brackets [] are placed at the end of the function's signature where, once again, commas are used to indicate the number of dimensions. Because the function is defined as returning a two-dimensional array of Int32 values, attempting to receive anything else (such as an array holding different types or of different dimensions) will result in a compile error.

One last point to make here is that I'm using Int32 types here as opposed to int types. Int32 is just a managed version of the int type that you've always worked with. Using it here implicitly tells the compiler that I want a managed array without having to explicitly specify the __gc[] keyword.

C++ typedefs to Allow Native Array Syntax with Managed Arrays

As the syntax for managed arrays is a bit odd, I've included a header file in the Tools folder of the book's CD that, among other goodies, also defines a typedef that enables you to use native array syntax when passing or returning managed arrays to and from functions:

```
#define MCArray(type) type##Array
typedef int intArray __gc[];
```

Compare the following three functions that each return a managed array of int types.

```
int Foo1() __gc[]   // Managed C++ syntax
intArray Foo2()     // Using the intArray typedef
MCArray(int) Foo3() // Using both #define and intArray typedef
```

Compare the following three functions that each take as a parameter a managed array of int types.

```
void Bar1 ( int (__gc *outArray) __gc[] ) // Managed C++ syntax
void Bar2 ( intArray* outArray ) // Using intArray typedef
void Bar3 ( MCArray(int) ) // Using both #define and intArray typedef
```

As you can see, using either the `intArray` typedef or the `MCArray` #define provide for a more native C++ syntax, leading to fewer mistakes and greater readability.

Destructors

One often overlooked difference between unmanaged and managed C++ classes is in the way destructors are called. Since the programmer is responsible for deleting dynamically allocated objects in unmanaged C++, we know exactly when the destructor is called. However, since the GC manages the deletion of dynamically allocated managed objects, we cannot determine exactly when object destruction occurs. Here's an example:

```
__gc class ReferenceClass
{
public:
 ReferenceClass(String* s)
 {
    this->s = s;

    TRACE1(_T("\tReferenceClass::ReferenceClass - %s\n"),
           (CString)this->s);
 }
 ~ReferenceClass()
 {
    TRACE1(_T("\tReferenceClass::~ReferenceClass - %s\n"),
           (CString)this->s);
 }

protected:
 String* s;
public:
 String* GetString() { return this->s; }
};
```

```
class CNativeClass
{
public:
 CNativeClass(CString str)
  {
    m_str = str;

    TRACE1(_T("\tCNativeClass::CNativeClass - %s\n"), m_str);
  }
 ~CNativeClass()
  {
    TRACE1(_T("\tCNativeClass::~CNativeClass - %s\n"), m_str);
  }
protected:
  CString m_str;
};
```

In the preceding code, I've defined two classes—a reference type and an unmanaged class aptly named `ReferenceClass` and `CNativeClass`, respectively. Both classes use MFC TRACE statements in their constructors and destructors, so that we can see when they're being allocated and deallocated relative to the scope of the application. The following code does just that:

```
void AllocateObjects()
{
    TRACE(_T("AllocateObjects - Entering function\n"));

#pragma push_macro("new")
#undef new
    TRACE(_T("\tAllocateObjects - Dynamically "
        "allocating CNativeClass\n"));
    CNativeClass* pNativeClass = new CNativeClass(_T("C++ heap"));

    TRACE(_T("\tAllocateObjects - Dynamically "
        "allocating ReferenceClass\n"));
    ReferenceClass* referenceClass = new ReferenceClass(S"CLR heap");

    TRACE(_T("\tAllocateObjects - Stack allocating CNativeClass\n"));
    CNativeClass nativeClass(_T("Stack"));

    TRACE(_T("\tAllocateObjects - Manually deleting "
        "CNativeClass object\n"));
```

```
    delete pNativeClass;
#pragma pop_macro("new")

    TRACE(_T("AllocateObjects - Exiting function\n"));
}
```

When a call to the `AllocateObjects` function was placed in the `OnInitDialog` function of a dialog-based application's main dialog class, it resulted in the following output:

```
AllocateObjects - Entering function
    AllocateObjects - Dynamically allocating CNativeClass
    CNativeClass::CNativeClass - C++ heap
    AllocateObjects - Dynamically allocating ReferenceClass
    ReferenceClass::ReferenceClass - CLR heap
    AllocateObjects - Stack allocating CNativeClass
    CNativeClass::CNativeClass - Stack
    AllocateObjects - Manually deleting CNativeClass object
    CNativeClass::~CNativeClass - C++ heap
AllocateObjects - Exiting function
    CNativeClass::~CNativeClass - Stack
```

Object Construction : Standard C++ vs. Managed Extensions

The construction order of a reference (`__gc`) class hierarchy is different than the standard C++ construction order that most of us are familiar with. For example, In standard C++, the top base class is always constructed first, with each succeeding base class being constructed next. However, in Managed Extensions C++, this order is reversed, with the furthest derived class constructor being called first. Also, virtual functions called in the constructor always forward to the furthest derived class. Additionally—in contrast to standard C++—constructors are not used in conversions. Finally, the `ICloneable::Clone` interface replaces the need to define copy constructors for each class.

As you can see, all the constructors behave exactly as expected, and nothing is out of the ordinary here. Similarly, the dynamically allocated unmanaged object was destroyed when the `delete` operator was explicitly called for it, and the stack-allocated unmanaged object was destroyed when it went out of scope (when its allocating function—`AllocateObjects`—ended). However, notice that our reference object was not destroyed—at least not that we can see. As you might suspect, the reason for this is that

the GC hasn't freed the memory yet and won't until it needs to. In a simple application that contains only these few lines of code, that isn't likely to happen until the application terminates. However, the point being made here is that you cannot accurately predict when garbage collection will destroy your objects.

Of course you can manually force a call to the destructor by explicitly calling the `delete` operator on the reference object. However, while the destructor method will execute (which may be enough in certain cases where your main concern is performing clean-up chores), the memory allocated for the object will not actually be freed until the next garbage-collection cycle. In fact we can verify this as follows:

```
. . .

ReferenceClass* referenceClass = new ReferenceClass(S"CLR heap");

. . .

delete referenceClass;
TRACE1(_T("referenceClass::m_s = %s",
      (CString)referenceClass->GetString()));

. . .
```

This would surely result in a memory-type exception using a native class. However, with a reference object, this *might* actually execute. The reason is that even after explicitly calling the `delete` operator (which resulted in the destructor being executed), the memory won't be freed until the next garbage-collection cycle. Obviously, you shouldn't do this, as you don't know when the memory will be freed. In addition, any referenced objects from within the object may have already been destructed (finalized). However, this does serve to illustrate that garbage collection is out of your hands—and, therefore, not something you can predict in terms of precisely when it will deallocate specific objects.

So how does the GC know when an object is no longer being used? Unlike COM, the CLR doesn't use reference counting to control an object's lifetime. Rather, the GC monitors object references and identifies objects that can no longer be reached by runtime code. Once an object can no longer be actively referenced and its scope has ended, it is eligible for destruction.

If you use the ILDASM utility mentioned earlier to examine the MSIL generated for a reference class with a defined destructor, you'll see that the destructor has been renamed `Finalize`. In addition, the compiler will automatically create a function called `_dtor` that first calls the `GC::SuppressFinalize` method and then calls `Finalize` (your destructor code). The `GC::SuppressFinalize` method tells the GC that it need not call the `Finalize` method on this object, thereby preventing situations where an object might be destructed multiple times.

Summary

This chapter has covered just enough Managed Extensions and .NET basics to ensure that you get the most out of the following chapters without encountering repetition of the basics in each chapter. You began by seeing the canonical "Hello World" application in order to understand the key differences between a Managed Extensions application and an MFC application. I provided broad overviews of such .NET topics as *assemblies, manifests,* and *references.* From there, you examined a basic demo illustrating how to mix MFC and .NET code. The following part introduced some important mixed-mode programming issues and how to deal with them. Then you began to explore the subject of working with .NET types—focusing on their definition and usage in your MFC applications. That part covered the topics of defining and using managed types, boxing and unboxing, managed and unmanaged pointers, managed arrays, and destructors.

Now that you have the basics under your belt, we can get started with the fun stuff—seeing how the BCL can increase your productivity in the real world. Chapter 2 begins this journey with a topic that very few MFC developers have had the chance to add to their toolkit, as it's not inherently supported in the MFC library—Regular Expressions.

Regular Expressions

Introduction

Regular Expressions provides a means for matching strings based on a format that includes normal alpha-numeric characters and *metacharacters* (characters that allow you to define complex patterns). For many years, Perl and awk programmers have enjoyed the ability to use Regular Expressions to parse and manipulate text in a powerful manner that developers in other languages could only dream of. In fact, we Visual C++ developers have always had to either painstakingly write our own string-parsing code—usually rewriting it for every project—or turn to third-party products. While MFC still doesn't implement this wonderfully robust text-processing capability, the .NET BCL does! In fact, not only does the `System::Text::RegularExpressions` namespace provide a very rich set of Regular Expressions classes, but these classes also define unique abilities such as *right-to-left pattern matching* and *on-the-fly compilation*. In this chapter, we focus on the most common text-processing tasks and provide numerous code snippets to illustrate just how to accomplish those tasks using the BCL implementation of Regular Expressions.

We begin by learning how to use the `Regex` class to split strings into substrings by specifying patterns (delimiters) to look for. Then you'll see how to parse strings in order to perform such tasks as extracting or isolated target substrings within a larger set of text. Next I introduce the world of groups and captures and see how they enable you to specify submatch criteria within your match criteria to perform such tasks as parsing for a phone number and then capturing within that phone number the area code and exchange. One of the most popular uses of Regular Expressions involves its ability to parse through textual data using a supplied pattern and modify isolated text with a supplied value (which can also be based on the input

pattern). The subsequent sections entitled "Parsing and Replacing Strings" and "Replacing Matches Using Groups and Substitution Patterns" cover this topic. The chapter's final part combines several aspects of what you'll learn throughout this chapter by illustrating a completely functional e-mail parsing function.

Using the `Regex` Class to Split Strings

The `Regex` class—defined in the `System::Text::RegularExpressions` namespace—encapsulates an immutable expression. By that, I mean that an instantiated `Regex` object can only represent the expression that was used to construct it. As a result, if you want to modify an expression once it's been used in the construction of a `Regex` object—or if you want to use multiple expressions—you'll need to construct a distinct `Regex` object for each desired modification or new expression. While at first blush this may seem like a limitation, as you work your way through the chapter, I'll bring this point into play at the appropriate times so that you see that the immutability of the `Regex` class is not a problem. I mention it at the outset of our discourse on the `Regex` class because many people new to the .NET implementation read about this issue and assume that it's a show-stopper for them. It's not. That said, let's examine two main string processing tasks: splitting a string according to a specified pattern and searching a string for a defined pattern.

Splitting a String

A very common programming task is the parsing of a string—splitting it into substrings—according to a specified delimiter or set of delimiters. For many years, this has been accomplished in C and C++ with the `strtok`, `wcstok`, and `_mbstok` functions. Since the principal point of this book is to illustrate how selectively incorporating various BCL classes can make you a more productive C++ programmer, let's dig that `strtok` function out of the closet, dust it off, and see how it would be used to handle a simple case of parsing a string.

The following example borrows the opening lines of Virgil's *Aeneid*, where the text is parsed searching for occurrences of semicolons and new-line characters.

```
void StrTokTest()
{
  char sz[] = _T("Of arms I sing; and of the man"
              "who first\nFrom Trojan shores "
              "beneath the ban of fate");
  char delimiters[] = _T(";\n");
  char* token;

  CString strResults = _T("Using strtok to delimit the following for "
                          "new-line and semicolons\r\n\r\n");

  /* Establish string and get the first token: */
  token = strtok(sz, delimiters);
  while (NULL != token)
  {
    /* While there are tokens in sz */
    strResults += token;
    strResults += _T("\r\n");

    /* Get next token: */
    token = strtok( NULL, delimiters );
  }

  // Display results
  AfxMessageBox(strResults);
}
```

The output of this simple string parsing is shown in Figure 2–1.

Now I have to be honest about something. I've used this `strtok` function for more years than I'd like to admit, and yet its syntax is so unlike anything I use in my everyday C++ and C# coding life that I have to look up

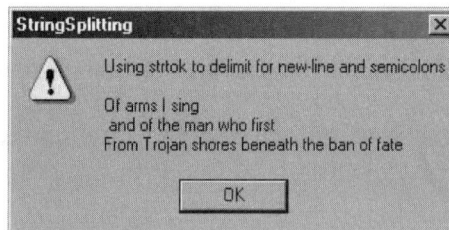

Figure 2-1 Simple `strtok` usage

the online-help sample each time I need to use it! Now compare the `strtok` example with the following example:

```
void RegexSplitTest()
{
#pragma push_macro("new")
#undef new
 try
   {
      CString str = _T("Of arms I sing; and of the man who "
                       "first\nFrom Trojan shores "
                       "beneath the ban of fate");

      CString strResults = _T("Using the Regex class to delimit for "
                             "new-line and semicolons\r\n\r\n");

      Regex* rx = new Regex(S";|\n");
      String* words[] = rx->Split(str);
      for(int i = 0; i < words->Length; i++)
      {
         strResults += (CString)words[i];
         strResults += _T("\r\n");
      }

      AfxMessageBox(strResults);
   }
   catch(Exception* e)
   {
      AfxMessageBox((CString)e->Message);
   }
#pragma pop_macro("new")
}
```

Like most current class libraries, the majority of the BCL classes (including the Regular Expressions classes) will throw exceptions when unexpected events occur. As a result, you should always wrap your Regular Expressions processing in try/catch statements.

This produces exactly the same output (see Figure 2–2), but it is much easier to use and certainly more intuitive. In order to use the Regex class for such a simple chore, we needed only to instantiate the Regex class with

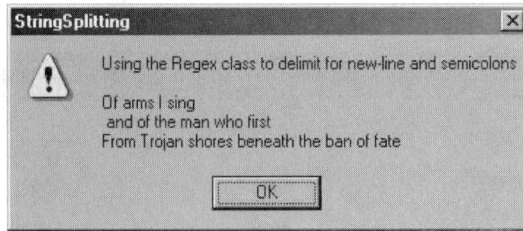

StringSplitting   [X]

⚠ Using the Regex class to delimit for new-line and semicolons

Of arms I sing
and of the man who first
From Trojan shores beneath the ban of fate

                    [OK   ]

Figure 2-2 Simple `Regex::Split` usage

the desired pattern to use in delimiting the string that is subsequently passed to the `Regex::Split` method. The `Split` method then returns an array of `String` objects that can be iterated with a `for` loop.

Note the use of the vertical bar within the pattern string passed to the `Regex` constructor. Similar to the C++ logical-OR operator, the vertical bar enables you to separate distinct characters that represent the delimiters to search for in the string. The following expression being passed to the `Regex` constructors states "find all occurrences of a semicolon *or* a new-line character."

```
Regex* rx = new Regex(S";|\n");
```

While these two examples show how easy it is to delimit strings with the `Regex` class, the use of Regular Expressions is obviously about much more than an object-oriented replacement for the `strtok` function. I used that as an example only because it's one of the most basic Regular Expressions tasks and has a C/C++ equivalent against which to compare the `Regex` class. The real power of Regular Expressions—even when doing something as simple as splitting a string—is in the ability to specify a pattern using special characters—*metacharacters.*

You can think of metacharacters as being somewhat similar to the DOS wildcards (`?` and `*`) that are used in the command shell when performing commands on files and directories. However, a glance at some of the more commonly used metacharacters in Table 2–1 shows that these metacharacters go far beyond the two simple wildcards that DOS provides.

Along with the version of the overloaded `Split` method that you've seen, where a single parameter indicates the string to be split, the `Split` method can also take an additional parameter—a number indicating the maximum number of splits performed by the Regular Expressions engine on a given string. (Recall from Chapter 1 that the `__gc[]` keyword at the

Table 2-1 Common Regular Expressions Metacharacters

Expression	Description
.	Matches any character except \n
[characters]	Matches a single character in the list
[^characters]	Matches a single character not in the list
[charX-charY]	Matches a single character in the specified range
\w	Matches a word character; same as [a–zA–Z_0–9]
\W	Matches a nonword character
\s	Matches a whitespace character; same as [\n\r\t\f]
\S	Matches a non-whitespace character
\d	Matches a decimal digit; same as [0–9]
\D	Matches a nondigit character
^	Match the beginning of a line
$	Match the end of a line
\b	On a word boundary
\B	Not on a word boundary
*	Zero or more matches
+	One or more matches
?	Zero or one matches
{n}	Exactly n matches
{n, }	At least n matches
{n, m}	At least n but no more than m matches
( )	Capture matched substring
(?<name>)	Capture matched substring into group name
|	Logical-OR

end of the method declaration indicates that the method will return a managed array.)

```
String* Split(
   String* str,
   int num_splits
) __gc[];
```

In the following example, our *Aeneid* string is being split a maximum of three times. (The differences between this function and the `RegexSplit Test` function are highlighted.)

```
void Split3Times()
{
#pragma push_macro("new")
#undef new
    CString str = _T("Of arms I sing; and of the man who "
                     "first\nFrom Trojan shores "
                     "beneath the ban of fate");

    CString strResults = _T("Splitting for spaces 3 times\r\n\r\n");

    Regex* rx = new Regex(S" ");
    String* words[] = rx->Split(str, 3);
    for(int i = 0; i < words->Length; i++)
    {
        strResults += (CString)words[i];
        strResults += _T("\r\n");
    }

    AfxMessageBox(strResults);
#pragma pop_macro("new")
}
```

As you can see, almost nothing changes in your code; Figure 2–3 shows the results.

You will often find that you have a string of text from which you want to remove only the text data, ignoring commas and spaces. The following example uses a new string to highlight a common problem that arises for

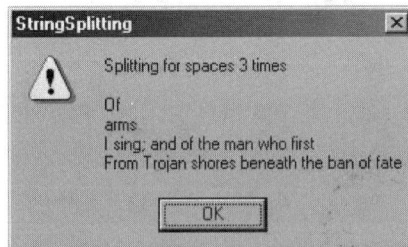

Figure 2-3 Setting the maximum number of times a string should be split

newcomers to Regular Expressions who attempt this operation for the
first time.

```
void RemoveSpacesAndCommas()
{
#pragma push_macro("new")
#undef new
  CString str = _T("Apples  Oranges  ,Bananas Plums");

  CString strResults = _T("Note the blank lines when I wanted all "
                          "items except spaces and commas\r\n\r\n");

  Regex* rx = new Regex(S" |,");
  String* words[] = rx->Split(str);
  for(int i = 0; i < words->Length; i++)
  {
    strResults += (CString)words[i];
    strResults += _T("\r\n");
  }

  AfxMessageBox(strResults);
#pragma pop_macro("new")
}
```

Once again, in the `RemoveSpacesAndCommas` function, I've used the vertical bar separator to indicate that I want the code to delimit the list with
commas and spaces. The results shown in Figure 2–4, however, reveal that
this code did not produce exactly what we wanted.

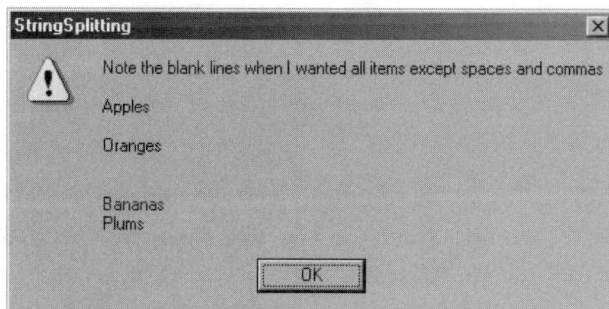

Figure 2-4 Attempting to split when consecutive instances of a pattern-
specified character exist in the string

If you were to examine the words array in the debugger, you'd see that its Count property indicates that it has seven elements and not the four we desire. The basic issue is with our expression |, which does not accommodate contingencies such as consecutive spaces or commas in a string. Thus the error here is not due to incorrect usage of the Regex class or its Split method; rather the problem is with the expression being employed. The following code corrects that problem (it includes a commented-out version of the original constructor for comparison purposes.)

```
// Regex* rx = new Regex(S" |,");
Regex* rx = new Regex(S"[ |,]+");
```

As you can see, we've now introduced two new elements into our bag of pattern tricks: the square bracket and the + sign. The square brackets simply allow us to cluster characters in the pattern so that we can indicate that another special character can be applied against it. Look back at the metacharacters defined in Table 2–1 and note that I'm essentially telling the Regular Expressions engine, "Delimit, for one or more (the + sign) characters in this group of characters, the space and comma (enclosed in square brackets)."

One thing to note is that you'll often hear the word "group" associated with Regular Expressions, and it has a very special meaning that I'll cover later in this chapter. To avoid confusion, the generic usage of the word appears in lowercase. When it is used in the sense of the BCL class that encapsulates the Regular Expressions definition, it will appear capitalized.

If you replace the RemoveSpacesAndCommas function's Regex constructor call with this one, your output will be as expected.

One common usage of the Regex::Split; method is when separating the columns of comma-delimited file input. Figure 2–5 shows an example running of the string-splitting demo (StringSplitting) that I've included on the book's CD, where my contact information has been entered in comma-delimited format and then split, with each column displayed as a separate entry in a list box.

One last thing I'll mention here before we move on to discuss Regular Expressions *matches* is that so far, we've essentially told the Regular Expressions engine to search through the strings for given patterns and return the strings not containing the characters in the patterns. However, what if you wanted to retrieve those characters? This is done via the parentheses

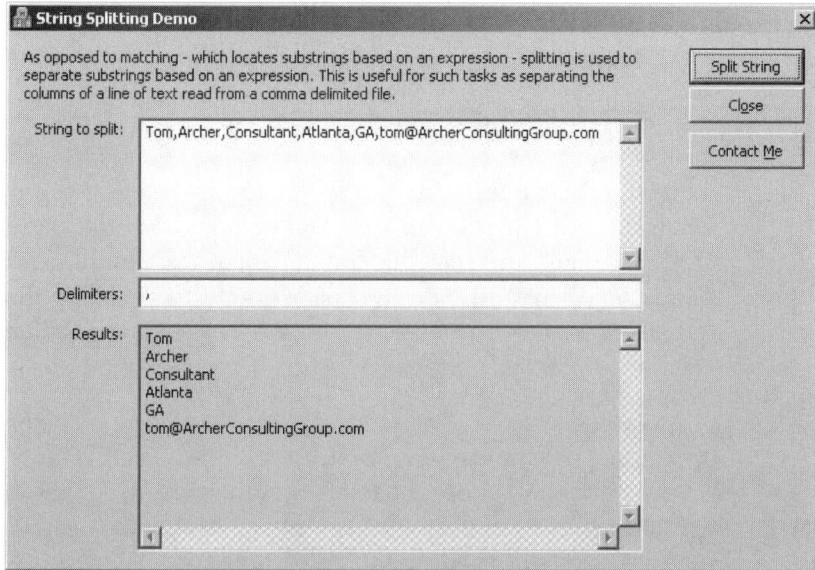

Figure 2-5 The CD's `StringSplitting` demo enables you to test various Regular Expressions patterns in splitting a string into its desired components. (Note that there is a space after the first square bracket in the `Regex`-pattern-edit field.)

and would result in a separate string in the `String` array returned by `Split` for every pattern matched. For example, using the following `Regex` construction in the `RemoveSpacesAndCommas` function would yield not only the four fruits from the string, but also `String` objects for each space and comma as well. This isn't necessarily something you'll need all the time when splitting strings, but it's good to know it's possible if you need to do it. (For those of you familiar with Regular Expressions and who know that the parentheses are used for "captures," that subject is covered later in the chapter under the heading "Processing Groups and Captures.")

```
// When used with Split method returns both the items
// being delimited as well as the delimiters in a
// String array
([ |,])+
```

Searching Strings with the `Match` and `MatchCollection` Classes

The `Match` class encapsulates information about a Regular Expressions pattern match—including a property to determine if the match was successful, the relative index into the string where the pattern was found (matched), and the matched value. This works by simply constructing a `Regex` object with the desired pattern and then calling the `Regex` class's `Match` method to return a `Match` object. The success or failure of the match can then be determined by evaluating the `Match::Success` Boolean property. For situations where a pattern might yield multiple matches, a collection object of type `MatchCollection` can be acquired via a call to the `Regex::Matches` method. As you might suspect, the `MatchCollection` is simply a collection of `Match` objects that can be iterated with a `for` loop (the `MatchCollection` has a `Count` property to control the loop). Before we get into some example code, take a peek at Table 2–2 where you'll find the most commonly used `Match` class members.

Table 2-2 Commonly Used `Match` Class Members

Expression	Description
`Success`	Boolean property that indicates whether a match was found. You should *always* test this property before processing a match.
`Index`	This numeric value indicates the starting position within a string where the match for the supplied pattern was found.
`Length`	This numeric value indicates the length of the string that matches the supplied pattern.
`Value`	This is the actual substring that matched the pattern.
`Groups`	This returns a collection of `Group` objects for the executed expression. Groups are used to identify subpatterns within an expression. For example, using groups would be perfect for a situation where you have an expression for a telephone number field and from within that expression you want to collect area codes. Groups are discussed shortly in the section entitled "Processing Groups and Captures."

Now let's see how to use this method to obtain a Match object using a code example. (The salient code is in bold to make it easier visually to skip over the display code.)

```
void FindSingleMatch()
{
#pragma push_macro("new")
#undef new
  CString strResult;

  CString strSource = _T("Tim, sometimes you have to "
                         "accept that times are changing");
  CString strPattern = _T("tim");

  Regex* rx = new Regex(strPattern);

  if (!rx->IsMatch(strSource))
  {
     strResult = _T("No matches found");
  }
  else
  {
    Match* pMatch = rx->Match(strSource);

    if (pMatch->Success)
    {
      strResult.Format(_T("Searching '%s' for substrings "
                          "containing the text '%s' yielded "
                          "a match at index %ld. It's value is '%s'"),
                       strSource, strPattern,
                       pMatch->Index, (CString)pMatch->Value);
    }
    else
    {
      strResult = _T("Regex class it had matches, but "
                     "no Match object returned!!");
    }
  }
  AfxMessageBox(strResult);
#pragma pop_macro("new")
}
```

Figure 2–6 shows the results of this simple test.

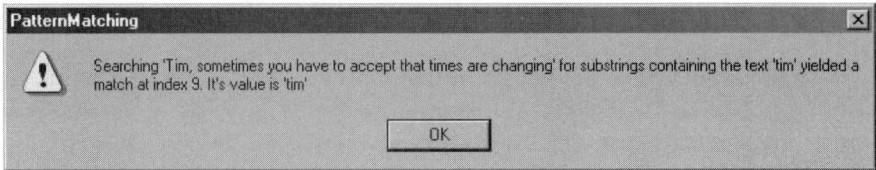

Figure 2-6 The `Match` class enables us to search through a string for a substring matching a given pattern.

As you see, our example has met with only limited success in this case, as the source string actually contains more than the single match returned. This situation is precisely why the `MatchCollection` class exists. The `MatchCollection` class—which implements both `ICollection` and `IEnumerable`—is basically an enumerable collection of `Match` objects. Just like the `Match` class, the `MatchCollection` class also has no public constructor, and a `MatchCollection` is obtained via a `Regex` method—`Matches`, in this case.

Let's examine an updated version of the previous function that returns all matches of the supplied pattern (again, the code of most interest is in bold).

```
void FindAllMatches()
{
#pragma push_macro("new")
#undef new
  CString strResults;

  CString strSource = _T("Tim, sometimes you have to "
                         "accept that times are changing");
  CString strPattern = _T("tim");

  Regex* rx = new Regex(strPattern);

  if (!rx->IsMatch(strSource))
  {
    strResults = _T("No matches found");
  }
  else
  {
    MatchCollection* pMatches = rx->Matches(strSource);

    strResults.Format(_T("Searching '%s' for substrings containing "
```

```
                               "the text '%s' yielded "
                               "%ld matches\r\n\r\n"),
                      strSource, strPattern, pMatches->Count);

   for (int i = 0; i < pMatches->Count; i++)
   {
      CString strTemp;
      strTemp.Format(_T("Match #%ld found at index %ld = %s\r\n"),
                    i,
                    pMatches->Item[i]->Index,
                    (CString)pMatches->Item[i]->Value);
      strResults += strTemp;
   }
 }
 AfxMessageBox(strResults);
#pragma pop_macro("new")
}
```

As the `MatchCollection` class implements the `ICollection` interface, we can use the `Count` property and a `for` loop to iterate through the collection. The `Item` property is an index property that returns individual `Match` objects. From there, it's pretty much like the previous example in that you can then use any needed `Match` members—such as `Index` `Value`. This new and improved match function yields the results shown in Figure 2–7.

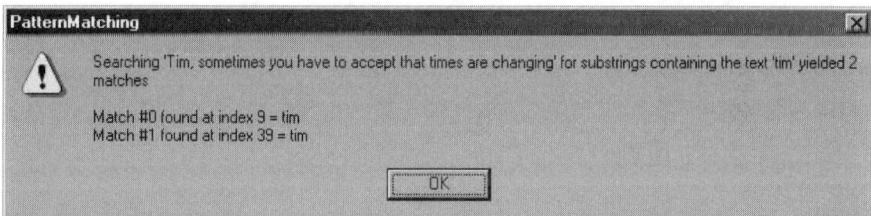

Figure 2-7 Using the `MatchCollection` class enables you to determine all matches found when searching a source string using a Regular Expressions pattern.

Noting that our function yields two matches, but the following string also has an instance of the substring "tim" in the very beginning, we might be a little puzzled. (The substrings that you would normally expect to have found with the previous example's code are in bold.)

Tim, sometimes you have to accept that **tim**es are changing

Looking at the differences between the substring that was not matched and the substrings that were matched, you might reasonably assume that the reason for the nonmatch had to do with capitalization, and you would be correct. What is needed is a means of specifying that the Regex object should ignore case, which is exactly what the following Regex constructor allows:

```
public Regex(
    string pattern,
    RegexOptions options
);
```

In this constructor, in addition to supplying the pattern, you can also specify an enumeration of options defined by RegexOptions that gives you even greater control over how the Regular Expressions engine parses the specified pattern. In this case, you would need only to change the construction of the Regex object in the FindAllMatches function as follows:

```
Regex* rx = new Regex(strPattern, RegexOptions::IgnoreCase);
```

Making this change will truly force the function to live up to its name, as it will then find all three occurrences of the "tim" substring. Table 2–3 lists the various RegexOptions enumeration values and their descriptions.

As you've seen, a Match object stores and provides access to all the substrings extracted by a given search. However, one interesting thing that I

Table 2–3 RegexOptions Enumerations

Enumeration	Description
Compiled	Specifies that the expression is compiled to an assembly. This yields faster execution but increases startup time.
CultureInvariant	Specifies that cultural differences in language are ignored.
ECMAScript	Enables ECMAScript-compliant behavior for the expression. This flag can be used only in conjunction with the IgnoreCase, Multiline, and Compiled flags. The use of this flag with any other flags results in an exception.
	(continued)

Table 2-3 `RegexOptions` Enumerations (*cont.*)

Enumeration	Description
`ExplicitCapture`	Specifies that the only time you want to capture is if you use the explicit group/capture syntax of (`?<name>`...). As a result, it isn't necessary to use the syntactically clumsy format (`?:`...)—called a capture-inhibitor operator—to simply dictate that you do not want to capture. This subject is covered under the heading "Processing Groups and Captures."
`IgnoreCase`	Specifies case-insensitive matching.
`IgnorePatternWhitespace`	Eliminates unescaped white space from the pattern and enables comments marked with #.
`Multiline`	Multiline mode. Changes the meaning of ^ and $ so they match at the beginning and end, respectively, of any line, and not just at the beginning and end of the entire string.
`None`	Specifies that no options are set. This equates to using the `Regex` constructor that only takes a single parameter (the expression).
`RightToLeft`	Specifies that the search will be from right to left instead of from left to right. This is useful in situations where you only care about the last occurrence of a substring. An example of that might be if you were attempting to determine the "first-level domain" of an e-mail address (e.g., .com, .org, .net, etc.).
`Singleline`	Specifies single-line mode. Changes the meaning of the dot metacharacter (.) so it matches every character (instead of every character except \n).

haven't broached until now is the fact that the `Match` object also keeps track of the string being searched and the expression being used, similar to reading a file with a file object and the file object keeping track of your current (cursor) position within the file for future reads and relative seeking (or repositioning). As a result, the `Match` object can be used to perform

continual searches that begin where the last search ended, as illustrated by the following code. Note the use of the `NextMatch` method.

```
void FindAllMatches2()
{
#pragma push_macro("new")
#undef new
  CString strResults;

  CString strSource = _T("Tim, sometimes you have to "
                        "accept that times are changing");
  CString strPattern = _T("tim");

  Regex* rx = new Regex(strPattern, RegexOptions::IgnoreCase);

  if (!rx->IsMatch(strSource))
  {
    strResults = _T("No matches found");
  }
  else
  {
    for (Match* pMatch = rx->Match(strSource);
        pMatch->Success;
        pMatch = pMatch->NextMatch())
    {
      CString strTemp;
      strTemp.Format(_T("Found match at index %ld = %s\r\n"),
                     pMatch->Index,
                     (CString)pMatch->Value);
      strResults += strTemp;
    }
  }
  AfxMessageBox(strResults);
#pragma pop_macro("new")
}
```

Since both the `FindAllMatches` and `FindAllMatches2` functions result in the same matches being realized, you might be tempted to ask if the choice of how to enumerate a `Regex` object's `MatchCollection` comes down to personal programming style. To a certain extent, yes. However, there is a real difference when choosing between either obtaining all the matches at once via the `Matches` property or acquiring one match at a time

via the NextMatch method—performance. The latter approach would be useful in situations where you don't want to wait for the collection to be completed before you have access to the search result or if you think there are going to be a huge number of matches and you're only interested in a subset.

Before moving on to consider groups and captures, let's look at one more example of an expression that is a bit more involved than you've seen. Let's say you had the following string and wanted to parse it for proper nouns (defined here as anything starting with a capital letter):

```
John and Harry are members of the Borbon club.
```

Using expressions, this could very easily be accomplished with the following expression:

```
[A-Z][a-z]+[ ]*
```

While this looks a bit strange at first, patterns become very easy to read with a little experience. We'll start from the left and read this just as the Regular Expressions engine sees it.

I'm searching for the following:

1. A single capital letter. The square brackets around the A–Z indicate a letter in this group. The A–Z indicates any letter within that range. The lack of a + sign after the right bracket indicates that I only want one leading capital letter.
2. Followed by any number of lowercase letters. Once again, the square brackets enclosing the a–z indicate a group of letters—in this case, a range of letters from a lowercase *a* to a lower case *z*. The + sign after the right bracket specifies that any number of lowercase characters is OK.
3. Terminated by zero or more spaces. The brackets are used to enclose a space character so that the parser doesn't get confused and combine the space with other parts of the pattern. The asterisk to the right of the right bracket indicates that zero or more of whatever is in the brackets will satisfy the search. Otherwise, if the last word in the sentence met the other parts of the pattern, but the string didn't end in a space, it would not be included in the matches found.

The results of this pattern would be the following proper nouns:

```
John
Harry
Borbon
```

However, what if the string were then changed to the following?

```
John and Harry are members of the Borbon club at IBM.
```

Looking at the expression, you can see that the IBM would not turn up in the Regex Matches collection because it doesn't match the pattern. The pattern specifically states "one capital letter followed by one-to-many lowercase letters followed by zero-to-many spaces." Updating the pattern to include any number of consecutive uppercase letters won't work because the substring would still need to include lowercase letters. The answer is to use the | operator with the second grouping of the pattern to allow for both capital and lowercase letters after the initial capital letter. Here's how that would look. (The change from the previous example pattern is in bold).

```
[A-Z][A-Z|a-z]+[ ]*
```

Now our expression will correctly yield the following:

```
John
Harry
Borbon
IBM
```

This works great until we realize that the expression will only find capitalized nouns that consist of two or more letters. This obviously omits the first person singular found in the following string:

```
John, Harry and I are members of the Borbon club at IBM.
```

The only change that needs to made here is to stipulate that there can be zero to many letters that follow the initial capital letter. Just as the asterisk was used to match on zero or more spaces, it can be used here as well for the final version of the expression.

```
[A-Z]([A-Z|a-z]*)+[ ]*
```

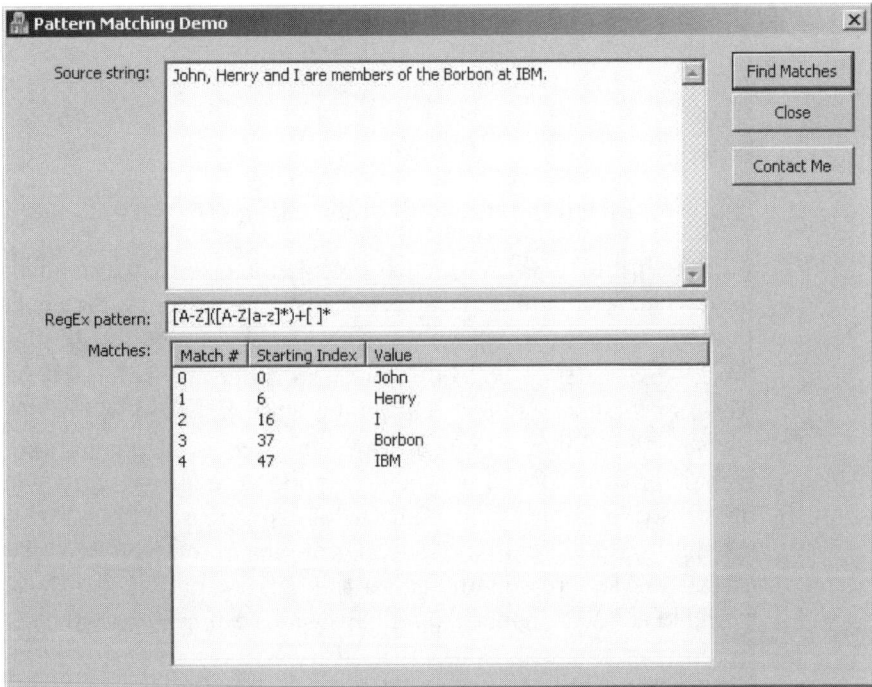

Figure 2-8 The `PatternMatching` demo enables you to quickly test your expressions.

In order to test this and other expressions, the CD includes a simple Regular Expressions matching demo called `PatternMatching`. Figure 2–8 shows that the last change to the pattern yields all the correct proper nouns.

Processing Groups and Captures

So far you've seen how to split strings and parse strings based on supplied patterns with specialized (meta-) characters. Now you'll take what you've learned a step further and discover the world of groups and captures. However, before we get into the technical details of how to work programmatically with groups and captures, let's cover what they are and what services they provide.

Grouping has two purposes. One is to designate captures—to specify a smaller part of a larger match that contains a string that you wish to extract

or isolate. In this vein, groups can be thought of as "sub-matches." An example of this would be in working with strings containing U.S. telephone numbers. Groups would come into play in this example if you needed to extract the area code from these telephone numbers. You'll soon see exactly how to accomplish this task. The second purpose of groups is to group several expressions together as a unit that can be operated on by other Regular Expressions modifiers, operators, or quantifiers.

Regarding the BCL Regular Expression classes, we've talked about the Regex class that contains a MatchCollection class that is a collection containing zero-to-many Match objects. Where do groups and captures fall into this hierarchy? Each Match within its MatchCollection parent object contains a GroupCollection object. This GroupCollection object, in turn, contains a collection of one-to-many Group objects. Each Group object then contains a CaptureCollection object and—as you might guess—each CaptureCollection object contains a collection of one-to-many Capture objects. While this seems almost as difficult to follow as a certain famous story involving the verb "beget," Figure 2–9 should illustrate how these objects relate to one another.

Let's now look at an example of using groups to extract the area code from an expression used to parse U.S. telephone numbers. We'll take one step at a time so that you can see the expression evolve from a simple pattern to a full-blown named group.

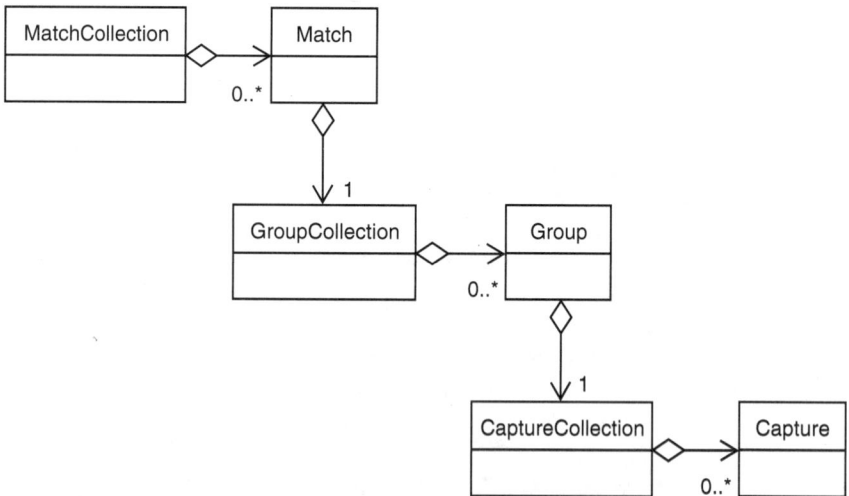

Figure 2-9 The relationship between matches, groups, and captures is much easier to understand when graphically illustrated.

The first thing we need in order to parse telephone numbers is a simple pattern. Since we know from Table 2–1 that \d is the metacharacter for any numeric value, we can start with the following pattern:

```
\d\d\d-\d\d\d\-\d\d\d\d
```

This pattern simply indicates that we're interested in anything that matches a contiguous string of "three numbers followed by a dash followed by three numbers followed by a dash followed by four more numbers." I know that sounds long-winded, but trust me. When you start getting into more complex expressions, you'll want to dissect them in your head like this. Looking at this, you might be thinking, "There's got to be a better way than all those repeated metacharacters." You'd be right. We can use the {n} metacharacter—where n represents the exact number of matches that must be met—as follows:

```
\d{3}-\d{3}-\d{4}
```

Now that we have a phone number pattern, let's look at some group/capture object code.

```
void DisplayGroups(CString const& strSource,
                   CString const& strPattern)
{
#pragma push_macro("new")
#undef new
  try
  {
    CString strResults;
    CString strTemp;

    Regex* rx = new Regex(strPattern);

    // for all the matches
    for (Match* pMatch = rx->Match(strSource);
         pMatch->Success;
         pMatch = pMatch->NextMatch())
    {
      strTemp.Format(_T("Match %s at %ld\r\n"),
                     (CString)pMatch->Value,
                     pMatch->Index);
      strResults += strTemp;
```

```
    // for all of THIS match's groups
    GroupCollection* gc = pMatch->Groups;
    for (int i = 0; i < gc->Count; i++)
    {
        strTemp.Format(_T("\tGroup %s at %ld\r\n"),
                        (CString)(gc->Item[i]->Value),
                        gc->Item[i]->Index);

        strResults += strTemp;
    }
  }
  AfxMessageBox(strResults);
}
catch(Exception* e)
{
  AfxMessageBox((CString)e->Message);
}
#pragma pop_macro("new")
}
```

The main points of interest are the two for loops used to iterate through all of the Match objects and, within those loops, all of the Group objects for each Match. Regarding the outer loop (the Match object loop), you saw this particular technique of using the NextMatch method) in the previous section. Here I'm simply updating a string (strResults) to be displayed at the end of the function with the current Match object's information (Value and Index). Once strResults has been updated, the function retrieves the current Match object's GroupCollection object (via the Match::Groups property) and enters a for loop to iterate through all of the Group objects within that collection, using the GroupCollection:: Count to control the loop. For each Group object found, the strResults is updated with the group information (Value and Index properties).

Calling the DisplayGroups function as shown in the following code would result in the output shown in Figure 2–10.

```
DisplayGroups(_T("770-555-1212"), _T("\\d{3}-\\d{3}-\\d{4}"));
```

The next step is grouping. How do we tell the Regular Expressions engine that we care about the area code within the phone number pattern? Simply by using parentheses. Every pair of parentheses sets up a submatch. Therefore, we only need to update the DisplayGroups client code as follows:

```
DisplayGroups(_T("770-555-1212"), _T("(\\d{3})-\\d{3}-\\d{4}"));
```

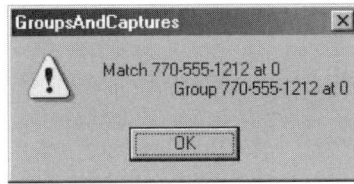

Figure 2–10 All matches have at least one `Group` object in their group collection.

That's it. Presto. I now have a group within a group (a submatch). Running the code again with this new expression yields the results you see in Figure 2–11.

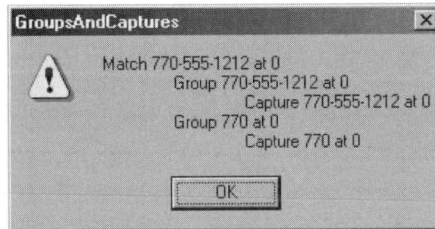

Figure 2–11 When groups (or submatches) are defined in an expression, they show up in the `Match` object's `GroupCollection`.

As the figure shows, we now have two groups: the entire match and the submatch (the area code). But we're not quite finished. It's not exactly enough to retrieve the area code separately; we need to know how to access it. The following `GetAreaCode` function, which takes as its only parameter a string phone number to parse, illustrates how to do that.

```
CString GetAreaCode(CString const& strPhoneNumber)
{
#pragma push_macro("new")
#undef new
  CString strAreaCode;

  try
  {
    CString strPhonePattern = _T("^(\\d{3})-\\d{3}-\\d{4}");
    Regex* rx = new Regex(strPhonePattern);
```

```
    Match* pMatch = rx->Match(strPhoneNumber);
    if (pMatch->Success)
    {
       strAreaCode = (CString)pMatch->Groups->Item[1]->Value;
    }
  }
  catch(Exception* e)
  {
    AfxMessageBox((CString)e->Message);
  }


  return strAreaCode;
#pragma pop_macro("new")
}
```

Most of this should be clear by now, but I'll walk through it in the name of completeness. The first thing the GetAreaCode function does is to pass the phone number pattern—^(\\d{3})-\\d{3}-\\d{4}—to the Regex constructor. You might notice one subtle addition to the pattern. It's the ^ character, which tells the Regular Expressions engine that you're only interested in matches *anchored* (or starting) at the beginning of the string. Obviously, you don't use this metacharacter unless you are certain that no other characters precede the beginning of the phone number, as is true in our case.

Once the phone number pattern is passed, the function retrieves the Match object via the Regex::Match property. After verifying that the match was good (by checking the Match::Success property), the function retrieves the second Group object from the Match object's Group Collection. We know the area code is in the second group because the entire match itself is the first group, and the area code grouping (the parentheses in the expression) is the first group defined going from left to right. (Besides, it's the only other group defined in this case.) The Value property of the Group object is then returned to the caller. Here's the client code for our GetAreaCode function.

```
CString strPhone = _T("770-555-1212");
CString strTemp;
strTemp.Format(_T("The area code for %s is %s"),
               strPhone,
               GetAreaCode(strPhone));
AfxMessageBox(strTemp);
```

Running this code yields the results shown in Figure 2–12.

Figure 2-12 Groups allow you to specify a smaller part (a submatch) of a larger match containing a string that you wish to extract or isolate.

Named Captures

Named groups are a powerful and very useful feature of Regular Expressions. By using named groups you can retrieve a Group object by its group name using the overloaded GroupCollection::Item property:

```
__property Group* get_Item(
  String* groupName
);
```

Looking back on the previous example, it was nice to be able to retrieve the phone number's area code so easily. However, it was not ideal to have to hard-code the index of the Match object's GroupCollection. Sure, we know today that the area code is the second group. However, put this code into production, and sure enough someone will come along and insert another group into the phone number expression, and the code starts to fail immediately. The solution is to simply name the group.

Naming a group is as simple as placing a ? after the open (left) parenthesis and then specifying a textual name between angle brackets. Updating the GetAreaCode gives us the following code (with the change shown in bold):

```
CString strPhonePattern = _T("^(?<areaCode>\\d{3})-\\d{3}-\\d{4}");
```

Now we just need to update the area code value extraction code as follows:

```
strAreaCode = (CString)pMatch->Groups->Item[S"areaCode"]->Value;
```

At this point, your code is protected from any attempt to change the order of the phone number expression's groups.

Let's look at another example to solidify your understanding of how groups work. In the following sample code we intend to separate currency symbols and amounts from a string and display them individually.

```
void DisplayCurrencyAmounts(CString const& strSource)
{
#pragma push_macro("new")
#undef new
  try
  {
    CString strResults;
    CString strTemp;

    Regex* rx = new Regex(S"(?<currency>[$|R])(?<amount>\\d\\w+)");
    MatchCollection* mc = rx->Matches(strSource);
    for(int i =0; i < mc->Count; i++)
    {
      GroupCollection* gc = mc->Item[i]->Groups;
      strTemp.Format(_T("Currency symbol = %s, Amount = %s\r\n"),
                      (CString)(gc->Item[S"currency"])->Value,
                      (CString)(gc->Item[S"amount"])->Value);
      strResults += strTemp;
    }
    AfxMessageBox(strResults);
  }
  catch(Exception* e)
  {
    AfxMessageBox((CString)e->Message);
  }
#pragma pop_macro("new")
}
```

If you were to execute this code and pass it a value of "I gave Robert $300 and he returned R100", you'd see a message box displaying the expected output:

```
Currency symbol = $, Amount = 300
Currency symbol = R, Amount = 100
```

Now let's dissect the expression to see why it worked the way it did.

```
(?<currency>[$|R])(?<amount>\d\w+)
```

Working from left to right, we can see two groups:

```
(?<currency>[$|R])
(?<amount>\d\w+)
```

We note that the first group is named "currency," and its pattern is the following:

```
[$|R]
```

So far, this is easy. As you've learned, the vertical bar represents the logical-OR operator between the characters represented in the brackets. Therefore, this line merely means "match on $ or R" and place in a group called `currency`.

The second group is right next to the first, so for a match, both expressions must match. That second group is

```
(?<amount>\d\w+)
```

The group is named `amount`, and a quick glance at Table 2–1 shows us that `\d\w+` represents a single numeric digit—`\d`—followed by any number of word characters—`\w+`. The latter part means "capture all alphanumeric characters from the current point to the next space or white-space or end-of-string.

Noncapturing Groups

You'll recall that the `DisplayGroups` function presented earlier allows a client to pass an expression, after which the function will display all the matches and groups associated with that expression. However, you might at times want to have groups in your expression that don't result in groups being created in the resulting `Match` object hierarchy. Such groups are called *noncapturing groups*.

Before presenting the syntax for defining a noncapturing group, I'll answer the logical question, "What is the purpose of grouping a subexpression if it's not going to be used?" Noncapturing groups are useful when you are using the OR (|) construct within an expression but don't want the overhead associated with creating `group` and `capture` objects.

Consider the following string:

```
An old man sat in the sun all day long.
```

Let's say we wanted to parse for all occurrences of "an," "on," and "in"—both lowercase and uppercase. Our expression might look like the following:

```
((A|a)|(O|o)|(I|i))n\s
```

If you were to run this expression using the GroupsAndCaptures demo from the book's CD, you'd see that this search results in five groups for the three matches! Obviously, if the code performing this expression doesn't need the groups, it's inefficient to let the Regular Expressions engine do the extra work. Therefore, we need a way to state that in some cases we don't want to capture groups. This can be done in one of two ways.

The first technique involves using the ?: syntax, as illustrated in the following updated expression:

```
(?:(?:A|a)|(?:O|o)|(?:I|i))n\s
```

Now if you tested this expression with the GroupsAndCaptures demo, it would result in only the three groups relating to the three matches.

The second technique is to specify the RegexOptions::Explicit Capture option when constructing the Regex object or calling a function that takes a RegexOptions enumeration.

Working with Captures

As you know, each time that a Match object is created it contains a Group Collection. As you've also seen, each Group object within that collection represents either a match or a submatch. But, what about Captures? Captures are the next level down. We haven't worried about captures yet because typically there is only one capture for each group; thus, most people use the terms *groups* and *captures* interchangeably. However, in some situations a given group may have many captures. Let's look at a couple of examples.

Suppose you are using the following expression to parse a string for a time construct:

```
(?<time>(\d|\:)+)
```

Nothing too difficult here. It's just a named group (time) that parses for continuous sequences of numbers and colons. Now imagine that expression is applied to the following time:

```
12:56:42
```

From what you've learned so far in this section, you can probably guess that this expression/string pair will yield three groups (the overall match) along with two sets of parentheses). It is not apparent, however, that the second group will contain eight captures! (This assumes that you have not specified the RegexOptions::ExplicitCapture enumeration.)

The reason for this is simple. The second group (the inner parenthesis in the expression) states that a submatch can be made on any single digit or colon. Since our source string contains eight such characters, we get eight captures! I've found it easier up to this point to use the term group to characterize the result of a submatch, but, technically speaking, submatches result in captures. As noted, however, since a group typically contains a single capture, using the terms interchangeably is also considered correct, unless you have a specific reason to differentiate between them—as in this case where a single group contains eight captures.

So how do you know how many capture objects a group contains? By enumerating the CaptureCollection of the Group object. We can insert the following code into the innermost for loop of the DisplayGroups function presented earlier in order to display the entire hierarchy of matches, groups and captures for a successful expression execution:

```
// inserted into innermost for loop of
// DisplayGroups function just after
// update of strTemp variable
...
// for all of THIS group's captures
CaptureCollection* cc = gc->Item[i]->Captures;
for (int j = 0; j < cc->Count; j++)
{
    strTemp.Format(_T("\t\tCapture %s at %ld\r\n"),
                   (CString)(cc->Item[j]->Value),
                   cc->Item[j]->Index);
    strResults += strTemp;
...
```

Running the GroupsAndCaptures application and providing the time expression with a time of 12:56:42 will result in the following output:

```
Match 12:56:42 at 0
      Group 12:56:42 at 0
            Capture 12:56:42 at 0
```

```
Group 2 at 7
       Capture 1 at 0
       Capture 2 at 1
       Capture : at 2
       Capture 5 at 3
       Capture 6 at 4
       Capture : at 5
       Capture 4 at 6
       Capture 2 at 7
Group 12:56:42 at 0
       Capture 12:56:42 at 0
```

As you learned in the preceding discussion entitled "Noncapturing Groups," to avoid the creation of unnecessary captures, simply use the ?: syntax as follows:

```
(?<time>(?:\d|\:)+)
```

Now the output from the `GroupsAndCaptures` application will change dramatically:

```
Group 12:56:42 at 0
       Capture 12:56:42 at 0
Group 12:56:42 at 0
       Capture 12:56:42 at 0
```

Parsing and Replacing Strings

One of the most powerful and popular features of Regular Expressions is its ability to locate substrings within a string using a supplied pattern and then replace that substring with another. Obviously, this ability is useful in a multitude of applications—whether in document-based systems such as editors, where the user needs a sophisticated search and replace capability, or in complex text parsers. With the BCL classes, this functionality is provided by the `Regex` class. In fact, the `Regex` class has several instance and static overloaded versions of the `Replace` methods from which to choose, depending on the exact nature of your needs.

The static versions of these methods are provided because we often do not need to instantiate a `Regex` object, and we sometimes need to process a

substantial number of patterns (remember that the `Regex` class is immutable), in which case it would be highly inefficient to construct a new `Regex` object for each new expression.

While we can't fake the age-old adage of someone forgetting to "dot their i's and cross their t's" in a demo, let's consider the following situation where someone has forgotten to capitalize their *i*'s:

```
Hi all. i    am Tom. i would like to thank everyone for buying my book
and supporting the cause
```

Let's look at a simple function that will allow us to parse this string for all occurrences of lowercase *i*'s and, using the `Replace` method, correct them with their uppercase equivalents.

```
void TheEyesHaveIt()
{
#pragma push_macro("new")
#undef new
  try
  {
    CString strResult;

    String* strSource = S"Hi all. i    am Tom. \ni would like to "
                        S"thank everyone for buying my "
                        S"book and supporting the cause.";

    Regex* rx = new Regex(S"\\si\\s");
    String* strDestination = rx->Replace(strSource, S" I ");

    strResult.Format(_T("%s\r\nconverted to\r\n%s"),
                     strSource,
                     strDestination);
    AfxMessageBox(strResult);
  }
  catch(Exception* e)
  {
    AfxMessageBox((CString)e->Message);
  }
#pragma pop_macro("new")
}
```

As you can see in our not-so-cleverly named `TheEyesHaveIt` function, I've chosen to employ the following expression:

```
\si\s
```

Looking back at Table 2–1, you can see that the `/s` metacharacter represents any whitespace—which basically represents spaces and nonprintable characters (such as new-lines and tabs). In addition, as you know by now, the + sign simply represents the fact that we don't care how many whitespace characters prepend or append the *i*. We still need it changed. Executing this function would simply cause the display of the source string with the lower-case *i*'s capitalized.

Let's examine the form of the `Replace` method that is being used here:

```
String* Replace(
   String* input,
   String* replacement
);
```

This particular instance overload takes two parameters. The first parameter (`input`) is a `String` object representing the source string that needs to be modified. The second parameter (`replacement`) is another `String` object—this one representing what strings to replace the located matches with —as located via the pattern used to construct the `Regex` object. As you know, the source string is not modified, because `String` objects are immutable, just as the `Regex` objects are. As a result, a new `String` object will be returned by the method.

Instead of instantiating a `Regex` object each time, we can also use one of the static overloads versions of the `Replace` method. Here's the syntax for one such static overload.

```
static string Replace(
    string input,
    string pattern,
    string replacement
);
```

The only difference syntactically between this method and the first instance method you saw is the inclusion of a `pattern` argument that will accept the same Regular Expression syntax as the `Regex` constructors.

Here's a simple (but useful) example illustrating how to replace a fully qualified local file path with a Web-based URL path:

```
void ConvertFilePathToWebPath()
{
#pragma push_macro("new")
#undef new
  try
  {
    CString strResult;
    CString strTemp;

    String* filepath = S"C:\\Inetpub\\wwwroot\\Abc.org.site"
                       S"\\music\\index.html";
    strTemp.Format(_T("Original Path: %s\r\n"), filepath);
    strResult += strTemp;

    String* webpath = Regex::Replace(filepath,
           S"C:\\\\Inetpub\\\\wwwroot\\\\Abc.org.site",
           S"http://www.abc.org");
    strTemp.Format(_T("After First Pass: %s\r\n"), webpath);
    strResult += strTemp;

    webpath = Regex::Replace(webpath,S"\\\\","/");
    strTemp.Format(_T("After second pass: %s"), webpath);
    strResult += strTemp;

    AfxMessageBox(strResult);
  }
  catch(Exception* e)
  {
    AfxMessageBox((CString)e->Message);
  }
#pragma pop_macro("new")
}
```

Note that in this example, expressions aren't needed—just simple strings to
search for. In addition, note that the code actually makes two passes at the
string before getting the value just right. After the first replacement, which
replaces the local drive and directory name with the Web domain name, we
still need to replace the backward slashes with the forward slashes via a
second call to Replace. As you can see in the second call, the webpath is

passed as the source string, and then the returned string is assigned again to webpath. Executing this application yields a correctly converted URL pathname.

Another example of a task where multiple replacement passes would be made against a single source string would be in converting similar languages—such as American English and British English—as in the following code snippet.

```
// convert to British spelling
String* s = S"My favorite color is gray";
String* newstr = Regex::Replace(s, S"favorite", S"favourite");
newstr = Regex::Replace(newstr, S"color", S"colour");
newstr = Regex::Replace(newstr, S"gray", S"grey");
```

As expected, the str variable would be modified from "My favorite color is gray" to "My favourite colour is grey."

Let's look at one more technique involving the Replace method known as *pattern stripping*. For example, we can easily write an expression to use in the Replace method that would strip all whitespace from a string; we simply specify a null string as the replacement string. Let's examine a small code snippet that uses this technique to extract individually the file name and directory path of an input file path string. In the case of strings containing file path information, you should use the BCL Path class (see Chapter 3) that we will use throughout this book. However, a fully qualified file name also serves as a good example of where to use pattern stripping.

```
String* fullpath = S"c:\\files\\backup\\file5.cpp";
String* fname = Regex::Replace(fullpath, S".*\\\\", S"");
String* path = Regex::Replace(fullpath, fname, S"");

AfxMessageBox((CString)fullpath);
AfxMessageBox((CString)fname);
AfxMessageBox((CString)path);
```

The ".*\\\\" will match any string that is terminated by a backslash, and in our case that will be the directory path portion of the file path. We replace it with an empty string, and the resulting string is obviously the bare file name of the input file path. Now getting the directory path alone is quite simple. We simply use the original path as input string and replace the file name string with an empty string; what remains will be the

directory path. Running the above snippet results in the following values being displayed in separate message boxes:

```
c:\files\backup\file5.cpp
file5.cpp
c:\files\backup\
```

Replacing Matches Using Groups and Substitution Patterns

In the previous two sections, you've learned how to group and capture submatches and how to replace text. This section shows you how to combine those two capabilities with substitution patterns to give you an even more powerful means of manipulating text. Substitution patterns are just special characters that tell the Regular Expression engine how you want text replaced. Table 2–4 shows the available substitution characters that can be used.

Of special importance in Table 2–4 are the first two patterns, which allow us to specify groups in the capture and then use those group names in the replacement. Let's see how to do that with an example that converts between U.S. (mm/dd/yyyy) and European (dd/mm/yyyy) date formats. We could go wild here and create a function to handle and convert between every conceivable date format, but the goal of this section is simply to give

Table 2–4 Regular Expression Substitution Patterns

Expression	Description
`$123`	Replaces the substring with the group at index 123.
`${groupName}`	Replaces the substring with the specified group name.
`$$`	Replaces a single "$" literal.
`$&`	Substitutes a copy of the entire match itself. This is generally used when you want to keep the previous match (not truly have it replaced) and then append other data to it.
`` $` ``	Substitutes all the text of the input string before the match.
`$'`	Substitutes all the text of the input string after the match.
`$+`	Substitutes the last group captured.
`$_`	Substitutes the entire input string.

you some examples of practical Regular Expression tasks and the patterns used to perform them. The following ConvertDate function takes as its only parameters a CString constant reference representing a source date and an enumeration value indicating the format of that date. ConvertDate then returns a CString object representing the alternative ("destination") date format. Here's the code to accomplish this task.

```
enum DateFormat {
  DateFormatUS,
  DateFormatEuropean
};


CString ConvertDate(CString const& strSourceDate, DateFormat
enumSourceDateFormat)
{
#pragma push_macro("new")
#undef new

  CString strDestinationDate;

  switch(enumSourceDateFormat)
  {
    case DateFormatUS :
    {
      String* regexp1 = S"(?<month>\\d{1,2})-"
                        S"(?<day>\\d{1,2})-"
                        S"(?<year>\\d{2,4})";
      String* regexp2 = S"${day}-${month}-${year}";
      strDestinationDate = (CString)Regex::Replace(strSourceDate,
                        regexp1, regexp2);

    }
    break;

    case DateFormatEuropean:
    {
      String* regexp1 = S"(?<day>\\d{1,2})-"
                        S"(?<month>\\d{1,2})-"
                        S"(?<year>\\d{2,4})";
      String* regexp2 = S"${month}-${day}-${year}";
      strDestinationDate = (CString)Regex::Replace(strSourceDate,
                        regexp1, regexp2);

    }
    break;
```

```
        default:
          throw _T("Invalid DateFormat specifier");
    }


    return strDestinationDate;
#pragma pop_macro("new")
    }
```

This first function should get you thinking about the possibilities that exist with regard to expressions. As mentioned, this function takes an enumeration value indicating whether the supplied date string is in U.S. or European format. Based on that value, a `switch` statement is used to control program flow. Let's examine how the function converts from U.S. format to European—converting from European to U.S. is just the same code in reverse.

First we build a `String` object (`regexp1`) containing the following expression:

```
(?<month>\d{1,2})-(?<day>\d{1,2})-(?<year>\d{2,4})
```

As you learned from the previous section, this expression simply creates three distinct named groups: `day`, `month`, and `year`. Each group is simply a match on a certain number of digits specified inside the curly braces. The numbers in those braces represent the number of valid digits: `day` and `month` can have one or two digits while `year` can have two or four digits. Obviously, `regexp1` string will hold the U.S. format.

Once the `regexp1` string is built, we specify the replacement pattern. This is where named groups come in handy. Instead of specifying the reversal of the previous expression, we can use the group names instead. Therefore, `regexp2` is as follows:

```
${day}-${month}-${year}
```

As you saw in Table 2–4, the `$` signifies that a group is being used here to substitute. In this case, the three groups being captured are all being used in the replacement pattern—just in a different order. Now all the code has to do is to call the static `Regex::Replace` method, passing it the `regex1` and `regex2` values. The `regex1` value is the expression; running it creates the specified named groups, which are then used in the replacement string. That's all there is to it!

Looking back again at Table 2–4, notice that we could have used the following replacement string as well, in which group index numbers are used:

```
$2-$1-$3
```

While this is certainly shorter and has the advantage that it would work for both conversions (as opposed to having a case statement for each), I personally prefer to use the group names because the resulting code is less subject to breaking should someone come along later and change the order of your expression groups.

Finally, suppose you wanted to see both the original value and add text around it—for instance, you might want to see the old value compared to the converted, or new, value. In this case, you would use the $& substitution pattern as follows:

```
[$& ==> ${day}-${month}-${year}]
```

In the case of a date such as 8/11/64, this would result in the text 8/11/64 ==> 11/8/64. You can test these and other replacement expressions by using the `ReplaceText` demo application on the CD that accompanies the book (Figure 2–13).

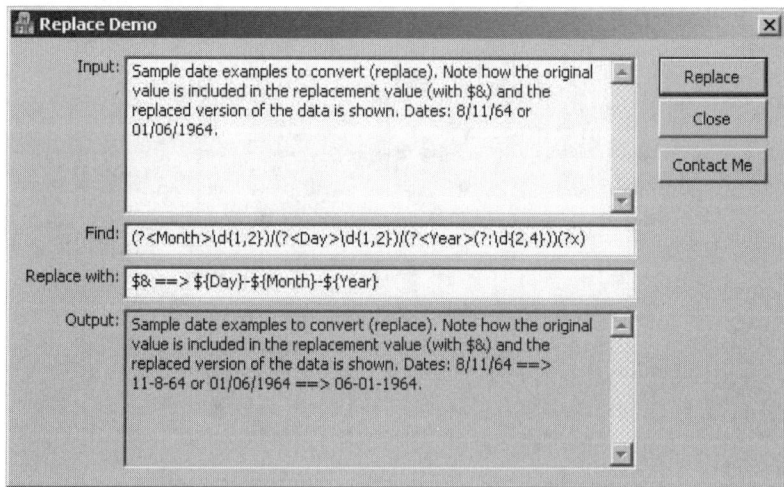

Figure 2–13 The combination of groups and substitution patterns provides a powerful means of replacing or modifying text.

Putting It All Together: Writing an E-mail Parser

A common task that utilizes the power of expressions is the extraction of e-mail addresses from Web pages and documents. For example, you might want to communicate *en masse* with the various people you met at a recent seminar by programmatically parsing a group of e-mail documents, filtering out all the e-mails, and sending a note to each address. Additionally, you might have the need to *harvest* e-mail addresses from a collection of Web pages or a complete domain for marketing purposes. Finally, you might need to verify that a given e-mail address is valid. (While you can't be sure of the validity of an e-mail address without sending and receiving data, you can at least verify the format of the address in situations where your application requires the input of an e-mail from a user.) Whatever your reasons, the ability to process e-mail addresses is certainly one of the most popular tasks for expressions.

Retrieving E-mails from Text

Let's start off with a function illustrating how to retrieve the e-mails from a string and store them in a string array that is then returned to the caller.

```
String* GetEmails(String* instring)[]
{
#pragma push_macro("new")
#undef new
  try
  {
    ArrayList* al = new ArrayList();

    MatchCollection* mc = Regex::Matches(instring,
                          S"[\\w]+@[\\w]+.[\\w]{2,3}");

    for (int i=0; i<mc->Count; i++)
      al->Add(mc->Item[i]->Value);

    String* strar[] = new String*[al->Count];
    al->CopyTo(strar);
    return strar;
  }
  catch(Exception* e)
```

```
    {
       AfxMessageBox((CString)e->Message);
    }
#pragma pop_macro("new")
}
```

The first thing this function does is to construct an `ArrayList` object. (The `ArrayList` resides in the `System::Collections` namespace, so you'll need to make sure you include a "using namespace" directive if you're playing along at home.) Next a `MatchCollection` object is instantiated using the static `Regex::Matches` method that takes both a source string and an expression as its parameters. Once that is done, a simple `for` loop is used (controlled by the `MatchCollection::Count` property), so that, one by one, each match (e-mail address) is added to the `ArrrayList` object. When the loop exits, the `ArrayList` objects are dumped into a newly created `String` array object (via `ArrayList::CopyTo`) and that `String` array object is returned to the caller.

Let's take a minute to look at the e-mail expression:

```
[\w]+@[\w]+.[\w]{2,3}
```

The expression first indicates that the match must contain one or more alphanumeric characters (this will be the username portion of the e-mail address), which should be followed by an @ character; then we look for one or more alphanumeric characters (this will be the domain-name portion of the e-mail address) followed by a . character. Finally, we look for two or three alphanumeric characters, which represent the first-level domain name like ".uk" or ".com." Certain e-mail addresses might not be correctly matched by this expression, but right now we simply want to demonstrate how this works, and this one is good enough to catch most valid e-mail addresses.

Doing Domain Counts

Speaking of domains, let's see how easy it is to retrieve just the domain information from an e-mail address. This will allow us to do things like reporting on the preponderance of our e-mails coming from the various "free" e-mail services such as Hotmail or Yahoo! The following function takes a `String` object representing an e-mail—for example an array

element returned from the previous section's GetEmails function—and returns the domain name in the form of a string.

```
String* GetDomain(String* email)
{
#pragma push_macro("new")
#undef new
  try
  {
    Match* m = Regex::Match(email,
                            S"@(?<domain>[\\w]+).");
    String* domstr = m->Groups->Item[S"domain"]->Value;
    return domstr;
  }
  catch(Exception* pe)
  {
    AfxMessageBox((CString)pe->Message);
  }

#pragma pop_macro("new")
}
```

Again, let's inspect the expression first:

```
@(?<domain>[\\w]+).
```

Basically this expression simply states that we wish to match on any string that has an @ character to its left and a . character to its right. This capture goes into a group called domain. As you learned in the discussion of groups and captures, naming groups just gives us an easier way to refer to sub-matches (for which the code is also easier to maintain).

Next, the named index property of the GroupCollection class is used to access the domain and assign its value to a String variable, which is then returned to the this function's caller. Here's an example using this function:

```
String* domain = GetDomain(S"abc@microsoft.org");
// Upon return domain would value 'microsoft' (without the quotes)
```

Demo Application to Parse for E-mail Information

In this example, which uses both the GetEmails and GetDomain functions, the function has a hard-coded sample e-mail header (instring) to parse for e-mail addresses. After initializing the instring variable, the function

initializes the local variables that will be used to keep track of the number of e-mails for the Mail.com and Yahoo.com as well as a bucket for tracking all other (miscellaneous) domain accounts.

From there, it's pretty straightforward. A call to `GetEmails` passing the `instring` variable nets an array of e-mail addresses (`strar`) that can then be enumerated with a `for` loop. As the function iterates through the `for` loop, it calls the `GetDomain` function (passing the e-mail address value) and, on the basis of the returned domain value, increments the appropriate domain counter value (`nMailAccounts`, `nYahooAccounts` or `nMisc Accounts`). As the function is doing all this, it updates a string value (`strResult`) that will be displayed in a message box at function end.

```
void ProcessEmails()
{
#pragma push_macro("new")
#undef new
  try
  {
    CString strResults;
    CString strTemp;

      String* instring = S" ----- Original Message ----- "
      S"From: \"Tom Archer\" <tom@ArcherConsultingGroup.com>"
      S"To: <chris@showprogramming.com>; "
      S"\"Erik\" <erik@function2form.net>;"
      S"\"Krista Crawley-Archer\" <kristalinski@yahoo.com>; "
      S" <tmacg@mail.com>"
      S"Sent: Monday, July 15, 2003 6:42 AM "
      S"Subject: Book is done!!";

      int nMailAccounts = 0;
      int nYahooAccounts = 0;
      int nMiscAccounts = 0;

      String* strar[] = GetEmails(instring);

      for (int i=0; i < strar->Length; i++)
      {
        CString strEmail = (CString)strar[i];
        CString strDomain = GetDomain(strar->Item[i]->ToString());

        strTemp.Format(_T("Email: %s\r\n"), strEmail);
        strResults += strTemp;
```

```
      if (0 == strDomain.CompareNoCase(_T("mail")))
      {
        nMailAccounts++;
      }
      else
      {
        if (0 == strDomain.CompareNoCase(_T("yahoo")))
          nYahooAccounts++;
        else
          nMiscAccounts++;
      }
    }

    // now tally domains
    strResults += _T("\r\n");

    strTemp.Format(_T("Total mail accounts = %ld\r\n"),
      nMailAccounts);
    strResults += strTemp;

    strTemp.Format(_T("Total yahoo accounts = %ld\r\n"),
      nYahooAccounts);
    strResults += strTemp;

    strTemp.Format(_T("Total other accounts = %ld\r\n"),
      nMiscAccounts);
    strResults += strTemp;

    AfxMessageBox(strResults);
  }
  catch(Exception* pe)
  {
    AfxMessageBox((CString)pe->Message);
  }
#pragma pop_macro("new")
}
```

Executing these functions yields results similar to those shown in Figure 2–14.

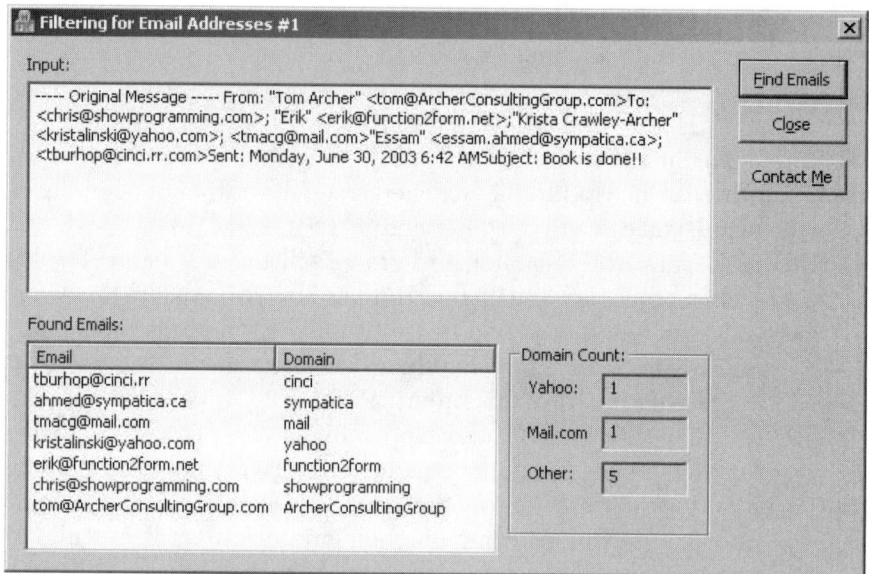

Figure 2-14 A great place to utilize the power of expressions is in parsing text values such as e-mail addresses.

A More Advanced E-mail Regular Expression

The GetEmails and GetDomain functions are definitely useful and will work for the majority of the e-mails you'll process, but they are not even close to perfect; a more complex expression will net a better result when filtering e-mails. In addition, a more advanced way of programming this would enable the e-mail expression to capture the domain as it's getting the e-mail.

Therefore, in this part we'll look at a much more powerful (all-inclusive) version of the original expression and examine in detail how it works. We begin with the following expression, which can be used to verify the format of an e-mail address. (Note that this expression is for exact matches only.)

```
^[^@]+@([-\w]+\.)+[A-Za-z]{2,4}$
```

This may seem a little overwhelming, so let's dissect this expression to better understand what is going on here.

```
^[^@]+@
([-\w]+\.)+
[A-Za-z]{2,4}$
```

The first portion, `^[^@]+@`, basically states, "Match everything to the left of the @ and include the @ itself." Table 2–5 gives a detailed explanation.

Now for the second portion: `([-\w]+\.)+`.

This is used to match everything in between the @ and the upper-level domain name (i.e., "com," "net," "de," etc.). As you learned earlier, parentheses will capture the matched text. If you will not need the captured text later, you can add the `?:` operator: `(?:[-\w]+\.)+`.

Conversely, naming the group will make it easier to obtain the text when processing the match. You saw how to do that in the Named Captures section. Since we would be capturing the top-level domain, this part of the expression could be something like `(?<domain>[-\w]+\.)+`. Table 2–6 details each part of this subexpression.

The remainder of the expression `[A-Za-z]{2,4}$` matches the terminating upper-level domain name. Table 2–7 details how this works.

If you were to run the `GroupsAndCaptures` demo application and input a document containing all sorts of different e-mail addresses, this expression would capture the overwhelming majority of them. Why not all? After quite a lot of investigation, it turns out that the most inclusive e-mail expression is over 6K in length!! Needless to say, that's a lot of expression to type, and most of it is not needed for 99% of the e-mail formats you'll ever see.

Now let's tweak the expression so that instead of working on direct matches (great for an e-mail validation program) it will work in parsing

Table 2–5 Explanation of Subexpression `^[^@]+@`

Expression	Description
`^`	Match the beginning of a line.
`[^@]+`	Character class that says, "Match at least one character that is NOT @." Like many of the metacharacters, the `^` has more than one meaning and is context-dependent. When it is used in a character class—as it is here—it negates the character class (i.e., "Match characters that are not the ones in the character class").
`+`	The + that follows the character class is a quantifier that says, "Match one or more."
`@`	Match the literal.

Table 2-6 Explanation of Subexpression `([-\w]+\.)+`

Expression	Description
`[-\w]+`	Says, "Match any amount of dashes (-) or word characters." Note that the square brackets (`[` and `]`) are used to delineate a character class definition in this context
`-`	Match the literal dash (-).
`\w`	This is a class shorthand that means "Match any word character." It is equivalent to `[a-zA-z0-9_]`.
`+`	Quantifier that says, "Match one more" and is applied to the entire character class definition.
`\.`	Says, "Match a literal period (.)."
`+`	As noted earlier, this is a quantifier meaning one or more and is applied to everything within the parentheses.

a string for all e-mail addresses. (Also note that we use the `?:` capture-inhibitor operator so as not to capture unneeded submatches.)

```
([-\.\w^@]+@(?:[-\w]+\.)+[A-Za-z]{2,4})+
```

The main differences here are the removal of the "match a line beginning with" and "end of line" metacharacters. Other than that, grouping is the main reason this new expression will find all e-mails in an arbitrary block of text.

Table 2-7 Explanation of Subexpression `[A-Za-z]{2,4}$`

Expression	Description
`[A-Za-z]`	This is a character class that says, "Match any capital letters (A–Z), or any lowercase letters (a–z)."
`{2,4}`	This is a quantifier that specifies a range of acceptable quantities, namely, 2, 3, or 4.
`$`	This is a metacharacter that denotes the end of a line (opposite of `^` that was mentioned earlier).

This new expression enables the following code to display every e-mail address in a specified block of text:

```
MatchCollection* mc = Regex::Matches(m_strInput,
    S"([-\\.\\w^@]+@(?<domain>[-\\w]+\\.)+[A-Za-z]{2,4})+");

for (int i=0; i<mc->Count; i++)
{
    CString email = mc->Item[i]->Value());

    // get domain name
    GroupCollection* gc = mc->Item[i]->Groups;
    CString domain = (CString)(gc->Item[S"domain"])->Value;
    CString domain = domain.Left(domain.GetLength()-1);
}
```

Note that the domain has to clip off a last character as the delimiting period was captured along with the domain name in the expression.

At this point, the power of expressions should be obvious, as we can now parse a document of any length, extract practically all legal e-mails and isolate the address and domain—all with a single expression and five lines of C++ code! As this task would take about a hundred lines of C++ code without expressions, this alone should convince you that a working knowledge of expressions is an indispensable addition to your programming toolkit.

Summary

The Regular Expressions classes add a lot of power to your MFC applications by giving you improved string-handling and manipulation techniques not found inherently in MFC. The classes themselves are few, and they are quite easy to use; the complexity involved depends only on the complexity of the expressions that you want to use. Irrespective of how simple or how complex your expressions are, the nature and behavior of the Regex and the other classes do not change.

During the course of this chapter, you saw how to use the Regex class, which represents an immutable Regular Expressions object, to split strings by specifying patterns (delimiters) to look for and by supplying expressions. After that, you discovered how to parse strings in order to perform such

tasks as extracting or isolated target substrings within a larger set of text. Then you were introduced to the world of groups and captures and learned how they enable you to specify submatch criteria within your match criteria to perform such tasks as parsing for a phone number and then capturing within that phone number the area code and exchange. One of the most popular uses of expressions is in its ability to parse through textual data using a supplied pattern and modify isolated text with a supplied value (which can also be based on the input pattern). These topics were discussed under the headings "Parsing and Replacing Strings" and "Replacing Matches using Groups and Substitution Patterns."

Finally, the chapter concluded by combining several aspects of what you've learned throughout this chapter in creating an e-mail parsing function. While the last e-mail parsing example used a hard-coded string as its input source, you would almost assuredly be using as input a file or stream containing the textual data to parse, which segues nicely into our next chapter's topics of file I/O. I just love it when a plan works out.

File I/O and Registry

Introduction

Almost all applications need to store and retrieve data. This can be in the form of structured data stored in a relational database management system (RDBMS) and accessed via an API (Application Programming Interface) such as ODBC (Open Database Connectivity) or ADO.NET (covered in Chapters 6 and 7). The data could be configuration information used to control the way in which the application works for a specific user or machine. This data is normally stored in the registry or in format-specific files (such as .INI or .CONFIG files) where a unique API exists to access the data. However, in many situations an application still needs to access a file directly on disk in either a generic manner or in a manner unique to the application. An example of the former would be a text editor application that needs the ability to read and write standard text files. An example of the latter would be an application whose data is stored in proprietary format so that the programmer needs to be able to open the file and read the data (in either text or binary format) as the application requires. For situations like these, .NET provides a set of classes defined in the System::IO namespace that enable the reading and writing of data to files and streams as well as access to basic file and directory support.

We begin by looking at streams and readers and writers under .NET. This section includes the topics of reading and writing data to disk using different formats as well as using memory and buffered streams. Then you'll learn how to perform various file and directory-level operations such as creating, moving, and deleting files; programmatically searching for files; and even implementing an event system to monitor for text-file changes within a specified directory. The chapter concludes with a section on using the registry classes provided by the BCL.

Stream, Readers, and Writers in .NET

The `System::IO` namespace provides a set of classes for file and file-content manipulation. Included in these types are classes that implement stream, writer, and reader behavior. As these concepts are new to most MFC developers, I'll first present an overview of these types and then move into a couple of demo applications in order to illustrate their use from MFC applications.

A stream is an abstraction of a sequence of bytes that might be represented by such entities as a disk file, an input/output device, or even a TCP/IP socket. (Where the data is actually stored is sometimes referred to as the stream's *backing store*.) The point is that the stream itself abstracts, or isolates, the programmer from the technical details of the actual storage specifics so that the programmer can focus on the data. To that end, the `System::IO` namespace defines the stream classes shown in Table 3–1.

The `CryptoStream` class is another important `Stream`-derived class that links data streams to cryptographic transformations. This class resides in the `System.Security.Cryptography` namespace (see Chapter 4, "Cryptography, Hash Codes, and Data Encryption."

Table 3-1 Stream Classes in the `System::IO` Namespace

Class	Description
Stream	The abstract base class `Stream` represents a sequence of bytes.
FileStream	Used to perform basic functions (open/close, read/write) to physical files. Also used to manipulate other file-related operating system handles such as pipes and standard input and output. Supports both synchronous and asynchronous functions.
MemoryStream	In contrast to the `FileStream`, the `MemoryStream` class creates streams that have memory as a backing store, as opposed to disk or a network connection.
BufferedStream	Represents a block of bytes in memory used to cache (or buffer) data, thereby reducing the number of calls to the operating system. Buffers improve read and write performance.

The abstract `Stream` class defines three main groupings of methods that are then overridden by any derived classes. These groupings provide the following capabilities:

- **Writing:** The output of a byte or array of bytes to a stream
- **Reading:** The input of a byte or array of bytes from a stream
- **Seeking:** The ability to query and modify the current position within a stream.

As not all `Stream`-derived classes may support these functions, the `Stream` base class defines properties to enable the user of a `Stream`-derived class to determine which of these functions is supported. These properties are `CanRead`, `CanWrite` and `CanSeek`. As a result, it's always a good idea to check these properties before attempting to call the corresponding function. That way, your code is protected in case you change the backing store for your stream in the future.

In contrast to MFC, which contains a `CFile` class for reading binary files and a `CStdioFile` class (derived from `CFile`) for reading text files a line at a time, .NET provides readers and writers that are attached to a given stream and define the manner in which the application deals with the stream's data. As you'll see while working through the chapter, this design is much more powerful and flexible than the older MFC design. Table 3–2 lists the available writers and readers that you can attach to the various .NET stream objects.

While the .NET stream classes do have methods for writing and reading data, they are designed for working with bytes and byte arrays and are therefore somewhat limited. As a result, when working with streams, you'll generally create a stream object and then attach a writer or reader object to it. Therefore, let's talk about writer and reader objects next. (Note that I'll speak specifically about files here, but the concept applies to all stream types regardless of the backing store.)

Many programmers think of files as either binary or text. However, in reality there is no such thing as a binary or text file because files are nothing more than a sequence of bytes persisted to disk. The designation of *binary* or *text* simply refers to how the file's contents are produced or consumed by an application.

To see what I mean, let's say you wish to output the contents of a Microsoft Word file (document) using the command shell's `type` command. The way the `type` command works is that it opens a specified file, reads its contents, translates each byte into its ASCII representation, and prints that

Table 3-2 Writer and Reader Classes in the `System::IO` Namespace

Class	Description
TextWriter	The abstract base class for several specific writer classes (such as `StreamWriter` and `StringWriter`). While the `Stream` class (and its derivatives) have methods for writing data, these methods are designed for byte arrays. In contrast, the TextWriter is designed for Unicode character input.
TextReader	The abstract base class for the `StreamReader` and `StringReader` classes.
StreamWriter	Used to write lines of data to a text files; similar to the output functions provided with the MFC `CStdioFile` class.
StreamReader	Where the `Stream` class is designed for byte input and output, the `StreamReader` class is designed for input in a particular encoding. As this class is typically used to read lines of data from a text file, it correlates nicely to the input functions provided by the MFC `CStdioFile` class.
StringWriter	The output reciprocal to the `StringReader` class, this class enables you to write data to a `String` (stored internally in a `StringBuidler` object) using streaming functions so that the application is abstracted from knowing the backing store.
StringReader	Used to read characters from a `String` so that standard stream functions can be used in situations where your data is in the form of a string rather than an actual file.
BinaryReader	Used to read binary data from a specified stream and encoding.
BinaryWriter	The output reciprocal of the `BinaryReader` class.

character to the standard output device. Word stores a lot of application-specific data in its files (such as text formatting options, page breaks, and so on) using values that don't translate to printable characters, so your output includes a lot of white space and strange characters as well as a few beeps (ASCII code 07 is a beep). Thus, in order to process this file (or simply view its contents), you would want to open it and view each byte without translation. However, that's just a matter of how you want to view the file and doesn't change the fact that the file is just a sequence of bytes.

As an even better example, let's look at the subject of *steganography*. Steganography is the ability to communicate something in a way that hides the existence of the communication. One very popular method is to place

the sensitive information in a file that most people will view as a text file. The hidden information is encoded using bytes that don't translate to ASCII characters, so that anyone opening and reading the file as text won't see the hidden message. However, someone who knows of the existence of a hidden message in the file can open and view the file as a stream of bytes and convert or decode those bytes to uncover the hidden message. My point is that the file isn't binary or text—it's just bytes; whether or not the data is read as binary or text is based solely on the file's producer or a consumer.

Now that you've learned a bit about streams, writers, and readers and have seen the available .NET class implementations, let's see how to use them in some MFC demo code.

Using the `FileStream` Class

As mentioned earlier, the `FileStream` class is used to perform basic functions (open/close, read/write) on physical files. As we've all worked with files before, let's move quickly into the syntax of how to perform the most common file functions using this class:

- Creating and opening streams
- Writing data
- Reading data
- Positioning within a stream

Creating and Opening Streams

Instead of explicit `Create` or `Open` methods, files are created and opened simply by instantiating a `FileStream` object, where you specify the mode of file access needed and the type of access desired to the underlying file. To that end, here is the list of available constructors for the `FileStream` class:

- `public: FileStream(String*, FileMode);`
- `public: FileStream(String*, FileMode, FileAccess);`
- `public: FileStream(String*, FileMode, FileAccess, File Share);`
- `public: FileStream(String*, FileMode, FileAccess, File Share, int bufferSize);`
- `public: FileStream(IntPtr, FileAccess);`

- ■ `public: FileStream(IntPtr, FileAccess, bool ownsHandle);`
- ■ `public: FileStream(IntPtr, FileAccess, bool ownsHandle, int bufferSize);`
- ■ `public: FileStream(IntPtr, FileAccess, bool ownsHandle, int bufferSize, bool isAsync);`

As you can see, the `FileStream` constructors can be split into two categories—those that take `String*` as the first parameter and those that take `IntPtr` as the first parameter.

While interesting, the latter category of constructors (that take the `IntPtr`) doesn't do much good for most MFC code bases for the following reasons. Typically in an MFC application file, access would be accomplished through a `CFile` (or derived) object. However, you can't pass the `CFile::m_hFile` as an `IntPtr` to the `FileStream` constructor because the `m_hFile` member represents a CRT (C Runtime) file descriptor, not the Win32 file handle that `FileStream` expects. Technically, you could use the `_get_osfhandle` function to get the OS handle from the CRT file descriptor, but this is not intended to be used for reading or writing into the handle, as the CRT file descriptor keeps additional state that will not be updated by `FileStream`. As a result, I'll focus on the set of `FileStream` constructors that take a `String` object as the first parameter.

The next set of parameters is the `FileMode` and `FileAccess` enumerations. The `FileMode` enumeration is used to specify whether a file is overwritten, created, or opened, or some combination thereof. Table 3–3 shows the available values and their impact on what can be done with the file.

Table 3-3 `FileMode` Enumeration Values

Enumeration Value	Description
`Append`	If the file exists, this enumeration will open it and seek to the end of the file so that any write operations will append data to the end of the file. If the file does not exist, it will be created. Note that `FileMode.Append` can only be used in conjunction with `FileAccess.Write`. Any attempt to read fails and throws an `ArgumentException`.
`Create`	Specifies that the file is to be created. If the file already exists, it will be overwritten.
`CreateNew`	Specifies that the file is to be created. If the file exists, an `IOException` will be thrown.

Table 3-3 `FileMode` Enumeration Values

Enumeration Value	Description
Open	Specifies that the operating system should open an existing file using the access specified with the `FileAccess` enumeration. If the file does not exist, `FileNotFoundException` is thrown.
OpenOrCreate	Specifies that the operating system should open a file if it exists; otherwise, a new file should be created. However, specifying the `FileMode::Create` along with the appropriate `FileAccess` enumeration for your code will accomplish the same thing.
Truncate	Opens a file, deletes its contents (truncates it), and moves the pointer to the first byte. Specifying `FileAccess::Read` with this `FileMode` enumeration results in an `InvalidArgument` exception. Attempting to read after opening a file simply returns no data because the pointer is at the end of the stream.

Once you've decided on the `FileMode` value, you need to specify the access level for the file: `Read`, `ReadWrite`, or `Write`. The `FileAccess` enumerations are fairly obvious, but I'll list them here for the sake of completeness.

- **`FileAccess::Read`**—Read-only access to the stream. Combine with `Write` for read/write access. *Note that read permission is always implied with the opening of a file.*
- **`FileAccess::ReadWrite`**—Both read and write access to the stream.
- **`FileAccess::Write`**—Write-only access to the stream.

Now, let's look a few basic examples of instantiating a `FileStream` object:

```
#pragma push_macro("new")
#undef new

// Create a new file with Write access
FileStream* pFS1 = new FileStream(S"test.fil",
                                  FileMode::Create,
                                  FileAccess::Write);
```

```
// Open a file -remember that FileAccess::Read is implied here
FileStream* pFS2 = new FileStream(strFileName,
                                  FileMode::Open);

// Open a file with explicit desire to read
FileStream* pFS3 = new FileStream(S"test.fil",
                                  FileMode::Open,
                                  FileAccess::Read);

// Create a new file and open it for write purposes
FileStream* pFS4 = new FileStream(S"test.fil",
                                  FileMode::Create,
                                  FileAccess::Write);

#pragma pop_macro("new")
```

Note that since most methods of the `Stream` class (and its derived classes) throw exceptions, you should always wrap your streaming code in `try/catch` statements. For the sake of brevity, I'm not doing it in these code snippets, but the demo applications provided with the book do follow this standard.

Writing Data

Before writing data, it's always a good idea to test the ability to write to the stream's underlying backing store with a call to the `CanWrite` property. The `CanWrite` property will return a Boolean value of `true` if data can be written to it and `false` if the stream is either closed or it was not opened with write access. For example, `CanWrite` would fail if an invalid path was used or the user account under which the application is executing had insufficient security privileges to write the data. Here's a simple example of making that call before attempting any write procedures.

```
...
if (pFS->CanWrite)
{
    // Write data
}
```

There are two basic ways of writing data directly to a `FileStream` object: by passing a byte array (`FileStream::Write`) or one byte at a time

(FileStream::WriteByte). While that might seem limited, remember that it makes sense once you take into consideration that a stream can be visualized as a contiguous sequence of bytes (a byte array) that is attached to a backing store (memory buffer, disk file, device, etc.). Also keep in mind that, as we progress through this chapter, you'll see other specialized streams that provide more flexible options in terms of overloaded read and write methods. Now let's look at the Write method syntax and usage:

```
public: void Write(unsigned char array __gc[], int offset, int count);
```

As you can see, the first parameter to the Write method is an (managed) unsigned char array, with the second parameter being the offset into that array (where the first byte of data should be read from) and the third parameter specifying the total number of bytes to read from that array and write to the stream.

If you're new to Managed Extensions, the syntax of the byte array may look a bit daunting. However, as the following code snippet illustrates, it's fairly straightforward when you realize that the only new part is the __gc modifier, which tells the compiler that the array is a managed array. Note that in the following code I've declared two buffers (one as a char array that will be implicitly converted to a Byte array and the other as an explicit Byte array). Both Write statements result in the same data being written to the stream.

```
#pragma push_macro("new")
#undef new

Byte buf1 __gc[] = {0xb4,0x02};
unsigned char buf2 __gc[] = {0xb4,0x02};

FileStream* pFS = new FileStream(S"test.fil",
                                 FileMode::Create,
                                 FileAccess::Write);
pFS->Write(buf1, 0, buf1->Length);
pFS->Write(buf2, 0, buf2->Length);

#pragma pop_macro("new")
```

As you might imagine, the WriteByte method is as simple as it gets; you only need to supply a single parameter:

```
// public: void WriteByte(unsigned char value);
```

Here are a couple of examples of writing single bytes to a stream:

```
Byte buf1 __gc[] = {0xb4,0x02};

// Write array one byte at a time
for (int i = 0; i < buf1->Length; i++)
  pFS->WriteByte(buf1[i]);

// Write 32 numeric values (0-31)
for (int i = 0; i < 32; i++)
  pFS->WriteByte(i);
```

Reading Data

As you might suspect, the reciprocal methods for the two write methods (Write and WriteByte) are Read and ReadByte, respectively, with the former taking a byte array, and the latter taking a single byte. Here is the syntax for these two methods:

```
public: int ReadByte();

public: int Read([In, Out] unsigned char array __gc[],
                 int offset,
                 int count);
```

In the following code snippet a file is opened, a byte array is declared (using the open file's length to determine the array size), and then the Read method is used to read the file's contents into that array.

```
#pragma push_macro("new")
#undef new

FileStream* pFS = new FileStream(S"test.fil", FileMode::Open);
Byte buffer __gc[] = new Byte __gc[pFS->Length];

pFS->Read(buffer, 0, buffer->Length);

#pragma pop_macro("new")
```

There probably aren't too many situations where you'd want to read a single byte at a time, but one example might be if each byte being read

needed to be translated into a character. In that case, the above example could be recoded as follows:

```
#pragma push_macro("new")
#undef new

CString str;

FileStream* pFS = new FileStream(S"Hello.txt", FileMode::Open);

byte ch;

for (int i = 0; i < pFS->Length;; i++)
{
    ch = pFS->ReadByte();
    str += (char)ch;
}

AfxMessageBox(str);

#pragma pop_macro("new")
```

Positioning with a Stream

Streams use the concept of an internal stream pointer: (much like the analogous database cursor) to keep track of the current position for reading and writing purposes. When you open a stream, the stream pointer is normally positioned at the first byte of the stream. (One exception would be when using the `FileMode::Append` enumeration; then the pointer is set to the end of the stream once the file is opened.) In order to manually position this pointer, most streams support *seeking,* which is the ability to move the internal stream pointer to an arbitrary position.

Before attempting to seek within a stream, it is definitely a good idea to call the `CanSeek` property. This method (whose syntax is shown below) returns a Boolean value of `true` if the stream's internal pointer can be repositioned and `false` if the stream's backing store is closed or if the stream was constructed from an operating system handle.

```
public: __property bool get_CanSeek();
```

Once you've determined if you can seek, you then call the `Seek` method, specifying an offset and an origin:

```
public: __int64 Seek(__int64 offset, SeekOrigin origin);
```

The offset is straightforward and is simply the number of bytes from the specified origin (the second parameter) to move the stream pointer. That second parameter is an enumeration of type SeekOrigin and is one of the following values: SeekOrigin ::Begin, SeekOrigin ::Current or SeekOrigin ::End. The main thing to keep in mind here is that whenever you want to move backward from a specified origin (such as when you pass a value of SeekOrigin::End), you must specify a negative value denoting the relative number of bytes to move the stream pointer.

You can also use the Position property to specify explicitly the exact position of the stream's pointer, with an exception being the result of any attempt to access an invalid position.

You can avoid closing and opening the same stream between writing and reading by seeking to the desired location when needed. This eliminates the classic create-write-close-open-read-close pattern, improving overall performance and making your code easier to work with. For example, the following code opens a file, writes data to it, tests to see whether the stream supports seeking, moves the internal stream pointer, and then writes additional data. (I've inserted comments to show what the values of the FileStream::Length and FileStream::Position properties would be at each juncture.) While this code snippet is basically the same as doing an append operation, these steps are the same regardless of whether you're attempting to seek to a specific position in order to read data, overwrite data, append data, or whatever.

```
#pragma push_macro("new")
#undef new

String* pstr = new String(S"Hello");

FileStream* pFS = new FileStream(S"hello.txt",
                                 FileMode::Create,
                                 FileAccess::Write);
// Length = 0, Position = 0

unsigned char buf1 __gc[] = {'H','e','l','l','o'};
pFS->Write(buf1, 0, buf1->Length);

// Length = 5, Position = 5

if (pFS->CanSeek)
{
    pFS->Seek(5, SeekOrigin::Begin);
```

```
    unsigned char buf2 __gc[] = {' ','W','o','r','l','d'};
    pFS->Write(buf2, 0, buf2->Length);
    // Length = 11, Position = 11
}
pFS->Close();


#pragma pop_macro("new")
```

One interesting last point about this code is that if you specify a value that is greater than the file's length, .NET will automatically increase the size of the file accordingly. Therefore, in the code above, if you were to specify a value of 15 for the Seek offset, .NET would write 10 bytes to the file (all having a value of 0) and then position the pointer at the end of the file.

FileStream Demo

At this point, you've seen how to perform the most common operations that involve the FileStream class. Now let's create a simple demo application to illustrate how to use that knowledge within the context of an MFC application. As mentioned earlier, the FileStream class's Read and Write methods are typically used when working with raw byte data. If your data is in another format (such as text), you would generally use a derived class that has overloaded methods for handling the desired format. Keeping that in mind, the demo allows the user to enter a string of data, and writes that data to a file along with the necessary DOS 16-bit assembler instructions to print that text on the screen. You'll probably never need to do this, but the code shown here is the same regardless of what type of data you're writing, and it's interesting to generate a .COM file that actually executes! Here are the steps:

1. Create a new dialog-based MFC application called FileStream.
2. Add the Managed Extensions option to the project.
3. Add the following directives to the stdafx.h file:

```
    #using <mscorlib.dll>
    #using <System.Windows.Forms.dll>
    using namespace System;
    using namespace System::Windows::Forms;
    using namespace System::IO;
```

4. Open the dialog template in the resource editor and add the controls, as shown in Figure 3–1.

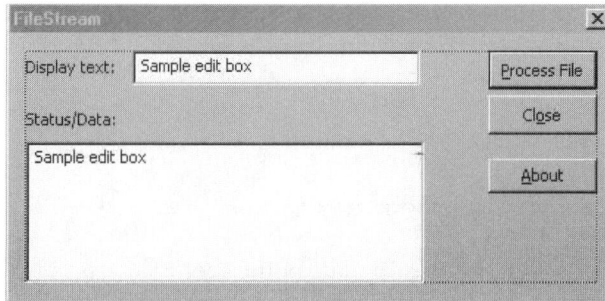

Figure 3-1 Dialog template for `FileStream` demo application

5. Add a DDX value variable for the input and status edit controls of type `CString`. Name them `m_strInput` and `m_strStatus`, respectively.
6. Add a DDX control variable for the status edit control called `m_edtStatus`.
7. Modify the status edit control's properties as follows:

   - `MultiLine = True`
   - `ReadOnly = False`
   - `VerticalScroll = True`

8. Add the following variable at the top of the `FileStreamDlg.cpp` file:

```
CString g_strFileName = _T("DisplayText.com");
```

9. Add an event handler for the Process File (OK) button and code it as follows:

```
void CFileStreamDlg::OnBnClickedOk()
{
  if (UpdateData())
  {
    if (0 < m_strInput.GetLength())
    {
      WriteFile();
```

```
        ReadFile();
    }
    else
    {
      AfxMessageBox(_T("Please enter a value before
                    "continuing."));
    }
  }
}
```

10. Add the `WriteFile` function as shown in the following code. As you can see, aside from including some (very) old 16-bit assembler instructions before and after the text, this code uses the same standard functions you've already read about. (The lines of code that specifically deal with the `FileStream` class are in bold.)

   First the function declares a few byte arrays for the various assembler instructions that are to be written. Then the assembler instructions are written out along with the text to be displayed once the file is executed in a command shell. Once that is done, the file is closed.

In case you're curious (or like me, haven't messed with assembly language in a long time), here's a quick breakdown on the assembler instructions being written to our demo file. The MOV AH, 02h places the value representing the "display output" function into the high part of the AX register. In the `for` loop, the insertion of the MOV DL instruction followed by the output of a character simply tells the assembler to move that character into the low part of the DX register. Then the INT 21 interrupt is raised, causing the OS to look at the AH register and call the "display output" function. When that is done, the "display output" function displays the character in the DL register. Finally, after all the characters have been processed, the INT 20 interrupt is raised, causing the application to exit.

```
void CFileStreamDlg::WriteFile()
{
#pragma push_macro("new")
#undef new
```

```
System::Byte MoveFunctionToAH __gc[] = {0xb4,0x02}; // MOV
  AH,02h
System::Byte MoveToDL __gc[] = {0xb2}; // MOV DL
System::Byte Int21 __gc[] = {0xcd,0x21}; // INT 21h
System::Byte Int20 __gc[] = {0xcd,0x20}; // INT 20h

try
{
  CString strTemp;
  strTemp.Format(_T("\r\nCreating file %s ... "),
    g_strFileName);
  UpdateView(strTemp);

  FileStream* pFS = new FileStream(g_strFileName,
                                     FileMode::Create,
                                     FileAccess::Write);

  UpdateView(_T("\r\nWriting data ... "));
  pFS->Write(MoveFunctionToAH, 0, MoveFunctionToAH->Length);

  for (int i = 0;
       i < m_strInput.GetLength();
       i++)
  {
    pFS->Write(MoveToDL, 0, MoveToDL->Length);

    pFS->WriteByte( m_strInput[i] );

    pFS->Write(Int21, 0, Int21->Length);
  }

  pFS->Write(Int20, 0, Int20->Length);

  UpdateView(_T("\r\nClosing file ... "));
  pFS->Close();

  UpdateView(_T("\r\nFile written successfully!\r\n"));
}
catch(Exception* pe)
{
  UpdateView((CString)pe->Message);
}
#pragma pop_macro("new")
}
```

11. Next, create the `CFileStreamDlg::ReadFile` function as follows. The first thing the code does is to instantiate a `FileStream` object on the `display.com` file. After that, it checks to see if the file can be read. If it can, a byte array of the same size as the file is dynamically allocated, and the entire file is read in to that array. From there, a simple `for` loop is used to iterate through the array to convert each byte into a string representation of its hex value, and to do some formatting to make the display look nice for the dialog. Finally, the dialog is updated with the results of the file read.

```
void CFileStreamDlg::ReadFile()
{
#pragma push_macro("new")
#undef new

  try
  {
    UpdateView(_T(""));

    CString strTemp;
    strTemp.Format(_T("\r\nOpening FileStream on file
      %s ... "),
                    g_strFileName);
    UpdateView(strTemp);

    FileStream* pFS = new FileStream(g_strFileName,
      FileMode::Open);

    UpdateView(_T("\r\nAttempting to read file ... "));
    if (pFS->CanRead)
    {
      UpdateView(_T("\r\nFILE CONTENTS:\r\n"));

      Byte buffer __gc[] = new Byte __gc[pFS->Length];
      pFS->Read(buffer, 0, buffer->Length);

      for (int i = 0; i < buffer->Length; i++)
      {
        CString c;
        c.Format(_T("%02x"), buffer[i]);

        if (0 == (i+1) % 8)
          c += _T("\r\n");
```

```
      else if (0 == (i+1) % 4)
        c += _T(" - ");
      else
        c += _T(" ");


      UpdateView(c);
    }
  }
  else UpdateView(_T("FileStream::CanRead returned
    FALSE!"));

  UpdateView(_T("\r\nClosing file ... "));
  pFS->Close();

  UpdateView(_T("\r\nFile read successfully!\r\n"));
}
catch(Exception* pe)
{
  UpdateView((CString)pe->Message);
}

#pragma pop_macro("new")
}
```

12. The last step is to add the CFileStreamDlg::UpdateView function. This function simply updates the status edit control and
scrolls to the bottom of the control.

```
void CFileStreamDlg::UpdateView(CString const& strNew)
{
  if (UpdateData())
  {
    m_strStatus += strNew;
    UpdateData(FALSE);

    int len = m_edtStatus.GetWindowTextLength();
    m_edtStatus.SetSel(len, len, FALSE);
  }
}
```

That's it! Now if you run the application, you should see results similar to
those shown in Figure 3–2.

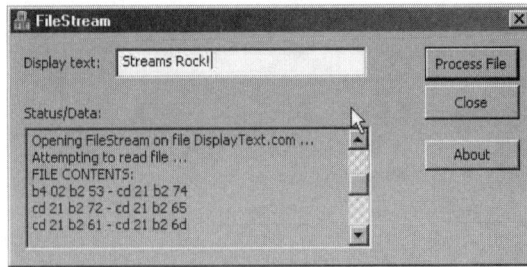

Figure 3-2 `FileStream` demo application at work

Now, for fun, open a command shell and run your application. Your assembler code should actually print out the specified string! As I mentioned earlier, it's not exactly every day that you'll need to output 16-bit DOS executable code, but what you've learned in this section will serve you regardless of the specifics of your binary data. In addition, even in its current limited incarnation, it's easy to see that with just a little more work you could actually turn this application into a full-blown hex editor. Now let's move onto the more specialized `MemoryStream` and `BufferedStream` classes.

Using the `MemoryStream` and `BufferedStream` Classes

Both the `MemoryStream` and `BufferedStream` classes derive from the abstract `Stream` class. While both classes are designed for streaming data into and out of memory, you would use the `BufferedStream` if you plan to write the data to disk, because that class defines constructors that take `Stream` objects as parameters. The `MemoryStream::Flush` method (overridden from its `Stream` base class) doesn't do anything, while the `BufferedStream::Flush` method (also overridden from the `Stream` base class) writes all unwritten data to the associated stream's backing store.

In the following code (taken from the `MemoryStream` demo on the CD that accompanies this book), after a `MemoryStream` object is instantiated, the `Length`, `Position`, and `Capacity` properties are retrieved and displayed. At that point, a `for` loop is used to write 24 bytes to the stream, and the same properties are displayed once again. Finally, the `Memory Stream::GetBuffer` is used (1) to retrieve the buffer into a locally allocated byte array, (2) to use a `for` loop to spin through the array and display each value of the array, and (3) to close the stream. (The salient `Memory`

Stream methods are in bold to separate them visually from the display and formatting code.)

```
void CMemoryStreamDlg::OnBnClickedOk()
{
#pragma push_macro("new")
#undef new
  m_strData = _T("");
  CString strTemp;

  MemoryStream* pMS = new MemoryStream(64);
  strTemp.Format(_T("Length: %02I64d Position:"
                    "%02I64d Capacity: %02ld"),
                    pMS->Length, pMS->Position, pMS->Capacity);
  m_strData += strTemp;
  m_strData += _T("\r\n");

  for (int i = 0; i < 24; i++)
  {
    pMS->WriteByte((byte)i);
  }
  strTemp.Format(_T("Length: %02I64d Position:"
                    "%02I64d Capacity: %02ld"),
                    pMS->Length, pMS->Position, pMS->Capacity);
  m_strData += strTemp;
  m strData +=  T("\r\n");

  m_strData += _T("\r\nContents:\r\n");

  byte buffer __gc[] = pMS->GetBuffer();
  for (int i = 0; i < buffer->Length; i++)
  {
    strTemp.Format(_T("%02ld"), buffer[i]);
    m_strData += strTemp;

    if (0 == (i+1) % 8) m_strData += _T("\r\n");
    else m_strData += _T(" ");
  }

  pMS->Close();

  UpdateData(FALSE);
#pragma pop_macro("new")
}
```

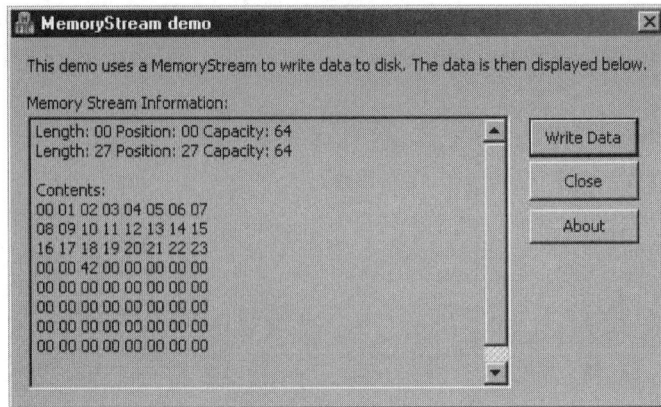

Figure 3-3 An example of writing to and reading from a `MemoryStream` object

Looking at Figure 3–3, you can see that each byte after the 24 explicitly assigned bytes has a value of 0; illustrating that upon allocation, the `Memory Stream` object is automatically initialized to zeroes.

Now what do you think would happen if we were to modify this code to write more than the originally allocated 64 bytes? Will it generate a runtime exception? Will it overrun into unallocated memory? Or will it just dynamically resize the buffer? One clue is to examine closely the previous code and Figure 3–3. You'll note that while the `MemoryStream` object's initial capacity was set to 64 in the constructor (confirmed by the call to the `Capacity` property), the `Length` property was still 0. It wasn't until after the writing of data that the stream was dynamically sized as needed. Therefore, we know that the `MemoryStream` object automatically resizes its attached buffer dynamically when the data being written to it exceeds its current length.

That brings up another interesting point. Our local byte array was allocated using the `MemoryStream::GetBuffer` method, and the code displayed its contents using a `for` loop (based on the array's `Length` property) resulting in 64 bytes being written—not the 24 we actually wrote. Therefore, the `GetBuffer` actually returns a buffer equal to the stream's defined capacity with the bytes written padded with 0s.

Obviously, the most useful feature of the `MemoryStream` object is its ability to execute the same standard stream methods you saw in the `FileStream` section on an array. For example, the following code snippet can be placed after the `for` loop that wrote the 24 bytes to the stream. As

you can see, the code simply moves the stream pointer forward two places and writes the value 42.

```
...
if (pMS->CanSeek)
{
  pMS->Seek(2, SeekOrigin::Current);
  pMS->WriteByte(42);
}
```

As you would assume, this results in the same display as Figure 3–3 except that the 27th byte of the stream (and local buffer) now contain the value 42, while the 25th and 26th bytes still contain the initialized value of 0.

Finally, let's look at one last important benefit of the MemoryStream object. If you need to persist the stream to disk, all you need to do is to instantiate a FileStream object and, when finished processing the Memory Stream, call the MemoryStore::WriteTo method, passing it the File Stream object. The following code snippet illustrates just that.

```
// Write to MemoryStream object (pMS)

...

FileStream* pFS = new FileStream(S"test.mem",
                                 FileMode::Create,
                                 FileAccess::Write);
pMS->WriteTo(pFS);
pMS->Close();
```

If you were to run this code, you would see that the resulting file would contain the data displayed in Figure 3–3 (plus the value 42 in the 27th position). However, one last thing to note is that the file's length would be 27 bytes as opposed to the 64 bytes that the GetBuffer method returns. This is because the MemoryStream::WriteTo method only writes out to the specified FileStream the actual bytes that were modified, whereas the GetBuffer method returns the entire capacity so that you have access to the entire buffer without having to write data first. An example of that is in the Seek example above, where I wrote 24 bytes but wanted to jump two unwritten (initialized) bytes before continuing the writing.

Following the footsteps of our MemoryStream demo code, let's perform the same tasks with the BufferedStream object. While these classes are very similar in functionality, there are subtleties that differentiate them:

- While a MemoryStream object can be instantiated in one of two main ways—with no parameters or by passing it a buffer to which the MemoryStream attaches—a BufferedStream can only be instantiated by initializing it with another Stream object.
- The BufferedStream object's internal buffer is managed differently, in that if an initialize size is not specified at object instantiation, it defaults to 4096 bytes.
- As opposed to the MemoryStream object, which defines Read, Read Byte, and GetBuffer methods, the BufferedStream only defines the overridden (from its base Stream class) Read and ReadByte methods.
- As a result of not being able simply to "dump" a stream into a buffer (using a method such as GetBuffer), care must be taken to position the pointer correctly when performing reads on a BufferedStream object.

Now let's look at some code (excerpted from the CD's Buffered Stream demo) illustrating how to use the BufferedStream class. Once again, focus on the code in bold.

```
void CBufferedStreamDlg::OnBnClickedOk()
{
#pragma push_macro("new")
#undef new

try
{
  m_strData = _T("");
  CString strTemp;

  FileStream* pFS = new FileStream(S"test.buf",
                                   FileMode::Create,
                                   FileAccess::ReadWrite);
```

```
BufferedStream* pBS = new BufferedStream(pFS);
strTemp.Format(_T("Length: %ld Position: %ld"),
                 pBS->Length, pBS->Position);
m_strData += strTemp;
m_strData += _T("\r\n");

for (int i = 0; i < 24; i++)
{
  pBS->WriteByte((byte)i);
}

strTemp.Format(_T("Length: %ld Position: %ld"),
                 pBS->Length, pBS->Position);
m_strData += strTemp;
m_strData += _T("\r\n");

m_strData += _T("\r\nContents:\r\n");

byte buffer __gc[] = new byte __gc[24];
pBS->Position = 0;
pBS->Read(buffer, 0, buffer->Length);
for (int i = 0; i < buffer->Length; i++)
{
  strTemp.Format(_T("%02ld"), buffer[i]);
  m_strData += strTemp;

  if (0 == (i+1) % 8) m_strData += _T("\r\n");
  else m_strData += _T(" ");
}

pBS->Close();
m_strData += _T("\r\nBuffer written disk (test.buf)");
}
catch(Exception* pe)
{
  CString strTemp;
  strTemp.Format(_T("Exception caught : %s"), pe->Message);
  m_strData += strTemp;
}

UpdateData(FALSE);
#pragma pop_macro("new")
}
```

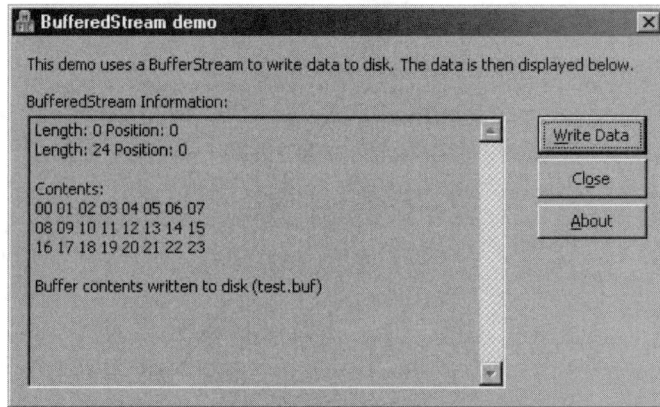

Figure 3-4 An example of writing to and reading from a `BufferedStream` object

Figure 3–4 shows the output from this code, which should not be too much of a surprise, given that it's almost identical to the `MemoryStream` demo code. However, there are a few subtleties to point out. First, the `BufferedStream` does not define a `Capacity` property. Second—and most important—the code doesn't explicitly write from the `BufferedStream` to the `FileStream` object as the `MemoryStream` demo did. This is because upon instantiation, the `BufferedStream` is associated with a stream object. Therefore, all writes are buffered (hence the name of the class) until the `BufferedStream::Close` is called, whereupon all data is flushed to the stream's backing store.

Text File Processing with `StreamReader` and `StreamWriter`

As you've now seen, the `FileStream` class is fine for reading and writing raw byte (binary) data. However, if your code needs to process character data, classes such as the `StreamWriter` and `StreamReader` (derived from `TextWriter` and `TextReader`, respectively) are far more suitable. These classes are still attached to an underlying stream class, but the overloaded read and write methods make it easier to use application-specific data without constantly having to convert from one type to another to satisfy the API. In fact, as you'll soon see, the combination of the `StreamWriter` and `StreamReader` classes should strike a chord with most MFC developers, as they closely mimic the `CStdioFile` write and read functions.

Writing Text Data

While there are quite a few constructor overloads for the StreamWriter class, most of them exist to allow the programmer to specify an encoding scheme for the stream's data. As encoding is a topic that we can't treat properly without straying too far from the topic of this chapter, I've chosen not to cover it here. Instead we'll be using the following constructors:

- StreamWriter(Stream* inputStream);
- StreamWriter(String* fileName);
- StreamWriter(String* fileName, bool append);

As you can see, the first StreamWriter constructor takes a Stream object as its only parameter. Therefore, you need only make sure that the Stream object to which you are attaching the StreamWriter has been opened with write access. The second and third constructors require a file name (in the form of a String object). As you are obviously opening the file for write purposes, the StreamWriter object will automatically instantiate the underlying FileStream object and attach it for you. Therefore, the only option you have is whether or not you want to overwrite the file if it exists or append to it.

In the following code snippet, the first line of code causes the file to be created if it does not exist and overwritten if it does. The second line does the same thing, as the second parameter indicates that the file is not to be appended to. The last line is used in situations where you want to create the file if it does not exist and append to it if it does exist.

```
#pragma push_macro("new")
#undef new

StreamWriter *pSW1 = new StreamWriter (S"test.txt");
StreamWriter *pSW2 = new StreamWriter (S"test.txt", false);
StreamWriter *pSW3 = new StreamWriter (S"test.txt", true);

#pragma pop_macro("new")
```

Now, let's look at how to actually write to text files, specifically where these StreamWriter methods improve upon their CStdioFile write function counterparts. To begin with, CStdioFile inherits from CFile the ability to write a block of data for a given count. From there, CStdioFile provides the WriteString function to write a provided line of text followed by a

carriage return/line feed combination. While this is certainly more convenient than manually converting textual data for use with the CFile class, it's
not nearly as flexible as the StreamWriter class, which has about a dozen
overloads for both the Write and WriteLine method, accepting almost any
conceivable type. Additionally, there are even overloads that take a formatted
String object followed by the values to use—similar to the C printf/
fprintf formatting options. Therefore, while at first glance the Stream
Writer object might seem to be a .NET equivalent of the CStdioFile, it's
actually much easier to work with. Let's now look at some code snippets that
use this class to create a text file and write text to a file.

This first example illustrates how easy it is to create a new file and write
some text to it:

```
void CreateTextFile()
{
#pragma push_macro("new")
#undef new

  StreamWriter *pSW = new StreamWriter(S"test.txt");

  pSW->WriteLine(S"Immature love says 'I love you because I need you'");
  pSW->WriteLine(S"Mature love says 'I need you because I love you.'");

  pSW->Close();

#pragma pop_macro("new")
}
```

That wasn't too complicated. Just instantiate a StreamWriter object
(covered earlier in the chapter), pass the full path of the file to create, and
then use WriteLine to write strings to the file. As mentioned earlier, the
WriteLine method will automatically add a carriage-return/line-feed pair
to the end of each line. If you do not want this behavior, you would simply
use the Write method, passing the same data.

This is OK, but what really makes the StreamWriter interesting is that
we can use it to write just about anything from any numeric value to .NET
objects such the FileStream object.

```
void WriteVariousBits()
{
#pragma push_macro("new")
#undef new
```

```
StreamWriter *pSW = new StreamWriter(S"test.txt");

pSW->WriteLine();
pSW->WriteLine(S"I can write dates and ints like this:");
pSW->WriteLine(S"As of now ({0}) and always, ",
                 __box(System::DateTime::Now.ToString));
pSW->Write(S"The Ultimate answer to Life, the Universe and "
             S"Everything is...");
pSW->WriteLine(42);

pSW->WriteLine();
pSW->WriteLine(S"I can write floats like {0} or any other "
                S"numeric type", __box(4.2.ToString));

pSW->WriteLine();
pSW->WriteLine(S"I can pass objects like this writer's
                S"base stream object: {0}", pSW->BaseStream);

pSW->Close();
#pragma pop_macro("new")
}
```

The output from this function follows:

```
I can write dates and ints like this:
As of now (6/17/2003 3:26:14 PM) and always,
The Ultimate answer to Life, the Universe and Everything is...42

I can write floats like 4.2 or any other numeric type

I can pass objects like this writer's base stream object:
   System.IO.FileStream
```

Note the last line where the FileStream object is output. Not exactly interesting information, is it? That's because internally the FileStream object's ToString method is being called, and it simply returns the string representation of the object type (System.IO.FileStream). However— you can probably guess where I'm headed with this—if .NET will automatically call the ToString method, that means that you can override the ToString method in your own reference (__gc) objects so that they

output whatever you want just by passing the object to the `Write/Write` `Line` method! Here's an example of just that:

```
__gc class Book
{
public:
  Book(String* title, String* author, String* publisher)
  {
    this->title = title;
    this->author = author;
    this->publisher = publisher;
  }
protected:
  String* title;
  String* author;
  String* publisher;

public:
  String* ToString()
  {
#pragma push_macro("new")
#undef new
    StringBuilder* pSB = new StringBuilder();
    pSB->AppendFormat(S"{0} by {1} published by {2}", title, author,
      publisher);
    return pSB->ToString();
#pragma pop_macro("new")
  }
};


void WriteBookInfo()
{
#pragma push_macro("new")
#undef new

  Book* pBook= new Book(
    S"Enhancing MFC Applications with the ".NET Framework",
    S"Tom Archer",
    S"Addison-Wesley");

  StreamWriter *pSW = new StreamWriter(S"test.txt");
```

```
pSW->WriteLine();
pSW->WriteLine(S"I can even override the ToString method "
  S"of my reference objects to work with StreamWriter!");
pSW->WriteLine();
pSW->WriteLine(S"Be sure and enjoy: {0}", pBook);

pSW->Close();
```

```
#pragma pop_macro("new")
}
```

This is the resulting output:

```
I can even override the ToString method of my reference objects to
  work with StreamWriter!

Be sure and enjoy: Extending MFC Applications with the .NET Framework
  by Tom Archer published by Addison-Wesley
```

The last thing I'll cover here regarding writing to a text file is appending text to the end of a file. As mentioned, you can instantiate a StreamWriter object with the desired file name and the second parameter set to true (for append). You could also simply open the file and seek to the end, as shown in the streaming examples earlier in the chapter. We can also use the File::AppendText method, which takes a file name as its only parameter and returns a StreamWriter object. Here's how that looks in code:

```
void AppendTextFile()
{
    StreamWriter* pSW = File::AppendText(S"test.txt");

    pSW->WriteLine(S"The End.");

    pSW->Close();
}
```

Reading Text Data

As with the StreamWriter class, the StreamReader has several constructors dealing with encoding. The two most commonly used constructors

simply take either a `Stream` object or a `String` representing the file to open in read mode.

- `StreamReader(Stream* inputStream);`
- `StreamReader(String* fileName);`

The following code illustrates how easy it is to instantiate such an object using either of these constructors:

```
#pragma push_macro("new")
#undef new

FileStream* pFS = new FileStream(S"test.txt",
                                 FileMode::Open,
                                 FileAccess::Read);
StreamReader* pSW = new StreamReader(pFS);

StreamReader* pSW2 = new StreamReader(S"test.txt");

#pragma pop_macro("new")
```

Aside from that, the `Read` and `ReadLine` methods mirror the `Write` and `WriteLine` methods in terms of being overloaded to satisfy just about any type your application may use. The following function is from the CD's `StreamWriterAndReaderDlg` demo, where a file is being read and its contents displayed in an edit control.

About the only thing to mention here about this otherwise self-explanatory code is the way in which I'm controlling the `while` loop. While the `ReadLine` method does return a null reference if the underlying stream's pointer is at the end of the data, you can also use the `Peek` method to determine if more data is available to read. The `Peek` method does not alter the stream's pointer; instead it returns the next character to be read. A value of –1 is returned when there is no more data to be read.

```
void CStreamWriterAndReaderDlg::OnBnClickedOk()
{
#pragma push_macro("new")
#undef new
  StreamReader* pStreamReader = new StreamReader(S"test.txt.cpp");

  CString strFileContents;
  CString strCurrLine;
```

```
   while (0 < pStreamReader->Peek())
   {
      strCurrLine = pStreamReader->ReadLine();
      strFileContents += strCurrLine;
      strFileContents += _T("\r\n");
   }
   pStreamReader->Close();

   m_edtContents.SetWindowText(strFileContents);
#pragma pop_macro("new")
}
```

String Processing with the `StringWriter` and `StringReader`

A common complaint of people new to .NET is that the .NET `String` object is immutable: it cannot be changed once instantiated. However, this is mitigated by the fact that the `StringBuilder` class is designed for purposes of building a string or dealing with a string whose contents may change. This is important to understand, because the functionality provided by the `StringWriter` and `StringReader` classes overlaps with that of the `MemoryStream` and `BufferedStream` classes as well as the `String` and `StringBuilder` classes. In fact, as you'll soon see, the `StringWriter` and `StringReader` classes actually use a `StringBuilder` object internally.

Let's look at some code to see how we can use the `StringWriter` object to create a mutable string in memory by providing both the `String Writer::Write` and `StringWriter::WriteLine>` methods with several input sources (e.g., literals, `StringBuilder`, `String::Format`, etc.), which all ultimately yield `String` objects.

```
#pragma push_macro("new")
#undef new
   StringWriter* pSW = new StringWriter();

   double moneyFound = 5.74;

   pSW->WriteLine(S"A bag of coins was found to contain ${0}",
                  __box(moneyFound));
   String* pstr = new String(S"There was the same number of "
                  S"pennies, nickels, dimes and quarters");
   pSW->Write(pstr);
```

```
pSW->Write(pSW->NewLine);
pSW->Write(new StringBuilder(S"How many of each?"));
pSW->WriteLine();
pSW->Write(S"\r\n");

double totalCoinValue = .25 + .10 + .05 + .01;

pSW->WriteLine(String::Format(S"Total coin face value : "
                S"25x + 10x + 5x + x = {0}x",
                    __box(totalCoinValue)));

pSW->WriteLine(String::Format(S"Finding for x : {0} / {1}",
                    __box(moneyFound),
                    __box(totalCoinValue)));

pSW->WriteLine(String::Format(S"x = {0}",
                    __box(moneyFound / totalCoinValue)));

pSW->Close();

AfxMessageBox((CString)pSW->ToString());
#pragma pop_macro("new")
```

Figure 3–5 shows the output from this simple math quiz.

One thing be wary of is that while the `StringWriter` class does define a `Flush` method, calling this method won't actually flush its underlying encoder unless you explicitly call `Close`. Closing the stream will automatically flush it and prepare it for destruction by calling `Dispose`.

In addition, look at Figure 3–6, in which a QuickWatch window in the Visual Studio .NET debugger shows that the `StringWriter` class (pSW)

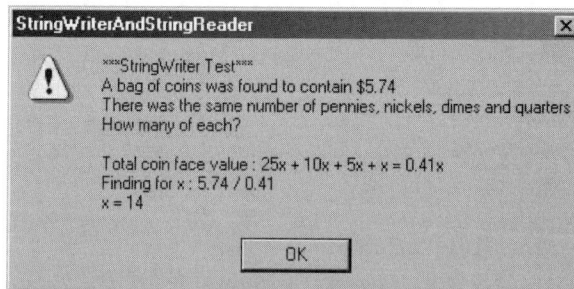

Figure 3–5 An example of writing to a `StringWriter` object

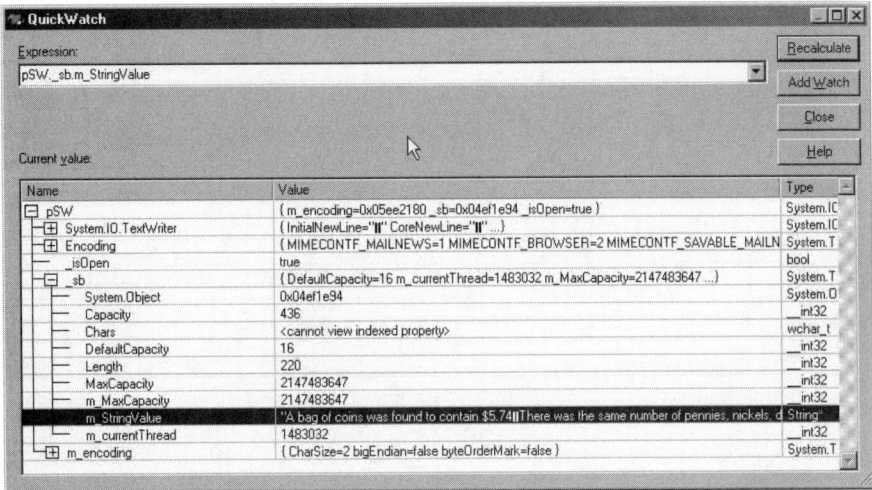

Figure 3-6 The `StringWriter` object contains an internal `StringBuilder` object.

contains an internal `StringBuilder` object (called `_sb`), which in turn contains a `String` field named `m_StringValue`. In fact, you can instantiate a `StringWriter` by passing it a `StringBuilder` object in the constructor—I just chose to use the default `StringWriter` constructor and have it create its own `StringBuilder` object.

Since a `StringBuilder` object exists within the `StringWriter` object, it's worth noting that you can perform the same string manipulation with a `StringBuilder` that you can perform with a `StringWriter`. In fact, it's easy to retrieve that `StringBuilder` object manually and directly call its methods. This is done via the `StringWriter::StringBuilder` method, as the following code snippet (appended to the previous example) illustrates.

```
// after last write of previous example

// now directly manipulate the StreamWriter's internal StringBuilder
StringBuilder* pSB = pSW->GetStringBuilder();
int i = pSB->Length;
pSB->AppendFormat(S"You scored {0}%", __box(100));
pSB->Insert(i, S"Congratulations! ");
pSB->Append(S" and are now a Mensa member!");

// before call to StreamWriter::Close
```

As you can see from the above, the `StringBuilder` class allows you either to insert data into the stream (at any given valid offset) or to append data to the end of the stream (using either the `Append` or `AppendFormat` methods).

Now let's turn our attention to the `StringReader` class. As is usually the case, once the writer class is understood, the reader class is far easier to work with. The following code (which can be added to the previous example) constructs a `StringReader` from the `StringWriter` object's internal string. A combination of `Read`, `ReadLine`, `ReadBlock`, and `Read ToEnd` method calls is used to illustrate the various ways in which data can be read using the `StreamReader` class. Note that while we used the `Read` method to read in a block of characters (just as with the `ReadBlock` method), you can also use it to read a single character at a time.

```
// after the previous example's StringBuilder usage
// make sure and move StringWriter close to end of this code

// now use the StringReader object
CString strData = _T("");

StringReader* pSR = new StringReader(pSB->ToString());

String* pstr2 = pSR->ReadLine();
strData += (CString)pstr2;

strData += (char)pSR->Read();

wchar_t buffer __gc[] = new wchar_t __gc[37];

pSR->Read(buffer, 0, 19);
for (int i = 0; i < 19; i++)
    strData += (char)buffer[i];

pSR->ReadBlock(buffer, 0, 37);
for (int i = 0; i < 37; i++)
    strData += (char)buffer[i];

strData += (CString)pSR->ReadToEnd();

pSR->Close();
pSW->Close();

AfxMessageBox(strData);
```

One question I hear often is "Why use the `StringWriter`/`StringReader` class pair to process a mutable string if the `StringBuilder` class provides the same functionality in a single class?" While one answer could be the age-old subjective one of personal taste, a more practical reason is that since both the `StreamWriter`/`StreamReader` and `StringWriter`/`StringReader` class pairs derive from `TextWriter`/`TextReader` classes, using the former allows you to write polymorphic code in which both sets of classes may be used interchangeably. In addition, the `StringWriter` class is useful when you need a writer of some kind (as with `XslTransform.Transform`). However, if all you need is the underlying `StringBuilder` capability, then the `StringBuilder` class is sufficient.

Using the `BinaryWriter` and `BinaryReader` Classes

Sometimes it's easiest to understand the functionality and reason for one class by comparing it with another. The `StreamWriter` class provides a text-interpolation layer on top of another stream—such as a `FileStream`. In order to write virtually any type of application-specific data, the `StreamWriter.Write` method is heavily overloaded to allow for the writing of value and reference types. By comparison, the `BinaryWriter` class provides the same overloaded means of writing almost any type of data to a stream, with the main difference being that the stream is written without this text interpolation, so that the resulting data is in binary form.

A demo illustrates this point well. In the following code, two files will be written: one using the `StreamWriter` class (test.txt) and one using the `BinaryWriter` class (test.dat).

```
void CBinaryWriterDlg::OnBnClickedOk()
{
#pragma push_macro("new")
#undef new
  try
  {
    double pi = 3.1415;

    CString strTextFileName = _T("test.txt");
    Stream* pStream1 = new FileStream(strTextFileName,
                                FileMode::Create);
    StreamWriter* pStreamWriter = new StreamWriter(pStream1);
```

```
    pStreamWriter->Write(String::Format(S"The answer to all "
                    S"life's questions: {0}",
                    __box(42)));
    pStreamWriter->Write(S" ");
    pStreamWriter->Write(String::Format(S"Pi = {0}", __box(pi)));
    pStreamWriter->Close();
    pStream1->Close();


    pStreamWriter->Close();
    pStream1->Close();


    CString strBinaryFileName = _T("test.dat");
    Stream* pStream2 = new FileStream(strBinaryFileName,
                    FileMode::Create);
    BinaryWriter* pBinaryWriter = new BinaryWriter(pStream2);
    pBinaryWriter->Write(String::Format(S"The answer to all "
                    S"life's questions: {0}", __box(42)));
    pBinaryWriter->Write(S" ");
    pBinaryWriter->Write(String::Format(S"Pi = {0}", __box(pi)));
    pBinaryWriter->Close();
    pStream2->Close();


    CString str;
    str.Format(_T("Data successfully written to the "
            "following files:\r\n\t%s\r\n\t%s"),
            strTextFileName,
            strBinaryFileName);
    AfxMessageBox(str);
  }
  catch(Exception* pe)
  {
    AfxMessageBox((CString)pe->Message);
  }
#pragma pop_macro("new")
}
```

The code writes exactly the same data (The answer to all life's
questions: 42 Pi = 3.1415) to both files using only the Write method
of the StreamWriter and BinaryWriter methods. Figure 3–7 shows
the resulting files.

As the figure shows, Visual Studio has determined that test.dat contains nonprintable characters and has opened the file in binary mode. It

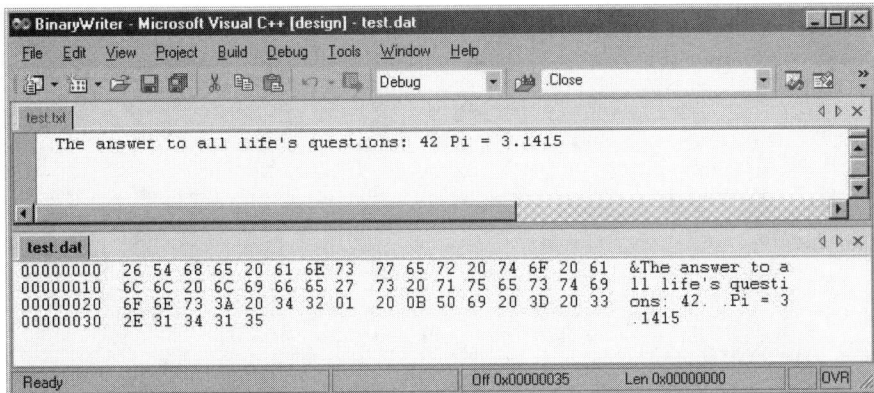

Figure 3-7 Comparing the use of the `StringWriter` and `BinaryWriter` classes

splits its view in half—the left side gives the hex representation of each byte, and the right side shows either printable characters or periods (for nonprintable characters).

The following code illustrates how to read the previous example's binary data using the `BinaryReader` class.

```
void CBinaryWriterDlg::OnBnClickedButton1()
{
#pragma push_macro("new")
#undef new
  try
  {
    CString strBinaryFileName = _T("test.dat");
    Stream* pStream = new FileStream(strBinaryFileName,
                                FileMode::Open,
                                FileAccess::Read);

    BinaryReader* pBinaryReader = new BinaryReader(pStream);

    CString strData;
    unsigned char ch;

    for (int i = 0; pBinaryReader->PeekChar() != -1; i++)
    {
      ch = pBinaryReader->ReadByte();
      strData += (CString)ch.ToString(S"X");
```

```
      if (0 == (i+1) % 8)
        strData += _T("\r\n");
      else
        strData += _T(" ");
   }

   pBinaryReader->Close();
   pStream->Close();

   AfxMessageBox(strData);
   }
   catch (Exception* pe)
   {
     AfxMessageBox((CString)pe->Message);
   }
#pragma pop_macro("new")
}
```

As you can see, once the code instantiates a FileStream on the desired file and passes that object to the BinaryReader constructor, all that is needed is a simple for loop, where a call to PeekChar prevents us from reading past the end of the stream, and ReadByte reads in the next character. This character is then converted to hex and added to a local CString object that will be displayed once the stream has been completely read. Figure 3–8 shows the output from this code snippet, which you can compare with what Visual Studio .NET displays when it opens the same file.

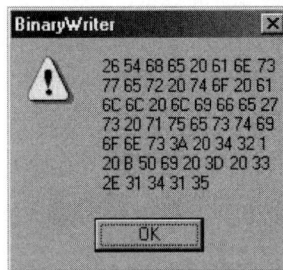

Figure 3-8 Using the BinaryReader class to read a small file

File System Classes

In addition to the Stream and Writer/Reader classes discussed so far, .NET Framework also provides a very powerful set of classes for encapsulating information and functionality that are suitable for processing files and directories. These classes, which also reside in the System.IO namespace, are listed in Table 3–4.

Now let's see just what these classes can provide for us and how best to utilize them.

Table 3–4 File System Classes in the System::IO Namespace

Class	Description
FileSystemInfo	The abstract base class for FileInfo and DirectoryInfo objects, this class contains methods that are common to both file and directory manipulation. As this is the base class for both, it is useful when writing polymorphic code that must handle a collection of both files and directories.
DirectoryInfo	While the majority of the information-level properties, such as directory attributes and various kinds of timestamp information, are inherited from FileSystemInfo, this class provides directory manipulation functionality such as create, delete, and move operations.
FileInfo	While the majority of the information level properties, such as file attributes and various kinds of timestamp information, are inherited from FileSystemInfo, this class provides file manipulation functionality such as create, delete, and move operations.
Directory	Provides numerous static methods for standard directory-level functions such as copying, moving, renaming, creating, and deleting directories. The Directory class can also be used to get and set timestamp information related to the creation, access, and writing of a directory.
File	Provides numerous static methods for standard file-level functions such as copying, moving, renaming, creating, and deleting files. The File class can also be used to get and set timestamp information related to the creation, access, and writing of a file.
Path	Provides a platform-independent means of processing a String that contains path information.

Acquiring File Information

The `FileInfo` class has only one constructor, and it takes a file path as its argument. You can pass either a fully qualified path name or a relative path based on the current directory. The following function illustrates how easy it is to acquire the information about a file by calling a few chosen `FileInfo` properties.

```
void DisplayFileInfo(CString const& strFileName)
{
#pragma push_macro("new")
#undef new

  try
  {
    FileInfo* pFileInfo = new FileInfo(strFileName);

    CString strInfo;

    strInfo += (CString)String::Format(S"Name : {0}", pInfo->Name);

    strInfo += (CString)String::Format(S"\r\nDirectory : {0}",
                    pInfo->Directory->ToString());

    strInfo += (CString)String::Format(S"\r\nExtension : {0}",
                    pInfo->Extension);

    strInfo += (CString)String::Format(S"\r\nLength : {0}",
                    __box(pInfo->Length));

    strInfo += (CString)String::Format(S"\r\nFull Name : {0}",
                    pInfo->FullName);

    strInfo += (CString)String::Format(S"\r\nCreation Time : {0}",
                    pInfo->CreationTime.ToString());

    strInfo += (CString)String::Format(S"\r\nLast Access Time : {0}",
                    pInfo->LastAccessTime.ToString());

    strInfo += (CString)String::Format(S"\r\nLast Write Time : {0}",
                    pInfo->LastWriteTime.ToString());

    // Display or use strInfo variable containing file
    // information for specified file.
```

```
    }
    catch(Exception* pe)
    {
        AfxMessageBox((CString)pe->Message);
    }
#pragma pop_macro("new")
}
```

Note especially things like how the various members need to be formatted to work with Visual C++. For example, the properties that return `String` objects don't need to be converted. The objects that can't be implicitly converted to `String` have their `ToString` method called. The properties that return numeric values are boxed (via __box). Figure 3–9 shows this code being used by the `DisplayFileInfo` demo application from the accompanying CD.

If you wanted to query information about a directory rather than a file, you'd need to have used the `DirectoryInfo` class instead of the `FileInfo` class. Anyway both the `FileInfo` class and the `DirectoryInfo` class are not just for querying information; they both provide various methods for copying, creating, enumerating, and deleting files and directories. I'll cover those methods next.

Enumerating Files and Directories

An MFC developer who needs to enumerate files and directories can use the `CFindFile` class. The Win32 SDK `FindFirstFile`, `FindNextFile` and related API calls may also be used. However, as you're about to see, neither technique is even close to being as easy to use as the `FileInfo` and `DirectoryInfo` classes. As an example, take a look at the following code,

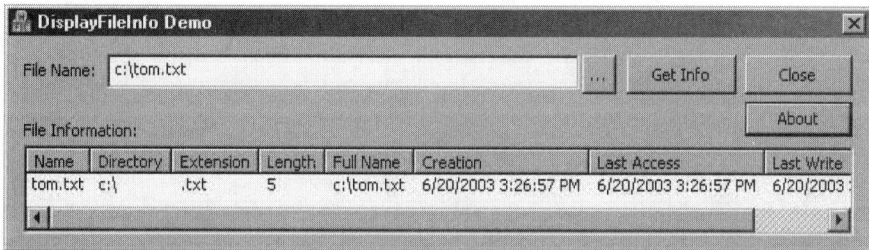

Figure 3–9 Example of the `DisplayFileInfo` demo acquiring and displaying file information

where just a few lines allow me to enumerate a directory's contents (the main points of interest are bold).

```
void ListFolder(CString const& strFolder, BOOL bDir = true)
{
#pragma push_macro("new")
#undef new

  CString strInfo;
  CString strTemp;

  try
  {
    DirectoryInfo* pDirInfo = new DirectoryInfo(strFolder);
    FileSystemInfo* aFileInfo[] = NULL;

    if (bDir)
    {
      strInfo = _T("Enumerating directories...\r\n");
      aFileInfo = pDirInfo->GetDirectories();
    }
    else
    {
      strInfo = _T("Enumerating files...\r\n");
      aFileInfo = pDirInfo->GetFiles();
    }

    for(int i=0; i<aFileInfo->Length; i++)
    {
      strTemp.Format(_T("%s\r\n"),
      (CString) aFileInfo[i]->FullName->ToString());
      strInfo += strTemp;
    }
  }
  catch(Exception* e)
  {
    AfxMessageBox((CString)e->Message);
  }

  AfxMessageBox(strInfo);
#pragma pop_macro("new")
}
```

As you can see, the first thing the `ListFolder` function does is to instantiate a `DirectoryInfo` object based on the specified directory. Once that is done, a Boolean value specifies whether or not directories or files will be enumerated. If the value is `true`, the `DirectoryInfo::GetDirectories` is called. If the value is `false`, the `DirectoryInfo::GetFiles` method is called. What's interesting, though, is that, while these methods return `DirectoryInfo` and `FileInfo` arrays, respectively, the fact that both the `DirectoryInfo` and `FileInfo` classes are derived from `FileSystemInfo` means that we can use the returned array polymorphically. In other words, the returned array is upcasted to an array of type `FileSystemInfo`. The array is then iterated, and its `FullName` property is called, resolving to a call to either `DirectoryInfo::FullName` or `FileInfo::FullName`, depending on the exact type at runtime.

Searching for Files

Let's look at a slightly more complex function that searches for a file in a directory and all of its subdirectories. The `FindMatchingFile` function takes two parameters, both of them `const` references to `CString` objects. The first parameter is the file name to search for, and the second parameter is the starting directory name in which to begin the search. The file will then return a `FileInfo` object if it finds the file (and a NULL value if the file is not found.) Now let's look at the implementation of this function.

```
FileInfo* FindMatchingFile(CString const& strFileName,
                           CString const& strFirstDir)
{
#pragma push_macro("new")
#undef new
  FileInfo* pFoundFile = NULL;

  DirectoryInfo* pDirInfo = new DirectoryInfo(strFirstDir);

  FileInfo* aFileInfo[] = pDirInfo->GetFiles();
  for(int i=0; i < aFileInfo->Length; i++)
  {
    CString strCurrName = aFileInfo->Item[i]->ToString();
    if (0 == strCurrName.CompareNoCase(strFileName))
    {
      pFoundFile = aFileInfo[i];
      return pFoundFile;
```

```
    }
  }

  DirectoryInfo* pSubDirInfo[] = pDirInfo ->GetDirectories();
  for(int i=0; i < pSubDirInfo->Length; i++)
  {
    pFoundFile = FindMatchingFile(strFileName,
    (CString)pSubDirInfo[i]->FullName->ToString());
    if (pFoundFile) break;
  }

  return pFoundFile;

#pragma pop_macro("new")
}
```

The function begins by initializing the value that will be returned
(pFileInfo) to NULL. A DirectoryInfo object is then instantiated using
the search directory. From there, an array of FileInfo objects is filled
using the DirectoryInfo object's GetFiles method. Once that is done,
a for loop is employed to iterate that array and, using the CString::
CompareNoCase member function, the string representation of each
FileInfo object's Item property is compared to the search file name. If a
match if found, pFoundFile variable is set to the current FileInfo object,
and the function returns.

If a match is not found, the DirectoryInfo::GetDirectories
method is called to load another local object array—this one is an array of
DirectoryInfo objects. Now the code simply calls itself recursively for
each directory in the array, checking the returned FileInfo object to
determine if a file was found. Using this technique, all subdirectories of
the originally passed directory are searched until a matching file name is
found.

The following example uses the FindMatchingFile function to search
for the file name FindMatchingFile.exe, starting at the current directory. The results are displayed, based on what the function returns.

Sampling Some Typical File Operations

There's really no need to go through each and every method in these
classes, so I'll show you some code snippets that perform the most basic file
operations typically needed in an application. As mentioned, the FileInfo

class can be used to copy, rename, and delete files using the `CopyTo`, `MoveTo`, and `Delete` methods, respectively. Here's some sample code illustrating those functions.

```
void PerformTypicalFileOps()
{
#pragma push_macro("new")
#undef new

  try
  {
    CString strOriginal = _T("c:\\TomArcher.txt");
    CString strClone = _T("c:\\TomsClone.txt");

    // Instantiate a new FileInfo object
    FileInfo* pFileInfo = new FileInfo(strOriginal);

    // Copy file
    pFileInfo->CopyTo(strClone);

    // Delete original
    pFileInfo->Delete();

    // Rename clone to original
    FileInfo* pFileInfo2 = new FileInfo(strClone);
    pFileInfo2->MoveTo(strOriginal);
  }
  catch(Exception* e)
  {
    AfxMessageBox((CString)e->Message);
  }

#pragma pop_macro("new")
}
```

One thing to notice about this code is that there's really no need for a method called `Rename`, as the `MoveTo` accomplishes the same thing. More important, note that the code instantiates two separate `FileInfo` objects in order to perform the tasks of copying, deleting, and renaming. While that's fine if your code already has these two objects, it's extremely inefficient to instantiate these objects if these are the only operations you need to perform. To that extent, the `File` class offers methods that do the same thing,

but they are static, so that your application does not need to incur the overhead of instantiating objects that are, in this case, overkill. Here's how to rewrite the preceding code snippet using those static File methods.

```
void PerformTypicalFileOps()
{
#pragma push_macro("new")
#undef new

    try
    {
        CString strOriginal = _T("c:\\TomArcher.txt");
        CString strClone = _T("c:\\TomsClone.txt");

        // Copy file
        File::Copy(strOriginal, strClone);

        // Delete original
        File::Delete(strOriginal);

        // Rename clone to original
        File::Move(strClone, strOriginal);
    }
    catch(Exception* e)
    {
        AfxMessageBox((CString)e->Message);
    }

#pragma pop_macro("new")
}
```

The DirectoryInfo class provides capabilities for directories similar to those we saw in the previous file-level maintenance code snippets. In the following PerformTypicalDirectoryOps function, a DirectoryInfo object is first instantiated, followed by calls to its Exists property to determine if the name passed to its constructor exists. If the directory does not exist, the DirectoryInfo::Create is called. Once that is accomplished, the DirectoryInfo::CreateSubdirectory is called, followed by a call to DirectoryInfo::Delete. Take note of that last function because the Delete function accepts a Boolean argument (set to true in our code snippet), which specifies how the object should react based on whether or not the directory is empty. If this value is true, the directory will be deleted

along with all contents (files and subdirectories). If the value is `false`, the directory will only be deleted if it is empty. In the case of an attempted deletion of a nonempty directory with a passed value of `false`, an `IOException` will be thrown—another example of why you should *always use exception handling in your file I/O code.*

```cpp
void PerformTypicalDirectoryOps()
{
#pragma push_macro("new")
#undef new

  try
  {
    CString strParentDir = _T("c:\\Test Parent Directory");
    DirectoryInfo* pDirInfo = new DirectoryInfo(strParentDir);

    if (pDirInfo->Exists)
    {
      CString str;
      str.Format(_T("The folder '%s' already exists."), strParentDir);
      AfxMessageBox(str);
    }
    else
    {
      pDirInfo->Create();

      CString str;
      str.Format(_T("'%s' created"), strParentDir);
      AfxMessageBox(str);
    }

    // Create a subdirectory within the directory wrapped
    CString strSubDir = _T("Test Child Directory");

    pDirInfo->CreateSubdirectory(strSubDir);
    CString str;
    str.Format(_T("'%s' created"), strSubDir);
    AfxMessageBox(str);

    // Delete the directory and subdirectories when finished
    pDirInfo->Delete(TRUE);
    AfxMessageBox(_T("Both directories deleted"));
```

```
   }
   catch(Exception* e)
   {
      AfxMessageBox((CString)e->Message);
   }

#pragma pop_macro("new")
}
```

Monitoring Directories and Files for Changes

One of the most interesting types in the System::IO namespace is the
FileSystemWatcher class. This .NET implementation of the publish-
subscribe design pattern as applied to the file space enables you to alert
your application that a watched directory or file has changed in some fashion. Basically, what is being provided here is an object-oriented wrapper for
the Win32 SDK ReadDirectoryChangesW function that is much easier to
use. Note that this Win32 call is only available in Windows NT, Windows
2000, and Windows XP. In other words, since the underlying Win32 function is not available in Windows 95, Windows 98 and Windows ME, the
.NET FileSystemWatcher does not work in those environments. In addition, while the FileSystemWatcher can be used on local directories as
well as remote directories or network shares, it cannot be used against
removable media such as CD and DVD drives.

The FileSystemWatcher is actually very easy to use once you know
the steps:

1. Instantiate a FileSystemWatcher object specifying either no
   parameters (default constructor), passing a single String object
   (indicating the directory to watch), or passing two String objects
   (indicating the directory to watch and types of files to monitor).
2. (Optionally) Set the FileSystemWatcher::Path property. This is
   the directory to be monitored. (As mentioned, you can specify this
   value in the constructor.)
3. (Optionally) Set the FileSystemWatcher::Filter property. This
   is the type of file to be monitored for changes. (As mentioned, you
   can specify this value in the constructor.)
4. Set the FileSystemWatcher::NotifyFilter property. This property takes a value of type FileSystemWatcher::NotifyFilters—
   an enumeration that specifies what changes to watch for. These
   enumerations are listed in Table 3–5.

5. Set the `FileSystemWatcher::IncludeSubdirectories` Boolean property to indicate whether or not your directory monitoring should extend from the specified directory to all subdirectories.
6. Set the `FileSystemWatcher::EnableRaisingEvents` Boolean property to indicate that the `FileSystemWatcher` object is either in an enabled (event-raising) state or disabled (non-event-raising) state.
7. Set the desired member events. These event members (listed in Table 3–6) are set to an instance of a `FileSystemEventArgs` object, which is constructed by specifying the object that will receive notification of the event(s) being raised and the method that is to be called when the specified event is raised.

Maybe it doesn't seem all that simple looking at this list of "to-do" items. However, take a peek at the `NotifyFilters` enumeration values and the

Table 3–5 `FileSystemWatcher::NotifyFilters` Enumeration Values

Member Name	Description
Attributes	The attributes of the file or directory have changed.
CreationTime	The creation time of the file or directory has changed, which usually means a file or folder got created.
DirectoryName	A directory got renamed.
FileName	A file got renamed
LastAccess	The last access time of the file or directory has changed.
LastWrite	The last write time of the file or directory has changed, which in the case of a file means it got modified.
Security	The security settings of the file or folder have changed.
Size	The size of the file or folder has changed.

Table 3–6 `FileSystemWatcher` Events

Event	Description
Changed	A watched file or directory has been modified.
Created	A new file or directory has been created.
Deleted	A file or directory has been deleted.
Renamed	A file or directory has been renamed or moved.

event types that are supported in Tables 3–5 and 3–6, respectively, and we'll jump into some code where you'll see that this is not as difficult as it seems.

As promised, let's look at a class that I put together for the TextFile Watcher demo application on the CD.

```
__gc class TextFileWatcher
{
public:
  // These functions -called when the associated event
  // is raised -are intentionally left blank, as
  // what you do here is application-specific.
  void OnChanged (Object* source, FileSystemEventArgs* e)
  {
  }
  void OnCreated (Object* source, FileSystemEventArgs* e)
  {
  }
  void OnDeleted (Object* source, FileSystemEventArgs* e)
  {
  }
  void OnRenamed(Object* source, RenamedEventArgs* e)
  {
  }

  void Start(CString const& strDirectory)
  {
#pragma push_macro("new")
#undef new

    // Instantiate the object
    FileSystemWatcher *pWatcher = new FileSystemWatcher();

    // Set the
    pWatcher->Path= strDirectory;
    pWatcher->Filter = S"*.txt";

    // Tell the watcher which changes I care about...
    pWatcher->NotifyFilter =
      static_cast<NotifyFilters>
        (NotifyFilters::FileName | NotifyFilters::Attributes |
        NotifyFilters::LastWrite | NotifyFilters::Size |
        NotifyFilters::DirectoryName);

    // Associate a method with each event
    pWatcher->Changed += new FileSystemEventHandler(this, OnChanged);
```

```
    pWatcher->Created += new FileSystemEventHandler(this, OnCreated);
    pWatcher->Deleted += new FileSystemEventHandler(this, OnDeleted);
    pWatcher->Renamed += new RenamedEventHandler(this, OnRenamed);

    // Indicate that changes to subdirectories should also be
    // monitored
    pWatcher->IncludeSubdirectories = true;

    // Enable the watcher!
    pWatcher->EnableRaisingEvents = true;
#pragma pop_macro("new")
  }
};
```

At this point, reading through the comments in this code should pretty much tell the whole story. And now with the `FileSystemWatcher` behavior encapsulated in our own class, starting it up from an MFC application is as easy as the following:

```
TextFileWatcher* pWatcher = new TextFileWatcher();
pWatcher->Start(_T("c:\\")); // Watch *all* .txt files
```

Figure 3–10 illustrates the CD's slightly more advanced `FileWatcher` `Demo` in action.

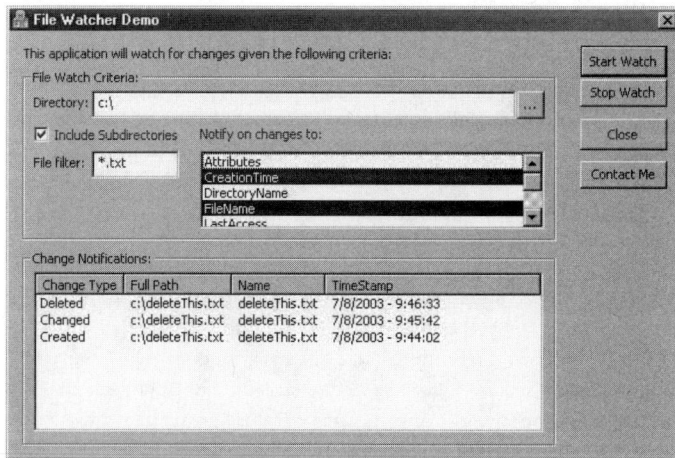

Figure 3–10 `FileWatcherDemo` application for monitoring file and directory changes

Handling Events in Mixed Mode Applications

When instantiating a delegate (such as `FileSystemEventHandler`), you must supply a method defined within a managed class. This is the main reason that the `Text FileWatcher` is used to encapsulate the `FileSystemWatcher`—so that its methods could be used to handle the various `FileSystemWatcher` events.

In terms of mixed-mode programming, this begs an important question. How hard is it to redirect the raised events from the managed class to unmanaged methods in a native class? It's not very hard at all, and there are actually many ways to do so.

One very simple way is to define the managed class (the `TextFileWatcher`, in our case) to take a pointer to an unmanaged class in its constructor. When the event methods are called, these methods would then call the appropriate unmanaged class member functions, passing the relevant data received from the `FileSystemEvent Args` when an event is raised.

Another way—one that leaves the `TextFileWatcher` class a little less coupled with its client—is to have the `TextFileWatcher` post a Windows message when a change occurs. The MFC client could then have standard message map entries for agreed-upon, user-defined messages.

Parsing Paths

When dealing with files, we commonly need to parse strings that contain path information. For example, determining whether a string contains a file name extension or combining multiple strings into a single, fully qualified path value has always been something that most developers "hand-roll" on an as-needed basis. The .NET BCL has a specialized class designed just for handling these situations—the `Path` class.

Note that when I use the term *path,* I'm referring to a string whose value includes the location of a file or directory, not an actual location on disk. It's important to make that distinction because most of the `Path` class members do not interact with the file system and therefore do not verify the existence of the file (or directory) specified. Therefore, `Path` methods that modify a path string only modify the `string` object and not any actual files or directories on disk. On the other hand, some `Path` class methods do validate that a specified path string has the correct form and will throw an exception if the string contains characters that aren't valid.

This utility class is definitely useful, and it's extremely easy to use. The following sample function illustrates how to retrieve and display path information from the path string provided by the `AfxGetAppName` function.

```
void DisplayCurrentExePathInfo()
{
```

```
#pragma push_macro("GetCurrentDirectory")
#pragma push_macro("GetTempPath")
#pragma push_macro("GetTempFileName")

#undef GetCurrentDirectory
#undef GetTempPath
#undef GetTempFileName

CString strInfo;
CString strModuleName = AfxGetAppName();

strInfo += (CString)String::Format(S"Path::GetFullPath = {0}\r\n",
  Path::GetFullPath(strModuleName));
strInfo += (CString)String::Format(S"Path::GetFileName = {0}\r\n",
  Path::GetFileName(strModuleName));
strInfo += (CString)String::Format(S"Path::GetFileNameWithoutExtension"
  S"= {0}\r\n", Path::GetFileNameWithoutExtension(strModuleName));
strInfo += (CString)String::Format(S"Path::GetDirectoryName = "
  S"{0}\r\n",
Path::GetDirectoryName(Directory::GetCurrentDirectory()));
strInfo += (CString)String::Format(S"Path::GetPathRoot = {0}\r\n",
  Path::GetPathRoot(Directory::GetCurrentDirectory()));
strInfo += (CString)String::Format(S"Path::GetTempPath = {0}\r\n",
  Path::GetTempPath());
strInfo += (CString)String::Format(S"Path::GetTempFileName = {0}\r\n",
  Path::GetTempFileName());

  AfxMessageBox(strInfo);

#pragma pop_macro("GetCurrentDirectory")
#pragma pop_macro("GetTempPath")
#pragma pop_macro("GetTempFileName")
}
```

Accessing the Registry

The Windows registry is perhaps the most abused feature in the various flavors of the Windows operating system. Just about any application that needs
to save some configuration value writes it to the registry and the rather awful
end result is that you end up with a massively bloated registry, which considerably and quite noticeably slows down your computer. That said, the

registry is really a nice centralized place to save data, and the use or abuse of it depends totally on your overall program design. As developers, we can think of the registry as a treelike structure consisting of registry-keys, where each key can have subkeys as well as values. Typically, we access the registry using the `Registry` access API functions provided in the Win32 API. Quite strangely, MFC does not provide any wrapper classes for accessing the registry, but most programmers use some kind of wrapper class of their own or from such popular MFC programming Web sites such as CodeProject. All of this brings us to the point of this section—the .NET Registry classes.

So what are the advantages of choosing the .NET equivalents of the Win32 API over your own hand-rolled (or borrowed) C++ classes? The first advantage is obvious: The .NET classes give you a much more flexible, object-oriented interface. Regarding having your own C++ registry classes, the advantage to using the .NET classes is that with the .NET classes you have a set of standard classes providing constant functionality, meaning that new developers coming into your shop will be able to work with familiar code—always a plus in terms of productivity. In this section, you'll see how to perform basic registry operations, such as how to enumerate and read subkeys and values, and how to create and modify existing keys and values.

Enumerating Registry Subkeys

Let's take as an example a situation where you would want to read all the subkeys for a given registry key; for instance, you might need to enumerate all the software that has been installed on a given machine. This can be done quite easily, as most applications update the Software subkey of the `HKEY_CURRENT_USER` root key at installation time. Needless to say, not all applications do this, but our main concern is with the ones that do in order to illustrate a common registry subkey that you would want to read.

The first class you'll need to know about for this task is the `Registry` class. The `Registry` class provides access to the basic root keys via public fields (not properties) that return `RegistryKey` objects. These `Registry Key` objects can then be used to create and open registry keys as well as read subkeys and values. Let's now look at the basic steps required to enumerate a registry subkey:

1. Call the appropriate `Registry` property to retrieve the desired the root key (in the form of a `RegistryKey` object). The list of available root keys and corresponding `Registry` public fields is shown in Table 3–7.

Table 3-7 Registry Root Keys and the `Registry` Class Methods to Retrieve Them

Windows Registry Key	.NET `Registry` Class Field Name
HKEY_CLASSES_ROOT	ClassesRoot
HKEY_CURRENT_CONFIG	CurrentConfig
HKEY_CURRENT_USER	CurrentUser
HKEY_DYN_DATA	DynData
HKEY_LOCAL_MACHINE	LocalMachine
HKEY_PERFORMANCE_DATA	PerformanceData
HKEY_USERS	Users

2. Once the root key's `RegistryKey` object has been acquired, use that object to open the desired subkey via a call to `RegistryKey::GetSubKey`. This method returns a `RegistryKey` object for the indicated subkey.

3. Now retrieve the values of the subkey by calling the `RegistryKey::GetSubKeyNames` method. This method will return an array of `String` objects, each containing the name of a subkey.

4. Using a `for` loop, iterate through the array, performing whatever application-specific logic you need to.

5. Close any open `RegistryKey` objects with a call to their `Close` methods.

Now let's look at the code implementation of how to enumerate subkeys:

```
void RetrieveSoftwareData()
{
  try
  {
    RegistryKey* keyCurrentUser = Registry::CurrentUser;
    RegistryKey* keySoftware =
      keyCurrentUser->OpenSubKey(S"Software");

    System::String* subkeys[] = keySoftware->GetSubKeyNames();
    for( int i = 0; i < subkeys->Length; i++)
    {
      CString strApplicationName = subkeys[i];
      // application specific logic for strApplicationName
```

```
        }
        keySoftware->Close();
   }
   catch(Exception* e)
   {
      AfxMessageBox((CString)e->Message);
   }
}
```

Figure 3–11 shows this code in practice in the CD's `RegistryOps` demo application.

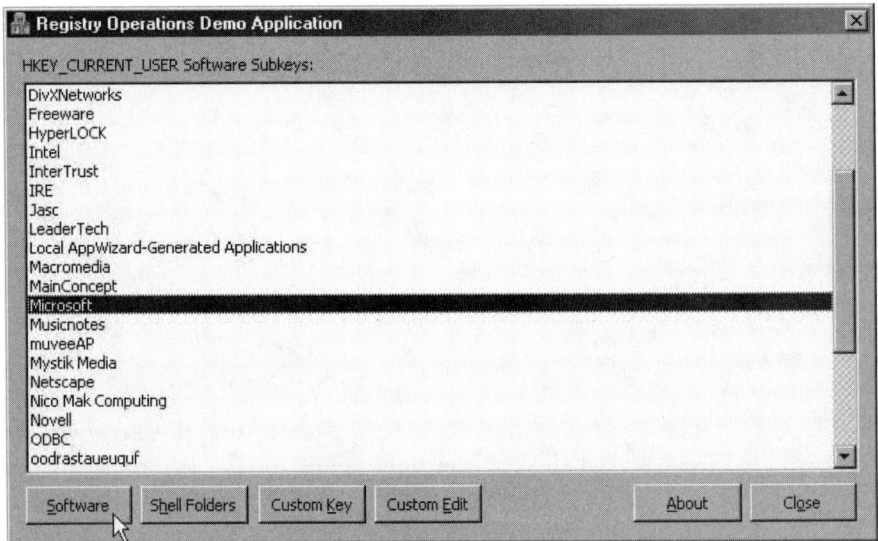

Figure 3-11 The `RegistryOps` demo application illustrates how easy it is to read from and manipulate the Windows registry using the .NET `Registry` classes.

Enumerating Values

The enumeration of registry values is very similar to the subkey enumeration code from the previous section except that instead of calling the `Get SubKeyNames` method, the `GetValueNames` method is called. This latter method returns an array of `string` objects, each of which represents the textual representation of a *value-name* stored under a given registry key. To then retrieve the actual value associated with each of the value-names, a call to the `GetValue` method is used.

```
void RetrieveShellFolderValues()
{
  try
  {
    RegistryKey* keyCurrentUser = Registry::CurrentUser;
    RegistryKey* keySoftware = keyCurrentUser->OpenSubKey(
      S"Software\\Microsoft\\Windows\\CurrentVersion\\"
      S"Explorer\\Shell Folders");
    System::String* valueNames[] = keySoftware->GetValueNames();
    for( int i = 0; i < valueNames->Length; i++)
    {
      CString strValue = static_cast<System::String*>
        (keySoftware->GetValue(valueNames[i]));
      // application-specific logic for strValue
    }

    keySoftware->Close();
    keyCurrentUser->Close();
  }
  catch(Exception* e)
  {
    AfxMessageBox((CString)e->Message);
  }
}
```

The `RegistryKey::GetValueNames` uses the `RegEnumValue` Win32 API call to get the list of value-names for the key that was opened when the `RegistryKey` was constructed. The `RegistryKey::GetValue` method then uses the `RegQueryValueEx` Win32 API call to get the value of the specified value-name. Note that the `GetValue` method returns a base `Object`, which must then be cast to the desired type. As a result, you need to know exactly what type you are expecting, because there is no way to find out if the underlying value in the registry is a string, numeric, or binary value. To see this code execute, simply run the `RegistryOps` demo on the book's CD and click the Shell Folders button, which will result in the display of all the `ShellFolder` values in the registry for the current user.

Creating New Registry Keys and Values

The last common registry task that we'll cover in this section is the ability to create new registry keys and their values. For this operation, the

RegistryKey class provides a method called CreateSubKey, which creates a new registry key using the RegCreateKeyEx Win32 API. From that point, creating new values is done via the RegistryKey::SetValue method, which can also be used to update existing values. The SetValue internally uses the RegSetValueEx Win API. In the following sample function, a new Software subkey is created with values for both a description and version number.

```
void CreateSoftwareSubKey(CString const& strDescription,
                          CString const& strVersion)
{
  try
  {
    RegistryKey* keyCurrentUser = Registry::CurrentUser;
    RegistryKey* keySoftware =
      keyCurrentUser->CreateSubKey(S"Software\\Our Custom Key");

    keySoftware->SetValue(S"Description",
                          (System::String*)strDescription);
    keySoftware->SetValue(S"Version",(System::String*)strVersion);

    keySoftware->Close();
    keyCurrentUser->Close();
  }
  catch(Exception* e)
  {
    AfxMessageBox((CString)e->Message);
  }
}
```

One important thing to realize is that methods like CreateSubKey do not throw exceptions if they fail because such a subkey already exists. The reason for this is philosophical. As an exception is technically an exceptional situation that the programmer is not anticipating (such as a hardware error), the fact that the key doesn't exist hardly qualifies. In these error (not exception) situations, both the CreateSubKey and OpenSubKey will simply return a value of null. In the following sample function, the code is attempting to open the subkey we wrote with the CreateSoftwareSubKey function. If it is found, the description and version will be updated. If not, the CreateSoftwareSubKey function will be called.

```
void CRegistryOpsDlg::UpdateSoftwareSubKey(
  CString const& strDescription,
  CString const& strVersion)
{
  try
  {
    RegistryKey* keyCurrentUser = Registry::CurrentUser;
    RegistryKey* keySoftware = keyCurrentUser->OpenSubKey(
      S"Software\\Our Custom Key");

    if (keySoftware)
    {
      keySoftware->SetValue(S"Description",
        (System::String*)strDescription);
      keySoftware->SetValue(S"Version", (System::String*)strVersion);

      keySoftware->Close();
      keyCurrentUser->Close();
    }
    else
    {
      CreateSoftwareSubKey(strDescription, strVersion);
    }
  }
  catch(Exception* e)
  {
    AfxMessageBox((CString)e->Message);
  }
}
```

Summary

This chapter began by introducing you to the BCL implementation of the concepts of streams, writers, and readers. Once the basics were out of the way, you explored the capabilities of the FileStream class and how to write binary data to a disk file. From there, you discovered the MemoryStream and BufferedStream classes. As you learned, both of these classes enable you to write to and read from memory to improve performance for file I/O–intensive applications. Additionally, since both classes derive from the base Stream class, you can polymorphically use the same code base

in reading and writing data without having to maintain two sets of I/O functions—one for performance-enhancing memory caches and one for disk files.

Once the subject of streams had been covered, you then moved into writers and readers. The first set covered was the `StreamWriter/Stream Reader` pair; you saw how they enable the ability to write and read text, respectively, to and from their backing stores. Once that was done, you learned about the `StringWriter/StringReader` classes (that enable you to use file I/O calls to manipulate strings) and the `BinaryWriter/Binary Reader` class pair (and how its many write and read methods enable you to write and read almost any type of data without writing conversion routines).

From there, you then discovered several of the file system classes—`FileSystemInfo`, `DirectoryInfo`, `FileInfo`, `File` and `Path`—that encapsulate the information and functionality suitable for processing files and directories. In fact, the extremely useful `Path` class (used to extract key information from and manipulate strings containing file path information) is used extensively throughout this book's code snippets and demos. Finally, the chapter ended with a section devoted to the BCL registry classes and how they provide you the ability to easily enumerate registry subkeys and values, and create new keys and values.

While MFC does provide decent file support—in the form of the `CFile` and `CStdioFile` classes—this chapter should have convinced you that the .NET implementation of streams, writers, and readers provide you a much more robust and flexible means of performing file I/O. In the next chapter, I'll continue this thread of showing you how the BCL classes can make you a more productive MFC developer by introducing a set of classes for a subject that has always been a difficult chore for MFC/Win32 developers until now—cryptography and data encryption.

Cryptography, Hash Codes, and Data Encryption

Introduction

The cryptography classes provided by the BCL should be reason enough for any programmer dealing with cryptography to add .NET support to their existing MFC applications. What in the past has taken you literally hundreds of lines of code and endless hours of debugging can now be achieved in a few minutes with a couple of lines of code using the various types found in the `System::Security::Cryptography` namespace. This namespace includes support for all the most common symmetric, asymmetric, and hash algorithms; providing a robust and flexible object-oriented interface to the powerful Microsoft CryptoAPI.

In this chapter, you will not only be introduced to several of these classes, but will also see complete hands-on demos for creating practical, usable applications. The chapter begins with an overview on the basics of cryptography. Specifically, the topics of data encryption (both symmetric and asymmetric), hash codes and digital signature are explained at a high level. From there we jump into a more detailed look at hash codes, where you'll learn that the previous difficult chore of generating hash codes using the CryptoAPI can now be done with a single line of code. There are two demos in that section—one to generate hash codes using various algorithms (such as MD5 and SHA1) and another to compare two string values using hash codes. The latter is especially significant, as you will see code that will allow you to do comparisons of large bodies of text in your own code with just a couple of lines of code!

After the hash code section, we dive into the world of data encryption. The section begins by talking about symmetric key encryption and provide a demo for both encrypting and decrypting files using DES. After that, you discover asymmetric key encryption and why it's usually combined with the former in real-world situations. There are two demos in that section: a complete class (TextCrypt) that encrypts text data to RSA (and vice versa) and a demo to illustrate encrypting and decrypting data using RSA. Finally the chapter ends with a section and demo on digital signatures.

Cryptography Basics

Cryptography is the science—or some would say, art—of protecting data by converting it (encrypting it) into a seemingly incomprehensible format (referred to as *cipher text*) so that unauthorized persons cannot gain access to the original data, or would probably have to work for years to unprotect it. Cryptography is particularly important in data transfers that take place over a network, such as the inherently unsafe Internet. Additionally, cryptography is not used only to secure data; it can also be used to verify that the data originated from a particular sender and not someone spoofing a valid sender's ID for malicious purposes. Having said that, here's a list of the most important cryptographic operations:

- **Data encryption:** Data encryption is used to protect the data and usually involves the conversion of the original data into an unreadable encrypted form—also known as cipher text—using an encryption key. There are two broadly different techniques used for data encryption:
  - **Symmetric key encryption:** Also known as private key encryption, this form of data encryption uses a single private key, which is shared by two parties. The same key is used both for encrypting the data and for decrypting it.
  - **Asymmetric key encryption:** Also known as public key encryption, this approach uses a public key/private key pair. Encryption is done with the public key of the recipient (which the recipient created and publicized for everyone to use), and decryption can be done only by using the recipient's private key (which the recipient keeps private).

> Symmetric key encryption is often referred to as private key encryption, just as asymmetric key encryption is often referred to as public key encryption. As a result, you'll see me use the terms interchangeably as well in this chapter.

- **Hash codes:** A hash code (sometimes called a *message digest* depending on the context) is a fixed-length byte sequence that is generated by a hash algorithm from input data of any length. I'll cover some of the more popular usages of hash codes later in this section and present two separate demo applications in the section entitled "Using Hash Codes."
- **Data signing:** This is a technique used to verify that the sender is who he or she claims to be. A digital signature is sent along with the data, usually a hash of the data that's encrypted using the sender's private key. Now the recipient can attempt to decrypt the signature using the sender's known public key and then compare it with a generated hash of the received data.

In real-life scenarios, data encryption, digital signatures, and hashing are quite commonly used in conjunction with one another to ensure data security and to verify the integrity of both data and sender. In fact, the need to protect data being sent over the Internet is so prevalent that most current e-mail clients (including Microsoft Outlook and Microsoft Outlook Express) have built-in support for data encryption and digital signatures. Now let's start getting into the details of the above-mentioned cryptography concepts.

Symmetric Key Encryption

As mentioned, symmetric key encryption is an encryption system in which both the sender and the receiver share a single private key that is used both to encrypt and decrypt the message. Symmetric key encryption is very fast compared to asymmetric key encryption and thus is useful for encrypting large amounts of data. Some of the most popular symmetric key encryption algorithms are DES (Data Encryption Standard), TripleDES, and RC2.

Symmetric key encryption algorithms are referred to as *block ciphers* because they encrypt data a block at a time, where the block is of length n bytes and n varies according to the algorithm used. Usually the data would

be much larger than the block length and is broken down into blocks and encrypted. The resultant number of encrypted bytes may not be equal to the number of input bytes and will vary according to the algorithm used and the data in the block.

One small issue with this is that an intruder who has somehow gained unauthorized access to data might be able to cause harm using a partial knowledge of the actual contents. The problem is that two identical blocks always become identically encrypted blocks. Now assuming that some particular text appears several times within a message, then its encrypted output will also appear several times in the encrypted message, and this gives the intruder a very good starting point from which to attack the encryption. Since this should obviously be avoided, the algorithm designers devised a technique called *cipher block chaining*.

Cipher block chaining involves encrypting each block of data by using both the private key and information from the previous block. Since the first block of data obviously doesn't have a previous block, an initialization vector is used. This technique ensures that identical input blocks will not produce identical output (encrypted) blocks, and thus intruders will not be able to identify common phrases and use them in their attacks on the encrypted message.

The main disadvantage of symmetric key encryption is that the security of the encryption depends entirely on the security of the private key; once the private key falls into the wrong hands, the encrypted message can be decrypted with zero effort. Obviously, this means the private key must never be sent along with a message, and in fact it's best to never send it across any network in plain-text (unencrypted) format. Thus both sender and receiver must have been able to agree upon the same private key via some secure means such as direct physical contact.

Fortunately there is a more feasible way of establishing and exchanging private keys across a network using a technique called asymmetric key encryption. The symmetric key is encrypted using the recipient's public key (which he makes public) and sent to the recipient, who can now decrypt it safely using his private key (which is safe with him). The next section goes into that in a bit more detail.

Asymmetric Key Encryption

Asymmetric key encryption is an encryption system in which two keys are used: a public key, which is known to everyone, and its corresponding private key, known only to the recipient. A message you want to send to the

recipient is encrypted using the recipient's public key and can only be decrypted with the recipient's private key. You might hear this kind of encryption being referred to as *Diffie-Hellman encryption* after its inventors, Whitfield Diffie and Martin Hellman, who invented this encryption system in 1976.

The private and public keys are generated as a pair so that there is a mathematical link between them, but it is an extraordinarily remote contingency that the private key might be derived using a public key, and we can essentially consider that eventuality as impossible. Any data that is encrypted using a given public key can be decrypted using its matching private key, and any data that has been signed using a private key can be verified using its matching public key. Along with its increased security, another advantage to asymmetric key encryption is that a sending party cannot deny sending the data. In the jargon of cryptography, this is referred to as assuring *nonrepudiation*—an essential element of e-commerce.

While there are extreme advantages to using asymmetric key encryption, there is one very real disadvantage—it is extremely slow. In fact, this technique is so slow that it is a given in the world of cryptography that it is not usable when working with large amounts of data. In real-world situations data is encrypted using a private key. The private key as well as the initialization vector (IV) are then signed and encrypted using a public key algorithm. These elements are then sent to the recipient. The recipient verifies the signature, decrypts the encrypted private key, and uses the private key to decrypt the encrypted data.

Two very popular public key cryptography algorithms are DSA (Digital Signature Algorithm) and RSA algorithms. (RSA is named after the engineers who invented the algorithm in 1978, Ron Rivest, Adi Shamir, and Leonard Adleman). Both of these algorithms have undergone rigorous public scrutiny and have been proven mathematically secure. You'll see them both used in demo applications later in the chapter.

Hash Codes

As hash codes were defined at the beginning of this section, I'll talk about some usages of hash codes here.

- **Password hashing:** One popular usage of hash codes is in situations where you either do not want to—or legally cannot—store sensitive information such as passwords. In these scenarios users are often allowed to create a password and a user-defined hint to help remind

them of their password. A hash code is then generated from the password, and both the password hash code and the hint are persisted (typically either to a file or a database). When a user attempts to log in, the user-entered password is hashed, and that value is compared against what was persisted for the specified user. If the two hash codes match, the user is authenticated. If the user cannot remember the password, the hint can be emailed or displayed to the user. If the user still cannot remember the password, a new one is generated. If you've ever used a system that warned that it could not recover your password should you lose it and you wondered how they were able to authenticate your password without storing it, now you know one technique for accomplishing this.

- **Database hash addressing:** Hash codes are also used in databases where the hash value of a key field is used as the index of a record. Thus, to access a record that has a matching key field, a hash code is generated and used as the index to retrieve the record.

- **Data (text) comparison:** An extremely common usage of hash codes is that they provide a very quick and easy means of comparing two sets of data. The `HashDemo` application that you'll see shortly illustrates this usage as it allows you to enter two text values and then compare the two values using hash codes. The application generates a hash code for each value and displays a message indicating whether the two hash codes (and thus, the two input fields) are identical.

- **Data integrity:** Since hash codes can be used to compare values by comparing the generated hash codes from two or more inputs, this works well in determining if data has been modified while in transit from a sender to a receiver. The sender generates a hash code on the data that will be sent to the receiver. This hash code is encrypted using the (more secure) public key encryption. The data is then encrypted using the (less secure, but faster) private key encryption. The sender then sends both the encrypted hash code and encrypted data to the receiver. Once these two items are received, the receiver decrypts both the hash code and the data, generates a hash code from the data (using the same hash algorithm that was used to generate the first hash code) and compares the two hash codes. If the two hash codes match, the receiver knows that the data has not been modified in transit. If the two hash codes do not match, it indicates that the data has been altered while it was in transit. I'll go through a more complete example of combining several of these cryptographic

techniques in the section entitled "Combining Symmetric Encryption, Asymmetric Encryption, and Hash Codes."

Data Signing

One very important security issue in Internet data transfers is to verify the source of the data. Public key encryption algorithms are used to generate digital signatures that can be used to verify the authenticity of a message (confirm that a message comes from a particular source). If the amount of data to be sent is small, the sender can simply encrypt the entire message using a private key, and thus any recipients can verify that it's actually from the expected sender by decrypting it with the sender's public key. Part of the message should be prefixed text so that the recipient has something with which to make comparisons. But in real-world situations the data to be sent cannot be assumed to be small, and since public key encryption is very much slower than private key encryption, it does not make sense to encrypt the entire message using a public key algorithm.

Here's an example of how digital signatures are employed, using Krista as the sender of data and Tom as the receiver: Krista generates a hash code for the data to be sent, and then signs the hash code using the private key of a public/private key pair. This encrypted code is called the digital signature. Now when Tom receives the data along with the digital signature, he can verify that Krista sent the message by first decrypting it using Krista's public key and then comparing it with a hash code based on the received message. If the hash values match, then Tom can be sure that the sender is indeed Krista.

The signature can be verified by anyone who has the sender's public key, and this is usually just about everyone, because public keys are always made publicly available. Remember that signatures only ensure the data came from a particular sender and that the actual data must be encrypted using other means—typically, the same private key encryption algorithm that was used to create the digital signature. In the section entitled "Using Digital Signatures," you'll see a demo application of how to create and verify digital signatures.

Using Hash Codes

If you are going to be dealing with cryptography applications, you're going to have to know how to generate hash algorithms and hash codes at some point. (In fact, as you saw in the previous section hash codes are useful in

many noncryptographic scenarios.) Hash codes are generated using hash algorithms, which may be written using any of several algorithms such as MD5, SHA1, and SHA256, to name a few. Each hashing algorithm has its benefits and drawbacks, but they all essentially operate in the same way—the key difference is that differing hashing algorithms produce hash values of different length. In some cases, you may need a very short hash value, so using SHA256 (which produces a hash value that's 256 bytes long) may not be practical.

Irrespective of the actual algorithm used, a hash value generates a fixed-length byte sequence for an input binary string of arbitrary length. While hash codes are not unique in the sense that two (or more) different input strings can produce the same hash codes, each of these input strings would be so different as to mean that no two semantically related strings would produce the same hash code. This gets a bit complicated, so for purposes of this discussion if you're comparing two meaningful input sources (and not just random data), the hash codes generated from the two will be different if the two strings are different and will be the same if the two strings are the same. The sidebar "Are Hash Codes Unique?" explains this in a bit more detail.

For working with hash codes, the BCL provides an abstract class— HashAlgorithm (defined in the System::Security::Cryptography namespace)—from which all other cryptographic hash algorithm classes must derive. As the HashAlgorithm is abstract, it cannot be instantiated or directly used; you must always use one of the derived classes. To that extent, the BCL provides the following HashAlgorithm-derived classes, where each class implements a different algorithm for generating hash values. (It's fairly easy to discern which particular hash function algorithm each class encapsulates merely by looking at the class's name.)

- System::Security::Cryptography::KeyedHashAlgorithm
- System::Security::Cryptography::MD5
- System::Security::Cryptography::SHA1
- System::Security::Cryptography::SHA256
- System::Security::Cryptography::SHA384
- System::Security::Cryptography::SHA512

Note that each of the classes above are themselves abstract base classes. This is done because many of the algorithms have two implementations in the BCL—a CryptoAPI wrapper and a pure managed implementation. As an example, the SHA1 base class defines members that are

common to the SHA1 specification, regardless of specific implementation. The two derived classes—SHA1CryptoServiceProvider and SHA1 Managed—then define the members that are specific to a given implementation. The base classes SHA1CryptoServiceProvider and SHA1Managed then provide the CryptoAPI and managed implementation, respectively.

In addition, you are free to derive your own classes from the above base classes if you have the requirement to use variations of one of the provided algorithms. In fact you may derive from the HashAlgorithm class if you want to write a completely new hash algorithm of your own.

Are Hash Codes Unique?

While you will hear people state that hash codes generate a unique value for a given input, the fact is that, while difficult to accomplish, it is technically feasible to find two different data inputs that hash to the same value. However, the true determining factors regarding the effectiveness of a hash algorithm lie in the length of the generated hash code and the complexity of the data being hashed.

Let's first talk about the hash algorithms themselves. As stated earlier, a hash algorithm generates a fixed-length hash code regardless of the length of the input. For example, the MD5 hash function always generates hash codes that are 16 bytes in length, the SHA1 hash function generates 20-byte hash codes, SHA256 generates 256-byte hash codes, and so on. Therefore, since there are a limited range of possible values for a given hash code and an unlimited range of values to hash, it stands to reason that the length of the hash code generated with a given hash algorithm is directly related to the difficulty of finding two inputs that will generate the same hash code.

This is easy to prove. If n is the number of possible hash codes, we only need $n + 1$ distinct input values to prove that there is an overlap. Granted that for most hash functions (including all the hash algorithms mentioned in this chapter) n is rather large, we can probably assume that for any meaningful input values, you would have a hard time finding another meaningful input that would give the same hash.

That brings me to the second point—the input value itself. It is assured that if the input is anything more involved than random data, then even if you were to find two inputs that generated the same hash value, the two inputs would have no semantic relationship to one another. This is especially evident if you are hashing text, because there are many more ways to produce gibberish than there are ways to produce meaningful words. In other words, it's extremely difficult to create random words, even harder to form sentences, and very unlikely that those sentences will form a paragraph or a document and still have it make sense.

As an example, let's say you're generating a hash code for a top-secret document that details the route of nuclear warheads for disposal. The chances of someone

(continued)

Are Hash Codes Unique? (*cont.*)

randomly generating words until they matched the hash code generated for the original document *and* producing a document that makes any sense are so small that you could accurately say it would be impossible.

Therefore, the fact that each possible hash code doesn't uniquely match an input only comes up when you are dealing with a small amount of simple (nonstructured) data, such as a password. (In the previous section I spoke briefly of the concept of password hashing.) To enable hash functions to work even for this type of data, a random piece of data (referred to as a *salt*) is usually added to the password before hashing. This accomplishes two things: First, even if someone were to obtain a list of hashed passwords, even if two passwords were the same, the hash codes would be different. Second, the addition of the salt greatly increases the difficulty in cracking the codes, because now the unauthorized person must determine the salt that gets added to the input and how it's added to the input before even attempting to generate the passwords.

Therefore, while there isn't a one-to-one correlation between every possible hash code and every possible input value such that all combinations are guaranteed to be unique, hash codes are an extremely reliable means of protecting data integrity.

Create Hash Code Demo

As an MFC developer, if you've ever had to use the CryptoAPI to access hash functions, you're probably excited to get started, as this promises to provide a level of productivity one could only dream of with the CryptoAPI. To give you an example of what I'm talking about, consider this: Using the BCL classes, you can generate a hash code from just about any standard hash function algorithm by calling a single method! Before you ask me if I'm on any sort of (heavy) medication, take a look at the following (overloaded) method—`HashAlgorithm::ComputeHash`:

```
unsigned char ComputeHash(unsigned char __gc[]) __gc[]
unsigned char ComputeHash(unsigned char __gc[], int, int) __gc[]
unsigned char ComputeHash(Stream*) __gc[]
```

Each of these overloaded methods returns a hash code in a byte array. The only difference is the flexibility of the input. In the first version, the input can be a byte array. In the second version, you also pass a byte array along with offset and length information to define the subset of the byte

array that will be used to compute the hash code. Finally, the third the version of the ComputeHash method enables you to specify a Stream object as containing the input data.

Up to now, I've studiously avoided a lot of step-by-step tutorials in this book for the simple reason that the topics of the two previous chapters—regular expressions and file streaming—required the explanation of so many details and examples that to go through a full demo application with each would have bloated each of those chapters unnecessarily. Therefore, in those chapters, I presented a lot of code snippets and made references to the various demo applications on the book's CD. However, with cryptology, you do need to roll up your sleeves and get your hands dirty a bit to see what is really going on. Therefore, let's do just that in creating the chapter's first demo—one that will accept input and produces three hash codes (SHA1, MD5, and MD5 using a subset of the input data).

1. Create a dialog-based application called CreateHashCode.
2. Update the Project properties to support Managed Extensions.
3. Open the stdafx.h file and add the following .NET support directives to the end of the file:

```
#using <mscorlib.dll>
#using <System.Windows.Forms.dll>
#using <System.dll>
using namespace System;
using namespace System::Windows::Forms;
using namespace System::Security::Cryptography;
#undef MessageBox
```

4. Open the dialog template resource and add the controls as you see them in Figure 4–1.
5. Set the ReadOnly property for the three bottom hash code value edit controls to True.
6. For the input edit control, set its AutoHScroll property to False and its Multiline property to True.
7. Create the DDX member variables as shown in Table 4–1.

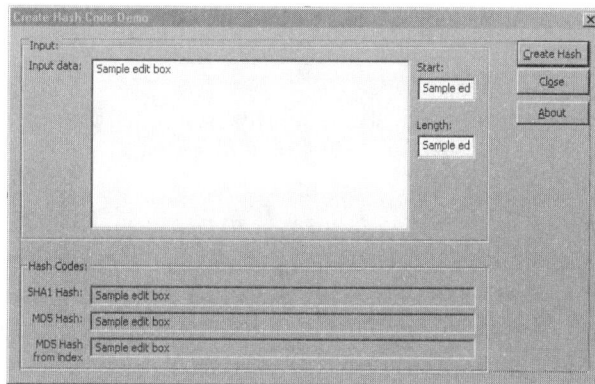

Figure 4-1 Dialog resource for the CreateHashCode demo application

Table 4-1 DDX Variables for the CreateHashCode Demo

Control	Variable Type	Variable Name
Input data edit	Cstring	m_strInputData
Start edit	Int	m_iStart
Length edit	Int	m_iLength
SHA1 edit	Cstring	m_strSHA1Hash
MD5 edit	Cstring	m_strMD5Hash
MD5 edit (substring)	CString	m_strMD5HashSubstring

8. At this point, double click the OK (Create Hash) button to create a handler for its BN_CLICKED message and code it as follows.

```
void CCreateHashCodeDlg::OnBnClickedOk()
{
#pragma push_macro("new")
#undef new
```

```
try
{
  if (UpdateData())
  {
    try
    {
      m_strSHA1Hash = _T("");
      m_strMD5Hash = _T("");
      m_strMD5HashSubstring = _T("");
      UpdateData(FALSE);

      Byte barr[] = new Byte[m_strInputData.GetLength()];
      for(int i = 0; i < m_strInputData.GetLength(); i++)
        barr[i] = static_cast<Byte>(m_strInputData [i]);

      SHA1CryptoServiceProvider* csp1 =
        new SHA1CryptoServiceProvider();

      MD5CryptoServiceProvider* csp2 =
        new MD5CryptoServiceProvider();

      Byte hash[];

      //generate the SHA1 hash value
      hash = csp1->ComputeHash(barr);
      m_strSHA1Hash = (CString)BitConverter::ToString(hash);

      //generate the MD5 hash value
      hash = csp2->ComputeHash(barr);
      m_strMD5Hash = (CString)BitConverter::ToString(hash);

      //generate MD5 hash for a substring
      hash = csp2->ComputeHash(barr, m_iStart, m_iLength);
      m_strMD5HashSubstring = (CString)BitConverter::
        ToString(hash);
    }
    catch(Exception e)
    {
      MessageBox::Show(e->Message);
    }

    UpdateData(FALSE);
  }
```

```
    }
    catch(Exception* e)
    {
        MessageBox::Show(e->Message);
    }

    #pragma pop_macro("new")
    }
```

As you can see, acquiring a hash code is as easy as I claimed it to be! After initializing the dialog's hash code value edit controls, the function constructs a `Byte` array of the same length as the data entered into the dialog's Input Data edit control. As mentioned in the sidebar entitled "Using `CString` Objects with the Cryptography Classes," there is no direct way to convert a `CString` to a `Byte` array that the `ComputeHash` requires. Therefore, a simple `for` loop is employed to accomplish the task.

From there, the function constructs both a `SHA1Crypto ServiceProvider` and `MD5CryptoServiceProvider` object. Finally, the `Compute` method is called for the two objects—once for the `SHA1CryptoServiceProvider` object and twice for the `MD5 CryptoServiceProvider` (the second time using the starting offset into the input buffer and length specified on the dialog). The results are then displayed by way of updating the three hash code value edit controls.

9. That's all there is to it. Figure 4–2 shows an example of running this application and supplying it with some test input data.

Text String Comparer Demo

Now that you've seen how easy it is to create a hash code using the algorithm of your choice, let's take the next step and write a very practical demo application. As mentioned, you can compare the hash codes of two input sources to test whether the two hash codes were generated from identical input data. This generally leads to the assumption that hash codes are only useful in cryptographic scenarios. However, there are many noncryptographic situations where you might want to compare two sets of data and not have to write and debug hundreds of lines of C++ code to do it.

One obvious example is a version-control system where you would need to compare the version of code being checked in with the last version to determine if changes were made. Another example comes up in an

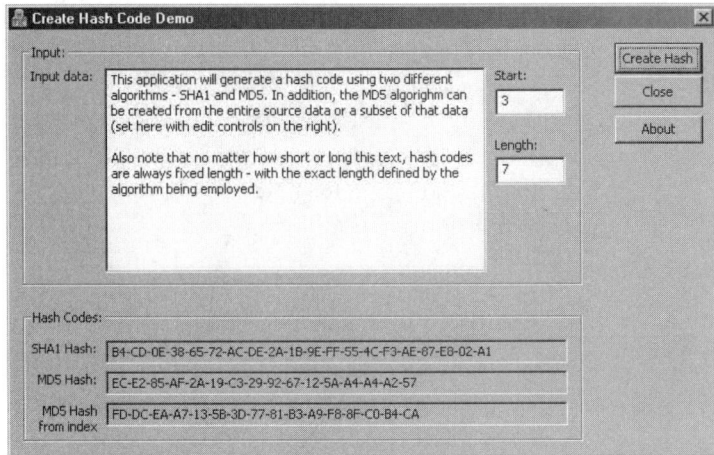

Figure 4-2 The `CreateHashCode` demo application clearly illustrates how easy it is to perform cryptographic functions using the .NET classes—especially in comparison to directly interfacing to the CryptoAPI.

Using `CString` Objects with the Cryptography Classes

As an MFC developer, you'll frequently encounter the need to convert `CString` objects and `CString` pointers to arrays of type `System::Byte` for use with the cryptography classes. While, there is no single conversion method that will accomplish this task for us, it's still relatively simple to do ourselves with code like the following:

```
//convert String* to byte array
String* str = S"This is a line of text that is to be hashed";
Byte barr[] = new Byte[str->Length];
for(int i=0; i<str->Length; i++)
  barr[i] = static_cast<Byte>(str->ToCharArray()[i]);


//convert CString to byte array
CString str = _T("This is a line of text that is to be hashed");
Byte barr[] = new Byte[str.GetLength()];
for(int i = 0; i < str.GetLength(); i++)
  barr[i] = static_cast<Byte>(str [i]);
```

editor-like application where you wish to know if the document is *dirty*, or changed since its last save. There are countless scenarios, but you can easily see why an extremely fast means of comparing text would be beneficial to a host of applications. This section's demo introduces an application that

allows you to enter two strings (of practically any length) and have them compared via hash codes.

1. Create a dialog-based application called HashDemo.
2. Update the Project properties to support Managed Extensions
3. Open the stdafx.h file and add the following .NET support directives to the end of the file:

```
#using <mscorlib.dll>
#using <System.Windows.Forms.dll>
#using <System.dll>
using namespace System;
using namespace System::Windows::Forms;
using namespace System::Security::Cryptography;
#undef MessageBox
```

4. Open the dialog template resource and add the controls shown in Figure 4–3.
5. Create the DDX member variables as shown in Table 4–2.
6. Set the ReadOnly properties for the results edit control and the two Hash Code edit controls to True.

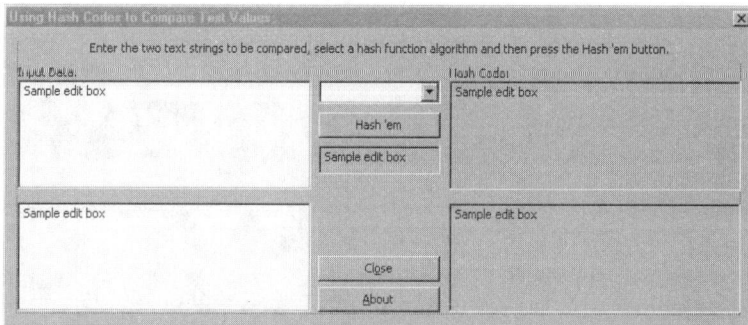

Figure 4–3 Dialog resource for the HashDemo application

Table 4-2 DDX Variables for the `HashDemo` application

Control	Variable Type	Variable Name
Top input edit	CString	m_strInput1
Bottom input edit	CString	m_strInput2
Top hash edit	CString	m_strHashCode1
Bottom hash edit	CString	m_strHashCode2
Combo box	CComboBox	m_cbxHashAlgorithms
Result edit	CString	m_strResults

7. Set the `Multiline` and `Vertical` properties for the two input edits and two hash code value edits to `True`.
8. Add the following code to the end of the `OnInitDialog` (just before the return statement) to populate the combo box with the available hash function algorithms that this application will support. (In a production system, I would associate enumeration values with the choices and store them in the combo box's item data, but we'll just keep the demo simple and compare on the text from the combo box when it's time to determine which algorithm the user selected.)

```
BOOL CHashDemoDlg::OnInitDialog()
{
 CDialog::OnInitDialog();

 //...

  // Initialize combo box with available
  // hash algorithms
  m_algorithm.AddString(_T("MD5"));
  m_algorithm.AddString(_T("SHA1"));
  m_algorithm.AddString(_T("SHA256"));

  m_cbxAlgorithms.SetCurSel(0); // select first item

  return TRUE;
}
```

9. Double click the Hash 'em button to create a BN_CLICKED handler for that button. In this function, you'll insert the actual comparison code where both hash code values are generated for the two input strings. Once generated, this function will simply compare the two hash code values. If the two values are identical, we know that the input strings are identical, and the read-only results edit will reflect that. Otherwise, we know the input sources are distinct, and the results edit will reflect that accordingly.

```cpp
void CHashDemoDlg::OnHashEm()
{
#pragma push_macro("new")
#undef new

 if (UpdateData(TRUE))
 {
   try
   {
    Byte baTop[], baBottom[];

    /*
      We need to first convert the CStrings into Byte arrays
    */
    baTop = new Byte[m_strInput1.GetLength()];
    for (int i = 0; i < m_strInput1.GetLength(), i++)
       baTop[i] = static_cast<Byte>(m_strInput1[i]);

    baBottom = new Byte[m_strInput2.GetLength()];
    for (int i = 0; i < m_strInput2.GetLength(); i++)
       baBottom[i] = static_cast<Byte>(m_strInput2[i]);

    //Get the strSelectedAlgorithm selected
    CString strSelectedAlgorithm;

    m_cbxAlgorithms.GetLBText(m_cbxAlgorithms.
                              GetCurSel(),
                              strSelectedAlgorithm);

    HashAlgorithm *md5csp;
```

```
/*
Based on the strSelectedAlgorithm selected, instantiate
the corresponding hash strSelectedAlgorithm class
*/
if (strSelectedAlgorithm == _T("MD5"))
    md5csp = new MD5CryptoServiceProvider();
if (strSelectedAlgorithm == _T("SHA1"))
    md5csp = new SHA1CryptoServiceProvider();
if (strSelectedAlgorithm == _T("SHA256"))
    md5csp = new SHA256Managed ();

//get the hash values using ComputeHash
Byte baHashCode1[] = md5csp->ComputeHash(baTop);
Byte baHashCode2[] = md5csp->ComputeHash(baBottom);

//Convert the Byte arrays to CStrings
m_strHashCode1 = BitConverter::ToString(baHashCode1);
m_strHashCode2 = BitConverter::ToString(baHashCode2);

/*
If the hash values match, then the original strings
  match too
*/
m_strResults.Format(_T("They are %s"),
                    (m_strHashCode1 == m_strHashCode2
                    ? _T("the same") : _T("different"))));
}
catch(Exception* e)
{
    MessageBox::Show(e->Message);
}
UpdateData(FALSE);
    }
#pragma pop_macro("new")
}
```

10. The last step is obviously to build, run, and test the application. When you do, you'll find that the hash values generated are identical for identical strings and are also of a fixed length. You'll also note that even the slightest change to the text changes the hash code value dramatically. The results of my test are shown in Figure 4–4.

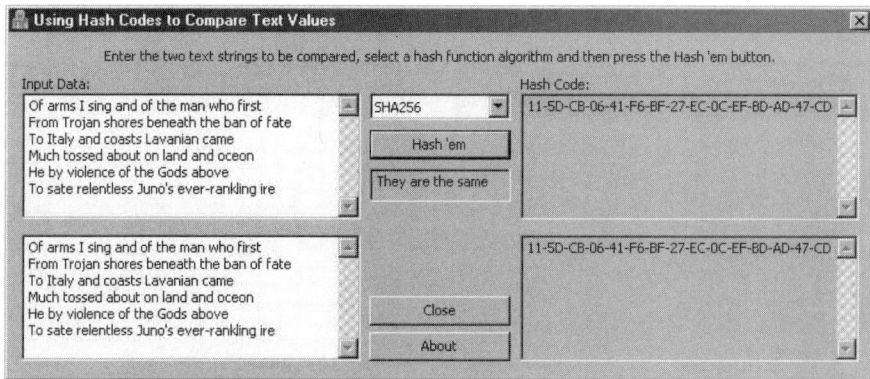

Figure 4-4 The HashDemo application illustrates how easy it is to generate and compare hash values to ensure that the data from which the hash values were created are identical.

Using Symmetric (Private) Key Encryption

As mentioned previously symmetric key encryption uses a single private key for both encryption and decryption. To provide support for this type of encryption, the BCL defines the System::Security::Cryptography:: SymmetricAlgorithm abstract class, which is the base class for all symmetric algorithm classes. There are wrapper classes for several popular symmetric cryptographic algorithms, all of which derive from this Symmetric Algorithm class:

- System::Security::Cryptography::DES
- System::Security::Cryptography::RC2
- System::Security::Cryptography::Rijndael
- System::Security::Cryptography::TripleDES

Essentially each of those classes are abstract base classes for their respective algorithms, and any class that implements any of those algorithms must derive from these classes. For example, a class called DESCrypto ServiceProvider (derived from the DES class) provides a wrapper for the DES implementation provided by the Cryptographic service provider.

Private Key Encrypting/Decrypting Demo Using DES

The CryptoStream class is used to perform cryptographic transformations on any data stream and is under the System::Security::Cryptography namespace. It derives from System::IO::Stream, and thus we can call any Stream methods on a CryptoStream object just as if it were a network or file stream object. The CryptoStream constructor is as follows:

```
CryptoStream(
  Stream* stream,
  ICryptoTransform* transform,
  CryptoStreamMode mode
);
```

The first parameter is a stream object that may be a file stream, a memory stream, or a network stream, and it's on this stream that the cryptographic transformation is performed.

The second argument is an ICryptoTransform object that defines the cryptographic transform that is to be performed on the stream. Any class that derives from the SymmetricAlgorithm has a CreateEncryptor method that returns an ICryptoTransform object. Thus, to perform a DES transform on the data stream, instantiate a DESCryptoService Provider object and call CreateEncryptor on it, or, if you want to perform a Rijndael transformation, then instantiate a RijndaelManaged object and call CreateEncryptor. In fact you may also pass any class derived from the HashAlgorithm class, because they will all be implementing the ICryptoTransform interface, though the issue would be that hash algorithms are not key-based, and thus the security of the encryption would be considerably reduced.

The third and last parameter for the CryptoStream constructor is a CrytopStreamMode enumeration value that specifies either read mode (CryptoStreamMode::Read) or write mode (CryptoStreamMode::Write).

Now let's look at what it takes to encrypt a file. We'll look at a function with the following syntax:

```
void EncryptFile(
  String* inputFile,
  String* outputFile
);
```

As you can see, this file takes only two parameters—an input file name and an output file name. However, before I throw some code at you, let's look at the generic steps involved in encrypting a file and the details of how the `EncryptFile` accomplished each given step.

1. **Acquire a key value:** One of the first tasks is to obtain a key value used to encrypt the file. In the case of the `EncryptFile` function, this value is hard-coded in the function in the form of a `String` object. Since the `Cryptography` class works with byte arrays, this value is then converted to a byte array using the incredibly useful `System::Text::Encoding` class. (You should note that while this function has a hard-coded key value for the sake of simplicity, you might want to add this value to the parameter list.)

2. **Construct the `FileStream` objects for input and output:** As you learned in Chapter 3, the `FileStream` object allows a generic means of reading and writing data to physical files. The `Encrypt File` function first constructs a `FileStream` object for the specified input file with read access. Then the function constructs a `FileStream` object for the specified output file name with create access.

3. **Construct a `SymmetricAlgorithm`-derived object:** While you can obviously select whatever algorithm you need for your particular application, I've chosen the `DESCryptoServiceProvider` object to support the DES algorithm. (As you might guess from the class name, the `DESCryptoServiceProvider` is an encapsulation of the CryptoAPI support for DES.)

4. **Construct a symmetric encryptor object with the specified key and IV:** All `SymmetricAlgorithm`-derived classes must implement the pure virtual `CreateEncryptor` function that takes key and IV parameters. This function returns an `ICryptoTransform` interface.

5. **Construct a `CryptoStream` object:** At this point, we have two of the main ingredients needed to encrypt a file: a `FileStream` object for the output and an `ICryptoTransform` interface. These two objects (along with a `CryptoStreamMode` enumeration value of `Write`) are then passed to the `CryptoStream` constructor. As you might imagine from the discussion of streams in Chapter 3, the `CryptoStream` class handles all the details of how to output the data (in the specified encrypted form) to the indicated output file name.

6. **Read input data:** Obviously this is application-specific, but, as mentioned in Step 2, the `EncryptFile` function presented here attaches a `FileStream` to the specified input file name. After construction of the `CryptoStream` object, the function simply reads through the input `FileStream` object, reading a block of 128 bytes at a time.

7. **Write data to the `CryptoStream`:** As each of the 128-byte blocks are read, they are output to the `CryptoStream` using the `Crypto Stream::Write` method. Internally, the `CryptoStream` object encrypts the data in the chosen format and outputs it to the underlying physical files.

8. **Close all opened `stream` objects:** In the case of the `Encrypt File` function, those would be the input and output `FileStream` objects and the `CryptoStream` object.

That's it. Eight simple steps (about 50 lines of C++ code) to do something as advanced as taking an input file, opening it, reading every byte, encrypting every byte, and then writing the encrypted data to disk. At this point, let's see the actual C++ code for the `EncryptFile` (Listing 4–1). (Note that the comments in the code are in bold to make it easier to tie the aforementioned file-encryption steps to the actual implementation code.)

Listing 4-1 The `EncryptFile` function outputs encrypted data from a specified input file to a specified output using a supplied private key.

```
void EncryptFile(String* inputFileName, String* outputFileName)
{
#pragma push_macro("new")
#undef new
  CryptoStream *cryptoStream;
  FileStream* outputFileStream;
  FileStream* inputFileStream;

  try
  {
    // Step #1 : Acquire a keyByteArray value - in this case hard-coded.
    String* keyString = S"KeyAbcGG";
    Byte keyByteArray[] = Text::Encoding::Default->GetBytes(keyString);
```

```
// Step #2: Construct the FileStream objects for input and output
inputFileStream = new FileStream(inputFileName,
                                    FileMode::Open,
                                    FileAccess::Read);

outputFileStream  = new FileStream(outputFileName,
                                       FileMode::Create,
                                       FileAccess::Write);

// Step #3: Construct a SymmetricAlgorithm-derived object
DESCryptoServiceProvider *serviceProvider =
      new DESCryptoServiceProvider();

// Step #4: Construct a symmetric encryptor object with the
// specified key and IV
ICryptoTransform* encryptor =
   serviceProvider->CreateEncryptor(keyByteArray,keyByteArray);

// Step #5: Construct a CryptoStream object
cryptoStream = new CryptoStream(outputFileStream,
                                   encryptor, CryptoStreamMode::Write);

Byte bytesread[] = new Byte[129];

// Step #6: Read input data
while(int n = inputFileStream->Read(bytesread, 0, 128))
{
    // Step #7: Write data to the CryptoStream
    cryptoStream->Write(bytesread, 0, n);
}
}
catch(Exception* e)
{
  throw e;
}
__finally
{
  // Step 8: Close all opened Stream objects
  cryptoStream->Close();
  outputFileStream->Close();
  inputFileStream->Close();
```

```
   }
#pragma pop_macro("new")
}
```

Now, let's turn our attention to the task of decrypting data. For this operation, I've included a function called `DecryptFile`:

```
void DecryptFile(
   String* encryptedfile,
   String* decryptedfile
)
```

Once again, I'll first list the basic steps of decrypting a file and outputting the results to another file and then present an implementation of that list in the form of the `DecryptFile` function.

1. **Acquire a key value:** As with encrypting a file using symmetric key encryption, a private key is needed to decrypt the file. Obviously, the key used to decrypt the file must match exactly the one that was used to encrypt. As with the `EncryptFile` function, I've hard-coded the value of the key in the `DecryptFile` for the sake of simplicity. (You'll want to change this key in your production environment.) This value is converted from its `String` object form to a byte array suitable for consumption by the BCL cryptography classes.

2. **Construct the `FileStream` objects for input and output:** As mentioned earlier in this section the `CryptoStream` object that will be used to decrypt the data can be constructed from a `Stream`—including a file-based `FileStream`. Since our input is a physical file, we must first construct a `FileStream` using the specified input file name. At the same time, another `FileStream` object is created for the output.

3. **Construct a `SymmetricAlgorithm`-derived object:** Obviously the algorithm used when decrypting a file must match the algorithm used when that file was encrypted. Since the `Encrypt File` function used the BCL wrapper for the CryptoAPI implementation of DES encryption, the `DecryptFile` function utilizes the `DESCryptoServiceProvider` here.

4. **Construct a symmetric decryptor object with the specified key and IV:** All `SymmetricAlgorithm`-derived classes must

implement the pure virtual `CreateDecryptor` function that takes key and IV parameters. This function returns an `ICryptoTransform` interface that will do the actual data decryption.

5. **Construct a `CryptoStream` object:** At this point, we have two of the main ingredients needed to decrypt a file—a `FileStream` object for the input and an `ICryptoTransform` interface. These two objects (along with a `CryptoStreamMode` enumeration value of `Read`) are then passed to the `CryptoStream` constructor.

6. **Read input data:** Utilizing the `CryptoStream::Read` method, data is read 128 bytes at a time into a `Byte` array. Once again, there is no decryption work on your end, as the transformer interface is doing the heavy lifting for you.

7. **Write data to the `CryptoStream`:** This is obviously application-specific, but in the case of the `DecryptFile` function, each 128-byte block that is read via the `CryptoStream` is written to a `FileStream` attached to a physical file whose name was passed as a parameter to the function.

8. **Close all opened `stream` objects:** In the case of the `Decrypt File` function, those would be the input and output `FileStream` objects and the `CryptoStream` object.

Once again, we have eight easy-to-follow steps for reading data from a file, decrypting it, and writing it to another file. Listing 4–2 shows the code for the `DecryptFile` (again, the comments are in bold so that you can match these steps with the actual C++ implementation).

Listing 4-2 The `DecryptFile` function takes input from a physical file and outputs DES decrypted data to a specified file name using a supplied private key.

```
void DecryptFile(String* inputFileName, String* outputFileName)
{
#pragma push_macro("new")
#undef new
  CryptoStream *cryptoStream;
  FileStream* outputFileStream;
  FileStream* inputFileStream;

  try
  {
```

```
// Step #1: Acquire a keyByteArray value - in this case hard-coded.
String* keyString = S"KeyAbcGG";
Byte keyByteArray[] = Text::Encoding::Default->GetBytes(keyString);


// Step #2: Construct the FileStream objects for input and output
inputFileStream = new FileStream(inputFileName,
                                 FileMode::Open,
                                 FileAccess::Read);

outputFileStream = new FileStream(outputFileName,
                                  FileMode::Create,
                                  FileAccess::Write);


// Step #3: Construct a SymmetricAlgorithm-derived object
DESCryptoServiceProvider* serviceProvider =
   new DESCryptoServiceProvider();


// Step #4: Construct a symmetric decryptor object with the
// specified key and IV
ICryptoTransform* decryptor =
   serviceProvider->CreateDecryptor(keyByteArray, keyByteArray);


// Step #5: Construct a CryptoStream object
cryptoStream = new CryptoStream(inputFileStream,
                                decryptor,
                                CryptoStreamMode::Read);


Byte bytesRead[] = new Byte[129];


// Step #6: Read input data
while(int n = cryptoStream->Read(bytesRead, 0, 128))
{
    // Step #7: Write data to the CryptoStream
    outputFileStream->Write(bytesRead,0,n);
  }
}
catch(Exception* e)
{
  throw e;
}
```

```
__finally
{
    // Step 8: Close all opened Stream objects
    cryptoStream->Close();
    outputFileStream->Close();
    inputFileStream->Close();
}
#pragma pop_macro("new")
}
```

It's one thing to present an allegedly helpful function (or two in this case), but let's take another step and see these functions at work in a demo application that allows you to both encrypt and decrypt files using a private key encryption algorithm—specifically, DES.

1. Create an MFC dialog-based application called DESDemo.
2. Update the Project properties to support Managed Extensions.
3. Open the stdafx.h file and add the following .NET support directives to the end of the file:

   ```
   #using <mscorlib.dll>
   #using <System.Windows.Forms.dll>
   #using <System.dll>
   using namespace System;
   using namespace System::Windows::Forms;
   using namespace System::IO;
   using namespace System::Security::Cryptography;
   #undef MessageBox
   ```

4. Open the dialog template resource and add the controls as you see them in Figure 4–5. (Note that the control on the bottom—below the "Status:" static text—is a list box control.)
5. For the larger edit control (below the "Result:" static text), set the ReadOnly property to True and the AutoHScroll property to False.
6. Create the DDX member variables as shown in Table 4–3.
7. Define the following member variable in the CDESDemoDlg class.

   ```
   class CDESDemoDlg : public CDialog
   {
       . . .
       CString m_strWorkingDir;
   ```

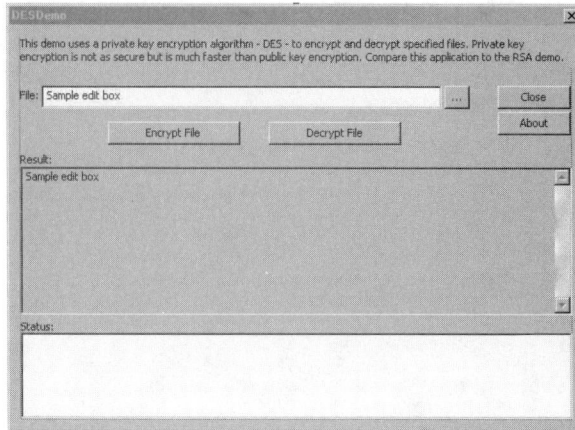

Figure 4-5 Dialog resource for the DESDemo demo application

Table 4-3 DDX Variables for the DESDemo Demo

Control	Variable Type	Variable Name
File name edit	CString	m_strFileName
File name edit	CEdit	m_edtFileName
Result edit	CString	m_strFileContents
Status list box	CList box	m_lbxStatus

8. Add the following code to the OnInitDialog function just before the return statement.

```
BOOL CDESDemoDlg::OnInitDialog()
{
    ...
    TCHAR buff[MAX_PATH];
    GetModuleFileName(NULL, buff, MAX_PATH);
```

```
m_strWorkingDir = System::IO::Path::GetDirectoryName(buff);
return TRUE;   // return TRUE  unless you set the focus to a
control
}
```

9. At this point, code the button with the ellipsis ("...") on it. This will allow users to use the Windows File Open dialog box to locate their files to encrypt and decrypt:

```
void CDESDemoDlg::OnBnClickedFindFile()
{
UpdateData();

CFileDialog dlg(TRUE);
dlg.m_ofn.lpstrInitialDir = m_strWorkingDir;
dlg.m_ofn.lpstrFilter =
  _T("Text Files (*.txt)\0*.txt\0"
  "Encrypted Files (*.enc)\0*.enc\0"
  "Decrypted Files (*.dec)\0*.dec\0"
  "All Files (*.*)\0*.*\0"
);

if (IDOK == dlg.DoModal())
{
  // Update the dialog with the selected file.
  m_strFileName = dlg.GetPathName();
  UpdateData(FALSE);

  // Common trick to move the cursor to the end
  // of the edit control so that the user can
  // see the file name in a long name.
  m_edtFileName.SetSel(m_strFileName.GetLength(),
                       m_strFileName.GetLength(),
                       FALSE);
}
}
```

10. Insert the `EncryptFile` and `DecryptFile` functions from Listings 4–1 and 4–2, respectively. These functions have been intentionally coded in such a way that they can be dropped into any mixed mode (MFC/MC++) application.

11. Now it's time to use these functions. First add a handler for the Encrypt File button and code it as follows. In a full-production

system, you would want to add more error-handling and centralize things like the hard-coded extensions I've chosen for this application (.enc and .dec for encrypted and decrypted files, respectively). However, I've tried to keep things as simple as possible here while still providing a practical, usable demo.

As you can see, this function's main purpose in life is to determine the output file name *(the original file is never overwritten)* from the input file name supplied by the user, to call the Encrypt File function, and, upon return, to open and display the encrypted file. In addition, status messages are inserted into the list box in reverse order.

Note that the input file name is automatically changed to the output name, as the next logical step for the user might be to decrypt the newly created file to test the encryption/decryption round trip.

```
void CDESDemoDlg::OnBnClickedEncryptFile()
{
#pragma push_macro("new")
#undef new
  CWaitCursor wc;


  UpdateData();
  if (0 < m_strFileName.GetLength())
  {
    FileStream* pFS;
    try
    {
      CString strStatus;
      m_strFileContents = _T("");

      // If extension ends in .enc, replace with .dec
      // to prevent continual enc/dec of same file resulting
      // in names like file.txt.enc.dec.enc.dec...
      // Otherwise just add .dec
      CString strOutputFile = m_strFileName;
      if (0 == strOutputFile.Right(4).CompareNoCase(
        _T(".dec")))
      {
        strOutputFile =
```

```
      strOutputFile.Left(strOutputFile.GetLength() - 4)
      + _T(".enc");
}
else
{
  strOutputFile += _T(".enc");
}


// Call EncryptFile function passing just-formatted
// output name
strStatus.Format(_T("Encrypting file %s"),
  m_strFileName);
EncryptFile(m_strFileName, strOutputFile);


strStatus.Format(_T("%s successfully encrypted"
                    "to file %s"),
                  m_strFileName,
                  strOutputFile);
m_lbxStatus.InsertString(0, strStatus);


// Using a FileStream open, read and display newly
// encrypted file in hex format.
m_lbxStatus.InsertString(0, _T("Attempting to"
  "read newly encrypted file"));


pFS = new FileStream(strOutputFile,
                     FileMode::Open,
                     FileAccess::Read);


if (pFS->CanRead)
{
  Byte buffer __gc[] = new Byte __gc[pFS->Length];
  pFS->Read(buffer, 0, buffer->Length);


  for (int i = 0; i < buffer->Length; i++)
  {
    CString c;
    c.Format(_T("%02x "), buffer[i]);
    m_strFileContents += c;
  }
  m_lbxStatus.InsertString(0, _T("Data read back"
                                 "and displayed"));
```

```
        m_strFileName = strOutputFile;
      }
    }
    catch(Exception* e)
      {
        m_lbxStatus.InsertString(0, (CString)e->Message);
      }
      __finally
      {
        pFS->Close();
      }

      UpdateData(FALSE);
    }
    else
    {
      MessageBox::Show(S"You must supply an input file name.");
    }
#pragma pop_macro("new")
}
```

12. Now, add a handler for the Decrypt File button and code it as follows. Much like the reciprocal Encrypt File button handler, this function's tasks are to determine the output file name from the input file name supplied by the user (so that the original file is not lost), call the DecryptFile function and, upon returning, open and display the decrypted file. In addition, status messages are inserted into the list box in reverse order.

```
void CDESDemoDlg::OnBnClickedDecryptFile()
{
#pragma push_macro("new")
#undef new
  CWaitCursor wc;

  UpdateData();
  if (0 < m_strFileName.GetLength())
  {
    StreamReader* pSR;
    try
    {
      CString strStatus;
      m_strFileContents = _T("");
```

```
// If extension ends in .enc, replace with .dec
// to prevent continual enc/dec of same file resulting
// in names like file.txt.enc.dec.enc.dec...
// Otherwise just add .dec
CString strOutputFile = m_strFileName;
if (0 == strOutputFile.Right(4).CompareNoCase(
  _T(".enc")))
{
  strOutputFile =
    strOutputFile.Left(strOutputFile.GetLength() - 4)
    + _T(".dec");
}
else
{
  strOutputFile += _T(".dec");
}

// Call DecryptFile function passing just-formatted
// output name
strStatus.Format(_T("Decrypting file %s"),
  m_strFileName);
DecryptFile(m_strFileName, strOutputFile);

strStatus.Format(_T("%s successfully decrypted to file
  %s"),
                m_strFileName,
                strOutputFile);
m_lbxStatus.InsertString(0, strStatus);

m_lbxStatus.InsertString(0, _T("Attempting to read "
  "newly decrypted file"));

// Using a StreamReader - open, read and display
// newly decrypted file.
pSR = new StreamReader(strOutputFile);

CString strCurrLine;

while (0 < pSR->Peek())
{
  strCurrLine = pSR->ReadLine();
  m_strFileContents += strCurrLine;
```

```
          m_strFileContents += _T("\r\n");
       }

       m_lbxStatus.InsertString(0, _T("Data read back"
                                      "and displayed"));
       m_strFileName = strOutputFile;
    }
    catch(Exception* e)
    {
       m_lbxStatus.InsertString(0, (CString)e->Message);
    }
    __finally
    {
       pSR->Close();
    }
    UpdateData(FALSE);
  }
  else
  {
    MessageBox::Show(S"You must supply an input file name.");
  }
#pragma pop_macro("new")
}
```

13. That's all there is to writing a simple encryption/decryption file using these powerful classes. Now run the application and supply an input file name to encrypt. Once you've done that, decrypt the same file to see the original file's contents. You can use the status list box in case of a problem—for example, the specified input file does not exist or you don't have write privileges for the output file. Your results should be similar to those shown in Figures 4–6 and 4–7.

Now that you've seen how private key encryption works, let's have a look at public key encryption.

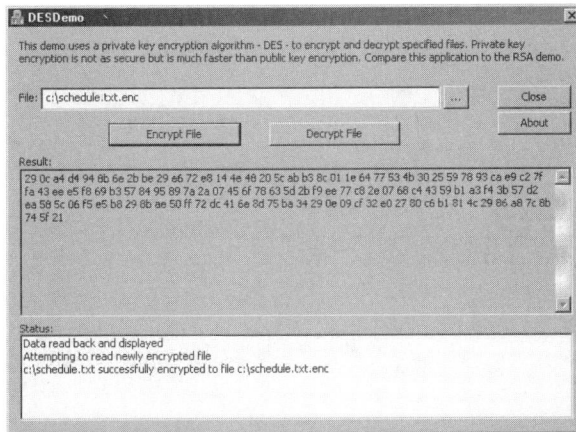

Figure 4–6 Example of encrypting a file using the CD's `DESDemo` demo

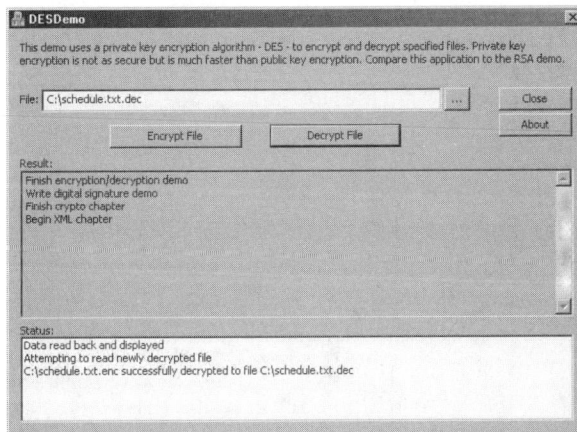

Figure 4–7 Example of decrypting an encrypted file using the CD's `DESDemo` demo

Using Asymmetric (Public) Key Encryption

Asymmetric key encryption makes use of two distinct keys: a private key and a public key. Both keys are generated at the same time and are mathematically linked. The private key is kept with the key's owner (typically the data's intended recipient), and the public key is given out to entities having authorization to encrypt and send data to the recipient. As the two keys are

intrinsically matched, any data encrypted using the private key can only be decrypted using its corresponding public key, and any data encrypted using the public key can only be decrypted using the private key. Two of the most popular public key algorithms are RSA and DSA, of which DSA is used for making digital signatures and not for encryption/decryption. RSA is used both for digital signatures as well as for data encryption/decryption.

The BCL provides the abstract `AsymmetricAlgorithm` class, which is the base class for all public key algorithm classes. The BCL also provides two more abstract classes—both of which derive from the `Asymmetric Algorithm` class: the DSA class and the RSA class, which provide abstract definitions for DSA and RSA algorithms. Any class that implements one of the two algorithms must derive from the corresponding class. Finally, the framework library provides us with the DSA-derived `DSACryptoService Provider` class as well as the RSA-derived `RSACryptoServiceProvider` class—both of which wrap the corresponding CryptoAPI implementation. Since the DSA algorithm is used only for data signing I'll cover that class in the next section, "Using Digital Signatures." Therefore, in this section I'll be using the `RSACryptoServiceProvider` to demonstrate how to encrypt and decrypt data using a public key algorithm.

One major difference between public key cryptography and private key cryptography is that in public key cryptography, streams cannot be used as input or output by the BCL classes. The input for an encryption routine is always a block of data that has a maximum length—depending on the algorithm used—and the returned value is also a fixed-length block of encrypted data. In the case of the RSA algorithm this length is the size of the encryption key divided by 8. Therefore, to encrypt a large amount of data, we must break it up into blocks whose length does not exceed the maximum allowed input block length. Similarly, while decrypting, we need to read blocks of data whose length is equal to the key size divided by 8.

As mentioned earlier, public key encryption/decryption is much slower than private key encryption/decryption. However, the following uses of private key encryption are widely employed (sometimes in conjunction with one another):

- If a large amount of data is being sent, that data is encrypted using a private key algorithm. The private key is then encrypted using a public key algorithm.
- A hash code is generated from the data. The hash code is encrypted using a public key. Both are sent to the recipient. After the recipient decrypts the hash code, another hash code (using the same hash

code algorithm) is generated from the data. The two hash codes are then compared. If they are equal, the data arrived unmodified. If not, we have trouble right here in River City. (For those of you not familiar with the musical *The Music Man*, this means it's not a good thing.)

In the process of constructing a RSACryptoServiceProvider object, a random public/private key pair is automatically generated, which should be persisted (typically to a disk file) in order to later decrypt data that has been encrypted using this key pair. Of course, the logical question would be, "If I lose the key pair, can I restore my data?" The only way to do that would be to either crack the RSA encryption algorithm or run a brute-force tool on the data for countless years. So, essentially the answer is that the data would be lost—which is how secure you want your data to be from someone who doesn't have the proper key anyway.

The RSACryptoServiceProvider class overrides two principal methods derived from its RSA base class—ToXmlString and FromXmlString. These methods enable us to acquire an XML string representation of the RSA object and reconstruct an RSA object from XML, respectively. We can extract either the public key alone, which we can then distribute to other people; or we can extract a private/public key pair, which we must keep private to our machine.

Writing a Class to Encrypt and Decrypt Data Using RSA

Mainly because the RSA class members don't take Stream objects as parameters, there's a lot of data munging that needs to take place in order to encrypt and decrypt data. There's also the issue of creating, saving, and restoring keys, and, of course, maintaining state for an application using any of the RSA classes. Therefore, in this section you'll see how to write a fully functional RSA encrypting/decrypting class capable of saving either its public key or a public/private key pair, reconstructing itself from the saved keys, encrypting data using the RSA algorithm, and decrypting using the same.

1. Create a new file called TextCrypt.h.
2. Within the TextCrypt.h file, declare the following directives. Inserting these directives into this class's header file makes it more mobile—enabling you to insert this class easily into any of your MFC projects.

```
#pragma once

#using <mscorlib.dll>
#include <tchar.h>
#include <atlstr.h>

using namespace System;
using namespace IO;
using namespace Security::Cryptography;
using namespace System::Text;
```

3. Now define a managed class called TextCrypt as follows. At this point, we're simply defining the class, its main member (the RSA CryptoServiceProvider), and the public constructor. Remember that the mere act of instantiating the RSACryptoService Provider will generate the private/public key pair.

```
__gc class TextCrypt
{
public:
  TextCrypt()
  {
    rsaServiceProvider = new RSACryptoServiceProvider();
  }

protected:
  RSACryptoServiceProvider* rsaServiceProvider;
};
```

4. Insert the following (public) method into the TextCrypt class's definition. As mentioned before the start of this demo, the RSA ServiceProvider provides a means of retrieving either the public key or the public/private key pair (you cannot retrieve only the private key). In order to retrieve only the public key, the ToXML String method is called with a parameter of false. This method returns a String object containing an XML representation of the public key value. At that point, the key is converted to a Byte array that can be written to a file using the FileStream class.

```
//Saves the public key to a specified file.
void SavePublicKey(CString fileName)
{
    FileStream* stream;
    try
```

```
    {
        // Retrieve key value in XML format
        String* publicKey = rsaServiceProvider->
            ToXmlString(false);

        // Convert String to byte array for FileStream::Write
        // method
        Byte publicKeyBytes[] =
            Encoding::Default->GetBytes(publicKey);

        // Construct a FileStream object and write out public
        // key
        stream = new FileStream(fileName,
            FileMode::OpenOrCreate);
        stream->Write(publicKeyBytes, 0,
                        publicKeyBytes->Length);
    }
    catch(Exception* e)
    {
        throw e;
    }
    __finally
    {
        stream->Close();
    }
}
```

5. The next obvious `TextCrypt` method to implement is the public `SaveKeys` method to save both the private and public keys of the class's `RSAServiceProvider` member. The only code difference between this method and the `SavePublicKey` method is that the `SaveKeys` method passes a value of `true` to the `RSAService Provider::ToXMLString` method so that both keys are returned in XML format. These keys are converted to a `Byte` array and saved to the specified file.

```
//Saves both the public and private key to a specified file.
void SaveKeys(CString fileName)
{
    FileStream* stream;
    try
```

```
    {
        // Retrieve key value in XML format
        String* keys = rsaServiceProvider->ToXmlString(true);

        // Convert String to byte array for FileStream::Write
        // method
        Byte keyBytes[] = Encoding::Default->GetBytes(keys);

        // Construct a FileStream object and write out public
        // key
        stream = new FileStream(fileName, FileMode::Create);
        stream->Write(keyBytes, 0, keyBytes->Length);
    }
    catch(Exception* e)
    {
        throw e;
    }
    __finally
    {
        stream->Close();
    }
}
```

6. At this point, our nifty RSA-based text encryptor class can save its generated keys. Now let's add a method to this class to generate an RSA object from the XML-formatted keys. This method is called after data has been encrypted and the keys saved in order to reconstruct the appropriate RSA object to decrypt the data.

```
 //Construct RSA object from key file
void FromKeys(CString fileName)
{
    FileStream* stream;
    try
    {
        // Construct a FileStream object based on the file
        // that contains the saved key(s)
        stream = new FileStream(fileName, FileMode::Open);

        // Read the key data (saved in XML) format
        // into a local buffer
        Byte keyBytes[] = new Byte[(int)stream->Length];
        stream->Read(keyBytes, 0, (int)stream->Length);
```

```
    // Convert the buffer to a String object
    String* keys = Encoding::Default->GetString(
      keyBytes);
    rsaServiceProvider->FromXmlString(keys);
}
catch(Exception* e)
{
  throw e;
}
  __finally
{
  stream->Close();
}

}
```

7. Now code the `TextCrypt` method to encrypt data and store the resulting cipher text in a specified file. As you can see, our `EncryptStringToFile` method takes as a `String` object the data to be encrypted and another `String` object representing the file name that upon return from this method will contain the encrypted data.

For encryption purposes you would only use the public key supplied by the recipient, but the recipient would obviously need to have both private and public keys in order to decrypt the data successfully.

Another thing to note here is all the conversions taking place. These conversions are one reason that this algorithm is much slower than a symmetric key encryption algorithm.

```
// Encrypt data and store in specified file
void EncryptStringToFile(CString str, CString fileName)
{
    FileStream* stream;
    try
    {
      // Convert data to a byte array
      String *s1 = new String(str);
      Byte inputBytes[] = Encoding::Default->GetBytes(str);

      // Construct a FileStream object that will
      // contain the newly encrypted data
      stream = new FileStream(fileName, FileMode::Create);
```

```
// Define two byte arrays.
//   #1: used to send 8 bytes at a time to the RSA
//   object
//   #2: used to write to the FileStream object
Byte outputBytes[];

// Read through input byte array 8 bytes at a time
for(int i = 0; i < inputBytes->Length; i += 8)
{
   // Verify that we set the length properly
   // when retrieving the last bytes from the array
   // which might be less than 8 bytes
    int len = 8;
    if ((inputBytes->Length - i) < 8)
       len = inputBytes->Length - i;

   // Copy the data into the byte array to pass
   // to the encryption method
    Byte blockToEncrypt[] = new Byte[len];
    Array::Copy(inputBytes, i, blockToEncrypt, 0, len);

   // Call the encryption method. Returns the encrypted
   // data
    outputBytes = rsaServiceProvider->
       Encrypt(blockToEncrypt,false);

   // Write the encrypted data to the FileStream
    stream->Write(outputBytes, 0, outputBytes->Length);
  }
}
catch(Exception* e)
{
   throw e;
}
__finally
{
   // Close the FileStream
   stream->Close();
}
}
```

8. One last method and our `TextCrypt` class is complete. Now implement the following decrypt method (`DecryptFileToString`). As

you can see from the method signature, the method takes two parameters. The first is a String object containing the file name— this file should contain the cipher text you're attempting to decrypt. The second parameter is a String object that, upon return from this method, will contain the decrypted data.

Since the TextCrypt::EncryptStringToFile encrypts bytes in 8-byte blocks, we know that each decrypted block will be 8 bytes or less. Once again, note all the conversions taking place. If we had tried to mix this code with unmanaged code (and incurred the performance penalty of yet more conversions) the extra data conversions required would slow down an already slow algorithm to unacceptable levels.

```
// Decrypt data from the specified file into a String
   void DecryptFileToString(CString fileName, CString& str)
{
  FileStream* stream;
  try
  {
    // Initialize the string that will hold the end
    // result of this operation
    str = S"";

    // Determine the byte size of the key. As the KeySize
    // property returns the size in bits, we need to divide
    // that number by 8
    int blockSize = rsaServiceProvider->KeySize / 8;

    // Construct a FileStream object that will
    // contain the newly encrypted data
    stream = new FileStream(fileName, FileMode::Open);

    // Declare byte array capable of holding
    // FileStream object's data
    Byte inputBytes[] = new Byte[(int)stream->Length];

    // Read data into input buffer and close
    // FileStream
    stream->Read(inputBytes, 0, (int)stream->Length);
```

```
            // Acquire an array of bytes equal to the block size
                    Byte inputBlock[] = new Byte[blockSize];

            // for every <blockSize> amount of data in the
            // input buffer...
            for(int i=0; i < inputBytes->Length; i += blockSize)
            {
                // . . . copy that amount of data from the
                // input byte array into the input block to
                // be decrypted
                Array::Copy(inputBytes, i, inputBlock, 0, blockSize);

                // Call the Decrypt method, which returns the
                // decrypted data in a byte array
                Byte decryptedBytes[] =
                  rsaServiceProvider->Decrypt(inputBlock, false);

                // Convert the decrypted byte array to a string value
                String* s1 = Encoding::Default->
                  GetString(decryptedBytes);

                // Update the string containing the total bytes
                // decrypted
                str = str + s1;
            }
        }
        catch(Exception* e)
        {
                    throw e;
        }
        __finally
        {
                    stream->Close();
        }
    }
```

That's all there is to it. You now have a fully functional RSA class that can be used in any of your applications. We'll put it to the test in the next section's demo—RSADemo.

Public Key Encrypting/Decrypting Demo Using RSA

As you've seen throughout this chapter, nothing beats rolling up your sleeves and jumping into a demo when dealing with something like cryptography. Therefore, in this section we'll create an application to show the RSA algorithm in action and, at the same time, test our TextCrypt class. It will also allow you to compare the speed of this public key encryption technique with the speed of the private key encryption technique used in the earlier DESDemo application. As far as trying to crack the two encrypted data sets to prove that public key encryption is more secure—I'll leave that one as an exercise for the reader.

1. Create an MFC dialog-based application called RSADemo.
2. Update the Project properties to support managed extensions
3. Open the stdafx.h file and add the following Managed Extensions directives: You'll note that I'm not including the System:: Security::Cryptography namespace as in the previous demos. That is because the cryptography (encryption and decryption) for this demo is isolated to the TextCrypt class you saw in the previous section, which will be inserted into this project shortly.

```
#using <mscorlib.dll>
#using <System.Windows.Forms.dll>
#using <System.dll>
using namespace System;
using namespace System..Windows::Forms;
using namespace System::IO;
#undef MessageBox
```

4. Open the dialog template resource and add the controls as you see them in Figure 4–8. (Note that the control on the bottom—below the "Status:" static text—is a list box control.)
5. Set the Sort property of the ListBox control to False.
6. Create the DDX member variables as shown in Table 4–4.
7. Add the TextCrypt.h file to the RSADemo project and insert an include directive for the header file in the CRSADemoDlg.h file.

In the chapter's DESDemo, the functions that perform the encryption and decryption contain a mixture of both managed and unmanaged code. However, in the RSADemo, a managed class (TextCrypt) is solely responsible for these tasks. The benefit to isolating the RSA code in a managed class is that it avoids a lot of managed/unmanaged data conversions and results in a more efficient application—something desperately needed when using a slow algorithm such as RSA.

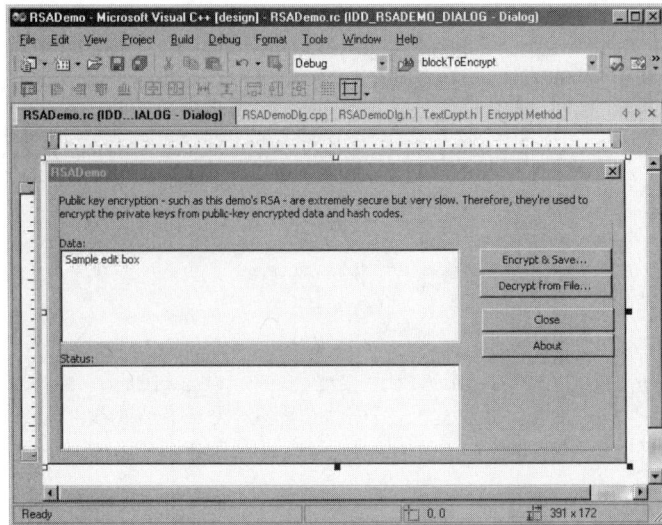

Figure 4-8 Dialog resource for the RSADemo application

8. As you learned in Chapter 1, you must use the gcroot template in order to define a managed member inside an unmanaged class. Therefore, open the RSADemoDlg.h file and declare the following CRSADemoDlg member.

```
class CRSADemoDlg : public CDialog
{
    . . .

protected:
    gcroot<TextCrypt*> m_pTextCrypt;

    . . .
};
```

Table 4-4 DDX Variables for the RSADemo demo

Control	Variable Type	Variable Name
File name edit	CString	m_strInput
Status list box	CList box	m_lbxStatus

9. Add the following members to the CRSADemoDlg class. The first is the declaration for an initialization function that will be called to construct the dialog's embedded TextCrypt object. The second is simply the working directory of the application that is used as the default directory in saving and reading the encrypted files.

```
protected:
    void CRSADemoDlg::InitializeTextCryptObject();
    CString m_strWorkingDir;
    CString m_strPublicKeyFile;
    CString m_strKeyPairFile;
```

10. Add a call to the InitializeTextCryptObject to the end of the CRSADemoDlg::OnInitDialog function just before the return statement.

```
BOOL CRSADemoDlg::OnInitDialog()
{

    . . .



    InitializeTextCryptObject();
    return TRUE;   // return TRUE  unless you set the focus to a
      control
}
```

11. Implement the InitializeTextCryptObject function as follows. After constructing the dialog's TextCrypt object, this function uses a couple of BCL classes covered in Chapter 3, Path and File (see Table 3–4).

First the function uses the global AfxGetAppName, which returns the application's fully qualified name. That value is then passed to the Path class's GetDirectoryName method, which strips out the directory name for us. Next, the Path::Combine method is used to concatenate the key file names with the path. The benefit of using this method is that we don't need to parse for ending backslashes and append only if needed, and so on. The Path class takes care of all that.

Once the fully qualified key file names have been determined, the static File::Exists method is called to see if the key file exists. If it does not, new keys are generated.

```
void CRSADemoDlg::InitializeTextCryptObject()
{
```

```
#pragma push_macro("new")
#undef new
 try
 {
    //Instantiate the TextCrypt class
    m_pTextCrypt = new TextCrypt();
    // Check and see if the key files are found.
    TCHAR buff[MAX_PATH];
    GetModuleFileName(NULL, buff, MAX_PATH);
    m_strWorkingDir = System::IO::Path::GetDirectoryName(
      buff);

    m_strPublicKeyFile =
      (CString)Path::Combine(m_strWorkingDir,
        S"publickey.xml");
    m_strKeyPairFile =
      (CString)Path::Combine(m_strWorkingDir,
        S"keypair.xml");

    // If keys were not found...
     if (!File::Exists(m_strKeyPairFile))
    {
        // . . . create and save them
        m_pTextCrypt->SavePublicKey(m_strPublicKeyFile);
        m_pTextCrypt->SaveKeys(m_strKeyPairFile);
    }
 }
 catch(Exception* e)
 {
    MessageBox::Show(e->Message);
 }
#pragma pop_macro("new")
 }
```

12. Implement an event handler for the Encrypt button and code it as
    follows. After making sure that the user entered some data to be
    encrypted, this function displays a Save File common dialog box
    to allow the user to specify where and in what file the data will
    be encrypted. Once the user specifies a file name, the RSA object
    is constructed from the publickeys.xml file (created in the
    InitializeTextCryptObject function) and calls the TextCrypt::

EncryptString method, passing it the user's input data and the
file name. Status messages are inserted into the ListBox to keep
the user informed of what transpired.

```
void CRSADemoDlg::OnBnClickedEncryptFile()
{
  UpdateData();
  if (0 < m_strInput.GetLength())
  {
    try
    {
     CFileDialog dlg(FALSE);
     dlg.m_ofn.lpstrInitialDir = m_strWorkingDir;
     dlg.m_ofn.lpstrDefExt = _T("rsa");
     dlg.m_ofn.lpstrFilter = _T("RSA Encrypted Files "
       "(using TextCrypt) (*.rsa)\0*.rsa\0\0");
     dlg.m_ofn.lpstrTitle = _T("Save RSA Encrypted File");
     if (dlg.DoModal()==IDOK)
     {
       m_pTextCrypt->FromKeys(m_strPublicKeyFile);

       m_lbxStatus.InsertString(0, _T("Constructed RSA "
         "object from public key file"));

       CString strFileName = dlg.GetPathName();
       m_pTextCrypt->EncryptStringToFile(m_strInput,
         strFileName);

       CString strTemp;
       strTemp.Format(_T("Encrypted data to %s"),
                    Path::GetFileName(strFileName));
       m_lbxStatus.InsertString(0, strTemp);
     }
    }
    catch(Exception* e)
    {
      MessageBox::Show(e->Message);
    }
  }
  else
  {
    MessageBox::Show(S"Please enter the data you wish to
      encrypt.");
```

```
        }
    }
```

13. Finally, add an event handler for the Decrypt button. This function will display the Open File common dialog box; allowing the user to specify the file to be decrypted. Once the user has specified the file, the function constructs the RSA object from the previously saved key pair. The `TextCrypt::DecryptFileToString` function is then called, which takes the file name and returns the string value of its decrypted data. That data is then displayed on the dialog box. Status messages are also inserted in the dialog box's `ListBox` to keep the user informed of success or failure.

```cpp
void CRSADemoDlg::OnBnClickedDecryptFile()
{
  m_strInput = _T("");
  UpdateData(FALSE);

  CFileDialog dlg(TRUE);
  dlg.m_ofn.lpstrInitialDir = m_strWorkingDir;
  dlg.m_ofn.lpstrDefExt = _T("rsa");
  dlg.m_ofn.lpstrFilter = _T("RSA Encrypted Files "
    "(using TextCrypt) (*.rsa)\0*.rsa\0\0");
  dlg.m_ofn.lpstrTitle = _T("Open RSA Encrypted File");
  if (IDOK == dlg.DoModal())
  {
    try
    {
    CString strFileName = dlg.GetPathName();

    m_pTextCrypt->FromKeys(m_strKeyPairFile);
    m_lbxStatus.InsertString(0, _T("Constructed RSA object "
      "from public/private key pair file"));

    m_pTextCrypt->DecryptFileToString(strFileName,
      m_strInput);
    UpdateData(FALSE);

    m_lbxStatus.InsertString(0, _T("Decrypted and"
                                "displayed data"));
    }
    catch(Exception* e)
    {
```

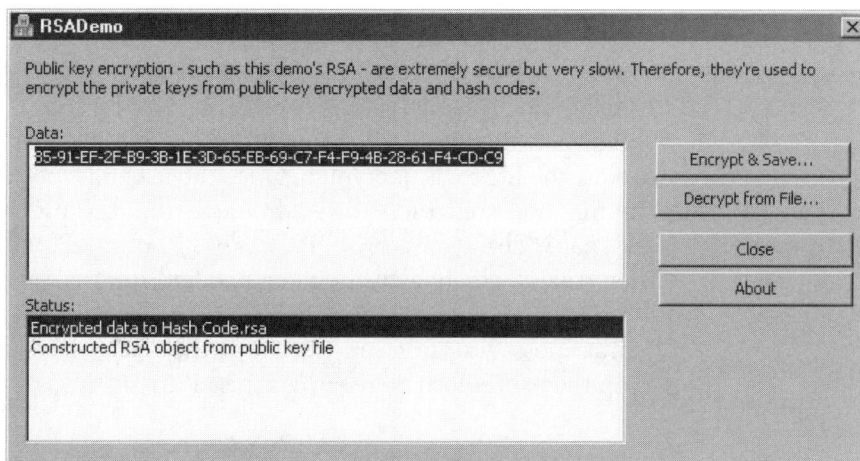

Figure 4–9 Public key algorithms such as RSA are very secure, but also very slow. As a result they are generally used to protect private keys to larger data sets and hash codes used to validate that received data has not been modified in transit.

```
        MessageBox::Show(e->Message);
    }
  }
}
```

14. At this point, you can build and run your RSA application. As you can see, it will allow you to enter data and have that data encrypted to the file of your choosing. You can then decrypt any file this application encrypts. Figure 4–9 shows an example of this application decrypting the hash code generated with this book's HashDemo application.

Combining Symmetric Encryption, Asymmetric Encryption, and Hash Codes

At this point, I've spoken generically about how symmetric and asymmetric encryption can be combined to present an efficient yet secure means of encrypting data. However, because the demos are written for a single user,

it might not be obvious *exactly* how this all ties together. Therefore, let's review some of the things you've read in this chapter:

- With *symmetric* encryption two parties each share a private key that is used both to encrypt and decrypt data. Algorithms using this technique are very fast, but not as secure as with asymmetric encryption. In addition, with symmetric encryption alone, there is no secure means by which the person generating the key can get it to the other person.
- With *asymmetric* encryption one person (typically the future recipient of the data) generates a pair of keys (one public and one private). The public one is sent out to anyone who would need to encrypt and send data to the recipient. The recipient's private key can then decrypt the file that was encrypted with the sender's public key. The problem is that this technique is too slow to be feasible for large amounts of data.

So now let's look at a sample scenario using the characters from before, Tom and Krista. Tom has a large amount of data to send to Krista. Krista needs this data sent in a secure fashion. They can't use symmetric encryption because Krista is at a client site in Australia, and Tom in the United States, so there's no secure way for one of them to generate a private key and get it to the other person. As you know by now, they can't use asymmetric encryption because it's too slow for a large amount of data.

In this section, I'll list the steps that are needed for Krista and Tom to exchange data in an efficient yet secure manner. First, take a look at Figure 4–10 which provides a UML Sequence Diagram of these steps. After the diagram, I'll list each step along with the optional ways of performing the steps. Where possible, I'll relate the step to the exact class or method you learned about in this chapter so that, when finished, you can easily play the role of either Tom or Krista in the real-life scenario of needing to exchange your own sensitive data with another party.

1. Since Krista will be receiving the data, she generates a public/private key pair using the asymmetric algorithm of her choice. We'll say she chose RSA. In order to use this algorithm, she could use the `RSACryptoServiceProvider` class you've learned about in this chapter to generate the needed keys.

   She could also plug this chapter's `TextCrypt` class into her code and call only two methods that write the key values in XML

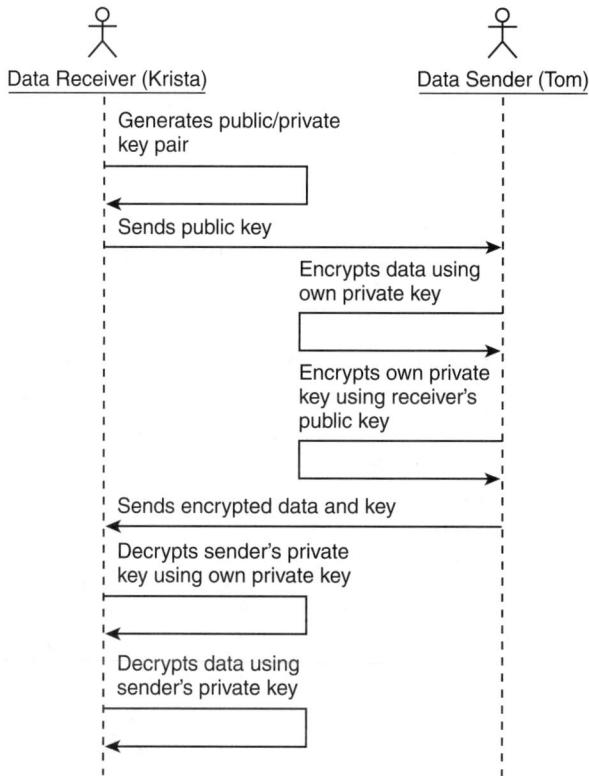

Figure 4-10 One technique used to ensure a secure, yet efficient exchange of sensitive data

format to disk—`TextCrypt::SaveKeys` outputs both the public and private key pair, and `SavePublicKey` outputs just the public key.

Finally, she could simply run this book's `GenerateAsymmKeys` application, which uses the `TextCrypt` class to create and output keys using any of several asymmetric algorithms.

2. Next Krista needs to send the public key to Tom. Once she has the public key in an XML formatted file, she can simply e-mail that to Tom. In fact, the `GenerateAsymmKeys` application has an E-mail button that uses MAPI so that you can e-mail the public key immediately after creating it.

There's no security concern because the only thing someone can do with this key is to encrypt data—not decrypt it. (Only

Krista's private key can decrypt the data.) It's analogous to someone having the number of a bank account that only accepts deposits.

3. Now it's Tom's turn. He needs to encrypt the large body of data to send to Krista. He does so by choosing a symmetric algorithm such as DES with a private key that he generates. I'll refer to this private key as Tom's private key so as not to confuse this key with Krista's private key.

   Tom could directly use the `DESCryptoServiceProvider` class himself or this chapter's `DESDemo` that will encrypt the file using a supplied key. (While the demo uses a hard-coded key, it would be trivial to add an edit control to the application's dialog box so that the application's users could specify any key value they desire.) The `DES Demo` on my Web site (`www.ArcherConsultingGroup.com`) actually allows you to generate random private keys that are GUIDs.)

4. At this point, Tom has encrypted the data he wants to send to Krista with his own private key (Tom's private key). However, he needs a secure way to get that private key to Krista. This is where Krista's public key comes in. Tom uses the public key to encrypt "Tom's private key." This can be done in one of at least three ways.

   He could directly use the `RSACryptoServiceProvider` class (remember that Krista chose RSA when generating the public key), or he could use the chapter's `TextCrypt` class. He only needs to call the `TextCrypt::FromKeys` method (passing it the name of the file containing the public key that Krista sent) followed by calling the `TextCrypt::StringToFile`. This will encrypt Tom's private key using Krista's public key.

   Finally, of course, he could simply run the chapter's `RSADemo`, copy and paste the public key into the dialog's box's edit control, click the Encrypt button, and specify an output file. It really just comes down to whether you want to use the utilities provided here or insert the code into your own application.

5. At this point, Tom has two encrypted files. One is a data file encrypted in Step 3 using Tom's private key. The other is an encrypted file containing Tom's private key. This file was encrypted using the public key provided by Krista (Steps 1 and 2). Now he needs only to send the data to Krista using whatever protocol they choose: FTP, SMTP, and so on.

6. Krista receives the data and decrypts Tom's private key using her private key.

    7. Once she has Tom's private key, she can finally decrypt the actual data file.

Now, we have a fairly secure system here until one day Krista states, "I'm really concerned about packet sniffing and the possibility that someone might intercept my data and change it before it arrives. How can I prevent that?" The answer is that you can't prevent it, but you can detect it using hash codes.

As you'll recall from the section entitled "Using Hash Codes," a hash code is a fixed-length byte sequence for an input binary string of arbitrary length. No two input strings will give the same hash value unless they are identical. Using the hash function of their choosing (such as MD5, SHA1, SHA256), Tom and Krista would now perform the following steps in conjunction with the above steps:

    1. Tom would generate a hash code based on the data (before encrypting it in Step 3). He could do this using the appropriate `HashAlgorithm`-derived class, or he could simply run this chapter's `CreateHashCode` application.

    2. During Step 4, he would also encrypt the hash code using Krista's public key in the same fashion as he encrypted Tom's private key.

    3. When he sends the other files in Step 5, he would also send this encrypted hash code file.

    4. Krista would now have a new step, in which she would decrypt the hash code file using her private key just as she decrypted the encrypted the file containing Tom's private key.

        After decrypting the data, she would generate her own hash code using the same algorithm Tom chose. The chapter's `HashDemo` application would allow her to compare the hash codes side by side and let her know if the data had been modified in any way after the original hash code was generated. If the two are equal, her data is safe. If not, the data's integrity cannot be assured.

In summary, using public key encryption gives two or more parties sharing data a level of confidence they wouldn't have otherwise. As authors H. X. Mel and Doris Baker state in their book *Cryptography Decrypted* (Boston: Addison-Wesley, 2000), the public key encryption is arguably the most outstanding cryptographic achievement of the twentieth century. Now let's look at yet another layer of security—that of digital signatures.

Using Digital Signatures

In addition to the already-stated benefits of public key encryption, another huge benefit is derived: digital signatures. Digital signatures ensure the e-commerce requirements of authentication, integrity, and nonrepudiation. One method of creating digital signatures is for the sender to encrypt the message using a private key. Anyone can then verify the authenticity of the sender by decrypting the message using the sender's known public key. In fact, the RSA algorithm is used to make digital signatures in this manner, though instead of signing the entire message, a hash of the message is created and this hash is then encrypted. The DSA algorithm does not encrypt data, but it can be used to create digital signatures. This might puzzle you a bit, and you might wonder how it can create a signature without being able to encrypt data. The secret here is that a unique signature is generated using the data and the private key.

The actual details are a bit more involved than this: The message is first hashed using SHA1, and then a random number k and the private key are also applied mathematically to generate two numbers—referred to as r and s—which together form the digital signature. During verification the SHA-1 hash is calculated, and then, using the public key, r, and s, some mathematical operations are performed to produce an output. If this output matches t, the signature is verified. Of course the signature algorithm used is a matter of choice. Typically the sender uses the same algorithm for encrypting the data and signing it. However, if you use two distinct algorithms—such as DSA or RSA—for the two tasks, you're assured that the recipient will be able to verify the signature, as both of these algorithms are quite popular and included in almost any encryption/decryption software.

Since the previous public key encryption algorithm chosen was RSA, I'll use DSA, just to expand the types of algorithms used in this chapter. You can, of course, use either one.

Signature Demo

We can use the DSACryptoServiceProvider class provided by the .NET Framework library. We use the SignData method of the class to create the signature and VerifyData to verify the signature.

1. Create an MFC dialog-based application called DSADigital Signature.

2. Update the Project properties to support Managed Extensions.

3. Open the `stdafx.h` file and add the following .NET support directives to the end of the file:

```
#using <mscorlib.dll>
#using <System.Windows.Forms.dll>
#using <System.dll>
using namespace System;
using namespace System::Windows::Forms;
using namespace System::IO;
using namespace System::Security::Cryptography;
using namespace System::Text;
#undef MessageBox
```

4. Open the dialog template resource and add two buttons to the dialog box. One button should have a caption like "Sign File . . ." and the second "Verify File . . ."

5. Add the following member variables to the `DSADigital SignatureDlg` class. The first—`m_dsaServiceProvider`—is a BCL wrapper for the CryptoAPI implementation of the DSA cryptography standard. The three `CString` members are used to determine the application's working directory and the file names of the public key and the public/private key pair.

```
class CDSADigitalSignatureDlg : public CDialog
{
    . . .


protected:
    gcroot <DSACryptoServiceProvider*> m_dsaServiceProvider;
protected:
    CString m_strWorkingDir;
    CString m_strPublicKeyFile;
    CString m_strKeyPairFile;
```

6. Insert a call to the `InitializeDSAServiceProvider` just before the return statement in the `CDSADigitalSignatureDlg::OnInit Dialog` function.

```
BOOL CDSADigitalSignatureDlg::OnInitDialog()
{
    . . .


    InitializeDSAServiceProvider();
```

```
      return TRUE;
}
```

7. Add the following DSA object initialization function. As you can
see, the main thing here is to make sure that the files are created
for the public and public/private keys.

```
void CDSADigitalSignatureDlg::InitializeDSAServiceProvider()
{
#pragma push_macro("new")
#undef new
    // Instantiate the member DSACryptoServiceProvider
    // object
    m_dsaServiceProvider = new DSACryptoServiceProvider();

    // Check and see if the key files are found.
    TCHAR buff[MAX_PATH];
    GetModuleFileName(NULL, buff, MAX_PATH);
    m_strWorkingDir = System::IO::Path::GetDirectoryName(
      buff);

    // Save public and public/private key pair file names
    m_strPublicKeyFile = (CString)Path::Combine
      (m_strWorkingDir,
       S"publickey.xml");
    m_strKeyPairFile = (CString)Path::Combine(m_strWorkingDir,
                                              S"keypair.xml");

    // If key keys were not found...
    if (!File::Exists(m_strKeyPairFile))
    {
      FileStream* fsPublicKey;
      FileStream* fsKeyPair;
      try
      {
        // . . . create and save them
        Byte baKey[];

        // Create public key
        String* strPublicKey = m_dsaServiceProvider->
          ToXmlString(false);
        baKey = Encoding::Default->GetBytes(strPublicKey);
        fsPublicKey = new FileStream(m_strPublicKeyFile,
          FileMode::Create);
        fsPublicKey->Write(baKey, 0, baKey->Length);
```

```
            // Create public/private key pair
            String* strKeyPair = m_dsaServiceProvider->
              ToXmlString(true);
            baKey = Encoding::Default->GetBytes(strKeyPair);
            fsKeyPair = new FileStream(m_strKeyPairFile,
              FileMode::Create);
            fsKeyPair->Write(baKey, 0, baKey->Length);
        }
        catch(Exception* e)
        {
          MessageBox::Show(e->Message);
        }
        __finally
        {
          fsPublicKey->Close();
          fsKeyPair->Close();
        }
      }
    #pragma pop_macro("new")
    }
```

8. Add an event handler for the Sign File button. This function first
   displays a File Open common dialog box so that the user can specify what file to sign. Digital signatures are generated from the private key of a public/private key pair, so the private key is read from
   disk. (This file was created by the `InitializeDSAService`
   `Provider` function.) A DSA object is then constructed using
   the private key. The input file is read. The `DSACryptoService`
   `Provider::SignData` method is called—passing it the input file.
   The `SignData` function returns a digital signature for the input
   data. This signature is written to a file whose name is the same as
   the input file name with a `.sig` extension appended.

```
void CDSADigitalSignatureDlg::OnBnClickedSignFile()
{
#pragma push_macro("new")
#undef new
  CFileDialog dlg(TRUE);
  dlg.m_ofn.lpstrInitialDir = m_strWorkingDir;
  dlg.m_ofn.lpstrDefExt = _T("*");
  dlg.m_ofn.lpstrFilter = _T("All files (*.*)\0*.*\0\0");
  dlg.m_ofn.lpstrTitle = _T("Open file to sign");
```

```
if (IDOK == dlg.DoModal())
{
    FileStream* fsKey;
   FileStream* fsInput;
   FileStream* fsout;

   try
   {
     CWaitCursor wc;

     // Determine input file name, signed file name and
     // key pair file name
     CString strInputFile = dlg.GetPathName();
     CString strSignedFile = strInputFile + _T(".sig");

     // Open key pair file as dig sig is created with
     // the private part of the public/private key pair.
     // Read data (keys) to byte array,
     // convert the byte array to a string and
     // construct a DSA object from the keys
     fsKey = new FileStream(m_strKeyPairFile,
                            FileMode::Open);
     Byte baKey[] = new Byte[(int)fsKey->Length];
     fsKey->Read(baKey,0,(int)fsKey->Length);
     String* strKey = Encoding::Default->GetString(baKey);
     m_dsaServiceProvider->FromXmlString(strKey);

     // Open user-specified input file, read data into
     // byte array
     fsInput = new FileStream(strInputFile, FileMode::Open);
     Byte baInput[] = new Byte[(int)fsInput->Length];
     fsInput->Read(baInput,0, (int)fsInput->Length);

     // Sign the data (via the service provider), take
     // returned signed byte array and write it to disk
     Byte baSignedData[] =
      m_dsaServiceProvider->SignData(baInput);
     fsout = new FileStream(strSignedFile, FileMode::Create);
     fsout->Write(baSignedData, 0, baSignedData->Length);
   }
   catch(Exception* e)
   {
```

```
    MessageBox::Show(e->Message);
  }
  __finally
  {
    fsKey->Close();
    fsInput->Close();
    fsout->Close();
  }
 }
#pragma pop_macro("new")
}
```

9. Add an event handler for the Verify File button. Once again a File
   Open common dialog box is displayed. The user selects a file for
   which they have a signature file. As digital signatures are verified
   with the public key of a public/private key pair, the public key is read
   from disk (where it was stored by the `InitializeDSAService`
   `Provider` function). A DSA object is constructed using the public
   key. The input data is read. The signature file (whose name is
   determined by appending `.sig` to the input file name) is read. The
   input data and signature data are passed to the `DSACryptoService`
   `Provider::VerifyData` method. Using the public key information it was constructed from, the DSA object verifies that the signature was one that would have been computed for the specified
   data. A return of `true` indicates a successful verification. Obviously,
   a return of `false` indicates failure.

```
void CDSADigitalSignatureDlg::OnBnClickedVerifyFile()
{
#pragma push_macro("new")
#undef new
 CFileDialog dlg(TRUE);
 dlg.m_ofn.lpstrInitialDir = m_strWorkingDir;
 dlg.m_ofn.lpstrDefExt = _T("*");
 dlg.m_ofn.lpstrFilter = _T("All files (*.*)\0*.*\0\0");
 dlg.m_ofn.lpstrTitle = _T("Open file for which you have "
                           "a signature file");

 if (IDOK == dlg.DoModal())
 {
    FileStream* fsPublicKey;
    FileStream* fsInput;
    FileStream* fsSigned;
```

```
try
{
  CWaitCursor wc;

  // Determine input file name, signed file name and
  // key pair file name
  CString strInputFile = dlg.GetPathName();
  CString strSignedFile = strInputFile + _T(".sig");

  // Open public key file as dig sig is verified with
  // the public part of the public/private key pair.
  // Read data (keys) to byte array,
  // convert the byte array to a string and
  fsPublicKey = new FileStream(m_strPublicKeyFile,
    FileMode::Open);
  Byte baKey[] = new Byte[(int)fsPublicKey->Length];
  fsPublicKey->Read(baKey, 0, (int)fsPublicKey->Length);
  String* strPublicKey = Encoding::Default->
    GetString(baKey);
  m_dsaServiceProvider->FromXmlString(strPublicKey);

  // Open user-specified input file, read data into
  // byte array
  fsInput = new FileStream(strInputFile, FileMode::Open);
  Byte baInput[] = new Byte[(int)fsInput->Length];
  fsInput->Read(baInput,0,(int)fsInput->Length);

  fsSigned = new FileStream(strSignedFile,
    FileMode::Open);
  Byte baSigned[] = new Byte[(int)fsSigned->Length];
  fsSigned->Read(baSigned, 0, (int)fsSigned->Length);

  BOOL success =
   m_dsaServiceProvider->VerifyData(baInput,
                                    baSigned);

  String* str =
    String::Format(S"File '{0}' {1} its"
                   S"verification based "
                   S"on its signature file - '{2}'",
                   Path::GetFileName(strInputFile),
                   success ? S"PASSED" : S"FAILED",
```

```
                              Path::GetFileName(strSignedFile));
              MessageBox::Show(str);
          }
          catch(Exception* e)
          {
              MessageBox(e->Message);
          }
          __finally
          {
              fsPublicKey->Close();
              fsInput->Close();
              fsSigned->Close();
          }
      }
      #pragma pop_macro("new")
      }
```

10. Build and run the demo. Click the Sign File button and select any file (I've chosen `stdafx.h`).

11. Once signed, click the Verify File button. As you can see in Figure 4–11 my `stdafx.h` file matched the signature.

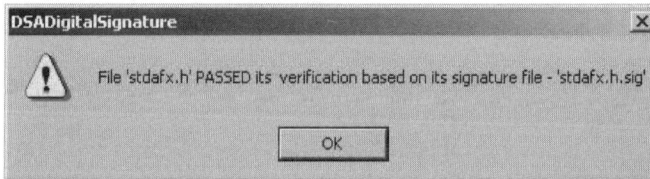

Figure 4-11 The DSADigitalSignature demo verifying a file/signature combination

12. Now modify the `stadfx.h` file and then attempt to verify it again. Since the signature no longer matches the data, you will receive an error message similar to that in Figure 4–12.

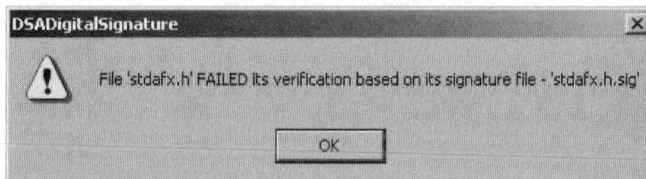

Figure 4-12 An example of a failed verification attempt: The file was modified after the signature was generated.

Summary

In this chapter we have seen how to use the various cryptography classes to perform data hashing, symmetric key encryption, and public key encryption, as well as to create digital signatures. Most of the classes are wrappers around the underlying CryptoAPI. But the classes have been so well designed that we can use them without needing to be cryptography gurus. If on the other hand, you've enjoyed what you've read here and would like to learn more, I highly recommend *Cryptography Decrypted*, by H. X. Mel and Doris Baker.

In this chapter, you discovered how the `HashAlgorithm`-derived classes can be used to generate hash values for a wide variety of hash algorithms. While many of the classes are CryptoAPI wrapper classes, there are also a few pure managed classes that wrap algorithms not currently supported by the CryptoAPI. For symmetric key cryptography, we saw how the `SymmetricAlgorithm`-derived classes can be used together with the `CryptoStream` class to allow us to use crypto streams to read decrypted data and write encrypted data to a file or network streams directly. The `CryptoStream` class will take care of all required buffering, and thus we need not be bothered with the details.

Similarly, you were introduced to how the `AsymmetricAlgorithm`-derived classes can be used to perform RSA and DSA cryptographic operations on data. The `RSACryptoServiceProvider` class can be used to encrypt/decrypt data using public key cryptography and can also be used to create digital signatures by encrypting a hash of the message using the private key. The `DSACryptoServiceProvider` class is exclusively used for creating digital signatures, and we saw how digital signatures are very useful in determining whether a user is who he or she claims to be.

For MFC programmers, the best thing about all these classes is the absolutely monumental productivity gains to be had both in doing cryptography and in using hash codes. In the next chapter, we'll look at yet another topic where the BCL provides a set of classes to give the MFC developer another tremendous boost in productivity: XML.

XML and the DOM

Introduction

In my MFC book, *Visual C++ .NET Bible* (New York: John Wiley & Sons, 2002), I included a chapter that illustrated how to write an MFC application to read and display the nodes of an XML file to a tree view control using the COM interfaces to the Microsoft XML DOM (Document Object Model). In that chapter's demo I illustrated that this seemingly trivial task required that you initialize COM, acquire an `IXMLDOMDocumentPtr` interface, call that interface's `load` method, and acquire the root element interface pointer (`IXMLDOMElementPtr`) via the `documentElement` member. Starting at the root element, you iterate through each node acquiring `IXMLDOMNodePtr` interface pointers and finally—on the basis of the node type—extract the name and value of the desired elements and attributes. When you think about having to do all that—along with the required conversions between C++ types and Automation types—it hardly seems worth it! However, that was the only standard means of interfacing Visual C++/MFC and XML until the release of the .NET Framework and the BCL XML-related classes. In fact, it was the rich support provided by these very classes that originally spawned the idea for this book. While fully covering all or even a few of these classes would take a complete book, I'll use this chapter to illustrate how much more quickly you can add support for XML processing to your MFC applications using the BCL.

This chapter is evenly divided into two sections. The first section illustrates how the `XmlTextWriter` and `XmlTextReader` classes are used to process XML documents in a sequential, fast-forward manner. In this section, you'll discover how to read and write the various nodes of an XML document—including the declaration node, elements (using data of various application-specific types), and attributes. The section concludes

with a demo illustrating maintaining a single-entity XML file using the `XmlTextWriter` and `XmlTextReader` classes. The second section dives into the more elaborate world of the DOM and the XML classes that encapsulate it—`XmlDocument` and `XmlNode`. There, I first provide a short introduction to the DOM itself and then present several task-oriented subsections that illustrate such common functions as loading and saving documents, and enumerating XML elements. From there, you will learn how to read and edit nodes from and to memory, how to retrieve and set the XML and text values of entire node branches, and how to search for specific nodes or groups of nodes based on tag names or values using XPath. You'll even see how to define an element attribute as a unique identifier in your document so that a specific nodes can be located when their element values are duplicated. This section (and the chapter) conclude with a pair of demos. The first illustrates how to parse and read any XML document and is included as a comparison point between using the (older) XML DOM COM interfaces and the BCL `XmlDocument` and `XmlNode` classes. The second demo illustrates a basic maintenance application using these classes.

Assumptions about the Reader

This chapter assumes that the reader is at least familiar with XML in terms of its definition and purpose. If you are new to XML, you will find a very basic overview and syntax in Appendix A.

Writing and Reading XML Documents

The easiest and quickest means of writing and reading XML documents is via the `XmlTextWriter` and `XmlTextReader` classes. I'll first explain how to create and write XML documents and then how to read those documents—parsing for the desired node types and values. Once that is done, you'll see a demo illustrating how to use these classes in writing and reading a simple XML file containing book-detail information.

Writing XML Documents

The first thing to understand about the `XmlTextWriter` class is that it encapsulates a fast, noncached, forward-only way of generating streams or files containing XML data that conforms to the W3C standards for XML.

Therefore, it can only be used to write from beginning to end an XML document—as opposed to the XmlDocument class, which allows you to position dynamically within the document according to your application's needs—such as in searching and modifying specific node values. (The XmlDocument class is covered in the section entitled "The DOM and the XmlDocument Class.") Additionally, the XmlTextWriter class assumes that the data you are writing is well-formed; therefore, it performs no validation. (The XmlValidatingReader—not covered in this chapter—performs validation while reading.) Therefore, the XmlTextWriter class is most useful when you've already validated your data and are looking for the most efficient means of outputting that data to an XML file.

Having said that, let's look at the steps required to write data to a new XML file using the XmlTextWriter class.

1. Construct an XmlTextWriter object.
2. Write the standard XML declaration node.
3. Write the desired elements and their attributes.
4. Close the document and perform clean-up.

Constructing an *XmlTextWriter* Object

There are three distinct versions of the XmlTextWriter class:

```
XmlTextWriter(TextWriter* textWriter);
XmlTextWriter(Stream* inputStream, Encoding*);
XmlTextWriter(String* fileName, Encoding*);
```

The first two constructors should be familiar, as they relate to what you learned about streams and writers/readers in Chapter 3, "File I/O and Registry." You would use the first constructor if you had already instantiated a TextWriter object in order to write to a text file.

The second constructor has two parameters: a System::IO::Stream object and a System::Text::Encoding object. Being able to pass a Stream object enables you to use any of the streaming classes—as illustrated in Chapter 3—as your XML document target. The Encoding object enables you to specify a character encoding such as Unicode, UTF7, or UTF8.

The last constructor enables you to simply specify a file name to be written. The XMLTextWriter takes care of any file level operations. You need only focus on the XML that you want to write to the document.

Writing the Standard XML Declaration Node

The first node of a well-formed XML document must be the declaration node, which contains information such as the XML version number and character encoding method. To that end, the XmlTextWriter provides a WriteStartDocument method that *must be the first XmlTextWrite method you call after constructing the XmlTextWriter object.* Calling any other write method before calling the WriteStartDocument causes an InvalidOperationException to be thrown.

The WriteStartDocument method is overloaded as follows:

```
void WriteStartDocument();
void WriteStartDocument(bool standAlone);
```

As you'll see in this chapter's first demo, the first (parameter-less) method writes out the following declaration node, where only the version attribute is specified.

```
<?xml version="1.0" ?>
```

That leaves us with the second overload, which accepts a Boolean parameter. This method will output the declaration node with the aforementioned version attribute as well as the standalone attribute, where its value corresponds to Boolean value passed. If you specify a value of true, then the following declaration node is written.

```
<?xml version="1.0" standalone="yes"?>
```

A value of false would result in the following:

```
<?xml version="1.0" standalone="no"?>
```

Writing Elements and Attributes

Once you've constructed an XmlTextWriter and written the declaration node, it's time to write your data. There are two basic ways of doing this, depending on the format of your element values and whether or not you have attributes to write.

In the simplest situation, you're writing elements (with no attributes) from string values. In this case, you would use the `WriteElement String`:

```
void WriteElementString(
   String* elementName,
   String* value
);
```

You would use the `WriteElementString` method, for example, if you wanted to write an XML element such as `<book>Extending MFC Apps with .NET</book>`. In this case, you would only need to call the `Write ElementString` method, passing it the name of the element and the value.

```
// Write an element from a String; no element attributes
xmlwriter->WriteElementString(S"book",
                              S"Extending MFC Apps with .NET");
```

The `WriteElementString` works well in situations where you don't have any attributes to define for the element you're writing. However, if you also need to write attributes for a given element, then you need a means of associating the element to its attributes. Luckily, the `XmlText Writer` class makes this very easy by providing a method to indicate that you're starting the definition of an element (`WriteStartElement`) and ending the definition of an element (`WriteEndElement`). Between these two methods, you can call the `WriteAttributeString` method to write as many attributes as your element needs along with a method to write the element's value. Note that the `XmlTextWriter` maintains an internal state machine, so that if you call `WriteAttributeString` outside of a `Write StartElement/WriteEndElement` pair, an `InvalidOperationException` will be thrown. Here's how you would add a language attribute (`lang`) to the previous element example:

```
// XmlTextWriter methods to output:
// <book lang="EN">Extending MFC Apps with .NET</book>

xmlwriter->WriteStartElement(S"book");
   xmlwriter->WriteAttributeString(S"lang", S"EN");
   xmlwriter->WriteString(S"Extending MFC Apps with .NET");
xmlwriter->WriteEndElement();
```

As you can see, the XmlTextWriter object will automatically enclose the attribute value you pass within the quotations. Also note that, once again, if you incorrectly call a method out of order—such as attempting to write the element value before the attribute—an InvalidOperation Exception will be thrown.

So far you've seen examples of writing string data to the XML file (via the WriteElementString and WriteString methods). This will suffice in most cases, as your data will generally be either string data or data you can easily convert to a string. However, there are times when you'll have data that is in an application-specific format that must be written exactly in that format. The XmlTextWriter offers several methods to handle such scenarios. While I won't go into much detail on every one of these methods—doing so quickly starts down a slippery slope of documenting tons of methods that would only be useful in very rare circumstances—I have listed several of the more interesting write methods along with descriptions and examples of their usage in Table 5–1.

Closing the Document and Performing Clean-Up

Once you've finished writing the various elements and attributes that will constitute your XML document, all you need to do is perform a couple of

Table 5–1 Examples of Various XMLTextWriter Methods to Write Data

Method	Usage
WriteBase64	Used to encode and output binary data from a byte buffer. You would use this, for example, when writing out encrypted data (see Chapter 4) or binary data read from a file (via the BinaryReader covered in Chapter 3). Using this method, you can output your binary data to XML format without unwanted conversions taking place.
WriteBinHex	Used to encode and output binary data in hex format. As an example, Chapter 3's FileStream demo application output an assembly program's opcodes in binary to a file. Using that demo's data (hard-coded into byte arrays), you could write the entire array (or a selected subset of the array) to an XML file using the WriteBinHex method without having to first convert the data from the byte array to a string.

Method	Usage
WriteCData	Writes out a block of data within the `<![CDATA[` and `]]>` delimiters. Normally all data within an XML document is parsed by the XML parser. In order to insert data that will not be parsed, you would place that data inside a CDATA tag via the `WriteCData` method. In fact, my Web site (www.TomArcherConsultingGroup.com) has an errata page for this book where the data is stored in an XML file. Since this data includes corrected code snippets that would have been parsed by the XML parser, these code snippets have been defined in the XML file using the CDATA tag.
WriteChars	As opposed to the `WriteString` method, which is intended to write a small amount of data, the `WriteChars` method is designed for situations where you have large amounts of text—allowing you to write that text one (character) buffer at a time.
WriteString	The most commonly used `Write` method, this allows you to output the value of an element from a `String` object. Note that the `WriteString` method automatically does simple formatting, such as replacing the reserved XML formatting characters `&`, `<`, and `>` with `&amp;`, `&lt;`, and `&gt;`, respectively.
WriteRaw	Used in situations where you wish to have complete control over the data being written, the `WriteRaw` method enables you to write the raw markup to the XML document. Obviously, no formatting or conversion is done, and thus you're responsible for ensuring the validity of the data being written.
WriteWhitespace	This method enables you to output whitespace so that you can manually control how the XML is formatted. Compare this with the `Formatting` method, which allows you to define how you want to format the document. Once the formatting is defined, the `XMLTextWriter` object will format the document automatically per your specifications.

trivial clean-up procedures. The first among those is to call the `WriteEnd Document` method (if you've previously called `WriteStartDocument`). `WriteEndDocument` will close any open elements or attributes and initialize the internal state of the `XmlTextWriter` to its "start" state. After that, you need only call the `Flush` and `Close` methods as you would with any BCL writer object. Once you've performed these clean-up duties, your

XML document is complete and should be viewable by any XML parser (such as Internet Explorer or Visual Studio .NET).

```
// write elements and attributes

xmlwriter->WriteEndDocument();
xmlwriter->Flush();
xmlwriter->Close();
```

Reading XML Documents

Like the XmlTextWriter class, the XmlTextReader provides a fast, non-cached, forward-only reader for XML data. The reader reads one node at a time (using the Read method), with the object's properties reflecting the various values of the current node. It's also important to realize that the XmlTextReader class does not provide data validation. Because of this, the XmlTextReader is ideally suited for parsing scenarios in which the data is known to be valid, and optimal performance is needed. In this section, I'll first talk about the different ways to construct an XmlTextReader object and then present several common ways to read various data elements from an XML document.

Constructing an *XmlTextReader* Object

There are 12 overloaded public constructors for the XmlTextReader. The three most commonly used correlate directly with the three XmlText Writer constructors I spoke of in the previous section.

```
XmlTextReader(TextReader* textReader);
XmlTextReader(Stream* outputStream);
XmlTextReader(String* fileName);
```

The first constructor is used in situations where you've already instantiated a TextReader object in order to read from a text file. Similarly, the second constructor is used if you've already attached a Stream object to an XML document. The last constructor enables you to simply specify a file name to be read.

Parsing XML Nodes

The XmlTextReader does not provide a mechanism for randomly reading a given element. Therefore you must read each node of an XML document,

processing either the desired node types or elements. This is accomplished by calling the `XmlTextReader::Read` method in a loop until the method returns `false`, indicating that end of the document has been reached. Within the `read` loop, you are able to determine the node type of the current node by checking the `XmlTextReader::NodeType` property. Table 5–2 lists the different `XmlNodeType` enumerations that you can filter.

Table 5-2 `XmlNodeType` Enumerations Used to Determine the Current Node Type

`XmlNodeType` Enumeration	Description
Attribute	An XML attribute defined within an element
CDATA	A CDATA section is a block of data that will not parsed by the XML parser.
Comment	An XML comment
Document	A root document object node
DocumentFragment	A document fragment
DocumentType	This represents the document type declaration.
Element	A complete element
EndElement	The end element tag, as in </book>
EndEntity	This represents the end of an entity.
Entity	This represents an entity declaration.
EntityReference	An entity reference
None	There is no node associated with the `XmlText Reader`, as when the `Read` method has not been called yet.
Notation	A notation node
ProcessingInstruction	An XML processing instruction
SignificantWhitespace	Whitespace between markup tags in a mixed content model
Text	The text content of an XML node
Whitespace	Whitespace between markup tags
XmlDeclaration	The XML node declaration

While the `XmlNodeType` enumeration defines values for `Document`, `Document Fragment`, `Entity`, `EndEntity`, and `Notation`, these values are never used by the `XmlNodeType::NodeType` property.

After the `NodeType`, the `XmlTextReader::Name` and `XmlTextReader::Value` properties are the most significant properties that you'll use in your code. Therefore, understanding which `NodeType` property correlates to which XML node and when the `Name` and `Value` properties are set is paramount to properly coding your `read` loop. To illustrate this, consider the following XML document:

```
<?xml version="1.0"?>
<!DOCTYPE BookDetail [
<!ELEMENT BookDetail (Title, Publisher, Authors, ISBN, Unformatted)>
<!ELEMENT Title (#PCDATA)>
<!ATTLIST Title Lang CDATA #REQUIRED>
<!ELEMENT Publisher (#PCDATA)>
<!ELEMENT Authors (Author+)>
<!ELEMENT Author (#PCDATA)>
<!ELEMENT ISBN (#PCDATA)>
<!ELEMENT Unformatted (#PCDATA)>
]>
<BookDetail>
  <Title Lang="EN">Extending MFC Apps with .NET</Title>
  <Publisher>Addison-Wesley</Publisher>
  <Authors>
    <Author>Tom Archer</Author>
    <Author>Nishant S.</Author>
  </Authors>
  <ISBN>032117352x</ISBN>
  <Unformatted><![CDATA[<456>]]></Unformatted>
</BookDetail>
```

As you can see, I've inserted a few basic node types such as a declaration node, document type node, some elements, and some attributes. Now consider the following method that takes as parameters an XML file to open and a reference to a `CListCtrl` in which to display the nodes of the XML file:

```
void ListAllNodes(String* fileName, CListCtrl& lstNodes)
{
```

```
#pragma push_macro("new")
#undef new
  try
  {
    lstNodes.DeleteAllItems();

    XmlTextReader* xmlreader = new XmlTextReader(fileName);

    // Read each node
    while (xmlreader->Read())
    {
      // Get NodeType enumeration's symbolic name
      String* s = String::Format(S"XmlNodeType::{0}",
                                 __box(xmlreader->NodeType));

      // Insert the NodeType enum name into the list control
      int idx = lstNodes.InsertItem(lstNodes.GetItemCount(),
                                    (CString)s->ToString());

      // Also insert the current node's Name and Value properties
      lstNodes.SetItemText(idx, 1, (CString)xmlreader->Name);
      lstNodes.SetItemText(idx, 2, (CString)xmlreader->Value);
    }
  }
  catch(Exception* e)
  {
    AfxMessageBox((CString)e->Message);
  }
#pragma pop_macro("new")
}
```

Now take a look at Figure 5–1 (from the CD's ListXMLNodes demo application). If you skip over the node declaration and whitespace nodes, you'll see that when the NodeType is set to XmlNodeType::Element, the Value is not set to the element value. Instead the Name property is set to the element name, and the Value property is set to blank. The Value property for a given element isn't set until the Text node type is read following the Element node type. (In the next section, I'll explain how to read element values.)

Another thing to notice in Figure 5–1 is that there are quite a few node types that you might not care about (such as XmlNodeType::Xml Declaration and XmlNodeType::XmlWhitespace) if you're only interested

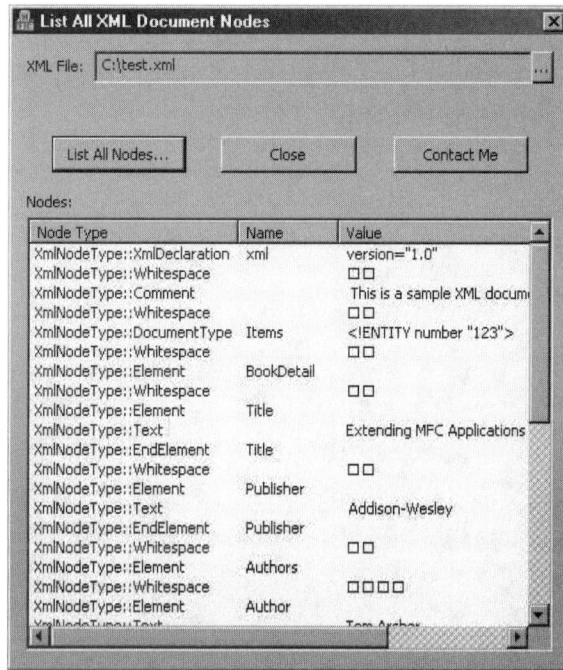

Figure 5-1 The `ListXMLNodes` demo lists all node types read by the
`XmlTextReader` object.

in retrieving data (elements and attributes). While you could simply ignore
these in your `read` loop, there are more efficient means of filtering these
items out.

   If your application doesn't need to process document-level information node types such as `XmlDeclaration`, `ProcessingInstruction`,
`DocumentType` or wishes to ignore any `Whitespace` or `Significant`
`Whitespace` nodes that appear before the start of the document's content,
you can use the `MoveToContent` method.

```
// Moves to first node of type CDATA, Element, EndElement,
// EntityReference, or EndEntity

public: XmlNodeType MoveToContent();
```

   While the `MoveToContent` seems fairly straight-forward, there is
one subtle issue that trips people up when they first use this method. If the
current node is not content (isn't of type `CDATA`, `Element`, `EndElement`,

EntityReference, or EndEntity) then the MoveToContent will move to the next node that is content. This means that you can't just modify the while loop above with the following:

```
XmlTextReader* xmlreader = new XmlTextReader(fileName);

// This will not work!
while (xmlreader->MoveToContent())
{
   . . .
```

The reason this won't work is that once the method moves to the first content node, it will not move from that node until you perform a read operation. Remember that if the current node is a content mode, MoveTo Content will not do anything. Therefore, you would need to use the MoveToContent in conjunction with the Read method in order to retrieve only content nodes. Here's one way of coding a content-only read loop:

```
XmlTextReader* xmlreader = new XmlTextReader(fileName);

for (XmlNodeType nodeType = xmlreader->MoveToContent();
     NULL != nodeType;
     xmlreader->Read(), (nodeType = xmlreader->MoveToContent()))
{
   . . .
```

With this loop, the first read is done via the MoveToContent method. This will jump nodes such as the XML declaration node and any other non-content node until it arrives at the first content node. The *loop-expression* (third part) of the for statement calls the Read method and the MoveTo Content method. If the Read method has read a noncontent node, the MoveToContent method will move to the next content node. Otherwise, if the Read method has read a node containing content, the MoveToContent method won't do anything.

The last thing we'll look at before getting into the subject of how to read elements and attributes is how to filter out whitespace. As you've seen, MoveToContent will effectively filter out any whitespace along with document-level nodes. However, there might be times when the only thing you want to filter out is whitespace. This is where the XmlTextReader:: WhitespaceHandling property comes in.

The WhitespaceHandling property accepts an enumeration value of type System::Xml::WhitespaceHandling whose values are All, None,

and `Significant`. The `All` and `None` enumeration values obviously refer to either wanting all whitespace nodes (the default) or no whitespace nodes, respectively. The `Significant` enumeration refers to nodes of type `SignificantWhitepsace`, which are only returned within the an `xml:space='preserve'` scope.

Now that you've seen how to filter out noncontent nodes that your application may not need to handle, Figure 5–2 shows you a version of the `ListXMLNodes` application that makes use of the `MoveToContent` and `WhitespaceHandling` members to display only content nodes.

Reading Elements and Attributes

Now that you know how to use the various `XmlTextReader` class methods and properties to read an XML document, parsing it for the desired node types, let's see how to read elements. If you look at either Figure 5–1 or Figure 5–2, you'll notice that when the current node is of type `Element`, the `Name` property is set to that element's name, but the `Value` property is

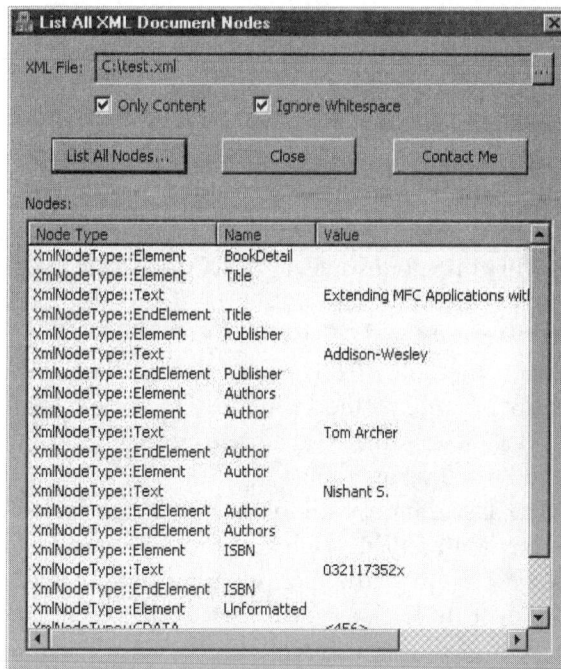

Figure 5–2 The `XmlTextReader` class defines methods and properties to filter out just the nodes your application needs.

blank. It's not until a subsequent read that the `Text` node is returned and the `Value` property is set to the current element's value. As a result, there are at least two ways to read elements.

The first way entails coding your `read` loop based on your knowledge that the `Text` node will be read after the `Element` node. As an example, let's say you have a dialog that has two fields—title and publisher—that you wish to populate from an XML document. As an MFC developer, you would normally create a DDX value member variable for each field. However, when the `read` loop has the `Element` node, you won't have the element's value, and then when you have the `Text` node, you won't know for which element the value applies. Therefore, one method involves defining DDX control variables for each control (in this case, the title and publisher controls), declaring a `CWnd` pointer in the same function that will perform the `read` loop, setting the `CWnd` pointer to the address of the appropriate edit control when an `Element` node type is returned, and finally calling the `CWnd::SetWindowText` when the `Text` node type is returned. This would be implemented as follows:

```
// m_edtTitle and m_edtPublisher are DDX control variables associated
// with the title and publisher edit controls, respectively

CWnd* pEdit;

while (xmlreader->Read())
{
  switch (xmlreader->NodeType)
  {
    case XmlNodeType::Element :
      if (0 == strElement.CompareNoCase("Title"))
        pEdit = &m_edtTitle;
      else if (0 == strElement.CompareNoCase("Publisher"))
        pEdit = &m_edtPublisher;
      else pEdit = NULL;
    break;

    case XmlNodeType::Text:
      if (pEdit)
        pEdit->SetWindowText((CString)xmlreader->Value);
    break;
  }
}
```

Coding your loop in this fashion is beneficial primarily if you need to read each node in exactly the order in which it is stored in the XML document—such as for an XML parser application that will read the data and then display that data in some graphical format.

However, if you're reading the XML document in search of element and attribute values, then the following loop is more appropriate:

```
// m_strTitle and m_strPublisher are DDX value variables associated
// with the title and publisher edit controls, respectively
Cstring strElement;

while (xmlreader->Read())
{
  switch (xmlreader->NodeType)
  {
    case XmlNodeType::Element :
      strElement = (Cstring)xmlreader->name;
      if (0 == strElement.CompareNoCase(S"Title"))
        m_strTitle = (CString)xmlreader->ReadString();
      else if (0 == strElement.CompareNoCase(S"Publisher"))
        m_strPublisher = (CString)xmlreader->ReadString();
      break;
  }
}
```

As you can see, in this second code snippet, I'm only checking for `Element` types and then calling the `XmlTextReader::ReadString` method to retrieve the element's value. You should note here that the `ReadString` method (as well as any of the other data read methods) will return the value of the current `Element` or `Text` node. Speaking of data read methods, the `XmlTextReader` class supports reciprocal read methods for each of the write methods you saw in Table 5–1.

Now that you've seen how to read elements, the next (and last) logical item type to learn how to read from an XML document is attributes. However, if you once again look back at Figure 5–1 or Figure 5–2, you'll see that there is no node type called "attribute." This is because the XML specification doesn't consider an attribute of an element to be its own node type. Rather, attributes are considered to be part of the element. As a result, when you read an `Element` node type, an `XmlTextReader` property called `Has Attributes` can be used to determine if the current element has any defined attributes. (Similarly, you can also check the `AttributeCount` property.)

As with elements, there are two distinct ways to read attributes depending on what type of application you're writing. If you're writing a parser-like application that needs to read the attribute names and values in the order in which they appear in the document, you could use code like the following:

```
while (xmlreader->Read())
{
  switch (xmlreader->NodeType)
  {
    case XmlNodeType::Element :
      if (xmlreader->HasAttributes)
      {
        for (int i = 0; i < xmlreader->AttributeCount; i++)
        {
          xmlreader->MoveToAttribute(i);
          // xmlreader::Name contains the attribute name
          // xmlreader::Value contains the attribute value
        }
      }
      // Get element value using one of the read methods
      // This should always follow any attribute processing
      break;
      ...
```

One thing to always keep in mind when using the various read and move methods is the impact the given method will have on the reader's current position within the document. For example, let's say that you're attempting to read both an element and that element's attributes. If you were to first use one of the XmlTextReader class's read methods (such as ReadString or ReadChars), the current node would become the End Element node type. At that point, attempting to determine if the element has attributes (via the HasAttributes method) would fail because the reader has moved the beyond the attributes, and, being a forward-only reader, it cannot backtrack. The correct way , therefore, to read an element that contains attributes is to read the attributes and then read the element's value.

The second way to read an element's attributes is to simply call the XmlTextReader::GetAttribute method, passing either the attribute name or the relative index of the attribute. For applications where you're

filtering for specific attributes and know their names (or relative positions), this is typically the method you would use. Referring back to the example XML element that defined a title and a language attribute, the following code snippet illustrates how those two values would be read.

```
while (xmlreader->Read())
{
  if (XmlNodeType::Element == xmlreader->NodeType)
  {
    CString name = (CString)xmlreader->Name;
    if (0 == name.CompareNoCase(_T("Title")))
    {
      String* lang = xmlreader->GetAttribute(S"Lang");
      String* title =xmlreader->ReadString();
      ...
```

Demo Maintenance Application Using XML

At this point, it's time to see a demo application implementing much of what you've learned up to this point. However, before we get started, it's important to keep in mind that the XmlTextWriter and XmlTextReader classes both provide sequential, forward-only access to an XML document. As a result, there are certain situations where these classes are appropriate, and there are situations where other System::Xml namespace classes that you'll see later in this chapter are the better choice.

In my experience, you should use these classes whenever you don't mind reading and writing every node of a document. For example, an XML parser specifically needs to read and write every node in the exact order in which it appears in the document. Other examples include applications that use small, "single-entity" XML documents, such as an application that writes its configuration information to an XML file, or a maintenance application, where each entity being maintained is persisted to its own individual file.

Having said that, let's look at this section's demo application, which illustrates how to use the XmlTextWriter and XmlTextReader classes to implement an application that allows the adding and updating of book information—such as author, ISBN, title, and so on—to an XML file.

1. To get started, create an MFC dialog-based application called `XMLSingle`.
2. Update the Project properties to support Managed Extensions.
3. Open the `stdafx.h` file and add the following .NET support directives to the end of the file. (The inclusion of the `Microsoft.VisualBasic` assembly will certainly throw many of you for a loop, but you'll see why this assembly provides a very cool little function shortly.)

```
#using <mscorlib.dll>
#using <System.Windows.Forms.dll>
#using <System.dll>
#using <System.xml.dll>
#using <Microsoft.VisualBasic.dll>

using namespace System;
using namespace System::Windows::Forms;
using namespace System::Xml;
using namespace System::IO;
using namespace Microsoft::VisualBasic;
```

4. Open the dialog template resource and add the controls as you see them in Figure 5–3.
5. For purposes of this demo, hard-code two entries into the language combo box control's `Data` property—EN and SP (separated by semicolons).

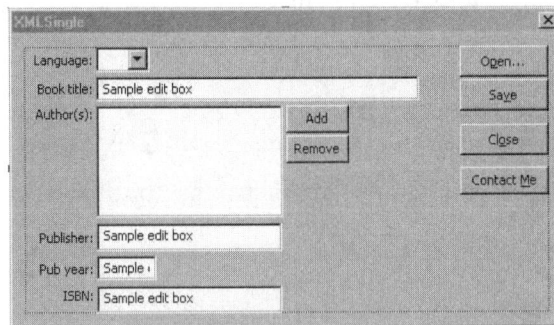

Figure 5-3 Dialog resource for the `XMLSingle` demo application

Table 5-3 DDX Variables for the XMLSingle Demo

Control	Variable Type	Variable Name
Language edit	CString	m_strLanguage
Book title edit	CString	m_strTitle
Author list box	CListBox	m_lbxAuthors
Publisher edit	CString	m_strPublisher
Pub year	int	m_iPubYear
ISBN	CString	m_strISBN

6. Set the Sort property for the author list box to False.
7. Create the DDX member variables as shown in Table 5–3.
8. The first task is to add a member CString variable to the dialog class called m_strWorkingDir

```
class CXMLSingleDlg : public CDialog
{
...
protected:
    CString m_strWorkingDir;
```

9. Update the OnInitDialog function by inserting the following code just before the return statement. Here we're simply saving the application's directory.

```
// Save application directory
TCHAR buff[MAX_PATH];
GetModuleFileName(NULL, buff, MAX_PATH);
m_strWorkingDir = Path::GetDirectoryName(buff);
```

10. As you'll want to initialize the controls when saving and opening a new XML file, add the following member function to the dialog class:

```
void CXMLSingleDlg::InitControls()
{
    m_strLanguage = _T("");
    m_strTitle = _T("");
    m_lbxAuthors.ResetContent();
    m_strPublisher = _T("");
    m_iPubYear = 0;
```

```
    m_strISBN = _T("");
    UpdateData(FALSE);
}
```

11. Now, let's add a function to allow users to add authors to the list box. One of the very few advantages that Visual Basic programmers have over us is that they've always been able to display a dialog box with a single input edit control simply by calling `InputBox`. We MFC developers, of course, have had to create and code a dialog for such a trivial task. However, since Microsoft was kind enough to define `InputBox` for backwards compatibility between Visual Basic 6 and Visual Basic .NET, that means that we also have access to it from MFC! Here I'm using that `InputBox` method to retrieve the author name from the user and (if it is not already in the list box) to insert the value into the list box.

```
void CXMLSingleDlg::OnBnClickedAddAuthor()
{
  CRect rect;
  GetWindowRect(&rect);
  String* author = Interaction::InputBox(S"Author:",
                                         S"New Author",
                                         S"",
                                         rect.left +
                                         (rect.Width() / 2),
                                         rect.top +
                                         (rect.Height() / 2));
  if (0 < author->Length)
  {
    CString strAuthor = (CString)author;
    if (LB_ERR == m_lbxAuthors.FindStringExact(-1, strAuthor))
      m_lbxAuthors.AddString((CString)author);
    else
      AfxMessageBox(_T("That author is already listed"));
  }
}
```

12. Now add the following "remove author" function. While these last two functions don't have anything to do with XML, they are small in scope and help to create a more practical application from which to learn XML. Here, I'm simply deleting the currently selected list box entry (author).

```
void CXMLSingleDlg::OnBnClickedRemoveAuthor()
{
  int iCurrSel;
  if (LB_ERR == (iCurrSel = m_lbxAuthors.GetCurSel()))
  {
    AfxMessageBox(_T("You must first select an author to"
                     "remove"));
  }
  else
  {
    m_lbxAuthors.DeleteString(iCurrSel);
  }
}
```

13. At this point, add an event handler for the Save button. After calling UpdateData, the function first determines the name of the new
file by combining the application's directory, the book title, and the
book language. Obviously, in a real-world application you would do
some validation here, but we'll assume valid data in order to keep
the focus on XML.

Once that is done, an XMLTextWriter object is instantiated,
and several of its methods that you learned about in this section are
put to use to start the document, add the appropriate elements and
attributes, and then close the file (causing it to be written to disk).

Of particular note here is the fact that I used the Write
StartElement/WriteEndElement method pair for the Book
Details as I need to write the Lang attribute followed by its value.
Similarly, I used the WriteStartElement/WriteEndElement
method pair for the Authors element because between those
methods, I need to write multiple Author elements (each retrieved
from the author list box). The remaining elements are output using
the WriteStringElement method. Finally, the document is ended
(via WriteEndDocument), flushed, and closed.

```
void CXMLSingleDlg::OnBnClickedSave()
{
#pragma push_macro("new")
#undef new
  XmlTextWriter* xmlwriter;

  try
  {
    UpdateData();
```

```
// In a real-world app, you would verify that
// all fields were completed
CString strNewFileName = m_strTitle + _T("-")
                            + m_strLanguage + _T(".xml");
strNewFileName = Path::Combine(m_strWorkingDir,
                            strNewFileName);

xmlwriter = new XmlTextWriter(strNewFileName,
                            Text::Encoding::ASCII);
xmlwriter->WriteStartDocument(true);

xmlwriter->WriteStartElement(S"BookDetails");
  xmlwriter->WriteAttributeString(S"Lang", m_strLanguage);

  xmlwriter->WriteElementString(S"Title", m_strTitle);

  xmlwriter->WriteStartElement(S"Authors");
    for (int i = 0; i < m_lbxAuthors.GetCount(); i++)
    {
      CString strAuthor;
      m_lbxAuthors.GetText(i, strAuthor);
      xmlwriter->WriteElementString(S"Author", strAuthor);
    }
  xmlwriter->WriteEndElement(); // end of Authors

  xmlwriter->WriteElementString(S"Publisher",
                            m_strPublisher);

  xmlwriter->WriteElementString(S"YearPublished",
                            m_iPubYear.ToString());

  xmlwriter->WriteElementString(S"ISBN", m_strISBN);

 xmlwriter->WriteEndElement(); // end of BookDetails

 xmlwriter->WriteEndDocument();
xmlwriter->Flush();
xmlwriter->Close();

 InitControls();
 AfxMessageBox("Book successfully saved");
}
```

```
catch(Exception* e)
{
  if (xmlwriter)
  {
    xmlwriter->Flush();
    xmlwriter->Close();
  }

  AfxMessageBox((CString)e->Message);
}
#pragma pop_macro("new")
}
```

14. Now let's see how to modify the code to read an XML file that meets our application's needs. Add an event handler for the Open . . . button as follows.

    Basically this function works like this. Once the user selects an XMLSingle-formatted file to read (from a displayed File Open common dialog box), the function instantiates an XMLTextReader object based on that file name. The function then uses a while loop to iteratively call the XMLTextReader::Read method until end of file. As you learned in the section entitled "Reading XML Documents," the Read method will read one node at a time, and its NodeType property indicates the type of node read.

    In that while loop, the function checks the NodeType looking for XmlNodeType::Element types. (The NodeType enumerations are listed in Table 5–2.) If the node type is an element, the function uses an if/else statement to determine if the current element is one whose value needs to be retrieved for display on the dialog box. For each such element, the function calls the ReadString method to retrieve the corresponding value and sets the appropriate DDX value member variable. The only element that is treated a bit differently is the BookDetails element. When that element is read, the GetAttribute is used to retrieve and display the Lang attribute.

```
void CXMLSingleDlg::OnBnClickedOpen()
{
#pragma push_macro("new")
#undef new
  CFileDialog dlg(TRUE);
  dlg.m_pOFN->lpstrInitialDir = m_strWorkingDir;
```

```
      dlg.m_pOFN->lpstrTitle = _T("Open an XMLSingle XML File");
      dlg.m_pOFN->lpstrFilter = _T("XML Files (*.xml)\0*.xml\0");

      if (IDOK == dlg.DoModal())
      {
        m_lbxAuthors.ResetContent();

        XmlTextReader* xmlreader  =
          new XmlTextReader(dlg.GetPathName());

        while (xmlreader->Read())
        {
          if (XmlNodeType::Element == xmlreader->NodeType)
          {
            CString name = (CString)xmlreader->Name;

            if (0 == name.CompareNoCase(_T("BookDetails")))
              m_strLanguage = (CString)xmlreader->
                GetAttribute(S"Lang");

            else if (0 == name.CompareNoCase(_T("Title")))
              m_strTitle = (CString)xmlreader->ReadString();

            else if (0 == name.CompareNoCase(_T("Author")))
              m_lbxAuthors.AddString((CString)xmlreader->
                ReadString());

            else if (0 == name.CompareNoCase(_T("Publisher")))
              m_strPublisher = (CString)xmlreader->ReadString();

            else if (0 == name.CompareNoCase(_T("YearPublished")))
              m_iPubYear= Convert::ToInt32(xmlreader->
                ReadString());

            else if (0 == name.CompareNoCase(_T("ISBN")))
              m_strISBN = (CString)xmlreader->ReadString();
          }
        }
        xmlreader->Close();
        UpdateData(FALSE);
      }
#pragma pop_macro("new")
}
```

15. Now build and run the application. Figure 5–4 shows an example of running this application with this book's details used as data.

Figure 5–5 shows the XML file created using this book's details and the XMLSingle application.

As I mentioned at the outset of the demo, the XmlTextWriter and XmlTextReader classes work well for situations where you don't mind reading and writing every node of an XML document in a forward-only manner. However, if your application needs a more efficient means of reading or writing specific elements or the ability to seek both forward and backward from a given point, then the subject of the next section—the XmlDocument class—might be just the answer.

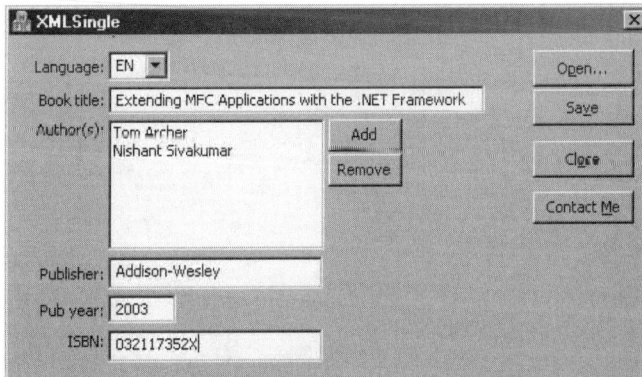

Figure 5–4 The XmlTextWriter and XmlTextReader classes provide a very easy means of writing and reading simple, single-entity XML files.

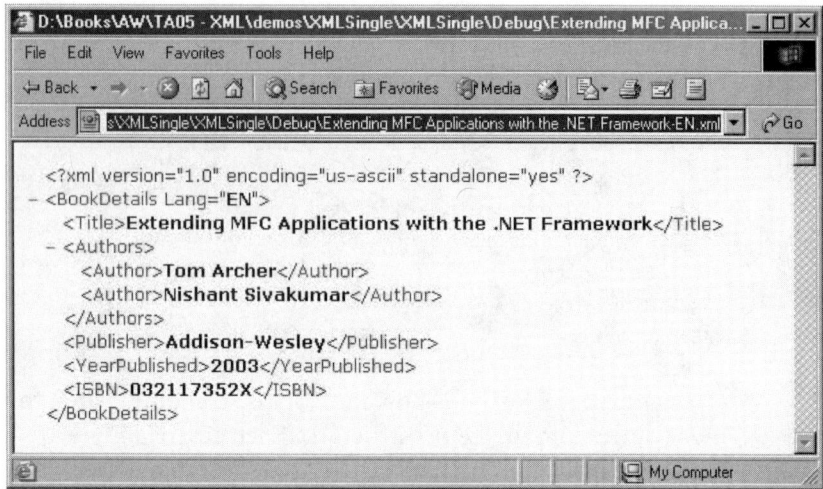

Figure 5–5 The `XmlTextWriter` and `XmlTextReader` classes enable you to write and read any standard XML nodes—including multiple levels of nested elements.

The DOM and the `XmlDocument` Class

As the `XmlDocument` class is an encapsulation of the DOM (Document Object Model), let's first get a basic overview of what the DOM is and how the capabilities it provides compare with what you've learned up to this point.

The DOM is an in-memory representation of an XML document that allows you to read, edit, and save XML data. However, unlike the `XmlText Writer/XmlTextReader` class pair, the DOM doesn't view the XML data in a linear fashion. Instead, when the DOM loads an XML document, the document is parsed, and the data is represented as a tree of related nodes. As an example, take a look the following, more expanded, version of our earlier `BookDetail` XML document:

```
<?xml version="1.0"?>
<BookDetail>
  <Book ISBN="032117352X">
    <Title Lang="EN">Extending MFC Apps with .NET</Title>
    <Authors>
      <Author>Tom Archer</Author>
```

```
    <Author>Nishant S.</Author>
  </Authors>
</Book>
<Publisher Name="Addison-Wesley">
  <ProjectEditor>Sondra Scott</ProjectEditor>
  <TechEditors>
    <TechEditor>Dmitri Riz</TechEditor>
    <TechEditor>Mark Burhop</TechEditor>
  </TechEditors>
</Publisher>
</BookDetail>
```

Figure 5–6 illustrates how memory is structured when this XML document is read into the DOM structure. Each circle in Figure 5–6 represents a node (encapsulated by the XmlNode class); both element names and values are defined as nodes. (The XmlNodeType enumeration—whose members are shown in Table 5–2—is used to distinguish nodes types.) In addition, note that attributes are also nodes, but they are treated a bit differently by the DOM. (I'll get into that shortly.) It is this treelike architecture—where nodes are arranged in a parent/child/sibling relationship—that enables the DOM to allow direct access to specific elements as well as access to a node's related nodes (such as its parent or any siblings).

As you can tell at this point, the two classes that are used most often when accessing the DOM are the XmlDocument and XmlNode classes. Therefore, before I get into the specifics of how to perform some common tasks using these classes, it might help to further define them in terms of their respective capabilities.

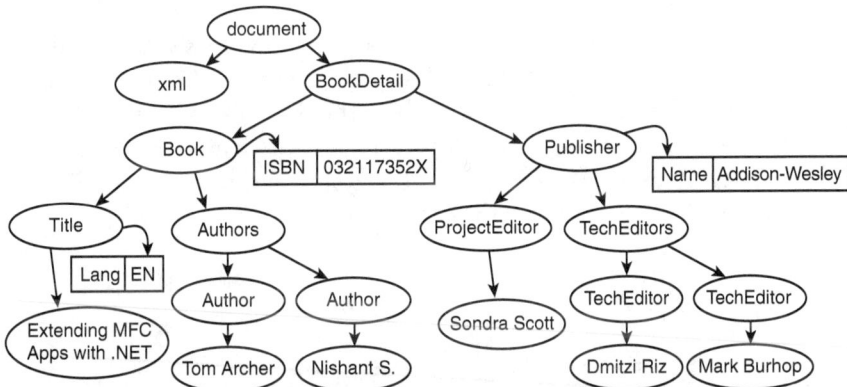

Figure 5-6 An example of XML DOM structure

The XmlDocument class is actually derived from the XmlNode class and supports both the inherited node-level functions (such as reading and modifying node data) as well as document-level functions (such as loading and saving XML). Having said that, let's look at the characteristics that define an XmlNode and put it into perspective with our previously illustrated BookDetail XML document.

- All nodes have a single parent node with the exception of the Document node. This can be seen in Figure 5–6, as each parent node appears directly above its child node.
- The Document node is encapsulated by the XmlDocumentType class (derived from XmlNode) and defines document-level properties such as the document's namespace, the complete XML of the document (the InnerXml property), and various properties that define how the document is to be processed.
- Attributes are also nodes, but they are treated differently; because they are not part of the parent, child, sibling relationship, you cannot call the same methods to retrieve them as you would call for other node types (such as elements). This will become clearer in the section entitled "Reading and Writing Attributes."
- Some node types can be parent nodes, while others cannot. The nodes that can be parent nodes are Document, DocumentFragment, EntityReference, Element, and Attribute. The node types that cannot have child nodes are XmlDeclaration, Notation, Entity, CDATASection, Text, Comment, ProcessingInstruction, and DocumentType.
- Nodes that are at the same level on the hierarchy are called *siblings*. In Figure 5–6, Book and Publisher are both sibling children of the BookDetails node. Likewise, Title and Authors are siblings, as are both Author nodes, and so on.

Now that you've seen how the DOM represents an XML document and that both the XmlDocument and the XmlNode class encapsulate that functionality, let's move into the sections on how to use these classes to read and write XML data.

Loading and Saving XML Documents

One of the most helpful capabilities of the XmlDocument class is that it allows you to load an XML document from either a source that is external

to your code—such as a file—or directly from a string containing XML markup. This gives you the ability to do things like build a framework for your XML document using the String (or StringBuilder class), load the XML string into an XmlDocument object, and then call any of the various XmlDocument and XmlNode methods and properties in order to further manipulate the data before persisting it. This functionality is provided via the XmlDocument::LoadXml method and has the following syntax:

```
void LoadXml(string xmlString);
```

Of note here is that the LoadXml method will not preserve whitespace and will not perform any DTD or Schema validation. However, if the XML is not well-formed, the LoadXml method will throw an XmlException. The following code snippet loads a newly instantiated XmlDocument object with a simple XML document and then calls the XmlDocument::Save. method.

Note that the Console::Out property returns a Stream object representing the standard output device. Since the Save method can write to any Stream object, this is a very quick means of verifying your XML by writing it to console (instead of writing to a file and having to open and view the file.)

```
#pragma push_macro("new")
#undef new

// Create the XmlDocument.
XmlDocument* xmlDocument = new XmlDocument();
xmlDocument->LoadXml(S"<Test><Element1>Text Value</Element1></Test>");

xmlDocument->Save(System::Console::Out);

#pragma pop_macro("new")
```

The alternative to the LoadXml method is the XmlDocument::Load method, which defines the following overloads:

```
void Load(string fileName);
void Load(Stream inputStream);
void Load(TextReader textReader);
void Load(XmlReader xmlReader);
```

The first method takes, as its only parameter, a String object containing the file name to open and parse. The second and third methods take

instances of the `Stream` and `TextReader` classes, respectively, that you learned about in Chapter 3. The last override is very interesting because it allows you to pass an `XmlReader` object. If `XmlReader` has not been used yet to read any nodes, then the `XmlDocument` will load the entire document. If the `XmlReader` is at a given node, then only that node (and its siblings and child nodes) will be loaded.

Once you've loaded a DOM structure in memory and have made the needed changes to the document's nodes, you can persist the entire structure with a single method call—`XmlDocument::Save`. I won't go into a lot of detail here because there's really no detail to cover. The `Save` method is incredibly easy to use, and its overloaded methods directly mirror the `Load` method overloads so that you have complete flexibility in saving your data:

```
void Save(string fileName);
void Save(Stream outputStream);
void Save(TextWriter textWriter);
void Save(XmlWriter xmlWriter);
```

Enumerating the Elements of an XML Document

As you know, every well-formed XML document has a root node, so that's where we'll start. The `XmlDocument::DocumentElement` property is used to retrieve the root node of the currently loaded XML document. Once you have the root node, you can code a recursive function using the `XmlNode::FirstChild` and `XmlNode::NextSibling` methods to retrieve each child. The following generic functions will print all the node names of any well-formed XML file:

```
void DisplayChild(XmlNode* node)
{
  Console::WriteLine(node->Name);
}

void DisplayChildren(XmlNode* parent)
{
  DisplayChild(parent);

  // simple for loop to get all children
  for (XmlNode* child = parent->FirstChild;
       NULL != child;
       child = child->NextSibling)
```

```
    {
        // for each child, call this function so that we get
        // its children as well
        DisplayChildren(child);
    }
}

void DisplayAllNodes(String* fileName)
{
#pragma push_macro("new")
#undef new
   try
   {
     XmlDocument* xmlDocument = new XmlDocument();
     xmlDocument->Load(fileName);

     XmlNode* rootNode = xmlDocument->DocumentElement;

     // Pass root to recursive DisplayChildren function
     DisplayChildren(rootNode);
   }
   catch(Exception* e)
   {
   }
#pragma pop_macro("new")
}
```

As you can see, the DisplayAllNodes function loads the specified XML file, retrieves the document's root node, and calls the DisplayChildren function. The DisplayChildren function calls DisplayChild to display the name of the current node and then begins a for loop that starts by calling the current node's FirstChild property. Within the loop, the DisplayChildren function is called in case the current node has any children. Finally, on the iteration cycle of the for loop, the current node's next sibling is retrieved. This manner of programming ensures us that we traverse the entire document starting with whatever node is originally passed to the DisplayChildren function.

If you were to run this code against the BookDetail XML file shown earlier, you would see the following output (focus on the bold text).

```
BookDetail
Book
```

```
Title
#text
Authors
Author
#text
Author
#text
Publisher
ProjectEditor
#text
TechEditors
TechEditor
#text
TechEditor
#text
```

As you can see from this output, some of the nodes have names of #text. As you might guess, these are the XmlNodeType::Text nodes and represent element values. One way to retrieve their values is to check the current node's type and display the desired property. The following modification to the DisplayChild function accomplishes this:

```
// Display either Name or Value depending
// on node type
String* value;

if (XmlNodeType::Text == parent->NodeType)
    value = parent->Value;
else
    value = parent->Name;
Console::WriteLine(value);
```

At this point, the DisplayChild function will output both the elements and their values (as opposed to #text). (The element values are bold for comparison purposes.)

```
BookDetail
Book
Title
Extending MFC Apps with .NET
Authors
Author
```

Tom Archer
Author
Nishant S.
Publisher
ProjectEditor
Sondra Scott
TechEditors
TechEditor
Dmitri Riz
TechEditor
Mark Burhop

Editing Element Values

One way to modify an element's value is to use the read/write capable `XmlNode::Value` property (the `XmlNode::Name` only readable) to modify an element's value by simply passing the new value in the right-hand side of an assignment operator. To illustrate this, consider the following change to the previously displayed `DisplayChild` function.

```
void DisplayChild(XmlNode* node)
{
...

case XmlNodeType::Text:
  node->Value = node->Value->ToUpper();
  value->Append(node->Value);
break;
```

Inserting this single line change into the `DisplayChild` function and then calling the `XmlDocument::Save` when the `DisplayChildren` function returns converts every element value in the document to upper case and saves the resulting document to disk. In addition to this method, you will often see element values written using the `XmlNode::InnerText` property. I'll cover that in the section entitled, "Getting and Setting the Text and XML of a Node or Node Branch."

Reading and Writing Attributes

Looking at the output generated from the `DisplayChildren/Display Child` functions, you can see that the `XmlTextReader` class (used in the

ListXMLNodes demo application) doesn't retrieve the element attributes as nodes, and neither does the XmlDocument. This is why I mentioned earlier that while XML attributes are encapsulated with the XmlNode class, they are not retrieved in the same fashion as other nodes because they do not participate in the parent/child/sibling hierarchy in the same manner as other node types. Instead, when an XML document is loaded, the Xml Document object stores each element's attributes in an XmlAttribute Collection object belonging to that element object. You can retrieve this collection simply by using the XmlNode::Attributes property.

The following code snippet illustrates how to enumerate the attributes of an element. As the attribute is of type XmlNode, its Name and Value properties can be used to retrieve the attributes name/value pair, respectively.

```
if (XmlNodeType::Element == node.NodeType)
{
  // Read each attribute's name and value pair
  XmlAttributeCollection* attributes = node->Attributes;
  for (int i = 0; i < attributes->Count; i++)
  {
    String name = attributes->Item(i)->Name;
    String value = attributes->Item[i]->Value;
  }
}
```

Now, let's make one file change to our DisplayChild function so that it will also retrieve the attributes of each element. As you can see, I'm doing a little fancy formatting with the StringBuilder object and passing the parameters to it as an array, so I made the lines of code that are specific to the retrieval of the attribute name and value strings bold so that they stand out a little more.

```
void DisplayChild(XmlNode* node)
{
#pragma push_macro("new")
#undef new
  StringBuilder* value = new StringBuilder();
  switch(node->NodeType)
  {
    case XmlNodeType::Element:
    {
      value->Append(node->Name);
```

```
XmlAttributeCollection* attributes = node->Attributes;
if (attributes)
{
  StringBuilder* attrString = new StringBuilder();
  for (int i = 0; i < attributes->Count; i++)
  {
    String* params[] = new String*[4];
    params[0] = String::Format(S"{0}", (0 == i ? S" [" : S"") );
    params[1] = attributes->Item(i)->Name;
    params[2] = attributes->Item(i)->Value;
    params[3] = String::Format(S"{0}",
      ((i + 1) == attributes->Count) ? S"]" : S", ");
    attrString->AppendFormat(S"{0}{1} = \"{2}\"{3}", params);

    value->Append(attrString->ToString());
  }
}
}
break;

case XmlNodeType::Text:
  value->Append(node->Value);
break;

default:
  value->Append(node->Name);
break;
}
Console::WriteLine(value->ToString());
#pragma pop_macro("new")
}
```

Our completed `DisplayChildren` function now enumerates any XML file—displaying its elements and attributes as in the following output from the `BookDetail` XML.

```
BookDetail
Book [ISBN = "032117352X"]
Title [Lang = "EN"]
Extending MFC Apps with .NET
Authors
Author
Tom Archer
```

```
Author
Nishant S.
Publisher [Name = "Addison-Wesley"]
ProjectEditor
Sondra Scott
TechEditors
TechEditor
Dmitri Riz
TechEditor
Mark Burhop
```

Modifying Attribute Values

The `Value` property of an attribute can also be used to modify an attribute's value just as you saw in the previous section, where it was used to modify an element's value.

Getting and Setting the Text and XML of a Node or Node Branch

In addition to providing the `Name` and `Value` properties for reading and writing node information, Microsoft has also added several extensions to DOM in the form of properties that can prove quite useful from time to time. Three of those extensions—`InnerText`, `InnerXml` and `OuterXml`— allow you to retrieve the text or XML within not only the current node, but all of its child nodes as well. The first two properties (`InnerText` and `InnerXml`) also allow you to modify the text and XML, respectively. To illustrate this, consider the following XML and code snippet. (The `xml` string is formatted to make the separate nodes clearer.)

```
#pragma push_macro("new")
#undef new
  String* xml = S"<Authors>"
                S"<Author>Tom</Author>"
                S"<Author>Nish</Author>"
                S"</Authors>";

  XmlDocument* doc = new XmlDocument();
  doc->LoadXml(xml);
  XmlNode* authors = doc->DocumentElement;

  ...
```

Now let's look at the aforementioned properties and discuss their impact on this XML. (Note that each code snippet assumes starting over with the initialized xml string and is not a continuation of the previous snippet.)

The InnerText property allows you to retrieve or set the concatenated values of the node object for which it called (as well as all child nodes). These values do not include the XML tags themselves.

```
// Displays the concatenated value: TomNish
Console::WriteLine( authors->InnerText );
```

Setting the InnerText property will therefore replace any child nodes with plain text:

```
// Replaces ...
// <Authors><Author>Tom</Author><Author>Nish</Author></Authors>
// With ...
// <Authors>TomNish</Authors>
node->InnerText = S"TomNish";
```

Additionally, you cannot use the InnerText property to write XML tags because any included markup characters will be escaped (as shown in Table A–1 of Appendix A).

```
// This produces the following escaped markup...
/*
<Authors>&lt;Author&gt;Tom's Clone&lt;/Author&gt;&lt;Author&gt;Nish's
Clone&lt;/Author&gt;</Authors>
*/
authors->InnerText = S"<Author>Tom's Clone</Author>"
                     S"<Author>Nish's Clone</Author>";
```

As you can see, this is probably not what you would want most of the time, as the escaped markup obliterated the child nodes. In order to replace a branch with a single line of code, you would use either the InnerXml or OuterXml properties, which I'll get to shortly. The Inner Text property is most often used when the element node is a *leaf node* (contains no child nodes), in which case the InnerText property would be used to either retrieve or set that element's value:

```
// Results in the following XML
/*
```

```
<Authors><Author>Tom's Clone</Author><Author>Nish's
Clone</Author></Authors>
*/


XmlNode* author = authors->FirstChild;
author->InnerText = S"Tom's Clone";

author = author->NextSibling;
author->InnerText = S"Nish's Clone";
```

Using the `Value` property is certainly convenient if you've already instantiated an `XmlNode` object representing a given element. However, if the code has the element node's parent node and only needs to update the element's value, it can use the parent node's `Item` property to retrieve a child node by name. As an example, consider the following XML fragment:

```
...
<Name>
  <First>Tom</First>
  <Last>Archer</Last>
</Name>
```

If your code has an `XmlNode` object representing the `Name` node and needed to update the `First` and `Last` nodes, it isn't very efficient to instantiate an `XmlNode` object for each of these elements. Instead, you could simply write code like the following (where it's assumed that you've already acquired the `Name` node in the form of `pNameNode`):

```
// Update first and last name using previously acquired Name node
pNameNode->Item[S"First"]->InnerText = S"New First";
pNameNode->Item[S"Last"]->InnerText = S"New Last";
```

As opposed to the `InnerText` property, which returns the concatenated text of an entire branch, the `InnerXml` returns the XML (all text including the XML tags) within a branch. The closely related `OuterXml` property returns the XML for the entire branch, including the node's own tags:

```
// Print branch XML within current tags
// <Author>Tom</Author><Author>Nish</Author>
Console::WriteLine( authors->InnerXml );
```

```
// Print branch XML including encompassing tags
// <Authors><Author>Tom</Author><Author>Nish</Author></Authors>
Console::WriteLine( authors->OuterXml );
```

As you can see, setting the InnerXml or OuterXml properties would enable you to replace entire branches with a single line of code:

```
// Results in the following XML
/*
<Authors><Author>Tom's Clone</Author><Author>Nish's
Clone</Author></Authors>
*/
authors->InnerXml = S"<Author>Tom's Clone</Author>"
                    S"<Author>Nish's Clone</Author>";
Console::WriteLine( authors->OuterXml );
```

One very powerful way to use this is if you need to rename an XML tag or search for and replace all occurrences of a string (XML tag or element/attribute value). As an example, let's say we wanted to change the Authors and Author tags to Writers and Writer, respectively.

```
// BEFORE:
<Authors><Author>Tom</Author><Author>Nish</Author></Authors>
Console::WriteLine( doc->InnerXml );

// Replace Author with Writer
String* authorsBranch = doc->InnerXml;
doc->InnerXml = authorsBranch->Replace(S"Author", S"Writer");

// AFTER:
<Writers><Writer>Tom</Writer><Writer>Nish</Writer></Writers>
Console::WriteLine( doc->InnerXml );
```

One thing you might have noticed is that I used the document node and not the root node. I did this because I also wanted to change the root node's tag, and since the OuterXml property is read-only, I couldn't do that with the root node. Instead I used the document node's InnerXml property, which allows me access to the entire document's XML.

That leads me to one last interesting note here. The OuterXml property of both the document node and the root node return the same content. This is because the document node has no tag associated with it. Therefore, you receive the document node's InnerXml content, which is the same as

the root node's `OuterXml` content. To illustrate this, the following code will output both the `OuterXml` and `InnerXml` properties of the document and root nodes.

```
Console::WriteLine( S"Entire XML\n{0}\n", xml);

Console::WriteLine( S"Document OuterXml\n{0}\n", doc->OuterXml );
Console::WriteLine( S"Root OuterXml\n{0}\n", authors->OuterXml );

Console::WriteLine( S"Document InnerXml\n{0}\n", doc->InnerXml );
Console::WriteLine( S"Root OuterXml\n{0}\n", authors->InnerXml );
```

This snippet yields the following output:

```
Entire XML
<Authors><Author>Tom</Author><Author>Nish</Author></Authors>

Document OuterXml
<Authors><Author>Tom</Author><Author>Nish</Author></Authors>

Root OuterXml
<Authors><Author>Tom</Author><Author>Nish</Author></Authors>

Document InnerXml
<Authors><Author>Tom</Author><Author>Nish</Author></Authors>

Root OuterXml
<Author>Tom</Author><Author>Nish</Author>
```

At this point, you should see why these properties are so useful. Now think of combining the string-parsing and replacement capabilities of Regular Expressions (covered in Chapter 2)! This combination would provide a very powerful means of manipulating either the data only (`InnerText`) or the data and tags (`InnerXml`) of either an entire XML document or select branches and nodes.

Searching for Specific Nodes

As mentioned at the outset of this section, one of the biggest advantages to using the `XmlDocument` over the `XmlTextReader` class is that it enables you to programmatically search for a given node without having to parse the entire document manually, doing string comparisons until you find the

target node. In that regard, there are three sets of methods that I'll cover in this section. The first set—XmlNode::SelectSingleNode and XmlNode::SelectNodes—is used to retrieve a single node and a group of nodes, respectively, using a supplied XPath expression. After that, you'll see the XmlDocument::GetElementById function, which allows you to retrieve elements by an attribute's value. Finally, you'll see how the XmlDocument::GetElementsByTagName method enables you to search for elements via their tag names.

What Is XPath?

As a discussion on XPath can get quite involved and would distract from the purpose of this section, I've included—in Appendix B—an overview of XPath in terms of what it is, what it's designed for, and how to use it. The text and code snippets of this chapter assume that you are familiar with XPath syntax.

For purposes of the following code snippets I'll be referring to the XML shown in Listing 5–1. (The blank lines are intended to make it easier to view the individual book nodes.)

Listing 5–1 Example of an XML document used in the Searching for Specific Nodes code snippets

```
<?xml version="1.0"?>

<!DOCTYPE root [
  <!ELEMENT Book ANY>
  <!ATTLIST Book ISBN ID #REQUIRED>
]>

<BookInventory>

  <Book ISBN="032117352X">
    <Title Lang="EN">Extending MFC Apps with .NET</Title>
    <Authors>
      <Author>Tom Archer</Author>
      <Author>Nishant S.</Author>
    </Authors>
    <MSRP Curr="USD">44.99</MSRP>
  </Book>
```

```
<Book ISBN="0201616475">
  <Title Lang="EN">Cryptography Decrypted</Title>
  <Authors>
    <Author>X.X.Mel</Author>
    <Author>Doris Baker</Author>
  </Authors>
  <MSRP Curr="USD">34.99</MSRP>
</Book>

<Book ISBN="0201734117">
  <Title Lang="EN">Essential .NET, Volume I: The CLR</Title>
  <Authors>
    <Author>Don Box</Author>
  </Authors>
  <MSRP Curr="USD">49.99</MSRP>
</Book>

</BookInventory>
```

As the name suggests, the `XmlNode::SelectSingleNode` is used for selecting single nodes based on an XPath expression. In the following example, I'm using the `SelectSingleNode` both to search for a book entitled *Cryptography Decrypted* and, once that node is located, to search for its title and price nodes. The title and price are then printed. (Note that I'm also accessing the price node's attributes collection for the currency symbol.)

```
// bookInventory is the root node from the XML in Listing 5-1:

String* search = S"//Book[Title='Cryptography Decrypted']";
XmlNode* book = bookInventory->SelectSingleNode(search);

if (book)
{
    XmlNode* title = book->SelectSingleNode(S"Title");
    XmlNode* price = book->SelectSingleNode(S"MSRP");
    if (price)
    {
        Console::WriteLine( S"Title: '{0}'\nMSRP: {1} {2}",
                            title->InnerText,
                            price->InnerText,
                            price->Attributes->Item(0)->Value);
```

```
      }
  }
```

This snippet results in the following output:

```
Title: 'Cryptography Decrypted'
MSRP: 34.99 USD
```

While the `SelectSingleNode` is great for retrieving single nodes, it will only return the first item found that matches your criteria. In cases where the search might return more than one match, you can use the `SelectNodes` method. This method also takes an `XPath` parameter that defines the search, but returns an `XmlNodeList` object containing all the matching `XmlNode` objects. In the following code snippet, I'm searching for all books that have an MSRP of less than 45.00. (The code to print the found book title, price, and currency symbol is the same as in the previous example; the new code that specifically deals with retrieving and handling multiple matches is bold.)

```
// bookInventory is the root node from the XML in Listing 5-1:

String* search = S"//Book[MSRP<45.00]";
XmlNodeList* books = bookInventory->SelectNodes(search);

if (books)
{
  Console::WriteLine(S"{0} books found\n", books->Count.ToString());
  for (int i = 0; i < books->Count; i++)
  {
    XmlNode* book = books->Item(i);
    if (book)
    {
      XmlNode* title = book->SelectSingleNode(S"Title");
      XmlNode* price = book->SelectSingleNode(S"MSRP");
      if (price)
      {
        Console::WriteLine( S"Title: '{0}'\nMSRP: {1} {2}\n",
                            title->InnerText,
                            price->InnerText,
                            price->Attributes->Item(0)->Value);

      }
    }
```

```
    }
}
```

As you can see, once the SelectNodes method returns the XmlNode List object, that object's Count property can be used to determine how many matches were located. The Item method is then used to fetch each node from the node list and print the following output:

```
2 books found

Title: 'Extending MFC Apps with .NET'
MSRP: 44.99 USD

Title: 'Cryptography Decrypted'
MSRP: 34.99 USD
```

In XML, attributes are commonly used to uniquely identify elements. For example, in this section's sample XML I have a list of books, and since the standard unique identifier for a book is its ISBN, that is defined as an attribute of the Book element. There are at least two ways to search for a node by its attribute value:

- Call the SelectSingleNode method
- Call the XmlDocument::GetElementById method

Using the SelectSingleNode method, you would simply specify the special XPath notation—@<attributeName>. Note the following example, where I've specified that within the Book node, I'm searching for an ISBN attribute (@ISBN) that has a value of 032117352X.

```
String* search = S"Book[@ISBN='032117352X']";
XmlNode* book = bookInventory->SelectSingleNode(search);
```

The second means of searching by attribute is via the GetElementById method. Here the Id part of this method name refers to a DTD ID. If you look back at Listing 5–1, you'll see the following DOCTYPE DTD at the top of the XML file:

```
<!DOCTYPE root [
  <!ELEMENT Book ANY>
  <!ATTLIST Book ISBN ID #REQUIRED>
]>
```

This basically defines an element named `Book` and states that the `Book` element has an attribute list consisting of an ISBN, which is required. Also, note the `ID` part of the `ATTLIST` tag. The `GetElementById` method will search through the document until it finds the first node containing an attribute of type `ID` and the specified value. Therefore, using this DTD, we can write the following:

```
// Because ISBN is defined as an ID, this will return the first
// element having an ISBN attribute with the value 032117352X
XmlNode* book = doc->GetElementById(S"032117352X");
```

The last search method I'll cover here is the `XmlDocument::Get ElementsByTagName` method. The `GetElementsByTagName` is designed to search an XML document for all nodes whose `Name` property matches the parameter passed. All matching nodes are returned in an `XmlNodeList` object. (Passing the special value `*` returns all nodes in the document.) This method can be used for many purposes, but is generally used to retrieve the top-level nodes of a document.

The following call would retrieve all the `Book` nodes from the sample XML in Listing 5–1:

```
// Get all book nodes
XmlNodeList* books = doc->GetElementsByTagName(S"Book");
```

Now that you've seen how to read, search, and edit XML using the `XmlDocument` and `XmlNode` classes, I'll present two demos. The first allows you to open and display (in a tree view control) any XML file. The second is a simple student maintenance application.

Demo to Display XML Files

At the beginning of this chapter, I alluded to an XML demo I wrote for another book that used the COM interface to the MS XML DOM. That application allowed users to open any valid XML file, iterated through its nodes, and displayed its elements and attributes in a tree view. In this section, you'll see that same demo accomplished with half the code, thanks to the `XmlDocument` class.

1. After creating an MFC dialog-based application called `Xml DocumentDisplayFile`, update the Project properties to support Managed Extensions.

2. Open the `stdafx.h` file and add the following .NET support directives and `struct` definition to the end of the file. (In the 1.1 release of the .NET Framework, the `_TREEITEM` was inadvertently omitted from the Platform SDK header file. Therefore, this `struct` definition is needed in order to use the tree view control in mixed-mode applications.)

```
#using <mscorlib.dll>
#using <System.dll>
#using <System.xml.dll>

using namespace System;
using namespace System::Xml;
using namespace System::IO;

struct _TREEITEM {};
```

3. Open the dialog template resource and add the controls as you see them in Figure 5–7.

4. Set the tree view control's `Has Buttons`, `Has Lines`, and `Lines At Root` properties to `True` and set the static control's `Path Ellipsis` property to `True`.

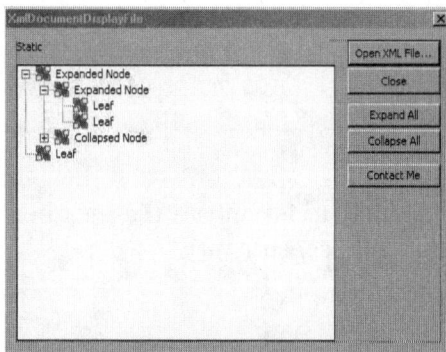

Figure 5-7 Dialog resource for the `XmlDocumentDisplayFile` demo application

Table 5–4 DDX Variables for the `XmlDocumentDisplayFile` Demo

Control	Variable Type	Variable Name
Static control	CString	m_strFileName
Tree view	CTreeCtrl	m_treeFileContent

5. Create the DDX member variables as shown in Table 5–4.
6. Now that the UI is taken care of, add the following member variable to the dialog class's header file. As in previous demos, this value will be used to store the application's current directory for use in setting the File Open common dialog box's initial directory.

```
protected:
    CString m_strWorkingDir;
```

7. Update the `OnInitDialog` with the following code just prior to the return statement.

```
TCHAR buff[MAX_PATH];
GetModuleFileName(NULL, buff, MAX_PATH);
m_strWorkingDir = Path::GetDirectoryName(buff);
```

8. Define a pointer to an `XmlDocument` object in the dialog class header file. (You'll remember from Chapter 1 that the `gcroot` template allows you to define a managed type within an unmanaged type.)

```
protected:
    gcroot<XmlDocument*> m_plDomDocument;
```

9. Update the `OnInitiDialog` function (just before the `return` statement) to instantiate the member `XmlDocument` object using the default constructor:

```
#pragma push_macro("new")
#undef new
    m_plDomDocument = new XmlDocument();
#pragma pop_macro("new")
```

10. Now let's add the code to open the XML file. Start by coding an event handler for the Open XML File . . . button as follows. As you can see, this function simply allows the user to specify an XML file using the File Open common dialog box. Once the user does that,

the specified file name is used in a call to the XmlDocument::Load method, the root node is retrieved (via the XmlDocument:: DocumentElement property, and a recursive function—Display Children—is called to display the nodes of the XML file in the tree view.

```
void CXmlDocumentDisplayFileDlg::OnBnClickedOpenXMLFile()
{
  // Let user specify XML file to display
  CFileDialog dlg(TRUE);
  dlg.m_pOFN->lpstrInitialDir = m_strWorkingDir;
  dlg.m_pOFN->lpstrTitle = _T("Open an XML File");
  dlg.m_pOFN->lpstrFilter = _T("XML Files (*.xml)\0*.xml\0");

  if (IDOK == dlg.DoModal())
  {
    m_strFileName = dlg.GetPathName();
    m_treeFileContent.DeleteAllItems();
    UpdateData(FALSE);

    // specify xml file name
    CString strFileName(m_strFileName);

    try
    {
      // Load the XML file
      m_plDomDocument->Load(m_strFileName);

      // Get the root node of the XML file
      XmlNode* pRootNode = m_plDomDocument->DocumentElement;

      // Pass root to recursive DisplayChildren function
      DisplayChildren(TVI_ROOT, pRootNode);
    }
    catch(Exception* e)
    {
      AfxMessageBox((CString)e->Message);
    }

    UpdateData(FALSE);
  }
}
```

11. Declare in the dialog class and code the following `Display Children` function. As you can see, this function uses the same technique for reading through a document's node structure that you saw earlier in the section entitled "Enumerating the Elements of an XML Document."

```
void CXmlDocumentDisplayFileDlg::DisplayChildren(
    HTREEITEM hParent, XmlNode* pParent)
{
    // display the current node's name
    HTREEITEM hItem = DisplayChild(hParent, pParent);

    // simple for loop to get all children
    for (XmlNode* pChild = pParent->FirstChild;
        NULL != pChild;
        pChild = pChild->NextSibling)
    {
        // for each child, call this function so that we get
        // its children as well
        DisplayChildren(hItem, pChild);
    }
}
```

12. As you saw in the previous step, the `DisplayChildren` function calls the `DisplayChild` function to displays the contents of each node. This function should be coded as follows, where a switch statement is used to determine the node type and set a local `CString` variable that will be inserted into the tree view control. After that, the current node's `XmlAttributeCollection` is enumerated, where each attribute name/value pair is formatted and appended to the element string, which is then inserted into the tree view control.

```
HTREEITEM CXmlDocumentDisplayFileDlg::DisplayChild(
    HTREEITEM hParent, XmlNode* pChild)
{
    // Determine the node type and, based on that type,
    // retrieve the value we want in the tree view.
    CString strElement;
    switch (pChild->NodeType)
    {
        case XmlNodeType::Element:
            strElement = (CString)pChild->Name;
        break;
```

```
case XmlNodeType::Text:
  strElement = (CString)pChild->Value;
break;


default:
  strElement = (CString)pChild->Name;
break;
}

// Get all the attributes for this node.
XmlAttributeCollection* pAttributes = pChild->Attributes;
if (pAttributes)
{
  CString strAttrs;
  for (int i = 0; i < pAttributes->Count; i++)
  {
    strAttrs.Format(_T("%s%s = \"%s\"%s"),
              (0 == i ? _T(" [") : _T("")),
              (CString)pAttributes->Item(i)->Name,
              (CString)pAttributes->Item(i)->Value,
              ((pAttributes->Count == (i + 1)) ? _T("]") :
               _T(", "))
             );
    strElement += strAttrs;
  }
}


// Add the element's node name to the tree view
// and return the HTREEITEM.
return m_treeFileContent.InsertItem(strElement, hParent);
}
```

Once you've finished coding the application, build and run it. Figure 5–8 shows an example of running this application.

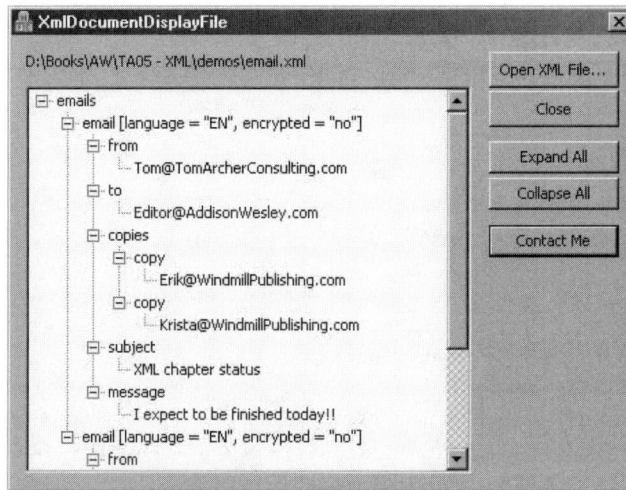

Figure 5-8 The XmlDocument class enables you to parse XML documents in half the time it takes with the COM interface to the XML DOM.

Simple Maintenance Applications Using the DOM

In the previous XmlDocumentDisplayFile demo, you saw how to load, parse, and display a complete XML document. In this demo, I'll illustrate how to write a basic maintenance application using the XmlDocument and XmlNode classes. This application will load a combo box control from an XML file containing students and grades for various programming courses. Upon selection of a student, the student's information will be read from the DOM and the dialog populated. When the user has finished editing the student's information, clicking the Save to Memory button will write the changes back to the DOM. Clicking the Commit All Changes button will write the DOM to disk.

While the scope of this application is intentionally small in order to focus on XML, the basic functions employed here are the same as those of any maintenance application. To that end, the following code will implement many of the things you've learned in this section—including loading an XML file, enumerating a key value and its attribute (the student and student ID), and saving the document's changes to disk. As the major benefit of using the DOM is its ability to allow you to dynamically search for specific nodes, the application makes use of the following methods: GetElementsByTagName, SelectSingleNode, GetElementById and SelectNodes.

Before we get started, look at Listing 5–2, which shows the XML file that this application is designed to maintain. This file (`students.xml`) is also included on the CD in the `xmlfiles` folder.

Listing 5-2 Sample XML document in `XmlDocumentMaintenance` demo

```
<?xml version="1.0"?>
<!DOCTYPE root [
  <!ELEMENT Student ANY>
  <!ATTLIST Student ID ID #REQUIRED>
]>
<Students>
  <Student ID="100">
    <Name>Tom Archer</Name>
    <Grades>
      <MFC>96</MFC>
      <ATL>66</ATL>
      <Winsock>78</Winsock>
    </Grades>
  </Student>
  <Student ID="101">
    <Name>Karen Colley</Name>
    <Grades>
      <MFC>94</MFC>
      <ATL>91</ATL>
      <Winsock>92</Winsock>
    </Grades>
  </Student>
  <Student ID="103">
    <Name>Krista Crawley-Archer</Name>
    <Grades>
      <MFC>88</MFC>
      <ATL>78</ATL>
      <Winsock>65</Winsock>
    </Grades>
  </Student>
  <Student ID="104">
    <Name>Vicki Tyson</Name>
     <Grades>
        <MFC>48</MFC>
        <ATL>63</ATL>
        <Winsock>72</Winsock>
     </Grades>
```

```
    </Student>
</Students>
```

Now let's see how to write an application to maintain this student XML file:

1. After creating an MFC dialog-based application called `Xml DocumentMaintenance`, update the Project properties to support Managed Extensions

2. Open the `stdafx.h` file and add the following .NET support directives and `struct` definition to the end of the file.

```
#using <mscorlib.dll>
#using <System.dll>
#using <System.xml.dll>

using namespace System;
using namespace System::IO;
using namespace System::Xml;
```

3. Open the dialog template resource and add the controls as you see them in Figure 5–9.

4. Set the combo box control's `Type` property to `Drop List`. Set the Student ID edit control's `Read Only` property to `True`.

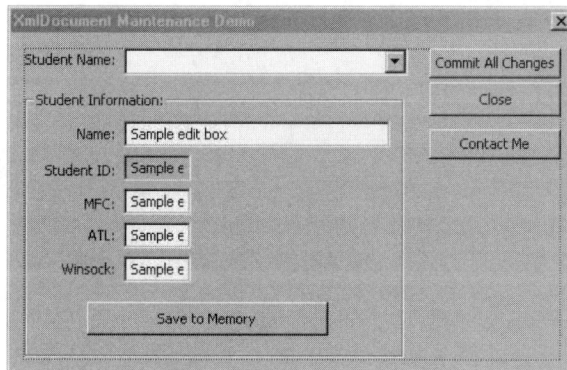

Figure 5–9 Dialog resource for the `XmlDocumentMaintenance` demo application

Table 5-5 DDX Variables for the `XmlDocumentMaintenance` Demo

Control	Variable Type	Variable Name
Student combo box	CComboBox	m_cbxStudents
Student name edit	CString	m_strName
Student ID edit	int	m_iID
MFC Grade edit	int	m_iGradeMFC
ATL Grade edit	int	m_iGradeATL
Winsock Grade edit	int	m_iGradeWinsock

5. Create the DDX member variables as shown in Table 5–5.

6. Add an `XmlDocument` member variable named `m_pDocument` to the dialog class:

```
class CXmlDocumentMaintenanceDlg : public CDialog
{
...
protected:
  gcroot<XmlDocument*> m_pDocument;
```

7. Instantiate the `m_pDocument` object and load the `students.xml` file in the `OnInitDialog` function just before the `return` statement. (Note that the `students.xml` file must be in the application directory for this application to perform correctly.)

```
BOOL CXmlDocumentMaintenanceDlg::OnInitDialog()
{
...

#pragma push_macro("new")
#undef new
  m_pDocument = new XmlDocument();

    TCHAR buff[MAX_PATH];
    GetModuleFileName(NULL, buff, MAX_PATH);
  String* sFileName =
    Path::Combine(Path::GetDirectoryName(buff),
    S"students.xml");
```

```
    try
    {
      m_pDocument->Load(sFileName);
      LoadStudents();
    }
    catch(Exception* pe)
    {
      m_pDocument = NULL;
      AfxMessageBox((CString)pe->Message);
    }
  #pragma pop_macro("new")

   return TRUE;
  }
```

8. Add the following helper member function to the dialog. This
function will read the student from the DOM and populate the dialog box's combo box control. As you can see, the function uses the
GetElementsByTagName to retrieve all of the Student nodes. As
each Student node is extracted from the student list (in a for
loop), the student's ID is saved and the student's Name value is
retrieved. Once that is done, the Name node value is added to the
combo box and that combo box element's item data is set to the student ID (so that we can easily retrieve the student's information
when selected).

```
void CXmlDocumentMaintenanceDlg::LoadStudents()
{
  // Get student node list
  XmlNodeList* pStudentNodes =
    m_pDocument->GetElementsByTagName(S"Student");

  XmlNode* pStudentNode;
  int idx;
  int iStudentId;
  CString strName;

  // For each student node...
  for (int i = 0; i < pStudentNodes->Count; i++)
  {
    pStudentNode = pStudentNodes->Item(i);
    if (pStudentNode)
```

```
        {
          // ...retrieve its id attribute
          iStudentId =
            Convert::ToInt32(pStudentNode->Attributes->Item(0)->
              Value);

          // ...retrieve its name
          strName = (CString)pStudentNode->Item["Name"]->
            InnerText;

          // ...insert student name into combo box
          idx = m_cbxStudents.AddString(strName);

          // ...and set its item data to the student id.
          m_cbxStudents.SetItemData(idx, (DWORD_PTR)iStudentId);
        }
      }
    }
```

9. Add an event handler for the combo box control's CBN_SELCHANGE notification message. This function first retrieves the item data (representing the student ID) for the currently selected student. The GetElementById method is then utilized to retrieve the Student node by this value. Once the Student node has been retrieved, its Grade node is retrieved—via a call to Select SingleNode—and finally the grade values are obtained.

```
void CXmlDocumentMaintenanceDlg::
  OnCbnSelchangeComboStudentNames()
{
  int iCurrSel = m_cbxStudents.GetCurSel();
  if (CB_ERR != iCurrSel)
  {
    // Get the student id from cbx item data
    m_iID = (int)m_cbxStudents.GetItemData(iCurrSel);

    // Retrieve the Student node using the student id
    // attribute
    XmlNode* pStudentNode =
      m_pDocument->GetElementById(m_iID.ToString());

    if (pStudentNode)
    {
```

```
// Retrieve the student name
m_strName = (CString)pStudentNode->Item[S"Name"]->
  InnerText;

// Acquire the Grades node
XmlNode* pGradesNode = pStudentNode->
  SelectSingleNode(S"Grades");
if (pGradesNode)
{
  // Retrieve each grade value
  m_iGradeMFC =
    Convert::ToInt32(pGradesNode->Item[S"MFC"]->
      InnerText);

  m_iGradeATL =
    Convert::ToInt32(pGradesNode->Item[S"ATL"]->
      InnerText);

  m_iGradeWinsock =
    Convert::ToInt32(pGradesNode->Item[S"Winsock"]->
      InnerText);
}
}
}

UpdateData(FALSE);
}
```

10. At this point, the application can load the students.xml file and retrieve student information as students are selected in the combo box. Now to save the data (in memory), implement a handler for the Save to Memory button. For the most part, this function is just the reverse of the student selection process.

```
void CXmlDocumentMaintenanceDlg::OnBnClickedSaveToMemory()
{
  UpdateData();

  int iCurrSel = m_cbxStudents.GetCurSel();
  if (CB_ERR != iCurrSel)
  {
    // Get the student id from cbx item data
    m_iID = (int)m_cbxStudents.GetItemData(iCurrSel);
```

```
// Retrieve the Student node using the student id
    attribute
XmlNode* pStudentNode = m
  _pDocument->GetElementById(m_iID.ToString());
if (pStudentNode)
{
  // Save the student name
  pStudentNode->Item[S"Name"]->InnerText = m_strName;

  // Acquire the Grades node
  XmlNode* pGradesNode =
    pStudentNode->SelectSingleNode(S"Grades");
  if (pGradesNode)
  {
    // Save each grade value
    pGradesNode->Item[S"MFC"]->InnerText =
      m_iGradeMFC.ToString();
    pGradesNode->Item[S"ATL"]->InnerText =
      m_iGradeATL.ToString();
    pGradesNode->Item[S"Winsock"]->InnerText =
      m_iGradeWinsock.ToString();

    // Did name change?
    CString strPreviousName;
    m_cbxStudents.GetWindowText(strPreviousName);
    if (0 != strPreviousName.Compare(m_strName))
    {
      // Name change!
      m_cbxStudents.DeleteString(iCurrSel);
      int idx = m_cbxStudents.AddString(m_strName);
      m_cbxStudents.SetItemData(idx, (DWORD_PTR)m_iID);
      m_cbxStudents.SetCurSel(idx);
    }

    AfxMessageBox(_T("Student information saved"));
  }
 }
}
}
```

11. Now that the students' data can be retrieved, modified, and saved to memory, it's time to persist that data. To do this, simply imple-

ment a handler for the Commit All Changes button as follows. As
you can see, the function is very straightforward—it simply determines the file name and passes that name to the m_pDocument
object's Save method.

```
void CXmlDocumentMaintenanceDlg::OnBnClickedCommitAllChanges()
{
  try
  {
    TCHAR buff[MAX_PATH];
    GetModuleFileName(NULL, buff, MAX_PATH);
    String* sFileName =
      Path::Combine(Path::GetDirectoryName(buff),
        S"students.xml");

    m_pDocument->Save(sFileName);

    AfxMessageBox(_T("File saved successfully"));
  }
  catch(Exception* pe)
  {
    AfxMessageBox((CString)pe->Message);
  }
}
```

12. That's all there is to writing a maintenance application using the
DOM! Obviously, a real application would have more error checking, but the various XML class methods and properties used will
remain largely the same regardless of the overall complexity of the
application. Figure 5–10 shows the application in use.

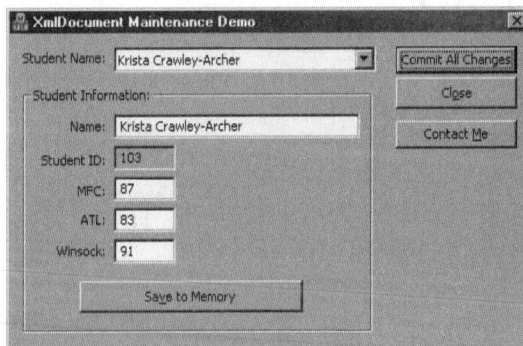

Figure 5-10 The XmlDocumentMaintenance demo application illustrates how
easy it is to read and write data using the DOM.

Summary

In the first section of this chapter, I illustrated how the `XmlTextWriter` and `XmlTextReader` classes are used to process XML documents in a sequential, fast-forward manner. In that section, you discovered how to read and write the various nodes of an XML document—including the declaration node, elements (using data of various application-specific types), and attributes. The section concluded with a demo illustrating the maintenance of a single-entity XML file using the `XmlTextWriter` and `XmlText Reader` classes. The second section then dove into the more elaborate world of the DOM and the XML classes that encapsulate it—`XmlDocument` and `XmlNode`. In that section, I first presented a short introduction to the DOM itself, followed quickly by several task-oriented subsections that illustrated such common functions as loading and saving documents and enumerating XML element. From there, you learned how to read and edit nodes from and to memory, how to retrieve and set the XML and text values of entire node branches, and how to search for specific nodes or groups of nodes based on tag names or values using XPath. You even saw how to define an element attribute as being a unique identifier in your document (using DTD) so that specific nodes can be located when their element values are duplicated. That section (and the chapter) concluded with a pair of demos. The first illustrated how to parse and read any XML document and was included as a comparison point between using the (older) XML DOM COM interfaces and the BCL `XmlDocument` and `XmlNode` classes. The second demo illustrated a basic maintenance application using these classes.

Even in more than 40 pages full of class descriptions, code snippets and demo applications, this chapter only barely scratched the surface of the incredible array of XML classes provided by the BCL. However, with what you've seen here, you should now realize just how easily you can incorporate these classes into your MFC applications. In the next chapter, I'll continue presenting BLC classes that help to both augment your knowledge and make you a more productive MFC developer by covering one of most exciting aspects of .NET—the ADO.NET classes.

Disconnected Data via ADO.NET and DataSets

Introduction

During the beta for Windows 2000, Microsoft released a very exciting part of COM+ that would allow for the manipulation of disconnected data in memory. This technology—called IMDB (in-memory database)—promised to solve a host of problems associated with distributed database applications, as it meant that an application could connect to a data source, retrieve into memory the necessary data, disconnect from the data source, and then read from and manipulate the in-memory representation of the data without incurring constant round-trips to the data store or keeping the connection alive for long periods of time, thereby locking crucial data. Once finished with its work, the application could then connect the IMDB back to its data store and synchronize any changes that were made. Unfortunately, just as Release Candidate 2 for the Windows 2000 operating system was being released, Microsoft announced that for various reasons IMDB just wasn't ready for prime time and would not ultimately be a part of the Windows 2000 final release.

While many of us were very disappointed by this decision, what we could not have known at the time was that IMDB was taking a slightly different—and more powerful—form and being included into something then called NGWS (Next Generation Windows Services)—what we now know as .NET. This .NET incarnation of an in-memory, disconnected data model is only one part of the overall ADO.NET data access layer. However, as MFC developers, we have plenty of data access technologies to pick from (MFC database classes, ODBC, OLEDB, ADO, DAO, etc.). Therefore,

this chapter will focus on the aspects of ADO.NET that we do not have—specifically the various classes that support the original concepts of IMDB.

The chapter begins with a section illustrating the basic terminology and classes specific to ADO.NET and disconnected data as well as the basics of constructing those objects and connecting to a data source. From there, you'll begin a section illustrating how to perform basic database operations (creating, reading, updating, and deleting of records). Included in that section are tips on handling the common issues of dealing with disconnected data and auto-increment primary keys, and how to specify primary key information for untyped datasets. The section ends with a fully functional MFC maintenance application using the ADO.NET against a sample database. At this point, you'll be fairly comfortable with the basics of reading and writing data to a disconnected data table. I'll then illustrate several techniques for reading and writing data in batch in order to keep round trips to the data server to a minimum and dramatically speed up insert operations. One of the key benefits of an IMDB is its support of the ability to search, sort, and filter data once it's been retrieved. Therefore, the next section illustrates how to accomplish these tasks as well as how to create multiple filtered views on a single dataset. Finally, the chapter ends with a section on generating strongly typed datasets, using them in an MFC application, and the pros and cons of using untyped vs. typed datasets.

ADO.NET Basics

As programmers, many of us like to jump right in and start using a technology with so much promise. However, logic dictates that we take a minute to familiarize ourselves with the terms and classes that implement this technology. Therefore, this section is meant as a primer for the rest of the chapter. I'll begin by quickly going over the terminology and main classes associated with disconnected data and then introduce a few tasks and concepts that will be used or referenced throughout the chapter.

ADO.NET Terminology and Main Classes

The first new term you'll hear quite often regarding ADO.NET is that of a *managed provider.* This is simply the .NET equivalent of terminology that was originally introduced with OLEDB (and later used by its COM

interface, ADO). In OLEDB, code that provides a generic interface to data is referred to as a *provider.* Therefore, since code written to run on top of the CLR is called "managed," we are given yet another new database term to remember. As of the time of this writing, the .NET Framework defines five managed providers:

- **OLEDB:** Supports data stores that have an OLEDB provider.
- **ODBC:** Supports data stores that have an ODBC driver.
- **Oracle:** A set of classes optimized for the Oracle database product.
- **SQL CE:** A .NET Compact Framework managed provider that supports Microsoft SQL Server CE.
- **SQL Server:** A set of classes that are optimized to support the Microsoft SQL Server database product.

While we're on the topic, I'll also be using the familiar terms *data source* and *data consumer.* (*Data source* is the generic name for the data being provided for consumption by the consumer.) Obviously, the consumer is any code that retrieves, stores, and manipulates data represented by the managed provider.

Like many other frameworks that you've seen throughout this book, ADO.NET is comprised of many classes. However, this chapter will focus on the following classes:

- **Connection**: Functions much like the ADO object of the same name and represents a connection to a data source.
- **Command**: Another holdover from ADO, the Command object represents a query or a command that is to be executed by a data source.
- **CommandBuilder**: Used to automatically generate the insertion, update, and delete commands for the data adapter object based on the select command. It is also used to provide optimistic concurrency for disconnected DataSet objects.
- **DataSet**: One of the key elements with ADO.NET is the DataSet. A little too involved to be defined with a single sentence, the DataSet represents an in-memory model of disconnected data and has built-in support for XML serialization. That latter capability is covered in Chapter 8, "Combining ADO.NET and XML."
- **DbDataAdapter**: The abstract base class for all data store–specific classes such as SqlDataAdapter, OracleDataAdapter, OleDbDataAdapter, and so on.
- **DataAdapter**: The base class for the DbDataAdapter class.

- **Data adapter**: Not really a class, but a generic designation for one of the `DbDataAdapter`-derived classes.
- **`DataView`**: This class is most easily defined to MFC developers as the data equivalent of a `CView` class for data. For example, in a standard MFC document/view class you can build multiple views that are built on—but work with different parts of—the same data. Likewise, multiple `DataView` objects represent different views on the same `DataSet`.
- **`XmlDataDocument`**: Enables you to treat `DataSet` data as XML data in order to support things like `XPath` search expressions, XSL (eXtensible Stylesheet Language) transformations, and so on.

Now that you've been introduced to the terms, it's easier to define a managed provider as a group of classes that interface to the generic `DataSet` class to abstract you from the specifics of the data you are reading or modifying. For example, the `System::Data::SqlClient` namespace defines about 15 classes and several delegates that are optimized for use with the SQL Server database product. Among these classes are derived types of the base classes I mentioned in the previous list: `SqlConnection`, `SqlDataAdapter`, `SqlCommand`, and `SqlCommandBuilder`.

Let's now look at the `DataSet` class a bit more closely. The `DataSet` class is a collection of data structures (other classes) that are used to model relational data. The following list details the main classes that comprise either the `DataSet` class or one of its member classes:

- **`DataTable`**: If you're familiar with ADO, then at first glance you might be tempted to think of a `DataSet` class as being comparable to souped-up ADO `Recordset` objects. However, datasets are so encompassing that there is no equivalent in ADO for them. The `DataTable` class, on the other hand, is a more true ADO.NET equivalent of the ADO `Recordset` object, as it encapsulates a two-dimensional array (rectangle) of data organized into columns and rows.
- **`DataColumn`**: Within the `DataTable` class are a collection of `DataColumn` definitions. As the `DataRow` class (described next) defines actual data, the `DataColumn` class defines the data store column definitions. Example members of this class are `ColumnName` and `DefaultValue` as well as Boolean properties such as `AllowDBNull`, `AutoIncrement` and `ReadOnly`.

- **DataRow**: The `DataRow` class encapsulates the data for a given `DataTable` object in addition to defining many members that support the disconnected capabilities of the `DataSet/DataTable`. These members include support for tracking the current and original values of each column, the current state of the row (a `DataRow` `State` enumeration with such values as `Added`, `Deleted`, `Detached`, `Modified`, and `Unchanged`) and a connection to the parent table to support `DataRelation` via the `GetParentRows` and `GetChildRows` methods
- **DataRelation**: `DataRelation` objects are used to define how multiple `DataTables` are associated. For example, it is quite common to use this feature when dealing with tables that have a parent/child relationship, such as order header and order detail tables. Using this feature, you can more easily navigate the related data of these two tables. This class is covered in more detail in the next chapter.
- **Constraint**: Each `DataTable` defines a collection of constraints that specify rules for maintaining data integrity. For example, when you delete a value that is used in one or more related tables, a `ForeignKeyConstraint` determines whether the values in the related tables are also deleted, set to null values, set to default values, or whether no action occurs.

Constructing and Filling `DataSet` Objects

Now that that you've been introduced to the main ADO.NET classes that will be used throughout this chapter, let's take a look at a code snippet that illustrates how to connect to and retrieve data from a data source. After the code snippet, I'll provide a walkthrough of the various classes that are being used here as well as a lot of not-so-obvious tasks that are being performed for us in order to facilitate a disconnected dataset.

```
SqlConnection* conn =
  new SqlConnection(S"Server=fantine;"
                    S"Database=Northwind;"
                    S"Integrated Security=true;");
SqlDataAdapter* adapter =
  new SqlDataAdapter(S"SELECT * FROM Employees", conn);
SqlCommandBuilder* cmd = new SqlCommandBuilder(adapter);
conn->Open();
```

```
DataSet* dataset = new DataSet();
  adapter->Fill(dataset, S"AllEmployees");
conn->Close(); // No longer needed
DataTableCollection* tables = dataset->Tables;
employeesTable = tables->Item[S"AllEmployees"];
// ... Use employees table as needed.
```

While this code looks pretty straightforward, there's much more going on here than meets the eye.

- The first thing the code snippet does is to connect to the SQL Server sample database, Northwind, using the `SqlConnection` class.

Specifying Connection Strings for Different Database Products

For this chapter, I chose to use the SQL Server database as it's the most commonly used database among Visual C++/MFC professionals. In addition, while much of the code that you'll see in this chapter can easily be massaged to work with any managed provider, the initialization of the `Connection` object is data source–specific. Therefore, if you are using another product, such as Oracle or Microsoft Access, or want to use the OLEDB or ODBC interfaces to these or other databases, the http://www.connectionstrings.com Web site is an invaluable resource, as it contains connection strings for virtually every data store.

- Once that is done, the code uses a `DbDataAdapter`-derived class (`SqlDataAdapter`) designed specifically for SQL Server access. As mentioned in the previous section, the data adapter is what connects a dataset to the underlying data store. However, what's really interesting here is that while I'm passing a "select" value to the `SqlData Adapter` class' constructor, the various data adapter classes define four distinct commands (in the form of `SqlCommand` classes): `Select Command`, `InsertCommand`, `UpdateCommand`, and `DeleteCommand`. (From here on, the latter three commands will be referred to en masse as *action commands*.) One extremely important note to make here is that *the data adapter does not automatically generate commands to reconcile changes made to a dataset based on the `select` statement used to construct the adapter.* You must either set these commands yourself or use a command builder class, which segues nicely into the next items of interest from the code snippet.

- Once the data adapter has been constructed with the desired select command, an `SqlCommandBuilder` object is instantiated and associated with the data adapter. The command builder automatically generates the appropriate action commands (complete with the underlying SQL code, ADO.NET `Command` objects, and their associated `Parameters` collections) based on the adapter's `Select Command`.

- Next, the connection is opened. One thing to note here is that the data adapter is designed to minimize the time a connection stays open. As you see more code in this chapter, take note that the data adapter's associated connection is never explicitly opened or closed. Instead the adapter knows when it needs to connect and disconnect. For example, when calling the data adapter object's `Update` method in order to commit changes to the dataset, the data adapter will automatically use an already open connection to the data store or make the necessary connection and automatically disconnect when finished.

- After that, we're finally down to the `DataSet` object itself. To construct and fill the `dataset`, you can simply use the `DataSet` class's default constructor and then call the data adapter object's `Fill` method, passing the constructed `DataSet` object as the first parameter. The `Fill` method retrieves data from the underlying data store based on the data adapter's `SelectCommand` value. (In this example, that value was set when the `SqlDataAdapter` was constructed.) You'll also notice that I specified a literal value of `"AllEmployees"` for the second parameter to the `Fill` method. This value specifies the name that I wish to give the `DataTable` that will be constructed with the returned data. If I had not named the dataset's data table, it would have been named "Table" automatically. (When more than one data table are generated and not specifically named, they are assigned the names Table1, Table2, and so on.)

Creating Multiple `DataTables` in a `DataSet`

While most of the chapter's code snippets and demo applications will only read and modify a single table, there might be times when you'll want a `DataSet` to contain multiple `DataTable` objects. The section entitled "Creating Multiple `DataTables` in a `DataSet`" will illustrate how to do this both by using multiple data adapters and also by combining multiple `SELECT` statements in a single data adapter in order to reduce round trips to the server.

- At this point, the requested data is in the `DataRow` members of the `DataTable` members of the `DataSet` object. Therefore, the code can safely disconnect from the data source and continue working until it wants to commit any changes made to the data!
- The last thing I'll illustrate here before moving on to the next code snippet is how to retrieve the desired `DataTable` objects from the `DataSet` object. As you can see from the code, the `DataSet` class has a public property called `Tables` that is simply a collection of `Data Table` objects. As with accessing any other .NET collection with Managed Extensions, you can use one of two overloaded `Item` indexers—one accepts the relative index and the other the named entity. Therefore, as the data adapter in this code snippet only constructed a single `DataTable` object that was named `"All Employees"` in the `Fill` method, it can be retrieved either by name or by passing its index value of 0.

Different Ways to Construct Datasets

There are three distinct methods to constructing and filling datasets. One way—used in this chapter—is from a data adapter (which is typically associated with a database). You can also construct a dataset programmatically from any data your application has access to, either read from another source or generated within the application. This technique—while not overly difficult—is not used very often and is beyond the scope of this chapter. Finally, you can also construct a `DataSet` object from an XML document in situations where you wish to treat XML data as you would any other data format. The topic of mixing ADO.NET and XML is covered in Chapter 8.

Untyped vs. Typed Datasets

There are two basic ways to use the `DataSet` objects: *untyped* and *typed*. When using untyped datasets, you use the base BCL-provided `DataSet` objects and pass the relevant information that specifies which table, column, row, and so on that you're working with. For example, let's say that you're working with a row of data (represented by a `DataRow` object) for a table that contains a column named `FirstName`. For each row, you could access and modify the `FirstName` column as follows:

```
// row is a DataRow object
// Retrieve value
```

```
String firstName = row->Item[S"FirstName"]->ToString();
// Set value row->Item[S"FirstName"] = S"Krista";
```

The DataRow—needing to be a generic interface for all data—provides methods for reading and updating column values, respectively, where you're responsible for specifying the column name and—if updating—an Object representing the value. This generic approach, which makes you responsible for the specifics, is used throughout all the DataSet classes. Therefore, the main drawback to untyped datasets is that the code is not type-safe. In other words, mistakes made in your code, such as misspelling the column name or passing an incompatible data type, will only be realized at runtime.

Typed datasets, on the other hand, are classes that are generated from a specified data store. It's important to realize that these classes are still directly derived from the ADO.NET base classes. However, they contain members specific to the data store schema and, as such, allow for compile-time error checking. To continue our Employees table example, a typed DataSet would include a DataRow-derived class called EmployeesRow. This class would then define members for each column in the Employees table, as shown in the following excerpt.

```
public: EmployeesDataSet::EmployeesRow*  AddEmployeesRow
(
   System::String*  LastName,
   System::String*  FirstName,
   System::DateTime HireDate,
   System::Byte Photo[],
   System::String*  Notes,
   System::Int32 ReportsTo
);
```

Using the typed dataset, our read and update code becomes the following:

```
// row is an EmployeesRow object
// Retrieve value
String firstName = row->FirstName;
// Set value row->FirstName = S"Krista";
```

As you can see, the main benefits to typed datasets are better readability and compile-time type checking—as each column is a class member

that is associated with its correct type within the class. To draw a parallel between typed datasets and our MFC world, you could say that typed datasets are analogous to using the MFC ODBC Consumer Wizard to generate a `CRecordSet` class. The main difference is that while the various ADO.NET classes can be bound to .NET Windows Forms controls, they were designed for a managed world; thus, there's nothing akin to RFX that will automatically bind the data to our MFC dialogs/views and controls. That we have to do manually.

I'll get into more of the advantages and disadvantages of using typed datasets in the section entitled "Working with Typed Datasets." However, I at least wanted you to know at this point that they both exist and to understand the main differences between them. Also note that while typed datasets have some obvious advantages, this chapter will use mostly untyped datasets for the following reasons:

1. Untyped datasets allow you to see more easily what is really going on in code snippets as the client code explicitly states table and column names, store-procedure parameter names, and so on, as opposed to the actual database entity names being hidden in a class.
2. Untyped datasets allow for shorter, more focused code snippets and demos. Otherwise, each demo would require extra steps to create the typed datasets and then would require a lot of cross referencing between the main code and the typed `DataSet` class code.

Basic Database Operations with ADO.NET

Whether you're working with a connected or disconnected data store, the majority of database operations involve **NURD** work—**N**ew, **U**pdate, **R**ead, **D**elete. However, as this section will illustrate, many of the sometimes very tedious database operations are made much easier with the help of the various ADO.NET classes.

Quick Note on This Section's Examples

This section's code snippets are all freestanding functions that can be plugged directly into your own test applications. They make the sole assumption that the `DataSet`, `DataAdapter`, `DataTable`, and `CommandBuilder` objects have all been properly

constructed. For example, in an SDI application you might declare each of these as member variables of the view and instantiate them (as illustrated in the section entitled "Constructing and Filling DataSet Objects") in the view's `OnInitialUpdate` function. This way, you can follow along, trying the various code snippets without having to see the same connection code repeated over and over in each code example.

Reading Data

The first "real" task that many people new to a given database access layer want to explore is that of reading data. In the previous sections, you learned how to connect to a data store and construct a `DataSet` object that in turn contains a collection of `Table` objects. That is where we'll pick up here—the objects within the `Table` object.

The `Table` class contains two integral collections that you'll use most often: one for holding all columns (`DataColumnCollection`) and one for holding all returned rows from a query or command (`DataRowCollection`). Within these collections are held the `DataColumn` and `DataRow` objects, respectively. Figure 6–1 shows the relationship between these classes.

Let's start with the `DataRowCollection` and `DataRow` classes. The `DataRowCollection` is simply the collection of `DataRow` objects returned from the query or command executed by the adapter. This class is used to

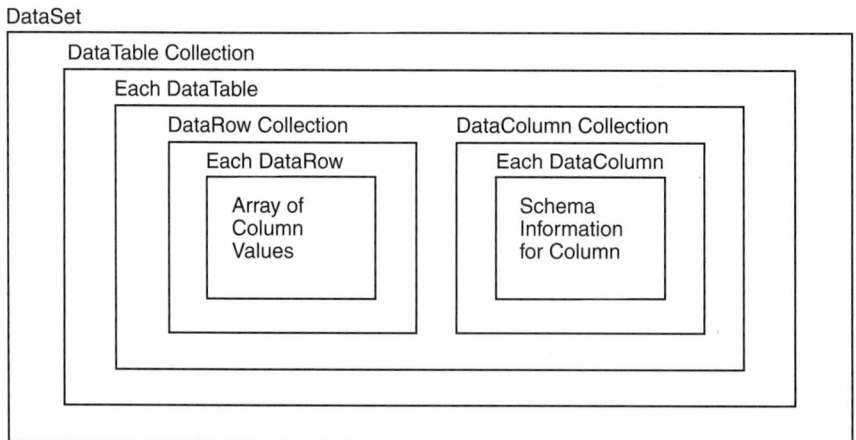

Figure 6–1 Basic relationship between `DataSet`, `DataTable`, `DataRow` `Collection`, `DataRow`, `DataColumnCollection`, and `DataColumn` classes

perform operations on the entire set of rows, such as inserting new rows, deleting rows, and searching for rows. Each of these tasks will be covered shortly. For now, we'll just be enumerating this collection in order to get at the row objects. To see how these two classes are used to read elements from a DataSet, take a look at the following function (ListAllEmployees) that connects to the sample SQL Server database Northwind, enumerates its Employees table, and retrieves the FirstName and LastName columns for each row:

```
void ListAllEmployees()
{
#pragma push_macro("new")
#undef new
  try
  {
    // Assumes employeesTable has already been filled by an adapter
    DataRowCollection* rows = employeesTable->Rows;
    DataRow* row;
    String* id;
    String* firstName;
    String* lastName;

    for (int i = 0; i < rows->Count; i++)
    {
      // Get DataRow object
      row = rows->Item[i];

      id = row->Item[S"EmployeeID"]->ToString();
      firstName = row->Item[S"FirstName"]->ToString();
      lastName = row->Item[S"LastName"]->ToString();
    }
  }
  catch(Exception* e)
  {
    MessageBox::Show(String::Format(S"Exception : {0}", e->Message));
  }
#pragma pop_macro("new")
}
```

As you can see, once the ListAllEmployees function retrieves the DataRowsCollection object from the Employees DataTable object (via the Rows method), it then employs a for loop to enumerate the collection.

Within the loop, the `DataRowCollection::Item` property is called in order to retrieve each `DataRow` object from the collection. Once the `DataRow` object has been secured, the `DataRow::Item` property is used to retrieve the desired column's data (by column name). As you can see, once the higher-level objects (such as `DataSet` and the data adapter) have been constructed, the database code resembles any other—except that here you have the power of disconnected data.

Now let's look at one way the `DataColumnCollection` and `Data Column` classes can help in reading data. As the `DataColumnCollection` object contains an array of every `DataColumn` object for a given `Data Table`, you can easily determine column-level information when needed—such as when coding an agnostic client—or one that has no knowledge of the data store's schema. By interrogating the `DataColumn` object, you can determine many important aspects of the column's definition, including column name, data type, whether or not the column is read-only, has a default value, and so on.

Take a look now at the following method, which illustrates how the `DataColumnCollection` and `DataColumn` classes can be used to dynamically determine column information and retrieve all data from a given table (the Employees table, in this case).

```
void DumpEmployeeTable()
{
  try
  {
    // Assumes employeesTable has already been filled by an adapter
    // Get all column names and column types...
    String* columnName;
    String* columnType;
    for (int i = 0; i < employeesTable->Columns->Count; i++)

    {
      // Get column name
      columnName = employeesTable->Columns->Item[i]->ColumnName;
      // Get column data type
      columnType =
        employeesTable->Columns->Item[i]->DataType->ToString();
      // Display column information
    }

    // Get all rows and within each row, all column data
    DataRowCollection* rows = employeesTable->Rows;
```

```
     DataRow* row;
     for (int i = 0; i < rows->Count; i++)
     {
       row = rows->Item[i];
       for (int j = 0; j < row->ItemArray->Count; j++)
       {
         String* value = row->Item[j]->ToString();
         // Display column data for current row
       }
     }
   }
   catch(Exception* e)
   {
     MessageBox::Show(String::Format(S"Exception : {0}", e->Message));
   }
 }
```

The DumpEmployeeTable function begins by enumerating the employeeTable object's DataColumnCollection with a for loop. Within that loop, each column is retrieved with a call to the DataColumn Collection::Item property. The ColumnName and DataType properties are then used to retrieve those values.

Once the column information has been acquired, the function loops through the employeeTable object's DataRowCollection in similar fashion to what you saw in the ListAllEmployees function. The main difference here being that instead of hard-coding the desired columns, the loop contains an inner loop to enumerate each row's columns. Within the inner loop, the overloaded DataRow::Item property that takes an array index value is used.

As you can see, a few trivial tweaks and this function could be modified to dump both the complete schema information and data of any table.

Inserting and Updating Rows

Having seen a bit of the DataRowCollection class, you might imagine that adding rows to a DataTable is easy, and you would be correct. There are two distinct means of inserting new rows into a DataTable, each facilitated by an overload of the DataRowCollection::Add method:

```
virtual DataRow* Add(Object* valueArray[]);
void Add(DataRow* newRow);
```

The first overload takes an `Object` array of values and returns a `DataRow` object representing the new data row. Here's an example of using this method. (Notice that in the code I'm not defining the first value of the array; I'll explain why shortly.)

```
try
{
  // Assumes employeesTable DataTable has already been constructed
  // Allocate enough elements for all the columns
  Object* values[] = new Object*[employeesTable->Columns->Count];
  // Populate the array
  // Intentionally skipping first element of array
  values[1] = S"Tom";
  values[2] = S"Archer";
  // ...

  // Add the row to the DataTable object
  DataRow* newRow = employeesTable->Rows->Add(values);
  // Must call DataAdapter::Update when ready to commit changes
  // to disconnected data source
  // adapter->Update(dataset, S"AllEmployees");
  //

  MessageBox::Show(S"Record added successfully");
}
catch(Exception* e)
{
  MessageBox::Show(String::Format(S"Exception : {0}", e->Message));
}
```

As you can see, this code snippet first allocates an array of `Object` types using the Managed C++ syntax for allocating an array of reference types, as discussed in Chapter 1. The `DataTable::Count` property is used to ensure that the proper number of elements is allocated (although in this example I only output a couple of values for example purposes). From there, the code populates the array and calls the `Add` method, with the new `DataRow` object being returned.

However, there a couple of key issues to cover here. First, note the comment regarding the data adapter object's `Update` method. Calling the `Update` method will obviously cause the data adapter to connect to the underlying data store in order to reconcile changes in the adapter's specified `DataTable`. Therefore, where you place this logic will be

application-specific. As an example, let's say you have a distributed application where you want to keep connections to the remote data store to a minimum. Instead of calling Update on every data change, you could place a UI element on the application (such a Commit Changes menu item) that calls the Update method and causes all updates, inserts, and deletes to be reconciled en masse.

The second issue to take note of is that of skipping the first element of the array. Typically as C++ programmers we would cringe to see someone allocate an array, not initialize the first element, and then pass that array to another function for further processing. However, in this case I know that the first element will not be used in the insert of the new record. This can be verified by inspecting the CommandBuilder object's InsertCommand (via the GetInsertCommand method). As shown here, note that the EmployeeID column is not specified in the SQL INSERT command.

```
"INSERT INTO Employees( LastName , FirstName , Title ,
TitleOfCourtesy , BirthDate , HireDate , Address , City , Region ,
PostalCode , Country , HomePhone , Extension , Photo , Notes ,
ReportsTo , PhotoPath ) VALUES ( @p1 , @p2 , @p3 , @p4 , @p5 , @p6 ,
@p7 , @p8 , @p9 , @p10 , @p11 , @p12 , @p13 , @p14 , @p15 , @p16 ,
@p17) "
```

However, that does bring up an interesting issue that often plagues users of tables with a primary key column that is defined as auto-increment. Specifically, the problem is determining how to insert a new record and then retrieve its auto-incremented primary key. For example, you'd want to do this if you had related tables where the primary key for one table was to be used as part of the key for another table. Also, you might need the primary key in your code in order to programmatically keep track of the records, or you might even need to return that value to the user. I'll cover a common technique for handling this situation in the section entitled "Disconnected Data and Auto-Increment Primary Keys."

Now, let's look at a code snippet that inserts new records into a Data Table object by first creating and populating a DataRow object. As a DataRow can be populated using either the DataRow:: Item property (one element at a time) or by using the ItemArray property (where an array of Object types can be specified), I'll illustrate both techniques.

```
try
{
    // Assumes employeesTable DataTable has already been constructed
```

```
    // Construct a DataRow object using an Object array
    DataRow* newRow1 = employeesTable->NewRow();
    Object* values[] = new Object*[employeesTable->Columns->Count];
    values[1] = S"Tom";
    values[2] = S"Archer";
    newRow1->ItemArray = values;
    employeesTable->Rows->Add(newRow1);
    // Construct a DataRow object one element at a time
    DataRow* newRow2 = employeesTable->NewRow();
    newRow2->Item[S"FirstName"] = S"Nishant";
    newRow2->Item[S"LastName"] = S"Sivakumar";
    employeesTable->Rows->Add(newRow2);
    MessageBox::Show(S"Records added successfully");
}
catch(Exception* e)
{
    MessageBox::Show(String::Format(S"Exception : {0}", e->Message));
}
```

With three row-insertion techniques to choose from, it's logical at this point to question which one is best. My personal recommendation is to use the NewRow method and then fill the returned DataRow object via the overloaded Item that allows you to specify the column name. This way, you eliminate common errors associated with assigning a value to an incorrect column. In addition, explicitly naming the column to which a value is being assigned results in code that is more readable—and therefore, more maintainable.

Deleting and Removing Rows

There are actually three distinct ways to delete (or remove) a row depending upon your needs. Here are the three methods and a brief explanation of their differences, specifically with regard to when you would use one over the other:

- DataRowCollection::Remove
- DataRowCollection::RemoveAt
- DataRow::Delete

First off, the only difference between the Remove and RemoveAt methods is that the former takes a DataRow object as its only parameter, and the latter takes an index value, or ordinal, as its sole parameter. I would

encourage you to be very careful when using the RemoveAt method, since adding rows to and removing rows from the DataRowCollection will have an impact on the ordinal value of the current rows in the collection. Therefore, unless you are very sure of the index of the row you wish to delete, I recommend using one of the search methods (see "Searching, Sorting, and Filtering Data") to acquire the desired DataRow object and then passing that object to the Remove method.

```
// Assumes employeesTable DataTable has already been constructed
// and that you have acquired the desired row to delete
employeesTable->Rows->Remove(row);
// Assumes employeesTable DataTable has already been constructed
// and removes the first row in the table's row collection
employeesTable->Rows->RemoveAt(0);
```

One important fact to keep in mind here is that the Remove/RemoveAt methods do not actually delete the row from the underlying data store upon calling the data adapter object's Update. Instead, the Remove/RemoveAt methods simply set the DataRow::State to DataRowState::Detached and remove the row from the row collection. In other words, calling the data adapter Update method to delete rows only works for rows that are still in the collection and marked as DataRowState::Deleted. This is accomplished via the DataRow::Delete method:

```
// Assumes employeesTable DataTable has already been constructed
// and that you have acquired the desired row to delete
row->Delete(); // delete row
```

It's worth reiterating that since the data is disconnected, the deletion of the row from the data store will not occur until you call data adapter's Update. Also, if you call the Delete method and then query either the DataRowCollection::Count property or enumerate the DataRow Collection object, it will appear as though the record has not been deleted. This is because when you directly access the DataRowCollection, you see all records in that collection regardless of their State value (Added, Changed, Deleted, etc.).

So how do we know if a record has been deleted so that it is not included in the enumeration of a row collection? This is done via the DataView object. I'll discuss the usage of DataView objects in more detail in the section entitled "Sorting, Searching, and Filtering Data with DataView Objects"). For now, however, I'll describe how the DataView

object allows you to filter data rows based on their `DataRowState` property. In fact, every `DataSet` has a "built-in" default data view that's accessible via the `DataTable::DefaultView` property. The default view's `RowState` `Filter` is set to `DataViewRowState::CurrentRows` so that the view includes unchanged, new, and modified rows, but excludes deleted rows. You can see this in the following code snippet and comments:

```
// Assumes employeesTable DataTable has already been constructed
// and that you have acquired the desired row to delete
row->Delete(); // delete row
// Will be the same as the number before the delete because
// no view is being applied
int dataRowCount = employeesTable->Rows->Count;
// Will be the one less than the count before the delete
// as the default view is set to DataViewRowState::CurrentRows
int defaultViewCount = employeesTable->DefaultView->Count;
```

So the question of which approach you should use really comes down to what capabilities your application will provide to the user. For example, if the application will allow the user to undo delete operations, then you should use the `Delete` method to delete rows and *not* use the `Remove`/ `RemoveAt` method, as it would permanently remove the row from the collection. You could then use two `DataView` objects to view your data—one using a filter that ignores rows marked as deleted, and one using a filter that only includes deleted rows. I'll go into this in more detail in the section entitled "Sorting, Searching, and Filtering Data with `DataView` Objects."

If, on the other hand, you have no need to allow for the revocation of delete operations, then you can either remove or delete the row(s). Just make sure that you understand that removing a row from the rows collection means that no matter what filter you set for your view, you will not be able to view the row. Finally, if you are going to delete *and* remove the row, you must call the `Delete` method first.

Disconnected Data and Auto-Increment Primary Keys

There might be times when you work with a table that contains a primary key column defined as *auto-increment* (also known as *auto-number*, or *IDENTITY*). This can be especially problematic for disconnected data scenarios where the application needs to determine the auto-increment value once the record has been inserted into the data store. Here are two examples of such scenarios:

- An application where the record's primary key is used to programmatically keep track of records. For example, a maintenance application might display records in a list and need to retrieve the user-selected record from the data row collection. A newly created record could be inserted into the list with a special value indicating that it was a new record. However, once the user chooses to update the data store, the application-supplied primary key would be incorrect, and the application would need a means of retrieving the new row's actual auto-incremented primary key from the data store.
- An application that uses multiple tables that have a parent-child relationship—such as order header and order detail tables. This is especially problematic if the parent table has an auto-incremented primary key that must be used in inserting rows into the child table.

In both cases, the technique used to determine the auto-incremented value of a newly inserted row is the same. In fact, the first scenario is the subject of the next section's demo application. The second scenario is a bit more complex and brings into play a special ADO.NET class called a `DataRelation`, which enables an application to define a programmatic relation between two tables—a parent and child—such that the data can be navigated in a much more intuitive manner. I'll get into using the `DataRelation` class in the next chapter.

To understand how you can retrieve the auto-incremented primary key value from the data store, it's important to understand the complete order of execution that occurs when a row is updated against a data store.

- The `DataRow` object's values are moved to the adapter's parameter members. These members govern the SQL statement executed against the data store.
- **The data adapter's `OnRowUpdating` method fires the `Row Updating` event.**
- The command is executed against the data store.
- The `DataRow` object is updated accordingly if the `FirstReturned Record` property is set or if any output parameters are specified.
- **The data adapter's `OnRowUpdated` method fires the `RowUpdated` event.**
- The `DataRow::AcceptChanges` method is called to commit the changes to the row.

As you can see, I've made two lines of interest here bold, where an event is fired both before and after the command is executed against the data store. As we're looking for a value returned as a result of executing a command, we're only interested in the second event; therefore, we need only handle the data adapter's `RowUpdated` event.

In .NET you handle an event by registering, for that event, a method that conforms to a *delegate*. A delegate is a signature for a method and is used in cases where one piece of code (a server) needs to define the signature for a method that will be passed to it from another piece of code (a client). Typical uses of delegates include callback scenarios or asynchronous event-handling situations. As a result, the event-handling method must be a managed method.

A common way of writing event-handling methods in mixed-mode applications is to define a managed class that specifies (and implements) the method that's called when the event is fired. Here are the steps involved in defining such a class to handle the `RowUpdated` event and retrieve the auto-incremented primary key for a newly inserted row.

1. Create a managed class that will register and implement the delegates that will handle the desired data adapter events.

```
__gc class SqlEventHandler
{
};
```

2. Implement a class constructor that takes a data adapter as a parameter and register (with that adapter) the method to be called for the desired events. Since we're handling the `RowUpdated` event, that means the code needs to call the data adapter's `add_RowUpdated` method and pass it an `SqlRowUpdatedEventHandler` object. The `SqlRowUpdatedEventHandler` constructor takes an instance of an object that will handle the event and the method name to call when the event is fired.

```
. . .
public:
  SqlEventHandler(SqlDataAdapter* adapter)
  {
    adapter->add_RowUpdated(
      new SqlRowUpdatedEventHandler(this,
        SqlEventHandler::OnUpdated));
  }
```

3. Implement the event-handling method. Since the `SqlRowUpdated` `EventHandler` delegate defines the signature for the method, the method you define must match it perfectly. In this case, the `Sql` `RowUpdatedEventHandler` signature states that the method must accept two parameters: an `Object` representing the source of the event (the data adapter object) and an `SqlRowUpdatedEventArgs` object, which will be passed when the event fires and the method is called.

Now let's look at an example method. The first thing you must do is verify that the row causing the event to be fired is a row that was added (as opposed to updated). Once you've determined that the row is a newly added row, construct a command object specifying the SQL `SELECT` `@@IDENTITY` command to query the data store for the new row's auto-incremented value. Calling the command object's `ExecuteScalar` method executes the command. Note that `ExecuteScalar` returns the first column of the first row of a result set that's enough for our needs, since we're only retrieving one value. With the new auto-increment value at hand, update the row's appropriate column. In the example's case, that is the `EmployeeID` column. Finally, call the `AcceptChanges` method to commit the changes to the `DataRow` object.

```
...
public:
  void OnUpdated(Object* obj, SqlRowUpdatedEventArgs* e)
  {
    if (StatementType::Insert == e->StatementType)
    {
      SqlCommand* command =
        new SqlCommand(S"SELECT @@IDENTITY",
                          e->Command->Connection);
      e->Row->Item[S"EmployeeID"] = command->ExecuteScalar());
      e->Row->AcceptChanges();
    }
  }
```

4. That's it for the event-handling side of things. Now you simply construct the new class—typically just after constructing the data adapter:

```
...
SqlConnection* conn =
```

```
new SqlConnection(S"Server=localhost;"
                  S"Database=Northwind;"
                  S"Integrated Security=true;");
adapter = new SqlDataAdapter(S"SELECT * FROM Employees", conn);
eventHandler = new SqlEventHandler(adapter, this);
...
```

Now the `SqlEventHandler::OnUpdated` gets called for each updated or inserted row anytime you call the adapter's `Update` method to synchronize your local in-memory changes with the data store.

Filling in Missing Schema and Primary Key Information for Untyped Datasets

When using untyped datasets, certain schema information—such as primary key information is not available. The following code snippet illustrates this point. If you want to test this code, you'll need to create the second dataset (`EmployeesDataSet`) using the steps listed in the section entitled "Generating a Typed DataSet").

```
SqlConnection* conn =
  new SqlConnection(S"Server=localhost;"
                    S"Database=Northwind;"
                    S"Integrated Security=true;");
adapter* = new SqlDataAdapter(S"SELECT * FROM Employees", conn);
conn->Open();
// Untyped DataSet
DataSet* untyped = new DataSet();
adapter->Fill(untyped, S"AllEmployees");
int count1 =
  dataset->Tables->Item["AllEmployees"]->PrimaryKey->Length;
// count1 will be 0
// Typed DataSet
EmployeesDataSet* typed = new EmployeesDataSet();
adapter->Fill(typed->EmployeesDataTable);
int count2 = typed->EmployeesDataTable->PrimaryKey->Length;
// count2 will be 1
conn->Close();
```

As the comments indicate, the data table will only return the correct count of primary keys if the `DataSet` is typed. This can cause a problem if you wish to use certain ADO.NET functionality, which relies on the data

table properly representing the data store. For example, the `DataRow Collection::Find` method allows you to search through a row collection based on a primary key. However, attempting to call the `Find` method using the untyped dataset, as illustrated in the previous code snippet, results in a `System::Data::MissingPrimaryKeyException` because no primary key has been defined—at least as far as the data table object is concerned. There are two solutions to this problem.

The first solution is simply to tell the data adapter to gather the schema information when filling the dataset. This is accomplished via the data adapter's `MissingSchemaAction` property, which accepts a `Missing SchemaAction` enumeration value that tells the adapter what action to take when the `DataSet` schema doesn't match the incoming data. For example, suppose you programmatically define a `DataTable` object that doesn't match the incoming data in terms of the number of columns. Let's say you've defined a `DataTable` that has only two columns, and you want to read the `Employees` table data (which contains many more columns) into that `DataTable`. Since the schemas don't match, the data adapter has to be told how to handle that situation, and that's exactly what the `Missing SchemaAction` enumeration is typically used for. By specifying a value of `MissingSchemaAction::Add`, the adapter will add any necessary columns to complete the schema. Specifying a value of `MissingSchemaAction. Ignore` results in the data adapter ignoring any extra columns (and not downloading that data) and `MissingSchemaAction Error` results in an exception if the schemas don't match. The default `MissingSchemaAction` property value is `Add`.

So, as you can see, the `MissingSchemaAction` property is mainly used to map dissimilar schemas to one another, which doesn't seem to be what we're after. However, there is one last `MissingSchemaAction` enumeration value—`AddWithKey`—that does exactly what we want. This value is similar to the `Add` value except that it also adds the primary key information to complete the schema. As a result, simply setting the `MissingSchema Action` property to `AddWithKey` just before the `Fill` method call will result in the `DataTable` being properly constructed with the data source's primary key information:

```
SqlConnection* conn =
  new SqlConnection(S"Server=localhost;"
                    S"Database=Northwind;"
                    S"Integrated Security=true;");
```

```
adapter = new SqlDataAdapter(S"SELECT * FROM Employees", conn);
adapter->MissingSchemaAction = MissingSchemaAction::AddWithKey;
conn->Open();
dataset = new DataSet(); adapter->Fill(dataset, S"AllEmployees");
// Untyped DataSet
DataSet* untyped = new DataSet();
adapter->Fill(untyped, S"AllEmployees");
int count1 =
   dataset->Tables->Item["AllEmployees"]->PrimaryKey->Length;
// count1 will now be 1
```

One important thing to note is that when a data adapter whose
`MissingSchemaAction` property is set to `AddWithKey` creates a `Data
Column` object for a column defined as a primary key in the data source,
that `DataColumn` object is marked as read-only (`DataColumn::ReadOnly`
is set to `true`). Obviously, if you do not need to alter the primary key value,
this solution will work fine for you.

However, there are plenty of scenarios where you'll need to modify the
local value for a primary key. For example, in the previous section you saw
that the local value for an auto-increment primary key of a new row can't be
realized until the row is inserted and that value is retrieved from the data
store. The local value is then updated to properly reflect the data store's
value. This can't be done if the `DataColumn` is set to read-only. Therefore,
another mechanism must be used to indicate that a given data source column is a primary key. We can do this via the `DataTable::PrimaryKey`
property.

```
SqlConnection* conn =
   new SqlConnection(S"Server=localhost;"
                     S"Database=Northwind;"
                     S"Integrated Security=true;");
adapter = new SqlDataAdapter(S"SELECT * FROM Employees", conn);
conn->Open();
dataset = new DataSet(); adapter->Fill(dataset, S"AllEmployees");
DataTableCollection* tables = dataset->Tables;
employeesTable = tables->Item[S"AllEmployees"];
DataColumn* primaryKeys[] = new DataColumn*[1];
primaryKeys[0] = employeesTable->Columns-> Item[0];
employeesTable->PrimaryKey = primaryKeys;
// Untyped DataSet
DataSet* untyped = new DataSet();
```

```
adapter->Fill(untyped, S"AllEmployees");
int count1 =
  dataset->Tables->Item["AllEmployees"]->PrimaryKey->Length;
// count1 will now be 1
```

After the DataSet object is filled, a DataColumn array is allocated, and
the first DataColumn object from the employees DataTable is inserted
into it, as the first column is the EmployeeId primary key. The PrimaryKey
property is then set to the DataColumn array, thereby accomplishing what
we need.

You could also set the MissingSchemaAction property and then, after
the DataSet is filled, set the desired DataColumn object's ReadOnly property to false. However, it's my opinion that if you're going to manually override what the adapter does anyway, you might as well save yourself the
overhead of requesting that the schema information be retrieved during
the DataSet fill and simply override the entire process.

So, in summary, I recommend using the MissingSchemaAction property if your application will not need to modify the DataColumn object corresponding to the primary key; otherwise use the PrimaryKey property.

Demo—Simple Maintenance Application

Let's test what you've learned to this point with a semi-realistic MFC SDI
demo application that lists all the employee records from the sample SQL
Server Northwind database's Employees table. The application will allow
you to create new records as well as edit and delete existing records using
the classes and techniques you've learned about up to this point. While this
demo contains a few more steps than I normally prefer to include in a book
demo, there are a lot of little things you have to do to make a realistic
ADO.NET application, and seeing how everything fits together is paramount to understanding how to use ADO.NET in a real-world application.

1. To get started, create a new MFC project called Employee
   Maintenance—where the application type is SDI and the view
   class is a CListView—and update the Project properties to support Managed Extensions.
2. Open the stdafx.h file and add the following .NET support directives to the end of the file.

   ```
   #using <mscorlib.dll>
   #using <system.dll>
   ```

```
#using <system.data.dll>
#using <system.xml.dll>
#using <system.windows.forms.dll>

using namespace System;
using namespace System::Data;
using namespace System::Data::SqlClient;
using namespace System::Xml;
using namespace System::Windows::Forms;
#undef MessageBox
```

3. In the view class's `PreCreateWindow` function, set the view window's style to "report" and single selection:

```
BOOL CEmployeeMaintenanceView::PreCreateWindow(CREATESTRUCT&
cs)
{
  cs.style |= LVS_REPORT;
  cs.style |= LVS_SINGLESEL;
  return CListView::PreCreateWindow(cs);
}
```

4. Add the following code to the end of the view's `OnInitialUpdate` function to initialize the list view.

```
void CEmployeeMaintenanceView::OnInitialUpdate()
{
    CListView::OnInitialUpdate();

    ...

  CListCtrl& lst = GetListCtrl();
  // All full row selection
  LONG lStyle =
  (LONG)lst.SendMessage(LVM_GETEXTENDEDLISTVIEWSTYLE);
  lStyle |= LVS_EX_FULLROWSELECT;
  lst.SendMessage(LVM_SETEXTENDEDLISTVIEWSTYLE,   0,
    (LPARAM)lStyle);
  // Add columns to listview lst.InsertColumn(0, _T("ID"));
  lst.InsertColumn(1, _T("First Name"));
  lst.InsertColumn(2, _T("Last Name"));
}
```

5. Now that the main UI is done, let's focus on the view-level ADO. NET objects and their construction. First, define the following

ADO.NET objects in the `CEmployeeMaintenanceView` class that will be used throughout the application.

```
class CEmployeeMaintenanceView : public CListView
{
. . .
protected:
  gcroot<DataSet*>dataset;
  gcroot<SqlDataAdapter*>adapter;
  gcroot<DataTable*>employeesTable;
  gcroot<SqlCommandBuilder*>commandBuilder;
. . .
```

6. Add the following code to the end of the view's `OnInitialUpdate` function to initialize the ADO.NET objects. As you can see, the first thing that is done is to make the connection to the sample SQL Server Northwind database. From there, the code constructs an `SqlDataAdapter` object (`adapter`) with an SQL SELECT statement that retrieves all records from the Employees table.

Once the adapter has been constructed, an `SqlCommand Builder` object (`eventHandler`) is instantiated, and the connection is opened. A `DataSet` object (`dataset`) is then constructed and filled with employee records via the `SqlDataAdapter::Fill` method, and the resulting `DataTable` object is named "All-Employees." At this point, we have the data in memory, and so the connection to the data store is closed. A `DataTable` object (`employeesTable`) is then allocated and points to the "All-Employees" `DataTable` created during the `Fill` method. (Note that declaring a `DataTable` object is done just for convenience, as the application could retrieve the table from the `DataSet` object's `DataTable` collection each time it needs access to the table.)

The `employeesTable` primary key is defined as the table's first column via the `PrimaryKey` property. This is done so that the `DataRowCollection::Find` method can be used to locate records by their primary key. Finally, a helper function called `ReadAll Employees` is called.

```
void CEmployeeMaintenanceView::OnInitialUpdate()
{
  CListView::OnInitialUpdate();

  . . .
```

```
#pragma push_macro("new")
#undef new
try
  {
    SqlConnection* conn =
      new SqlConnection(S"Server=localhost;"
                         S"Database=Northwind;"
                         S"Integrated Security=true;");
    adapter = new SqlDataAdapter(S"SELECT * FROM Employees",
      conn);
    commandBuilder = new SqlCommandBuilder(adapter);
    conn->Open();
    dataset = new DataSet();
    adapter->Fill(dataset, S"AllEmployees");
    conn->Close(); // No longer needed
    DataTableCollection* tables = dataset->Tables;
    employeesTable = tables->Item[S"AllEmployees"];
    // Set the table's primary key column(needed for
    // DataRowCollection::Find method) Can't use
    // DataAdapter::MissingSchemaAction because that
    // would make the EmployeeID readonly and I need to
    // set that value after inserts are realized against
    // the data store.
    DataColumn* primaryKeys[] = new DataColumn*[1];
    primaryKeys[0] = employeesTable->Columns->Item[0];
    employeesTable->PrimaryKey = primaryKeys;
    employeesTable->Columns->Item[0]->ReadOnly = false;
    ReadAllEmployees();
  }
  catch(Exception* e)
  {
    MessageBox::Show(e->Message, S".NET Exception Thrown",
                     MessageBoxButtons::OK,
                     MessageBoxIcon::Error);
  }
#pragma pop_macro("new")
}
```

7. Implement the following `ReadAllEmployees` member function. After initializing the list view, this function enumerates the `employeesTable` object's row collection, retrieving each record's

EmployeeId, FirstName and LastName values and inserting them into the list view.

```
void CEmployeeMaintenanceView::ReadAllEmployees()
{
  try
  {
    CWaitCursor wc;
    CListCtrl& lst = GetListCtrl(); lst.DeleteAllItems();
    DataRowCollection* rows = employeesTable->Rows;
    DataRow* row;
    String* firstName;
    String* lastName;
    String* id;
    for (int i = 0; i < rows->Count; i++)
    {
      row = rows->Item[i];
      id = row->Item[S"EmployeeID"]->ToString();
      firstName = row->Item[S"FirstName"]->ToString();
      lastName = row->Item[S"LastName"]->ToString();
      int idx = lst.InsertItem(i, (CString)id);
      lst.SetItemText(idx, 1, (CString)firstName);
      lst.SetItemText(idx, 2, (CString)lastName);
    }
  }
  catch(Exception* e)
  {
    MessageBox::Show(e->Message, S".NET Exception Thrown",
                     MessageBoxButtons::OK,
                     MessageBoxIcon::Error);
  }
}
```

8. At this point, you should be able to build and run the application and see the employee records in the list view. Now, let's add the ability to create, edit, and delete records. Begin by adding an Employees pop-up menu as shown in Figure 6–2.

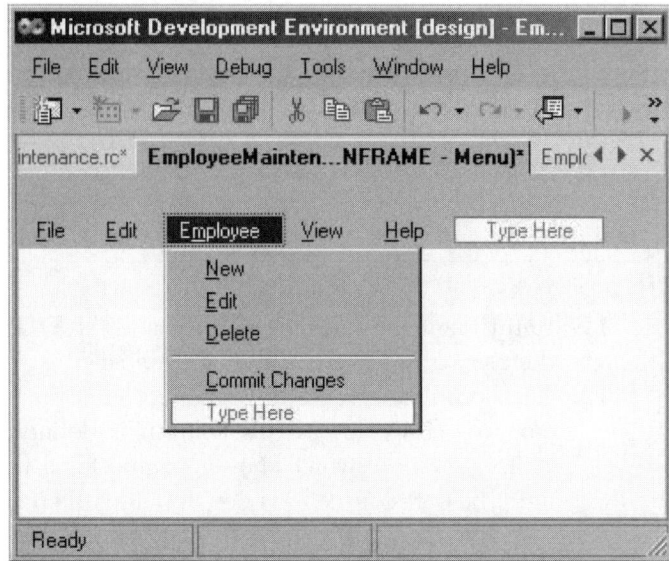

Figure 6-2 The Employee Maintenance menu for adding, editing, and deleting employee records

9. Now let's implement the ability to create new employee records. Start by adding the dialog shown in Figure 6–3.

10. Add a `CDialog`-based class called `CEmployeeDlg` for the dialog resource and include its header file in the view's implementation file (`EmployeeMaintenanceView.cpp`).

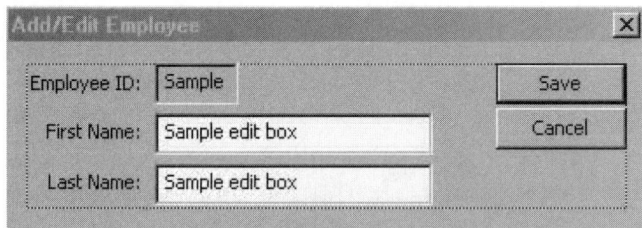

Figure 6-3 The dialog for adding and editing employee records

Table 6-1 DDX variables for the `EmployeeMaintenance` demo

Control	Variable Type	Variable Name
Employee ID	int	m_iEmployeeId
First Name	Cstring	m_strFirstName
Last Name	Cstring	m_strLastName

11. Returning to the resource editor, add the DDX value variables for the `CEmployeeDlg` as shown in Table 6–1.

12. Each employee record is uniquely identified by its `EmployeeId` column. However, as this column is defined as auto-increment (`IDENTITY`), we won't know its value for a newly created record until the data source has been updated, the database has generated the value, and that value is retrieved using a technique such as that described in the section entitled "Disconnected Data and Auto-Increment Primary Keys." However, we need a way to uniquely identify newly created records before the data store is updated. Therefore, we need to associate any newly created records with an application-generated key that will not conflict with any possible real values. One way to accomplish this is to create an application variable that starts out as a negative value and decrements with each new record. This will work because database auto-increment values start at 0 and increment. Therefore, define the following static variable (`newRowId`) to the top of the EmployeeMaintenanceView.cpp file.

```
// Used as a temporary id for new records not yet added to the
database
static newRowId = -1;
```

13. The next thing we'll want to do is to track which records in the list view are new records. We'll do this so that when the data store is updated, it's easy to locate the affected records and update their `EmployeeId` value with the data store generated value. We'll use a simple MFC map collection to do this. Define the following `CMapPtrToPtr` member in the view class.

```
class CEmployeeMaintenanceView : public CListView
{
...
```

```
protected:
  CMapPtrToPtr mapIdToLVIndex;
```

14. Now that everything's in place, add the following #define directive and event handler for the New option of the Employee menu to the CEmployeeMaintenanceView class. The function first displays the CEmployeeDlg. If the user clicks the Save (IDOK) button, the function constructs a new DataRow object and initializes it by setting the EmployeeId to the value of the newRodId and the First Name and LastName values to those of the dialog's member variables (representing the data entered by the user).

    The data is then added to the list view. You'll notice that I opted to set the list view's employeeID column to NEW_RECORD ("NEW") instead of the newRowId value. This was simply a choice of aesthetics. However, since I need to keep track of the row's newRowId value, I stuff that value into the item's item data for later retrieval. The map is then updated where the newRowId maps to the index of the newly added item in the list view. Finally, the newRowId value is decremented for the next new record.

```
#define NEW_RECORD "NEW"


...


void CEmployeeMaintenanceView::OnEmployeeNew()
{
  try
  {
    CListCtrl& lst = GetListCtrl();
    CEmployeeDlg dlg; if (IDOK == dlg.DoModal())
    {
      CWaitCursor wc;
      DataRow* newRow = employeesTable->NewRow();
      newRow->Item[S"EmployeeID"] = __box(newRowId);
      newRow->Item[S"FirstName"] =
        (String*)dlg.m_strFirstName;
      newRow->Item[S"LastName"] = (String*)dlg.m_strLastName;
      employeesTable->Rows->Add(newRow);
      int idx = lst.InsertItem(lst.GetItemCount(),
        NEW_RECORD);
      lst.SetItemData(idx, (int)newRowId);
      lst.SetItemText(idx, 1, dlg.m_strFirstName);
      lst.SetItemText(idx, 2, dlg.m_strLastName);
```

```
      mapIdToLVIndex.SetAt((void*)newRowId, (void*)idx);
      // when finished, decrement static new row counter
      newRowId -;
    }
  }
  catch(Exception* e)
  {
    MessageBox::Show(e->Message, S".NET Exception Thrown",
                     MessageBoxButtons::OK,
                     MessageBoxIcon::Error);
  }
}
```

15. We can add records to the list view, but that's all. Therefore, let's implement the Commit Changes menu item so that we can see the data store get updated. As you can see, there's really nothing much to do here besides call the data adapter's Update method—specifying which table to update—and handle any potential exceptions.

```
void CEmployeeMaintenanceView::OnEmployeeCommitChanges()
{
  try
  {
    CWaitCursor wc;
    adapter->Update(employeesTable);
    MessageBox::Show(S"Changed committed", S"Information",
                     MessageBoxButtons::OK,
                     MessageBoxIcon::Information);
  }
  catch(Exception* e)
  {
    MessageBox::Show(e->Message, S".NET Exception Thrown",
                     MessageBoxButtons::OK,
                     MessageBoxIcon::Error);
  }
}
```

16. At this point, building and running the code results in new records being added to the data store. However, we still have one hurdle to clear. We need to update the DataRow (and list view) with the data store generated EmployeeId. If we skip this, attempts to edit or delete the row after updating the data store will result in an exception, as our application-supplied temporary value is invalid. To start

with, forward-declare the following class before the CEmployee
MaintenanceView class. We have to do this because—as you'll soon
see—there's a circular reference between the SqlEventHandler
class and the CEmployeeMaintenanceView class.

```
__gc class SqlEventHandler;
class CEmployeeMaintenanceView : public CListView
{
...
```

17. Now, implement the SqlEventHandler class. This class will be
used to handle the data adapter's RowUpdated event so that we can
determine the new EmployeeId value for newly inserted records.
The class's constructor takes both a pointer to an adapter object as
well as a pointer to the view object. The latter is needed because
the object needs to tell the view when it has handled an event.
The OnUpdated method—what actually gets called when the data
adapter's RowUpdated event fires—saves the previous EmployeeId
value and then uses the SELECT @@IDENTITY command to retrieve
the new EmployeeId value. Both values are then passed to the
view's OnRowInserted function.

```
__gc class SqlEventHandler
{
public:
  SqlEventHandler(SqlDataAdapter* adapter,
                  CEmployeeMaintenanceView* parentView)
  {
    adapter->add_RowUpdated(new
      SqlRowUpdatedEventHandler(this,
      SqlEventHandler::OnUpdated));
    this->parentView = parentView;
  }
protected:
  CEmployeeMaintenanceView* parentView;
public:
  void OnUpdated(Object* obj, SqlRowUpdatedEventArgs* e)
  {
    if (StatementType::Insert == e->StatementType)
    {
      SqlCommand* command =
        new SqlCommand(S"SELECT @@IDENTITY",
                       e->Command->Connection);
```

```
      int previousId =
        *dynamic_cast<__box int*>(e->Row->
          Item[S"EmployeeId"]);
      e->Row->Item[S"EmployeeID"] = command->ExecuteScalar();
      int newId =
        *dynamic_cast<__box int*>(e->Row->
          Item[S"EmployeeId"]);
      e->Row->AcceptChanges();
      parentView->OnRowInserted(previousId, newId);
    }
  }
};
```

18. Now, implement the view's (public) OnRowInserted function as follows. Here the function is simply using the previous EmployeeId value to search the mapIdToLVIndex collection for the list view index of the just-inserted row. The list view item is then updated with the new EmployeeId value, and the entry is removed from the mapIdToLVIndex collection, as this is the only function that uses it, and it's no longer needed once the EmployeeId is updated.

```
void CEmployeeMaintenanceView::OnRowInserted(int iPrevId, int
iNewId)
{
  void* iListViewIdx;
  mapIdToLVIndex.Lookup((void*)iPrevId, (void*&)iListViewIdx);
  ASSERT(-1 < (int)iListViewIdx); if (-1 < (int)iListViewIdx)
  {
    CListCtrl& lst = GetListCtrl();
    CString strNewId;
    strNewId.Format(_T("%ld"), iNewId);
    lst.SetItemText((int)iListViewIdx, 0, strNewId);
    mapIdToLVIndex.RemoveKey((void*)iPrevId);
  }
}
```

19. Define a CEmployeeMaintenanceView member variable called eventHandler that is of type SqlEventHandler This is why we had to forward-declare the SqlEventHandler class before the view class's definition.

```
class CEmployeeMaintenanceView : public CListView
{
  ...
```

```
public:
    gcroot<SqlEventHandler*>eventHandler;
```

20. Now that the code is in place to handle the data adapter's Row
Updated event, we need only subscribe to the event. Since the sub-
scription to the event takes place in the SqlEventHandler object's
construct, we just need to construct the SqlEventHandler at the
appropriate time. The perfect place to do this is in the view's
OnInitialUpdate function immediately after the data adapter
object has been instantiated:

```
void CEmployeeMaintenanceView::OnInitialUpdate()
{
...
adapter = new SqlDataAdapter(S"SELECT * FROM Employees",
    conn);
eventHandler = new SqlEventHandler(adapter, this);
```

21. At this point, you can now add records to the data store! Let's finish up this demo by implementing the edit and delete functions.
Start by adding the following helper function to the CEmployee
MaintenanceView class, which will return the currently selected
item index of the list view.

```
int CEmployeeMaintenanceView::GetSelectedItem()
{
    int iCurrSel = -1;
    CListCtrl& lst = GetListCtrl();
POSITION pos = lst.GetFirstSelectedItemPosition();
    if (pos)
        iCurrSel = lst.GetNextSelectedItem(pos);
    return iCurrSel;
}
```

22. Implement the following event handler for the Edit option of the
Employee menu. As you can see, the function attempts to determine the EmployeeId for the row by first looking at the first column of the list view. However, if that value is equal to "NEW", then
the function retrieves the EmployeeId value from the list view
item's item data. From there, the CEmployeeDlg object's member
variables are initialized, and the dialog is displayed.

   If the user enters data and clicks the Save button (IDOK),
the DataRow for the edited row is located by specifying the
EmployeeId to the DataRowCollection::Find method. Being

able to search the `DataRowCollection` is another reason why the application needed to maintain a temporary `EmployeeId` for new rows. Once the row is retrieved, it's `FirstName` and `LastName` columns are updated and so is the list view.

```
void CEmployeeMaintenanceView::OnEmployeeEdit()
{
  try
  {
    CListCtrl& lst = GetListCtrl();

    int currSel = GetSelectedItem();
    if (-1 < currSel)
  {
    CEmployeeDlg dlg;

    CString strId = lst.GetItemText(currSel, 0);
    Int32 id;
    if (0 == strId.Compare(NEW_RECORD))
      id = (int)lst.GetItemData(currSel);
    else
      id = atoi(strId);

    dlg.m_iEmployeeId = id;
    dlg.m_strFirstName = lst.GetItemText(currSel, 1);
    dlg.m_strLastName = lst.GetItemText(currSel, 2);

    if (IDOK == dlg.DoModal())
    {
      CWaitCursor wc;
      DataRow* row = employeesTable->Rows->Find(__box(id));
      if (row)
      {
        row->Item[S"FirstName"] = (String*)dlg.m_strFirstName;
        row->Item[S"LastName"] = (String*)dlg.m_strLastName;

        lst.SetItemText(currSel, 1, dlg.m_strFirstName);
        lst.SetItemText(currSel, 2, dlg.m_strLastName);
      }
    }
  }
    else
    {
      MessageBox::Show(S"You must first select an employee "
```

```
                       S"to perform this operation.",
                       S"Alert",
                       MessageBoxButtons::OK,
                       MessageBoxIcon::Error);
    }
  }
  catch(Exception* e)
  {
    MessageBox::Show(e->Message, S".NET Exception Thrown",
              MessageBoxButtons::OK,
              MessageBoxIcon::Error);
    }
  }
```

23. Finally, implement the following event handler for the Delete option of the Employee menu. The function starts out much like the Edit menu event handler by first retrieving the DataRow for the currently selected employee. Once that is done, the DataRow::Delete method is called, and the row is removed from the list view.

```
void CEmployeeMaintenanceView::OnEmployeeDelete()
{
  try
  {
    CListCtrl& lst = GetListCtrl();
    int currSel = GetSelectedItem(); if (-1 < currSel)
    {
      CWaitCursor wc;
      CString strId = lst.GetItemText(currSel, 0);
      Int32 id;
      if (0 == strId.Compare(NEW_RECORD))
        id = (int)lst.GetItemData(currSel);
      else
        id = atoi(strId);
      DataRow* row = employeesTable->Rows->Find(__box(id));
      if (row)
      {
        row->Delete();
        lst.DeleteItem(currSel);
      }
    }
    else
    {
```

```
        MessageBox::Show(S"You must first select an employee "
                         S"to perform this operation.",
                         S"Alert",
                         MessageBoxButtons::OK,
                         MessageBoxIcon::Error);
    }
}
catch(Exception* e)
{
    MessageBox::Show(e->Message, S".NET Exception Thrown",
                     MessageBoxButtons::OK,
                     MessageBoxIcon::Error);
}
}
```

Finally we're done! As I mentioned in the beginning of this chapter, while disconnected data is a wonderful thing for certain scenarios, it also increases the complexity of even something as simple as a maintenance application. However, just saying (or writing) that doesn't have the same impact as actually walking through a step-by-step demo such as this EmployeeMaintenance application and seeing for yourself the various issues that arise in a disconnected setting and how to deal with them one by one. Figure 6–4 shows an example of the application being run.

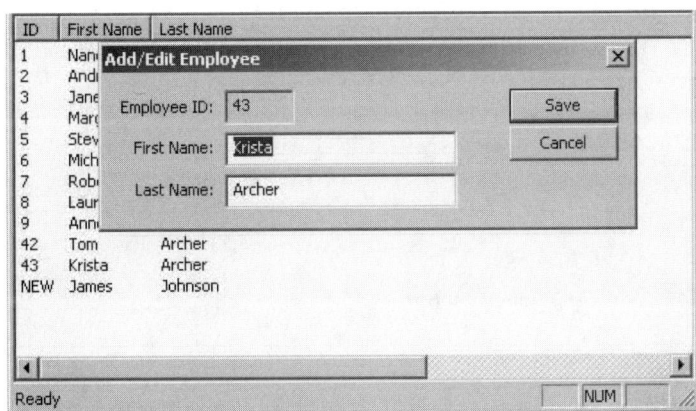

Figure 6-4 The EmployeeMaintenance demo at work

Working in Batch

There are times when it's advantageous to perform certain operations in batch. For example, multiple read operations can be performed with a single `Fill` method, thereby causing the generation of more than one local `DataTable` while incurring the cost of only one round trip between the client and the server. In addition to that, this section will also illustrate how to turn off `DataTable` constraints so that multiple changes can be made to a `DataRow` in batch and then either all accepted or canceled with a single method call. Finally, we'll also look at a way to turn index maintenance off to improve the performance associated with loading large amounts of data into a `DataTable`.

Creating Multiple `DataTables` in a `DataSet`

The principal benefits of using disconnected data are to prevent sustained client connections to the data source and to minimize round trips between client application(s) and the data server. To that end, there are times when it's advisable to batch your reads or queries against the data source so that you can fill more than one `DataTable` with a single round trip. Take the following example, where I'm creating a `DataSet` object with two `DataTable` objects—one filled with records from the Employees table and one filled with records from the Products table.

```
void MultipleRoundTrips()
{
#pragma push_macro("new")
#undef new
  try
  {
    SqlConnection* conn =
      new SqlConnection(S"Server=localhost;"
                        S"Database=Northwind;"
                        S"Integrated Security=true;");
    // Construct the employees data adapter
    SqlDataAdapter* employeesAdapter =
      new SqlDataAdapter(S"SELECT * FROM Employees", conn);
    // Construct the products data adapter
    SqlDataAdapter* productsAdapter =
```

```
          new SqlDataAdapter(S"SELECT * FROM Products", conn);
      conn->Open();
      DataSet* dataset = new DataSet();
      // Get all the employees - FIRST TRIP TO SERVER
      employeesAdapter->Fill(dataset, S"AllEmployees");
      // Get all the products - SECOND TRIP TO SERVER
      productsAdapter->Fill(dataset, S"AllProducts");
      conn->Close(); // No longer needed
      DataTableCollection* tables = dataset->Tables;
      DataTable* table;
      for (int i = 0; i < tables->Count; i++)
      {
        table = tables->Item[i];
        Console::WriteLine(String::Format(S"{0} table has {1} rows",
                      table->TableName,
                      __box(table->Rows->Count)));

      }
   }
   catch(Exception* e)
   {
      Console::WriteLine(e->Message);
   }
#pragma pop_macro("new")
}
```

As the comments indicate, two round trips are being made here—one for each `DataTable`. In situations like this—where you have all the information you need to specify your SELECT statement—it's more efficient to specify a SELECT statement in the data adapter's constructor that will cause each table to be retrieved in single round trip. Simply delimiting each SELECT statement with a semicolon accomplishes this:

```
String* allEmployees = S"SELECT * FROM Employees";
String* allProducts = S"SELECT * FROM Products";
String* select = String::Format(S"{0};{1}", allEmployees,
                             allProducts);
SqlDataAdapter* adapter = new SqlDataAdapter(select, conn);
```

At this point, you might be wondering how to name the `DataTable` when using this technique. After all, the `Fill` method takes the name of a

`DataTable`, but now we're reading two tables. The following function answers that question:

```
void BatchRead()
{
#pragma push_macro("new")
#undef new
  try
  {
    SqlConnection* conn =
      new SqlConnection(S"Server=localhost;"
                        S"Database=Northwind;"
                        S"Integrated Security=true;");
    String* allEmployees = S"SELECT * FROM Employees";
    String* allProducts = S"SELECT * FROM Products";
    String* select = String::Format(S"{0};{1}",
                                    allEmployees,
                                    allProducts);
    // Read two tables at once to reduce round trips
    SqlDataAdapter* adapter = new SqlDataAdapter(select, conn);
    conn->Open();
    DataSet* dataset = new DataSet();
    // Fill dataset. Don't bother naming the table because
    // the second table has to be renamed anyway.
    adapter->Fill(dataset);
    conn->Close(); // No longer needed
    DataTableCollection* tables = dataset->Tables;
    DataTable* table;
    for (int i = 0; i < tables->Count; i++)
    {
      table = tables->Item[i];
      Console::WriteLine(table->TableName);
    }

    tables->Item[0]->TableName = S"AllEmployees";
    tables->Item[1]->TableName = S"AllProducts";
    for (int i = 0; i < tables->Count; i++)
    {
      table = tables->Item[i];
      Console::WriteLine(String::Format(S"{0} table has {1} rows",
                         table->TableName,
                         __box(table->Rows->Count)));
    }
```

```
    }
    catch(Exception* e)
    {
        Console::WriteLine(e->Message);
    }
#pragma pop_macro("new")
}
```

This function produces the following output:

```
Table
Table1
AllEmployees table has 11 rows
AllProducts table has 77 rows
```

As mentioned earlier in the chapter, if you call the `Fill` method and do not specify a name for the `DataTable` that will be generated, it defaults to Table. If more than one `DataTable` is created, they are named Table1, Table2, and so on, as you can see in the first two lines of the output. However, what happens if you do specify a name? In this case, the first table gets the specified name and the remaining tables are simply named the same thing with a sequential numeric value (starting with 1) affixed to the end. Therefore, in order to have symbolic `DataTable` names, you'll need to set the `DataTable::TableName` property manually after a read that returns more than one table. As the last two lines of the output indicate, the two `DataTable` objects were renamed, and their row counts reflect the fact that we did indeed perform two separate queries with one round trip.

Batch Row Changes

The term "batch row changes" or "batch row updates" can be a bit misleading; it simply refers to the ability to change multiple properties (including column values) of a `DataRow` object before committing those changes. This can be extremely useful in several scenarios. For example, because of certain constraints on a given table, you might need to make two or more changes to a row before the row is valid, yet the first change results in an exception because you are in violation of the constraint. Let's say you're working with the Employees table and you want to store either a location to the employee's picture (such as a URL) or the actual binary data that comprises the picture. Since these are two drastically different data types (string and binary, respectively) you could create a column for each possibility and

set up a constraint that states that one of the columns must have a value and the other must be null—your standard XOR situation. Now let's say that you add a record with the picture data column set and the picture location column null, but later want to reverse that. Because of the aforementioned constraint, this would be extremely difficult.

As another practical example of needing to process multiple row changes in parallel, or batch, suppose you had a wizard-like interface that allowed the user to modify different columns of a row through a series of dialogs. As the user might cancel the operation at any point, you would need the ability to reverse the changes the user did make to the `DataRow`.

Both of these scenarios can be handled very simply by using the `Data Row` class's `BeginEdit`, `EndEdit`, and `CancelEdit` methods. When the `BeginEdit` method is called, all events are temporarily suspended for the row so that any changes made do not trigger validation rules (constraints). Calling `EndEdit` then commits these changes and brings back into play any events and constraints defined for the table/row. If you decide that you do not want to commit the changes, you can simply call the `CancelEdit` method and any changes made to the row since the `BeginEdit` method was called will be lost. Using the picture example from above, the code to change the values could now be written as follows:

```
// row is a DataRow having a column for picture data,
// which is currently set and a column for picture location,
// which needs to be set - all without breaking the constraint
// that one of these columns needs to be null at all times
// Start the edit process - turning off constraints
row->BeginEdit();
// Make changes row->Item[S"PictureData"] = DBNull::Value;
row->Item[S"PictureLocation"] = S"http://somelocation.com";
// Commit the changes
row->EndEdit();
```

Batch Table Updates

Much as the `DataRow` class allows you to suspend validation rules in order to perform certain operations in batch, the `DataTable` also provides the same capability at the table level, using the `BeginLoadData` and `EndLoad Data` method pair. In fact, because we are talking about the table level, index maintenance is also suspended, providing a much faster means of loading already validated data into a `DataTable`. In fact, my own benchmarking—using some of the performance and benchmarking techniques

discussed in Chapter 10—indicate up to a 30% increase in speed simply by calling `BeginLoadData`, loading the rows via `LoadDataRow`, and then calling `EndLoadData` when finished!

As an example of a realistic situation where you'd want to use these methods, let's say you were loading a large amount of data into a `Data Table` from another data source—such as a text file. Listing 6–1 shows the partial listing of the `rockets.txt` file included on the book's CD for the `BatchTableLoad` demo application.

Listing 6-1 Partial listing of sample comma-delimited text file (`rockets.txt`)

```
"Yao","Ming",13.5,"Dynasty"
"Steve","Francis",21.0,"Franchise"
"Cuttino","Mobley",17.5,"Cat"
...
```

Now let's see how we could read this data and batch load it into a `DataTable`.

```
void BatchLoad()
{
#pragma push_macro("new")
#undef new
  try
  {
    // Define a DataTable
    DataSet* dataset = new DataSet();
    DataTable* table = new DataTable(S"Players");
    dataset->Tables->Add(table);
    // Define the table's columns
    table->Columns->Add(S"FirstName", __typeof(String));
    table->Columns->Add(S"LastName", __typeof(String));
    table->Columns->Add(S"PointsPerGame", __typeof(Double));
    table->Columns->Add(S"Nickname", __typeof(String));
    // Open the csv file and read entire file into a String object
    StreamReader* reader = new StreamReader(S"rockets.txt");
    String* rocketStats = reader->ReadToEnd();
    reader->Close();
    String* pattern = S"\"(?<fname>[^\"]+)\","
                      S"\"(?<lname>[^\"]+)\","
                      S"(?<ppg>[^,]+),"
```

```
                        S"\"(?<nickname>[^\"]+)\"";
    Regex* rx = new Regex(pattern);
    // Disable notifications, constraints and index maintenance
    table->BeginLoadData();
    // for all the matches
    for (Match* match = rx->Match(rocketStats);
        match->Success;
        match = match->NextMatch())
    {
      Object* columns[] = new Object*[table->Columns->Count];
       GroupCollection* groups = match->Groups;
      columns[0] = groups->Item[S"fname"]->Value;
      columns[1] = groups->Item[S"lname"]->Value;
      columns[2] = groups->Item[S"ppg"]->Value;
      columns[3] = groups->Item[S"nickname"]->Value;
      // Load all fields with one call
      table->LoadDataRow(columns, true);
    }

    // Re-enable notifications, constraints and index maintenance
    table->EndLoadData();
  }
  catch(Exception* e)
  {
    Console::WriteLine(e->Message);
  }
#pragma pop_macro("new")
}
```

Once the DataTable is constructed and its schema defined, the text file is read using the StreamReader class covered in Chapter 3. Specifically, the ReadToEnd method is called so that the entire file is read into a String object. At that point, a regular expression is used, in which each Match object represents a row in the text file and each named Group object within the Match object represents a column of that row. (Regular expressions, matches and groups are covered in Chapter 2.)

Now that we have the data in a format that can be easily enumerated, the function calls BeginLoadData to disable notifications, constraints, and index maintenance while the rows are loaded into the DataTable. This loading is done via a call to the LoadDataRow method. As you can see, the first parameter to this method is an array of objects, each of which represents a column.

The second parameter is a Boolean value that indicates if the `DataTable::AcceptChanges` should be called after the table is loaded.

There are also a couple of other things you should note here. First, if a column in the `DataRow` is auto-generated, or if you want the default value used, you should simply set the object corresponding to that column to `System::Object::Empty`. However, keep in mind that the new row will only be appended to the `DataTable` if the key field does not already exist. If you're attempting to add a row with a duplicate primary key, the first row will be overwritten with the second. Therefore, while I didn't use a primary key in this example in order to keep things simple (and since I've already shown how to use primary keys in previous examples), you should take care to avoid duplicate keys. Finally, when the data has been loaded, a call to `EndLoadData` re-enables the `DataTable` object's notifications, constraints, and index maintenance.

While the batch loading of records is the most common use for the `BeginLoadData` and `EndLoadData` methods, they are also used to handle another common problem—that of swapping the primary keys of two rows. In fact, I've seen numerous times on various newsgroups that someone has attempted to use the `DataRow` class's `BeginEdit` and `EndEdit` methods to accomplish this to no avail.

```
// Disable constraint checking for both rows
row1->BeginEdit();
row2->BeginEdit();

Int32 temp = *dynamic_cast<__box int*>(row1->Item[S"ID"]);
row1->Item[S"ID"] = row2->Item[S"ID"];
row2->Item[S"ID"] = __box(temp);
row1->EndEdit();   // ERROR! - WILL FAIL HERE!
row2->EndEdit();
```

As the comment indicates, this technique will fail. The calls to the `BeginEdit` method for the two `DataRow` objects result in the disabling of constraints for the respective rows, as we want. However, once `EndEdit` is called for the first `DataRow`, the primary key value for that first row conflicts with the primary key value for the not-yet-committed second `DataRow` object. Once again, the `BeginLoadData` and `EndLoadData` methods save the day, as illustrated in the following example:

```
void SwapPrimaryKeyValues()
{
```

```
#pragma push_macro("new")
#undef new
  try
  {
    DataSet* dataset = new DataSet();
    DataTable* table = new DataTable(S"Players");
    dataset->Tables->Add(table);
    table->Columns->Add(S"ID", __typeof(Int32));
    table->Columns->Add(S"Name", __typeof(String));
    DataColumn* primaryKeys[] = new DataColumn*[1];
    primaryKeys[0] = table->Columns->Item[0];
    table->PrimaryKey = primaryKeys;
    // Create row #1
    DataRow* row1 = table->NewRow();
    row1->Item[S"ID"] = __box(1);
    row1->Item[S"Name"] = S"Foo";
    table->Rows->Add(row1);
    // Create row #2
    DataRow* row2 = table->NewRow();
    row2->Item[S"ID"] = __box(2);
    row2->Item[S"Name"] = S"Bar";
    table->Rows->Add(row2);
    // Turn off notifications, constraints and
    // index maintenance.
    table->BeginLoadData();
    // Pull the primary key switch-a-roo while nobody's
    // watching.
    Int32 temp = *dynamic_cast<__box int*>(row1->Item[S"ID"]);
    row1->Item[S"ID"] = row2->Item[S"ID"];
    row2->Item[S"ID"] = __box(temp);
    // Turn back on notifications, constraints and
    // index maintenance.
    table->EndLoadData();
  }
  catch(Exception* e)
  {
    Console::WriteLine(e->Message);
  }
#pragma pop_macro("new")
}
```

Working with Row State Information

As mentioned earlier in the chapter and illustrated in the section entitled "Deleting and Removing Rows," the `DataRow` class defines a property called `RowState` that is a `DataRowState` enumeration with the values shown in Table 6–2.

Table 6-2 `DataRowState` Enumeration Values

Member Name	Description
`Added`	The `DataRow` object has been added to the `DataRowCollection` object, but `DataRow::AcceptChanges` has not been called.
`Deleted`	The `DataRow` object—belonging to a `DataRowCollection`— has been deleted using the `DataRow::Delete` method.
`Detached`	Either the `DataRow` object has not been added to the collection or it has been removed via either the `DataRowCollection::Remove` or `DataRowCollection::RemoveAt` method.
`Modified`	The `DataRow` object—belonging to a `DataRowCollection`—has been edited, but `DataRow::AcceptChanges` has not been called.
`Unchanged`	The `DataRow` object—belonging to a `DataRowCollection`—has not changed since the last time `DataRow::AcceptChanges` was called.

As you can you can see, these enumeration values cover the various states of a `DataRow`. In fact, you can use the Visual Studio .NET debugger with the `EmployeeMaintenance` demo application and easily see the various states of the respective `DataRow` objects as you add, edit, and delete employee records. The following example illustrates the manipulation of a `DataRow` object and the impact of those changes on the object's `RowState` property. (As you've already seen examples of connecting to and reading existing data into `DataSet` objects, this code also illustrates how to manually define your own `DataSet` and `DataTable` objects from application data. It's not something you'll use every day, but it's definitely useful at times.)

```
void PrintRowState(String* currentAction, DataRow* row)
{
```

```
    Console::WriteLine(String::Format(S"{0,-20}\t\tDataRowState::{1}",
                    currentAction,
                    __box(row->RowState)));
}


void CreateAndQueryRowStates()
{
#pragma push_macro("new")
#undef new
  try
  {
    // Define a DataSet
    DataSet* dataset = new DataSet();
    DataTable* table = new DataTable(S"Players");
    dataset->Tables->Add(table);

    table->Columns->Add(S"ID", __typeof(Guid));
    table->Columns->Add(S"FirstName", __typeof(String));
    table->Columns->Add(S"LastName", __typeof(String));
    table->Columns->Add(S"Nickname", __typeof(String));
    // Create new row
    DataRow* row = table->NewRow();
    row->Item[S"ID"] = Guid::NewGuid().ToString();
    row->Item[S"FirstName"] = S"Yao";
    row->Item[S"LastName"] = S"Ming";
    row->Item[S"Nickname"] = S"";
    PrintRowState(S"Created new row", row);
    // Add row table->Rows->Add(row);
    PrintRowState(S"Added new row", row);
    // Accept changes        row->AcceptChanges();
    PrintRowState(S"Accepted changes", row);
    // Edit row
    row->Item[S"Nickname"] = S"Dynasty";
    PrintRowState(S"Edited row", row);

    // Delete row
    row->Delete();
    PrintRowState(S"Deleted row", row);
  }
  catch(Exception* e)
  {
    Console::WriteLine(e->Message);
  }
```

```
#pragma pop_macro("new")
}
```

Plugging these two functions into a Win32 console application with Managed Extensions support and the appropriate using directives would result in the following output:

```
Created new row        DataRowState::Detached
Added new row          DataRowState::Added
Accepted changes       DataRowState::Unchanged
Edited row             DataRowState::Modified
Deleted row            DataRowState::Deleted
```

While most of this is what you'd expect after seeing the `DataRowState` enumeration descriptions in Table 6–2, there are a couple of things of note. First, a newly created `DataRow` has its `RowState` set to `Detached` until it is added to the `DataRowCollection` via the `DataRowcollection::Add` method. At that point, the `RowState` is initialized to `Added` until the `DataRow::AcceptChanges` method is called, at which point the `RowState` becomes `Unchanged`. Finally, note that if you call `DataRow::Delete`, the row's `RowState` becomes `Deleted`, and subsequently calling `DataRow Collection::Remove` (or `DataRowCollection::RemoveAt`) will not affect its `RowState`. The opposite is also true: Removing a `DataRow` from a `Data RowCollection` sets the row's `RowState` to `Detached`, but subsequently deleting it will not alter its `RowState` value.

Searching, Sorting, and Filtering Data

So far, you've learned how to connect to a data store, retrieve the desired data, close the connection, modify the data locally (including adding, editing, and deleting records), and then update the data store at will—once again only holding the connection long enough to synchronize your in-memory representation of the data with the data store. As you'll discover in this section, the ADO.NET classes also support the ability to search, sort, and filter data once it's been retrieved. In fact, being able to perform these operations against your disconnected data without making continual round trips to the server is one of the strongest arguments for using disconnected data to begin with.

We'll begin by looking at how to search the `DataTable` and `DataRow Collection` objects using both primary keys and column data. After that, you'll see how to create `DataView` objects to represent different views on the same `DataTable` object where you can filter on both column values and row state information, which we covered in the previous section. Finally, you'll discover how to sort and search through these filtered `DataView` objects.

Searching `DataTable` and `DataRowCollection` Objects

There are two basic ways to search a `DataTable` and its internal `DataRow Collection`—by primary key and by column value. We'll start with the primary key search, as that technique was used in the `EmployeeMaintenance` demo application.

Searching on Primary Key

The `DataRowCollection::Find` method is used to retrieve a `DataRow` from a `DataRowCollection` by its primary key value. This method has overloaded versions that return a `DataRow` instance. A value of `null` is returned if a row containing the specified primary key value(s) cannot be located in the collection.

```
public: DataRow* Find(Object*);
public: DataRow* Find(Object*[]);
```

The first `Find` method enables you to specify a single primary key value, while the second enables you to specify an array of values if the primary key has more than one column. Note that passing an array whose length is greater than the number of defined primary keys columns results in an `ArgumentException`, where exception object's `Message` property will indicate both how many values were expected and how many were specified in the `Find` method.

Also keep in mind that in order to use the `Find` method with an untyped dataset, you'll need to either set the data adapter object's `Missing SchemaAction` property to `MissingSchemaAction::AddWithKey` (before the call to the adapter's `Fill` method) or manually set the `DataTable` object's `Primary` key property by passing it an array of `DataColumn` objects that correspond to the table's primary key columns. Both of these techniques are described in more detail in the section entitled "Filling in Missing Schema and Primary Key Information for Untyped Datasets."

In the following example, I'm using the `Find` method to locate an employee record with a primary key value of 1. As the Employees table's `EmployeeID` column is defined as its only primary key column, then this search is basically the equivalent of an SQL statement like `SELECT` * where `EmployeeID = 1`.

```
// Search for a primary key made up of a single column
// Assumes a connection to the Employees table
...

DataRow* row = employeesTable->Rows->Find(__box(1));
if (row)
{
  // Have DataRow to use
  Console::WriteLine(String::Format(S"Found record  -> {0} {1}",
                                    row->Item[S"FirstName"],
                                    row->Item[S"LastName"]));
}
else
  Console::WriteLine(S"Could not locate record");
```

The sample Employee table contains a record for a "Nancy Davolio" that has an `EmployeeID` value of 1, so her name should print out. Now let's see how to perform a `Find` operation with more than one primary key column. As the Employees table has only a single column defined as its primary key column, the following sample will also include code to override that fact and tell the `DataTable` that the `FirstName` and `LastName` columns are now the primary key columns.

```
// Search for a primary key made up of multiple columns
// Assumes a connection to the Employees table
...

// Tell the DataTable that the FirstName and LastName
// columns are now the primary key columns
employeesTable->PrimaryKey = NULL;
DataColumn* keys[] = new DataColumn*[2];
keys[0] = employeesTable->Columns->Item[S"FirstName"];
keys[1] = employeesTable->Columns->Item[S"LastName"];
employeesTable->PrimaryKey = keys;
// Build the primary key search value array
Object* search[] = new Object*[2];
```

```
search[0] = S"Nancy";
search[1] = S"Davolio";

DataRow* row = employeesTable->Rows->Find(search);
if (row)
{
  // Have DataRow to use
  Console::WriteLine(String::Format(S"Found record --> {0}",
                                    row->Item[S"EmployeeId"]));
}
else
  Console::WriteLine(S"Could not locate record");
```

In addition to the `Find` method, the `DataRowCollection` also includes a method called `Contains`, which allows you to perform the same exact search as the `Find` method, with the difference being that instead of a `DataRow` being returned, a Boolean value is returned indicating the success or failure of the search.

Searching (and Sorting) on Column Values

In order to search a `DataTable` for a given column value, you can use the `DataTable::Select` method and specify search criteria much like those of the SQL `SELECT` statement. The `Select` method has the following overloads.

```
DataRow* Select() [];
DataRow* Select(String* criteria) [];
DataRow* Select(String* criteria, String* sort) [];
DataRow* Select(String* criteria, String* sort, DataViewRowState) [];
```

The first (parameter-less) version of this method simply returns all rows in the `DataTable` object's `DataRowCollection` in a `DataRow` array sorted by primary key. If the table does not contain a primary key, then the rows are sorted in arrival sequence (the order in which they were added to the collection).

The second—and most used—version allows you to specify the search criteria. As mentioned, this criterion is much like specifying the value of an SQL `SELECT` statement without the SQL verbs. As with the parameter-less `Select` method, the resulting `DataRow` array is either sorted by primary key or, in the absence of a defined primary key column, arrival sequence.

The third version of the Select method enables you to specify search criteria as well as the columns to use in sorting the returned DataRow array. You can also append the standard SQL ASC and DESC to the column to indicate an ascending or descending sort, respectively.

Finally, the last Select method overload enables you to specify search criteria using the sort columns and a DataViewRowState enumeration value. In the section entitled "Working with Row State Information," you learned that the DataRow object maintains a state property indicating such things as whether or not the row has been added or changed (but not committed), deleted, or unchanged. The valid DataViewRowState enumeration values and their descriptions are listed and explained in the section entitled "Filtering on Row State Information."

Here are several examples of how to use the Select method that illustrate how robust this feature is. For example, aside from the standard comparison operators such as =, <, >, and <>, you also have the ability to use the SQL wildcards * and % with the LIKE predicate, user-defined values for comparisons on types such as dates, and special built-in functions such as SUBSTRING.

```
void PrintResults(DataRow* rows[])
{
  Console::WriteLine();
  String* format = S"{0,-7} {1, -10} {2}";
  Console::WriteLine(format, S"Record", S"First Name", S"Last Name");
  for (int i = 0; i < rows->Count; i++)
  {
    Console::WriteLine(format,
                       __box(i),
                       rows[i]->Item[S"FirstName"],
                       rows[i]->Item[S"LastName"]);
  }
}

void Select()
{
#pragma push_macro("new")
#undef new
  try
  {
    DataRow* rows[];
    // Search on a single column
    rows = employeesTable->Select(S"LastName = 'Archer'");
```

```
      PrintResults(rows);
      // Search on multiple columns
      rows = employeesTable->Select(S"FirstName = 'Tom' "
                                    S"AND LastName = 'Archer'");
      PrintResults(rows);
      // Search using the SQL LIKE predicate
      rows = employeesTable->Select(S"Title LIKE 'Sales*'");
      PrintResults(rows);
      // Search and Sort on two columns
      rows = employeesTable->Select(S"City = 'London'",
                                    S"LastName,FirstName");
      PrintResults(rows);
      // Search and Sort in DESCending order
      rows = employeesTable->Select(S"Region = 'WA'", S"BirthDate
        DESC");
      PrintResults(rows);
      // Search using a user-defined value
      rows = employeesTable->Select(S"BirthDate > #01/01/1964#");
      PrintResults(rows);
      // Search using the SUBSTRING expression
      rows = employeesTable->Select(S"SUBSTRING(HomePhone,2,3) =
        '206'");
      PrintResults(rows);
   }
   catch(Exception* e)
   {
      Console::WriteLine(e->Message);
   }
#pragma pop_macro("new")
}
```

Sorting, Searching, and Filtering Data with `DataView` Objects

When programming "true" .NET applications—as opposed to mixed mode applications— `DataView` objects are used to provide data binding between an application's data and Windows Forms (and Web Forms) controls. However, as this chapter is about using ADO.NET in the context of an MFC application, I think of the `DataView` being associated with the `DataTable` much as a `CView` is associated with a `CDocument`. Therefore, for our purposes `DataView` objects allow you to create multiple concurrent views on the same data so that any change to the data is realized by all the views— depending on their filter.

Table 6-3 `DataViewRowState` Enumeration Values

Member Name	Description
`Added`	Returns new rows.
`CurrentRows`	All current rows—including new, changed, and unchanged rows.
`Detached`	Rows marked as deleted, but not yet removed from the collection.
`ModifiedCurrent`	Each `DataRow` object maintains two versions of its data—an "original" version and a "modified" version. This facilitates such things as the batch operations where you can cancel edits, as seen in "Batch Row Changes." The `ModifiedCurrent` enumeration simply specifies that you want the current versions of modified rows. (Compare with `ModifiedOriginal`.)
`ModifiedOriginal`	Specifies that you want the original values of any modified rows. Compare with `ModifiedCurrent`, which allows you to obtain the current values.
`None`	No rows are returned.
`OriginalRows`	This will return all of the original rows, including those that have been deleted or modified.
`Unchanged`	Only returns rows that have not been modified.

While it's not always obvious, each `DataTable` has a built-in `DataView` object associated with it. This view can be accessed via the `DataTable::DefaultView` property. Additionally, you can create as many `DataView` objects on your `DataTable` as your application needs. Typically these `DataView` objects are created in order to provide a filtered view of your data. Therefore, in this section you'll first see how to create filtered views—both by column value and by row state—and then how to sort and search your view's data. Note that I'll be defining new `DataView` objects in the code snippets in this section, as most times when a `DataView` is used it is because the application needs multiple views on the same data. However, you can apply the same techniques to the `DataTable::DefaultView` if your application has no need for a second view.

Filtering on Row State Information

In the section entitled "Working with Row State Information" you learned about the `DataRow::RowState` property and how various operations affect

its value. In addition to being able to create views that are filtered on column values, you can also create filtered views specific to rows having a given RowState value. In fact, implementing this feature can be a very powerful addition to your application. For example, let's say that in your disconnected application you wish to provide an Undo feature for any changes the user has made to the DataTable. As you'll soon see, you could easily construct a view and display a dialog with all modified, added, or deleted rows so that the user could choose which ones were to be undone. You could also use the same strategy in presenting a dialog that would allow the user to confirm which data to actually update with the server data source. This is all done simply by instantiating a DataView object and setting its RowStateFilter property to one of the DataViewRowState enumeration values shown in Table 6–3.

The following code was once again run against the Employees table, with the difference being that this time I specified in the data adapter's constructor that I only wanted records with EmployeeID values less than 4. (This was done merely to produce a shorter printout from the example code.)

```
void DataViewFilterByRowState()
{
#pragma push_macro("new")
#undef new
  try
  {
    // Add rows
    DataRow* row1 = employeesTable->NewRow();
    row1->Item["FirstName"] = S"Fred";
    row1->Item["LastName"] = S"Flintstone";
    employeesTable->Rows->Add(row1);
    // Edit row
    employeesTable->Rows->Item[0]->Item[S"FirstName"] = S"Test";

     // Delete row
     employeesTable->Rows->Item[1]->Delete();
    PrintDataViewInfo(DataViewRowState::OriginalRows);
    PrintDataViewInfo(DataViewRowState::CurrentRows);
    PrintDataViewInfo(DataViewRowState::Added);
    PrintDataViewInfo(DataViewRowState::Deleted);
    PrintDataViewInfo(DataViewRowState::ModifiedCurrent);
    PrintDataViewInfo(DataViewRowState::ModifiedOriginal);
    PrintDataViewInfo(DataViewRowState::Unchanged);
```

```
  }
  catch(Exception* e)
  {
     Console::WriteLine(e->Message);
  }
#pragma pop_macro("new")
}
```

As you can see, the first thing I did was simply to add a new row, edit an existing row, and then delete a row. The `PrintDataViewInfo` function (shown next) is then called to test how these three changes (four if you include the initial reading of the data) would impact a `DataView` using each of the various `DataViewRowState` enumerations.

```
void PrintDataViewInfo(DataViewRowState rowStateFilter)
{
#pragma push_macro("new")
#undef new
  try
  {
     DataView* view = new DataView(employeesTable);
     view->RowStateFilter = rowStateFilter;
     Console::WriteLine(S"Using RowStateFilter = {0}",
                        __box(view->RowStateFilter));
     Console::WriteLine(S"Row count = {0}", __box(view->Count));
     String* format = S"{0,-3} {1, -10} {2}";
     Console::WriteLine(format, S"Row", S"First Name", S"Last Name");
     for (int i = 0; i < view->Count; i++)
     {
        Console::WriteLine(format,
                           __box(i),
                           view->Item[i]->Item[S"FirstName"],
                           view->Item[i]->Item[S"LastName"]);
     }
     Console::WriteLine();
  }
  catch(Exception* e)
  {
     Console::WriteLine(e->Message);
  }
#pragma pop_macro("new")
}
```

As you can see, the `PrintDataViewInfo` function constructs a `DataView` object based on the Employees table, sets the `RowStateFilter` to the passed `DataViewRowState` value, and then simply enumerates and outputs the records for that newly created view. Here is the output from running these two functions:

```
Using RowStateFilter Added
Row count = 1
Row First Name Last Name
0    Fred         Flintstone

Using RowStateFilter OriginalRows
Row count = 3
Row First Name Last Name
0    Nancy        Davolio
1    Andrew       Fuller
2    Janet        Leverling

Using RowStateFilter Deleted
Row count = 1
Row First Name Last Name
0    Andrew       Fuller

Using RowStateFilter CurrentRows
Row count = 3
Row First Name Last Name
0    Test         Davolio
1    Janet        Leverling
2    Fred         Flintstone

Using RowStateFilter ModifiedOriginal
Row count = 1
Row First Name Last Name
0    Nancy        Davolio

Using RowStateFilter ModifiedCurrent
Row count = 1
Row First Name Last Name
0    Test         Davolio
```

As you can see, with the `DataView` you can retrieve the original rows that were read into the `DataTable`, filter only added or deleted rows, view

all current rows (which is what the `DataTable::DefaultView` does), or even specify that you want to see only modified rows (either the original values or the current values).

Filtering on Column Values

Much as you can use the `DataTable::Select` method to search for rows giving a search criteria, you can also search `DataView` objects via the `DataView::RowFilter` property. The main difference is that the `Select` method returns an array of `DataRow` objects, while the `RowFilter` property simply changes the records that the `DataView` presents when enumerating the row collection or requesting a specific row (via the `Item` property or one of the search methods).

In addition, the `RowFilter` takes a `String` parameter that acts as the search criterion that allows the same expression values as the `Select` method. Therefore, I'll refer you back to the "Searching (and Sorting) on Column Values" section in order to see examples of what types of expressions you can pass to the `RowFilter` property.

Sorting

Along with filtering `DataView` objects, you can also sort them via the `DataView::Sort` property. In order to set this value, simply specify the name of one or more columns separated by commas and include either ASC for (ascending sort) or DESC (for descending sort). Note that if you do not specify ASC or DESC for a given column, then its sort order defaults to ascending.

. . .

```
DataView* view = new DataView(employeesTable);
// Sort by a single column in ascending order
view->Sort = S"LastName";
// Sort by birthdate in descending order
view->Sort = S"BirthDate DESC";
// Sort by multiple columns in various orders
view->Sort = S"LastName, BirthDate DESC, Title";
```

Working with Typed Datasets

As mentioned earlier in the chapter, typed datasets are not a separate set of ADO.NET classes. Instead, they are derived directly from the standard ADO.NET classes and define extra members that are specific to your database schema. To understand the main benefit of typed datasets, think of how data has been accessed throughout this chapter using the generic `DataRow::Item` property, where column name or index value is specified. Because the `Item` property has no intimate knowledge of the underlying data schema, any mistakes made when using it—such as such as misspelling the column name or using an incompatible data type—aren't realized until runtime. Conversely, typed datasets are generated directly from the database schema and thus produce members that allow for compile-time checking of these common mistakes.

In this section, I'll first illustrate—step-by-step—how to generate a typed dataset for the SQL Server sample Northwind database's Employees table. After that, I'll introduce an MFC SDI application that uses a typed dataset to list all records from the Employees table. Finally, I'll wrap up the section—and chapter—with a listing of the pros and cons of using untyped vs. typed datasets.

Generating a Typed Dataset

Now let's see how to use Visual Studio .NET to generate a typed dataset. As stated at the outset of this chapter, the demo will be using the SQL Server sample Northwind database.

1. Either create a new project with support for Managed Extensions, or open an existing Managed Extensions project.
2. Select the Add New Item option from the Project menu (or from the project's context menu in the Solution Explorer).
3. When the Add New Item dialog appears (Figure 6–5), click the Data folder (category) on the left and then the `DataSet` icon (template) on the right.
4. Enter a name for the header file that will contain the definitions of the typed dataset and any included `DataTable` objects. While this demo will only use the Northwind Employees table, you might want add other `DataTable` definitions later, so name the file `Northwind` and click the Open button.

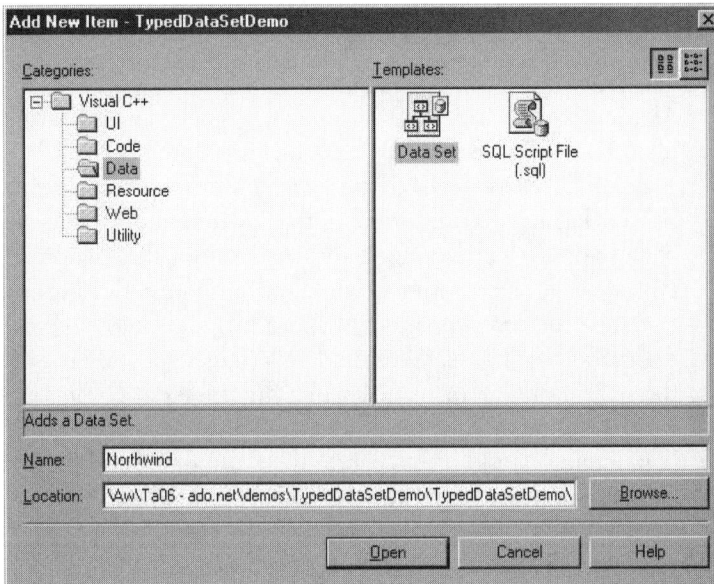

Figure 6-5 Visual Studio .NET provides a template for generating typed datasets.

5. At this point a script will run that will generate the file containing the typed dataset information. *It's important to realize that this script might trigger a response from your anti-virus software.* If that is the case, just select the option to allow the script to run. The Visual Studio .NET Output window will display the progress of the script.

6. When a typed dataset file is generated, an XSD file (XML Schema Definition) file is created to describe the dataset. This file is automatically opened upon the creation of the typed dataset (Figure 6-6) in a window that is sometimes called a *canvas*. Database entities can be added to the dataset by dragging them from the Server Explorer onto the canvas, so at this point, click the canvas's Server Explorer link (also available from the Project menu).

7. If the desired database is on your local machine, simply browse to that database using the Server Explorer tree view. However, if your database is on a remote machine, you'll need to click the Connect to Server button at the top of the Server Explorer dialog. When the Add Server dialog is displayed, simply type the name of the server. (There's no need to add the UNC path information.)

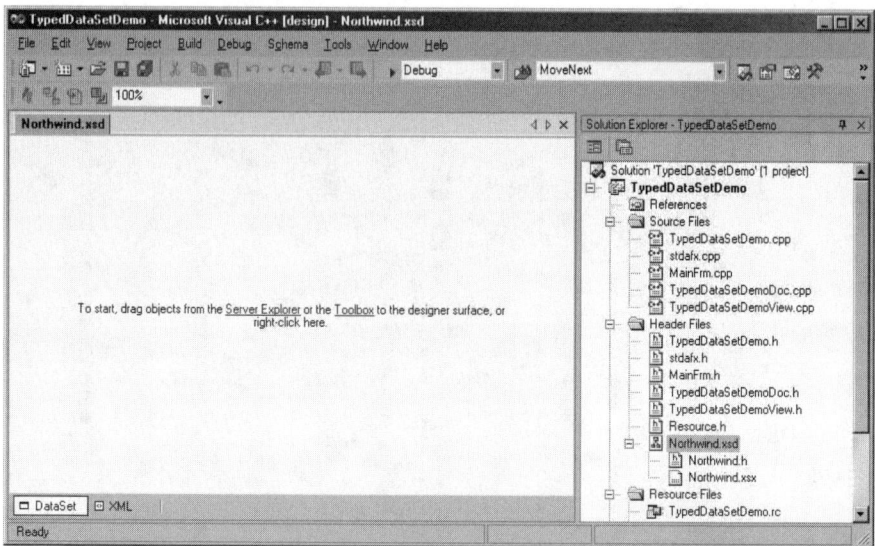

Figure 6-6 A drop-target canvas is used to visually represent a typed dataset.

8. At this point, you should have a database displayed in the Server Explorer—local or remote. Figure 6–7 shows my particular configuration. As you can see, JEANVALJEAN is the name of my programmer machine, and it has no SQL Server databases. Therefore, I added (connected to) a server named FANTINE to my Server Explorer so that I could access its SQL Server databases.

9. While the icons next to the various databases include an "x", once you expand the item the Server Explorer connects to the database and retrieves the selected database's information—such as its Tables, Views, Stored Procedures, and so on. Here you would simply drag the desired database entity to the dataset palette. For purposes of this demo—and so we can make an "apples and apples" comparison between the untyped datasets you've been using throughout the chapter and a typed dataset—drag the Employees table from the Northwind database to the typed dataset canvas.

10. If you open the `Northwind.h` file, you won't see any Employee `DataTable` information. That information isn't generated until you build the project. Building the project now will result in Visual Studio .NET generating the expected classes and members in the `Northwind.h` file for the Employees table.

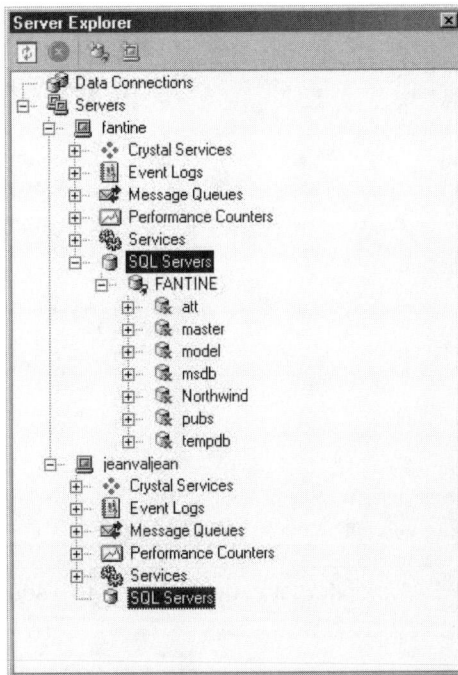

Figure 6–7 The Server Explorer allows you to drag database entities—such as tables, views, and stored procedures—onto the typed dataset view.

A tremendous amount of code is actually generated for a typed dataset, so we'll just briefly look at some of it—but enough so that you get an idea of what all has been done for you. When you open the `Northwind.h` file, the first thing you notice is that the file is free-standing in that the appropriate `#using` and `using` namespace directives have been inserted. Also note that the typed dataset classes have been defined within a namespace of the same name as the current project. That means you'll have to either qualify all uses of the types defined in this file or insert a `using` namespace directive into any of your typed dataset client code.

The first class that you encounter will be the `DataSet`-derived Northwind class. This class also defines nested classes for the Employees table that include a `DataTable`-derived class, a `DataRow`-derived class, and a delegate for subscribing to row-change events.

```
public __gc class Northwind : public System::Data::DataSet {
  public : __gc class EmployeesDataTable;
```

```
public : __gc class EmployeesRow;
public : __gc class EmployeesRowChangeEvent;
...
```

You'll note that the file does not include any connection-specific information as you might expect. This is so that the typed dataset can be used against any connection. Therefore, the typed dataset only refers to the data entity's schema—not to where the underlying data store is located or how to connect to it. Therefore, the way you would use these types is to connect and build a data adapter just as you would with an untyped dataset, and then use the typed `DataSet`-derived and `DataTable`-derived classes as follows.

```
#include "Northwind.h"
using namespace System::Data::SqlClient;
using namespace TypedDataSetDemo;


...


SqlConnection* conn =
  new SqlConnection(S"Server=FANTINE;"
                    S"Database=Northwind;"
                    S"Integrated Security=true;");
SqlDataAdapter* adapter =
  new SqlDataAdapter(S"SELECT * FROM Employees", conn);
SqlCommandBuilder* commandBuilder = new SqlCommandBuilder(adapter);
conn->Open();
Northwind* northwind = new Northwind();
adapter->Fill(northwind, S"Employees");
conn->Close(); // No longer needed
Northwind::EmployeesDataTable* employees = northwind->Employees;


...
```

Let's look at this code—especially in comparison to the connection code used previously in this chapter using untyped datasets. Obviously, I need to include the `Northwind.h` file, as it defines the typed dataset types. After that, I need to specify the `System::Data::SqlClient` namespace, as the `Northwind.h` file is using generic base classes and the `SqlClient` namespace is specific to SQL Server ADO.NET types. Finally, the `Typed DataSetDemo` namespace is referenced so that we don't have to qualify the references to the various typed dataset types.

After the standard connection object creation and instantiation of a data adapter object, I declare an instance of the DataSet-derived Northwind class. From there, I call the adapter's Fill method. However, note that I've changed my table name from the "AllEmployees" value that I've used throughout the chapter to "Employees". While seemingly trivial, I'm actually forced to do this because the Northwind::Employees property is hard-coded to return an EmployeesDataTable object with a name of "Employees".

One last thing that I'll point out before you see a demo illustrating how to use a typed dataset in an MFC application is that no DataRow Collection-derived class is generated for us. This is because the Employees DataTable implements the IEnumerable interface and acts as a collection for its rows. The various collection classes, the IEnumerable interface, and even creating your own enumerable classes are covered in Chapter 11. The EmployeesDataTable also implements the Count and Item properties so that an EmployeesDataTable object can be iterated like an array.

```
public :
[System::Diagnostics::DebuggerStepThrough]
__gc class EmployeesDataTable :
  public System::Data::DataTable,
  public System::Collections::IEnumerable
{

...

__property System::Int32 get_Count();

...

public:
__property TypedDataSetDemo::Northwind::EmployeesRow* get_Item(
  System::Int32 index

);
```

Because these properties have been implemented, the Employees DataTable can be enumerated as simply as this:

```
...

// For every employee record
for (int i = 0; i < employees->Count; i++)
```

```
{
    int id = *dynamic_cast<__box int*>(employees->Item[i]->EmployeeID);
    String* firstName = employees->Item[i]->FirstName;
    String* lastName = employees->Item[i]->LastName;
}
```

As you can see, each row is retrieved via the Item property, and, as the row is a Northwind::EmployeesRow, there are properties with the same names and types as the actual Employees table:

```
public : [System::Diagnostics::DebuggerStepThrough]
__gc class EmployeesRow : public System::Data::DataRow {
```

    . . .

```
public: __property System::Int32 get_EmployeeID();
public: __property  void set_EmployeeID(System::Int32 value);
public: __property System::String*  get_LastName();
public: __property  void set_LastName(System::String*_u32 ?value);
public: __property System::String*  get_FirstName();
public: __property  void set_FirstName(System::String*  value);
```

    . . .

Using Typed Datasets

Now let's see how to use typed dataset types in an MFC demo. We'll keep the application simple, as it will connect to the Employees table, read all the employee records, and display them in a list view. However, combining what you'll see here with what you've learned throughout the chapter should make it easy for you to use typed datasets for all your dataset needs should you decide to go that route.

1. To get started, create and MFC SDI application called Typed DataSetDemo, setting the view base class to CListView and updating the Project settings to support Managed Extensions.
2. As explained in the previous section, create a typed dataset called Northwind and add to it the Employees table.
3. Add the following directives to the stdafx.h file:

```
#using <mscorlib.dll>
#using <system.dll>
```

```
#using <system.windows.forms.dll>
using namespace System;
using namespace System::Windows::Forms;
#undef MessageBox
```

4. Open the `TypedDataSetDemoView.h` file and add the following directives just before the `CTypedDataSetDemoView` class declaration.

```
#include "Northwind.h"
using namespace System::Data::SqlClient;
using namespace TypedDataSetDemo;
```

5. Declare the following ADO.NET objects in the `CTypedDataSet DemoView` class.

```
class CTypedDataSetDemoView : public CListView
{
...
protected:
  gcroot<Northwind*>northwind;
  gcroot<SqlDataAdapter*>adapter;
  gcroot<Northwind::EmployeesDataTable*>employees;
  gcroot<SqlCommandBuilder*>commandBuilder;
...
```

6. Add the following code to the view's `OnInitialUpdate` function to initialize the list view.

```
void CTypedDataSetDemoView::OnInitialUpdate()
{
  CListView::OnInitialUpdate();
  // Add columns to listview
  CListCtrl& lst = GetListCtrl();
  lst.InsertColumn(0, _T("ID"));
  lst.InsertColumn(1, _T("First Name"));
  lst.InsertColumn(2, _T("Last Name"));
...
```

7. In the view class's `PreCreateWindow` function, set the view window's style to "report":

```
BOOL CTypedDataSetDemoView::PreCreateWindow(CREATESTRUCT& cs)
{
  cs.style |= LVS_REPORT;
  return CListView::PreCreateWindow(cs);
}
```

8. Insert the following code just before the end of the `OnInitial`
`Update` function.

```
#pragma push_macro("new")
#undef new
try
    {
        SqlConnection* conn =
            new SqlConnection(S"Server=localhost;"
                              S"Database=Northwind;"
                              S"Integrated Security=true;");
        adapter = new SqlDataAdapter(S"SELECT * FROM Employees",
                                     conn);
        commandBuilder = new SqlCommandBuilder(adapter);
        conn->Open();
        northwind = new Northwind();
        adapter->Fill(northwind, S"AllEmployees");
        conn->Close(); // No longer needed
        employees = northwind->Employees;
        ReadAllEmployees();
    }
    catch(Exception* e)
    {
        MessageBox::Show(e->Message,
                         S".NET Exception Thrown",
                         MessageBoxButtons::OK,
                         MessageBoxIcon::Error);
    }
#pragma pop_macro("new")
```

9. Finally, implement the `ReadAllEmployees` function as follows:

```
void CTypedDataSetDemoView::ReadAllEmployees()
{
  try
  {
    CWaitCursor wc;
    CListCtrl& lst = GetListCtrl(); lst.DeleteAllItems();
    for (int i = 0; i < employees->Count; i++)
    {
      int idx = lst.InsertItem(i,
        (CString)__box(employees->Item[i]->EmployeeID)->
          ToString());
      lst.SetItemText(idx, 1,
        (CString)employees->Item[i]->FirstName);
```

```
        lst.SetItemText(idx, 2,
            (CString)employees->Item[i]->LastName);
    }
}
catch(Exception* e)
{
    MessageBox::Show(e->Message, S".NET Exception Thrown",
                    MessageBoxButtons::OK,
                    MessageBoxIcon::Error);
}
}
```

Building and running the application will result in something similar to what you see in Figure 6–8. The code in this section was really not much different than what you've seen throughout this chapter, with the principal difference being that the use of typed dataset types enabled you to have the compiler check for common errors as opposed to having them realized at runtime. This segues nicely into the next part of this chapter—weighing the pros and cons of using typed datasets.

Weighing the Pros and Cons of Typed Datasets

Covering the entirety of typed datasets would take an entire chapter by itself. However, you've learned enough about typed datasets in this section to realize some of the positive points regarding using them in your ADO.NET code. Let's now briefly look at some of the advantages and disadvantages of using typed datasets (vs. using untyped datasets).

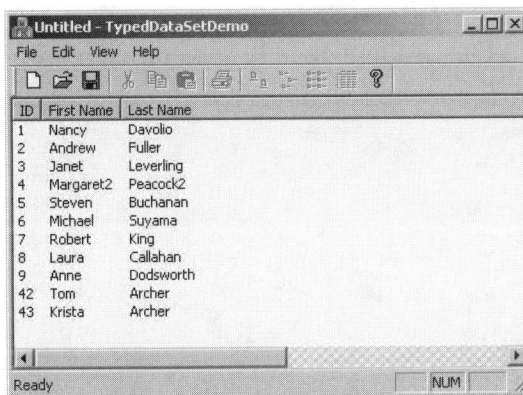

Figure 6-8 Example of running the `TypedDataSetDemo` application

Advantages of typed datasets:

- **Compile-time type checking**—Reduces runtime errors by having members based on the data's actual schema as opposed to untyped datasets, where you call a generic function and can pass an object of any type.
- **Schema-specific members**—Typed datasets define properties for getting and setting values where the property name is the same as the underlying column name. They also define properties for determining if the column is null and methods for searching the table via primary key(s).
- **Data binding support in VS.NET**—Only useful with Windows Forms applications, but bears mentioning if you plan on doing development based entirely on .NET in addition to the mixed-mode programming that is the focus of this book.
- **Intellisense support**—When using untyped datasets, you have to know beforehand the names of the columns and the types that the respective columns work with. With typed datasets, as soon as you enter the name of the type, Intellisense displays its members, thereby saving you development and debugging time.

Disadvantages of typed datasets:

- **Versioning**—Typed datasets can actually increase development time in situations where your schema changes, because you'll need to update the typed dataset information manually. This is obviously the same problem we've battled for years with CRecordset classes—having to modify them manually when the underlying schema changes.
- **Tightly coupled**—In its current design, typed datasets are difficult to extend and can't be modified (as they're auto-generated each time the project is built). In addition, they force a tight coupling of client to data access code, which might not be best for all situations.

As mentioned earlier in the chapter, I use untyped datasets in code snippets and demos, as this provides for shorter, to-the-point code listings. However, I personally choose to use typed datasets in my production code, as once I'm to the implementation stage of a project, I'm finished with any major changes to the database schema, and with the introduction of dataset

annotations, the only two real disadvantages for me are moot. (Annotations are extensions to the XSD format used to define typed datasets that allow some changes such as the renaming of classes and properties as well as defining how to deal with null values.)

Summary

The chapter began with a section illustrating the basic terminology and classes specific to ADO.NET and disconnected data as well as the basics of constructing those objects and connecting to a data source. The next section illustrated how to perform basic database operations (creating, reading, updating, and deleting records). Included in that section were tips on handling the common issues of dealing with disconnected data and auto-increment primary keys, and how to specify primary key information for untyped datasets. The section ended with a fully functional MFC maintenance application using the ADO.NET against a sample database. Once that section was completed, I then illustrated several techniques for reading and writing data in batch in order to keep round trips to the data server to a minimum and dramatically speed up insert operations. As one of the key benefits of an IMDB is its support of the ability to search, sort, and filter data once it's been retrieved, the next section illustrated how to accomplish these tasks as well as how to create multiple filtered views on a single dataset. Finally, the chapter ended with a section on generating strongly typed datasets, using them in an MFC application, and the pros and cons of using untyped vs. typed datasets.

While disconnected data can make applications more complex, it also dramatically improves the versatility and design of client/server and *n*-tier systems. For example, disconnected data means that systems with thousands of remote clients do not have to worry about how the server is going to attempt to handle so many concurrent connections to the database. In addition, because you can design your system so that the majority of the database operations are done in-memory before updating the database server, you can greatly reduce the number of network round trips. Using disconnected data also enables you to fill a `DataTable` with data taken from virtually any data source (or even application-generated) and process it with the same routines, regardless of the data's origin. In addition, the decoupled architecture of the `DataSet`- and `DataAdapter`-derived classes makes it possible to read data from one source and send updates to another

source. These are just a few of the possible scenarios where disconnected data can be a boon for software developers.

Please note that I'm not saying disconnected data is a silver bullet by any means. Every solution to a problem has its own drawbacks. In fact, using disconnected data is flat-out a very bad idea in certain circumstances. Specifically, just as with any other solution that introduces a certain level of complexity to a system, you must weigh the added complexity against the derived benefits. However, for systems—especially *n*-tier solutions— where both database connections and round trips must kept to a minimum, disconnected data via the ADO.NET class can significantly improve the overall design and execution of the system.

Finally, there's a lot more even to the parts of ADO.NET that support disconnected data than I could possibly convey in a single chapter. However, hopefully you've learned enough in this chapter to decide for yourself if and when ADO.NET and disconnected data are right for you. Now that you know these basics, in the next chapter I'll delve into the more advanced areas of ADO.NET such as merging `DataSet` objects, reading binary data (especially BLOBs), dealing with concurrency in a disconnected environment, and so on.

Advanced ADO.NET

Introduction

As mentioned in the summary of the previous chapter, there are a quite a few ADO.NET topics that simply couldn't all fit into a single chapter. Obviously, since entire books are written on the subject of ADO.NET, I'm not going to be able to cover it all in two chapters either. However, this chapter continues where the previous chapter left off with a number of ADO.NET topics that I'm frequently queried about once people get past the "gee whiz" factor of disconnected data and start implementing production systems. Among these topics are learning how to read and write binary data, such as photos; associating related tables with the `DataRelation` class for easier and more efficient navigation; and handling concurrency issues with multiuser, disconnected data systems.

The chapter begins by illustrating how to use the `DataReader` class's `GetBytes` method to read and write binary image data and then write it to disk using the `FileStream` class. As the `DataReader` only provides a means of reading data, the section then illustrates how to manually set up `SqlCommand` and `SqlParameter` objects to write binary data to a database. By way of comparison, the `DataSet` is then used to both read and write the same binary data. The section concludes with a demo maintenance application that uses the `DataSet` class to load data from a provided sample table that contains a photo column and data. The demo allows you to both select and view an image where the combination of GDI+ and the `MemoryStream` class is used to display an image without having to write it first to a temporary file on disk. The demo also allows you to select a new image for any record from disk and then update the database with that image.

The next section illustrates how the `DataRelation` class gives you the ability to define relationships between related tables such that both tables can be navigated more easily and efficiently. Using this class, you can

download two related tables, specify which columns create the parent/child linkage between the tables, and then, when in possession of a row from one of the tables, use either the `GetParentRow` or `GetChildRows` methods to retrieve the related rows from the other table. While the `DataRelation` class is extremely useful, there is the issue of inserting a parent table row that contains an auto-increment (IDENTITY) primary key and then needing to insert that value into the child rows. Using stored procedures with return parameters, you can easily solve this problem. The chapter then wraps up with a section on concurrency. This section explains what concurrency is, the various levels supported with ADO.NET, and the roles played by the data adapter and command builder classes. After a demo application illustrates an example of a concurrency conflict, the section presents techniques for handling the errors and advice on where in your code to handle them.

Creating the Sample Database for SQL Server or MSDE

You can, of course, use the techniques shown in this chapter with any data store supported by the ADO.NET classes. However, I've also provided a SQL Server database script—`ExtendingMFCWithDotNet.sql`—to create a sample database for use with this chapter's various demo applications. This script automatically generates a database named `ExtendingMFCWithDotNet` along with three tables (`Contributors`, `OrderHeaders`, and `OrderDetails`) and a stored procedure (`proc_InsertOrder`).

 If you're using SQL Server, open the SQL Query Analyzer application and log onto the desired SQL Server instance (either remote or local), supplying the appropriate user and password information. Once connected, open the SQL script file from the book's CD and execute it by pressing the F5 key. Once the script is running, you should see a couple of diagnostic messages indicating the status of running the script.

 Once the database has been generated, you'll need to import the sample data for the demo applications. For that purpose, I've included a `.dat` file for each table as well as an `import.cmd` file that calls the SQL Server BCP command line utility to import the data from each of these files into the appropriate table.

 Please note that the `.cmd` file might have to be tweaked for your environment. Specifically, the call to the BCP utility includes several switches such as the -S switch, which signifies the machine name. Either omitting the -S switch or specifying a value of (`local`)—including the parenthesis—indicates the local computer. The -U and -P switches indicate user and password, respectively. I'm using the system-supplied default values of `sa` for the user and a blank password, as this is a test, and not a production, database.

 If you're using the MSDE (Microsoft Developer's Edition) version of SQL Server, you also have access to BCP, but not to the graphical Query Analyzer. However, you

> should be able to use the OSQL, which is basically a command-line version of Query Analyzer, and be able to execute the scripts like this:
>
> ```
> osql -I ExtendingMFCWithDotNet.sql -S(local) -Usa -P
> ```
>
> Note that if you're using Windows Authentication, you'll need to use the -E switch instead of the -U and -P switches.

Working with Binary (BLOB) Data

A common task I'm often asked about is how to read and write binary data—typically representing a graphical image—using the ADO.NET classes. As most ADO.NET articles and books don't seem to cover this important task, I'll use this chapter's first section to illustrate two distinct ways to work with stored binary data—one with the `DataReader` class and the other with the `DataSet` class. I'll also explore the use of "chunking" to facilitate working with very large amounts of binary data in an efficient manner. The section concludes with a demo application that allows you to view the images stored for a sample database and update the database with any other image stored in the file system.

Using the `DataReader` Class

Most managed providers define a data reader object that allows for reading a forward-only stream of rows from a specific data store type. For example, there is a data reader class for SQL Server (`SqlDataReader`), Oracle (`OracleDataReader`), OLE DB provider support (`OleDbDataReader`), and ODBC driver support (`OdbcDataReader`). Up to this point, I've ignored the data readers for the simple reason that—for the most part—they don't provide us MFC developers with anything that we don't already have, since we can pick from a plethora of native database access technologies. However, the data reader does make the task of reading and writing binary data very simple; hence its inclusion in this section.

The first step in constructing a data reader object is to connect to a data store using one of the managed data connection objects (such as `Sql Connection`). Once that's done, you then construct a command object (such as `SqlCommand`) specifying the query to run against the data store. If

the command will yield a result set, you can then call the command object's `ExecuteReader` method, which returns a data reader object.

```
SqlConnection* conn =
  new SqlConnection(S"Server=localhost;"
                    S"Database=NorthWind;"
                    S"Integrated Security=true");
SqlCommand* cmd =
  new SqlCommand(S"SELECT * FROM Employees", conn);

conn->Open();

SqlDataReader* reader = cmd->ExecuteReader();

...

conn->Close(); // Don't call until finished using the data reader!
```

Note that, unlike the disconnected nature of the `DataSet`, the data reader does not read all of the data in a result set into memory at once. Instead, the data reader is an object that keeps a database connection open and basically behaves like a connected, server-side, read-only cursor. It does this by reading only as much of the complete result set as necessary, thereby saving memory, especially in cases where you expect a large result set. As a result, the act of closing the connection also closes the data reader.

Reading Binary Data with a Data Reader Object

Once the data reader object has been constructed, call its `Read` method to advance to the result set's next record. Simply call this method successively until it returns a Boolean value of `false`, indicating that there are no more records to read.

```
while (reader->Read())
{
    // data accessed via overloaded Item indexer
    // Object* o1 = reader->Item[S"columnName"];
    // Object* o2 = reader->Item[columnIndex];
}
```

While most data can be obtained through the overloaded `Item` indexer (into an `Object`), the data reader classes also provide a number of

data-type-specific methods such as GetBoolean, GetChars, GetDateTime, and GetGuid. The data-type-specific method we're most interested in here is the GetBytes method. The GetBytes method is used to read binary data and enables you to specify the binary data's column index, the buffer into which to read the data, the index of the buffer where writing should begin, and the amount of data to copy.

```
public: virtual __int64 GetBytes(
  int columnIndex,
  __int64 dataIndex,
  unsigned char buffer __gc[],
  int bufferIndex,
  int bytesToCopy
);
```

Here's an example of allocating and then populating a Byte array with binary data from a data reader object:

```
Byte image[] =
  __gc new Byte[Convert::ToInt32((reader->GetBytes(columnIndex,
                                   0, 0, 0,
                                   Int32::MaxValue
                                   )))];
reader->GetBytes(columnIndex, 0, image, 0, image->Length);
```

Note that the code snippet calls GetBytes twice. This is done because GetBytes can't be called to retrieve the data until a receiving buffer of the required length is allocated. Therefore, the first call is made with the buffer (third parameter) set to a null reference in order to determine the number of bytes to allocate. Once the image buffer has been allocated, the second call to GetBytes results in the population of the buffer.

Most image classes are designed to read the graphic from a specified file in order to display the graphic. Therefore, you could write the image data to disk using the FileStream object covered in Chapter 3. Here's a snippet illustrating the writing of a Byte buffer to disk.

```
FileStream* stream = new FileStream(destination,
                                   FileMode::Create,
                                   FileAccess::Write);
stream->Write(image, 0, image->Length);
stream->Close();
```

Now that you've seen the individual steps involved in using the `DataReader` class to read a binary value from a database and, optionally, save it to a file, here's a generic function (`GetPictureValue`) that takes a `SqlDataReader` object, column index value, and destination file name. Using what you've just learned, this function first ensures that the column value is not null (`GetBytes` will throw an exception if the column is null), allocates a `Byte` buffer, reads the data into that buffer, and then saves the data to the specified destination file name.

```cpp
void GetPictureValue(
  SqlDataReader* reader,
  int columnIndex,
  String* destination)
{
#pragma push_macro("new")
#undef new
  FileStream* stream;
  try
  {
    if (!reader->IsDBNull(columnIndex))
    {
      // Allocate a byte array
      Byte image[] =
        __gc new  Byte[Convert::ToInt32((reader->GetBytes(columnIndex,
                                    0, 0, 0, Int32::MaxValue)))];

      // Read the binary data into the byte array
      reader->GetBytes(columnIndex, 0, image, 0, image->Length);

      // Open FileStream and write buffer to file.
      stream = new FileStream(destination,
                              FileMode::Create,
                              FileAccess::Write);
      stream->Write(image, 0, image->Length);
      Console::WriteLine(S"{0} written", destination);
    }
    else
    {
      Console::WriteLine(S"Image column is null");
    }
  }
  catch (Exception* e)
```

```
  {
    // Handle exception
  }
  __finally
  {
    stream->Close();
  }
#pragma pop_macro("new")
}
```

You can now write something like the following where the code reads every photo in the Employees table and writes that photo out to a file named using the EmployeeID value. (The value 1 being passed to the Get PictureValue function refers to the second column—Photo—specified in the command object's constructor.)

```
SqlConnection* conn =
  new SqlConnection(S"Server=localhost;"
                    S"Database=NorthWind;"
                    S"Integrated Security=true");
SqlCommand* cmd =
  new SqlCommand(S"SELECT EmployeeID, Photo FROM Employees", conn);

conn->Open();

SqlDataReader* reader  = cmd->ExecuteReader();
while (reader->Read())
{
  String* fileName = String::Format(S"{0}.jpg", reader->Item[0]);
  GetPictureValue(reader, 1, fileName);
}

conn->Close();
```

Writing Binary Data with a Command and Parameter Object

As the data reader objects are read-only, they can't be used to insert into or update a data store. Instead, you would use a command object. You can't insert the binary data into an SQL statement; however, each command object contains a collection of parameter objects that serve as placeholders in the query. Parameters are used in situations where either the data cannot be passed in the query (as in our case) or in situations where the data to be

passed won't be resolved until after the command has been constructed. To specify that a command has a parameter, you simply use the special designation @parameterName in the query passed to the command object's constructor. In the following example, I'm specifying that once executed, the command object's parameter collection will include a parameter object named Photo that will contain the data to be used in this SQL UPDATE statement.

```
SqlCommand* cmd = new SqlCommand(S"UPDATE Employees SET Photo=@Photo",
                                 conn);
```

The next thing you would do is to construct the parameter object. Most managed providers define a parameter class that is specific to a given data store. For the SQL Server managed provider, this class is called Sql Parameter. Here's an example of setting up an SqlParameter object and then adding it to the command object's parameter collection, where the constructor is being used to specify such values as parameter name, data type, data length, parameter direction (input, in this case, as we're setting a database value), and the data value.

```
SqlParameter* param = new SqlParameter(S"@Photo",
                                       SqlDbType::VarBinary,
                                       image->Length,
                                       ParameterDirection::Input,
                                       false,
                                       0, 0, 0,
                                       DataRowVersion::Current,
                                       image);
cmd->Parameters->Add(param);
```

Once the parameters that you've specified for a given command have been constructed and added to the command object, you can then execute the command. As I presented a GetPictureValue in the previous section, here's a SetPictureValue function that illustrates how to use parameter objects for both the image data as well as the EmployeeID column value. This allows you to set up the connection and command one time and then SetPictureValue for each row whose image you wish to set.

```
void SetPictureValue(SqlCommand* cmd, String* fileName, int id)
{
```

```
#pragma push_macro("new")
#undef new
  FileStream* stream;
  try
  {
    // Read image into buffer.
    stream = new FileStream(fileName,
                            FileMode::Open,
                            FileAccess::Read);
    int size = Convert::ToInt32(stream->Length);
    Byte image[] = __gc new  Byte[size];
    stream->Read(image, 0, size);


    // Create parameter object named @Photo for the image data.
    SqlParameter* paramPhoto =
      new SqlParameter(S"@Photo",
                       SqlDbType::VarBinary,
                       image->Length,
                       ParameterDirection::Input,
                       false,
                       0, 0, 0,
                       DataRowVersion::Current,
                       image);
    cmd->Parameters->Add(paramPhoto);


    // Create parameter object named @ID for the passed employee id
    SqlParameter* paramID = new SqlParameter(S"@ID", __box(id));
    cmd->Parameters->Add(paramID);


    // Execute the query and report number of rows updated
    int rowsUpdated = Convert::ToInt32(cmd->ExecuteNonQuery());
    Console::WriteLine(S"{0} rows updated with specified image",
                       __box(rowsUpdated));


    // Remove parameters when finished
    cmd->Parameters->Remove(paramPhoto);
    cmd->Parameters->Remove(paramID);
  }
  catch(Exception *e)
  {
    // Handle exception
  }
```

```
  __finally
  {
    stream->Close();
  }
#pragma pop_macro("new")
}
```

Note that you could also have the client set up the command's parameters so that it's only done once and then have the `SetPictureValue` update the various parameter object members such as value and length. However, I like to minimize the work required by the client, and in the case of allocating an object such as the parameter object, the trade-off of performance vs. client work is such that I don't mind instantiating the parameter object each time. Obviously, this is reversed with regard to the connection because—depending on the environment—a connection may take some time to establish. Therefore, with this function, the client is responsible for creating the connection. In this simple test, I'm calling `SetPictureValue` for two employees. Also note the two parameters specified in the command object's constructor.

```
SqlConnection* conn =
  new SqlConnection(S"Server=localhost;"
                    S"Database=NorthWind;"
                    S"Integrated Security=true");

// Construct SQL command object specifying that a
// parameter named @Photo will be used
// (constructed and added to the command shortly).
SqlCommand* cmd = new SqlCommand(S"UPDATE Employees "
                                 S"SET Photo=@Photo "
                                 S"WHERE EmployeeID=@ID",
                                 conn);

// Open connection
conn->Open();

// Set two employee photos
SetPictureValue(cmd, S"c:\\sample.jpg", 1);
SetPictureValue(cmd, S"c:\\sample.jpg", 2);

// Our work is done here. Close the connection.
conn->Close();
```

Using the `DataSet` Class

In contrast to the `DataReader` class, the `DataSet` class is much easier to use—with the trade-off being much less control than the `DataReader` offers. As you already saw how to connect to a data source and fill a `DataSet` in Chapter 6, the following code assumes that you have a `DataRow` object containing a column (named `Photo`) that contains binary data. The following code is all you need in order to read a binary value from a `DataRow` object and then output that data to a file.

```
// Assumes that you have a DataRow object called row

// get photo from row
Byte pictureData[] = (Byte[])(row->Item[S"Photo"]);

// write data (image) to disk
int size = pictureData->GetUpperBound(0);
if (-1 < size)
{
  String* fileName = S"Test.jpg";

  FileStream* stream = new FileStream(fileName,
                              FileMode::OpenOrCreate,
                              FileAccess::Write);
  stream->Write(pictureData, 0,size);
  stream->Close();
}
```

As you can see, all that is required is to simply cast the data returned from the `Item` property! The reason this is so much simpler is that the `DataReader` gives you much more control in reading data in terms of specifying data type, precision, size, scale, and so on. This paradigm is fine for the `DataReader`, since with it you read one column of data at a time for each row. However, the `DataSet`'s internal `DataTable` object(s) is/are filled with data with a single call to the data adapter's `Fill` method, so that ADO.NET is obligated to look at the schema of the data store and download all the data at once, pursuant to the type of data. You can verify this by enumerating the column objects defined by each `DataSet` object's `Data Table` object, where you'll see that the schema information for that column was also downloaded and, ostensibly, used in determining how to read the data. Therefore, with the column data already downloaded and accessible

via the `Item` property, moving the data into a variable is much easier, albeit at the cost of flexibility.

That said, writing binary data is just as easy. The following code snippet first constructs a `FileStream` object encapsulating the file that will contain the image data for a given record. A `Byte` array is then allocated for the length of the file, and the data is read into that array. A `DataRow` object representing the record is then updated. As with any update to disconnected data, you then need to call the adapter's `Update` method to commit the changes to the data store.

```
// read specified file
FileStream* stream = new FileStream(S"Test.jpg",
                                    FileMode::Open,
                                    FileAccess::Read);
Byte pictureData[] = new Byte[stream->Length];
stream->Read(pictureData, 0, System::Convert::ToInt32(stream->
  Length));
stream->Close();

// Acquire row object
row->Item[S"Photo"] = pictureData;
```

DataReader vs. DataSet When Reading Binary Data

As you've seen, the `DataReader` and command object provide a much finer level of control when reading and writing binary data. However, the `DataSet` is much easier to use and has the benefit of being usable in a disconnected environment. What you must keep in mind is that using the `DataSet` (and associated data adapter) results in all the data—corresponding to the query when you call the data adapter's `Fill` method—being retrieved at once. Therefore, in some situations a "blended" approach might work best.

For example, let's say that you want to take advantage of the `DataSet` and design your system to be disconnected from the data source. However, one of the columns contains extremely large amounts of binary data that the user rarely needs (either for display or update). Instead of taking the performance hit of including such a large, rarely used column in the `DataSet`, you could exclude it from the `DataSet`. Then if the user wanted to view or update that column, you could use the `DataReader` and command objects to synchronously perform these tasks. This would give you the best of both worlds in terms of functionality and performance.

Demo Application to Read and Write Image Data

Assuming you've run the SQL Server script provided for the chapter's demo applications and imported the necessary data (as shown in the sidebar entitled "Creating the Sample Database for SQL Server or MSDE" at the beginning of this chapter), you should have a database named ExtendingMFCWithDotNet that includes a table called Contributors. This table defines four columns.

- ID—An IDENTITY column used to uniquely identify each row.
- Name—The full name of a contributor that is displayed in the list view.
- Role—The role of the contributor—such as Author, Co-Author and Tech Editor—that is also displayed in the list view.
- Photo—The image column that will contain a picture representing the contributor.

As mentioned, my main focus in these ADO.NET chapters is on the disconnected side of things. Therefore, this demo will illustrate using the DataSet class to load the Contributors table, displaying the Name and Role values of each record in a list view. When a particular record is selected, the associated image data (Photo value) will be displayed on the dialog. As an added bonus, you'll also see how to display an image from memory using GDI+—as opposed to saving to a temporary file first. The demo will also allow the user to select—from the file system—a different image file for the record. The record can then be saved with this new image data.

1. To get started, create an MFC dialog-based application called BLOBData and update the Project properties to support Managed Extensions.
2. Open the stdafx.h file and add the following .NET support directives to the end of the file. You've seen most of these namespaces used in the previous chapters. The drawing namespaces are used in order to facilitate the displaying of the image.

```
#using <mscorlib.dll>
#using <system.dll>
#using <system.data.dll>
#using <system.xml.dll>
#using <system.drawing.dll>
#using <system.windows.forms.dll>
```

```
using namespace System;
using namespace System::Data;
using namespace System::Data::SqlClient;
using namespace System::Xml;
using namespace System::Windows::Forms;
using namespace System::IO;
using namespace System::Drawing;
using namespace System::Drawing::Drawing2D;

#undef MessageBox
```

3. Update the project's main dialog as shown in Figure 7–1. The control below the list view is a `Picture` control with its `Type` property set to `Frame`.
4. Set the list view control's `View` property to `Report` and the picture control's `Sunken` property to `True`.
5. Add the DDX value variables for this dialog as shown in Table 7–1.

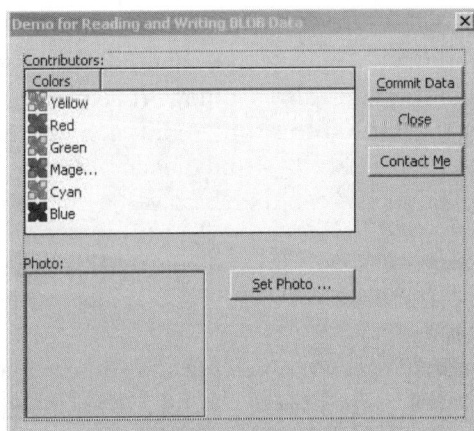

Figure 7-1 The `BLOBData` demo application's main dialog

Table 7-1 DDX Variables for the `BLOBData` Demo

Control	Variable Type	Variable Name
List view control	CListCtrl	m_lstContributors
Picture control	CStatic	m_wndPhoto

6. Define the following ADO.NET member variables for the dialog class.

```
class CBLOBDataDlg : public CDialog
{
...
protected:
  gcroot<DataSet*>dataset;
  gcroot<SqlDataAdapter*>adapter;
  gcroot<DataTable*>contributors;
  gcroot<SqlCommandBuilder*>commandBuilder;
...
```

7. Add the following list control initialization code to the end of the dialog's `OnInitDialog` function.

```
// All full row selection
LONG lStyle =
  (LONG)m_lstContributors.SendMessage(LVM_
    GETEXTENDEDLISTVIEWSTYLE);
lStyle |= LVS_EX_FULLROWSELECT;
m_lstContributors.SendMessage(LVM_SETEXTENDEDLISTVIEWSTYLE,
                          0, (LPARAM)lStyle);

// Add columns to listview
m_lstContributors.InsertColumn(0, _T("ID"));
m_lstContributors.InsertColumn(1, _T("Name"));
m_lstContributors.InsertColumn(2, _T("Role"));

TCHAR buff[MAX_PATH];
GetModuleFileName(NULL, buff, MAX_PATH);
m_strWorkingDir = System::IO::Path::GetDirectoryName(buff);
```

8. Add the following code to the end of the dialog's `OnInitDialog` function (just before the `return` statement) to initialize the ADO.

NET objects. The first thing that is done is to make a connection to the sample database (`ExtendingMFCWithDotNet`). A `SqlData Adapter` object (`adapter`) is then constructed with an SQL SELECT statement that retrieves all records from the Contributors table. Once the adapter has been constructed, its `Missing SchemaAction` property is set so that when the `Fill` method is called, primary key information will also be acquired. This is needed so that the `DataRowCollection::Find` method will allow us to search for records by primary key.

An `SqlCommandBuilder` object (`commandBuilder`) is then instantiated, as this demo will also update the data source. After that, the connection is opened, and the `SqlDataAdapter::Fill` method is called, with the resulting `DataTable` object being named `AllContributors`. At this point, we have the data in memory, so the connection to the data store is closed.

A `DataTable` object (`contributors`) is then allocated and points to the `AllContributors` `DataTable` created during the `Fill` method. Finally, a helper function (`ReadAllRows`) is called to read and display the downloaded records.

```
#pragma push_macro("new")
#undef new
  try
  {
    SqlConnection* conn =
      new SqlConnection(S"Server=localhost;"
                        S"Database=ExtendingMFCWithDotNet;"
                        S"Integrated Security=true;");

    adapter = new SqlDataAdapter(S"SELECT * FROM"
                                 "Contributors",
                                 conn);
    adapter->MissingSchemaAction =
      MissingSchemaAction::AddWithKey;

    commandBuilder = new SqlCommandBuilder(adapter);

    conn->Open();

    dataset = new DataSet();

    adapter->Fill(dataset, S"AllContributors");
```

```
    conn->Close(); // No longer needed

    DataTableCollection* tables = dataset->Tables;
    contributors = tables->Item[S"AllContributors"];

    ReadAllRows();
  }
  catch(Exception* e)
  {
    MessageBox::Show(e->Message, S".NET Exception Thrown",
                     MessageBoxButtons::OK,
                     MessageBoxIcon::Error);
  }
#pragma pop_macro("new")
```

9. At this point, let's implement the ReadAllRows methods. This function enumerates the contributors DataRowCollection member, inserting the ID, name, and role of each record into the list view. Note that the ID is shown in the first column so that it can later be used to search the DataRow record in the DataRow Collection. You could also simply store this value as item data if you didn't want to display it in a production system.

```
void CBLOBDataDlg::ReadAllRows()
{
  try
  {
    CWaitCursor wc;

    m_lstContributors.DeleteAllItems();

    DataRowCollection* rows = contributors->Rows;
    DataRow* row;

    String* id;
    String* name;
    String* role;
    for (int i = 0; i < rows->Count; i++)
    {
      row = rows->Item[i];

      id = row->Item[S"ID"]->ToString();
      name = row->Item[S"Name"]->ToString();
      role = row->Item[S"Role"]->ToString();
```

```
      int idx = m_lstContributors.InsertItem(i, (CString)id);
      m_lstContributors.SetItemText(idx, 1, (CString)name);
      m_lstContributors.SetItemText(idx, 2, (CString)role);
    }
  }
  catch(Exception* e)
  {
    MessageBox::Show(e->Message, S".NET Exception Thrown",
                     MessageBoxButtons::OK,
                     MessageBoxIcon::Error);
  }
}
```

10. Implement the following helper function that will return the index of the currently selected list view item.

```
int CBLOBDataDlg::GetSelectedItem()
{
  int iCurrSel = -1;

  POSITION pos =
    m_lstContributors.GetFirstSelectedItemPosition();
  if (pos)
    iCurrSel = m_lstContributors.GetNextSelectedItem(pos);

  return iCurrSel;
}
```

11. When the user clicks on a given list view item (representing a record in the Contributors table) we want that record's photo displayed. Therefore, implement the following handler for the list view's LVN_ITEMCHANGED message. After retrieving the record ID for the selected item (from the first column), the DataRow Collection::Find method is called to retrieve the DataRow object encapsulating that record. If the row contains data for the Photo column, a Byte array is then allocated and filled with that image data. This buffer is then passed to the DisplayImageFile function. If the Photo column is blank, then the DisplayImage File is called with a value of NULL to indicate that the static control used to display the image should be erased (in case it's displaying the value from a previously selected record).

```
void CBLOBDataDlg::OnLvnItemchangedList1(
  NMHDR *pNMHDR, LRESULT *pResult)
```

```
{
#pragma push_macro("new")
#undef new
  try
  {
    CWaitCursor wc;

    LPNMLISTVIEW pNMLV = reinterpret_cast<LPNMLISTVIEW>
      (pNMHDR);

    int iCurrSel = GetSelectedItem();
    if (-1 < iCurrSel)
    {
      int id = atoi(m_lstContributors.GetItemText(iCurrSel,
        0));

      DataRow* row = contributors->Rows->Find(__box(id));
      if (row)
      {
        // get photo from row
        Byte pictureData[] = (Byte[])(row->Item[S"Photo"]);

        // display photo or erase photo based
        // on if we found image data
        int size = pictureData->GetUpperBound(0);
        if (-1 < size)
        {
          // display image
          DisplayImageFile(pictureData);
        }
        else
        {
          // Clear the photo display
          DisplayImageFile(NULL);
        }
      }
    }
  }
  catch(Exception* e)
  {
    MessageBox::Show(e->Message, S".NET Exception Thrown",
                     MessageBoxButtons::OK,
                     MessageBoxIcon::Error);
```

```
      }

      *pResult = 0;
#pragma pop_macro("new")
}
```

12. Now for the `DisplayImageFile` function. This function uses the
GDI+ `Bitmap` and `Graphic` objects in order to display an image
file. If data is passed (indicating an image to display), the function
first copies the image data buffer into a `MemoryStream` object,
which is then passed to the `Bitmap` constructor. The device context
of the static control is then used to draw the image. If a value of
NULL is passed as the image data, then the static control's client area
is erased.

```
void CBLOBDataDlg::DisplayImageFile(Byte pictureData __gc[])
{
#pragma push_macro("new")
#undef new

    // Forward-declare the following objects so they can be
    // released in the __finally statement
    System::IO::MemoryStream* dataStream = NULL;
    Bitmap* image = NULL;
    Graphics* graphics = NULL;
    CClientDC dc(&m_wndPhoto);

    try
    {
      CWaitCursor wc;

     if (NULL != pictureData)
     {
        // Wrap the pictureData in a Stream object so as to
        // avoid creating a temporary file on disk
        dataStream = new System::IO::MemoryStream(pictureData);

        // Create the GDI+ Bitmap and Graphic objects
        image = new Bitmap(dataStream);
        graphics = Graphics::FromHdc(dc.GetSafeHdc());

        // Erase static control's client area
        RECT clientRect;
        m_wndPhoto.GetClientRect(&clientRect);
```

```
        graphics->FillRectangle(SystemBrushes::Control,
            clientRect.left, clientRect.top,
            clientRect.right - clientRect.left,
            clientRect.bottom - clientRect.top);


        // Draw image on static control's client area
        graphics->DrawImage(image, 0, 0, image->Width,
                            image->Height);
    }
    else // erase static control's client area
    {
        // Create GDI+ Graphics object from HDC
        graphics = Graphics::FromHdc(dc.GetSafeHdc());

        // Draw background
        RECT clientRect;
        m_wndPhoto.GetClientRect(&clientRect);

        graphics->FillRectangle(SystemBrushes::Control,
            clientRect.left, clientRect.top,
            clientRect.right - clientRect.left,
            clientRect.bottom - clientRect.top);
    }
}
catch(Exception* e)
{
  MessageBox::Show(e->Message, S".NET Exception Thrown",
                MessageBoxButtons::OK,
                MessageBoxIcon::Error);
}
__finally
{
 // Important! Clean up GDI+ and IO resources!
 if (graphics) graphics->Dispose();
 if (image) image->Dispose();
 if (dataStream) dataStream->Dispose();
}
#pragma pop_macro("new")
}
```

13. At this point, the application should display any photos stored for a given record. Now let's add the ability to change the photo so as to illustrate how to write image data. To do that, implement an event handler

for the Set Photo . . . button. After displaying the File Open common dialog, the function opens the selected image file using a `FileStream` object. The file's data is read into a `Byte` array and, after locating the `DataRow` being updated (via the `DataRowCollection::Find` method), its `Photo` column is set to the `Byte` array. Finally, the `DisplayImageFile` presented in the previous step is called.

```
void CBLOBDataDlg::OnBnClickedPhotoSet()
{
#pragma push_macro("new")
#undef new
  FileStream* stream;
  try
  {
    CWaitCursor wc;

    int iCurrSel = GetSelectedItem();
    if (-1 < iCurrSel)
    {
      CFileDialog dlg(TRUE);
      dlg.m_pOFN->lpstrInitialDir = m_strWorkingDir;
      dlg.m_pOFN->lpstrTitle = _T("Open an image file");
      dlg.m_ofn.lpstrFilter = _T("Image"
        "Files\0*.bmp;*.gif;*.jpg;\0"
        "All Files (*.*)\0*.*\0\0");
      if (IDOK == dlg.DoModal())
      {
        // read specified file
        CString strFileName = dlg.GetPathName();
        stream = new FileStream(strFileName,
                                FileMode::Open,
                                FileAccess::Read);
        Byte pictureData[] = new Byte[stream->Length];
        stream->Read(pictureData, 0,
                System::Convert::ToInt32(stream->
                Length));

        // update DataTable
        int id = atoi(m_lstContributors.GetItemText(
          iCurrSel, 0));
        DataRow* row = contributors->Rows->Find(__box(id));
        if (row)
```

```
        {
            row->Item[S"Photo"] = pictureData;
        }

        // display image
        DisplayImageFile(pictureData);
        }
    }
    else MessageBox::Show(S"You must first select a
        contributor");
    }
    catch(Exception* e)
    {
        MessageBox::Show(e->Message, S".NET Exception Thrown",
                        MessageBoxButtons::OK,
                        MessageBoxIcon::Error);
    }
    __finally
    {
        stream->Close();
    }
#pragma pop_macro("new")
}
```

14. The last step is to implement a handler for the Commit Data button so that, at any time, the user can commit the changes made in the application to the DataSet object's underlying data source. As you saw in the previous chapter, this is accomplished via calling the data adapter's Update function, and it will only work if a command builder has been set up to automatically generate the appropriate UpdateCommand.

```
void CBLOBDataDlg::OnBnClickedCommitData()
{
    try
    {
        CWaitCursor wc;

        adapter->Update(contributors);
        MessageBox::Show(S"Data successfully saved", S"Success",
                        MessageBoxButtons::OK,
                        MessageBoxIcon::Information);
```

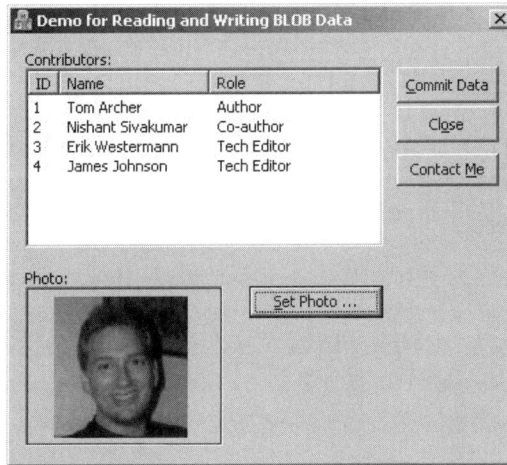

Figure 7-2 The `BLOBData` demo application illustrates how to read and write image data.

```
     }
     catch(Exception* e)
     {
        MessageBox::Show(e->Message, S".NET Exception Thrown",
                    MessageBoxButtons::OK,
                    MessageBoxIcon::Error);
     }
  }
```

At this point, you should have a fully functional application that allows you to both read and write image data to a SQL Server database. Running this application should provide results similar to those shown in Figure 7–2.

Associating Related Tables with the `DataRelation` Class

One of the most powerful aspects of ADO.NET—and, specifically, of the `DataSet` class—is the ability to define relationships between related tables such that both tables can be navigated more easily and efficiently. For example, let's say you have a typical order entry system with both an Order-Headers and OrderDetails table. In fact, the sample database (`Extending MFCWithDotNet`) provided on the CD with this book contains two such tables whose schemas are shown in Tables 7–2 and 7–3.

Table 7-2 Sample OrderHeaders Table Schema

Primary Key (y/n)	IDENTITY (y/n)	Column Name	Type
YES	YES	OrderId	int
		CompanyName	varchar
		ShippingAddress	varchar

Table 7-3 Sample OrderDetails Table Schema

Primary Key (y/n)	IDENTITY (y/n)	Column Name	Type
YES	YES	OrderId	int
		ProductId	int
		ProductName	varchar
		UnitPrice	decimal
		Quantity	int

As you can see, these tables would be related via the two `OrderId` columns. Normally, you would join these two tables using this primary key/foreign key combination, as with the following SQL statement.

```
SELECT * from OrderHeaders as oh, OrderDetails as od
WHERE oh.OrderId = od.OrderId
```

However, doing so yields a result set with duplicate data, since each detail record will also contain the data of its parent header record, as shown in Figure 7-3.

If you're developing a standalone application, the extra bandwidth to transfer the data might not be a big issue. However, if you're reading these ADO.NET chapters with the principal focus being on disconnected data, chances are that decreasing the amount of data transferred is indeed a big issue. Therefore, we turn to the `DataRelation` class.

The `DataRelation` class enables you to relate two `DataTable` objects to each other through specified `DataColumn` objects. The result is that both tables are downloaded and exist in a `DataSet` independently while still allowing you the ability to navigate up and down the hierarchy for related

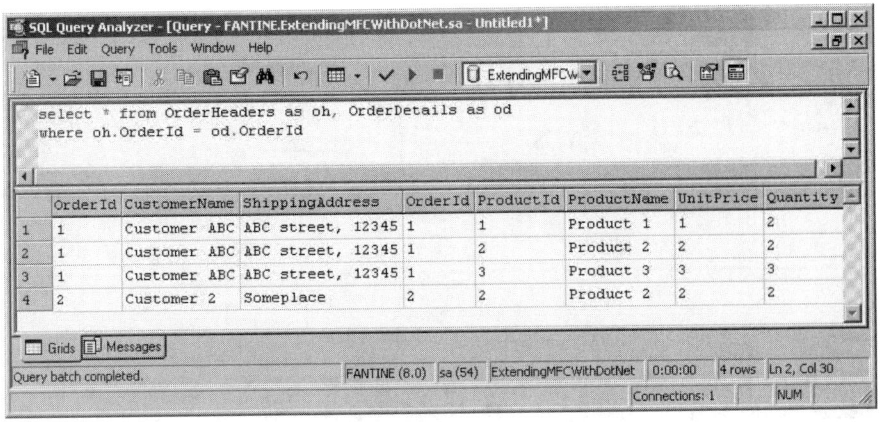

	OrderId	CustomerName	ShippingAddress	OrderId	ProductId	ProductName	UnitPrice	Quantity
1	1	Customer ABC	ABC street, 12345	1	1	Product 1	1	2
2	1	Customer ABC	ABC street, 12345	1	2	Product 2	2	2
3	1	Customer ABC	ABC street, 12345	1	3	Product 3	3	3
4	2	Customer 2	Someplace	2	2	Product 2	2	2

Figure 7-3 Example of result set with duplicate data caused by traditional join of header and detail tables

records. Here's an example of how to construct a `DataRelation` object using our OrderHeaders and OrderDetails tables:

```
// Assumes that the dataset contains two DataTable objects -
// one each the OrderHeaders and OrderDetails database tables

// Get the DataSet relations collection
DataRelationCollection* relations = dataset->Relations;

// Create a new DataRelation
DataRelation* relation =
  new DataRelation(S"OrderHeaderToOrderDetail",
                   orderHeaders->Columns->Item[S"OrderId"],
                   orderDetails->Columns->Item[S"OrderId"])

// Add the new relation to the DataSet relations collection
relations->Add(relation);
```

As you can see from this code, the `DataSet` object contains a collection (`DataRelationCollection`) of all defined `DataRelation` objects for its tables. To create the new `DataRelation` object, the code simply passes a string value naming the relation (used later when navigating the related tables) and specifies the two `DataColumn` objects that form the linkage from one table to the other. Finally, the newly constructed `DataRelation`

object is added to the DataSet relations collection. Now that you've seen how to define the relationship, let's look at both navigating related tables as well as inserting data into tables that contain auto-increment columns.

Navigating Hierarchical Data

In the previous chapter, you learned how to navigate the DataRow objects in a DataTable object's row collection. In situations where a DataRelation object has been defined, the DataRow class defines two methods that allow for the ability to navigate to the related records of other tables. These two methods are called GetChildRows and GetParentRow.

When calling the GetChildRows method, you can either specify a DataRelation object or the name used when the desired DataRelation object was constructed. As the DataRelation object defines which columns in the two tables form the relationship, ADO.NET can then return only the appropriate rows from the child table based on the value of the current row's column value.

```
// From the current header record (parent)
// get all of its detail records (children)
// using a DataRelation named OrderHeaderToOrderDetail
DataRow* details[] = header->GetChildRows(
   S"OrderHeaderToOrderDetail");
```

With regard to going in the other direction—retrieving a record's parent—there are actually two methods: GetParentRow and GetParent Rows. The only difference between the two is whether or not the child row has multiple parents or not. In our order example, the tables are defined such that an order header record can have any number of detail records, but a detail record can only belong to one order header record. Therefore, in our situation we would use the GetParentRow method as follows:

```
// From the current detail record (child)
// get its header record (parent)
// using a DataRelation named OrderHeaderToOrderDetail
DataRow* header = detail->GetParentRow(S"OrderHeaderToOrderDetail");
```

Now that you've seen the basics, let's look at a simple function to enumerate all of the order header and detail records in the sample database. For this function to work, you must first follow the steps covered in the

sidebar "Creating the Sample Database for SQL Server or MSDE" at the beginning of the chapter.

```
void DisplayAllOrders()
{
#pragma push_macro("new")
#undef new
  try
  {
    SqlConnection* conn =
      new SqlConnection(S"Server=localhost;"
                        S"Database=ExtendingMFCWithDotNet;"
                        S"Integrated Security=true;");
    conn->Open();

    DataSet* dataset = new DataSet();

    SqlDataAdapter* adapter =
      new SqlDataAdapter(S"SELECT * FROM OrderHeaders; "
                         S"SELECT * FROM OrderDetails",
                         conn);

    // Get all data with a single round trip
    adapter ->Fill(dataset);

    // Retrieve both data tables
    DataTable* orderHeaders = dataset->Tables->Item[0];
    DataTable* orderDetails = dataset->Tables->Item[1];

    // Create the relationship between the two tables
    DataRelation* relation =
      new DataRelation(S"OrderHeaderToOrderDetail",
                       orderHeaders->Columns->Item[S"OrderId"],
                       orderDetails->Columns->Item[S"OrderId"]));
    dataset->Relations->Add(relation);

    // get all headers
    DataRowCollection* headers = orderHeaders->Rows;
    DataRow* header;
    for (int i = 0; i < headers->Count; i++)
    {
      header = headers->Item[i];
      Console::WriteLine(String::Format(S"Order: {0}\t"
```

```
                                 S"Customer: {1}\t"
                                 S"Shipping Address: {2}",
                                 header->Item["OrderId"],
                                 header->Item["CustomerName"],
                                 header->Item[
                                 S"ShippingAddress"]));

      // get all detail lines
      String* format = S"{0,-5} {1,-5} {2,-10} {3,-10} {4,-10}";
      Object* values[] = new Object*[5];
      values[0] = S"Order";
      values[1] = S"PID";
      values[2] = S"Product";
      values[3] = S"Price";
      values[4] = S"Qty";
      Console::WriteLine(String::Format(format, values));

      DataRow* details[] =
        header->GetChildRows(S"OrderHeaderToOrderDetail");
      for (int j = 0; j < details->Count; j++)
      {
        Object* values[] = new Object*[5];
        values[0] = details[j]->Item[S"OrderId"];
        values[1] = details[j]->Item[S"ProductId"];
        values[2] = details[j]->Item[S"ProductName"];
        values[3] = details[j]->Item[S"UnitPrice"];
        values[4] = details[j]->Item[S"Quantity"];
        Console::WriteLine(String::Format(format, values));
      }
      Console::WriteLine();
    }

    conn->Close();
  }
  catch(Exception* e)
  {
    Console::WriteLine(e->Message);
  }
#pragma pop_macro("new")
}
```

As you can see, the DisplayAllOrders function is using several techniques covered in this and the previous chapter. First both tables are

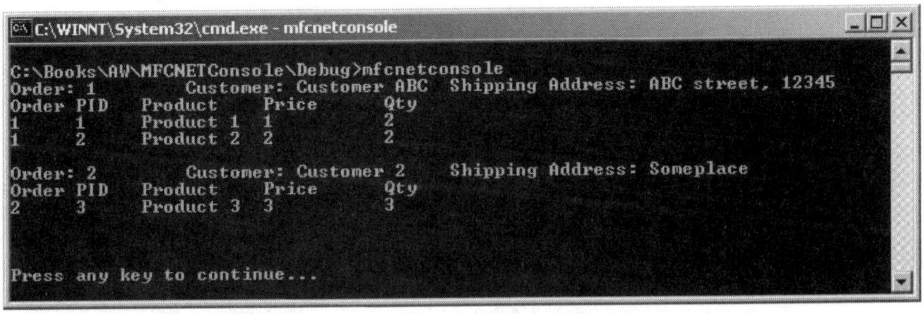

Figure 7-4 Example of using the `DataRelation` class and `DataRow::GetChildRows` to navigate header and detail (parent and child) data

specified in the data adapter's constructor so that when the adapter's `Fill` method is called, only a single round trip is made. From there, both `Data Tables` are retrieved from the `DataSet`, and a `DataRelation` called `Order HeaderToOrderDetail` is constructed between the `OrderId` columns of both tables. The `orderHeaders` row collection is then enumerated, where, for each header record, the header table's data is displayed, and the `DataRow::GetChildRows` is called. This results in the related order detail records being returned in a `DataRow` array, which are then also enumerated and displayed. Figure 7-4 shows the results of running this code against some sample data on my machine.

Using Stored Procedures to Insert Hierarchical Data with Auto-Increment Columns

Inserting data into or updating a `DataRow` that has a `DataColumn` involved in a defined `DataRelation` is typically no different than modifying any other `DataRow`—unless one of the tables has a column defined as an auto-increment (also known as IDENTITY, or auto-number) column and you need to use that value to also insert data into the second related table. Continuing our order header and order detail example, take a look back at Tables 7-2 and 7-3. As you can see, the OrderHeaders table defines a primary key called `OrderId`, which is also used as the foreign key for the OrderDetails table. As a result, when an order is created, we need a way of first inserting the `OrderHeaders` (parent) record and then retrieving the auto-incremented value of the `OrderId` column to use in inserting the associated `OrderDetails` (children) records. This section addresses just this issue.

In the section entitled "Disconnected Data and Auto-Increment Primary Keys" in Chapter 6, I illustrated how to respond to the data adapter's OnUpdated event to retrieve the auto-incremented value of a newly inserted row's primary key. This technique works great for many situations where retrieving the value in an asynchronous fashion doesn't present a problem. However, when writing a header record, it would be much more convenient to be able to know immediately the value of the auto-incremented column for use in writing the detail records. One way to handle such scenarios is with a stored procedure that automatically returns the auto-incremented value and updates the local primary key column for use in inserting associated child records. For example, the stored procedure for inserting rows into this chapter's OrderHeaders table looks like the following:

```
CREATE PROCEDURE proc_InsertOrder
(@OrderId int output,
 @CustomerName varchar(50),
 @ShippingAddress varchar(50)
)
AS
INSERT INTO OrderHeaders (CustomerName, ShippingAddress)
VALUES (@CustomerName, @ShippingAddress)
SELECT @OrderId=@@IDENTITY
GO
```

If you're new to stored procedures, this might look a bit strange. However, it's really easy to understand with a little guidance. First, the name of the procedure is proc_InsertOrder. Its parameters then follow and have names that are always prefixed with an at-sign: @. Each parameter designation also includes its type and, optionally, its direction. For example, the @OrderId parameter is defined as output because it will be used to return a value only. By default, parameters are defined as input only. Everything after the AS specifies what the procedure will do when called. Here the procedure simply uses an SQL INSERT statement to insert the values contained in the @CustomerName and @ShippingAddress parameters into the OrderHeaders table. As the OrderId column is defined as auto-increment, the database will set its value for us. Finally, the procedure sets the @OrderId parameter to the special @@IDENTITY value, which returns the new auto-increment value for the newly inserted row. Now you know how calling this procedure will automatically return the new OrderId value.

Now let's look at the generic steps to use a stored procedure for inserting data into a parent table so that the auto-incremented primary key value

is returned to insert associated child rows. Once you've seen the steps, I'll then illustrate a function that implements these steps to insert a rows into both the sample database's OrderHeaders and OrderDetails tables.

1. Create the stored procedure that returns the `@@IDENTITY` value in one of the parameters that is defined as an output parameter.
2. Once the parent table's data adapter object has been constructed, instantiate a command object that will call the stored procedure.
3. Set the adapter's `InsertCommand` to the newly created command object.
4. Construct the necessary parameter objects for the stored procedure. Ensure that the parameter that will contain the returned auto-increment value has its `Direction` member set to `Parameter Direction::Output`.
5. Either call the data adapter's `Fill` method to download any desired data or, if no data is to be downloaded, call the data adapter's `FillSchema` method to download the schema information for the parent table into a `DataSet` object.
6. If you're using the `FillSchema` method (which doesn't allow you to name the `DataTable` that will hold the downloaded schema), you'll want to rename the `DataTable`. This is because the name defaults to Table, and if you don't rename it, when you download the schema for the child table, the schemas will be merged instead of a second `DataTable` being created.
7. Construct the data adapter object for the child table.
8. Construct a command builder object for the child table. This step wasn't necessary for the parent table because a stored procedure is being used to insert data into that table for the sole purpose of retrieving the auto-increment value. With the child table, you can simply allow the command builder object to automatically generate the proper `SQL INSERT` statement.
9. As with the parent data adapter, either call the child adapter's `Fill` method to download desired data or, if no data is to be downloaded, call its `FillSchema` to download the schema information.
10. Construct the `DataRelation` object to link the desired `Data Column` objects of the parent and child `DataTable` objects.
11. When adding a parent row and its associated child records, call the `DataRow::SetParentRow` for each child row, passing the parent row objects.

12. Finally, when you're ready to commit the new records to the data store, call the `Update` method for both the parent and child adapters. The former will result in the stored procedure being called for the parent row, while the latter will result in the command builder–generated `InsertCommand` being used for each of the child rows.

That's a lot of steps, so let's see how this all looks in actual C++ code. In the following function (`InsertOrderDataWithIdentity`), I'm simply following the steps you just read and inserting a row into both the Order-Headers and OrderDetails tables with hard-coded test data.

```
void InsertOrderDataWithIdentity()
{
#pragma push_macro("new")
#undef new
  try
  {
    // Connect to the data store
    SqlConnection* conn =
      new SqlConnection(S"Server=localhost;"
                        S"Database=ExtendingMFCWithDotNet;"
                        S"Integrated Security=true;");

    // Open the connection
    conn->Open();
    DataSet* dataset = new DataSet();

    // Construct an adapter for the parent table.
    SqlDataAdapter* headersAdapter =
      new SqlDataAdapter(S"SELECT * FROM OrderHeaders", conn);

    // Construct a command object passing the name of the
    // stored procedure
    headersAdapter->InsertCommand =
      new SqlCommand(S"proc_InsertOrder", conn);

    // Set the parent adapter's Insert command to this newly
    // created command object
    SqlCommand* cmdInsert = headersAdapter->InsertCommand;

    // Specify that this command is a stored procedure command
    cmdInsert->CommandType = CommandType::StoredProcedure;
```

```
// Construct the parameters expected by the stored procedure
SqlParameter* param = new SqlParameter(S"@OrderId",
                                       SqlDbType::Int);


// The OrderId value will be returned so its direction
// must be set to Output
param->Direction = ParameterDirection::Output;
param->SourceColumn = S"OrderId";
cmdInsert->Parameters->Add(param);


param = new SqlParameter(S"@CustomerName",
                         SqlDbType::VarChar,
                         50,
                         S"CustomerName");
cmdInsert->Parameters->Add(param);


param = new SqlParameter(S"@ShippingAddress",
                         SqlDbType::VarChar,
                         50,
                         S"ShippingAddress");
cmdInsert->Parameters->Add(param);


// No data will be read, so we must download the schema
headersAdapter->FillSchema(dataset, SchemaType::Source);


// Save a pointer to the OrderHeaders schema and then
// change the table name so that the schema is not merged with
// the OrderDetails table schema, which would produce
// a single table schema for both tables
DataTable* headersTable = dataset->Tables->Item[S"Table"];
headersTable->TableName = S"OrderHeaders";


// Construct the child data adapter
SqlDataAdapter* detailsAdapter =
  new SqlDataAdapter(S"SELECT * FROM OrderDetails", conn);


// We'll let a command builder generate the InsertCommand
// for the OrderDetails table
SqlCommandBuilder* cmd = new SqlCommandBuilder(detailsAdapter);


// Download the OrderDetails schema
detailsAdapter->FillSchema(dataset, SchemaType::Source);
```

```
    DataTable* detailsTable = dataset->Tables->Item[S"Table"];
    detailsTable->TableName = S"OrderDetails";

    // Create the relationship between the two tables
    DataRelation* relation =
      new DataRelation(S"OrderHeaderToOrderDetail",
                     headersTable->Columns->Item[S"OrderId"],
                     detailsTable->Columns->Item[S"OrderId"]);
    dataset->Relations->Add(relation);

    // Create an order header row
    DataRow* header = headersTable->NewRow();
    header->Item[S"CustomerName"] = S"Customer ABC";
    header->Item[S"ShippingAddress"] = S"ABC street, 12345";
    headersTable->Rows->Add(header);

    // Create an order detail row
    DataRow* detail = detailsTable->NewRow();
    detail->Item[S"ProductId"] = __box(1);
    detail->Item[S"ProductName"] = S"Product 1";
    detail->Item[S"UnitPrice"] = __box(1);
    detail->Item[S"Quantity"] = __box(2);

    // Associate the two rows by calling SetParentRow
    detail->SetParentRow(header);
    detailsTable->Rows->Add(detail);

    // Update the data source.
    headersAdapter->Update(headersTable);
    detailsAdapter->Update(detailsTable);

    // Close the connection
    conn->Close();
  }
  catch(Exception* e)
  {
    Console::WriteLine(e->Message);
  }
#pragma pop_macro("new")
}
```

Understanding Concurrency

The most challenging part of working with disconnected data is that each user is working with a snapshot of the data at the point that their respective dataset was filled. This is fine if you only have a handful of users who never work on the same data concurrently. However, in the real world, this isn't always the case. For example, let's take a standard distribution application where inventory can be updated by several modules. In our example, Bob works in receiving and has just received a shipment of four widgets. He pulls up the widget record, which shows a total of two units on hand. When he updates the record to the data store, it will reflect a total of six widgets on hand. However, before he clicks the Save button, an order entry user enters an order to reflect that someone has just purchased the only two widgets. Now there are actually four widgets on hand. However, when Bob hits that Save button, the application will attempt to write that a total of 6 widgets exist. If you're new to disconnected data, your first response might be to attempt to lock the data. However, this completely nullifies many of the benefits of disconnected data. After all, one of the biggest advantages of disconnected data is to allow the user to work on a local snapshot of data without maintaining the data store connection. This is where handling concurrency comes in. Put simply, concurrency is the ability to synchronize changes made by multiple users.

ADO.NET defines several types of concurrency:

- **Optimistic**—With optimistic concurrency, rows are not locked when they are read. However, when an attempt is made to update a given row, the application determines whether another user has changed the row since it was read. In other words, optimistic concurrency is generally used in disconnected environments where there is low contention for the data, and you never want to overwrite newer updates made by other users. As you'll see shortly, ADO.NET employs this concurrency level by default.
- **Pessimistic**—Pessimistic concurrency is really row-level locking. In a pessimistic model, when a user performs an action that causes a lock to be applied, other users cannot perform actions that would conflict with the lock until the lock owner releases it. This model is primarily used in environments where there is heavy contention for data, and the cost associated with locking the data is outweighed by the cost of dealing with conflicts when they occur.

■ **Destructive**—I'm really only including this type of concurrency for completeness because technically it simply involves a "last in, wins" methodology, where the last person writing the data overwrites whatever is currently in the database. You obviously would only use this type of concurrency in very select situations.

Demo Application to Test Various Concurrency Issues

There are several options when working with concurrency. These include understanding how the DataSet handles concurrency by default as well as understanding the role of the data adapter and command builder objects. There are also several options for handling concurrency conflicts as well as implementing the different levels of concurrency. Therefore, at this point, let's see a simple maintenance application that will be used to illustrate these various points throughout this section.

1. Create an MFC dialog-based application called Concurrency and update the Project properties to support Managed Extensions.
2. Open the stdafx.h file and add the following .NET support directives to the end of the file. You've seen most of these namespaces used in the previous chapters. The drawing namespaces are used in order to facilitate the displaying of the image.

```
#using <mscorlib.dll>
#using <system.dll>
#using <system.data.dll>
#using <system.xml.dll>
#using <system.drawing.dll>
#using <system.windows.forms.dll>

using namespace System;
using namespace System::Data;
using namespace System::Data::SqlClient;
using namespace System::Xml;
using namespace System::Windows::Forms;
using namespace System::IO;
using namespace System::Drawing;
using namespace System::Drawing::Drawing2D;

#undef MessageBox
```

3. Update the project's main dialog as shown in Figure 7–5.

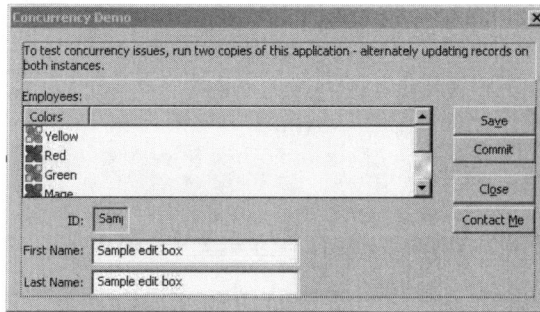

Figure 7-5 The Concurrency demo application's main dialog

4. Set the list view control's View property to Report and add the DDX value variables for this dialog as shown in Table 7–4.

Table 7-4 DDX variables for the Concurrency demo

Control	Variable Type	Variable Name
List view control	CListCtrl	m_lstEmployees
ID edit control	int	m_iId
First Name edit control	CString	m_strFirstName
Last Name edit control	CString	m_strLastName

5. Define the following ADO.NET member variables for the dialog class.

```
class CConcurrencyDlg : public CDialog
{
...
protected:
  gcroot<DataSet*>dataset;
  gcroot<SqlDataAdapter*>adapter;
  gcroot<DataTable*>employees;
```

```
gcroot<SqlCommandBuilder*>commandBuilder;
...
```

6. Add the following list control initialization code to the end of the dialog's `OnInitDialog` function to initialize the list control and call the `GetEmployees` helper function.

```
// All full row selection
LONG lStyle =
   (LONG)m_lstEmployees.SendMessage(LVM_
      GETEXTENDEDLISTVIEWSTYLE);
lStyle |= LVS_EX_FULLROWSELECT;
m_lstEmployees.SendMessage(LVM_SETEXTENDEDLISTVIEWSTYLE,
                           0, (LPARAM)lStyle);

// Add columns to listview
m_lstEmployees.InsertColumn(0, _T("ID"));
m_lstEmployees.InsertColumn(1, _T("First Name"));
m_lstEmployees.InsertColumn(2, _T("Last Name"));

// Read and display all Employee records
GetEmployees();
DisplayEmployees();
```

7. Now code the `GetEmployees` function as follows. This code will first initialize the various ADO.NET objects—connected to the Northwind sample database and filling the previously declared `DataSet` object with all records from the Employees table.

```
void CConcurrencyDlg::GetEmployees()
{
#pragma push_macro("new")
#undef new
  try
  {
    SqlConnection* conn =
      new SqlConnection(S"Server=localhost;"
                        S"Database=Northwind;"
                        S"Integrated Security=true;");

    adapter = new SqlDataAdapter(S"SELECT * FROM Employees",
                        conn);
```

```
        commandBuilder = new SqlCommandBuilder(adapter);

        conn->Open();

        dataset = new DataSet();

        adapter->MissingSchemaAction =
          MissingSchemaAction::AddWithKey;
        adapter->Fill(dataset, S"AllEmployees");

        conn->Close(); // No longer needed

        DataTableCollection* tables = dataset->Tables;
        employees = tables->Item[S"AllEmployees"];
      }
      catch(Exception* e)
      {
        MessageBox::Show(e->Message, S".NET Exception Thrown",
                         MessageBoxButtons::OK,
                         MessageBoxIcon::Error);
      }
#pragma pop_macro("new")
    }
```

8. Now, implement the `DisplayEmployees` method as follows to enumerate the `DataRowCollection` collection of the `employees` object filled in the `GetEmployees` method.

```
void CConcurrencyDlg::DisplayEmployees()
{
  try
  {
    CWaitCursor wc;

    m_lstEmployees.DeleteAllItems();

    DataRowCollection* rows = employees->Rows;
    DataRow* row;

    String* id;
    String* firstName;
    String* lastName;
    for (int i = 0; i < rows->Count; i++)
```

```
                {
                    row = rows->Item[i];

                    id = row->Item[S"EmployeeId"]->ToString();
                    firstName = row->Item[S"FirstName"]->ToString();
                    lastName = row->Item[S"LastName"]->ToString();

                    int idx = m_lstEmployees.InsertItem(i, (CString)id);
                    m_lstEmployees.SetItemText(idx, 1, (CString)firstName);
                    m_lstEmployees.SetItemText(idx, 2, (CString)lastName);
                }
            }
            catch(Exception* e)
            {
                MessageBox::Show(e->Message, S".NET Exception Thrown",
                                 MessageBoxButtons::OK,
                                 MessageBoxIcon::Error);
            }
        }
```

9. Implement the following helper function that will return the index of the currently selected list view item.

```
int CConcurrencyDlg::GetSelectedItem()
{
    int iCurrSel = -1;

    POSITION pos =
        m_lstEmployees.GetFirstSelectedItemPosition();
    if (pos)
        iCurrSel = m_lstEmployees.GetNextSelectedItem(pos);

    return iCurrSel;
}
```

10. Implement the following handler for the list view's LVN_ITEM CHANGED message to update the dialog with the selected employee record's data.

```
void CConcurrencyDlg::OnLvnItemchangedList1(
    NMHDR *pNMHDR,
    LRESULT *pResult)
{
    try
```

```
    {
      CWaitCursor wc;

      LPNMLISTVIEW pNMLV =
        reinterpret_cast<LPNMLISTVIEW>(pNMHDR);

      int iCurrSel = GetSelectedItem();
      if (-1 < iCurrSel)
      {
        int id = atoi(m_lstEmployees.GetItemText(iCurrSel, 0));

        DataRow* row = employees->Rows->Find(__box(id));
        if (row)
        {
          m_iId = Convert::ToInt32(row->Item[S"Id"]);
          m_strFirstName = (CString)row->Item[S"FirstName"]->
            ToString();
          m_strLastName = (CString)row->Item[S"LastName"]->
            ToString();
          UpdateData(FALSE); // update dialog
        }
      }
    }
    catch(Exception* e)
    {
      MessageBox::Show(e->Message, S".NET Exception Thrown",
                       MessageBoxButtons::OK,
                       MessageBoxIcon::Error);
    }

    *pResult = 0;
}
```

11. Add an event handler for the Save button.

```
void CConcurrencyDlg::OnBnClickedSave()
{
  try
  {
    CWaitCursor wc;

    UpdateData();

    int iCurrSel = GetSelectedItem();
```

```
if (-1 < iCurrSel)
{
  int id = atoi(m_lstEmployees.GetItemText(iCurrSel, 0));

  DataRow* row = employees->Rows->Find(__box(id));
  if (row)
  {
    row->Item[S"FirstName"] = (String*)m_strFirstName;
    m_lstEmployees.SetItemText(iCurrSel, 1,
                                 m_strFirstName);

    row->Item[S"LastName"] = (String*)m_strLastName;
    m_lstEmployees.SetItemText(iCurrSel, 2,
                                 m_strLastName);
  }
}
}
catch(Exception* e)
{
  MessageBox::Show(e->Message, S".NET Exception Thrown",
                   MessageBoxButtons::OK,
                   MessageBoxIcon::Error);
}
}
```

12. Finally, add an event handler for the Commit button. Note that the number of successfully updated rows is displayed. This will come into play shortly.

```
void CConcurrencyDlg::OnBnClickedCommit()
{
  try
  {
    CWaitCursor wc;

    adapter->Update(employees);
    int rowsUpdated = adapter->Update(employees);
    MessageBox::Show(String::Format(S"{0} rows"
                     S"successfully updated",
                     __box(rowsUpdated)),
                     S"Information",
                     MessageBoxButtons::OK,
                     MessageBoxIcon::Information);
```

```
    }
    catch(Exception* e)
    {
        MessageBox::Show(e->Message, S".NET Exception Thrown",
                         MessageBoxButtons::OK,
                         MessageBoxIcon::Error);
    }
}
```

Once you've finished coding the `Concurrency` demo, build and run *two instances* of the application. When both instances execute the `Get Employees` method, each will have a snapshot of the data. Update an existing row in the first demo. *Note that when I refer to "updating" a row, I'm referring to clicking both the Save button and the Commit button after modifying either the `FirstName` or `LastName` values.* Now, modify and update the same record with the second instance of the application. This should result in an `DBConcurrencyException` being thrown as a result of a concurrency conflict similar to that shown in Figure 7–6.

The conflict you see in Figure 7–6 occurs because of the way the data adapter and command builder implement optimistic concurrency.

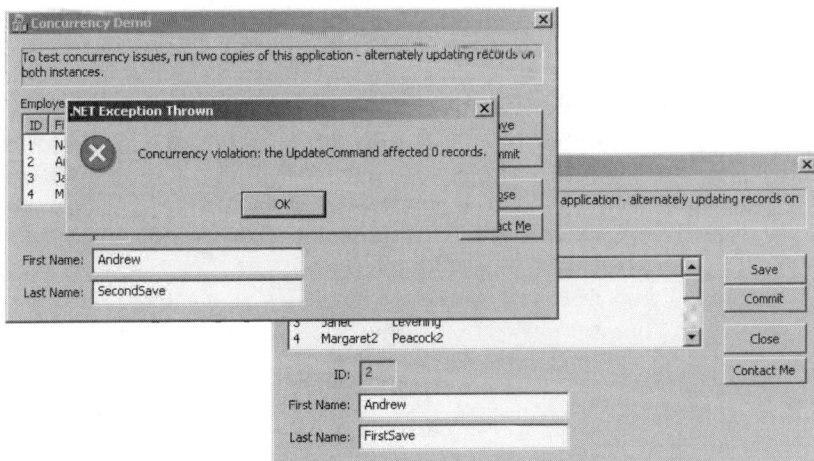

Figure 7–6 The `DataSet` class uses optimistic concurrency by default, which means that you'll need to resolve situations where multiple users attempt to update the same row(s).

Therefore, before we can get into the issue of how to handle concurrency conflicts, we need to cover the basics of the roles the data adapter and command builder each play regarding updating a data store with the changed data in a dataset.

The Roles of the Data Adapter and Command Builder in Concurrency

In the previous chapter, you learned that the data adapter's `Update` method is used to commit changes made to rows held in a data table to that table's underlying data store. However, for purposes of understanding concurrency, it's necessary to delve just a bit deeper and look at what is really going on when `Update` is called. First, there are several overloaded versions of the `Update` method. For example, a `DataTable`, `DataSet`, and even a `DataRow` array can be passed. If a `DataSet` is passed, it must contain a `DataTable` named Table (the default name given a `DataTable` when constructed as a result of a `DataSet` being filled from a data adapter). The most common means of calling the `Update` method is as you've seen in this book—specifying a given `DataTable` object.

The next thing you should realize is that the `Update` method can only update one `DataTable` at a time. You'll recall that in the section entitled "Creating Multiple `DataTables` in a `DataSet`" in Chapter 6 I illustrated how to minimize round trips by having a single data adapter read multiple tables into a single dataset. However, those tables were being read for input purposes only. When accessing data that will be updated, each dataset should only contain one data table and be associated with only one data adapter. The reason for this is that the data adapter contains command objects for each of the three action commands—`InsertCommand`, `Update Command`, and `DeleteCommand`. Since there's only one set of these commands per adapter, it would be impossible to use the same set for multiple tables without a lot of manual work on your end in continually changing those commands to work properly with a given table.

You can always manually set any of the data adapter's action commands. In fact, earlier in this chapter (see "Using Stored Procedures to Insert Hierarchical Data with Auto-Increment Columns") you saw a demo application that set the `InsertCommand` to call an SQL Server stored procedure. However, for most standard, noncomplex queries, constructing a command builder object and associating it with a given data adapter will automatically result in the creation of the data adapter's action commands. This is done via the command builder's determining of the data store's schema from the

data adapter's select command. As you've seen in the past two chapters, this is all usually done as follows:

```
SqlConnection* conn =
   new SqlConnection(S"Server=localhost;"
                       S"Database=Northwind;"
                       S"Integrated Security=true;");

adapter = new SqlDataAdapter(S"SELECT * FROM Employees", conn);

commandBuilder = new SqlCommandBuilder(adapter);

. . .
```

When the data adapter's `Update` method is called, it first determines the rows that have been modified, inserted, or deleted by checking each row's `RowState` property. For each row, the command associated with its state is called to update the data store. If the row is new, the data adapter's `InsertCommand` is used. If the row has been modified, the `UpdateCommand` is used, and if the row has been deleted, the `DeleteCommand` is used. Also note that the data adapter makes a round trip for each and every row change. Unfortunately, there is no ability to batch command updates with the adapter object.

At this point, we've arrived at the crux of the issue—the automatic construction of the command objects by the command builder. In order to see this in action, you could simply insert the following code in the demo application's Commit button handler.

```
MessageBox::Show(commandBuilder->GetUpdateCommand()->CommandText,
                 S"CommandBuilder UpdateCommand for Employees Table");
```

If you run this code, you'll see the results in Figure 7–7, showing the automatically generated `UpdateCommand` from our initial select value passed to the data adapter's constructor (`"SELECT * FROM Employees"`).

This is where and how ADO.NET supports optimistic concurrency. As you can see, the SQL simply verifies that—for each column—both the original value and the current value are identical. (While each managed provider's action command SQL might differ slightly, the SQL will be almost identical across providers.) If you were to output the `InsertCommand` and `DeleteCommand` values (via `GetInsertCommand` and `GetDeleteCommand`, respectively), you'd find similar code to ensure optimistic concurrency.

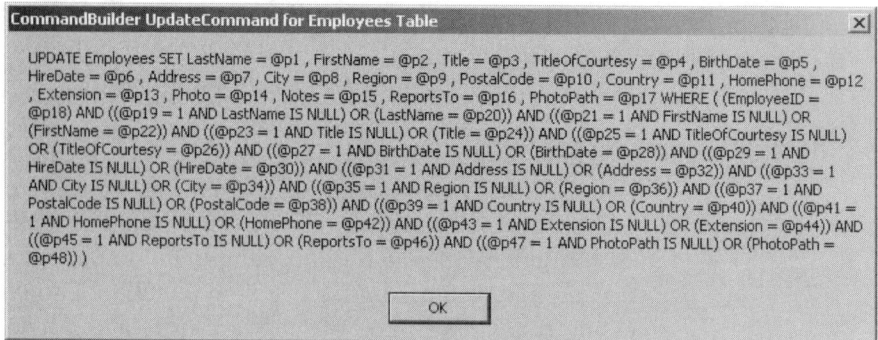

```
CommandBuilder UpdateCommand for Employees Table                                    ✕

UPDATE Employees SET LastName = @p1 , FirstName = @p2 , Title = @p3 , TitleOfCourtesy = @p4 , BirthDate = @p5 ,
HireDate = @p6 , Address = @p7 , City = @p8 , Region = @p9 , PostalCode = @p10 , Country = @p11 , HomePhone = @p12
, Extension = @p13 , Photo = @p14 , Notes = @p15 , ReportsTo = @p16 , PhotoPath = @p17 WHERE ( (EmployeeID =
@p18) AND ((@p19 = 1 AND LastName IS NULL) OR (LastName = @p20)) AND ((@p21 = 1 AND FirstName IS NULL) OR
(FirstName = @p22)) AND ((@p23 = 1 AND Title IS NULL) OR (Title = @p24)) AND ((@p25 = 1 AND TitleOfCourtesy IS NULL)
OR (TitleOfCourtesy = @p26)) AND ((@p27 = 1 AND BirthDate IS NULL) OR (BirthDate = @p28)) AND ((@p29 = 1 AND
HireDate IS NULL) OR (HireDate = @p30)) AND ((@p31 = 1 AND Address IS NULL) OR (Address = @p32)) AND ((@p33 = 1
AND City IS NULL) OR (City = @p34)) AND ((@p35 = 1 AND Region IS NULL) OR (Region = @p36)) AND ((@p37 = 1 AND
PostalCode IS NULL) OR (PostalCode = @p38)) AND ((@p39 = 1 AND Country IS NULL) OR (Country = @p40)) AND ((@p41 =
1 AND HomePhone IS NULL) OR (HomePhone = @p42)) AND ((@p43 = 1 AND Extension IS NULL) OR (Extension = @p44)) AND
((@p45 = 1 AND ReportsTo IS NULL) OR (ReportsTo = @p46)) AND ((@p47 = 1 AND PhotoPath IS NULL) OR (PhotoPath =
@p48)) )

                                         OK
```

Figure 7-7 There's more going on in a command builder's automatically generated action command than one would assume.

Deciding Where to Catch Concurrency Conflicts

Obviously, you can write code to handle any concurrency conflicts in the catch block associated with the try block that contains the call to the data adapter's Update method. However, in addition to that, there are two more common places in your code to handle these errors: in an event handler that is called for each updated row or after the call to the Update method returns, where all errors can be handled at once. Obviously, which one you choose will be specific to the design of your application, but I'll give you examples of how to implement each here.

Catching Concurrency Conflicts with the RowUpdated Event

The data adapter's RowUpdated event will fire for every row that the adapter attempted to update. That much you learned in the previous chapter, as implementing a handler for this event was used to illustrate one way to retrieve the auto-increment value for a newly added record's primary key column. However, another common use of this event is in determining the success or failure of an update attempt. To use this technique with the Concurrency demo, simply perform the following steps:

1. Define the following managed class that will represent our event handler. This class contains both a constructor and the event handler that will be called when the data adapter's RowUpdated event is raised. There are a couple of important things to note here. First, the SqlRowUpdatedEventArgs has a property called

`RecordsAffected` whose value can be used to determine if the update was successful or not. A value of 0 indicates failure. (In addition, the `Status` property will be equal to `UpdateStatus::ErrorsOccurred`.)

The second thing to note is that the default behavior of ADO.NET is to stop processing whenever a concurrency conflict occurs, and any already saved data is not rolled back. By setting the `SqlRowUpdatedEventArgs::Status` to `UpdateStatus::SkipCurrentRow`, you are basically telling ADO.NET to continue processing other rows once the method exits and not to attempt to update or report errors (throw exceptions) on this. You can then log the error or alert the user to the error. In the next section, I'll also cover a few options that you can employ to handle the conflict.

```
__gc class SqlEventHandler
{
public:
  void OnUpdated(Object* obj, SqlRowUpdatedEventArgs* e)
  {
    if (0 == e->RecordsAffected)
    {
      e->Status = UpdateStatus::SkipCurrentRow;
    }
  }
};
```

2. You can either construct the event-handler object (`SqlEventHandler`) after the data adapter object construction and store it as a member variable or simply construct it just before the call to the data adapter's `Update` as follows:

```
void CConcurrencyDlg::OnBnClickedCommit()
{
  ...

#pragma push_macro("new")
#undef new
  SqlEventHandler* eventHandler = new SqlEventHandler();
  adapter->add_RowUpdated(
    new SqlRowUpdatedEventHandler(eventHandler,
                                  SqlEventHandler::
                                  OnUpdated));
#pragma pop_macro("new")
```

```
. . .

    int rowsUpdated = adapter->Update(employees);

. . .
```

Catching Optimistic Concurrency Errors in Batch

In addition to handling each conflict as it occurs, you can also wait until all rows have been processed and then decide what to do with each conflict on a case-by-case basis. To do that, you would simply set the data adapter's ContinueUpdateOnError to true (by default it is set to false), check the DataTable::HasErrors property (set to true if any of the updates failed), and then enumerate through the rows that encountered errors upon update (retrieved via the DataTable::GetErrors method). Once again, using the Concurrency demo application's CConcurrencyDlg::OnBn ClickedCommit method, implementing this technique would look like the following:

```
void CConcurrencyDlg::OnBnClickedCommit()
{
  . . .

  adapter->ContinueUpdateOnError = true;

  int rowsUpdated = adapter->Update(employees);

  if (employees->HasErrors)
  {
    DataRow* rowsInError[] = employees->GetErrors();
    for (int i = 0; i < rowsInError->Count; i++)
    {
      // handle conflict
    }
  }
}

. . .
```

Deciding How to Handle Concurrency Conflicts

I intentionally used the phrase "catching concurrency conflicts" in the previous section because until now, I've only shown you how (and where in

code) to determine if an error has occurred; not how to actually handle that error. While there are many ways to deal with concurrency errors, I'll cover the most common methods here. The first thing you'll do is to see how to update the `Concurrency` demo to display the following versions of any record whose update fails:

- The original record that the user worked with
- The current record in the data store
- The user's proposed changes (that failed in the initial update attempt)

Users will then be presented with a choice of overwriting the data store with their current changes or canceling their current update attempt and refreshing their data from the data store. To begin, implement the code from the previous section, "Catching Optimistic Concurrency Errors in Batch." Once you've done that, insert the following call in the `for` loop.

```
// Handle each conflict on a case-by-case basis
HandleConcurrencyError(rowsInError[i]);
```

As you can see, the `HandleConcurrencyError` function takes a single parameter—a `DataRow` object. Now implement this function in the demo as follows. (While most of the code is probably self-documenting to you by now, a code explanation follows the listing.)

```
void CConcurrencyDlg::HandleConcurrencyError(DataRow* rowInError)
{
#pragma push_macro("new")
#undef new
  try
  {
    String* connString = S"Server=localhost;"
                         S"Database=Northwind;"
                         S"Integrated Security=true;"

    // Get current values from database
    SqlConnection* conn = new SqlConnection(connString);
    conn->Open();
    String* sql =
      String::Format(S"SELECT FirstName, LastName "
                     S"FROM Employees WHERE EmployeeId = {0}",
                     rowInError->Item["EmployeeId"]);
    SqlCommand* command = new SqlCommand(sql, conn);
```

```
SqlDataReader* results =
  command->ExecuteReader(CommandBehavior::CloseConnection);
results->Read();
String* databaseFirstName = results->GetString(0);
String* databaseLastName = results->GetString(1);
results->Close();

String* format = S"{0}: FirstName='{1}' LastName='{2}'\n";

StringBuilder* message = new StringBuilder();

message->Append(S"Another user has modified a record "
                S"you wish to update!\n\n"
                S"Do you want to overwrite the "
                S"current record in the database "
                S"with your proposed change? "
                S"(Selecting 'No' will update your local data "
                S"with the values in the database.\n\n");

message->AppendFormat(format,
  S"Original",
  rowInError->get_Item(S"FirstName",
                       DataRowVersion::Original)->ToString(),
  rowInError->get_Item(S"LastName",
                       DataRowVersion::Original)->ToString());

message->AppendFormat(format,
  S"Database",
  databaseFirstName,
  databaseLastName);

message->AppendFormat(format,
  S"Proposed",
  rowInError->get_Item(S"FirstName",
                       DataRowVersion::Current)->ToString(),
  rowInError->get_Item(S"LastName",
                       DataRowVersion::Current)->ToString());

DialogResult response = MessageBox::Show(message->ToString(),
                                         S"Concurrency Conflict",
                                         MessageBoxButtons::YesNo,
                                         MessageBoxIcon::Stop);
switch(response)
```

```
{
  case DialogResult::Yes:
  {
    // Overwrite database values
    SqlConnection* conn = new SqlConnection(connString);
    conn->Open();
    String* sql =
      String::Format(S"UPDATE Employees "
                     S"SET FirstName='{0}', LastName='{1}' "
                     S"WHERE EmployeeId = {2}",
      rowInError->get_Item(S"FirstName",
                      DataRowVersion::Current)->ToString(),
      rowInError->get_Item(S"LastName",
                      DataRowVersion::Current)->ToString(),
      rowInError->Item["EmployeeId"]);

    SqlCommand* command = new SqlCommand(sql, conn);
    int rowsAffected = command->ExecuteNonQuery();
    conn->Close();

    rowInError->AcceptChanges();

    MessageBox::Show(String::Format(S"{0} rows "
                       "successfully updated",
                       __box(rowsAffected)),
                     S".NET Exception Thrown",
                     MessageBoxButtons::OK,
                     MessageBoxIcon::Information);
  }
  break;

  case DialogResult::No:
  {
    // Set local changes to that of database
    rowInError->Item[S"FirstName"] = databaseFirstName;
    rowInError->Item[S"LastName"] = databaseLastName;
    rowInError->AcceptChanges();

    // Update UI
    LVFINDINFO info;
    info.flags = LVFI_STRING,
    CString strId =
      (CString)rowInError->Item[S"EmployeeId"]->ToString();
```

```
            info.psz = strId;
            int idx = m_lstEmployees.FindItem(&info);
            if (-1 < idx)
            {
              m_lstEmployees.SetItemText(idx, 1,
                 (CString)rowInError->Item[S"FirstName"]->ToString());
              m_lstEmployees.SetItemText(idx, 2,
                 (CString)rowInError->Item[S"LastName"]->ToString());
            }
          }
        break;
      }
    }
    catch(Exception* e)
    {
      MessageBox::Show(e->Message, S".NET Exception Thrown",
                       MessageBoxButtons::OK,
                       MessageBoxIcon::Error);
    }
    #pragma pop_macro("new")
  }
```

In a production system I would certainly recommend localizing things like the connection string and breaking this rather long function into a couple of generic functions that would accomplish the type of concurrency conflict resolution you decide to implement for your particular system. However, as this definitely gets into issues that are specific to your system design, this function gives you all the pieces of the puzzle, which you're free to assemble as you need.

Now, let's look at what the code does. The first thing the function does is to connect to the data store and—using an SqlCommand and DataReader—retrieve the current values of the FirstName and LastName columns from the database. The function then presents the user with a dialog that displays three versions of values for these two columns—the original values that the user saw before modifying them, the values currently in the database, and the values the user entered. The user then has a choice of overwriting the database values or canceling the update and having the local values overwritten with the database values. Figure 7–8 shows an example of this message box.

If the user chooses to overwrite what's in the database (by clicking the Yes button), a connection is made to the data store, an SqlCommand object

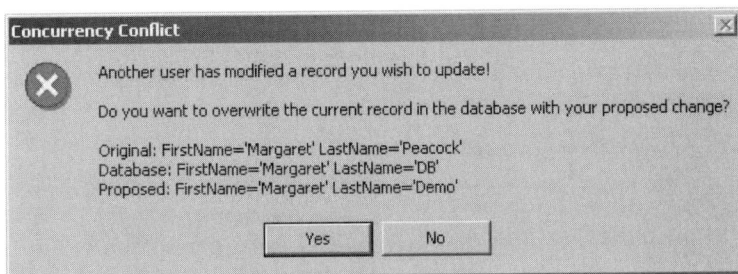

Figure 7-8 Example of allowing the user to specify how to handle a concurrency conflict

is constructed with the necessary SQL UPDATE syntax, and the SqlCommand:: ExecuteNonQuery method is called. As you can discern from the code, ExecuteNonQuery is used for situations where a result set will not be returned. Instead a value indicating the number of rows affected by the command is returned, which is then displayed to the user. Finally, the DataRow::AcceptChanges is called so that the state of the row is initialized.

If the user chooses to overwrite the local values with what's in the database (by clicking the No button), the database values retrieved when displaying the dialog are used. Once again, DataRow::AcceptChanges is called to finalize the changes and initialize the state of the DataRow object. Finally, the function locates the updated row in the list view and modifies its columns to reflect the database values.

Summary

As you've seen in the past two chapters, disconnected data can open your system design to an entirely new set of possibilities. However, there are some aspects of working with disconnected data and the DataSet class that aren't obvious. This chapter addressed several of these issues, including reading and writing binary data, associating related tables with the DataRelation class for easier and more efficient navigation, and handling concurrency issues with multiuser, disconnected data systems. The chapter began by illustrating how to use the DataReader class to read binary data. As DataReader only provides a means of reading data, the section then illustrated how to manually set up SqlCommand and SqlParameter objects

to write binary data to a database. By way of comparison, the DataSet class was then used to both read and write the same binary data. The section then concluded with a demo maintenance application that employed the DataSet class to load data from a provided sample table containing a photo column and data.

In the next section, you discovered how the DataRelation class gives you the ability to define relationships between related tables so that both tables can be navigated more easily and efficiently. Using this class, you can download two related tables, specify which columns create the parent/child linkage between the tables, and then, when in possession of a row from one of the tables, use either the GetParentRow or GetChildRows methods to retrieve the related rows from the other table. The section wrapped up by showing how to use stored procedures with return parameters to solve the issue of inserting parent rows that contain an auto-increment primary key so that the value of that key can be retrieved and used in inserting the child rows.

Finally, the chapter ended with a section on concurrency. This section explained what concurrency is, the various levels supported with ADO. NET, and roles played by the data adapter and command builder classes. After a demo application was presented to illustrate an example of a concurrency conflict, the section presented several techniques for handling concurrency conflicts in a graceful manner.

In Chapter 5, you learned a good deal about XML, and in the past two chapters, you've been exposed to quite a bit of ADO.NET. This is definitely not by chance, as Microsoft sees XML and ADO.NET as two of the cornerstone technologies for the future of .NET and data access in general. Therefore, in the next chapter, we'll bring all this full circle with a complete chapter on combining ADO.NET and XML.

CHAPTER 8

Combining ADO.NET and XML

Introduction

In Chapter 4, you learned a good deal about using the various classes in the `System::Xml` namespace in order to work with and manipulate XML documents. In Chapters 6 and 7, you were exposed to quite a bit of ADO.NET—with the focus on the topics most likely to impact and interest MFC developers. The fact that four of this book's eleven chapters involve either XML or ADO.NET is definitely not a coincidence, as Microsoft views XML and ADO.NET as two of the cornerstone technologies for the future of .NET and data access in general. Therefore, since you've already seen how to use these technologies independently, the only task left at this point regarding using XML and ADO.NET is learning how to move easily from one format to the other. Note that when I say that, I'm being a little liberal in my verbiage, as technically XML is a data format, and ADO.NET is a data access layer. However, in the context of this particular chapter, I'm referring to being able to take data currently held in an ADO.NET object—usually a `DataSet`—and then either serialize it to XML or deserialize XML documents into ADO.NET objects.

The chapter begins by illustrating how to serialize a `DataSet` containing a single `DataTable`—or multiple unrelated `DataTable` objects—to XML. From there, a section illustrates how to perform this same task with related tables. Once you've seen how to output your ADO.NET data to XML, you'll then be presented with a couple of sections that will demonstrate how to format the serialized XML in terms of the generated XML node types and indentation options. After a section that covers writing the schema (XSD) information for `DataSet` objects, you'll see how to fill a

`DataSet` object from XML documents—both with and without included schema information. The chapter then concludes with a section detailing how DiffGrams enable you to generate an XML file containing both the current and original values of modified data for more efficient transport of changed data across a network connection.

Prerequisites for This Chapter

As the focus of this chapter is about serializing `DataSet` data to XML and filling `DataSet` objects with XML, this chapter makes the assumption that you're comfortable with both topics. If not, I would recommend reading the previous three chapters in order to get the most out of this chapter.

If you want to follow along in this chapter or run the demo applications, you'll need to create and populate the `ExtendingMFCWithDotNet` sample SQL Server database using the instructions provided at the beginning of Chapter 7.

Serializing XML from a Single `DataTable` or Multiple Unrelated `DataTables`

The first thing we'll look at is how to generate XML from a `DataSet` that contains a single table. In fact, in doing so, you'll quickly see how much attention the .NET development team has paid to integrating ADO.NET and XML. There are two main methods for serializing a `DataSet` to XML: `WriteXml` and `GetXml`. The `WriteXml` method is generally used when you want to persist the XML representation of a `DataSet` object's `DataTable` objects. This method has a number of overloads that allow you to use `Stream` and `Writer` object types as well as a file name and a value (`Xml WriteMode`) indicating the type of schema information you want included with the XML data. Several sections covering how to control this output will be presented once we get past these first sections that cover the basics of generating XML from `DataSet` objects. The second method—`GetXml`—returns a string representation of the `DataSet` object's data. Note that since this method must first allocate and fill a `string` object, it has more overhead associated with it than the `WriteXml` method. In addition, the `GetXml` method does not produce any schema information—instead returning only the data. This is the equivalent of calling the `WriteXml` method with a value of `XmlWriteMode::IgnoreSchema`.

The following code reads all rows from the Contributors table, generates the XML representation of those rows, and displays them. Note that I'm omitting the `Photo` column for output so that it is easier to read. The inclusion of a binary column is no different than any other column types.

```
void DisplayContributorsInXml()
{
#pragma push_macro("new")
#undef new
  try
  {
    SqlConnection* conn =
      new SqlConnection(S"Server=localhost;"
                         S"Database=ExtendingMFCWithDotNet;"
                         S"Integrated Security=true;");
    SqlDataAdapter* adapter =
      new SqlDataAdapter(S"SELECT Name, Role FROM"
                         S"Contributors", conn);

    DataSet* dataset = new DataSet();

    adapter->Fill(dataset, S"Contributors");

    conn->Close();

    Console::WriteLine(dataset->GetXml());
  }
    catch(Exception* e)
    {
      Console::WriteLine(e->Message);
    }
#pragma pop_macro("new")
}
```

As you can see, the majority of this code implements what you've learned throughout the previous three chapters, which is exactly the point I wanted to make here. For the relatively simple tasks involving converting ADO.NET data into XML—and vice-versa—the Framework provides so much support that many of these tasks can be done with a few lines of code. The sample function produces the following output:

```
<NewDataSet>
  <Contributors>
```

```
    <Name>Tom Archer</Name>
    <Role>Author</Role>
  </Contributors>
  <Contributors>
    <Name>Nishant Sivakumar</Name>
    <Role>Co-author</Role>
  </Contributors>
  <Contributors>
    <Name>Erik Westermann</Name>
    <Role>Tech Editor</Role>
  </Contributors>
  <Contributors>
    <Name>James Johnson</Name>
    <Role>Tech Editor</Role>
  </Contributors>
</NewDataSet>
```

There are a few things to note here.

- The XML document's root node uses the `DataSet` object's `DataSet Name` property. If a name is not specified in the `DataSet` constructor, the default value is `NewDataSet`—the value you see in this XML listing.
- By default, all table columns are inserted as elements instead of attributes. I'll illustrate how to change that behavior shortly.
- By default, all columns in all tables in the `DataSet` are included in the serialization process. I'll also illustrate how to omit columns shortly.
- Any time you call the `DataSet::GetXml` method, all of the `Data Table` objects in that `DataSet` will be output in a flat format. For example, if the sample database included Customers and Vendors tables—which would not normally share a relationship—the resulting XML from the `GetXml` method would look something like the following:

```
<NewDataSet>
  <Customers>
    <Name>Customer #1</Name>
    ... Other Customers columns
  </Customers>
  <Customers>
    <Name>Customer #2</Name>
```

```
          ... other Customers columns
       </Customers>


       ...

       <Vendors>
         <Name>Vendor #1</Name>
         ... other Vendors columns
       </Vendors>
       <Vendors>
         <Name>Vendor #2</Name>
         ... other Vendors columns
       </Vendors>
     </NewDataSet>
```

This format is perfect for situations where the DataTable objects are not related. However, for related data—such as the sample database's OrderHeaders and OrderDetails tables, the following flat output would not be appropriate.

```
<NewDataSet>
  <OrderHeaders>
    <OrderId>1</OrderId>
    <CustomerName>Customer ABC</CustomerName>
    <ShippingAddress>ABC street, 12345</ShippingAddress>
  </OrderHeaders >

  ... other order headers

  <OrderDetails>
    <OrderId>1</OrderId>
    <ProductId>1</ProductId>
    <ProductName>Product 1</ProductName>
    <UnitPrice>1</UnitPrice>
    <Quantity>2</Quantity>
  </OrderDetails>

  ... other order details
</NewDataSet>
```

As you can see, there's no connection between the parent and child data, so that the order detail rows for each order header appear within that order header. In order to get the desired hierarchical output, you'll need to manually specify the DataTable relationships, as shown in the next section.

Serializing Datasets Containing Related Data Tables

In the previous chapter you learned how to construct `DataRelation` objects to define the linkage, or point of relationship, between the two `DataTable` objects contained within a `DataSet`. In that chapter, `Data Relation` objects were employed to more efficiently retrieve data (since duplicate data in the result set containing joined tables is eliminated) and more easily navigate to the related rows of data (via the `GetParentRow` and `GetChildRows` methods). `DataRelation` objects are also used internally by the `DataSet::GetXml` method to output XML that correctly generates the nested nature of related data. Consider the following function that generates an XML document based on the sample database's OrderHeaders and OrderDetails tables.

```
void DisplayRelationalDataInXml()
{
#pragma push_macro("new")
#undef new
  try
  {
    // Connect to data store
    SqlConnection* conn =
      new SqlConnection(S"Server=localhost;"
                        S"Database=ExtendingMFCWithDotNet;"
                        S"Integrated Security=true;");

    // Batch two reads in a single adapter
    SqlDataAdapter* adapter =
      new SqlDataAdapter(S"SELECT * FROM OrderHeaders;"
                         S"SELECT * FROM OrderDetails",
                         conn);

    DataSet* dataset = new DataSet();

    // Fill both tables with a single round trip
    // to the data store
    adapter->Fill(dataset);

    // Close the connection
    conn->Close();
```

```
  // Tables are named Table and Table1 by default.
  // Change them for more readable code.
  DataTableCollection* tables = dataset->Tables;
  tables->Item[0]->TableName = S"Orders";
  tables->Item[1]->TableName = S"OrderDetails";

  // Get DataTable objects for convenience
  DataTable* ordersTable = dataset->Tables->Item[S"Orders"];
  DataTable* detailsTable = dataset->Tables->Item[S"OrderDetails"];

  // Create DataRelation defining parent/child relationship
  // between OrderHeaders table and OrderDetails table
  DataRelation* ordersToOrderDetails =
    new DataRelation(S"OrdersToOrderDetails",
                       ordersTable->Columns->Item[S"OrderId"],
                       detailsTable->Columns->Item[S"OrderID"]);

  // IMPORTANT! Specify that the relationship should be nested
  // so that the XML generation correctly generates hierarchical
  // data
  ordersToOrderDetails->Nested = true;

  // Add the DataRelation to the DataSet
  dataset->Relations->Add(ordersToOrderDetails);

  // Retrieve XML for tables in the dataset using
  // the defined relations
  Console::WriteLine(dataset->GetXml());
}
catch(Exception* e)
{
  Console::WriteLine(e->Message);
}
#pragma pop_macro("new")
}
```

As you can see, once the DataRelation object has been constructed—by linking the two tables by their respective OrderId columns—the DataSet::Nested property must be set to true and the DataRelation is added to the DataSet. Setting the Nested property ensures that the XML serialization mechanism generates the XML document so that the data is

related in a hierarchical manner. Running this function produces the following XML document (snipped for brevity).

```
<NewDataSet>
  <Orders>
    <OrderId>1</OrderId>
    <CustomerName>Customer ABC</CustomerName>
    <ShippingAddress>ABC street, 12345</ShippingAddres>
    <OrderDetails>
      <OrderId>1</OrderId>
      <ProductId>1</ProductId>
      <ProductName>Product 1</ProductName>
      <UnitPrice>1</UnitPrice>
      <Quantity>2</Quantity>
    </OrderDetails>
    <OrderDetails>
      <OrderId>1</OrderId>
      <ProductId>2</ProductId>
      <ProductName>Product 2</ProductName>
      <UnitPrice>2</UnitPrice>
      <Quantity>2</Quantity>
    </OrderDetails>
  </Orders>

  ... other orders

</NewDataSet>
```

Mapping `DataTable` Columns to XML Node Types

As you've seen in the various sample functions, when a `DataSet` is serialized to XML, all of the columns (`DataColumn` objects) for each of the set's `DataTable` objects are output as XML elements. This might be fine for simple demos, but in a practical setting it probably won't be what you want all the time. To aid in controlling how `DataColumn` information is serialized to XML (in terms of the node type used), the `DataColumn` has a property called `ColumnMapping` that is set to any of the `MappingType` enumeration values shown in Table 8–1.

Table 8-1 MappingType Enumeration Values

Member Name	Description
Attribute	Maps the column to an attribute of the element having the same name as the column parent's table name. For a DataTable named "Employees," setting the MappingType to Attribute for a Data Column named EmployeeId (with a value of 1) would result in XML similar to the following: `<Employees EmployeeId="1">` . . .
Element	This is the default column mapping and simply states that the DataColumn will be mapped as a nested element under its parent element. For a DataTable named "Employees," setting the MappingType to Attribute for a DataColumn named EmployeeId (with a value of 1) would result in XML similar to the following: `<Employees>` `<EmployeeId>1</EmployeeId>` . . .
Hidden	This simply means to omit the column from being serialized to XML.
SimpleContent	This value cannot be used if any columns are using the Element mapping or with defined (parent/child) relationships. The resulting XML for columns defined as SimpleContent is text within the parent table element.

In the following modification to the DisplayContributorsInXml function, all of the columns of the Contributors DataTable are having their ColumnMapping property set to MappingType::Attribute.

```
void DisplayContributorsInXml()
{
...

    adapter->Fill(dataset, S"Contributors");
```

. . .

```
DataTable* contributors = dataset->Tables->Item[S"Contributors"];
DataColumn* column;
for (int i = 0; i < contributors->Columns->Count; i++)
{
   column = contributors->Columns->Item[i];
   column->ColumnMapping = MappingType::Attribute;
}
```

. . .

```
   Console::WriteLine(dataset->GetXml());
}
```

This yields the following XML. Note how the serialization mechanism creates empty element tags as none of the elements contain other elements or values.

```
<NewDataSet>
  <Contributors ID="5" Name="Tom Archer" Role="Author"/>
  <Contributors ID="6" Name="Nishant Sivakumar" Role="Co-author"/>
  <Contributors ID="7" Name="Erik Westermann" Role="Tech Editor"/>
  <Contributors ID="8" Name="James Johnson" Role="Tech Editor"/>
</NewDataSet>
```

As I mentioned back in Chapter 5, when deciding what values to represent using an attribute and what to represent using an element, I generally choose to represent anything that uniquely identifies the parent element as an attribute and represent everything else as elements. In the `Display ContributorsInXml` function, you'll notice that I omit the `Photo` column—I'm doing this for the simple reason that we haven't reached the discussion on column mapping yet. As you now know, you can simply omit columns from being serialized by setting the `DataColumn::Column Mapping` to `MappingType::Hidden`.

Having said that, I'll consider the `DisplayRelationalDataInXml` function as an example. Here, I want the Orders `OrderId` as an element, and since the data is already hierarchically defined, I don't need to include the `OrderId` in the child `OrderDetails` elements, so I omit that column from getting serialized by setting its `DataColumn::ColumnMapping`

property to MappingType::Hidden. With those changes incorporated, the
DisplayRelationalDataInXml function now looks like this:

```
void DisplayRelationalDataInXml()
{
  ...

  adapter->Fill(dataset);

  ...

  // Map the Orders OrderId column to an XML attribute type
  column = ordersTable->Columns->Item["OrderId"];
  column->ColumnMapping = MappingType::Attribute;

  // Omit the OrderDetails OrderId column when serialized
  column = detailsTable->Columns->Item["OrderId"];
  column->ColumnMapping = MappingType::Hidden;

  ...

  Console::WriteLine(dataset->GetXml());
}
```

This now produces the following updated XML representation of the
OrderHeaders/OrderDetails sample data:

```
<NewDataSet>
  <Orders OrderId="1">
    <CustomerName>Customer ABC</CustomerName>
    <ShippingAddress>ABC street, 12345</ShippingAddres
    <OrderDetails>
      <ProductId>1</ProductId>
      <ProductName>Product 1</ProductName>
      <UnitPrice>1</UnitPrice>
      <Quantity>2</Quantity>
    </OrderDetails>
    <OrderDetails>
      <ProductId>2</ProductId>
      <ProductName>Product 2</ProductName>
      <UnitPrice>2</UnitPrice>
      <Quantity>2</Quantity>
```

```
    </OrderDetails>
    </Orders>

    ... other orders

</NewDataSet>
```

Saving Formatted XML

Earlier in the chapter, I mentioned that the WriteXml method is used to persist the XML representation of a DataSet and that the GetXml method is used to simply retrieve that XML in a String object. Since then, you've seen how to control the structure of serialized XML via mapping columns to XML node types, how to specify which columns to omit, and how to define hierarchical relationships between elements. At this point, it's time to turn our attention to the issue of controlling the format of the XML when it's persisted.

You can use the members of the XmlWriter class as well as the Xml WriteMode enumeration to exercise the utmost control over the way the XML serialized. Consider the following code to control the indentation of an XML file:

```
// Assumes that the dataset has already been set up
// such as in the DisplayContributorsInXml or
// DisplayRelationalDataInXml functions

...

// Retrieve XML for tables in the dataset using
// the defined relations
// Console::WriteLine(dataset->GetXml());
StreamWriter* streamWriter = new StreamWriter(S"C:\\temp.xml");
XmlTextWriter* writer = new XmlTextWriter(streamWriter);

writer->Formatting = Formatting::Indented;
writer->Indentation = 5;

dataset->WriteXml(writer);
```

As you can see, specifying the indentation is a matter of setting the Xml TextWriter::Formatting to the desired Formatting enumeration value.

The only two values are `Formatting::Indented` and `Formatting::None` (the default). The `XmlTextWriter::Indentation` property is then used to specify the desired number of spaces to indent.

Here's an example of using this code with the Contributors table (instead of the `GetXml` method call), where I've omitted the `Photo` column, set the `ID` column to a mapping of `Attribute`, and let the remaining columns default to a mapping of `Element`.

```
<NewDataSet>
    <Contributors ID="5">
        <Name>Tom Archer</Name>
        <Role>Author</Role>
    </Contributors>
    <Contributors ID="6">
        <Name>Nishant Sivakumar</Name>
        <Role>Co-author</Role>
    </Contributors>
    <Contributors ID="7">
        <Name>Erik Westermann</Name>
        <Role>Tech Editor</Role>
    </Contributors>
    <Contributors ID="8">
        <Name>James Johnson</Name>
        <Role>Tech Editor</Role>
    </Contributors>
</NewDataSet>
```

Note that many XML parsers will apply their own indentation when displaying an XML file. For example, Internet Explorer (which has its own XML parser) will display the file the same way regardless of the value you specify for the `Indentation` property. Therefore, you'll need to view the file in Visual Studio or some other editor that doesn't change your indentation in order to test your code.

Writing Schema Information

When you use either the `WriteXml` or `GetXml` methods, the resulting XML is the complete data contained within the `DataSet`, but without the schema information. The schema defines the structure of the data in terms of things like column names and types, primary keys, and any defined relations. You

need to generate schema information as either a standalone file or in addition to the XML data if you want to have third-party applications make sense of the DataSet-generated XML data. In this section, we'll look at how to accomplish both these goals.

In order to write just the schema of an XML file, you can use the DataSet::WriteXmlSchema method. Like most methods in the BCL that allow you to persist something, the WriteXmlSchema method has several overloads:

```
void WriteXmlSchema(Stream*);
void WriteXmlSchema(String*);
void WriteXmlSchema(TextWriter*);
void WriteXmlSchema(XmlWriter*);
```

To test the type of information that can be generated let's look at simply replacing the GetXml call in the DisplayRelationalDataInXml function with the following:

```
void DisplayRelationalDataInXml()
{
   . . .

   dataset->Relations->Add(ordersToOrderDetails);

   String* filename =
     String::Format(S"{0}.xsd",
                    ordersToOrderDetails->RelationName);
   dataset->WriteXmlSchema(filename);
}
```

The code creates a file that's based on the value of the RelationName property along with an XSD extension (the file is called OrdersToOrder Details.XSD), since schemas are generally stored in a file with an XSD extension.

This code now generates the following schema information for both the OrderHeaders and OrderDetails tables. (Some elements have been broken into multiple lines for book-formatting purposes only.)

```
<?xml version="1.0" standalone="yes"?>
<xs:schema id="NewDataSet" xmlns=""
xmlns:xs="http://www.w3.org/2001/XMLSchema"
          xmlns:msdata="urn:schemas-microsoft-com:xml-msdata">
```

```
<xs:element name="NewDataSet" msdata:IsDataSet="true">
  <xs:complexType>
    <xs:choice maxOccurs="unbounded">
      <xs:element name="Orders">
        <xs:complexType>
          <xs:sequence>
            <xs:element name="CustomerName" type="xs:string"
minOccurs="0" msdata:Ordinal="1" />
            <xs:element name="ShippingAddress" type="xs:string"
minOccurs="0" msdata:Ordinal="2" />
            <xs:element name="OrderDetails" minOccurs="0"
maxOccurs="unbounded">
              <xs:complexType>
                <xs:sequence>
                  <xs:element name="ProductId" type="xs:int"
minOccurs="0" msdata:Ordinal="1" />
                  <xs:element name="ProductName" type="xs:string"
minOccurs="0" msdata:Ordinal="2" />
                  <xs:element name="UnitPrice" type="xs:decimal"
minOccurs="0" msdata:Ordinal="3" />
                  <xs:element name="Quantity" type="xs:int"
minOccurs="0" msdata:Ordinal="4" />
                </xs:sequence>
                <xs:attribute name="OrderId" type="xs:int"
use="prohibited" />
              </xs:complexType>
            </xs:element>
          </xs:sequence>
          <xs:attribute name="OrderId" type="xs:int" />
        </xs:complexType>
      </xs:element>
    </xs:choice>
  </xs:complexType>
  <xs:unique name="Constraint1">
    <xs:selector xpath=".//Orders" />
    <xs:field xpath="@OrderId" />
  </xs:unique>
  <xs:keyref name="OrdersToOrderDetails" refer="Constraint1"
msdata:IsNested="true">
    <xs:selector xpath=".//OrderDetails" />
    <xs:field xpath="@OrderId" />
  </xs:keyref>
</xs:element>
</xs:schema>
```

As you can see from the listing, I've used bold for a few key points. First, the top-level element is the Orders table, with its columns and the OrderDetails table being nested under it as child elements. The OrderDetails table's columns are then nested under the OrderDetails element. The constraints are then listed, where you can see that the OrderId column is used as a unique identifier (primary key) for the Orders table. Finally, the relation between the Orders and OrderDetails tables is described by linking the OrderID of OrderDetails with Orders using a unique constraint (all OrderIDs must be unique).

In order to serialize both the data and the schema of a DataSet, you would simply specify the XmlWriteMode enumeration value of Write Schema when calling the DataSet::WriteXml method. The valid values for this enumeration are listed in Table 8–2. The following code produces both the XML and schema for the data contained in a DataSet object.

```
void DisplayRelationalDataInXml()
{
   . . .

   dataset->Relations->Add(ordersToOrderDetails);

   String* filename =
      String::Format(S"{0}.xml",
                     ordersToOrderDetails->RelationName);
   dataset->WriteXml(filename, XmlWriteMode::WriteSchema);
}
```

Table 8–2 XmlWriteMode Enumeration Values

Member Name	Description
DiffGram	Writes the entire DataSet as a DiffGram, including original and current values. DiffGrams are covered in the section entitled "Working with DiffGrams."
IgnoreSchema	The default and results in only the contents of the DataSet being written. If the DataSet is empty, nothing is written.
WriteSchema	Writes the XSD schema of the DataSet followed by the current contents of the DataSet. If the DataSet has only a schema with no data, only the inline schema is written. If the DataSet does not have a current schema, nothing is written.

Generating XML from a `DataTable`

One question I receive quite often regarding serializing ADO.NET objects to XML format is how to generate XML from a standalone `DataTable`, or one that is not included in a parent `DataSet` object. If you look up the `GetXml` or `WriteXml` method in online help, you'll find that while both methods are implemented by the `DataSet` class, neither is implemented by the `DataTable` class. This is a bit confusing at first because, as you've seen, calling and using these methods against a `DataSet` containing multiple `DataTable` objects does result in each table being processed. After exchanging a couple of e-mails with friends on the .NET development team, I've found that the `DataSet` does define methods to serialize a specified `DataTable`, but that these methods are not exposed. However, this limitation is very easy to circumvent with the following generic function—`WriteDataTableToXml`—which serializes a specified `DataTable` object to XML.

```
void WriteDataTableToXml(
   DataTable* table,
   String* filename,
   XmlWriteMode mode)
{
#pragma push_macro("new")
#undef new
   try
   {
      // Create a temporary dataset
      DataSet* dataset = new DataSet();

      // Does table belong to a dataset already?
      if (NULL == table->DataSet)
      {
         // Add it to the temporary DataSet
         dataset->Tables->Add(table);
      }
      else
      {
         // Make a copy of the DataTable and
         // add it to the temporary DataSet
         dataset->Tables->Add(table->Copy());
      }
```

```
    dataset->WriteXml(filename, mode);
  }
  catch(Exception* e)
  {
    Console::WriteLine(e->Message);
  }
#pragma pop_macro("new")
}
```

. The WriteDataTableToXml function takes three parameters—a DataTable object, a String that's used as a file name, and an XmlWrite Mode enumeration value. After the function constructs an empty DataSet, it checks to see if the DataTable already belongs to a DataSet by inspecting the DataTable::DataSet property. This is done due to the restriction that a single instance of a DataTable cannot simultaneously belong to more than one DataSet. If the DataTable *does not* belong to a DataSet, it is added to the temporary DataSet. If the table *does* belong to a DataSet, then a temporary copy of the DataTable is made and added to the temporary DataSet. Finally, the DataSet::WriteXml is called, passing it the specified file name and XmlWriteNode parameter values. The following test function programmatically constructs a DataTable, defining its columns and adding a couple of rows of test data before calling the Write DataTableToXml function.

```
void TestWriteDataTableToXml()
{
#pragma push_macro("new")
#undef new
  try
  {
    // Define a DataTable
    DataTable* table = new DataTable(S"Players");

    // Define the table's columns
    table->Columns->Add(S"FirstName", __typeof(String));
    table->Columns->Add(S"LastName", __typeof(String));

    DataRow* newRow;
    Object* values[] = new Object*[table->Columns->Count];

    // Add test rows to the DataTable object
    values[0] = S"Yao";
```

```
      values[1] = S"Ming";
      newRow = table->Rows->Add(values);

      values[0] = S"Steve";
      values[1] = S"Francis";
      newRow = table->Rows->Add(values);

      WriteDataTableToXml(table,
                          S"Players.xml",
                          XmlWriteMode::WriteSchema);
    }
    catch(Exception* e)
    {
      Console::WriteLine(e->Message);
    }
  #pragma pop_macro("new")
}
```

As the following output indicates, our little helper function now enables us to serialize `DataTable` objects with or without the accompanying schema information.

```
<?xml version="1.0" standalone="yes"?>
<NewDataSet>
  <xs:schema id="NewDataSet"
    xmlns=""
    xmlns:xs="http://www.w3.org/2001/XMLSchema"
    xmlns:msdata="urn:schemas-microsoft-com:xml-msdata">
    <xs:element name="NewDataSet" msdata:IsDataSet="true">
      <xs:complexType>
        <xs:choice maxOccurs="unbounded">
          <xs:element name="Players">
            <xs:complexType>
              <xs:sequence>
                <xs:element name="FirstName"
                    type="xs:string" minOccurs="0" />
                <xs:element name="LastName"
                    type="xs:string" minOccurs="0" />
              </xs:sequence>
            </xs:complexType>
          </xs:element>
        </xs:choice>
      </xs:complexType>
```

```
      </xs:element>
    </xs:schema>
    <Players>
      <FirstName>Yao</FirstName>
      <LastName>Ming</LastName>
    </Players>
    <Players>
      <FirstName>Steve</FirstName>
      <LastName>Francis</LastName>
    </Players>
  </NewDataSet>
```

Filling `DataSet` Objects from XML

So far, you've seen how to serialize a `DataSet` object's schema and contents to XML and the various issues regarding the formatting of the XML. In this section, we'll do the reverse—fill a `DataSet` from an XML document using the `DataSet::ReadXml` method.

The `ReadXml` method is a mirror-image of the `WriteXml` method to the extent that, much as the `WriteXml` method takes an `XmlWriteMode` enumeration value as a parameter, the `ReadXml` method takes an `XmlReadMode` enumeration value that specifies how to read XML data and a relational schema into the `DataSet`. At its most basic level, the `ReadXml` method is incredibly easy to use. For example, let's say that you had run the `Display RelationalDataInXml` function in order to serialize the sample Order-Headers and OrderDetails data to an XML file named `OrdersToOrder Details.xml` with schema information. This data could easily be read back into a new `DataSet` as follows:

```cpp
void ReadOrdersXmlWithSchema()
{
#pragma push_macro("new")
#undef new
  try
  {
    DataSet* dataset = new DataSet();
    dataset->ReadXml(S"OrdersToOrderDetails.xml",
                     XmlReadMode::ReadSchema);
```

```cpp
// Let's see if the data is there and properly related!
DataTable* orderHeaders = dataset->Tables->Item[0];
DataTable* orderDetails = dataset->Tables->Item[1];

// get all headers
DataRowCollection* headers - orderHeaders->Rows;
DataRow* header;
for (int i = 0; i < headers->Count; i++)
{
  header = headers->Item[i];
  Console::WriteLine(String::Format(S"Order: {0}\t"
                                    S"Customer: {1}\t"
                                    S"Shipping Address: {2}",
                                    header->Item["OrderId"],
                                    header->Item["CustomerName"],
                                    header->Item["ShippingAddress"
                                      ]));

  // get all detail lines
  String* format = S"{0,-5} {1,-5} {2,-10} {3,-10} {4,-10}";
  Object* values[] = new Object*[5];
  values[0] = S"Order";
  values[1] = S"PID";
  values[2] = S"Product";
  values[3] = S"Price";
  values[4] = S"Qty";
  Console::WriteLine(String::Format(format, values));

  DataRow* details[] =
    header->GetChildRows(S"OrdersToOrderDetails");
  for (int j = 0; j < details->Count; j++)
  {
    Object* values[] = new Object*[5];
    values[0] = details[j]->Item[S"OrderId"];
    values[1] = details[j]->Item[S"ProductId"];
    values[2] = details[j]->Item[S"ProductName"];
    values[3] = details[j]->Item[S"UnitPrice"];
    values[4] = details[j]->Item[S"Quantity"];
    Console::WriteLine(String::Format(format, values));
  }
  Console::WriteLine();
}
```

```
  }
  catch(Exception* e)
  {
    Console::WriteLine(e->Message);
  }
#pragma pop_macro("new")
}
```

Running this function produces the following output:

```
Order: 1    Customer: Customer ABC   Shipping Address: ABC street, 12345
Order PID   Product      Price        Qty
1    1      Product 1    1            2
1    2      Product 2    2            2


Order: 2    Customer: Customer 2     Shipping Address: Someplace
Order PID   Product      Price        Qty
2    3      Product 3    3            3
```

As you can see, everything worked perfectly in terms of both retrieving the correct data as well as the relationships between the tables! This is because the XML file contained the schema that defined everything we needed, including the nested relation between the Orders and Orders Details elements. Since the schema was included with the data, all the ReadXMLIntoDataSet had to do was call the DataSet::ReadXml method, specifying the file name and an XmlReadMode value of ReadSchema.

Note that if you specify the XmlReadMode::ReadSchema enumeration and the schema is not included in the XML file, then nothing will be read. This brings up an important point about filling DataSet objects from XML files. Even though the XmlReadMode enumeration includes a value called InferSchema, it's only really useful in very simplistic scenarios. For situations where the XML file does not include schema information, you're going to need to have some prior knowledge of that file in order to programmatically use that data in a meaningful manner.

To illustrate my point, consider the following example, where the DisplayContributorsInXml is modified very slightly so that instead of displaying the contents of the Contributors DataTable, it saves the data to a file named Contributors.xml—without schema information.

```
void DisplayContributorsInXml()
{
    ...
```

```
    // Don't display the data
    // Console::WriteLine(dataset->GetXml());

    // Instead save it to an XML file without schema
    dataset->WriteXml(S"Contributors.xml");
}
```

Now we could attempt to use the XmlReadMode::InferSchema enumeration to attempt to read and use the data with the following function:

```
void ReadContributorsWithoutSchema()
{
#pragma push_macro("new")
#undef new
  try
  {
    DataSet* dataset = new DataSet();
    dataset->ReadXml(S"Contributors.xml", XmlReadMode::InferSchema);

    DataTable* contributors = dataset->Tables->Item[0];

    DataRow* row;
    DataColumn* column;
    for (int r = 0; r < contributors->Rows->Count; r++)
    {
        row = contributors->Rows->Item[r];
        for (int c = 0; c < contributors->Columns->Count; c++)
        {
         column = contributors->Columns->Item[c];
         Console::WriteLine(String::Format(S"{0} = '{1}'",
                           column->ColumnName,
                           row->Item[c]));
        }
        Console::WriteLine();
    }
  }
  catch(Exception* e)
  {
    Console::WriteLine(e->Message);
  }
#pragma pop_macro("new")
}
```

This function produces the following output:

```
Name = 'Tom Archer'
Role = 'Author'
ID = '5'

Name = 'Nishant Sivakumar'
Role = 'Co-author'
ID = '6'

Name = 'Erik Westermann'
Role = 'Tech Editor'
ID = '7'

Name = 'James Johnson'
Role = 'Tech Editor'
ID = '8'
```

As you can see, filling a DataSet from a simple flat table with no relation to any other table and no constraints is very easy. However, strip the schema information from the OrdersToOrderDetails.xml file, modify the ReadOrdersXmlWithSchema to use the XmlReadMode::InferSchema value, and the output from that function becomes the following, where you can see that no child rows were read.

```
Order: 1   Customer: Customer ABC  Shipping Address: ABC street, 12345
Order PID  Product     Price       Qty

Order: 2   Customer: Customer 2    Shipping Address: Someplace
Order PID  Product     Price       Qty
```

Stepping into the debugger does confirm that the GetChildRows function is not returning any rows for the parent table. The reason for this is that the GetChildRows refers to a relation that, while it exists in the XML, it doesn't go by that name. In fact, there are several things in our code that will no longer work because of the way the ReadXml had to infer schema information when none was included with the XML data.

The first problem is that although the ReadXml method was smart enough to build a nested relation when it parsed the XML file, it could not know the relation's name as we had given it in our original code; therefore, it was given an automatically generated name. This name always takes the

format `<parent>_<child>` such that the generated name for this relation would be `Orders_OrderDetails`.

The second problem has to do with the hidden data of the child table. If you recall, we specified a `ColumnMapping` of `MappingType::Hidden` for the OrderDetails `OrderId` column. However, once again the `ReadXml` method is smart enough to see how the two tables are linked and includes that column in our `DataSet`. But, since it can't know the name of the column it is added to the OrderDetails `DataTable` with a generated name taking the form `<parent>_Id` such that it now has the name `Orders_Id`.

Now that you know these two bits of information, the following function will illustrate how to read the `OrdersToOrderDetails.xml` file and correctly display its nested information, even though the file doesn't contain schema information.

```
void ReadOrdersXmlWithoutSchema()
{
#pragma push_macro("new")
#undef new
  try
  {
    DataSet* dataset = new DataSet();
    dataset->ReadXml(S"OrdersToOrderDetails.xml",
                     XmlReadMode::InferSchema);

    // Let's see if the data is there and properly related!
    DataTable* orderHeaders = dataset->Tables->Item[0];
    DataTable* orderDetails = dataset->Tables->Item[1];

    // get all headers
    DataRowCollection* headers = orderHeaders->Rows;
    DataRow* header;
    for (int i = 0; i < headers->Count; i++)
    {
      header = headers->Item[i];
      Console::WriteLine(String::Format(S"Order: {0}\t"
                                        S"Customer: {1}\t"
                                        S"Shipping Address: {2}",
                                        header->Item["OrderId"],
                                        header->Item["CustomerName"],
                                        header->Item["ShippingAddress"
                                          ]));
```

```
    // get all detail lines
    String* format = S"{0,-5} {1,-5} {2,-10} {3,-10} {4,-10}";
    Object* values[] = new Object*[5];
    values[0] = S"Order";
    values[1] = S"PID";
    values[2] = S"Product";
    values[3] = S"Price";
    values[4] = S"Qty";
    Console::WriteLine(String::Format(format, values));

    DataRelation* relation = dataset->Relations->Item[0];
    DataRow* details[] = header->GetChildRows(relation);
    for (int j = 0; j < details->Count; j++)
    {
      Object* values[] = new Object*[5];
      values[0] = details[j]->Item[S"Orders_Id"];
      values[1] = details[j]->Item[S"ProductId"];
      values[2] = details[j]->Item[S"ProductName"];
      values[3] = details[j]->Item[S"UnitPrice"];
      values[4] = details[j]->Item[S"Quantity"];
      Console::WriteLine(String::Format(format, values));
    }
    Console::WriteLine();
  }
}

    catch(Exception* e)
    {
      Console::WriteLine(e->Message);
    }
#pragma pop_macro("new")
}
```

This function now outputs the correct order information as follows:

```
Order: 1   Customer: Customer ABC  Shipping Address: ABC street, 12345
Order PID  Product     Price      Qty
1     1    Product 1   1          2
1     2    Product 2   2          2

Order: 2   Customer: Customer 2    Shipping Address: Someplace
Order PID  Product     Price      Qty
2     3    Product 3   3          3
```

Working with DiffGrams

A DiffGram is an XML document that is used to identify current and original versions of data elements. Because a DiffGram can be defined to contain only modified data rows, it is the most efficient means of transferring DataSet objects for the purpose of updating a remote data store. In fact, the DataSet internally uses DiffGrams to load and persist its contents, and to serialize its contents for transport.

A DataSet can be written as a DiffGram by specifying a value of Xml WriteMode::DiffGram when calling the DataSet::WriteXml method. When this is done, the resulting XML file is populated with all the necessary information to accurately recreate the contents, though not the schema, of the DataSet, including column values from both the original and current row versions, row error information, and row order. In order to produce a DiffGram containing only changed records, first call the DataSet::GetChanges method. This method returns a copy of the DataSet object containing only those records that have been inserted, modified, or deleted since the DataSet was filled or the last time the AcceptChanges method was called.

DiffGram Format

The DiffGram format is divided into three sections: the current data, the original data, and an errors section, as shown in the following example.

```
<?xml version="1.0" standalone="yes"?>
<diffgr:diffgram
   xmlns:msdata="urn:schemas-microsoft-com:xml-msdata"
   xmlns:diffgr="urn:schemas-microsoft-com:xml-diffgram-v1">

   <DataInstance>
   </DataInstance>

   <diffgr:before>
   </diffgr:before>

   <diffgr:errors>
   </diffgr:errors>

</diffgr:diffgram>
```

Here's a rundown of each of these data blocks.

- **<DataInstance>**—This element is either the DataSet name or a DataTable row. For example, since all of the examples in this chapter have constructed DataSet objects without a name, this value would be NewDataSet if we had generated DiffGrams for those files. This block of the DiffGram format contains the current data, whether it has been modified or not. An element, or row, that has been modified is identified with the diffgr:hasChanges annotation. For example, the following indicates that the row has been modified since its changes were last committed. Elements would then follow that would include the column names and data values for this row. The diffgr::id annotation is how you map the original value row back to this row to compare the differences.

```
<AllEmployees
   diffgr:id="Employees1"
   msdata:rowOrder="0" diffgr:hasChanges="modified"
>
```

- **<diffgr:before>**—This block of the DiffGram format contains the original version of any row that has been modified. Elements in this block are matched to elements in the <DataInstance> block using the diffgr.id annotation. Continuing the previous example, the following element would contain the original values for the element defined in the <DataInstance> block with a diffgr::id of "Employees1".

```
<AllEmployees diffgr:id="AllEmployees1" msdata:rowOrder="0">
```

- **<diffgr:errors>**—This block of the DiffGram format is used to contain error information for a particular row in the <DataInstance> block. Once again, the diffgr:id annotation can be used to map rows across blocks.

Example of Creating DiffGrams

Now let's look at how to create a DiffGram of some changes to the Northwind Employees table. In the following function, once the Employee rows are read into a DataSet, a single row is added, and another is modified. A DiffGram is then generated containing all rows (both changed and unchanged) with the XML saved to a file called all.xml. A second DiffGram

is then generated containing only the changed rows and saved to a file
called changes.xml.

```
void TestDiffGram()
{
#pragma push_macro("new")
#undef new
  try
  {
    SqlConnection* conn =
      new SqlConnection(S"Server=localhost;"
                         S"Database=Northwind;"
                         S"Integrated Security=true;");

    SqlDataAdapter* adapter =
      new SqlDataAdapter(S"SELECT EmployeeId, FirstName, LastName "
                         S"FROM Employees", conn);

    SqlCommandBuilder* commandBuilder = new
      SqlCommandBuilder(adapter);

    conn->Open();

    DataSet* dataset = new DataSet();

    adapter->Fill(dataset);

    conn->Close(); // No longer needed

    // Add a new record
    DataTable* table = dataset->Tables->Item[0];
    DataRow* row;
    row = table->NewRow();
    row->Item[S"FirstName"] = S"TestFirst";
    row->Item[S"LastName"] = S"TestLast";
    table->Rows->Add(row);

    // Modify a row
    row = table->Rows->Item[0];
    row->Item["FirstName"] = S"ChangeTest";

    dataset->WriteXml(S"all.xml", XmlWriteMode::DiffGram);
```

```
    DataSet* changes = dataset->GetChanges();
    if (changes)
      changes->WriteXml(S"changes.xml", XmlWriteMode::DiffGram);

    adapter->Update(changes);
  }
  catch(Exception* e)
  {
    Console::WriteLine(e->Message);
  }
#pragma pop_macro("new")
}
```

One important note to make is that there must be a primary key designation when creating DiffGrams. Omitting the EmployeeId from the SELECT statement would result in an InvalidOperationException stating the following:

```
Dynamic SQL generation for the UpdateCommand is not supported against a
SelectCommand that does not return any key column information.
```

Now let's look at the all.xml file containing both changed and unchanged records. The bold elements indicate the modified rows. Note both the diffgr:id and diffgr:hasChanges values.

```
<?xml version="1.0" standalone="yes"?>
<diffgr:diffgram
  xmlns:msdata="urn:schemas-microsoft-com:xml-msdata"
  xmlns:diffgr="urn:schemas-microsoft-com:xml-diffgram-v1">
  <NewDataSet>
    <AllEmployees diffgr:id="AllEmployees1" msdata:rowOrder="0"
      diffgr:hasChanges="modified">
      <FirstName>ChangeTest</FirstName>
      <LastName>Davolio</LastName>
    </AllEmployees>
    <AllEmployees diffgr:id="AllEmployees3" msdata:rowOrder="2">
      <FirstName>Janet</FirstName>
      <LastName>Leverling</LastName>
    </AllEmployees>
    .. other unchanged rows
    <AllEmployees diffgr:id="AllEmployees10" msdata:rowOrder="9"
```

```
      diffgr:hasChanges="inserted">
      <FirstName>TestFirst</FirstName>
      <LastName>TestLast</LastName>
   </AllEmployees>
 </NewDataSet>
 <diffgr:before>
   <AllEmployees diffgr:id="AllEmployees1" msdata:rowOrder="0">
      <FirstName>Nancy</FirstName>
      <LastName>Davolio</LastName>
   </AllEmployees>
 </diffgr:before>
</diffgr:diffgram>
```

The only difference between the `all.xml` and `changes.xml` file is that the latter doesn't contain the unchanged rows in the `<DataInstance>` block. At this point, you can merge the `changes` `DataSet` with another `DataSet` or send it to a remote server for processing.

Summary

In the previous three chapters, you learned how to work with ADO.NET and XML independently. However, there might be times when you will want to migrate one format to the other. As ADO.NET uses XML internally as its serialization mechanism, this is very easy to do and has been the subject of this chapter. The chapter began by illustrating how to serialize a `DataSet` containing a single `DataTable`—or multiple unrelated `DataTable` objects—to XML. From there, a second section illustrated how to perform this same task with parent/child related tables, bringing into play the `DataRelation` class you learned about in Chapter 7. Once you had seen how to serialize ADO.NET data to XML, you were then presented with a couple of sections that demonstrated how to format the serialized XML in terms of the generated XML node types and indentation options. After a section that covered writing schema (XSD) information for `DataSet` objects, you then saw how to fill a `DataSet` object from XML documents—both with and without included schema information. The chapter concluded with a section detailing how DiffGrams enable you to generate an XML file—containing both the current and original values of modified data—for more efficient transport of changed

data across a network connection. In the next chapter, we'll switch gears a little bit and move into the exciting area of remoting and specifically, how it provides significant advantages over using DCOM when sending MFC objects over a network connection or even within processes on the same machine.

Remoting

Introduction

.NET remoting is a technology that can be used to develop distributed applications. Basically in remoting you have the remoting servers, which provide the remoted objects that can be accessed and used by remoting clients. The remoting clients and the remoting servers may be in the same process, on the same machine, in different processes on the same machine, or on two different machines altogether—as long as the machines can access each other through the network. The best thing about remoting is that it totally abstracts the remoting communications process. There are multiple communication protocols that you can use, different types of data formatters, and different types of serialization mechanisms. For example, when a .NET application wants to talk to a native application, it can use an XML formatter for the communication, but two .NET applications can communicate to each other using a binary formatter, which will be slightly more efficient than the XML formatter. Thus there is an amazing amount of flexibility involved in remoting.

Remoting does not insist on a particular type of programming model. You can have remoting servers and clients running as Web services on a Web server, as Windows Services on a network server, as simple desktop applications, or even as console applications. They can be used for interprocess communication; for network communication; and in complex multiserver, multiclient distributed environments. Object activation and life-time management are all done for you by the .NET Framework, so you don't have to worry about all that. Security is built into the remoting framework, so you are spared the pains of providing a custom security model to your remoting applications. In this chapter we'll discuss how remoting solves some of the

problems associated with DCOM, how remoting compares with .NET Web services, and a few examples of how to use remoting as an interprocess communication mechanism both locally and across the network.

Issues with DCOM

DCOM is very effective and efficient when the distributed application runs only on a network where all participating computers are of similar type and use the same communication protocols. While this is quite all right in a small network, on a bigger network with multiple platforms, DCOM suddenly proves to be a problem because it lacks any decent support for interoperability across platforms and protocols. With DCOM you rarely have any control over the format in which the data is transmitted or the channels that are used to transmit the data. .NET remoting supports various data transmission and communication protocols, which makes it easy for programs running on other OS platforms to communicate with the distributed application's services. Unlike DCOM, which uses a proprietary binary format for data transfer, .NET can use any of several formats, including binary formats and XML formats. And since XML is a language that is universally understood by any language running on any operating system or platform, there is absolutely no problem for interoperability. .NET remoting even allows you to roll your own type of data format to use for communication, if you feel the need to do so.

Another major issue people faced when they used DCOM was the problem associated with going through firewalls. Most firewalls won't allow traffic through the ports that DCOM natively tries to use, and of course reconfiguring those firewalls to allow traffic through these ports would totally render the firewalls useless, and defeat the very purpose of having those firewalls. Even assuming that you manage to redirect DCOM traffic through the HTTP ports, a good many decent firewalls won't allow binary traffic through the HTTP port. With .NET remoting, you can use HTTP listeners with XML data formatters, and the firewalls won't even notice what's passing through, underneath their noses. Thus, remoting is essentially the improved version of DCOM for the .NET Framework, and it completely overrides DCOM, which can now be safely said to be obsolete. The only advantage DCOM will have is that DCOM will probably deliver much better performance because of its more efficient binary format and fast native running environment. But interoperability is usually far more

deserving of attention than is efficiency, and on the Internet the tiny bottle-necks in speed introduced by using remoting will be too infinitesimally small to be even noticed. And by specifying binary formatters, remoting can be made nearly as fast as DCOM applications.

Remoting and Web Services

Web services are preferred for Internet applications rather than remoting because Internet traffic usually goes through rather secure firewalls. Remoting can technically be made suitable for Internet applications by using HTTP, SOAP, and XML formatters, but typically that's what Web services are meant for. Thus, while remoting is more popular for intranet solutions, Web services are preferred for Internet solutions. Remoting can use both HTTP and TCP (as well as custom communication protocols), while Web services are strictly HTTP. Web services require a Web server, typically the IIS Web server that comes with Windows server editions. The basic advantage with remoting is that you do not need a Web server. Remoting servers run their own HTTP or TCP server and act as listeners. But the basic advantage with Web services is that because they use open standards such as WSDL and UDDI, any kind of client can easily access these Web services without having to know very much about how they are implemented. (See the sidebar "SOAP, UDDI, and WSDL" for definitions of these technologies.)

SOAP, UDDI, and WSDL

SOAP (Simple Object Access Protocol) is a simple, XML-based messaging protocol that is used to encode request and response messages that are transmitted across a network, and it is typically associated with Web services. SOAP is OS- and protocol-independent, and can thus be used on any OS and with any protocol, including but not limited to HTTP, SMTP, POP3, and so forth.

UDDI (Universal Description, Discovery, and Integration) is an OS-independent directory framework for Web services, which allows businesses to register their Web services on the Internet and allows clients searching for a particular Web service to be able to locate it easily. It's quite similar to a regular telephone directory's yellow pages section.

WSDL (Web Services Description Language) is an XML-based language used to describe the services offered by a Web service. WSDL, which was jointly developed by Microsoft and IBM, is extensively used by UDDI.

The problem with remoting servers is that you'll need to provide remoting clients, or at least provide very detailed specifications about how the server can communicate with remoting clients. Web services are stateless, which means that every request requires a new object to be instantiated. On the other hand, remoting allows us control over state management, thereby enabling the correlation of requests from the same client. On a general note, remoting is used for intranet applications and Web services, especially because remoting is pretty much faster and more efficient than Web services. (See the sidebar entitled "Using `sproxy.exe`" for information on the sproxy tool that ships with VS.NET 7.0/2003.)

Using `sproxy.exe`

For native C++ projects that want to consume an XML Web service, there is a command-line tool called `sproxy.exe` that will automatically generate C++ client code to access the Web service, based on the WSDL description for the Web service. The `sproxy.exe` tool can be found in the `\vc7\bin` directory of your Visual C++ .NET installation folder.

Demo: **Writing a Demo Remoting Application**

In this section we'll write a simple remoting server that remotes an object, and a remoting client that accesses and uses this object. The remoting server and client can be in different AppDomains, in different processes on the same computer, on different computers on the same network, or on different computers that can connect to each other through the Internet. It does not make any difference at all, as the remoting layer will handle it for us. For the example we'll design a simple object that will have two methods: one that returns a simple .NET object and another that returns a user-defined complex object.

Designing and Coding the Interface

The reason we need an interface is simple, and its function is similar to the use of interfaces in developing COM components. Basically to be able to use the remote object properly, a client program must know the structure, properties, and methods of the remote object. Obviously having the remote object and its containing library on the client to ensure that the client has

access to the remote object's definition is ridiculous and would defeat the whole purpose of remoting.

The big advantage with an interface assembly is that you simply distribute the interface assembly with the remoting clients. All the client needs is to be able to get the type of the remote object, and it can easily do that by using the interface library. It simply treats the remote object as the interface object and makes calls on it likewise. But under certain unusual circumstances, like when you actually need to work with objects and not with interfaces, the .NET remoting will allow you to do that, and this is a distinct advantage over traditional COM.

For our example, we have two methods in our interface, one that returns a `String` and another one that returns a user-defined object. We have to derive the user-defined class from `MarshalByRefObject` to allow it to be transferred across a remoting medium. `MarshalByRefObject` allows an object to be accessed across application domain boundaries in remoting scenarios. An alternative is to use `MarshalByValueComponent`, where a copy of an object is passed, but obviously this is less efficient, and we would usually prefer to have a reference to the actual object than a copy; in addition, this will only work in the case of simple value types. `MarshalByValue Component` objects make sense in remoting when you are transmitting raw data rather than trying to invoke object methods on the remote server. You'll get a serialization error if you try to pass managed objects by value across a remoting boundary, unless you mark them as serializable, either by using the `Serializable` attribute or by implementing `ISerializable`.

```
public __gc class Marks : public MarshalByRefObject
{
public:
  Marks()
  {
    Math = Mfc = Atl = 0;
  }
  Marks(int m, int f, int a)
  {
    Math = m;
    Mfc = f;
    Atl = a;
  }
  char GetGrade()
  {
    char g = 'F';
```

```
    if( (Math > 50) && (Mfc > 50) && (Atl > 50) )
    {
      g = 'C';
      int tot = Math + Mfc + Atl;
      if(tot > 240)
        g = 'A';
    else if(tot > 210)
        g = 'B';
    }
    return g;
  }
private:
  int Math;
  int Mfc;
  int Atl;
};

public __gc __interface ISimpleInterface
{
  System::String* GetStudentName(int rollnumber);
  Marks* GetMarks(int rollnumber);
};
```

As you can see, we have the ISimpleInterface, which has the two methods I mentioned earlier, and the Marks object, which is derived from the MarshalByRefObject class. This is necessary because we will get a serialization exception if we try to invoke the method on a remoted object that returns this object. As a result, keep in mind that you must not only derive your remoted object from MarshalByRefObject, but you also need to derive all objects that are returned by any of the primary object methods from MarshalByRefObject. The interface and the definition of our custom class are compiled into a class library, which can then be distributed to machines hosting the remoting clients and machines hosting the remoting servers.

Implementing the Remoting Server

In the remoting server, the first thing we need to do is to add a reference to the interface library that defines the interface for the remoted object. Then we need to implement a class that derives from this interface. The class will also have to derive from MarshalByRefObject to allow the object to be transferred across remoting boundaries. Just in case you are wondering

how you can have multiple inheritance in a managed class, remember that only `MarshalByRefObject` is a class; `ISimpleInterface` is an interface and not a class.

```
public:
  System::String* GetStudentName(int rollnumber)
  {
    return String::Concat(S"Student ", rollnumber.ToString());
  }

  Marks* GetMarks(int rollnumber)
  {
    int math = 75;
    int atl = rollnumber < 10 ? 70 : 90;
    int mfc = 77;
    return new Marks(math,mfc,atl);
  }
};
```

Well, the functions are simply some dummy code to try to return some meaningful values to the remoting client. In a real-life scenario, the server would probably query some kind of database or even contact another remoting server and form a kind of remoting server chain before sending a reply back to the remoting client. Now, since we have implemented the object that will be remoted, we can set up the remoting server. If you think that's going to be a lot of code, you are in for a big surprise.

```
TcpChannel* TcpChan = new TcpChannel(9999);

ChannelServices::RegisterChannel(TcpChan);

RemotingConfiguration::RegisterWellKnownServiceType(
  Type::GetType(S"StudentInfo"),
  S"SimpleRemotingDemo",WellKnownObjectMode::SingleCall);

Console::WriteLine(S"Press any key to stop server");
getch();
```

Well, that wasn't very long now, was it? The `TcpChannel` creates a remoting channel in which data is transmitted using the TCP protocol with a binary formatter. At its core, it uses TCP sockets to send and receive data, and it uses SOAP for data serialization. We have passed 9999 to the

constructor, which means our remoting server will listen for TCP connections on port 9999. This is not exactly a firewall-friendly port, but this is OK for intranet-based remoting applications. Once we have created our `TcpChannel` object, we use `ChannelServices::RegisterChannel` to register our `TcpChannel` with the channel services. Until we register a channel, we cannot remote objects through the channel. Now we use `RemotingConfiguration::RegisterWellKnownServiceType` to register our object (that is to be remoted) as a well-known type.

```
RemotingConfiguration::RegisterWellKnownServiceType(
   Type::GetType(S"StudentInfo"),
   S"SimpleRemotingDemo",WellKnownObjectMode::SingleCall);
```

We use the following override of the `RegisterWellKnownServiceType` function:

```
static void RegisterWellKnownServiceType (
   Type* type,
   String* objectUri,
   WellKnownObjectMode mode
);
```

The first argument is the type of the object to be remoted, which in our case is `StudentInfo`, which implements the `ISimpleInterface` interface as well as the `MarshalByRefObject` class. The second parameter is a string that is used to form the URI that the remoting clients use to access the server. The third parameter is a `WellKnownObjectMode` enumeration member, which can be `SingleCall` or `Singleton` (as shown in Table 9–1).

We have specified `SingleCall`, which means every incoming message is serviced by a new instance of the object. If you specify `Singleton`, a single object is maintained at the server and is used to service every

Table 9-1 `WellKnownObjectMode` Enumeration Values

Member	Description
SingleCall	Every request results in the instantiation of a new object.
Singleton	A single object is instantiated at the first request and is persisted until a timeout occurs.

incoming message. When the first call is made, an object is instantiated and then proxied to the client caller. If the `WellKnownObjectMode` used is `SingleCall`, then the object is destroyed after the remote client has invoked the method, and a new object is created each time a method is invoked. If the `WellKnownObjectMode` used is `Singleton`, then the object is kept alive for further calls from the client. Internally there will be a maximum timeout value, which, when exceeded, will result in the object being destroyed. Once we have set up the remoting service as above, by registering it as a known service, any remoting client can access the object if it knows the named URI for the service. The URI will be in the following form:

```
tcp://[hostname]:[port]/[string name we used]
```

which in our case is:

```
tcp://earth:9999/SimpleRemotingDemo
```

Implementing the Remoting Client

Setting up a remoting client is easy. The first thing to do is to add a reference to the interface library DLL, so that the client understands the type of the object that is to be accessed remotely. Once we have done this, accessing the remote object is as simple as the following:

```
TcpChannel* TcpChan = new TcpChannel();
ChannelServices::RegisterChannel(TcpChan);

ISimpleInterface* RemoteObject =
  __try_cast<ISimpleInterface*>
    (Activator::GetObject(Type::GetType(S"ISimpleInterface,"
      S"SimpleInterface"),
    S"tcp://earth:9999/SimpleRemotingDemo"));
```

Basically we instantiate a `TcpChannel` object, but unlike in the case of the server, we do not need to specify a port. Just as in TCP socket-based client-server applications, the server will be listening on a specified port, but the clients will use any random available port to make the TCP connections as and when required. In this case a random free channel is chosen for us by the .NET Framework layer. Once we have instantiated the `TcpChannel` object, we need to register the channel with channel services using

`ChannelServices::RegisterChannel,` just as we did in the remoting server application. Now that we have registered the channel, we simply use `Activator::GetObject` to get a reference to the remote object:

```
ISimpleInterface* RemoteObject = __try_cast<ISimpleInterface*>
    (Activator::GetObject(Type::GetType(S"ISimpleInterface",
        S"SimpleInterface"),
    S"tcp://earth:9999/SimpleRemotingDemo"));
```

We use the following override of the `Activator::GetObject` method:

```
static Object* GetObject (
  Type* type,
  String* url
);
```

The first argument is the type of the well-known object that we want to access across the remoting boundary. As you can see, we specify the `ISimpleInterface` as the type of the object because the remote object implements this interface, and we don't have to specify the actual type of the remote object. The second argument is the URI for the remote object on the server, as mentioned earlier in the previous section. Once you have done this much, you now have a reference to the object, and you can make calls on the object just as if it was on your local machine:

```
Console::WriteLine(RemoteObject->GetStudentName(1));
getch();

Marks* m = RemoteObject->GetMarks(1);

Console::WriteLine(Char::ToString(m->GetGrade()));
```

Well, it couldn't have gotten any better and simpler than this, eh? Try to think about how much more complicated and long-winded this same application would have been if you had used DCOM. In real-life scenarios, the client can then make method calls that might update a database on the server. Typically, the remoting server will be written as a Windows Service and will run on a server machine, while the remoting clients are usually written as GUI applications that run on multiple client workstations. Well, that's that, and you have created your first remoting application with a remoted object that has methods that return both simple and complex objects.

Understanding When Objects Get Constructed

Well, basically you have seen that a remote client can request an object from a remoting server, and then the remoting server gives it to the client. All that is well and good, but obviously as a programmer you need to understand the life-time of the object. If you remember, this was how we set up the remoting server to listen for connections:

```
RemotingConfiguration::RegisterWellKnownServiceType(
     Type::GetType(S"StudentInfo"),
     S"SimpleRemotingDemo",WellKnownObjectMode::SingleCall);
```

The interesting bit is the last parameter, where we specify the `SingleCall` mode. When we configure the remoting server to use the `SingleCall` mode, the object is created each time the client invokes a method on the object. But an object is not created when the client makes a call to `Activator::GetObject`. In fact we can easily verify this now; first add a constructor to our remoted object on the server:

```
public __gc class StudentInfo :
  public ISimpleInterface, public MarshalByRefObject
{
public:
  StudentInfo()
  {
    Console::WriteLine(S"I just got constructed");
  }
```

Now make the following changes to the client code; basically we add a few `getch()` statements so we can trace through the program flow manually:

```
ISimpleInterface* RemoteObject =
static_cast<ISimpleInterface*>
     (Activator::GetObject(Type::GetType(S"ISimpleInterface,"
        S"SimpleInterface"),
        S"tcp://earth:9999/SimpleRemotingDemo")));

getch();

Console::WriteLine(RemoteObject->GetStudentName(1));
getch();
Marks* m = RemoteObject->GetMarks(1);
getch();
```

OK. Now run the server in a DOS window and run the client in another DOS window. You'll see that a new object gets constructed on the server twice, first when you call the `GetStudentName` method and then when you make the call to `GetMarks`. But you'll see that no object is constructed when you use `Activator::GetObject`. Try running multiple instances of the client program, and you'll see that a new object gets created for each method invocation you do.

Of course, sometimes this might not be what you want; you might actually prefer to persist the object on the server. In that case you must use `WellKnownObjectMode::Singleton` when you start the remoting server:

```
RemotingConfiguration::RegisterWellKnownServiceType(
  Type::GetType(S"StudentInfo"),
  S"SimpleRemotingDemo",WellKnownObjectMode::Singleton);
```

Now run one or more clients, and you'll see that only one object gets created. The object is created when the first client makes the first method call, and this object is maintained throughout the life-time of the server.

Wrapping MFC Objects for Remoting

One serious limitation when using remoting is that you cannot remote MFC objects across application domains. Consider the following two classes that we can attempt to remote:

```
namespace SecondDll
{
  public __gc class StrWrapper : public MarshalByRefObject
  {
  public:
    StrWrapper(CStringArray* list)
    {
      m_list = list;
    }
    CStringArray* m_list;
  };


  public __gc class WebList : public MarshalByRefObject
  {
```

```
private:
  CStringArray* m_list;
public:
  WebList()
  {
    m_list = new CStringArray();
    m_list->Add(S"Amazon");
    m_list->Add(S"Rediff");
    m_list->Add(S"Yahoo");
    m_list->Add(S"Google");
  }
  ~WebList()
  {
    delete m_list;
  }
  CStringArray* GetList()
  {
    return m_list;
  }
  StrWrapper* GetList2()
  {
    return new StrWrapper(m_list);
  }
};


}
```

Essentially we have two methods, one that returns a CStringArray* and another that returns a managed object that itself has a public CStringArray* member variable. Now assume we have a remoting server that remoted this object:

```
TcpChannel* TcpChan = new TcpChannel(9988);

ChannelServices::RegisterChannel(TcpChan);

RemotingConfiguration::RegisterWellKnownServiceType(
      Type::GetType(S"SecondDll.WebList,SecondDll"),
        S"SecondRemotingDemo",WellKnownObjectMode::SingleCall);

Console::WriteLine(S"Press any key to stop server");
getch();
```

And a client that accesses it:

```
TcpChannel* TcpChan = new TcpChannel();
ChannelServices::RegisterChannel(TcpChan);

#undef GetObject

WebList* RemoteObject = static_cast<WebList*>
   (Activator::GetObject(Type::GetType(S"SecondDll.WebList,
SecondDll"),S"tcp://earth:9988/SecondRemotingDemo")));

//CStringArray* sa = RemoteObject->GetList();
StrWrapper* sw = RemoteObject->GetList2();
//CString s = sw->m_list->GetAt(0);

Console::WriteLine(S"Press any key to quit");
getch();
```

Just try running it after uncommenting the call to GetList(). You will get a System.NotSupportedException because it is simply not possible to remote a CStringArray* across a remoting boundary. But you'll notice that the call to GetList2 works, which might make you feel that you have successfully remoted a managed object that wraps a CStringArray*. But the problem is not there at all. The basic issue is that the object you have is a reference object; thus, when you try to access the CStringArray* member, that's when the remoting framework attempts to transfer the requested object across the application domain boundaries. Uncomment the line of code that tries to access the CStringArray* object, and you'll get a System.Runtime.Serialization.SerializationException. Perhaps in future versions of .NET and MFC, Microsoft might add some features to the MFC classes that will allow us to remote them across application boundaries. Until then, what we can do is to write wrapper classes that not only wrap an MFC object but also every function in the MFC object that might be needed at the remote client. In our above example, supposing that the remote client needs to access the list members using an index, then we'd need to wrap the GetAt() method of the CStringArray object:

```
public __gc class StrWrapper : public MarshalByRefObject
{
public:
  StrWrapper(CStringArray* list)
  {
```

```
      m_list = list;
   }
   String* GetAt(int index)
   {
      return m_list->GetAt(index);
   }
private:
   CStringArray* m_list;
};
```

It's basically a matter of making a corresponding call to the private `CStringArray` object and returning a type that is convertible to the corresponding native type—in this case we have returned a `String` object, which implicitly converts to a `CString` object at the client side. Now the client can access this method in the following manner:

```
#undef GetObject
WebList* RemoteObject = static_cast<WebList*>

(Activator::GetObject(Type::GetType(S"SecondDll.WebList, SecondDll"),
    S"tcp://earth:9988/SecondRemotingDemo"));

StrWrapper* sw = RemoteObject->GetList2();
CString s1 = sw->GetAt(1);
_tprintf("%s\r\n",s1);
```

We only need to wrap those methods in the MFC object that we'd be accessing at the client code. An important point to note is that the MFC object exists only on the server, and any method you invoke on the wrapper object will eventually execute the corresponding method on the MFC object on the server side and not on the client side.

Selecting Remoting Channels

In our examples above we have been using the `TcpChannel` class for our remoting server and client. The `TcpChannel` is definitely a faster channel for remoting as it uses a binary formatter internally and you can also specify the TCP port to use for remoting. But obviously it has a problem when going through firewalls. As mentioned previously, however, in .NET remoting you are not restricted to any particular protocol. In this particular case

where you need to go through networks that are firewalled, you can use the HttpChannel class, which transports messages across remoting boundaries using the SOAP protocol, which means it basically uses XML formatted data over an HTTP connection. Obviously, this will not be as efficient as using a TCP channel, and Web services are more suitable for Internet-based distributed applications anyway. So for most purposes your best option is to use the TcpChannel class, unless you have some very strong requirements to do otherwise. Using an HttpChannel is almost identical to using a TcpChannel, as the following code snippet will show you, except that internally the framework will be using SOAP and XML to transport your remote objects across the boundaries.

```
__gc class RemObj : public MarshalByRefObject
{
public:
  String* GetName()
  {
    return S"Anakin Skywalker";
  }
};
```

This is a very simple object which we will use as our remoted object.

```
int _tmain(int argc, char**argv)
{
  if(argc == 2)
  {
    Console::WriteLine(argc.ToString());
    ClientCode();
    return 0;
  }
  HttpChannel* hchan = new HttpChannel(8080);

  ChannelServices::RegisterChannel(hchan);

  RemotingConfiguration::RegisterWellKnownServiceType(
    Type::GetType(S"RemObj"),
    S"HttpChanDemo",WellKnownObjectMode::SingleCall);

  Console.:WriteLine(S"Press a key to quit server");
  getch();
  return 0;
}
```

```
void ClientCode()
{
   HttpChannel* hchan = new HttpChannel();
   ChannelServices::RegisterChannel(hchan);

   RemObj* RemoteObject = static_cast<RemObj*>
      (Activator::GetObject(Type::GetType(S"RemObj"),
         S"Http://earth:8080/HttpChanDemo"));

      Console::WriteLine(RemoteObject->GetName());

      Console::WriteLine(S"Press a key to quit client");
      getch();
}
```

As you can see, I have put the server code and the client code in a single executable, as this saves the hassle of having to put the remoted class in a separate DLL. The problem is that to get the type of the object, both server and client will need to access the same assembly, else even if you use the same code byte by byte, the type of the class will be different in each file. Basically what this means is that two identical classes in two different executables or DLLs will not be the same type of object. Of course the right solution is to create an interface DLL and to use that, but for the sake of demonstration, I decided to keep the example simple and to the point. As you can see from the code, there is not much difference from creating a `TcpChannel`. Thus you can at any point in time during your development cycle decide to switch between TCP channels and HTTP channels; this really adds flexibility to your application. If your first client has an intranet-only requirement, you can go for speed and use the TCP channel, but if the second client wants the same application on the Internet, you can sacrifice a little speed and go for the HTTP channel. You'd only have to change a very small percentage of the total code in your application.

Using Asynchronous Callbacks

Most of you who have done any kind of socket programming will understand the problems associated with blocking sockets, and in our previous examples we face this exact problem, where a remote method might take quite a while to return, and then our program will simply freeze until the

remote call returns. This is not at all a favorable kind of scenario and one to be totally avoided. An idea that might strike you instantly is to start a worker thread to make remote calls, but as the number of remote calls increases, this soon becomes a major headache as far as maintenance and data synchronization goes. Luckily, remoting provides us with the ability to use asynchronous callbacks. Basically what this means is that you have a delegate that holds a pointer to the remote method, and you invoke the delegate asynchronously, passing a callback method that gets called when the remote method finishes execution. Thus your main thread is left free to continue whatever it is that it was doing before getting rudely interrupted by slow-executing remote calls.

Essentially, asynchronous remoting is nearly identical to asynchronous programming in a nonremoting scenario. The remoting framework does not do anything special to provide asynchronous remoting capabilities. The caller code takes all the required steps to establish asynchronous remoting. In fact the caller (the remoting client) can decide whether to make a call asynchronously or synchronously, and the server need not even know what mode the call is being made in.

The following sample application should make things clearer. First we define a simple remotable object to use in our sample and put this in a DLL. We avoid interfaces for the sake of keeping the sample simple.

```
namespace AsyncDemoDll
{
public __gc class TheClass : public MarshalByRefObject
  {
  public:
    String* GetHello()
    {
      return S"Hello World";
    }
  };
}
```

The server in our demo is no different from normal remoting servers. In fact the server need not even know whether the clients are accessing it asynchronously or not.

```
TcpChannel* TcpChan = new TcpChannel(9977);
ChannelServices::RegisterChannel(TcpChan);
```

```
RemotingConfiguration::RegisterWellKnownServiceType(
  Type::GetType(S"AsyncDemoDll.TheClass,AsyncDemoDll"),
    S"AsyncDemo",WellKnownObjectMode::SingleCall);
```

And this is how we write the asynchronous client:

```
__delegate String* RemoteMethodDelegate();
```

We first declare a delegate that matches the remote method we are planning to invoke.

```
int _tmain()
{
  TcpChannel* TcpChan = new TcpChannel();
  ChannelServices::RegisterChannel(TcpChan);


  TheClass* RemoteObject = static_cast<TheClass*>
  (Activator::GetObject(Type::GetType(S"AsyncDemoDll.TheClass",
    S"AsyncDemoDll"),
    S"tcp://earth:9977/AsyncDemo"));

  RemoteMethodDelegate* rmd = new
    RemoteMethodDelegate(RemoteObject,
      &TheClass::GetHello);

  rmd->BeginInvoke(new AsyncCallback(0, &CallBackClass::MyCallBack), 0);

  Console::WriteLine(S"Press any key to quit");
  getch();
  return 0;
}
```

Basically the code is identical to our previous remoting client up to the point where we call `Activator::GetObject`. After that, basically what we do is declare our delegate object and point it to the method on the remoted object that we need to call. And we asynchronously invoke the delegate, passing to it our custom callback function, which will get called once the remote method has finished.

```
__gc class CallBackClass
{
```

```
public:
  static void MyCallBack(IAsyncResult* ar)
  {
    //First we cast the IAsyncResult* to AsyncResult*
    AsyncResult* ares = __try_cast<AsyncResult*>(ar);
    // Now we obtain the RemoteMethodDelegate object
    // from the AsyncResult object using its AsyncDelegate
    // property
    RemoteMethodDelegate* rmd =
      __try_cast<RemoteMethodDelegate*>(ares->AsyncDelegate);
    //Obtaining the result of the delegate invocation
    //is now a matter of calling EndInvoke
    Console::WriteLine(rmd->EndInvoke(ar));
  }
};
```

We also need to define a method that will act as our callback function. As you can see the callback function receives an `IAsyncResult` object, and we get back our delegate object with some casting and by using the `AsyncDelegate` property, and then call `EndInvoke` on it. Keep in mind that the remote function has already terminated, and its termination has nothing to do with `EndInvoke`, which simply gives back the results of the operation. If you compile and run the server on one console and the client on another console, you'll see something like this on the client console:

```
Press any key to quit
Hello World
```

As you can see from the output, the main thread continues execution, which is why the "Press any key to quit" string gets printed first, and then the remote method returns, and we get the string that's returned from the remote method. You must note that this delay occurred despite the client and server running on the same machine, so you can just try to imagine how much slower this will be if the client and server are on a slow, congested network and the call involves database access. Obviously you'd be much better off with asynchronous calls in such situations.

Returning Objects to the Server

So far, we've seen situations where we get a remote reference to an object on the remote server and make calls on it. Now there might be times when you want to pass a local object to the remote server, so that the remote object can access the local object. We'll see such an example now, where we pass an object by reference to the server and the server objects manipulate this object.

The Remoted Objects

As usual we define our remotable objects in a library that both client and server can share. This time we have two objects: the object that is remoted by the server and the object that is passed from the client to the server. Let's look at the object that is passed by reference to the server:

```
public __gc class MyClass   : public MarshalByRefObject
{
public:
  MyClass(IntPtr h)
  {
    m_hWnd = h;
  }
  void MyMethod(int i)
  {
    PostMessage((HWND)m_hWnd.ToPointer(),
      WM_APP+1, NULL, (LPARAM)i);
  }
  IntPtr m_hWnd;
};
```

Note that the constructor uses an HWND instead of a CWnd pointer because the CWnd class is not designed to be thread-safe, and it would be dangerous to access a CWnd object from outside the thread that created it.

Also, at first glance this code might not seem to do much—simply posting a message to the specified window handle. And obviously this window handle is passed to it by the remoting client and not by the server. Basically this is just great, because we are actually constructing an object at the client side and passing it to the remote server, so whatever calls the remote server

makes on this object get executed at the client. Now let's look at the basic remoted object:

```
public __gc class CBDemo : public MarshalByRefObject
{
public:
  void CheckPass(MyClass* mc)
  {
    for(int i = 0; i < 40; i++)
    {
    if ( (i % 7) == 0 )
    mc->MyMethod(i);
    }
  }
};
```

As you can see, it has a method called CheckPass that accepts a MyClass object as argument; this object will be passed to it from the client, and it will be a client-side object. When it invokes the MyMethod call on the MyClass object, the call gets executed at the client and not on the server.

Implementing the Remoting Server

There are a few small changes to the server code, as you will soon see. The basic issue is that, starting with the .NET Framework 1.1 release, the security has been tightened so that you can't simply pass a client-side object to the server and expect it to work. You have to manually bring down the security level of the server to be able to pass remote objects by reference. The reason for beefing up security is that the remoting system relies on run-time type validation, and malicious programs can take advantage of this to pass in rogue objects that might eventually be able to take control of your server. Anyway, in our case, since we know exactly what the client code is going to do, we can safely bring down the security levels:

```
int _tmain()
{

  BinaryServerFormatterSinkProvider *provider =
    new BinaryServerFormatterSinkProvider();
  provider->TypeFilterLevel =
  System::Runtime::Serialization::Formatters::
    TypeFilterLevel::Full;
```

```
IDictionary *props = new Hashtable();
props->set_Item(S"port",__box(7777));

TcpChannel* TcpChan = new TcpChannel(props,0,provider);

ChannelServices::RegisterChannel(TcpChan);

RemotingConfiguration::RegisterWellKnownServiceType(
   Type::GetType(S"CallBackDll.CBDemo,CallBackDll"),
   S"CBDemo",WellKnownObjectMode::SingleCall);

Console::WriteLine(S"Press any key to stop server");
   getch();
   return 0;
}
```

Basically all we have to do is to change the serialization level of the `BinaryServerFormatterSinkProvider` object and use an overload of the `TcpChannel` constructor that accepts a `BinaryServerFormatterSink Provider` as argument. Now we can happily pass objects by reference from the clients to the server.

Implementing the Remoting Client

For the sake of the demo, I have created a simple MFC dialog-based application that will create a `MyClass` object and pass it to the remote server. We need to register the TCP channel in the `OnInitDialog` as shown below:

```
#undef new
   TcpChan = new TcpChannel(7766);
   ChannelServices::RegisterChannel(TcpChan);
```

As you can see there is slight difference in the code we used. This time we pass a port number to the `TcpChannel` constructor, because when the remote server tries to access the local object that we have passed to it, it needs a listening port on the client to be able to connect to it. Other than this there is not much difference in the way the remoting client is coded.

```
void CCallBackClientDlg::OnBnClickedStart()
{
#undef GetObject
```

```
m_listbox.ResetContent();

CBDemo* RemoteObject = static_cast<CBDemo*>
(Activator::GetObject(Type::GetType("CallBackDll.CBDemo,
   CallBackDll"),
   S"tcp://earth:7777/CBDemo")));

   MyClass* mc = new MyClass(m_hWnd);
   RemoteObject->CheckPass(mc);
}
```

We simply construct the MyClass subject and pass the HWND of the dialog to the constructor. Then we pass this object to the remote object's CheckPass method. Now we need to handle the message that will be posted to our window, so we add the following line to the message map:

```
BEGIN_MESSAGE_MAP(CCallBackClientDlg, CDialog)
      .  .  .
      ON_MESSAGE( WM_APP+1, OnCallBack )
      .  .  .
END_MESSAGE_MAP()
```

And we add the message handler:

```
LRESULT CCallBackClientDlg::OnCallBack( WPARAM wParam, LPARAM lParam)
{
   int y = (int) lParam;
   CString str;
   str.Format_T("%d has passed",y);
   m_listbox.AddString(str);
   return 0;
}
```

Basically, the message handler acts here as a callback function for us. The server code posts messages, and we handle them here at the client side in the message handler. This can be really useful in many situations where we will be able to set the properties of an object only from the client side. And of course the biggest advantage is that the calls execute on the client rather than on the server, allowing a sort of two-way traffic between the client and the server.

Using Remoting in Interprocess Communication

So far we have seen the use of remoting as a client-server technology, where there are remoting servers and remoting clients that connect to them and access objects remotely across a network on an intranet or even the Internet. But one very powerful use of remoting is an interprocess communication (IPC) mechanism. Now you actually have the ability to transfer a full object across application boundaries. In this section we'll do a simple application that demonstrates one way of doing this. Let's assume you have an MFC dialog-based program that asks for a user name before the program is run. Now suppose you have another MFC-based application that keeps logging all users who use the former application as well as the time at which they used it. We'll have the logging application remote an object on a TCP port, and we'll have the dialog applications all connecting to this remoting server and sending in the details of the user who just logged in.

The Remoted Object

Let's take a look at how we will implement the object that will be remoted.

```
#define WM_INFO WM_APP + 1
namespace MfcShare
{
public __gc class ProgInfo : public MarshalByRefObject
  {
  public:
    void Inform(String* User)
    {
      CString strUser(User);
        if(m_hWnd != NULL)
          SendMessage((HWND)m_hWnd.ToPointer(), WM_INFO,  NULL,
            (LPARAM)strUser.GetBuffer(0));
        strUser.ReleaseBuffer();
    }
    static IntPtr m_hWnd = NULL;
  };
}
```

As usual we have an assembly where we define our remoted object, with a single method called Inform. The class also has a static member m_hWnd, to

which we will later assign the window handle of our remoting server program's view window. We then use SendMessage to send the user's info to this window. We use SendMessage instead of PostMessage because we need to ensure that the CString variable does not get destroyed until the message has been handled. And if you are wondering whether the LPCSTR returned by GetBuffer will be valid, remember that this assembly will be loaded into the address space of the remoting server process, and thus the address of the string will be perfectly valid in the view class.

Implementing the Remoting Server

We will create a default MFC SDI application and select CListView as our view class, using the Application Wizard. We will then start the remoting server in the OnInitialUpdate method of the view class:

```
void CMfcServerView::OnInitialUpdate()
{
  CListView::OnInitialUpdate();

  ProgInfo::m_hWnd = m_hWnd;

  // TODO: You may populate your ListView with items
  // by directly accessing
  //   its list control through a call to GetListCtrl()

  GetListCtrl().ModifyStyle(0,LVS_REPORT);
  GetListCtrl().InsertColumn(0,_T("User"),LVCFMT_LEFT,250);
  GetListCtrl().InsertColumn(1,_T("Time"),LVCFMT_LEFT,750);

#undef new
  TcpChannel* TcpChan = new TcpChannel(8888);

  ChannelServices::RegisterChannel(TcpChan);

  RemotingConfiguration::RegisterWellKnownServiceType(
    Type::GetType("MfcShare.ProgInfo,MfcShare"),
    S"ProgInfo",WellKnownObjectMode::SingleCall);

}
```

We just change the list view to report mode and add the two columns we require. Then we start a TCP remoting server on port 8888. In addition

to starting the server, we also need to process the custom message that will be sent when a client connects to the server. So we add this entry to our message map:

```
BEGIN_MESSAGE_MAP(CMfcServerView, CListView)
  ON_MESSAGE( WM_INFO, OnProgInfo )
END_MESSAGE_MAP()
```

And we also write the handler as follows:

```
LRESULT CMfcServerView::OnProgInfo( WPARAM wParam, LPARAM lParam)
{
  CString strUser((LPTSTR)lParam);
  int i = GetListCtrl().InsertItem(0,strUser);
  CTime t = CTime::GetCurrentTime();
  CString strTime;
  strTime.Format(_T("%02d:%02d:%02d"),t.GetHour(),t.GetMinute(),
    t.GetSecond());
  GetListCtrl().SetItemText(i,1,strTime);
  return 0;
}
```

We just cast back the LPARAM to an LPTSTR and construct a CString, using that to get the user's details. Then we insert this into the first column and insert the current time into the second column. There is nothing out of the ordinary there—just nice, old-fashioned MFC code in there.

Implementing the Remoting Client

The client is a simple dialog-based application that pops up a child dialog that prompts the user for his or her name. This information is then sent to the logging application by remoting to it. Basically all the interesting code is in the OnInitDialog:

```
BOOL CMfcClientDlg::OnInitDialog()
{
  CDialog::OnInitDialog();

  // Set the icon for this dialog.  The framework does this
  // automatically
  //  when the application's main window is not a dialog
  SetIcon(m_hIcon, TRUE);            // Set big icon
  SetIcon(m_hIcon, FALSE);           // Set small icon
```

```
#pragma push_macro("new")
#pragma push_macro("GetObject")

#undef new
#undef GetObject

  TcpChannel* TcpChan = new TcpChannel();
  ChannelServices::RegisterChannel(TcpChan);

  ProgInfo* RemoteObject = static_cast<ProgInfo*>
  (Activator::GetObject(Type::GetType("MfcShare.ProgInfo",
    S"MfcShare"),
      S"tcp://earth:8888/ProgInfo"));

  CLoginDlg lDlg;
  if(lDlg.DoModal() == IDOK)
    RemoteObject->Inform(lDlg.m_user);
  else
    EndDialog(0);

#pragma pop_macro("GetObject")
#pragma pop_macro("new")

  return TRUE;
}
```

Now we'll compile and run the server (the CListVew-based SDI application), then start a few instances of the client (the Dialog application), and you'll see how the names entered into the Dialog application's login prompt get logged by the SDI application (see Figure 9–1). Of course in our example we used a rather simple object to transfer data, but you can basically remote any object that is remotable, and this makes it a very powerful and flexible mechanism for interprocess communication.

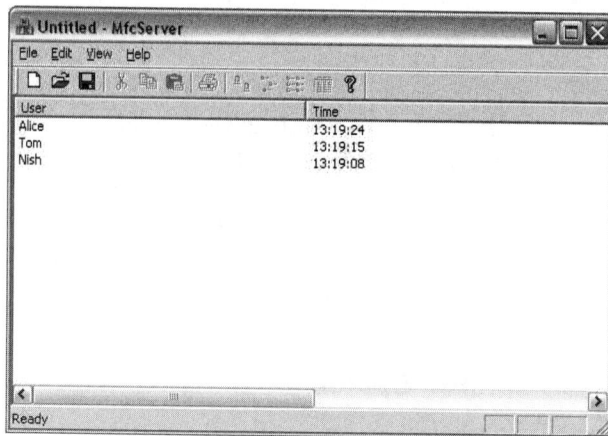

Figure 9-1 The SDI demo application lists all users who have logged in using the Dialog client.

Implementing a Remoting Server in a Windows Service

So far we've seen simple examples where the remoting server is in a console application or on an MFC GUI application. But for real-life purposes, it's best to put the remoting server in a Windows Service. Basically Windows Services start before the user is logged in; this comes in mighty handy, because remoting servers might usually be run on Network servers on which users typically don't remain logged in except for administrative purposes. In this section we'll see how to write a Windows Service and put a remoting server in it. We'll use the service classes provided by the .NET Framework to write our service. The demo program shows how to write a simple expiration service, where a program can run n number of times before its license expires. The example here simply writes a count value into the registry; thus, it is not a realistic protection for shareware and trialware authors, but is just used to demonstrate the running of a remoting server as a service.

We'll also see one other major use for employing a service: by having the service run as Local System account, we can have restricted user-level programs communicating to the service, using remoting and accomplishing tasks for which they would otherwise have required administrative privileges. This is one area where MFC programmers can save a lot of time by

using remoting. Programs often have a problem when the user is running in a restricted access level and is unable to do something on the system that requires administrative rights. The simple solution is to have a Windows Service that is running as Local System account that will have a remoting server listening; now all the restricted-user-owned programs have to do is to obtain a remote object through remoting from the Windows Service and make calls on it. The object executes remotely, and not at the client, so the object methods get invoked with the privileges of the Local System account, which means we now have full rights on the machine.

The Remoted Object

The remoted object is defined in a .NET class library DLL:

```
namespace ThirdDll
{
 public __gc class Expirer : public MarshalByRefObject
 {
 public:
  int IncrementRun(int MaxRuns)
  {
    int count = 1;
    RegistryKey* rkey = Registry::LocalMachine;
    RegistryKey* rkey1=rkey->OpenSubKey(
      S"SOFTWARE\\AbsoluteSoftware",true);
    if(rkey1 == 0)
    {
      rkey1 = rkey->CreateSubKey(S"SOFTWARE\\AbsoluteSoftware");
      rkey1->SetValue(S"Count",count.ToString());
    }
    else
    {
      String* tmp = static_cast<String*>(rkey1->GetValue(S"Count"));
      count = Convert::ToInt32(tmp);
      count++;
      rkey1->SetValue(S"Count",count.ToString());
    }
      rkey1->Close();
      if(count > MaxRuns)
        return -1;
      return count;
```

```
        }
    };
}
```

Basically what we do is to write a registry value into a subkey under HKEY_LOCAL_MACHINE. If the key does not exist, we create it, or else we read the current value of the run-count, increment it, write the new value, and return TRUE or FALSE, depending on whether the count has exceeded the maximum runs allowed. For registry access, I have made use of the Registry and the RegistryKey classes, both of which are defined under the Microsoft::Win32 namespace and make registry access and modification a real piece of cake. (Accessing the Windows Registry is covered in Chapter 3.)

Implementing the Remoting Server

Create a new project using the application wizard and select the Windows Service (.NET) project type (see Figure 9–2). Basically the wizard generates a class for you derived from System::ServiceProcess::ServiceBase.

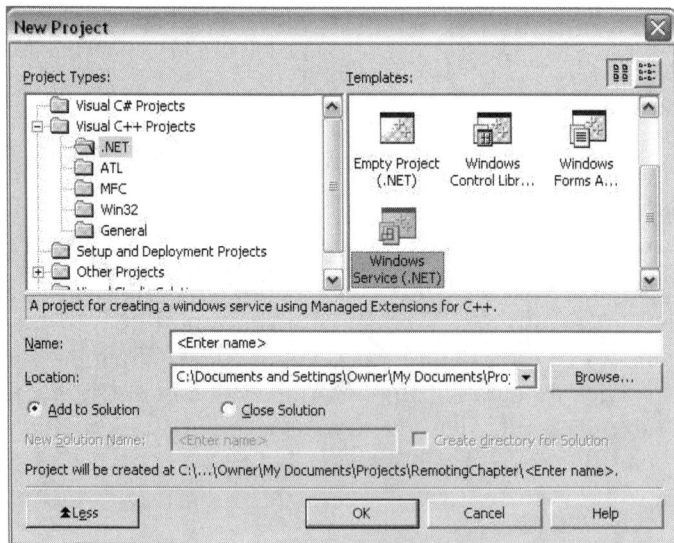

Figure 9-2 The VC++ .NET 2003 Windows Service (.NET) AppWizard option

Basically you have something like this:

```
__gc class ThirdServiceWinService: public ServiceBase
{
public:
  ThirdServiceWinService ()
  {
    //set some basic stuff here
    InitializeComponent();
  }
protected:
  void OnStart(String*[])
  {
    //start remoting
  }
};
```

You'll notice that the wizard has added a method for us called `Initialize Component()` and called it in the constructor:

```
void InitializeComponent(void)
{
  this->components =
    new System::ComponentModel::Container();
  this->CanStop = true;
  this->CanPauseAndContinue = true;
  this->AutoLog = true;
  this->ServiceName = S"ThirdServiceWinService";
}
```

You can change the various properties here if you want to. For our demo service, just leave the default settings as they are.

You also have an `OnStop` if you want to put in any clean-up code, but in our case we don't have anything to do. When the service is stopped, the service executable simply terminates, and our remoting server is stopped automatically. All we need to do is to put our remoting server listener setup code in the `OnStart` as shown below:

```
void OnStart(String* args[])
{
  TcpChannel* TcpChan = new TcpChannel(8989);
```

```
ChannelServices::RegisterChannel(TcpChan);
   RemotingConfiguration::RegisterWellKnownServiceType(
  Type::GetType(S"ThirdDll.Expirer,ThirdDll"),
  S"Expirer",WellKnownObjectMode::SingleCall);
}
```

Well, there is nothing special in the above code, except that by virtue of being in a service, the remoting server is started even before a user has logged into the machine. This is quite a good thing as far as we are concerned, especially if this is a network server and we have multiple clients all using this server to determine the number of runs left. So the end-user is not restricted to just one machine but is free to try out the program on any number of machines on the same network, while maintaining the total number of runs allowed.

Implementing the Service Installer

Let's quickly put together a small program that will install the service for us. We'll use just plain old API to do this so we can keep it nice and simple.

```
int _tmain(int argc, _TCHAR* argv[])
{
  if(argc !=2)
  {
    cout << "Usage:\r\n";
    cout << "       ThirdServiceInstaller
[Path]\\ThirdServiceWinService.exe\r\n";
    return 1;
  }

  SC_HANDLE ThirdService,hScm;
  hScm=OpenSCManager(0,0,SC_MANAGER_CREATE_SERVICE);
  if(!hScm)
  {
    cout << "Service Manager could not be opened\r\n";
    return 1;
  }

  ThirdService=CreateService(hScm,S"ThirdServiceWinService",
    S"ThirdServiceWinService",
  SERVICE_ALL_ACCESS,SERVICE_WIN32_OWN_PROCESS,
    SERVICE_DEMAND_START,
```

```
    SERVICE_ERROR_NORMAL,
    argv[1],
    0,0,0,0,0);
  if(!ThirdService)
  {
    CloseServiceHandle(hScm);
    cout << "Service could not be created\r\n";
    return 1;
  }
  CloseServiceHandle(ThirdService);
  CloseServiceHandle(hScm);

  cout << "Service created successfully\r\n";
  return 0;
}
```

We simply used the service API to install the service, which goes to prove that there is nothing special about a .NET service. It's just the same as any other service written using Win32 API.

Implementing the Remoting Client

The remoting client in this case is our sample fake-trialware program, which will execute only for *n* number of times, each time showing the number of runs left (see Figure 9–3). After that it will show an expiry notice (see Figure 9–4). For our demo, we just create a simple dialog-based application and add the following code to the `OnInitDialog`:

```
BOOL CThirdClientDlg::OnInitDialog()
{
  CDialog::OnInitDialog();

  . . .
  . . .

#undef new
#undef GetObject

  TcpChannel* TcpChan = new TcpChannel();
  ChannelServices::RegisterChannel(TcpChan);
```

```
    Expirer* RemoteObject = static_cast<Expirer*>
       (Activator::GetObject(
Type::GetType(S"ThirdDll.Expirer,ThirdDll"),
                   S"tcp://earth:8989/Expirer"));

    int y = -2;

    try
    {
      y = RemoteObject->IncrementRun(10);
    }
    catch(Exception*)
    {
      MessageBox(_T("There was a problem with your license"));
      EndDialog(0);
      return TRUE;
    }

    if( y == -1 )
    {
      MessageBox(_T("Your trial period has expired"));
      EndDialog(0);
      return TRUE;
    }

    CString str;
    str.Format(_T("This is run %d. You have %d runs left"),
       y,10-y);
    MessageBox(str);

    return TRUE;
}
```

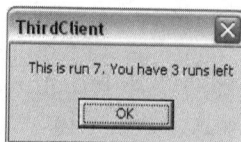

Figure 9-3 A dialog showing the number of runs left in the dummy trialware application

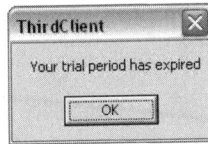

Figure 9-4 A dialog saying that the trialware application has expired

We connect to the remote server, obtain the remote object, and make calls on it. The big advantage here is that the user might be logged in as a restricted user, but since the remoting server is running as a Windows Service configured to run as a Local System account, we can actually read and write to HKEY_LOCAL_MACHINE, which would otherwise have been impossible for a non-administrator user. This is a very powerful feature that we can take advantage of, and is much easier than in the unmanaged days, when the closest analogous technique would have been to use pipes!

Summary

As you have seen from this chapter, .NET remoting gives you an easy and at the same time very powerful mechanism to develop truly distributed applications. Your client programs can use objects on other programs running on the same machine, on the same network, or even somewhere else on the Internet. It is much easier to use than DCOM, and it is a lot more flexible in that you can choose the communication protocol that you wish to use. And it is also much faster when compared to Web services by nature of its binary data transmission mechanism. You also have the option of asynchronously connecting to the remoting server, which might at times greatly improve the efficiency of your client programs.

Remoting can also be used as an excellent means of interprocess communication, since you can actually pass entire objects between communicating processes, regardless of whether these processes are running on the same machine or on two different machines in the same network. By hosting remoting servers inside Windows Services running with higher privileges, lower-privilege programs can actually access the remote objects to perform operations that they could not otherwise have performed due to lack of access rights. The ability to pass a local object by reference to the remote server, allowing the remote server to make calls on it, is another

extremely useful feature with a lot of potential. The remote server can actually make calls on the local object, which will run on the local client machine rather than on the remote server. But as with every other programming technology, the final benefits are derived by carefully deciding what exactly you are trying to achieve and making intelligent choices as to when to use remoting and when not to. If you are not careful, you may get carried away by this extremely powerful .NET feature.

Event Log, Process Control, and Benchmarking

Introduction

At first glance the subjects of the event log, process control, and benchmarking don't seem to have all that much in common, but they do all fit within the umbrella of diagnostics. Not so coincidentally, the types that support these topics are all defined in the `System::Diagnostics`. If you have ever used the Win32 API to access and modify the event log—along with message files and the message compiler—you'll really appreciate the first section of this chapter as I'll illustrate how the `EventLog` class (and related classes) make this task so much easier than ever before. In fact, not only will you learn the common operations of enumerating event logs, creating custom logs, writing event entries, and enumerating log entries, but you'll also see how to write an event handler that is notified whenever a new entry is added to a specified event log. From there, you'll discover the `Process` class, which encapsulates just about everything you'll need to start, control, monitor, and stop processes—both on your local machine as well as on remote servers. After that, the chapter will wrap up with a look at the `PerformanceCounter` class, where you'll see how we can use it to monitor system performance and even create new performance measurement categories.

Working with the Windows Event Log

The various Windows operating systems (with the exception of Windows 95, Windows 98, and Windows Millennium) provide a central repository that allows user-written applications (and the operating system itself) to record

487

errors, warnings, and informational messages—referred to as events. This repository is called the event log. Windows provides three event logs: Application, System, and Security. The System event log is used by Windows system applications and other kernel mode programs, including device drivers. The Security event log stores entries relating to successful and unsuccessful log-on attempts, suspicious port scans, and other security and resource-related events. The Security event log restricts normal-mode applications to read-only access for obvious reasons. That leaves the Application event log, which is used by the majority of Windows applications. In addition to the system-supplied event logs, application-specific logs—called *custom event logs*—can be created.

The event log is actually a Windows service that starts automatically when Windows is loaded. Along with this service, Windows also provides an interactive tool for viewing and modifying logs and event entries called the event viewer. I should note here that while all users are allowed to run the event viewer and can view the entries of both the application and system logs (as well custom logs), only system administrators can gain access to the security log.

Now let's briefly talk about the types of events that should be written to the event log. While you typically use other means to alert the user to "normal" events—such as displaying a dialog to indicate a data input validation error—the following are just a few examples of scenarios where you would use the event log to record application-level events:

- **Services**—By definition, services are meant to be run unattended and are typically used for low-level and/or crucial system-level operations. Therefore, any errors that occur during the running of one of your services should be written to the event log.
- **System startup problems**—These would include any missing files or applications that are needed for your users to correctly run either the operating system or your supplied system.
- **Resource problems**—These include running out of memory, hardware errors, network (including Internet) connectivity issues, and so on.
- **Corrupt application files**—Many applications depend on certain files in order to initialize and run successfully. Obviously, any error in loading or reading these files would qualify as an event that must be dealt with by the system or application administrator.
- **Registry**—Missing or corrupt registry keys or values typically indicate an application that was not installed (or uninstalled) successfully

or might even be installed on the machine illegally—such as being copied on the machine instead of using the application's installation utility.

You'll note that in each of these cases, the end-user would not have any control over solving the problem, and thus the information is put into this central repository so that further diagnostics can be performed or action taken to address the reported event.

Those of you unfamiliar with programmatically interfacing to the event log might wonder, "Why not simply write a log file?" That's certainly a fair enough question, as there are literally dozens of error-logging articles on the Internet that come complete with examples and source code. However, the event log has many advantages over the "roll your own" method. Here are just a few of those benefits:

- Standard means of recording messages. As a result, you can use the system-supplied event viewer to view your application events on any Windows machine.
- The event log's accompanying event viewer application allows you to sort, search, and filter messages by many factors—source applications, message category, user, system, and so forth.
- Application messages can be entered into one of the system-supplied logs (where you could later filter for just the messages from your application), thereby providing a central place to view all application events.
- Application-specific logs can be created—either interactively (via the event viewer) or programmatically—in order to store and track application events separately from other application events while at the same time enjoying the other benefits of using the event log.
- Event categories can be created to semantically group messages within a log.
- Event data can be stored in either ASCII or binary form.
- Events can be exported to a log file and then imported on other machines.
- The Win32 provides an API for creating and deleting logs as well as creating, modifying, and deleting events.

You can view event logs on remote systems using the event viewer—great for remote support scenarios. The point is that if your application needs to log crucial events in a standardized manner, you should use the

event log. That said, the next logical question might very well be, "If the event log is so great, why do so many people write application events to text files?"

Obviously, I can't speak for everyone or about every situation. However, it's been my experience that the answer to that question can be realized by simply looking at the cumbersome steps required for an application to write a simple message to the event log using the standard Win32 API. Specifically, you can't just call a function and pass a log name and event data. Instead you must first manually create a *message file* (with an extension of `.mc`). The message file contains the properties of every event and event category that the application will use. These properties include things such as a symbolic name, language, event severity and facility, description, parameter holders, and a mapping between the category and event. This file is then compiled using the message compiler (`mc.exe`), which generates a resource source file (`.rc`) as well as a header file containing `#define` directives for the symbolic names. Both of these files are then added to the project. These symbolic names in the generated header file can then be referred to in the `ReportEvent` Win32 function. However, before you can access the event log, you must create certain keys and values in the Registry that define things such as the event source that can write events, what log it can write to, and the types of events it can write. If you don't enter these keys, the event viewer will only display the numeric values of your events and not the actual intended event descriptions.

If you're interested in seeing what applications are set up to write to your event log, you can use your favorite Registry editor/viewer (such as `regedt32.exe`) and browse to the various `HKLM\SYSTEM\Current ControlSet\ServicesEventLog\<LogName>` (where `LogName` is either the value `Application`, `System`, `Security`, or the name of a custom log) and view its `Sources` value. This value is a space-delimited string defining every event source that can write to that particular log. In addition, each of these event sources defines a key within the log key that contains such values as `EventMessageFile` and `TypesSupported`. Figure 10–1 shows an example of the .NET runtime event source key on my system.

After all that, is it any wonder so many programmers forego using the event log? This is where the BCL `EventLog` class comes in. As you will soon see, the `EventLog` class—which resides in the `System::Diagnostics` namespace—makes the process of working with event logs dramatically easier. In addition to the `EventLog` class, which provides the methods to enumerate logs and to manipulate logs and events, this section will also

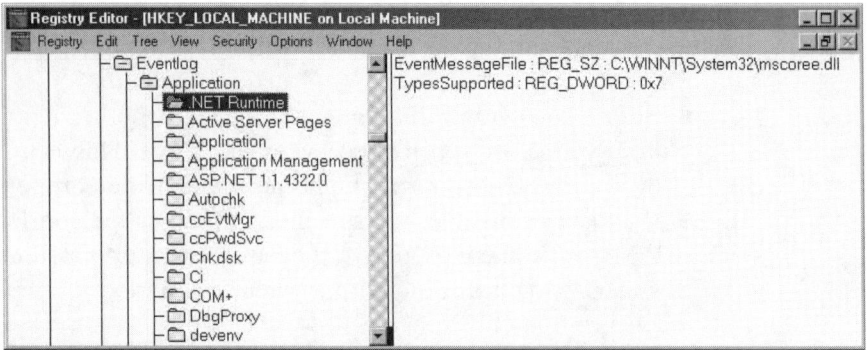

Figure 10-1 Example of the Registry entries that indicate "event sources" that are allowed to write to the event log

illustrate how the `EntryWrittenEventHandler` enables you to interact with event logs asynchronously so that your application can be notified if a specified log has been modified. Finally, the `EventLogPermission`, `EventLogEntryType`, and `EventLogEntryCollection` classes provide access to more detailed control, including permission restrictions, the ability to specify event log types (which control the type of default data that is written with an event log entry), and enumerate event logs.

Constructing `EventLog` Objects and Enumerating Event Logs

`EventLog` objects can be constructed in one of two ways—either by calling the static `EventLog::GetEventLogs` method (which returns all logs for which the user has permission) or by calling one of the overloaded `Event Log` constructors. In the latter case, the various overloads allow you to specify log name, machine name on which that log exists, and the event source name, as shown (used only when you want to write to the event log).

```
EventLog()
EventLog(String* logName)
EventLog(String* logName, String* machineName)
EventLog(String* logName, String* machineName, String* source)
```

Each constructor allows you to specify increasingly more detail about the event log that you are intending to use. However, there are some key

points to be made here about which of these values are needed, when they're needed, and the permissions involved.

- Before you can use the `EventLog` object (whether reading or writing), you must specify the event log name This can be done at `EventLog` object construction time or via the `Log` property.
- If a machine name is not specified (either via the constructor or via the `MachineName` property), the local computer is assumed. In addition, passing a `String` with a value of a period (`.`) represents the local computer.
- Regarding the `MachineName` syntax, if you want to specify another computer on the network, you do not need to include the UNC prefix characters—\\.
- The Windows account under which the application is being run must have the appropriate permissions to access the specified event log. These permissions are listed in Table 10–1.

Table 10-1 User Account Permissions Needed for Various Event Log Operations

User Account	Application Log			System Log			Security Log		
	Read	Write	Clear	Read	Write	Clear	Read	Write	Clear
LocalSystem	Yes	Yes	Yes	Yes	Yes	Yes	Yes	Yes	Yes
Administrator	Yes	Yes	Yes	Yes	Yes	Yes	Yes	Yes	No
ServerOp	Yes	Yes	Yes	Yes	No	No	No	No	No
Everyone Else	Yes	Yes	No	Yes	No	Yes	No	No	No

Here are some examples of constructing the `EventLog` object. Note that any checks on permissions or the machine name will not be performed until you attempt to write to or read from the event log.

```
#pragma push_macro("new")
#undef new

// Specifies that work will be done with the Security log
EventLog* log = new EventLog(S"Security");

// Specifies the Application log on the JEANVALJEAN server
EventLog* log2 = new EventLog(S"Application", S"JEANVALJEAN");
```

```
// Specifies the local System log and that the event source
// is named MyApp
EventLog* log3 = new EventLog(S"System", S".", S"MyApp");


#pragma pop_macro("new")
```

As mentioned, you can also call the static `EventLog::GetEventLogs` method, which returns an array of `EventLog` objects. You can then either enumerate all the logs or select the desired log to work with. The following generic function displays a message box listing all the event logs on a specified machine. Also note that the `EventLog::LogDisplayName` property is used to retrieve the event log's "friendly name."

```
void EnumerateLogs(String* machine)
{
#pragma push_macro("new")
#undef new
  try
  {
    StringBuilder* logNames = new StringBuilder();

    EventLog* logs[] = EventLog::GetEventLogs(machine);
    EventLog* log;
    for (int i=0; i < logs->Count; i++)
    {
      log = logs[i];
      logNames->AppendFormat(S"{0}\n", log->LogDisplayName);
    }

    MessageBox::Show(logNames->ToString(), );
  }
  catch(Exception* e)
  {
    Console::WriteLine(e->Message);
  }
#pragma pop_macro("new")
}
```

Creating Event Sources and Custom Event Logs

In addition to the three supplied event logs, you might also want to create your own custom logs. The most common reason for wanting to do this is if you simply want to separate your application's log entries from those of

other applications. This makes tasks such as enumerating and exporting entries easier. Create a custom log by calling the static `EventLog::Create EventSource` method:

```
void CreateEventSource(String* source, String* logName);
void CreateEventSource(String* source, String* logName, String*
machineName);
```

It might seem unusual to call a method called `CreateEventSource` to create an event log—the reason for this is that you can write to an event log only when you specify an event source. While any number of event sources can write to a given log, each event source can write to only one log. The following example illustrates the steps for creating a custom event log:

```
void CreateCustomEventLog(String* eventSource, String* logName)
{
#pragma push_macro("new")
#undef new

    // Does the Log already exist?
    if (!EventLog::Exists(logName))
    {
        // Does the event source already exist?
        if (EventLog::SourceExists(eventSource))
        {
            // Delete the event source as an event source
            // since it can only be associated with one log
            EventLog::DeleteEventSource(eventSource);
        }
        // Create the event source and associate it
        // with the new custom log.
        EventLog::CreateEventSource(eventSource, logName}
    }
}
#pragma pop_macro("new")


...


// Create the application's custom log
CreateCustomEventLog(S"ACME Killer Application 2.1", S"ACME Log");
```

This function first checks to see if the event log already exists. If the log does not exist, the function needs to determine if the event source already

exists. If the event source exists, it is deleted using the `EventLog::Delete EventSource` method. At that point, the new log can be created and associated with the event source.

Writing Event Log Entries

As mentioned earlier, you must specify the event source (`Source` property) of the `EventLog` object before you can write to a log. There are a couple of issues regarding the event source that you need to be aware of:

- Attempting to write a log entry specifying an event source that is already associated with another log results in an `ArgumentException`.
- .NET will automatically create the Registry entries that I described at the beginning of this chapter if the specified event source has not been associated with the log that you're attempting to write to.

When creating a new log entry, you have the option of specifying the *event type*. Each entry in an event log can have one of several possible entry types, as defined by the `EventLogEntryType` enumeration. This value is passed to the `EventLog::WriteEntry` method; it indicates the importance

Table 10-2 `EventLogEntryType` Enumeration Values

Member Name	Description
Error	This is used to tell the user that a serious error has occurred that could be critical to the proper functioning of the application. For example if a proxy server cannot start its service because the port it wants to use is already in use, then it might make an `Error` entry into the event log, since this situation affects the proxy server's functionality.
Warning	This indicates that something has happened that is not severe enough to be an error but is significant enough to warrant notifying the user.
Information	This indicates a normal operation. In general it is best to minimize logging of `Information` events, as it only slows down the application as well as the computer as a result of writing out a lot of informational messages to the event log.
SuccessAudit	This is used to log an audited security access attempt that succeeds—for example, successful reading of a file from a network share.
FailureAudit	This is used to log an audited security access attempt that fails—for example, a failed attempt to connect to a network share.

or significance of the event log entry and defines the icon that will be displayed by the event viewer when viewing the entry's details. Table 10–2 lists the available `EventLogEntryType` enumeration values.

The following code snippet shows several examples of the `WriteEntry` method. (Note that I've used a shorter version of the book's title as the event source.)

```
void WriteLogEntries()
{
#pragma push_macro("new")
#undef new
  try
  {
    EventLog* log = new EventLog(S"Application");

    log->Source = S"Extending MFC with .NET Book";

    // Text (type defaults to EventLogEntryType::Information)
    log->WriteEntry(S"Informational message");

    // Text + Type
    log->WriteEntry(S"Warning message",
                EventLogEntryType::Warning);

    // Text + Type + eventID
    Int32 eventId = 100;
    log->WriteEntry(S"Message with event id",
                EventLogEntryType::SuccessAudit,
                eventId);

    // Text + Type + eventID + category
    Int32 catId = 1000;
    log->WriteEntry(S"Message with event id and category id",
                EventLogEntryType::SuccessAudit,
                eventId,
                catId);

    // Text + Type + eventId + category + binaryData.
    Byte binaryData[] = new Byte[4];
    for (int i = 0; i < 4; i++)
    {
      binaryData[i] = i;
```

```
    }
    log->WriteEntry(S"Error with event id, category id and binary
        data",
        EventLogEntryType::Error,
        eventId,
        catId,
        binaryData);


    }
    catch(Exception* e)
    {
        Console::WriteLine(e->Message);
    }
#pragma pop_macro("new")
}
```

I first create an `EventLog` object for use with the Application event log and set the event source (`Source` property) to `"Extending MFC with .NET Book"`. I then create several new log entries, using the overloaded `WriteEntry` method. There are a few things to note here:

- The event ID is an application-specific number that can be associated with an entry. Once the entry is in the log, the event viewer allows you to filter events by this value.
- The category ID is also a way to classify your events. The difference between the category ID and the event ID is that the category ID can be associated with a textual description that is displayed in the event viewer. Unfortunately, the current version of the .NET Framework does not yet support this capability through the `EventLog` class. Therefore, the most you can do at this point is to specify a numeric value.
- The last example call to `WriteEntry` is especially interesting because it allows you to specify the data associated with the event. Because you can pass a binary value, this gives you a tremendous amount of flexibility in creating log entries. For example, you could dump a part of memory that has caused a given error message.

Figure 10–2 shows the result of adding these entries on my machine, and in Figure 10–3 I've opened the last entry, where you can see the binary data for that particular log entry.

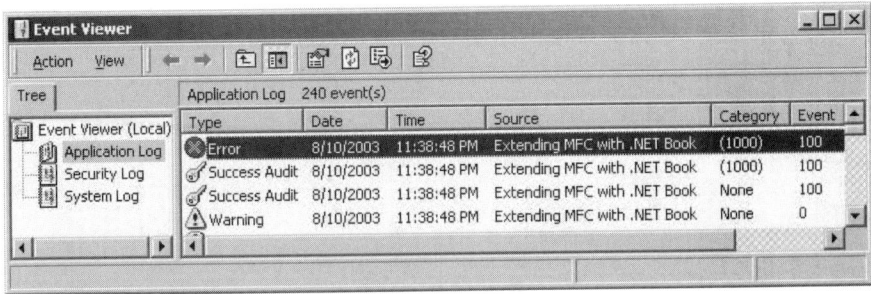

Figure 10-2 The `WriteEntry` method allows you to specify several different parameters, including event ID and category ID, that can then be filtered using the `Event Viewer` application.

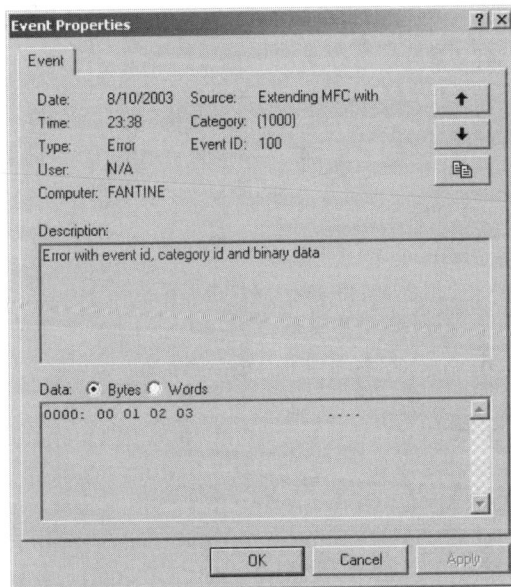

Figure 10-3 The ability to write binary data with a log entry allows you to save important data useful for diagnostic and debugging purposes.

Enumerating Event Log Entries

Now you've seen how to construct event logs, create custom logs, and write log entries, let's look at how to enumerate a log (either system-supplied or custom). This operation is accomplished via the `EventLog::Entries` property, which returns an `EventLogEntryCollection` object. The following

EnumerateLogEntries function takes a machine name and log name as its parameters and outputs the last five entries of the specified machine name and log combination.

```
void EnumerateLogEntries(String* machine, String* logName)
{
#pragma push_macro("new")
#undef new

  try
  {
    Console::WriteLine("SLast 5 entries for {0}/{1} event log",
                       (0 == machine->CompareTo(S".")) ?
                       Environment::MachineName :
                       machine,
                       logName);

    String* format = S"{0, -10} {1}";
    Console::WriteLine(format, S"DateTime", S"Message");

    EventLog* log = new EventLog(logName, machine);

    EventLogEntry* entry;
    Int32 nEntries = log->Entries->Count;
    for (int i = nEntries;
         i > max(0, nEntries - 5);
         i--)
    {
      entry = log->Entries->Item[i-1];
      Console::WriteLine(format,
                         entry->TimeGenerated.ToShortDateString(),
                         entry->Message);
    }
  }
  catch(Exception* e)
  {
    MessageBox::Show(e->Message);
  }
#pragma pop_macro("new")
}
```

As I only wanted the last five entries made to the specified log, and the EventLogEntryCollection object is ordered by the date/time stamp

associated with the event when it was inserted into the log (represented by the `EventLogEntry::TimeGenerated` property), I simply use a `for` loop that initializes the number of `EventLogEntry` objects in the `EventLog EntryCollection` collection and, decrementing the counter with each iteration, read the previous entry object until I reach a count of five or until all entries have been read.

Handling Event Log Notifications

The `EventLog` class defines several members that support the ability to receive event notifications whenever a specified event log has been modified with a new entry. In order to receive these events, you simply need to implement the following in your code:

1. Define a managed class and implement a member method that will be called when the event is raised. This method needs to have the same signature as the `EntryWrittenEventHandler` delegate. Here's an example of that:

```
// Example event handler managed class for new event log
entries
__gc class NewLogEntryEventHandler
{
public:
  NewLogEntryEventHandler() {}

public:
void OnNewLogEntry(
  Object* sender,
  EntryWrittenEventArgs* e)
{
  // Retrieve and work with the newly created entry...
  EventLogEntry* entry = e->Entry;
};
```

2. Instantiate the `EventLog` and set its `EnableRaisingEvents` property to `true`. This Boolean value controls whether or not events are raised when entries are added to the `EventLog` object's specified log.

```
EventLog* log = new EventLog("Application");
log->EnableRaisingEvents = true;
```

3. Call the `EventLog::add_EntryWritten` method, passing it the address of the managed event-handling class's method.

```
NewLogEntryEventHandler* handler = new
  NewLogEntryEventHandler();
log->add_EntryWritten(
  new EntryWrittenEventHandler(handler,

&NewLogEntryEventHandler::OnNewLogEntry));
```

The ease with which you can subscribe to this event is the good news. Now, let's talk briefly about a couple of limitations regarding handling event log notifications that you should be aware of:

- Only events for logs maintained on the local system can be handled.
- If a large number of events are written to the log in a short period of time, the system may not send event notifications for each of them.
- For an especially busy event log, the event will sometimes not signal immediately, resulting in a lag between the time of the event entries and then a sudden burst of notifications for the events that had been queued.

At this point, let's see a sample application that demonstrates how to handle event log write notifications. Actually, we'll write two very simple applications: one to write event entries and one to monitor for event log notifications.

1. Begin by creating an MFC dialog-based application called `Create LogEntries` that supports Managed Extensions.
2. Add the following directives to the project's `stdafx.h` file.

```
#using <mscorlib.dll>
#using <System.dll>
#using <System.Windows.Forms.dll>
using namespace System;
using namespace System::Windows::Forms;
using namespace System::Diagnostics;
#undef MessageBox
```

3. Open the dialog template resource and add the controls as you see them in Figure 10–4.

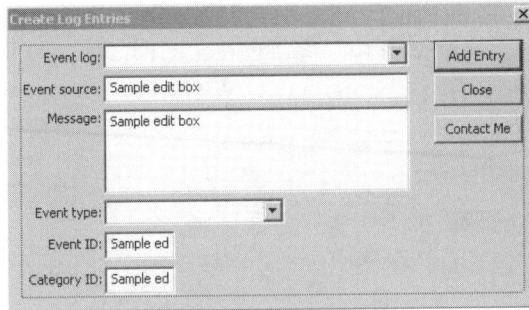

Figure 10-4 Dialog resource for the CreateLogEntries demo

4. Set the various control properties as shown in Table 10–3.
5. Create the DDX member variables as shown in Table 10–4.
6. Add the following code to the end of the dialog's OnInitDialog function (just before the return statement). The first part of this code enumerates the local machine's event logs and displays each log's name (EventLog::LogDisplayName) in the event log combo box. The code then enumerates all of the EventLogEntryType

Table 10-3 Control Property Settings for CreateLogEntries Demo

Control	Property	Value
Message edit	AutoHScroll	False
Message edit	AutoVScroll	True
Message edit	MultiLine	True
Message edit	VerticalScroll	True
Event log combo box	Type	Drop List
Event type combo box	Type	Drop List

Table 10-4 DDX Variables for the `CreateLogEntries` Demo

Control	Variable Type	Variable Name
Event log combo box	CComboBox	m_cboEventLogs
Event log combo box	CString	m_strEventLog
Event source edit	CString	m_strEventSource
Message edit	CString	m_strMessage
Event type combo box	CComboBox	m_cboEventTypes
Event ID edit	int	m_iEventId
Category ID edit	int	m_iCategoryId

enumerations, placing each of the enumeration value's symbolic names in the Event type combo box and setting that combo box item's data to the enumeration value.

```
BOOL CCreateLogEntriesDlg::OnInitDialog()
{
  CDialog::OnInitDialog();

  ...

#pragma push_macro("new")
#undef new
  try
  {
    // enumerate and display all event logs for local machine
    EventLog* logs[] = EventLog::GetEventLogs();
    for (int i = 0; i < logs->Count; i++)
      m_cboEventLogs.AddString((CString)logs[i]->
        LogDisplayName);

    // enumerate and display all event entry types
    int idx;
    int value;
    Object* eventEnum;

    Array* eventTypes =
      Enum::GetValues(__typeof(EventLogEntryType));
    for (int i = 0; i < eventTypes->Count; i++)
```

```
      {
        eventEnum = eventTypes->GetValue(i);
        idx = m_cboEventTypes.AddString((CString)eventEnum-
>ToString());

        value =
Convert::ToInt32(Enum::Format(__typeof(EventLogEntryType),
                               eventEnum,
                               S"D"));
        m_cboEventTypes.SetItemData(idx, (DWORD_PTR)value);
      }
    }
    catch(Exception* e)
    {
      MessageBox::Show(e->Message);
    }
#pragma pop_macro("new")

    return TRUE;
  }
```

As the comment indicates, I've left out error-checking to keep the demo as brief as possible and to keep the focus on the event log code. The function first constructs an `EventLog` object on the specified log name. From there, it determines (via the item data of the Event type combo box's currently selected item) the `Event LogEntryType` value to be used in creating the new event entry. The `EventLog::WriteEntry` method is then called, passing the information from the dialog. If all is successful (no exception is thrown), a success message box is displayed.

7. Add the following code to the dialog's `OnBnClickedOk` function:

```
void CCreateLogEntriesDlg::OnBnClickedOk()
{

// !!! Assuming that all needed data has been entered !!!

#pragma push_macro("new")
#undef new
  try
  {
    UpdateData();
    EventLog* log = new EventLog(m_strEventLog);
```

```
      if (0 < m_strEventSource.GetLength())
        log->Source = m_strEventSource;

      int idx = m_cboEventTypes.GetCurSel();
      EventLogEntryType eventType =
        (EventLogEntryType)m_cboEventTypes.GetItemData(idx);

      log->WriteEntry(m_strMessage,
                      eventType,
                      m_iEventId,
                      m_iCategoryId);

      MessageBox::Show(S"Event successfully written");

      // Initialize message
      m_strMessage = _T("");
      UpdateData(FALSE);
   }
   catch(Exception* e)
   {
      Console::WriteLine(e->Message);
   }
#pragma pop_macro("new")
   }
```

At this point, build and test the application. You should see that you can add entries at will to any event log on the local machine for which you have write access—including any custom event logs you've created.

Now let's build the event notification application that will subscribe to new log entry events for the Application log. For each item added on this application's watch, it will add that item's data to a list control.

1. Begin by creating an MFC dialog-based application called Event LogWatcher that supports Managed Extensions.
2. Add the following directives to the project's stdafx.h file.

```
#using <mscorlib.dll>
#using <System.dll>
#using <System.Windows.Forms.dll>
using namespace System;
using namespace System::Windows::Forms;
using namespace System::Diagnostics;
#undef MessageBox
```

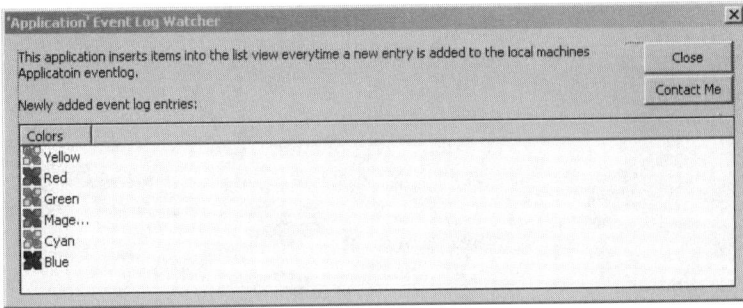

Figure 10-5 Dialog resource for the `EventLogWatcher` demo

3. Open the dialog template resource and add the controls as you see them in Figure 10–5.
4. Set the list control's `View` property to `Report`.
5. Create a DDX member variable of type `CListCtrl` called `m_lst NewEvents` for the list control.
6. Add the following code to the end of the dialog's `OnInitDialog` function to insert columns into the list control for the log entry data that the application will display.

```
BOOL CEventLogWatcherDlg::OnInitDialog()
{
    ...

    // Add columns to listview
    m_lstNewEvents.InsertColumn(0, _T("Data/Time"));
    m_lstNewEvents.InsertColumn(1, _T("Event Source"));
    m_lstNewEvents.InsertColumn(2, _T("Event Type"));
    m_lstNewEvents.InsertColumn(3, _T("Event ID"),
                        LVCFMT_RIGHT);
    m_lstNewEvents.InsertColumn(4, _T("Category ID"),
                        LVCFMT_RIGHT);
    m_lstNewEvents.InsertColumn(5, _T("Message"));
```

```
    return TRUE;
}
```

7. Now add the following managed class (NewLogEntryEvent Handler) to the bottom of the EventLogWatcherDlg.h file. This class implements the method that gets called when the new entry event is raised. While this class follows the general guidelines of the steps listed earlier in the section, there are a couple of important differences here. First, this class must be able to communicate the new entry's information to the dialog class. Therefore, this NewLog EntryEventHandler constructor takes an CEventLogWatcherDlg object as a parameter. Second, because NewLogEntryEvent Handler is defined using the gcroot template (so that it can be a member variable of the dialog class), an instance of it can't be passed to the EventLog::add_EntryWritten method. As a result, we have the NewLogEntryEventHandler constructor call the add_EntryWritten method, passing itself as the parameter. To allow this to happen, an EventLog object must be passed to the NewLogEntryEventHandler constructor.

```
__gc class NewLogEntryEventHandler
{
public:
 NewLogEntryEventHandler(CEventLogWatcherDlg* parent,
   EventLog* log)
 {
    this->parent = parent;
    log->add_EntryWritten(
      new EntryWrittenEventHandler(this, OnNewLogEntry));
 }

protected:
  CEventLogWatcherDlg* parent;

public:
   void OnNewLogEntry(Object* sender, EntryWrittenEventArgs* e)
   {
    // Pass entry to parent dialog for processing.
    parent->OnNewLogEntry(e->Entry);
```

```
    }
};
```

8. As mentioned, the `CEventLogWatcherDlg` will define the `NewLog
EntryEventHandler` as a member and, as you saw in the previous
step, will have an `OnNewLogEntry` function. Therefore, define the
following member variable in the `CEventLogWatcherDlg` class:

```
class CEventLogWatcherDlg : public CDialog
{
...
public:
  gcroot<EventLog*> log;
  void OnNewLogEntry(EventLogEntry* entry);
  gcroot<NewLogEntryEventHandler*>handler;
...
```

At this point, you can see that we have a circular reference, since
both classes have pointers to one another. Therefore, forward-
declare the `NewLogEntryEventHandler` just before the declaration of the `CEventLogWatcherDlg` class.

```
__gc class NewLogEntryEventHandler;

// CEventLogWatcherDlg dialog
class CEventLogWatcherDlg : public CDialog
{
...
```

9. We're almost there. Add the following initialization code to the
`OnInitDialog` function. Here I'm just constructing the member
`EventLog` object and setting its `EnableRaisingEvents` property
to `true`. I then construct the `NewLogEntryEventHandler` member object, which will subscribe to the passed `EventLog` object's
new entry event.

```
BOOL CEventLogWatcherDlg::OnInitDialog()
{
...

#pragma push_macro("new")
#undef new
  log = new EventLog(S"Application");
  log->EnableRaisingEvents = true;
```

```
    handler = new NewLogEntryEventHandler(this, log);
#pragma pop_macro("new")

return TRUE;
}
```

10. All that's left at this point is to actually implement the dialog's On
    NewEntry function. As you can see, this function merely receives
    the EventLogEntry object and adds its data to the list control.

```
void CEventLogWatcherDlg::OnNewLogEntry(EventLogEntry* entry)
{
    // Date/Time stamp
    int idx = m_lstNewEvents.InsertItem(0,
      (CString)entry->TimeGenerated.ToString());

    // Event source
    m_lstNewEvents.SetItemText(idx, 1, (CString)entry->Source);

    // Entry type
    m_lstNewEvents.SetItemText(idx, 2,
      (CString)__box(entry->EntryType)->ToString());

    // Event id
    m_lstNewEvents.SetItemText(idx, 3,
      (CString)entry->EventID.ToString());

    // Category id
    m_lstNewEvents.SetItemText(idx, 4,
      (CString)entry->CategoryNumber.ToString());

    // Message
    m_lstNewEvents.SetItemText(idx, 5, (CString)entry->Message);
}
```

At this point, you can run both applications side by side, and when you
add an entry to the Application log with the CreateLogEntries application, you should see that entry pop up in the list control of the EventLog
Watcher application (Figure 10–6).

Event logs are useful for determining the system's state in the past;
however, your applications will often need to take action as they execute.
This is where process control comes into play. Process control gives you
fine-grained control over running processes (applications or services) not
only on the local system, but also on remote systems.

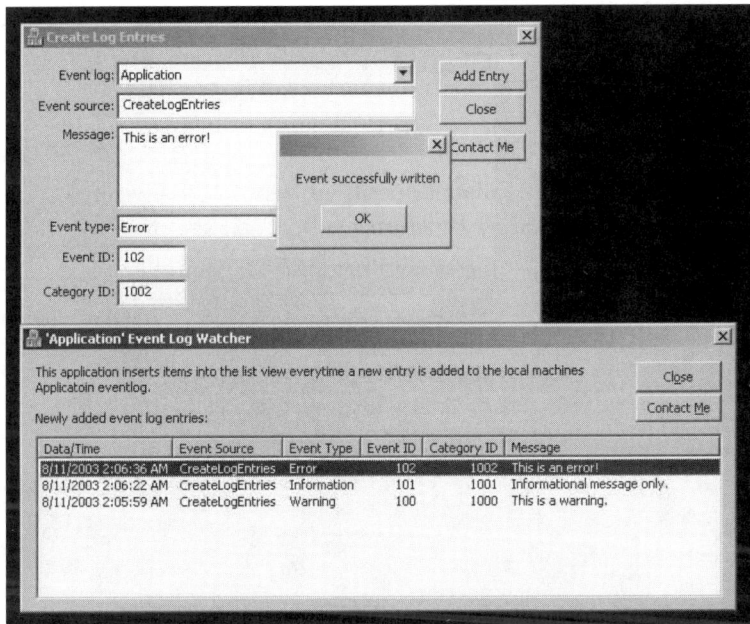

Figure 10-6 When the `EventLogWatcher` is signaled that a new event has been added to the Application log, it displays that entry's data in a list control.

Process Control

The `System::Diagnostics` namespace also includes the extremely useful `Process` class, which allows you to start and stop new processes, enumerate the running processes on the system, and retrieve information about any running process. You can also enumerate processes on a remote machine, but you may not start or stop processes on the remote machine. The `Process` class has several very useful properties, such as `Process::Id`, which exposes the unique process ID of a process; `Process::Handle`, which returns a process's native Win32 handle, and many more. You'll see some of the more useful methods and properties in action in this section.

Starting a New Process

Start a new process by calling the `Process::Start` method, which has four overloads. Of the four overloads, three are static, which means you don't need an instance of the `Process` class to start a new process. One of

the overloads, the most useful one, takes an instance of `ProcessStart Info`. The only member of `Process::StartInfo` that needs to be filled in is the `ProcessStartInfo::FileName` property; other useful properties include the `ProcessStartInfo::Arguments` property, which is used to set the command-line arguments passed to the process, and the `Process StartInfo::WorkingDirectory` property, which sets the process's initial directory. We can also use the `ProcessStartInfo::WindowStyle` property to configure the state of the process's window. The `WindowStyle` property takes a `ProcessWindowStyle` type, which is an enumeration type. The various members of the `ProcessWindowStyle` enumeration are listed below.

- **Hidden**—The process is started in a hidden state, just as if you were using `SW_HIDE` in a native Win32 application.
- **Normal**—The process is started in the normal state, just as if you had used the `SW_SHOW` parameter in a native Win32 application.
- **Minimized**—This starts the process in a minimized state, similar to using the `SW_MINIMIZE` in native Win32 applications.
- **Maximized**—This starts the process in a maximized state, similar to using `SW_MAXIMIZE` in a native Win32 application.

Instead of filling in the `StartInfo` property directly, you may also use an overload of `Process::Start`, which takes a `ProcessStartInfo` object as parameter. This overload is useful if you don't have an instance of a `Process` object, but you need to pass the required information to some method that does have an instance of a `Process` object. The code snippet below demonstrates how you can start a new process.

```
void StartProg1()
{
  Process* p = new Process();
  p->StartInfo->FileName = "Notepad.exe";
  p->StartInfo->Arguments = "abc.txt";
  p->Start();
  cout << "Process has been started with Id :";
  cout << p->Id << "\r\n";
  p->WaitForExit(9000);
  cout << "About to close main window\r\n";
  if(p->HasExited)
    cout << "User has closed process\r\n";
  else
```

```
{
    p->CloseMainWindow();
    cout << "Process has been closed\r\n";
}
}
```

The `Process::Start` is not a blocking call—it starts a new process and returns immediately. There are times when you might want to block execution of your application until the process has terminated. For example, you might start a new process that creates some kind of input file and then parse the input file. If you attempt to parse the file before the process has terminated, you are in danger of trying to open a nonexistent file or one that is still open in write-mode. Luckily, we can use `Process::WaitForExit`, which will block execution of the current thread until the process has returned or the specified time has elapsed. `Process::Wait ForExit` has two overloads:

- `void Process::WaitForExit()`—This will block forever until the process terminates.
- `bool Process::WaitForExit(int waitmilliseconds)`—This will block until either the process terminates or the interval specified by `waitmilliseconds` parameter expires. `WaitForExit` returns `true` if the process exits and `false` if the call returns because the wait interval expired. You can also use the `Process::HasExited` property to determine at any time if the process is currently running.

The sample code uses `Process::WaitForExit`'s overloaded version that specifies 9,000 for the `waitmilliseconds` parameter. Thus, the call blocks for 9 seconds or until the process terminates. If the `waitmilliseconds` interval expires and the process has not yet terminated, you can manually terminate the process by using the `Process::CloseMainWindow` method. `Process::CloseMainWindow` basically sends a `WM_CLOSE` message to the process's main window and has the same effect as clicking the process's [x] button. This is the best way to terminate GUI applications because it gives the application a chance to clean up before terminating.

For non-GUI applications you can use the more forceful `Process:: Kill` method, which immediately terminates a process. `Process::Kill` results in the abnormal and immediate termination of an application—thus the application never gets a chance to free any resources that are currently in

use. You should use `Process::Kill` only as a last measure if all else fails. For example, in our code snippet above, we start Notepad. If you type something into Notepad and then call `Process::CloseMainWindow`, Notepad will prompt you to save the file before exiting. If you use `Process::Kill`, the application will exit abruptly and any text you typed will be lost.

You might also have noticed that we have not passed the full path for the `FileName` property. This is because Windows will first search the Windows folder and the Windows System folder if we give it a file name without a fully qualified path. If you want to execute a program in a custom directory, then you must specify the fully qualified path to the file, as in `"C:\\YourFolder\\YourSubFolder\\YourProgram.exe"`.

Getting Exit Notification

Sometimes using `Process::WaitForExit` might not be suitable, and you want your application to continue execution. At the same time you might want to take some action once a particular process has terminated. The `Process` class has an `EnableRaisingEvents` property that, when set to true, raises an `Exited` event when a process terminates. We can add a delegate to handle this event. The `Exited` event's prototype is a simple `Event Handler` delegate. The following code snippet should make things clearer:

```
__gc class MyProcess
{
public:
  MyProcess(String* procfilename)
  {
    m_process = Process::Start(procfilename);
    m_process->add_Exited(new
      EventHandler(this,&MyProcess::MyEventHandler));
    m_process->EnableRaisingEvents = true;
  }
  void MyEventHandler(Object* sender, EventArgs* e)
  {
    Console::WriteLine(m_process->
    ExitTime.ToLongTimeString());
  }
private:
  Process* m_process;
};
```

```
// . . .

MyProcess* myp = new MyProcess("notepad");
```

Notice how we use the static overload of `Process::Start` to start a new process directly, which returns a `Process` object that is associated with the newly started process. Then we add a new event handler for the `Exited` event and set `Process::EnableRaisingEvents` to true. In the event handler, we simply display the time when the process was terminated. You should remember that at this stage the process has terminated, and thus most of the properties and methods are now invalid as far as this `Process` object is concerned. Calling a method would result in an `Invalid OperationException`. Some of the properties that you can use after a process terminates are `ExitCode`, `ExitTime`, `HasExited`, and `Handle`.

Enumerating Running Processes

You've probably seen some of those task-manager type programs that show a list of running processes in the system. Well, it is easy to write a program that not only lists all running processes but also shows information on each running process, using the `Process` class. The `Process` class has a static method called `GetProcesses` that returns an array of `Process` objects that are each associated with a running process. `GetProcesses` has two overloads: one for enumerating processes on the local system and the other for remote machines.

```
static Process* Process::GetProcesses() [];
static Process* Process::GetProcesses(String* machinename) [];
```

Creating a task-manager type application takes only a few steps:

1. Create a dialog-based MFC application and add CLR support to it. Add a list control to the dialog and associate a control variable with it called `m_list`.
2. Add the following code snippet to the `OnInitDialog` handler:

```
// Set up our list view control properly
m_list.InsertColumn(0, _T("Id"),LVCFMT_LEFT,60);
m_list.InsertColumn(1, _T("Process name"),LVCFMT_LEFT,280);

//Setting the full row select extended style
m_list.SetExtendedStyle(LVS_EX_FULLROWSELECT);
```

```
//Get list of running processes
Process* procs[] = Process::GetProcesses();

//Enumerate through the array
for (int i=0; i < procs->Length; i++)
{
 //Insert Process Id as column 1
 int ind = m_list.InsertItem(0,
                               (CString)procs[i]->
                               Id.ToString(),0);
 //Insert Process name as column 2
 m_list.SetItem(ind,1,LVIF_TEXT,
     (CString)procs[i]->ProcessName,0,0,0,0);
}

//Select the first item in the list
m_list.SetItemState(0,LVIS_SELECTED,LVIS_SELECTED);
```

3. Compile and run the program. You'll see a list of all running processes on your system, as shown in Figure 10–7. Try running multiple instances of a program, say Notepad, and rerun the program. You'll see how each instance is listed as a distinct process, but with unique process IDs. We have used the `Process::Id` and `Process::ProcessName` properties to get process information. The `Process` class has lots of other very useful properties, some of which we'll see later in this section.

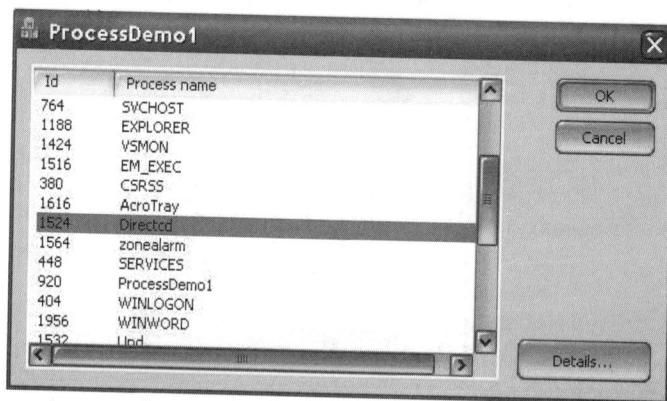

Figure 10-7 The `Process` class enables you to easily enumerate all running processes on a specified machine.

4. Add a new dialog to your project and then add a list box to this dialog. Using the Class Wizard, create a new class for the dialog named `CProcessInfoDlg` and associate a member variable named `m_lbox` to the list box control.

5. Add an `int` type member variable to this dialog named `pid`.

6. Add a button labeled Details to the main dialog box and add a button click handler that has the following code in it:

```
void CProcessDemo1Dlg::OnBnClickedButton1()
{
  int i = (int)m_list.GetFirstSelectedItemPosition();
  CString s = m_list.GetItemText(i-1,0);
  CProcessInfoDlg dlg;
  dlg.pid = atoi(s);
  dlg.DoModal();
}
```

Essentially we figure out the currently selected item in the list control and get the process ID of the selected item using `GetItem Text`. Then we bring up the `CProcessInfoDlg` dialog after setting the `pid` member variable to the process ID of the selected process.

Now the rest of our code is in the `OnInitDialog` of the `CProcessInfoDlg` class, where we populate the list box with information about the selected process (whose ID has been passed to the `pid` member variable).

7. Add the following code to the `OnInitDialog` function:

```
BOOL CProcessInfoDlg::OnInitDialog()
{
  CDialog::OnInitDialog();
  CString title = "Process Information - ";

  //Get the process that has the process id passed to pid
  Process* proc = Process::GetProcessById(pid);
  title += (CString)proc->ProcessName;
  //Putting the Process name in the window title
  SetWindowText(title);

  //We don't want to handle idle and system
  //which are not real processes
  if(pid == 0 || pid == 4)
      return TRUE;
```

```
//In case of any error, just return
//The process might not be running now
if(proc == NULL)
    return TRUE;


CString s;


//Populate the list box using some of the
//properties available to us through the Process class
s.Format(_T("Start time ==> %s"),proc->
  StartTime.ToLongTimeString());
m_lbox.AddString(s);


s.Format(_T("Process handle ==> %s"),proc->Handle.ToString());
m_lbox.AddString(s);


s.Format(_T("Number of threads ==> %d"),proc->Threads->Count);
m_lbox.AddString(s);


//Get the ProcessModule object associated with the
//main module of the given process
ProcessModule* pmod = proc->MainModule;
//If it does not have a main module, just return
if(pmod == NULL)
    return TRUE;


//Now use the ProcessModule object to further
//populate the list box
s.Format(_T("Module Name ==> %s"),pmod->ModuleName);
m_lbox.AddString(s);


s.Format(_T("Module FileName ==> %s"),pmod->FileName);
m_lbox.AddString(s);


s.Format(_T("Module Memory Size ==> %d bytes"),pmod->
  ModuleMemorySize);
m_lbox.AddString(s);


//Get a FileVersionInfo object from the FileVersionInfo
//property of the ProcessModule object
FileVersionInfo* fver = pmod->FileVersionInfo;
//If there is no version info, just return
```

```
if(fver == NULL)
    return TRUE;

//Add some version related information to the list box
//using some of the properties available through the
//FileVersionInfo object
s.Format(_T("File Version ==> %s"),fver->FileVersion);
m_lbox.AddString(s);

s.Format(_T("File description ==> %s"),fver->FileDescription);
m_lbox.AddString(s);

s.Format(_T("Company name ==> %s"),fver->CompanyName);
m_lbox.AddString(s);

s.Format(_T("Product name ==> %s"),fver->ProductName);
m_lbox.AddString(s);

return TRUE;
}
```

Running this code yields results similar to those shown in Figure 10–8.

We use the `Process::GetProcessById` method to get a `Process` object associated with the process with the specified process ID. Note that a new `Process` object is created here and associated with an existing process, whereas in earlier scenarios the new process itself was started using a `Process` object we had instantiated. This is sometimes referred to as binding a `Process` object to a running process. In addition to the `Process::GetProcessById` method, you can also use the `Process::Get ProcessesByName` method, which is similar except that it accepts a process

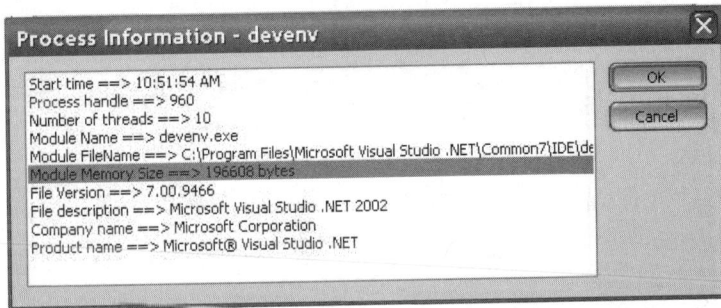

Figure 10–8 Almost any module information is retrievable using the `ProcessModule` class.

name string as parameter. Now there can be only one process with any given process ID, and thus `GetProcessById` returns a single `Process` object, but there can be multiple processes with a given name (for example you might have five instances of Notepad running), and thus `GetProcesses ByName` returns an array of `Process` objects. This behavior can be particularly useful if you want to close all instances of an application, as shown in the following listing:

```
void KillNotepads()
{
  Process* notepads[] = Process::GetProcessesByName(S"Notepad");
  for(int i=0; i<notepads->Length; i++)
  {
    notepads[i]->CloseMainWindow();
  }
}
```

Once we have the `Process` object bound to the running process, it's just a matter of using some properties to retrieve information about the process. In our sample program we first use three direct properties: `Start Time`, `Handle`, and `Threads` (which is a `ProcessThreadCollection` object on which we use the `Count` property to get the total number of threads in the process). Then we get a `ProcessModule` object, which is associated with the main module of the specified process, using the `Process::MainModule` property.

Now that we have the `ProcessModule` object, we can use some of its properties, such as `ModuleName`, `FileName`, and `ModuleMemorySize`. We then use the `ProcessModule::FileVersionInfo` property to get a `FileVersionInfo` object that contains version-specific information about the main module of the specified running process. Now it's just a matter of using some of the properties available through the `FileVersionInfo` object, such as `FileVersion`, `FileDescription`, `CompanyName`, and `ProductName`.

Determining If a Process Is Responding

Sometimes you have a situation where a process has crashed or its user interface frozen. If you are writing an application that starts several processes and then terminates them, you might want to determine whether each of those processes is in a responsive state before attempting to close them using `CloseMainWindow`. This is because `CloseMainWindow` won't

have much of an effect on a nonresponding process. For nonresponding processes, we need to use the `Kill` method. To determine if a process is responding or not, use the `Responding` property. For example our Notepad-killer sample should ideally be coded as follows:

```
void KillNotepads()
{
   Process* notepads[] = Process::GetProcessesByName(S"Notepad");
   for(int i=0; i<notepads->Length; i++)
   {
     if(notepads[i]->Responding)
         notepads[i]->CloseMainWindow();
     else
         notepads[i]->Kill();
   }
}
```

The code now checks if the Notepad is responding and uses the `Close MainWindow` method to terminate it. If the Notepad's GUI is frozen, the code uses the `Kill` method to forcefully terminate the process.

Enumerating Process Modules

The `Process::Modules` property makes it possible to enumerate the various modules that have been loaded by the specified process. The `Modules` property returns a `ProcessModuleCollection` object, which, as the name implies, is a collection of `ProcessModule` objects. Once we obtain the `ProcessModule` object for a module, we can retrieve information about that module using the various properties offered by the `ProcessModule` class, such as `FileName`, `ModuleName`, and `FileVersionInfo`. The following listing shows how you can enumerate the modules loaded by a running process:

```
void ListModules()
{
   //Start a new process
   Process* p = Process::Start(S"Notepad");

   //Make sure it's finished loading
   p->WaitForInputIdle();

   ProcessModuleCollection* mods = p->Modules;
   for (int i=0 ; i < mods->Count; i++)
```

```
    {
        ProcessModule* pm = mods->Item[i];
        Console::WriteLine(pm->ModuleName);
    }
    p->WaitForExit();
}
```

As you might observe, we call `WaitForInputIdle` on the `Process` object before enumerating the loaded assemblies. We do this because the process takes time to load all its required modules, and we run the risk of attempting to enumerate modules too early in the life-cycle of the process. `WaitForInputIdle` essentially waits until the process enters an idle state and we can be sure that it has finished loading all its required modules.

So far you've seen how to figure out what has happened on a system in the past using the event log and how to control running processes using the `Process` class. These two facilities are primarily useful for individual applications or processes. A larger view that encompasses the state of the entire system is often more useful, since you can see processes and events in the context of the system at large. The next section discusses how the .NET Framework makes it easy not only to monitor a system but also to add new monitoring information about your applications.

Monitoring System Performance

Windows supports `Performance` objects (also referred to as `Performance` categories), where each `Performance` object has one or more counters associated with it. Essentially a `Performance` object is a logically similar set of `Performance` counters that are grouped together; for example, there is a `Performance` object called `Processor` that has several counters like `Processor Time`, `User Time`, and so on. In addition to counters, `Performance` objects also have something called instances, which allow us to have multiple `Performance` counters of the same type under the same category. Thus, assuming there is a category called `CAT` with two counters—`CNTRA` and `CNTRB`, then if it supports two instances, as in `INSA` and `INSB`, you end up with four sets of data output: {CAT, CNTRA, INSA}, {CAT, CNTRA, INSB}, {CAT, CNTRB, INSA}, and {CAT, CNTRB, INSB}. Note that it is possible to have `Performance` counters that do not have instances (in which a single instance is always assumed). You can monitor

your system's performance using the `Performance Monitor` application that ships with Windows (click `Start > Run`, and then type `perfmon`).

For developers, the .NET BCL provides the `PerformanceCounter` and the `PerformanceCounterCategory` classes, both of which reside in the `System::Diagnostics` namespace. You can create and delete your own custom counters, write to custom counters, and you can read from both custom as well as local system counters. Writing to performance counters is referred to as publishing performance data.

Enumerate Existing `Performance` Objects and Counters

The `PerformanceCounterCategory` class has a static method called `Get Categories` that returns an array of `PerformanceCounterCategory` objects in the system (or on a remote computer, depending on the overload you use), as shown in the following listing:

```
PerformanceCounterCategory* cats[] =
   PerformanceCounterCategory::GetCategories();
for(int i=0; i<cats->Count; i++)
{
   Console::WriteLine( cats[i]->CategoryName );
}
```

Once we have the array of `PerformanceCounterCategory` objects, it's just a matter of using the `CategoryName` property to list the various `Performance` categories on the local machine. You might also want to enumerate the counters associated with each category as well as verify whether the category has more than one instance. Assume you want to do this for the `Processor` system performance object. The first thing you need to do is to construct a `PerformanceCounterCategory` object and associate it with the `"Processor"` category:

```
PerformanceCounterCategory* proccat =
   new PerformanceCounterCategory("Processor");
```

Alternatively, you could also reuse an instance of a `PerformanceCounter Category` object multiple times:

```
PerformanceCounterCategory* proccat =
   new PerformanceCounterCategory();
// . . .
   proccat->CategoryName = S"Processor";
```

Use the GetInstanceNames method first to determine if this category has multiple instances:

```
String* instnames[] = proccat->GetInstanceNames();
```

Now call the GetCounters method, which has two overloads, one of which takes an instance name argument. For a performance object with multiple instances, you must use the correct overload to avoid an exception, as shown in the following listing:

```
PerformanceCounter* pcarr[];
if(instnames->Count == 0)
  pcarr = proccat->GetCounters();
else
  pcarr = proccat->GetCounters(instnames[0]);
```

All instances of a performance object have the same set of counters, but you must be aware that the counters will be different objects for different instances. Since we only intend to enumerate the counters rather than access their data, we only need to use the first instance when we call Get Counters. The full code listing for enumerating the counters in the Processor performance object follows.

```
PerformanceCounterCategory* proccat =
  new PerformanceCounterCategory("Processor");
String* instnames[] = proccat->GetInstanceNames();
PerformanceCounter* pcarr[];
if(instnames->Count == 0)
  pcarr = proccat->GetCounters();
else
  pcarr = proccat->GetCounters(instnames[0]);
for(int i=0; i<pcarr->Count; i++)
{
  Console::WriteLine( pcarr[i]->CounterName);
}
```

Reading Performance Data

Reading performance counters is easy—here's an example.

1. As before, create a simple MFC dialog-based application with support for Managed Extensions.

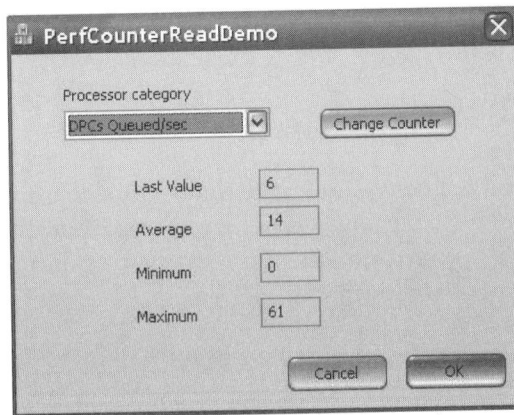

Figure 10-9 Example of running the `PerfCounterReadDemo` application

2. Add a combo box button and four text boxes to the dialog box as shown in the screen shot of the demo application (Figure 10–9).
3. Associate the `m_CountersCombo` member variable with the combo box, which we will fill in with the various performance counters available under the `"Processor"` category. Basically you need to end up with the following member variables in your `CDialog`-derived class:

```
CComboBox m_CountersCombo;
CString m_counter; //holds the selected counter name
gcroot<PerformanceCounter*> m_pc;
int m_lastvalue; //edit box DDX variable
int m_average; //edit box DDX variable
int m_minimum; //edit box DDX variable
int m_maximum; //edit box DDX variable
int count; //Total number of data request calls
int tot; //Sum of all values obtained
```

4. Add a handler for the `WM_TIMER` message (`OnTimer`) as well as an `ON_BN_CLICKED` handler for the button (`OnChangeCounter`).
5. Add some code to the `OnInitDialog` handler as shown below:

```
BOOL CPerfCounterReadDemoDlg::OnInitDialog()
{
  CDialog::OnInitDialog();

  // . . .
```

```
#pragma push_macro("new")
#undef new

    //Populate Combo box with all the counters
    //under the "Processor" category
    PerformanceCounterCategory* proccat =
      new PerformanceCounterCategory(S"Processor");
    String* instnames[] = proccat->GetInstanceNames();
    PerformanceCounter* pcarr[];
    if(instnames->Count == 0)
      pcarr = proccat->GetCounters();
    else
      pcarr = proccat->GetCounters(instnames[0]);
    for(int i=0; i<pcarr->Count; i++)
    {
      m_CountersCombo.AddString(
        (CString)pcarr[i]->CounterName);
    }
    if(pcarr->Count)
      m_CountersCombo.SetCurSel(0);

#pragma pop_macro("new")

    SetTimer(1000,1000,NULL);

    return TRUE;
}
```

As you might observe, we have used the `PerformanceCounter` category class to enumerate the counters associated with the `"Processor"` category, and we have populated the combo box with this list of counters. We have also started a timer at one-second intervals.

6. Add the following code to the timer handler function:

```
void CPerfCounterReadDemoDlg::OnTimer(UINT nIDEvent)
{
    if(m_counter != _T(""))
    {
        m_lastvalue = m_pc->NextValue();
        if(m_lastvalue < m_minimum)
            m_minimum = m_lastvalue;
```

```
if(m_lastvalue > m_maximum)
   m_maximum = m_lastvalue;
tot += m_lastvalue;
count++;
m_average = tot/count;
UpdateData(false);


}
   CDialog::OnTimer(nIDEvent);

}
```

The code first checks to see if m_counter contains anything valid before proceeding to do anything. As you can see, I have used the PerformanceCounter::NextValue method to acquire the performance counter calculated value. I then just fill in the various edit boxes with the maximum, minimum, average, and current values, which I compute using the existing data.

7. Add this code to the OnChangeCounter function:

```
void CPerfCounterReadDemoDlg::OnChangeCounter()
{
  m_lastvalue = m_minimum = m_maximum = m_average = tot =
    count = 0;
  UpdateData(false);
  int index = m_CountersCombo.GetCurSel();
  m_CountersCombo.GetLBText(index, m_counter);
#pragma push_macro("new")
#undef new
  m_pc = new PerformanceCounter(S"Processor", m_counter,
    S"_Total");
#pragma pop_macro("new")
}
```

The code just initializes all the variables to zero, gets the current selection from the combo box, and creates a new PerformanceCounter object for the selected performance counter. I have hard-coded "_Total" as the instance of the "Processor" category, but it would have been easy to dynamically choose the instance, in which case all I'd need to do is use the GetInstanceNames method of the PerformanceCounterCategory class and populate another combo box. For demonstration purposes, this has not been done to keep the code simple and to the point. Note that when you run the application, you must click the button after changing the selection in the combo box for the changes to take effect.

Publish Performance Data

It's also possible to publish custom data to the performance objects. To demonstrate this, I'll show you an MFC dialog-based application that creates a new `Performance` object, `BookExampleCategory`, with two counters called `"Basic Counter"` and `"Second Counter"`. This program sets a timer and publishes random data to the two counters. In a real-life situation, you'd publish some meaningful data, but nothing beats random data for demonstration purposes. The program itself shows the default dialog on screen, and to see all the action, you must use the Windows Performance monitor tool. Run `perfmon.exe` (our application should be up and running by now) and use the "new window from here" option from the Action menu. You'll end up with a blank screen. Click Add icon (the + icon), which brings up the Add Counters window as shown in Figure 10–10.

Add both our counters using the Add button and close this dialog box. Now sit back and watch the fun (you'll also realize what I meant when I said earlier that random data can be fun). Figure 10–11 shows an example of this running on my machine.

Now let's take a look at the program itself in more detail. The first thing we need to do is to create the `"BookExampleCategory"` object if it does not already exist. We can use `PerformanceCounterCategory::Exists` to determine if it has already been created (as when we run the program after

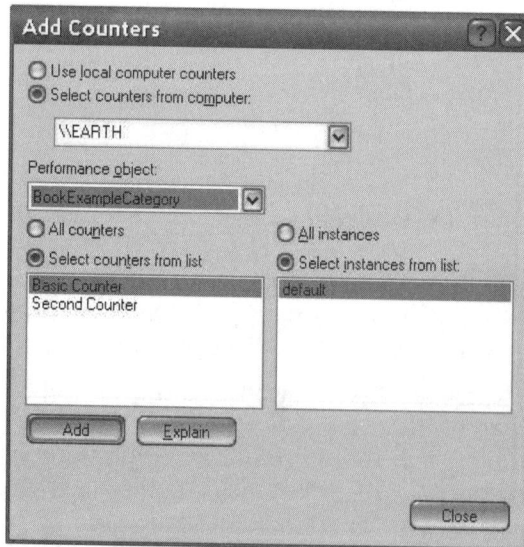

Figure 10–10 Example of running the Add Counters demo

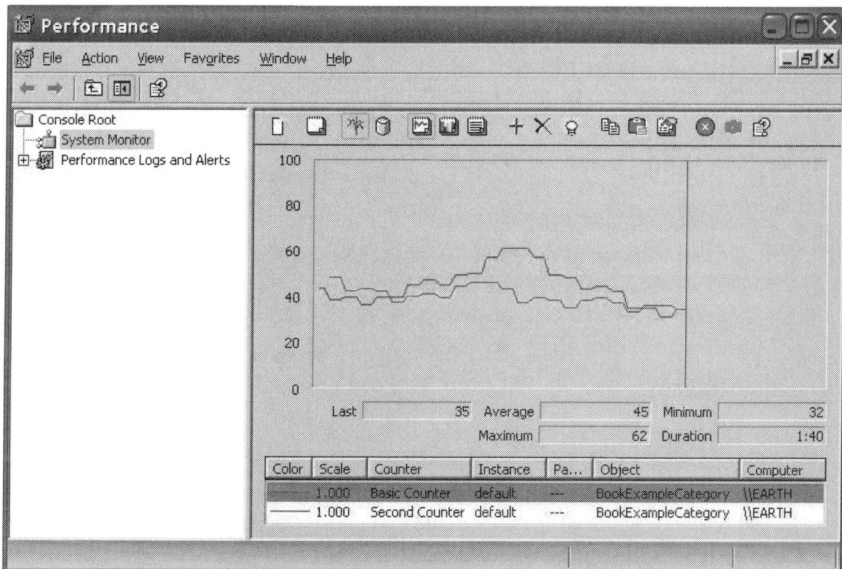

Figure 10–11 Example of inputting random data—representing real-time application benchmark data—in the `Performance` demo

the first time). If it has not been created, we proceed to create it and add the two counters. First we use the `CounterCreationData` class to define our two counters, as shown below.

```
CounterCreationData* ccd[] = new CounterCreationData*[2];
ccd[0] = new CounterCreationData(S"Basic Counter",
     S"Basic Counter", PerformanceCounterType::NumberOfItems32);
ccd[1] = new CounterCreationData(S"Second Counter",
     S"Second Counter", PerformanceCounterType::NumberOfItems32);
```

The first argument to the constructor is the name of the counter, which must be unique per category; the second argument is the counter's help text. The third argument is a `PerformanceCounterType` enumeration that describes the behavior of the counter. Various types of counters are possible, and I have used `NumberOfItems32`, which is an immediate-value type counter in the sense that it simply returns the current value. Other available types of counters include difference counters (which show the difference between the last two values), average counters (which show the average over a period of time), percentage counters (which calculate value as a percentage), and rate counters (which show the rate of change of

the counter values). You must choose your counter type according to your specific requirements. Once we have defined the two counters we'll be using, we create the category itself:

```
PerformanceCounterCategory::Create(
      S"BookExampleCategory",
      S"Book Example Object",
      new CounterCreationDataCollection (ccd));
```

We use the static `Create` method to create our `Performance` object, along with its two associated counters. Now we create `Performance Counter` objects for each of these counters, so that we can start writing data to them:

```
m_pc1 = new PerformanceCounter(S"BookExampleCategory",
      S"Basic Counter",S"Default",false);
m_pc2 = new PerformanceCounter(S"BookExampleCategory",
      S"Second Counter",S"Default",false);

m_pc1->RawValue = 50; //set default value
m_pc2->RawValue = 50; //set default value
```

We've used the following overloaded constructor:

```
PerformanceCounter(String* category, String* counter,
      String* instance, bool ReadOnly);
```

The category, counter, and instance arguments respectively represent the category name, the counter name, and the name of the instance. The `bool` flag specifies if this is a read-only counter object; in our case we set this argument to `false` because we intend to publish data and not just read it. The listing for the `OnInitDialog` method follows (missing AppWizard code):

```
BOOL CPerfCounterWriteDemoDlg::OnInitDialog()
{
      CDialog::OnInitDialog();

      // . . .

#pragma push_macro("new")
#undef new
```

```
//Check if it already exists
if(!PerformanceCounterCategory::Exists("BookExampleCategory"))
{
  //Create the two counters
  CounterCreationData* ccd[] = new CounterCreationData*[2];
  ccd[0] = new CounterCreationData(S"Basic Counter",
    S"Basic Counter",
    PerformanceCounterType::NumberOfItems32);
  ccd[1] = new CounterCreationData(S"Second Counter",
    S"Second Counter",
    PerformanceCounterType::NumberOfItems32);

  //Create the performance object with the two counters
  PerformanceCounterCategory::Create(
    S"BookExampleCategory",
    S"Book Example Object",
    new CounterCreationDataCollection (ccd));
}

//Construct the two performance counter objects
m_pc1 = new PerformanceCounter(S"BookExampleCategory",
  S"Basic Counter",S"Default",false);
m_pc2 = new PerformanceCounter(S"BookExampleCategory",
  S"Second Counter",S"Default",false);

//Set initial values
m_pc1->RawValue = 50;
m_pc2->RawValue = 50;

#pragma pop_macro("new")

//start a timer
SetTimer(1000,3000,NULL);
return TRUE;
}
```

Now let's see our timer-handler function:

```
void CPerfCounterWriteDemoDlg::OnTimer(UINT nIDEvent)
{
    //set seed
    srand((unsigned)time(0));
```

```
//increase randomness
for(int i=0; i<20; i++,rand());

int i1 = ((float)10*((float)rand()/(float)RAND_MAX));
int i2 = ((float)10*((float)rand()/(float)RAND_MAX));
int i3 = ((float)10*((float)rand()/(float)RAND_MAX));
int i4 = ((float)10*((float)rand()/(float)RAND_MAX));

m_pc1->IncrementBy(i1-i2);
m_pc2->IncrementBy(i3-i4);

CDialog::OnTimer(nIDEvent);
}
```

Basically I am generating random numbers between –10 and +10, and thus the two counters keep publishing values that either take them up or take them down in any graphical display. This accounts for the rather spectacular display when you run the Windows Performance Monitor tool. The counter value is changed using the `PerformanceCounter::IncrementBy` method.

Publish Differential Data

In the last example, we published data that can be retrieved as an instantaneous value; but sometimes we might want to measure a quantity's rate of change, as in disk interrupt calls issued per second. We'll now write a real-world program that measures the frequency of window activation on a given machine.

Essentially, we install a global CBT (computer-based training) hook. For global hooks, the hook-handler function must reside in a DLL, so that it can be loaded into the virtual address space of all running applications. I've written a simple Win32 DLL called `MonitorWindowHook.dll`. I'll quickly run through the DLL source code. First of all, we have some shared variables, which are put into a shared data section so that they can be accessed safely across processes.

```
#pragma data_seg(".SHRD")
HHOOK m_hook = NULL;
HANDLE ghModule = NULL;
HWND gParenthWnd = NULL;
#pragma data_seg()
#pragma comment(linker, "/section:.SHRD,rws")
```

In the `DllMain` function, we simply save the DLL's instance handle, which we will need later when we call `CallNextHookEx`.

```
BOOL APIENTRY DllMain( HANDLE hModule,
  DWORD  ul_reason_for_call,
  LPVOID lpReserved
    )
{
  ghModule =  hModule;
  return TRUE;
}
```

Then we have the `SetHook` and `UnSetHook` functions as follow .

```
__declspec( dllexport ) void SetHook(HWND hWnd)
{
  gParenthWnd = hWnd;
  m_hook = SetWindowsHookEx(WH_CBT,
    CBTProc ,(HINSTANCE)ghModule,0);
}


__declspec( dllexport ) void UnSetHook()
{
  UnhookWindowsHookEx(m_hook);
}
```

We save the `HWND` of the main application in the global `gParenthWnd` so that we can later post a message to the parent application. Now for the hook procedure.

```
__declspec( dllexport ) LRESULT CALLBACK CBTProc(int nCode,
  WPARAM wParam, LPARAM lParam )
{

  if(nCode<0)
    return CallNextHookEx(m_hook, nCode, wParam, lParam);

  UINT  wm_GotAHit = RegisterWindowMessage(
    "MonitorWindowHook" );

  if(nCode==HCBT_ACTIVATE)
  {
    PostMessage(gParenthWnd,wm_GotAHit,wParam,lParam);
```

```
    }

    return 0;
}
```

We first check if nCode equals HCBT_ACTIVATE, which indicates that a window is about to be activated. If so, then we post a user-defined message to the parent window using PostMessage, which is always safer than SendMessage when doing cross-process communication. Well, that's all to our DLL, and there was nothing .NETish about it at all.

Now let's take a look at our main program, which is again an MFC dialog-based application with CLR support. Add a list box to the dialog as shown in Figure 10–12 and associate a control member-variable to it.

Add the following to your dialog cpp file:

```
const UINT  wm_GotAHit = RegisterWindowMessage(
  "MonitorWindowHook" );
```

And add this to your message map:

```
ON_REGISTERED_MESSAGE( wm_GotAHit , OnNotification )
```

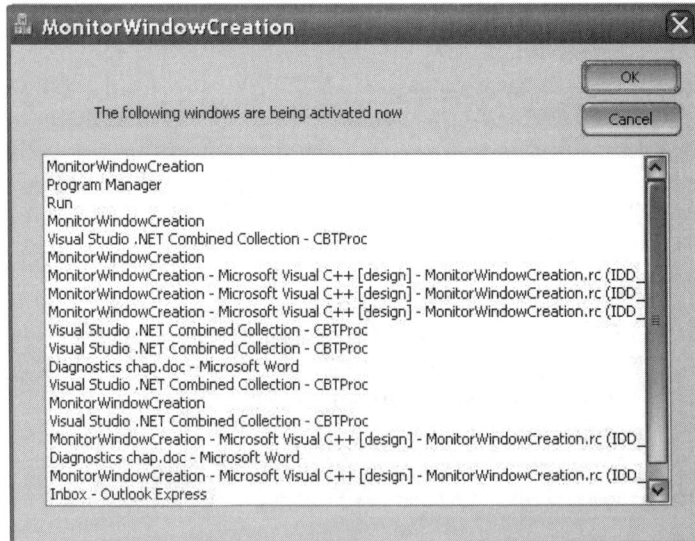

Figure 10–12 Example execution of the MonitorWindowCreation demo

Now add the following code snippet to your `OnInitDialog`:

```
BOOL CMonitorWindowCreationDlg::OnInitDialog()
{
  CDialog::OnInitDialog();

  SetIcon(m_hIcon, TRUE);
  SetIcon(m_hIcon, FALSE);

#pragma push_macro("new")
#undef new

  //Check if the category exists, and if
  //it does not, then create it
  if(!PerformanceCounterCategory::Exists(
    S"WindowActivationMeasure"))
  {
    //Create our counter first and set its type
    //to PerformanceCounterType::RateOfCountsPerSecond32
    CounterCreationData* ccd[] = new CounterCreationData*[1];

    ccd[0] = new CounterCreationData(S"MainCounter",
      S"This is the only counter in this category",
  PerformanceCounterType::RateOfCountsPerSecond32);

    //Now create the category as well as the counter
    PerformanceCounterCategory::Create(
      S"WindowActivationMeasure",
      S"A measure of the rate at which top level windows get
        activated", new CounterCreationDataCollection(ccd));
  }


  //Instantiate a new counter object and associate it
  //with the above created category
  m_pc = new PerformanceCounter(
    S"WindowActivationMeasure","MainCounter",
    S"DefaultInstance",false);

#pragma pop_macro("new")

  //Set our global hook
  SetHook(m_hWnd);
```

```
    return TRUE;
}
```

Well, there is not a lot of difference in the code compared to our previous example, except that this time we have created a counter of type `PerformanceCounterType::RateOfCountsPerSecond32`. And after we finish all that, we set the global hook. Remember to unset the global hook when we exit, and since there are only two ways to exit an MFC dialog-based application, we handle both those contingencies.

```
void CMonitorWindowCreationDlg::OnBnClickedOk()
{
  UnSetHook();
  OnOK();
}


void CMonitorWindowCreationDlg::OnBnClickedCancel()
{
  UnSetHook();
  OnCancel();
}
```

Now let's take a look at where all the action is taking place, namely, the handler for our user-defined message, which the hook procedure posts to our dialog window.

```
LRESULT CMonitorWindowCreationDlg::OnNotification(
  WPARAM wParam, LPARAM lParam)
{
  HWND h = (HWND)wParam;
  char buff[256];
  ::GetWindowText(h,buff,255);
  if(strlen(buff))
  {
    m_pc->Increment();
    m_listbox.InsertString(0,buff);
  }
  return 0;
}
```

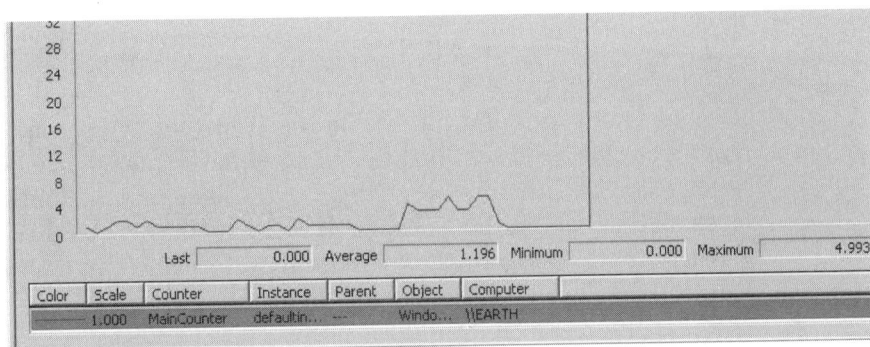

Figure 10-13 Sample running of the `Performance` demo

Basically there is only one line there that matters as far as the performance counter is concerned.

```
m_pc->Increment();
```

The rest of the code is just to add to the fun by populating the list box with the titles of the activated windows. To see the effect of everything we've done so far, simply run the program and then run the Windows Performance Monitor program (`perfmon.exe`). Just start a few programs and click around your desktop to simulate some window creation activity. In fact you may even write a small program that pops up several message boxes inside a loop. You'll see something similar to that shown in Figure 10–13.

Summary

In this chapter you first saw how to use the `EventLog` class and other related types to work with the event log. Specifically, you learned how these types enable applications to enumerate the available event logs, create custom logs, write log entries, enumerate log entries, and even subscribe to the `EntryWritten` event in order to be notified asynchronously whenever a new log entry is added to a specified log. From there, you discovered the `Process` class and how it can be used to start, control, and stop new processes. You saw that it also allows us to wait for a process to exit and to check whether a process is still running or not. We can also handle the

`Exited` event if we need to be asynchronously notified when a process terminates. Finally, you saw code that illustrated how to enumerate running processes and construct a `Process` object, attaching it to one of those processes. The benefit here is that once we accomplish this, we can control and monitor that process just as if the application had started the process itself. The chapter ended with a section displaying the capabilities and usefulness of the `PerformanceCounterCategory` and `PerformanceCounter` classes that are used to diagnose system performance. In that section, you learned that by using these classes, you can enumerate existing performance counters and read performance data, as well as create your own custom counters and publish data to them.

Managing Your Managed Objects

Introduction

Back in Chapter 1, I illustrated just enough .NET and Managed Extensions basics to get you to the point of being able to mix MFC and BCL code. In the ensuing chapters, I intentionally stayed away from allowing this book to become a "Managed Extensions book" and instead focused on the coverage of various BCL classes and how they could be used in MFC applications. Hopefully, since you've reached this point of the book, you've discovered what led me to pen this text to begin with—namely, that by combining the MFC and BCL class libraries, you end up with the best of both worlds and, as a result, become a more productive software developer. That leads me to the focus of this particular chapter.

While the majority of the previous chapters' code snippets and demo applications declared managed objects either as native class members or as local variables, you'll find that the more you mix MFC and managed code, the more you'll have a need to treat your managed objects as you do your native objects. For example, it's common in MFC applications to group semantically related objects in collections or arrays. Typically this is done with either the various map collections or their "templatized" cousins. Unfortunately, you can't store managed objects in the native collections that we're accustomed to using; therefore, one of the tasks that we'll cover here is how to use the various collection types in the BCL to group managed objects.

To that end, the chapter begins by introducing the basic interfaces that collection classes implement in order to be called collections. From there, you will see how several of the most commonly used collection classes

chose to implement those interfaces in order to give you a wide variety of collection types. The collection classes that you'll learn about are the `ArrayList`, `Hashtable`, `SortedList`, `Queue`, and `Stack` classes. Once you've seen the most commonly used collection classes and understand their interface underpinnings, you'll then learn how to design and code a custom collection class that can be treated (from a client code perspective) just as if it were one of the BCL-supplied collection classes. The example illustrated will allow for the enumeration and sorting of a collection of `Programmer` objects and will be completed with a step-by-step MFC demo application that shows how easy a custom collection can make client-side code that needs to use an enumerable, sortable collection for grouping managed objects.

Another basic task that is performed on native objects is that of *serialization* via the `CArchive` and related MFC classes. Once again, attempting to serialize managed classes using a native technique is far more trouble than it's worth. However, it's obviously very convenient to be able to persist your objects to and from disk in certain applications. Therefore, the last section of this chapter will introduce you to a very quick and easy technique of serializing entire managed objects to and from disk. After that you'll learn how to use attributes to specify which members of a class to serialize and even discover the technique of custom serialization, which gives you the most control over which members are serialized and how they're serialized.

Using .NET Collections

Collections in .NET are objects that implement various interfaces defined in the `System::Collections` namespace. This section will begin by giving an overview of these interfaces, their members, and their relations to one another. From there, you'll be introduced to some of the various BCL-supplied classes that implement these interfaces. For example, the first class you'll see is the `ArrayList` class and how it encapsulates the functionality of a dynamic array that automatically resizes as items are added. From there, you'll learn about the `Hashtable` class, which encapsulates a collection of key/value pairs that are organized on the basis of on the hash code of the key. (If you're new to hash codes, I would recommend reading the relevant sections of Chapter 4, "Cryptography, Hash Codes, and Data Encryption.") The next class to be covered is the `SortedList` class. As the name

implies, this class also implements a key/value pair collection. However, the SortedList class is a hybrid between the Hashtable class and Array class in that you can access collected objects using a key or an index value. The penultimate class that I'll cover in this section is the Queue class, which encapsulates a first-in, first-out (FIFO) collection of objects. Finally, the last class in this section is the Stack class, which represents the standard stack metaphor—a last-in-first-out collection of objects. The section winds up with a demo application illustrating how you can use the various collection interfaces to create your own enumerable managed collection types.

Collection Basics

Collections are groupings of semantically related objects that—depending on the collection type—can be enumerated, compared, and retrieved either by index or key. In the BCL, collections are defined by what interfaces they implement. With that in mind, I'll briefly cover the various collections interfaces in this section and explain how each helps to define the collection class that implements it.

- **IEnumerable**—The IEnumerable interface is what a collection implements to publicize that it is an enumerable entity. Its main job is to provide a layer of abstraction between the client and the collection's internal enumerator, represented by the IEnumerate interface. I'll get into this in a lot more detail as the chapter goes on and especially when you reach the section entitled "Implementing a Custom Collection Class," where you see firsthand how to create your own enumerable and sortable collection class.
- **ICollection**—The ICollection interface defines a few informational properties regarding the collection as well as a copy method that enables clients of the collection to copy the contents of one collection to another.
- **IList**—The IList interface adds support for the most basic index-based collection operations such as item retrieval, addition, and removal.
- **IDictionary**—Of course index-based collections aren't all that is supported by the framework: there are also interfaces for supporting dictionary (key/value) collections. The IDictionary interface adds support for the most basic key/value-pair-based collection operations, where values can be retrieved, inserted, and removed on the basis of a key value.

- **IDictionaryEnumerator**—The `IDictionaryEnumerator` class adds a more direct method of enumerating a collection based on key/value pairs. You'll see this interface used in the section entitled "Enumerating Hash Table Entries."

Also note that there are some dependencies regarding the implementation of, or derivation from, the various `System::Collections` namespace interfaces.

- The `IList` interface requires the implementation of both `ICollection` and `IEnumerable`.
- The `ICollection` interface requires the implementation of `IEnumerable`.
- The `IDictionary` interface requires the implementation of `ICollection` and `IEnumerable`. Also note that the `IDictionary` interface defines a method that returns an instance of a class implementing `IDictionaryEnumerator`.

The `IEnumerable` and `IEnumerator` Interfaces

In order to support the enumeration of its data items, a class implements two distinct interfaces: `IEnumerable` and `IEnumerator`. The `IEnumerable` interface exposes a single member—`GetEnumerator`—that acquires the collection's `IEnumerator` interface. It's the `IEnumerator` interface that is used to iterate over the data in the collection; as a result, when I use the generic term *enumerator*, I'm referring to the `IEnumerator` interface. The `IEnumerator` interface defines the three methods listed in Table 11–1.

Table 11–1 `IEnumerator` Members

Member name	Description
`__property Object* get_Current();`	Returns the current item in the collection.
`bool MoveNext();`	Moves to the next item in the collection, returning `true` if it successfully acquired the next item, and `false` if the enumerator has passed the end of the collection.
`void Reset();P`	Sets the enumerator to its initial position, which is before the first element in the collection.

It's important to understand that the IEnumerator interface doesn't do anything itself. It just defines the members listed in Table 11–1 such that any client working with a class that implements IEnumerator knows that the class's data can be enumerated using those members. Also note that while the class is free to store the enumerable data internally in any fashion, there are some basic guidelines regarding the implementation of the IEnumerator interface that should be followed so that clients of the class experience behavior that is consistent with all classes implementing the IEnumerator interface.

- For each instance of the class, the object should maintain an internal position, or cursor, value that enables it to keep track of where the client is in enumerating the object's data.
- The Current property should then return the data item (in the form of an object) currently referenced by the position value.
- The MoveNext method should increment the internal position value, but doesn't return an object. (Only the Current property returns data). MoveNext should also return a Boolean value indicating if there is more data to be enumerated.
- The Reset method should cause the class to initialize the position value to one less than the beginning of the data so that MoveNext will move to the first data item.
- While the IEnumerator instance is in use, any modifications to the collection should invalidate the enumerator, as the current position cannot be guaranteed. Therefore, the class should keep track of whether or not the enumerator is currently valid at all times. If the enumerator is not valid and the client calls Current or MoveNext, an InvalidOperationException exception should be thrown.
- For an invalidated enumerator, calling Reset reinitializes the enumerator so that the enumerator is valid again.
- Calling the Current property when the position value doesn't point to a valid data item should result in an InvalidOperation Exception. For example, calling Current after initializing the enumerator, but before calling MoveNext, is an error. Also, calling Current once the MoveNext has moved beyond the last data item is an error. (This is why it's important for MoveNext to return a value properly indicating if more data can be read).

This looks like a lot of rules, but as you'll see in the section entitled "Implementing a Custom Collection Class," it's really quite easy to implement a

customized collection, and adhering to these rules definitely makes coding easier on the client side.

The ICollection Interface

The ICollection interface is a key interface implemented by all collection types defined by the BCL. This interface defines three properties: Count, IsSynchronized, and SyncRoot, as well as a CopyTo method. You'll see how these properties are used as we get into the actual BCL classes that implement the interfaces. For now, property and method definitions are shown in Table 11–2.

IList Interface

The IList interface encapsulates a collection of objects that can be individually stored, retrieved, and removed by index or value. Table 11–3 shows the members of the IList interface. (Note that the IList interface implements both the ICollection and the IEnumerable interfaces. Therefore, while I've omitted those members in the table for the sake of brevity, those members are also inherited.)

Now we finally have something that looks like its collection as opposed to supporting a collection's definition! In fact, any class that implements the IList interface can be treated as if it were an array. Note that I said "as if it were an array." The underlying implementation does not have to be an

Table 11-2 ICollection Members

Member name	Description
`__property int get_Count();`	A read-only property, which returns the number of items added to the collection.
`__property bool get_IsSynchronized();`	Returns whether access to the underlying collection is synchronized, meaning that only one operation on the collection can occur at a time.
`__property Object* get_SyncRoot();`	When implemented, this property is used to synchronize access to a collection object.
`void CopyTo ( Array* array, int index );`	Copies the elements from the collection to a managed array.

Table 11-3 `IList` Members

Member name	Description
`__property bool` `  get_IsFixedSize();`	Returns whether the collection is of a fixed size.
`__property bool` `  get_IsReadOnly();`	Returns whether the collection is read-only; i.e., addition, removal, and modification of elements aren't allowed.
`__property Object*` `  get_Item( int index );`	Returns the item at the specified index.
`__property void set_Item(` `  int index, Object* );`	Sets the item at the specified index.
`int Add( Object* value );`	Adds an item to the collection, returning the items index.
`void Clear();`	Clears all items from the collection.
`bool Contains(` `  Object* value );`	Returns whether the specified value is in the collection.
`int IndexOf(` `  Object* value );`	Returns the index of the specified value; returns −1 if the value isn't in the collection.
`void Insert( int` `  index, Object* value );`	Adds an item at the specified index to the collection, pushing other elements down.
`void Remove( Object*` `  value );`	Removes an item from the collection; elements further down the collection are pulled up to fill the hole.
`void RemoveAt( int` `  index );`	Removes the item from the specified index; elements further down the collection are pulled up to fill the hole.

array at all, but that doesn't matter to the class's client code. This is why it's important to use the `IEnumerator` interfaces as opposed to directly accessing the object's data. In using the `IEnumerator` to traverse the elements of a given object (collection), you're using a standard interface and are abstracted from the object's internals (such as how the data is stored).

As an example of what I mean, let's say I were to create a linked-list class. I might be tempted to implement the `GetItem` method to traverse to the specified node each time it was called. This would work fine until the client wants to traverse the entire collection. With this (inefficient) design,

the class would end up traversing the early nodes multiple times (i.e., 0, 0 -> 1, 0 -> 1 -> 2, 0 -> 1 ...). A better design would be for the class to implement the IEnumerator interface so that the class can keep track of the current node as the client is enumerating the class's data. That way, the class doesn't start from the beginning of the internal data list each time the client wants to move to the next element.

Using the IList interface is fairly simple because the methods it exposes are like those you would find on any C++ array class wrapper. Here's the basic client code for using an IList-implementing class.

```
#using namespace System::Collections;

// Get an IList interface from an implementing class
// For example, the DataView and SqlParameterCollection classes
// (covered in Chapter 6) both implement IList.
IList* ilist = (IList*)...

// Construct item to add - specific to what the
// collection expects.
Object* item  = ...

// Add item to the collection
long index = ilist->Add(item),

// Is our item in the collection?
bool bContains = ilist->Contains(item);

// Retrieve item by index
item = ilist->Item[index];

// Construct a second item
Object* item  = ...

// Replace the value at index with the new item
ilist->Item[index] = item2;
```

Besides the two specialized classes I mentioned in the previous code snippet's comments, two very commonly used BCL types that implement the IList interface are the Array and ArrayList classes

- **Array**—The most basic type of collection is the array. As you learned in Chapter 1, arrays are represented in the BCL by the

`Array` class. As you might suspect by having worked with `Array` objects throughout this book and by comparing experience with the members listed in Table 11–3, the `Array` class implements the `IList` interface.

- **`ArrayList`**—The most common type of dynamic collection used is an `ArrayList`. This class creates an array that will dynamically size itself as items are added or removed. It also doubles the array size when it has run out of space in the array. It does expose a `Capacity` property, which will return the size of the array, and setting this value will reallocate the array to the specified number of elements. I'll cover the `ArrayList` in a bit more detail in the next section.

Throughout the BCL there are many different collection classes. Most of these collections are designed to be used specifically in concert with other classes. For example, the `System::Data::DataView` collection encapsulates an array of data rows and is really of no use unless you are specifically working with ADO.NET. However, there are a few collection classes that are specifically designed to be used in any situation where you need to collect generic managed objects. In the next few sections, I'll introduce several of the more commonly used of these collection classes: `ArrayList`, `Hashtable`, `SortedList`, `Queue` and `Stack`.

Notes on This Chapter's Code Snippets and Demos

The collection classes support dozens of methods and properties, and covering each one would be very tedious reading indeed. Therefore, I've chosen to cover only the most commonly used methods and group them by task. Even then, the majority of these tasks can be accomplished with a single line of code. As a result, these tasks are illustrated using short code snippets as opposed to the full demos that you've seen throughout this book, for the simple reason that complete MFC applications would be overkill for this particular topic matter. (The CD does contain demos for your convenience that illustrate using the collection classes presented in this chapter.)

In addition, as you probably won't read this section from beginning to end, dividing this section into task-related subsections enables you to use this section as a reference, so that you can easily jump to the specific class and task as needed.

Finally, the sections illustrating how to design and define your own custom collection and how to serialize objects contain complete demos, as those two subjects are more involved.

Using the `ArrayList` Class

At the end of this section, you'll see a demo where I illustrate how to create a simple custom collection to store numbers. In that demo, I specifically chose the `ArrayList` collection as I wanted a collection that automatically resized itself when needed. That is the key benefit of using the `ArrayList` over the `Array` object, and it is the subject of this section. Let's first look at some key tasks performed with this class and then a small demo application illustrating its use.

Adding and Inserting Items

As mentioned earlier, once objects are added to the `ArrayList` class, they can be searched, retrieved, or removed either by their index or by specifying the value inserted into the `ArrayList`. The simplest way to add an object is by using either the `Add` method or the `Insert` method:

```
int idx;
ArrayList* arrayList = new ArrayList();

idx = arrayList->Add(__box(42));          // index 0
idx = arrayList->Add(S"dummy entry");     // index 1
idx = arrayList->Insert(1, S"excuse me"); // index 1
```

As you can discern from the previous code example, calling the `Add` method appends the specified object to the end of the `ArrayList`, while the `Insert` method is used to pinpoint the exact index of the insertion.

In addition to these two methods, the `ArrayList` also supports the `AddRange` and `InsertRange` methods. These two methods enable you to add (or insert at a specified index) any object that implements the `ICollection` interface—in other words, any collection object. In the following code example, three distinct arrays are created: the second one is added to the first, and the third is inserted into the first at a specified index.

```
// Create and initialize three ArrayList objects
ArrayList* arrayList1 = new ArrayList();
arrayList1->Add(S"January");
arrayList1->Add(S"February");
arrayList1->Add(S"March");
arrayList1->Add(S"April");
```

```
ArrayList* arrayList2 = new ArrayList();
arrayList2->Add(S"September");
arrayList2->Add(S"October");
arrayList2->Add(S"November");
arrayList2->Add(S"December");
```

```
// Copy the arrayList2 elements to the END of arrayList1
arrayList1->AddRange(arrayList2);
```

```
ArrayList* arrayList3 = new ArrayList();
arrayList3->Add(S"May");
arrayList3->Add(S"June");
arrayList3->Add(S"July");
arrayList3->Add(S"August" ;
```

```
// Insert arrayList3 elements arrayList1 at index 4
arrayList1->InsertRange(4, arrayList3);
```

This code results in the following arrays:

```
// arrayList1
[0] = January
[1] = February
[2] = March
[3] = April
[4] = May
[5] = June
[6] = July
[7] = August
[8] = September
[9] = October
[10] = November
[11] = December
```

```
// arrayList2
[0] = September
[1] = October
[2] = November
[3] = December
```

```
// arrayList3
[0] = May
```

```
[1] = June
[2] = July
[3] = August
```

Removing Items and Item Ranges

As with adding items, you can remove an item from an ArrayList either by its index or its object value, using the RemoveAt and Remove methods, respectively. Additionally, you can remove an entire range of items at once via the RemoveRange method, where you specify a starting index and the number of elements to remove. Finally, removing all of an ArrayList object's elements can be accomplished by calling the Clear method, as shown in the following listing.

```
// Create and initialize an ArrayList object
ArrayList* arrayList1 = new ArrayList();
arrayList1->Add(S"January");
arrayList1->Add(S"February");
arrayList1->Add(S"March");
arrayList1->Add(S"April");
arrayList1->Add(S"May");
arrayList1->Add(S"June");

// Remove element at index 1 (February)
arrayList1->RemoveAt(1);

// Remove March element (currently index 1)
arrayList1->Remove(S"March");

// Remove everything after January/April
arrayList1->RemoveRange(2, arrayList1->Count - 2);

// Remove all remaining elements
arrayList1->Clear();
```

Retrieving and Setting Items by Index

The Item property enables you to retrieve an array list element by its index or even replace an element at a specified index.

```
// Create and initialize an ArrayList object
ArrayList* arrayList1 = new ArrayList();
```

```
arrayList1->Add(S"January");
arrayList1->Add(S"December");


// retrieve second element
String* s = arrayList1->Item[1]->ToString();


// replace second element
arrayList1->Item[1] = S"February";
```

Searching for Items by Value

There are two methods that provide the ability to search for an item in an
ArrayList. One method is the Contains method where the return
boolean value indicates the success or failure of the search. The second
method is the IndexOf method, which is more involved in that not only will
it return the index of a successful search, but it also allows you to search in a
given range of elements.

```
// Create and initialize an ArrayList object
ArrayList* arrayList1 = new ArrayList();
arrayList1->Add(S"January");
arrayList1->Add(S"February");
arrayList1->Add(S"March");
arrayList1->Add(S"April");
arrayList1->Add(S"May");
arrayList1->Add(S"June" ;


// This search would fail
if (arrayList1->Contains(S"December"))
   ;


int idx;


// Searches entire array and would return an index of 2
idx = arrayList1->IndexOf(S"March");


// Searches array starting at index 1 and would
// return an index of 2, as the returned index value
// is relative to the beginning of the array - not
// the beginning of the search
idx = arrayList1->IndexOf(S"March" , 1);
```

```
// Searches array starting at index 1 and searches
// only 2 elements. Therefore, the search would fail
// as March is the 3rd element (index 2)
idx = arrayList1->IndexOf(S"March", 1, 2);
```

Enumerating the `ArrayList`

There are two ways to enumerate an ArrayList object. The first would be the standard for loop; the second takes advantage of the fact that the ArrayList class implements the IEnumerable interface. Here are two examples of using the IEnumerator interface—one involving reference objects and one involving value objects that need to be unboxed.

```
ArrayList* arrayList1 = new ArrayList();
arrayList1->Add(S"January");
arrayList1->Add(S"February");
arrayList1->Add(S"March");

// Enumerate reference types
IEnumerator* enumerator1 = arrayList1->GetEnumerator();
while ( enumerator1->MoveNext() )
{
  String* value = enumerator1->Current->ToString();

  // ...
}

ArrayList* arrayList2 = new ArrayList();
arrayList2->Add( __box(6) );
arrayList2->Add( __box(42) );

// Enumerate value types
IEnumerator* enumerator2 = arrayList2->GetEnumerator();
while (enumerator2->MoveNext())
{
  int number = *dynamic_cast<__box int*>(enumerator2->Current);

  // ...
}
```

Using the `Hashtable` Class

The `Hashtable` class, which implements `IDictionary`, is the most commonly used BCL dictionary-based collection and is the standard sort taught in computer science courses. By the way, while the class name `Hashtable` does look like a typo, it isn't—and this will take some getting used to, as it's completely inconsistent with the naming conventions of other BCL classes: *t* in *table* is lowercase. A hash table is basically a key/value-pair collection, where each value is stored using an index, which is the hash of the key for that particular value. When an element is added to the hash table, it is saved so that future retrievals of that element based on its key will be performed as efficiently as possible. Hash tables internally consist of various subtables called *buckets*. When an element is added, on the basis of its key's hash value, a particular bucket is chosen to hold the element. Thus when the value is requested later, only the bucket needs to be searched instead of the entire hash table. Now let's see some code snippets illustrating how to use the `Hashtable` class.

Adding Entries

Keys (and their associated values) are added to a `Hashtable` via the `Add` method, where the first parameter is the key and the second parameter is the value to be associated with the key. Note that while the key cannot be NULL, the value can be NULL.

```
Hashtable* hashTable = new Hashtable();

hashTable->Add( S"January", S"Enero");
hashTable->Add( S"February", S"Febrero");
hashTable->Add( S"March", S"Marzo");
```

Removing Entries

The `Hashtable` class supports the removal of all entries at once (via the `Clear` method) or by key value (via `Remove`).

```
Hashtable* hashTable = new Hashtable();

hashTable->Add( S"January", S"Enero");
hashTable->Add( S"February", S"Febrero");
hashTable->Add( S"March", S"Marzo");

hashTable->Remove ( S"February" );
```

One quirk about the `Remove` method is that if the key does not exist, it will not throw an exception, and since it returns a `void`, the return value also does not give us any indication of what actually happened. Therefore it's always best to first check if the key exists using `ContainsKey` (shown next) if you are interested in knowing whether or not a deletion took place.

Retrieving and Setting Individual Entries

As with all other collections, you can use the `Item` property to retrieve or set the `Hashtable` entries, respectively. In fact, you can use the `Item` property to add entries to the `Hashtable`—with the caveat being that if the key already exists, you will be replacing it. Therefore, it is best to use the `Add` method when adding new entries and the `Item` property when replacing entries.

Searching Hash Tables

The `ContainsKey` method will return a Boolean value indicating if a particular key exists in the hash table. In addition, the `Hashtable` class also supports searching by value via the `ContainsValue` method. (Note that the `Hashtable` class also implements the `IDictionary::Contains` method, which is the functional equivalent of the `ContainsKey` method.)

```
Hashtable* hashTable = new Hashtable();

hashTable->Add( S"January", S"Enero");
hashTable->Add( S"February", S"Febrero");
hashTable->Add( S"March", S"Marzo");

// All of the following searches return true
hashTable->Contains ( S"February" ); // returns true
hashTable->ContainsKey ( S"February" ); // returns true
hashTable->ContainsValue ( S"Febrero" ); returns true;

// Returns false as February is the key and not the value
hashTable->ContainsValue ( S"February" );
```

Enumerating Hash Table Entries

As with any collection, you can use an enumerator to iterate over the `Hashtable` collection. However, in the case of the `Hashtable` class the

returned enumerator is an IDictionaryEnumerator interface. The significance is the IDictionaryEnumerator contains three members: Key,
Value, and Entry (which is a DictionaryEntry value-type class containing both the Key and Value members). Here's an example of enumerating
a Hashtable.

```
Hashtable* hashTable = new Hashtable();

hashTable->Add( S"January", S"Enero");
hashTable->Add( S"February", S"Febrero");
hashTable->Add( S"March", S"Marzo");

if (hashTable->Count)
{
  IDictionaryEnumerator *enumerator = hashTable->GetEnumerator();

  String* format = S"{0,-50} "
                   S"{1,-10} "
                   S"{2}";

  Console::WriteLine(format, S"Object Type", S"Key", S"Value");

  while ( enumerator->MoveNext() )
  {
    // Can now use enumerator->Key or enumerator->Value
    Console::WriteLine(format,
                       enumerator, enumerator->Key, enumerator->
                       Value);

    // Can also retrieve DictionaryEntry object
    // and use its Key and Value properties
    DictionaryEntry entry = enumerator->Entry();
    String* key = entry.Key->ToString();
    String* value = entry.Value->ToString();

    Console::WriteLine(format,
                       __box(entry)->GetType()->ToString(),
                       entry.Key->ToString(),
                       entry.Value->ToString());
  }
}
```

Note that I had to box the `DictionaryEntry` object in order to retrieve its type information as it's a value type. This code snippet produces the following output:

Object Type	Key	Value
System.Collections.Hashtable+HashtableEnumerator	March	Marzo
System.Collections.DictionaryEntry	March	Marzo
System.Collections.Hashtable+HashtableEnumerator	February	Febrero
System.Collections.DictionaryEntry	February	Febrero
System.Collections.Hashtable+HashtableEnumerator	January	Enero
System.Collections.DictionaryEntry	January	Enero

Using the `SortedList` Class

Despite its misleading name, the `SortedList` class does not implement the `IList` interface; rather, it is a dictionary-based collection (implementing the `IDictionary` interface) that behaves much like an array. In fact, the `SortedList` is a hybrid of the `ArrayList` and the `Hashtable` classes in that it supports an indexed sequence of elements as well a *sorted* key/value-pair collection. Note that an `ArrayList` or a `Hashtable` class is far more efficient unless you need a sorted collection or will need to the ability to both search by key and sequentially enumerate the collection. Let's now look at some typical operations using this class.

Adding Items

As with the `Hashtable` class, adding items to the `SortedList` class is done via the `Add` method.

```
SortedList* sorted = new SortedList();

sorted->Add(S"Tom Archer", S"Author");
sorted->Add(S"Nishant S.", S"Contributing Author");
sorted->Add(S"James Johnson", S"Technical Editor");
sorted->Add(S"Erik Westermann", S"Technical Editor");
```

As each element is added to the `SortedList` collection, it is inserted into the right position so that the list remains sorted on the key even after the insertion. As a result, you should never assume that an index value will remain constant.

Removing Items

As with all collections, the `Clear` method will remove all entries. In addition, you can remove items by specifying either the index or key value.

```
SortedList* sorted = new SortedList();

sorted->Add(S"Tom Archer", S"Author");
sorted->Add(S"Nishant S.", S"Contributing Author");
sorted->Add(S"James Johnson", S"Technical Editor");
sorted->Add(S"Erik Westermann", S"Technical Editor");

// Remove element at index 1 (Nishant S.)
sorted ->RemoveAt(1);

// Remove "James Johnson" element (currently index 1)
sorted ->Remove(S"James Johnson");

// Remove all entries
sorted->Clear();
```

Retrieving and Setting Items

There are two ways in which to retrieve and set items: by index or by key. The first technique is accomplished via the `GetByIndex` and `SetByIndex` method pair.

```
SortedList* sorted = new SortedList();

sorted->Add(S"Tom Archer", S"Author");
sorted->Add(S"Nishant S.", S"Contributing Author");
sorted->Add(S"James Johnson", S"Technical Editor");
sorted->Add(S"Erik Westermann", S"Technical Editor");

sorted->SetByIndex( 3, S"Contributing Author" );
String* erik = (String*) sorted->GetByIndex( 3 );
```

The second technique for retrieving or setting a `SortedList` object's items is via the dual-mode `Item` property.

```
SortedList* sorted = new SortedList();

sorted->Add(S"Tom Archer", S"Author");
sorted->Add(S"Nishant S.", S"Contributing Author");
```

```
sorted->Add(S"James Johnson", S"Technical Editor");
sorted->Add(S"Erik Westermann", S"Technical Editor");

String* position = sorted->Item[S"Erik Westermann"]->ToString();
sorted->Item[S"Nishant S."] = S"Nishant Sivakumar";
```

Enumerating a *SortedList*

You can enumerate a SortedList by taking into account the fact that it implements the IDictionary interface, just as you did with the Hashtable. In the following example, I'm enumerating the sorted list and retrieving the key and value for each entry into two String objects, which are then output to the screen.

```
. . .

IDictionaryEnumerator* enumerator = sorted->GetEnumerator();

String* format = S"{0,-20} {1}";
Console::WriteLine(format, S"Key", S"Value");

while (enumerator->MoveNext())
{
  String* key = enumerator->Key->ToString();
  String* value = enumerator->Value->ToString();

  Console::WriteLine(format, key, value);
}
```

This enumeration through our SortedList object results in the following output. Notice that the values are sorted according to the key, as opposed to the Hashtable collection, which provides no sorting capability.

```
Key                  Value
Erik Westermann      Technical Editor
James Johnson        Technical Editor
Nishant S.           Nishant Sivakumar
Tom Archer           Author
```

As mentioned earlier, a `SortedList` object can also be enumerated in sequential order (in the order in which the items were added) by treating it as an indexed list.

. . .

```
String* format = S"{0,-20} {1}";
Console::WriteLine(format, S"Key", S"Value");

for( int i = 0; i < sorted->Count; i++ )
{
   String* key = sorted->GetKey(i)->ToString();
   String* value = sorted->GetByIndex(i)->ToString();

   Console::WriteLine(format, key, value);
}
```

If you run the preceding code, you will see the same output as that in the previous example.

Using the `Queue` Class

The `Queue` class is slightly different than the preceding collection classes in that it does not implement either `IDictionary` or `IList`. However, as it implements both the `ICollection` and `IEnumerable` interfaces, it is technically a collection—in this case representing a first-in, first-out (FIFO) list, or queue, of items that automatically sizes itself as needed when items are added.

The `Queue` class is implemented as a circular array. This is, of course, the standard technique used in implementing a queue class, as it reduces the amount of required copying when the first item is dequeued (removed from the queue). The basic operations that can be done on a queue are adding an element to the end of the queue (*enqueuing*), removing an item from the beginning of the queue (*dequeuing*), and peeking at the first item in a queue without removing it. We'll see some sample code snippets on how to use the `Queue` class to perform these basic operations.

Adding Entries

Adding entries, or enqueuing, to a `Queue` object is done via the `Enqueue` method. You can add NULL values to the queue, and since the collection is

not a dictionary, you can also specify duplicate values. If the number of items in the queue equals the capacity of the circular array used to store the elements, then a new array is created with increased capacity, the existing array is copied to the new array, and then the new element is inserted. By default the initial capacity of a queue is 32, and the growth factor is 2.0. (You can override these default values by specifying them with one of the overloaded Queue class constructors.)

```
Queue* q = new Queue();

q->Enqueue(S"Mail 01");
q->Enqueue(S"Mail 02");
q->Enqueue(S"Mail 03");
q->Enqueue(S"Mail 04");
```

Removing Entries

Removing entries from a Queue collection, or dequeuing, is accomplished via the Dequeue method—keeping in mind that you can only remove the first entry from a queue. As a result, repeatedly calling the Dequeue method will empty the queue, as will simply calling its Clear method.

Peeking at the First Element

Peeking lets you examine the first element in the queue without having to remove that element. If you try to peek on an empty queue, the collection throws an InvalidOperationException exception. Thus it's safest to first check the Count property and make sure that it's not zero.

```
if (q->Count)
{
   Object* obj = q->Peek(); // peek but don't remove
}
```

Note that the Peek method always returns the queue's first element. Therefore, repeatedly calling Peek will not let you enumerate the queue.

Enumerating a Queue

Some would argue that theoretically you shouldn't enumerate a queue, as you should only be able to access the first element in the queue. But obviously real-world scenarios crop up where being able to enumerate over a

queue might be handy. As a result, you can always take advantage of the fact that while the `Queue` doesn't implement its own special enumerator type, it can still be enumerated like any other collection:

```
void DisplayQueue(Queue *q)
{
  Console::WriteLine(S"Total elements : {0}", __box(q->Count));

  IEnumerator* en = q->GetEnumerator();
  while (en->MoveNext())
  {
    Console::WriteLine(en->Current->ToString());
  }
}
```

This results in the following output:

```
Total elements : 4
Mail 01
Mail 02
Mail 03
Mail 04
```

Using the `Stack` Class

The `Stack` class is a last-in, first-out (LIFO) data structure. As with the `Queue` class, the `Stack` class only implements the `ICollection` and `IEnumerable` interfaces, and is mainly used when storing temporary data for later processing. Choosing between the `Queue` and `Stack` classes is therefore predicated by the desired order in which you want to retrieve the elements.

As with most stack implementations, the `Stack` class defines two methods—`Push` and `Pop`—that allow for the addition and removal of items to and from the collection, respectively. Additionally, the `Stack` class provides the nonstandard `Peek` method, which returns the item at the top of the stack without removing (popping) it. Now, let's look at the common operations through some simple code snippets.

Pushing Values onto the Stack

The `Push` method will add an element to the top of the stack. If the stack has reached its current capacity, a new internal buffer twice the size of the

existing buffer is created, and the old buffer is copied to the new buffer. The new element is then is inserted into this buffer.

```
Stack* st = new Stack();
st->Push( S"First" );
st->Push( S"Second" );
st->Push( S"Last" );
```

Popping and Peeking

The Pop method will return the top-most element in the stack and will then remove that element. If you do not want to remove the item, use the Peek method.

```
// Retrieve (and remove) top item off stack
Object* o1 = st->Pop();

// Retrieve (but leave) top item off stack
Object* o2 = st->Peek();
```

Implementing a Custom Collection Class

So far we've seen how to use some of the basic collection classes provided by the .NET Framework library. But obviously it would make good sense to be able to design your own custom collection class that can be treated just as if it were one of the BCL collection classes. One basic feature expected of any collection class is that it should be enumerable, and to achieve this, all we need to do is implement the IEnumerable interface in our collection class. Essentially this is all you need to do to have a legally valid collection class. In this section we'll see an example where we write our own collection class, and just to make it more interesting we'll also add sorting ability to the collection class by implementing the IComparable interface. We'll discuss each of these interfaces and other required interfaces further as we use them in our example project.

Designing the Collected Class

Obviously, if you're using generic strings or numbers and such, any BCL collection will suffice. However, the point of designing and coding your own collection class is that you have a design that is specific to a given type. In our example the collected type will be a class called Programmer. In fact,

the design of this class will also enable the direct comparison of two Programmer objects, thereby making it possible to sort a collection of Programmer objects.

To be able to compare two Programmer objects, we'll need to implement the IComparable interface for the Programmer class. The IComparable interface has just one member—a method called CompareTo.

```
// The IComparable interface defines only one member
int CompareTo(Object* obj);
```

The CompareTo method compares the current object instance with the specified object. Obviously, this object should be of the same type—otherwise an ArgumentException will be thrown by the runtime. Much like the various string-comparison functions in C and C++, the CompareTo method returns an int that determines the results of the comparison. A value of 0 indicates that the two objects were deemed "equal." A negative return value indicates that the current instance is "less than" the specified object, and a positive value indicates that the current instance is "greater than" the object. Obviously, the terms "equal to," "less than," and "greater than" can be class-specific and might refer to any desired member of the class that you wish to use for sorting purposes.

You'll recall that the IEnumerable interface is used to publicize the fact that an object can be enumerated, while the IEnumerator interface is what is used to perform the actual enumeration. Similar to that design, the IComparable interface has a closely related IComparer interface that is implemented by classes whose main purpose is to act as comparers. Examples of this are the Array and ArrayList classes, which both support overloaded Sort methods that take an IComparer object as an argument to use in sorting the collection. This enables us to have objects that can be sorted by any desired class member.

The IComparer interface defines just a single method called Compare, which takes two objects of the same type, compares them, and returns a value indicating whether they are equal or not.

```
// The IComparer interface defines only one member
int Compare( Object* obj1, Object* obj2 );
```

Just as in the case of the IComparable::CompareTo method, Compare returns 0 to show that the objects are identical, a negative number if the first object is less than the second object, and a nonzero positive number if the first object is greater than the second object.

Now let's see the basic framework of the `Programmer` class. (We'll fill in the methods as we go forward.)

```
__gc class Programmer : public IComparable
{
public:
  Programmer(String* fullName, int experienceInYears, int
    annualSalary)
  {
    this->fullName = fullName;
    this->experienceInYears = experienceInYears;
    this->annualSalary = annualSalary;
  }

protected:
  String* fullName;
  Int32 experienceInYears;
  Int32 annualSalary;

public:
  __property String* get_FullName() { return this->fullName; }

  __property Int32 get_ExperienceInYears()
  { return this->experienceInYears; }

    __property  Int32 get_AnnualSalary() { return this->annualSalary; }

public:
  int CompareTo(Object* obj)
  {
    /* implemented shortly */
  }

  __gc class ProgrammerSalaryComparer : public IComparer
  {
    /* implemented shortly */
  };

  __gc class ProgrammerExperienceComparer : public IComparer
  {
    /* implemented shortly */
  };
};
```

As you can see, we now have the beginnings of a class that implements the IComparable. Aside from that, all we've done is to define a constructor, a few protected member variables, and some access properties. Now we need to implement the CompareTo method.

You'll also notice that I defined two classes within the Programmer class— ProgrammerExperienceComparer and ProgrammerSalaryComparer—both of which implement the IComparer interface. As you probably surmised from the names, these two comparer classes give the class client code flexibility in terms of being able to sort on both the experience and salary members in addition to the default sorting by name, which we will implement now in the CompareTo method.

```
int CompareTo(Object* obj)
{
    Programmer* tmp = dynamic_cast<Programmer*>(obj);
    return (fullName->CompareTo(tmp->fullName));
}
```

Notice how I've delegated the actual comparison to a lower level by using the CompareTo method of the fullName member, which is a String object. As the String class implements IComparable, we are saved the trouble of actually comparing the two strings. Thus any collection of Programmer objects will—by default—get sorted on the basis of the Programmer::fullName member.

However, there may very well be situations where we would want the sorting to be done on the basis of another members such as Programmer::experienceInYears or Programmer::annualSalary. Let's see how the ProgrammerSalaryComparer and ProgrammerExperienceComparer could be coded to handle that.

```
__gc class Programmer : public IComparable
{
    . . .

    __gc class ProgrammerSalaryComparer : public IComparer
    {
    public:
      int Compare(Object* o1, Object* o2)
      {
        Programmer* p1 = dynamic_cast<Programmer*>(o1);
        Int32 p1Salary = p1->annualSalary;
```

```
      Programmer* p2 = dynamic_cast<Programmer*>(o2);
      Int32 p2Salary = p2->annualSalary;

      return -1 * p1Salary.CompareTo(__box(p2Salary));
   }
};


__gc class ProgrammerExperienceComparer : public IComparer
{
public:
  int Compare(Object* o1, Object* o2)
  {
     Programmer* p1 = dynamic_cast<Programmer*>(o1);
     Int32 p1Experience = p1->experienceInYears;

     Programmer* p2 = dynamic_cast<Programmer*>(o2);
     Int32 p2Experience = p2->experienceInYears;

     return -1 * p1Experience.CompareTo(__box(p2Experience));
  }
   };

 ...
} // end of Programmer definition
```

As both classes implement the IComparer interface, they must each provide an implementation of the Compare method. As you can see, they both do this; the only difference between the two is that the Programmer SalaryComparer class compares on the Programmer::annualSalary member, whereas the ProgrammerExperienceComparer class compares on the Programmer::experienceInYears member. Notice that, once again, we don't even have to do the comparison ourselves. Since the members being compared are both Int32 objects, and the Int32 class implements the CompareTo method, this method is called to perform the actual comparison in both cases.

Designing the Collection Class

We will now define a couple of classes and an enumeration type:

- **ProgrammerCollection class**—Implements the IEnumerable interface and acts as a collection for Programmer objects.

- **ProgrammerEnumerator class**—Defined within the `Programmer` `Collection` class, this class implements `IEnumerator` interface for allowing the actual enumeration of `Programmer` objects.
- **ProgrammerSortOption enumeration**—Enumeration that defines the three sort options (by name, experience, and salary).

As mentioned earlier in the chapter, every collection class that implements the `IEnumerator` interface needs to define a `GetEnumerator` method. This method returns an object that implements the `IEnumerator` interface. In our custom collection example, the `ProgrammerCollection::` `GetEnumerator` method will return an object of type `Programmer` `Enumerator`

As you saw in Table 11–1, the `IEnumerator` interface defines two methods and a property—`Current`, `MoveNext`, and `Reset`. Here's a more in-depth explanation of these members and how they'll be implemented in our demo collection.

- **Current**—`Current` is a read-only property that returns the current element in the collection. The `MoveNext` method must be called at least once before the `Current` property is accessed, because by default the enumerator is positioned one behind the first element in the collection. An attempt to access the `Current` property either when the enumerator is positioned before the first element in the collection or when the enumerator has gone past the end of the collection causes an `InvalidOperationException` exception. Accessing the `Current` property will not move the enumerator forward, and multiple calls to the `Current` property will return the same element.
- **MoveNext**—The `MoveNext` method is used to move the enumerator to the next element in the collection. The `MoveNext` method must be called at least once before the `Current` property can be used to access the elements in the collection. The `MoveNext` method returns `true` if the enumerator has moved to the next element in the collection and `false` if the enumerator has gone past the end of the collection. The first call to `MoveNext` will position the enumerator at the first element in the collection. Once the enumerator has been moved past the end of the collection, all further calls to `MoveNext` will return `false`, and any attempt to access the `Current` property will result in an `InvalidOperationException` being thrown. The moment the collection has changed (as when it has been sorted, or a

new member or members have been added), the enumerator should be invalidated. Once an enumerator has been invalidated, any further call to MoveNext will throw an InvalidOperationException, which basically means that if you change the collection after you've got an enumerator for it, then you'll need to get a new enumerator, as the current enumerator has been invalidated.

■ **Reset**—The Reset method will set the enumerator to a position that is one behind the first element in the collection. This is the initial position when an enumerator is first instantiated or obtained, and in this position the Current property is undefined. Access the Current property by first making at least one call to MoveNext. Just as in the case of the MoveNext method, the Reset method is valid only as long as the enumerator itself is valid. This means if any change has been made to the collection after the enumerator has been created, all further calls to the Reset method cause an InvalidOperationException exception.

Now let's see the actual implementation of our ProgrammerSort Option enumeration and ProgrammerCollection class:

```
__value enum ProgrammerSortOption
{
  ByName,
  ByExperience,
  BySalary
};

_gc class ProgrammerCollection : public IEnumerable
{
protected:
  ArrayList* programmers;
  bool dirty;

public:
  ProgrammerCollection()
  {
    programmers = new ArrayList();
    dirty = false;
  }

  __gc class ProgrammerEnumerator : public IEnumerator
  {
```

```
      // implemented shortly
   };

public:
   IEnumerator* GetEnumerator()
   {
      // implemented shortly
   }

   int Add(Object* value)
   {
      // implemented shortly
   }

protected:
   ProgrammerSortOption sortOption;
public:
   void Sort(enum ProgrammerSortOption sortOption)
   {
      // implemented shortly
   }

   int IndexOf(Programmer* programmer)
   {
      // implemented shortly
   }
};
```

As you can see, our collection is internally represented as an
ArrayList object (ProgrammerCollection::programmers). We also
have a member called dirty that represents a Boolean flag that is used to
verify if the collection has been changed. This flag is defined because when
the collection is altered in any way, we need to invalidate any enumerators
that might have been created before the collection was changed.

Now let's see how the individual ProgrammerCollection methods can
be implemented. We'll start with the GetEnumerator method. The method
first initializes the current collection instance's dirty flag. It then constructs a ProgrammerEnumerator object based on the current Programmer
Collection instance and returns that (cast to IEnumerator) to the caller.
(You'll see the ProgrammerEnumerator class shortly.)

```
IEnumerator* GetEnumerator()
{
```

```
    dirty = false;

    return dynamic_cast<IEnumerator*>(new ProgrammerEnumerator(this));
}
```

Now we'll look at the Add method. As mentioned, any time an object is added to the collection, the enumerator has to be invalidated.

```
int Add(Object* value)
{
  dirty = true;
  return programmers->Add(value);
}
```

Now for the Sort method. The first thing the method does is to save the current sort option. After that, the dirty flag is set to true. As the ArrayList::Sort method can take a IComparer interface as a parameter, this allows me to call the Sort method and pass either a Programmer:: ProgrammerExperienceComparer or Programmer::ProgrammerSalary Comparer object. The Sort method will then call the passed object's CompareTo method, which, as you saw earlier, does the actual comparison.

```
void Sort(enum ProgrammerSortOption sortOption)
{
   this->sortOption = sortOption;

   dirty = true;

   switch(sortOption)
   {
     case ProgrammerSortOption::ByName:
       programmers->Sort();
       break;
     case ProgrammerSortOption::ByExperience:
       programmers->Sort(new Programmer::ProgrammerExperienceComparer);
       break;
     case ProgrammerSortOption::BySalary:
       programmers->Sort(new Programmer::ProgrammerSalaryComparer);
       break;
   }
}
```

At times it might be necessary to ascertain at what index a given object has been stored relative to the current sort order. In fact the reason for the `ProgrammerSortOption` member of the `ProgrammerCollection` collection is specifically to keep track of the current sort order for just this purpose. As the `ArrayList` implements an `IndexOf` method, all I needed to do was to define my own `IndexOf` method that sorts the collection and then calls the `programmers::IndexOf` method. Very shortly—when we get to the MFC UI part of this demo—you'll see why this is important.

```
int IndexOf(Programmer* programmer)
{
  Sort(this->sortOption);
  return programmers->IndexOf(programmer);
}
```

Now let's take a look at the nested `ProgrammerEnumerator` class:

```
__gc class ProgrammerEnumerator : public IEnumerator
{
public:
  ProgrammerEnumerator(ProgrammerCollection* collection)
  {
    this->collection = collection;
    position = -1;
  }

protected:
  int position;
  ProgrammerCollection* collection;

  __property Object* get_Current()
  {
    // implemented shortly
  }

  void Reset()
  {
    // implemented shortly
  }

  bool MoveNext()
  {
```

```
    // implemented shortly
  }
};
```

We pass the `ProgrammerCollection` object to the `Programmer Enumerator` class so that the `ProgrammerEnumerator` will be able to access the `dirty` flag of the `ProgrammerCollection` for which it is acting as enumerator. It also has an index member (`ProgrammerEnumerator::position`), which is by default set to −1. This is done because, as mentioned earlier, enumerators should always initialize their internal position to one position behind the first element in the collection.

Now let's examine each of the methods in detail. First up is the `Current` property.

```
__property Object* get_Current()
{
  if (0 > position
  || position >= collection->programmers->Count
  || collection->dirty)
    throw new InvalidOperationException();

  return collection->programmers->Item[index];
}
```

As you can see, the `Current` property returns the current item in the collection and results in an `InvalidOperationException` if any of the following is true:

- The index is less than 0—meaning that the enumerator has not been positioned properly by first calling `MoveNext` before the call to the `Current` property.
- The enumerator has been moved beyond the last element in the collection.
- The dirty flag for the collection is set to `true`, which means that the collection has been changed.

If none of these conditions is true, then the method returns the element of its `ArrayList` (programmer collection) using its `position` member as the index.

The `Reset` method simply initializes the `dirty` and `position` members.

```
void Reset()
{
  collection->dirty = false;
  position = -1;
}
```

Note that the documentation suggests that if the collection has been changed, then the Reset method should throw an exception. However, as implementers we are free to make our changes to the working of the class without affecting general behavior. In this particular case we simply reset the enumerator to its initial position because we know that even if the collection has changed, resetting the enumerator to the default position is a perfectly valid operation.

The MoveNext method is the last method of our custom collection and completes the required IEnumerator members that must be implemented.

```
bool MoveNext()
{
  if (collection->dirty)
    throw new InvalidOperationException();

  position++;

  return (position < collection->programmers->Count);
}
```

After verifying that the dirty flag is not set, the MoveNext method increments the internal position member by 1 and returns a Boolean value indicating whether or not we've passed the end of the collection.

Using the Collection in an MFC Demo

At this point, let's see our brand-new, specialized collection class in action with a demo application. This application will allow the creation of Programmer objects that are inserted into a list view control. Upon clicking any of the list view header columns, the notification will be handled and the list view resorted.

1. Create an MFC dialog-based application called Custom CollectionsDemo that supports Managed Extensions.

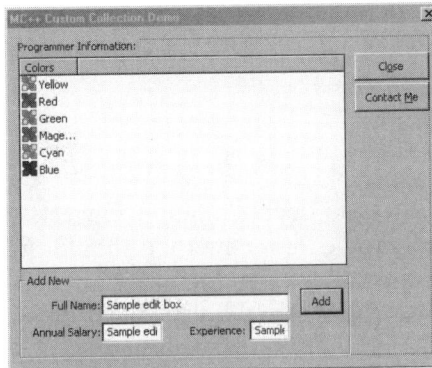

Figure 11-1 Dialog resource for the `CustomCollectionDemo` application

2. Open the dialog template resource and add the controls as you see them in Figure 11–1.
3. Set the list view control's `View` property to `Report`.
4. Create the DDX member variables as shown in Table 11–4.
5. As we've already walked through the definitions of the `Programmer`, `ProgrammerCollection`, and related classes, I won't rehash all that here. If you're following along, simply copy the `Programmer.h` file from the CD's version of this demo and add it to the current project.

Table 11-4 DDX variables for the `CustomCollectionDemo` demo

Control	Variable Type	Variable Name
List view	`CListCtrl`	`m_lstProgrammers`
Full name edit	`CString`	`m_strFullName`
Salary edit	`int`	`m_iAnualSalary`
Experience edit	`int`	`m_iExperienceInYears`

6. Insert an include directive for the `Programmer.h` file in the dialog's header file.

7. Declare a member variable of type `ProgrammerCollection` in the `CCustomCollectionDemoDlg` class.

```
class CCustomCollectionDemoDlg : public CDialog
{
  . . .
protected:
   gcroot<ProgrammerCollection*> m_collProgrammers;
```

8. Update the `CCustomCollectionDemoDlg::OnInitDialog` function as follows to construct the `ProgrammerCollection` object and to insert the list view columns.

```
BOOL CCustomCollectionDemoDlg::OnInitDialog()
{
  . . .

#pragma push_macro("new")
#undef new
   m_collProgrammers = new ProgrammerCollection();

   m_lstProgrammers.InsertColumn(0, _T("Full Name"));
   m_lstProgrammers.InsertColumn(1, _T("Experience"));
   m_lstProgrammers.InsertColumn(2, _T("Salary"));

#pragma pop_macro("new")

   return TRUE;
}
```

9. Implement the following Add button handler. This function begins by constructing a Programmer object using the values entered on the dialog. After that, the newly created `Programmer` object is added to the dialog's `ProgrammerCollection` member by calling the `ProgrammerCollection::Add` function. Now the collection's `IndexOf` function is called in order to determine where the `Programmer` object was inserted into the collection relative to its current sort option. The returned value is used to insert the

Programmer object's data into the list view at the same index. Finally, the dialog is initialized.

```
void CCustomCollectionDemoDlg::OnBnClickedAdd()
{
#pragma push_macro("new")
#undef new
      UpdateData(TRUE);

   // Create new Programmer object
   Programmer* p = new Programmer(m_strFullName,
                                    m_iExperienceInYears,
                                    m_iAnnualSalary);
   // Add to collection
   m_collProgrammers->Add(p);

   // Ask collection where new item was inserted
   int i = m_collProgrammers->IndexOf(p);

   // Based on item's insertion point, insert
   // item into list control at same index
   int idx = m_lstProgrammers.InsertItem(i, m_strFullName);

   CString strTemp;

   strTemp.Format(_T("%ld"), m_iExperienceInYears);
   m_lstProgrammers.SetItemText(idx, 1, strTemp);

   strTemp.Format(_T("%ld"), m_iAnnualSalary);
   m_lstProgrammers.SetItemText(idx, 2, strTemp);

   SizeAllColumns(m_lstProgrammers);

   // Init dialog
   m_strFullName = _T("");
   m_iExperienceInYears = 0;
   m_iAnnualSalary = 0;

   UpdateData(FALSE);
#pragma pop_macro("new")
}
```

10. Now, implement a handler for the list view header's HDN_ITEM CLICK message. All this function needs to do is to call the

ProgrammerCollection::Sort method, passing the Programmer
SortOption enumeration value that equates to the column header
that was clicked. Because our collection is sortable via a single
method call, that's all there is to providing a sortable list view
for it!

```
void CCustomCollectionDemoDlg::OnHdnItemclickList1(
  NMHDR *pNMHDR, LRESULT *pResult)
{
  LPNMHEADER phdr = reinterpret_cast<LPNMHEADER>(pNMHDR);
  switch( phdr->iItem )
  {
    case 0:
      m_collProgrammers->Sort(ProgrammerSortOption::ByName);
      break;
    case 1:
      m_collProgrammers->
        Sort(ProgrammerSortOption::ByExperience);
      break;
    case 2:
      m_collProgrammers->Sort(ProgrammerSortOption::BySalary);
      break;
  }

  RefreshList();

  *pResult = 0;
}
```

11. Finally, add the following RefreshList function that is called
    each time the sort order is changed so as to resort the list control's
    contents.

```
void CCustomCollectionDemoDlg::RefreshList()
{
  m_lstProgrammers.DeleteAllItems();

  int index = 0;

  IEnumerator* en = m_collProgrammers->GetEnumerator();
  while (en->MoveNext())
  {
    Programmer* programmer = dynamic_cast<Programmer*>(en->
      Current);
```

```
int idx =
  m_lstProgrammers.InsertItem(m_lstProgrammers
    .GetItemCount(),
    (CString)programmer->FullName);
m_lstProgrammers.SetItemText(idx, 1,
    (CString)programmer->ExperienceInYears.ToString());

m_lstProgrammers.SetItemText(idx, 2,
    (CString)programmer->AnnualSalary.ToString());
  }
}
```

12. Now build and run the application (Figure 11–2). You should see that each time you add a new programmer, the data is inserted into the list view according to the sort order and that you can change the sort order simply by clicking on the desired list view header column.

Well that's it, you have learned how to design and code a .NET custom collection class that behaves just like any other collection class available in the BCL but is specific to your application needs. In the next section, we'll look at serializing managed objects to and from disk.

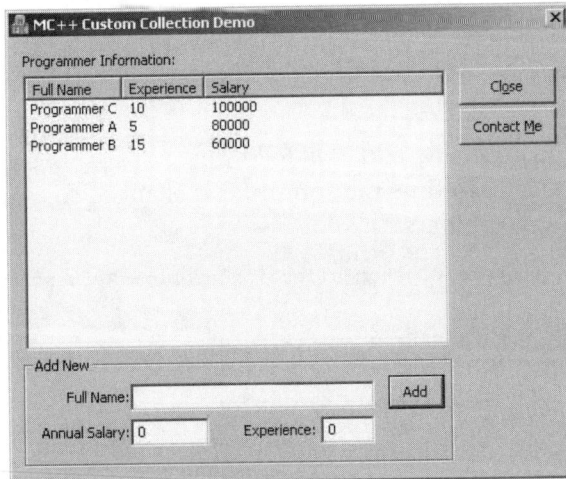

Figure 11–2 Demo application illustrating how to create your own enumerable, sortable collection classes

Serializing Managed Objects to and from Disk

Object serialization refers to the mechanism in which an object is converted to a form in which it can be saved on disk or transferred across a network and then de-serialized back to its original object form. The basic process of serialization involves converting all the members of the class along with the required metadata into bytes, which are then dumped to disk or to a network socket. The bytes are in stream form and are not necessarily read in the correct order all the time. The framework handles the actual serialization and de-serialization processes for you behind the scenes; therefore, as a developer, you won't need to bother with those details.

An important aspect of .NET serialization is that it cannot be applied to *unmanaged* objects. This limitation rules out any chance of using .NET serialization to serialize MFC objects. As a result, if your primary intention is to accomplish serialization for purely MFC objects, then you'd be much better off sticking to the serialization techniques available in the MFC framework, but if you need to serialize simple data to and from a disk or a network socket, .NET serialization is a quick and easy way of realizing this. In this section, you'll learn how to serialize both entire objects and selected members of a given managed object.

Serializing Entire Objects

Serializing an entire object to disk is just as easy as serializing an unmanaged MFC object. The first thing you need to do is to annotate the desired __gc class with the Serializable attribute. It's important to annotate all members that are to be serialized, even private members, using the Serializable attribute with the exception of basic types. Basic types are serializable by default, which means that the following class's members are already serializable without having to explicitly annotate each and every desired member.

In this class—Book—I've defined four simple members and the properties needed to get and set their respective values.

```
[Serializable]
__gc class Book
{
```

```
public:
    Book(String* title,
            String* author,
            String* publisher,
            Int32 year)
    {
      this->title = title;
      this->author = author;
      this->publisher = publisher;
      this->year = year;
    }

protected:
  String* title;
  String* author;
  String* publisher;
  Int32 year;

public:
    __property String* get_Title() { return this->title; }
    __property void set_Title(String* title) { this->title = title; }

    __property String* get_Author() { return this->author; }
    __property void set_Author(String* author) { this->author = author;
}

    __property String* get_Publisher() {return this->publisher; }
    __property void set_Publisher(String* publisher) { this->publisher
= publisher;  }

    __property Int32 get_Year() { return this->year; }
    __property void set_Year(Int32 year) { this->year = year; }
};
```

Fairly straightforward, right? Now let's see the client code needed to cause the serialization of a Book object's data to disk.

```
void WriteBookToDisk()
{
#pragma push_macro("new")
#undef new
    try
    {
```

```
    Book *pBook = new Book(
      S"Extending MFC Applications with the .NET Framework",
      S"Tom Archer",
      S"Addison Wesley",
      2003);

    FileStream *pFS = new FileStream(S"Book.dat",
                                     FileMode::Create,
                                     FileAccess::ReadWrite);

    BinaryFormatter *pBF = new BinaryFormatter();
    pBF->Serialize(pFS, pBook);

    pFS->Close();
  }
  catch(Exception* pe)
  {
    AfxMessageBox((CString)pe->Message);
  }
#pragma pop_macro("new")
}
```

As you can see from the WriteBookToDisk function, first the code instantiates a Book object. The code then constructs FileStream and Binary
Formatter objects. The BinaryFormatter class (defined in the System::
Runtime::Serialization::Formatters::Binary namespace), is used
because serialized data must be saved in binary format. Therefore, the
BinaryFormatter comes in handy here as it does all the heavy lifting for
us with regard to encoding our object in the proper format. Once instantiated, a call to the BinaryFormatter::Serialize method—passing the
FileStream object and the object to be serialized—is all we need!

If you insert these two functions into your code and place a call to the
WriteBookToDisk method, you could confirm that the Book object's data
was indeed written to disk by opening the Book.dat file in Visual Studio
.NET.

The obvious next step is to "deserialize"—or read—the data back from
the file. Here's the reciprocal function to the WriteBookToDisk function.

```
Book* ReadBookFromDisk()
{
#pragma push_macro("new")
#undef new
```

```
   Book* pBook = NULL;

   try
   {
     FileStream *pFS = new FileStream(S"Book.dat",
                                      FileMode::Open,
                                      FileAccess::ReadWrite);

     BinaryFormatter *pBF = new BinaryFormatter();
     pBook = static_cast<Book*>(pBF->Deserialize(pFS));

     pFS->Close();
   }
   catch(Exception* pe)
   {
     AfxMessageBox((CString)pe->Message);
   }

   return pBook;
#pragma pop_macro("new")
}
```

The `ReadBookFromDisk` function begins by initializing a local `Book` pointer to `NULL`. It then constructs a `FileStream` object on the `Book.dat` file. Once that is done, a `BinaryFormatter` object is instantiated, and its `Deserialize` method is called—passing it the `FileStream` object. As the `BinaryFormatter::Deserialize` method returns a deserialized object (in the form of a `System::Object`) read from the specified file, this value needs to be cast to the appropriate type (in our case, a `Book` class). The `FileStream` object is closed, and we're done.

Using this function is easy, as shown in the following code snippet:

```
   Book* pBook = ReadBookFromDisk();
   if (pBook)
   {
     CString str;
     str.Format(_T("Be sure and read the new book by %s "
                 "called '%s'. It is published by %s and due out "
                 "%ld"),
              pBook->Author,
              pBook->Title,
              pBook->Publisher,
```

```
        pBook->Year);
  AfxMessageBox(str);
}
else
{
  AfxMessageBox(_T("ReadBookFromDisk returned NULL"));
}
```

Figure 11–3 shows an example execution of the book's `SerializeObjects` demo.)

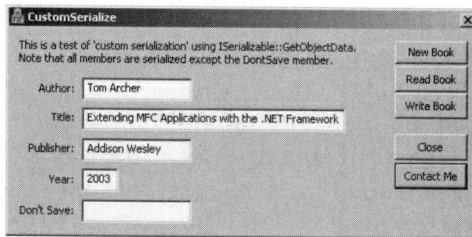

Figure 11-3 The `SerializeObjects` demo application illustrates how easy it is to serialize entire managed objects to disk.

Serializing Selected Members

Obviously, there are bound to be situations where you don't want to serialize an entire object. One example of such a situation is in an application using *remoting*, where bandwidth and performance would naturally be a concern. Therefore, in this section I'll cover a couple of ways in which to define your class so that only the desired members are serialized to disk.

As mentioned in the previous example, by default all basic types are serializable. What I mean by that is that the types defined by .NET are serializable—you have to annotate your own types with the `Serializable` attribute. You should also note that the `Serializable` attribute works regardless of the member's access modifier (public, protected, and private).

One technique to handle this situation is to simply annotate the member you do not want serialized with the `NonSerialized` attribute. Using the `Book` example from the previous section, you could simply annotate the attribute to the desired `Book` member as follows:

```
[Serializable]
__gc class Book
```

```
{
  . . .
protected:
  String* title;
  [NonSerialized] String* author;
  String* publisher;
  Int32 year;


  . . .
};
```

Now when the `ReadBookFromDisk` function is called, it would return a `Book` object, and since the `Author` property was not written, it would contain a value of `null`.

However, there is a far more powerful and flexible way of serializing only the selected members of a managed object—*custom serialization*. Custom serialization is realized via implementing the `ISerializable` interface (defined in the `System::Runtime::Serialization` namespace) in the managed class that will be serialized. This interface defines a single method called `GetObjectData`, which is called by the `Formatter` object during serialization. This gives you complete control over just how the serializing is done—including any need to encode the data or do any pre- or post-processing as needed. Here's the syntax for the `ISerializable::GetObjectData` method.

```
void ISerializable::GetObjectData(SerializationInfo* info,
                                  StreamingContext context);
```

The basic functionality of this method is to populate the `Serialization Info` object with the data needed to perform serialization, where the `StreamingContext` specifies the destination for the serialization. You typically won't need to manipulate the `StreamingContext` parameter, as it has already been initialized for you. It's passed to you for the rare occasion where you might need to know the information regarding the ultimate destination of the object's data.

Let's see a modification of the earlier class, where we have an extra data member that will not be serialized. (I've made the changes between this incarnation of the `Book` class and the previous one bold.)

```
[Serializable]
__gc class Book : public ISerializable
```

```
{
public:
  Book(String* title,
       String* author,
       String* publisher,
       Int32 year,
       String* dontSave)
  {
    this->title = title;
    this->author = author;
    this->publisher = publisher;
    this->year = year;
    this->dontSave = dontSave;
  }
protected:
    Book(SerializationInfo *si, StreamingContext sc)
    {
      this->title = si->GetString(S"title");
      this->author = si->GetString(S"author");
      this->publisher = si->GetString(S"publisher");
      this->year = si->GetInt32(S"year");
      // Note that this->dontSave is not read
      // because it was never written
    }

protected:
    String* title;
    String* author;
    String* publisher;
    Int32 year;
    String* dontSave;
public:
    __property String* get_Title() {  return this->title;  }
    __property void set_Title(String* title) { this->title = title; }

    __property String* get_Author() {  return this->author;  }
    __property void set_Author(String* author) { this->author =
      author; }

    __property String* get_Publisher() {  return this->publisher;  }
    __property void set_Publisher(String* publisher) {
      this->publisher = publisher; }
```

```
    __property Int32 get_Year() { return this->year; }
    __property void set_Year(Int32 year) { this->year = year; }

    __property String* get_DontSave() { return this->dontSave; }
    __property void set_DontSave(String* dontSave)
    { this->dontSave = dontSave; }

public:
    void GetObjectData(SerializationInfo *si, StreamingContext sc)
    {
      si->AddValue(S"title", this->title);
      si->AddValue(S"author", this->author);
      si->AddValue(S"publsher", this->publisher);
      si->AddValue(S"year", this->year);
      // Note that this->dontSave is not being saved here
    }
};
```

As you can see, a bit of work needs to be done, but not much and certainly not difficult. The first thing to notice is the addition of the DontSave property. While this is a simple String object here, it might very well represent a much more elaborate object that you do not need to have serialized with the rest of the managed object's members.

From there, notice how the GetObjectData method is implemented. All that is needed here is to populate the SerializationInfo object using the overloaded AddValue method—passing each member we desire to serialize to that method.

So what happens to the client code? Absolutely nothing. The serialization details of the object are fully encapsulated within the class's definition, so the client code doesn't need to know anything about how the object is serializing itself. The only change you would need to make to the Write BookToDisk function is obviously to instantiate the Book object with the new DontSave member, as the constructor now takes that additional parameter.

You'll note that the Book class now has an additional (protected) constructor. This constructor is used by the deserialization process. When you attempt to deserialize the object—via the BinaryFormatter:: Deserialize method that is called from our example ReadBookFromDisk function—the formatter reads the data from the passed FileStream object and calls this constructor, passing it the SerializationInfo and StreamingContext objects I covered earlier in this section. All the

managed class writer has to do is to populate the members from the
_SerializationInfo object via its various Get methods (e.g., GetString,
GetInt64, GetInt64). These functions take a String value representing
the name of the field to retrieve and return a desired type. Obviously, there
needs to be a distinct method for each type (as opposed to the overloaded
SerializationInfo::AddValue method) as overloaded methods cannot
differ by return type alone.

Once again the client code doesn't change at all, as the deserialization
details are fully encapsulated within the Book class. If you were to run this
code (supplied in the CD's CustomSerialize demo application), you
would see that all members are serialized except for the DontSave member, which is just what we wanted to accomplish.

One last point to make here is that while we only used Binary
Formatter objects for serialization, it's also possible to use XML formatters if you need to serialize data that will be deserialized by applications
running on other platforms.

In summary, the main limitation of using the .NET serialization techniques I've covered here is that you are restricted to using managed .NET
objects, and thus it won't be easy to port existing MFC serialization objects.
However, as you mix more and managed and unmanaged code, you will
undoubtedly need to serialize your managed objects from time to time, and
what you've learned here will help you immensely in that endeavor.

Summary

It is beyond the scope of this book to cover all the BCL collection classes
available—and would get quite boring anyway, as there is much overlap
between the various collection classes. However, in this chapter, we have
covered many of the more commonly used collection classes in terms of
both their syntax and the situations to which they're best suited. You started
out by learning the basic interfaces that collections implement in order to
be called collections. From there, you saw how to use the ArrayList class,
which implements the IList interface through a dynamically resized array.
You also saw how to perform typical operations using some rather handy
collection classes such as the Hashtable class, which represents a collection of key/value pairs stored on the basis of the hash of the key; the Sorted
List class, which is a key/value-pair collection that is sorted on the key and
is also accessible using the index of the key; and the Queue and the Stack

classes, which are implementations of the queue and stack data structures that all computer students and programmers eventually encounter somewhere in their coding career. The Stack class and the Queue class are implemented in similar fashion using an internal circular array to store the elements. The major difference between the Queue class and the Stack class is that the Queue class is a FIFO (first-in, first-out) structure, while the Stack class is a LIFO (last-in, first-out) structure.

Once the basics were covered of what underlying interfaces are implemented in order to define a collection class, and you saw several examples of the most commonly used collection classes and how they implemented those interfaces, a section was introduced that illustrated how to design and code a custom collection class that can be treated (from a client-code perspective) just as if it were one of the BCL-supplied collection classes. As one of the most basic features expected of any collection class is that its data be enumerable, the example custom collection implemented the IEnumerable and IEnumerator interfaces to serve just that purpose. In addition, the section also illustrated how to add sorting ability to the collection class by implementing the IComparable interface. The section wrapped up with an MFC dialog-based demo application that showed how easy a custom collection can make client-side code that needs to use an enumerable, sortable collection.

The last section of the chapter—"Serializing Managed Objects to and from Disk"—focused on an issue that most MFC developers are intimately familiar with—serialization. As I mentioned in the chapter's introduction, attempting to serialize managed classes using a native technique (such as the MFC CArchive class) would be far more trouble than it's worth. Therefore, this section showed a very quick and easy technique—using the BinaryFormatter class—to serialize entire objects to and from disk. Once that was done, you then saw how the Serializable and NonSerialized attributes can be used to define whether or not an entire class or specific members can be serialized. Finally, you discovered how to implement custom serialization to give you far greater control over which members are serialized and how they're serialized, using the ISerializable interface and SerializationInfo and StreamingContext classes.

XML Overview and Syntax

Intended Audience

This appendix is not meant as a full-fledged tutorial on XML or even the XML syntax. Rather, it is intended as a complement to Chapter 5 for those people who've never worked with XML before and want to have at least a firm understanding of the basics of XML and its syntax in order to get the most out of Chapter 5's material.

XML stands for Extensible Markup Language, and as the name implies, it is a bit like HTML in that both are standards used to mark up (or define) a document. However, there are some important distinctions between the two. Even if you don't specifically develop Web sites, chances are extremely high that you know about HTML and its purpose. Put simply, HTML is a markup language designed to describe how to display information. The various HTML tags allow you to specify exactly how an agent (typically a browser) should present that data. XML, on the other hand, has been designed to add structure to and describe data. That is, XML provides a means of creating self-describing data.

Another difference between HTML and XML is that HTML has predefined tags that you can use to mark up your Web pages. Using XML, you must define your data-specific tags. Also, XML uses something called a DTD (Document Type Definition) that is used to describe the data in terms of structure and type to clients (consumers) of that data. Therefore, XML is not a replacement for or competitor to HTML. In fact, XML is meant to complement HTML in that, while HTML is used to display information, XML is used to describe and project structure onto information. A

browser, like Microsoft's Internet Explorer, consumes HTML to display Web pages on your screen. HTML would be useless without the browser, since HTML marked-up documents are often difficult to read in their raw (source) form. Similarly, XML gets consumed by an XML parser—software that's capable of interpreting an XML document. Like browsers, there are many different types of XML parsers available, each providing features like varying support for current and new standards. So now that we know what XML is, what is it used for? Let's look at that now.

It's actually very easy to create a valid XML document, since the rules that define a properly formed XML document are simple, although they are very strict. Unlike HTML, where you can sometimes do things like incorrectly nest tags, try that with XML, and no XML parser is going to be able to read the document. Therefore, let's go over the terminology and syntax associated with XML documents and XML in general.

The first term we'll look at is *XML parser.* As the name implies, an XML parser is simply code that can read and understand XML-formatted documents. Both Internet Explorer and Visual Studio .NET provide XML parsers. Now let's look at a simple XML document and break down its components (nodes):

```
<?xml version="1.0" encoding="us-ascii" standalone="yes"?>
<email language="EN">
  <from>Tom@TomArcherConsultingGroup.com</from>
  <to>Editor@AddisonWesley.com</to>
  <subject>XML chapter status</subject>
  <message>I expect to be finished today!!</message>
</email>
```

- The first line of the document is called the XML node declaration and includes information such as the XML version, character encoding method, and standalone attribute.
- Each tag in XML—beginning with a < and ending with a >—denotes an element.
- In the XML example, the <email> line denotes the root element in this particular document, with all other elements being defined between the <email> start tag and the </email> end tag. It is like saying "this document is an email."
- A node that contains a start tag and an end tag is collectively referred to as an *XML element.*
- The next four elements ("from," "to," "subject," and "message") all describe child elements of the "e-mail" root element.

- You'll notice that the "email" element has a name/value pair within its start tag. This node is called an attribute node and is used to further define the element.
- Finally, the </email> tag terminates the scope of the <email> tag.

As you can see, there's really nothing to defining an XML document in terms of the tags you can use. Now let's look at some syntax rules regarding what is termed a well-formed XML document.

- All elements must have a closing tag. The only exception to this rule is the declaration node, and it doesn't count because technically it is not an XML element. (In fact, when you see the ListXMLNodes application shortly, you'll see that the XmlTextReader class refers to it as an "XmlDeclaration" node type as opposed to an "Element" node type.)
- Empty elements can be defined with a single tag. This is useful for situations where your document defines a given tag, but you don't have a value for that tag. For example, using the sample email XML document, let's say that no subject was supplied. In that case you could write the subject element using either of the following forms (where the first method is called an empty element):

```
<subject/>
<subject></subject>
```

- As mentioned, all XML documents have a single root element. Every other element in the XML document is then a child of this root. Therefore, the following XML would be incorrect, as we have defined more than one root:

```
<?xml version="1.0" encoding="us-ascii" standalone="yes"?>
<!-- INVALID! Multiple root elements -->
<email language="EN">
  <from>Tom@TomArcherConsultingGroup.com</from>
  <to>Editor@AddisonWesley.com</to>
  <copies>
    <copy>Erik@WindmillPublishing.com</copy>
    <copy>Krista@WindmillPublishing.com</copy>
  </copies>
  <subject>XML chapter status</subject>
  <message>I expect to be finished today!!</message>
</email>
```

```
<email language="EN">
  <from>Tom@TomArcherConsultingGroup.com</from>
  <to>Erik@WindmillPublishing.com</to>
  <copies/>
  <subject>I'm ready for my next book! &lt;<eom&gt;</subject>
  <message/>
</email>
```

In this case (where we have multiple "email" elements) you would need to define your root element as though it were a collection. A hint of how to do this is the "copies" element as it allows for multiple "copy" child elements. Here's the corrected XML.

```
<?xml version="1.0" encoding="us-ascii" standalone="yes"?>
<!-- Sibling elements can have the same name, but there -->
<!-- can only be one root -->
<emails>
  <email language="EN">
    <from>Tom@TomArcherConsultingGroup.com</from>
    <to>Editor@AddisonWesley.com</to>
    <copies>
      <copy>Erik@WindmillPublishing.com</copy>
      <copy>Krista@WindmillPublishing.com</copy>
    </copies>
    <subject>XML chapter status</subject>
    <message>I expect to be finished today!!</message>
  </email>
  <email language="EN">
    <from>Tom@TomArcherConsultingGroup.com</from>
    <to>Erik@WindmillPublishing.com</to>
    <copies/>
    <subject>I'm ready for my next book! &lt;eom&gt;</subject>
    <message/>
  <email>
</emails>
```

■ XML tags are case-sensitive. This is not a big deal to those of us who program in C++. However, it's worth mentioning because if you also do any Web "development," you may have become accustomed to tags being case-insensitive. As a result, an element named <cmail> is a completely different element than one named <Email>. In addition, the starting and ending tags for an element

must also match in terms of case. The following would be invalid, as they do not match.

```
<!-- INVALID! Start and End Elements do not match -->
<subject>XML chapter status</Subject>
```

- All elements must be properly nested, with end elements appearing in exact reverse of their counterpart start elements. The following XML is in error, as the "from" element precedes the "to" element, but the "to" end element appears before the "from" end element.

```
<from>
  <to>Editor@AddisonWesley.com
  </from>
</to>
```

- Attribute name/value must be defined using the syntax <name>= "value".

```
<-- Correct attribute syntax -->
<email language="EN">
```

- As you've noticed throughout these XML snippets, comments can be placed in XML documents using the <-- start and --> end syntax.
- XML defines few reserved characters that you must be careful with as you use them in your data. Each of these characters has a substitute that you should use in your data so that the XML parser doesn't mistake its usage. These are shown in Table A–1.

Table A–1 Reserved XML Characters and Their Substitutes

Reserved (or "markup") Character	Substitute (or "escape" value)
<	&lt;
>	&gt;
&	&amp;
"	&quot;
'	&apos;

Note that sometimes you will hear people refer to a markup character being "escaped." That simply means that the substitute was automatically inserted in place of the reserved character.

As mentioned at the outset of this section, a complete dissertation on the XML standard would take several chapters. However, if you're new to XML, what you've seen up to now should, in terms of terminology and syntax, be enough to help easily navigate through the remainder of the text and demo applications in Chapter 5.

XPath Overview and Syntax

Intended Audience

This appendix is not meant as a complete tutorial on XPath. Rather it is intended as a complement to Chapter 5 for those people who've never used the querying capabilities of the XPath language. It serves as a collection of sample XPath expressions that can be tested with some of Chapter 5's code snippets and demo applications. While not a prerequisite for using XML, understanding XPath will enable you to perform much richer and more complex queries against your data.

Before getting into the XPath syntax and examples, let's first talk about what it is. At its simplest, XPath is a W3C standardized syntax for accessing parts of an XML document and is typically used in building complex queries against that document. The name XPath is derived from the fact that, as XML is a data format whereby related data elements exist in a hierarchical fashion, XPath allows you to use search strings that resemble file path information whereby each level of the hierarchy is delimited by a forward slash character. In addition to allowing you to perform queries based on element values, XPath also defines a library of standard functions that can be used to locate specific nodes. Now let's look at a sample document that will be used throughout this text's examples:

```
<?xml version="1.0" encoding="ISO-8859-1"?>
<inventory>
  <sku id="111-006-116">
    <manufacturer>VOL</manufacturer>
    <description>Oil</description>
```

```
    <price>3.90</price>
  </sku>
  <sku id="111-106-126">
    <manufacturer>VOL</manufacturer>
    <description>Brake Fluid</description>
    <price>3.45</price>
  </sku>
  <sku id="211-222-336">
    <manufacturer>VOL</manufacturer>
    <description></description>
    <price>1.90</price>
  </sku>
</inventory>
```

Basic XPath Syntax

As explained in Appendix A, XML documents are most often represented as a hierarchy of nodes, with each node relating to another as either a parent, child, or sibling. An XPath expression is a slash-delimited list of child element names that describe a path through the XML document from a given point. The expression is said to locate, or select, elements that match the path indicated in the expression.

```
Locate all price elements of all sku elements of the inventory element
/inventory/sku/price
```

As you can discern from this, starting the pattern with a slash indicates that the path is an absolute (and not relative) path to an element. A relative path is one that does not start with a slash and is used in cases where you want the search to be conducted starting from the current node. Another note to make is that if the path starts with two slashes, all elements in the document that fulfill the criteria will be selected, regardless of level.

```
Locate all price elements in the document regardless of what sku the
price belongs to.
//price
```

XPath also supports the insertion of the asterisk character—*—as a wildcard in expressions where the asterisk represents any value. Consider the following examples:

```
Locate all child elements of all sku elements of the inventory element
/inventory/sku/*
```

```
Locate all price elements that are grandchild elements of the
inventory element regardless of their parent element
/inventory/*/price
```

```
Locate all price elements which have 2 direct ancestors
/*/*/price
```

```
Shorthand for selecting all elements in a document
//*
```

By using square brackets in an XPath expression, you can further refine your search:

```
Locate the first sku element of the catalog element:
/inventory/sku[1]
```

```
Locate the last sku child element
/inventory/sku[last()]
```

```
Locate all sku elements that have a price child element:
/inventory/sku[price]
```

```
Locate all sku elements of the inventory element that have a price
element with a value of 1.90:
/inventory/sku[price=1.90]
```

```
Locate all price elements of all the sku elements of the inventory
element that have a price element with a value of 1.90:
/inventory/sku[price=1.90]/price
```

Note that while there is a built-in function called last, there is no reciprocal first function. Instead, you would specify a hard-coded value of 1, as shown above.

XPath allows you to conduct a search based on attributes by using the at-sign—@—as in the following examples:

```
Locate all sku elements that have an attributed named id
//sku[@id]

Locate all sku elements that have any attributes at all
//sku[@*]

Locate
//sku[@id='111-006-116']
```

Expressions can be combined using the vertical bar character—|—as in the following. Note that the result will be the equivalent of running each expression independently.

```
Locate all manufacture and description elements that exist within
a sku element within an inventory element
/inventory/sku/manufacturer | /inventory/sku/description

Locate all manufacturer and description elements regardless of
where they exists in the hierarchy
//manufacturer | //description

Locate all manufacture elements that exist within
a sku element within an inventory element as well as all
description elements regardless of hierarchy
/inventory/sku/manufacturer | //description
```

Summary

As mentioned, this is not even close to a full-fledged tutorial on the XPath language, but I hope it has served you by illustrating the types of queries that are possible with this powerful language. In fact, I've covered just the very tip of what can be done with XPath, as this language supports a host of built-in functions and capabilities. If you're interested in learning more about XPath, I would suggest Erik Westermann's *Learn XML in a Weekend* (Cincinnati, OH: Premier Press, 2002), which has a great chapter on the XPath language.

Index

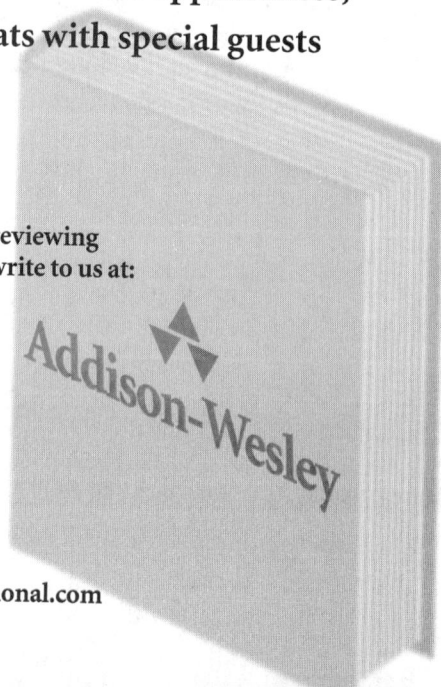

CD-ROM Warranty

Addison-Wesley warrants the enclosed CD-ROM to be free of defects in materials and faulty workmanship under normal use for a period of ninety days after purchase. If a defect is discovered in the CD-ROM during this warranty period, a replacement CD-ROM can be obtained at no charge by sending the defective CD-ROM, postage prepaid, with proof of purchase to:

Editorial Department
Addison-Wesley Professional
Pearson Technology Group
75 Arlington Street, Suite 300
Boston, MA 02116
Email: AWPro@awl.com

Addison-Wesley makes no warranty or representation, either expressed or implied, with respect to this software, its quality, performance, merchantability, or fitness for a particular purpose. In no event will Addison-Wesley, its distributors, or dealers be liable for direct, indirect, special, incidental, or consequential damages arising out of the use or inability to use the software. The exclusion of implied warranties is not permitted in some states. Therefore, the above exclusion may not apply to you. This warranty provides you with specific legal rights. There may be other rights that you may have that vary from state to state. The contents of this CD-ROM are intended for personal use only.

More information and updates are available at:
www.awprofessional.com